T0313149

IMAGE-GUIDED HYPOFRACTIONATED STEREOTACTIC RADIOSURGERY

IMAGE-GUIDED HYPOFRACTIONATED STEREOTACTIC RADIOSURGERY

A Practical Approach to Guide
Treatment of Brain and Spine Tumors

SECOND EDITION

edited by
Arjun Sahgal, MD
Simon S. Lo, MD
Lijun Ma, PhD
Jason P. Sheehan, MD

CRC Press
Taylor & Francis Group
Boca Raton London New York

CRC Press is an imprint of the
Taylor & Francis Group, an **informa** business

Second edition published 2022
by CRC Press
6000 Broken Sound Parkway NW, Suite 300, Boca Raton, FL 33487–2742

and by CRC Press
2 Park Square, Milton Park, Abingdon, Oxon, OX14 4RN

© 2022 Taylor & Francis Group, LLC

First edition published by CRC Press 2016

CRC Press is an imprint of Taylor & Francis Group, LLC

This book contains information obtained from authentic and highly regarded sources. While all reasonable efforts have been made to publish reliable data and information, neither the author[s] nor the publisher can accept any legal responsibility or liability for any errors or omissions that may be made. The publishers wish to make clear that any views or opinions expressed in this book by individual editors, authors or contributors are personal to them and do not necessarily reflect the views/opinions of the publishers. The information or guidance contained in this book is intended for use by medical, scientific or health-care professionals and is provided strictly as a supplement to the medical or other professional's own judgement, their knowledge of the patient's medical history, relevant manufacturer's instructions and the appropriate best practice guidelines. Because of the rapid advances in medical science, any information or advice on dosages, procedures or diagnoses should be independently verified. The reader is strongly urged to consult the relevant national drug formulary and the drug companies' and device or material manufacturers' printed instructions, and their websites, before administering or utilizing any of the drugs, devices or materials mentioned in this book. This book does not indicate whether a particular treatment is appropriate or suitable for a particular individual. Ultimately it is the sole responsibility of the medical professional to make his or her own professional judgements, so as to advise and treat patients appropriately. The authors and publishers have also attempted to trace the copyright holders of all material reproduced in this publication and apologize to copyright holders if permission to publish in this form has not been obtained. If any copyright material has not been acknowledged please write and let us know so we may rectify in any future reprint.

Except as permitted under U.S. Copyright Law, no part of this book may be reprinted, reproduced, transmitted, or utilized in any form by any electronic, mechanical, or other means, now known or hereafter invented, including photocopying, microfilming, and recording, or in any information storage or retrieval system, without written permission from the publishers.

For permission to photocopy or use material electronically from this work, access www.copyright.com or contact the Copyright Clearance Center, Inc. (CCC), 222 Rosewood Drive, Danvers, MA 01923, 978–750–8400. For works that are not available on CCC please contact mpkbookspermissions@tandf.co.uk

Trademark notice: Product or corporate names may be trademarks or registered trademarks and are used only for identification and explanation without intent to infringe.

ISBN: 9780367462789 (hbk)
ISBN: 9780367478728 (pbk)
ISBN: 9781003037095 (ebk)

Typeset in Garamond Pro
by Apex CoVantage, LLC

Contents

Contributors

Ahmed Abugharib, MD
Clinical Oncology Department
Sohag University Hospital
Sohag University
Sohag, Egypt

Justus Adamson, PhD
Duke University
Department of Radiation Oncology
Durham, North Carolina

John R. Adler, MD
Zap Surgical Inc.
San Carlos, California

Christopher Alvarez-Breckenridge, MD PhD
Department of Radiation Oncology
The University of Texas MD Anderson Cancer
    Center
Houston, Texas

Carey Anders, MD
Department of Radiation Oncology
Duke University
Durham, North Carolina

Lilyana Angelov, MD
Department of Neurosurgery and
Rose Ella Burkhardt Brain Tumor and
    Neuro-Oncology Center
Cleveland Clinic
Cleveland, Ohio

Josue Avecillas-Chasin, MD
Department of Radiation Oncology
The University of Texas MD Anderson Cancer
    Center
Houston, Texas

Steven Babic, PhD
Department of Medical Physics
Odette Cancer Centre
Sunnybrook Health Sciences Centre
Toronto, Ontario, Canada

Ehsan H. Balagamwala, MD
Department of Radiation Oncology
Taussig Cancer Institute
Cleveland Clinic
Cleveland, Ohio

Gene H. Barnett, MD
Department of Radiation Oncology
The University of Texas MD Anderson Cancer
    Center
Houston, Texas

Kathryn Beal, MD
Department of Radiation Oncology
Memorial Sloan Kettering Cancer Center
New York, New York

A.M. Bergman, MD
Department of Electrical Engineering and
    Computer Science
Lassonde School of Engineering York University
Toronto, Ontario, Canada

John M. Boyle, MD
Department of Radiation Oncology
Duke University
Durham, North Carolina

Paul D. Brown, MD
Department of Radiation Oncology
The University of Texas MD Anderson Cancer
    Center
Houston, Texas

Adomas Bunevicius, MD
Department of Neurological Surgery
University of Virginia
Charlottesville, Virginia

Aimee Chan, MS
Department of Medical Imaging
Toronto, Ontario, Canada

Michael Chan, MD
University of Toronto
Hospital for Sick Children
Toronto, Ontario, Canada

Eric L. Chang, MD
Department of Radiation Oncology
Keck School of Medicine and
    Norris Cancer Center
University of Southern California
Los Angeles, California

Joe H. Chang, MD
Radiation Oncology Centre
Peter MacCallum Cancer Centre
Melbourne, Australia

Samuel T. Chao, MD
Department of Radiation Oncology
Taussig Cancer Institute
and
Rose Ella Burkhardt Brain Tumor and Neuro-
    Oncology Center
Cleveland Clinic
Cleveland, Ohio

Steven J. Chmura, MD, PhD
Department of Radiation and Cellular Oncology
The University of Chicago
Chicago, Illinois

Cynthia Chuang, MD
Department of Radiation Oncology
Stanford University
Stanford, California

Jeffrey M. Clarke, MD
Department of Radiation Oncology
Duke University
Durham, North Carolina

Gregory J. Czarnota, PhD, MD
Department of Radiation Oncology and
    Imaging Research and Physical Sciences
Sunnybrook Health Sciences Centre
and
Department of Medical Biophysics
University of Toronto
Toronto, Ontario, Canada

Sunit Das, MD
St. Michael's Hospital
University of Toronto
Toronto, Ontario, Canada

Elizabeth David, MD, FRCPC
Interventional Radiology
Sunnybrook Health Sciences Centre
University of Toronto
Toronto, Ontario, Canada

N. Dea, MD
Department of Electrical Engineering and
    Computer Science
Lassonde School of Engineering York University
Toronto, Ontario, Canada

Martina Descovich, PhD
Department of Radiation Oncology
University of California, San Francisco
San Francisco, California

Jay Detsky, MD
Department of Radiation Oncology
Odette Cancer Centre
Sunnybrook Health Sciences Centre
University of Toronto
Toronto, Ontario, Canada

Kevin Diao, MD
Departments of Neurological Surgery and
    Radiation Oncology
University of Pittsburgh Medical Center
Pittsburgh, Pennsylvania

Emma M. Dunne, MD
Department of Radiation Oncology
BC Cancer, Vancouver Centre
Vancouver, British Columbia

Susannah Ellsworth, MD
Department of Radiation Oncology
UPMC Hillman Cancer Center
Pittsburgh, Pennsylvania

Peter E. Fecci, MD
Department of Radiation Oncology
Duke University
Durham, North Carolina

Joel Finkelstein, MD
Sunnybrook Health Sciences Centre
University of Toronto
Toronto, Ontario, Canada

John C. Flickinger, MD
Departments of Neurological Surgery and
    Radiation Oncology
University of Pittsburgh Medical Center
Pittsburgh, Pennsylvania

Scott R. Floyd, MD
Department of Radiation Oncology
Duke University
Durham, North Carolina

Matthew Foote, MD
Department of Radiation Oncology
Princess Alexandra Hospital and University of
    Queensland
Woolloongabba, Queensland, Australia

Peter C. Gerszten, MD, MPH
Departments of Neurological Surgery and
    Radiation Oncology
University of Pittsburgh Medical Center
Pittsburgh, Pennsylvania

Amol J. Ghia, MD
Departments of Neurological Surgery and
    Radiation Oncology
University of Pittsburgh Medical Center
Pittsburgh, Pennsylvania

Stefan Glatz, MD
Department of Radiation Oncology
University Hospital Zurich
Zurich, Switzerland

Matthias Guckenberger, MD
Department of Radiation Oncology
University Hospital Zurich
Zurich, Switzerland

Michael Hardisty, MD
Sunnybrook Research Institute
University of Toronto
Toronto, Ontario, Canada

Sara Hardy, MD
Department of Radiation Oncology
University of Rochester School of Medicine
Rochester, New York

Yaser Hasan, MD
Department of Radiation Oncology
The University of Texas MD Anderson Cancer Center
Houston, Texas

Ahmed Hashmi, MD
Department of Radiation Oncology
Odette Cancer Centre
Sunnybrook Health Sciences Centre
University of Toronto
Toronto, Ontario, Canada

Chris Heyn, MD
Medical Imaging Department
Sunnybrook Health Science Centre
Toronto University
Toronto, Ontario, Canada

Zhibin Huang, PhD
Department of Radiation Oncology
East Carolina University
Greenville, North Carolina

Zain A. Husain, MD
Radiation Oncology Department
Sunnybrook Health Science Centre
Toronto University
Toronto, Ontario, Canada

Seyed Ali Jalalifar, MSc, MD
Department of Electrical Engineering and
    Computer Science
Lassonde School of Engineering York University
Toronto, Ontario, Canada

Einsley-Marie Janowski, MD
Department of Radiation Oncology
University of Virginia
Charlottesville, Virginia

Jonathan P.S. Knisely, MD
Department of Neurological Surgery
Weill Cornell Brain and Spine Center
New York, New York

R. Kosztyla, MD
Department of Electrical Engineering and
    Computer Science
Lassonde School of Engineering York University
Toronto, Ontario, Canada

Robert Koucheki, MD
Interventional Radiology
Sunnybrook Health Sciences Centre
University of Toronto
Toronto, Ontario, Canada

Jeremie Larouche, MD
Orthopedic Division
Surgery Department
Sunnybrook Health Science Centre
Toronto University
Toronto, Ontario, Canada

David Larson, MD
Department of Radiation Oncology
University of California, San Francisco
San Francisco, California

Young Lee, PhD
Department of Medical Physics
Odette Cancer Centre
Sunnybrook Health Sciences Centre
Toronto, Ontario, Canada

Eric J. Lehrer, MD
Department of Radiation Oncology
Icahn School of Medicine at Mount Sinai
New York, New York

Paula Alcaide Leon, MD
University of Toronto
Hospital for Sick Children
Toronto, Ontario, Canada

Nir Lipsman, MD
Department of Radiation Oncology
The University of Texas MD Anderson Cancer Center
Houston, Texas

M. Liu, MD
Department of Electrical Engineering and
    Computer Science
Lassonde School of Engineering York University
Toronto, Ontario, Canada

Simon S. Lo, MD
Department of Radiation Oncology
University of Washington
Seattle, Washington

Melissa LoPresti, MD, MPH
Department of Radiation Oncology
The University of Texas MD Anderson Cancer
    Center
Houston, Texas

Lijun Ma, PhD
Department of Radiation Oncology
University of California, San Francisco
San Francisco, California

Sean S. Mahase, MD
Weill Cornell Brain and Spine Center
Department of Neurological Surgery
New York, New York

Pejman Maralani, MD
Medical Imaging Department
Sunnybrook Health Science Centre
Toronto University
Toronto, Ontario, Canada

Ariel E. Marciscano, MD
Department of Radiation Oncology
    and Molecular Radiation Sciences
Johns Hopkins University
Baltimore, Maryland

Kajsa Mayo, BS, MD
Department of Radiation Oncology
University of Rochester School of Medicine
Rochester, New York

Nina A. Mayr, MD
Department of Radiation Oncology
University of Washington
Seattle, Washington

Mary Frances McAleer, MD, PhD
Department of Radiation Oncology
The University of Texas MD Anderson Cancer
    Center
Houston, Texas

Christopher McGuinness, PhD
Department of Radiation Oncology
University of California, San Francisco
San Francisco, California

John T. McKenna, MD
Weill Cornell Brain and Spine Center
Department of Neurological Surgery
New York, New York

Pejman Jabehdar Maralani, MD, FRCPC
University of Toronto
Toronto, Ontario, Canada

Hatef Mehrabian, MD
University of Toronto
Toronto, Ontario, Canada

Harley Meirovich, MD
Interventional Radiology
Sunnybrook Health Sciences Centre
University of Toronto
Toronto, Ontario, Canada

Ying Meng, MD
Department of Radiation Oncology
The University of Texas MD Anderson Cancer Center
Houston, Texas

Michael T. Milano, MD, PhD
Department of Radiation Oncology
University of Rochester School of Medicine
Rochester, New York

Michael J. Moravan, MD
Department of Radiation Oncology
Duke University
Durham, North Carolina

Hima Bindu Musunuru, MD
Departments of Neurological Surgery and
    Radiation Oncology
University of Pittsburgh Medical Center
Pittsburgh, Pennsylvania

Sten Myrehaug, MD
Department of Radiation Oncology
Princess Margaret Hospital
University of Toronto
Toronto, Ontario, Canada

Mihir Naik, MD
Department of Radiation Oncology
Maroone Cancer Center, Cleveland Clinic Florida
Weston, Florida

Timothy K. Nguyen, MD
London Health Sciences Centre
Western University
London, Ontario, Canada

Alan Nichol, MD
BC Cancer Agency
Vancouver, British Columbia

Kevin Oh, MD
Department of Radiation Oncology
Massachusetts General Hospital
Boston, Massachusetts

Susan C. Pannullo, MD
Department of Neurological Surgery
Weill Cornell Brain and Spine Center
New York, New York

Phillip M. Pifer, MD
Departments of Neurological Surgery and
    Radiation Oncology
University of Pittsburgh Medical Center
Pittsburgh, Pennsylvania

Luke Pike, MD
Department of Radiation Oncology
Memorial Sloan Kettering Cancer Center
New York, New York

Christopher B. Pople, MD
Department of Radiation Oncology
The University of Texas MD Anderson Cancer Center
Houston, Texas

Richard Popple, PhD
Department of Radiation Oncology
University of Alabama at Birmingham
Birmingham, Alabama

Kristin J. Redmond, MD, MPH
Department of Radiation Oncology
    and Molecular Radiation Sciences
Johns Hopkins University
Baltimore, Maryland

Johannes Roesch, PhD
Department of Radiation Oncology
University Hospital Zurich
Zurich, Switzerland

Diana A. Roth O'Brien, MD
Clinical Oncology Department
Sohag University Hospital
Sohag University
Sohag, Egypt

Mark Ruschin, MD
Radiation Oncology Department
Sunnybrook Health Science Centre
Toronto University
Toronto, Ontario, Canada

Ali Sadeghi-Naini, PhD
Department of Electrical Engineering and
    Computer Science
Lassonde School of Engineering
York University
Toronto, Ontario, Canada

David Schlesinger, PhD
Departments of Radiation Oncology and
    Neurological Surgery
University of Virginia Health System
Charlottesville, Virginia

Deepa Sharma, PhD
Department of Radiation Oncology
Sunnybrook Health Sciences Centre
University of Toronto
Toronto, Ontario, Canada

Jason Sheehan, MD, PhD
Department of Neurological Surgery
University of Virginia
Charlottesville, Virginia

Hany Soliman, MD
Department of Radiation Oncology
Odette Cancer Centre
Sunnybrook Health Sciences Centre
University of Toronto
Toronto, Ontario, Canada

Gregory C. Stachelek, MD
Department of Radiation Oncology
    and Molecular Radiation Sciences
Johns Hopkins University
Baltimore, Maryland

John H. Suh, MD
Department of Radiation Oncology
Taussig Cancer Institute
and
Rose Ella Burkhardt Brain Tumor and Neuro-
    Oncology Center
Cleveland Clinic
Cleveland, Ohio

Suganth Suppiah, MD
Department of Radiation Oncology
The University of Texas MD Anderson Cancer
    Center
Houston, Texas

Claudio Tatsui, MD
Department of Radiation Oncology
The University of Texas MD Anderson Cancer Center
Houston, Texas

Bin S. Teh, MD
Department of Radiation Oncology
Houston Methodist Hospital
Weill Cornell Medical College
Houston, Texas

Isabelle Thibault, MD
Departement of Radiation Oncology
Centre Hospitalier Universitaire (CHU) de Québec
Université Laval
Québec, Canada

Jordan A. Torok, MD
Department of Radiation Oncology
Duke University
Durham, North Carolina

Daniel M. Trifiletti, MD
Department of Radiation Oncology
Mayo Clinic
Jacksonville, Florida

Chia-Lin Tseng, MD
Department of Radiation Oncology
Odette Cancer Centre
Sunnybrook Health Sciences Centre
University of Toronto
Toronto, Ontario, Canada

Arjun Sahgal, MD
Department of Radiation Oncology
Odette Cancer Center
Sunnybrook Hospital
Toronto, Ontario, Canada

April K.S. Salama, MD
Department of Radiation Oncology
Duke University
Durham, North Carolina

Joseph K. Salama, MD
Department of Radiation Oncology
Duke University
Durham, North Carolina

Arman Sarfehnia, MD
Radiation Oncology Department
Sunnybrook Health Science Centre
Toronto University
Toronto, Ontario, Canada

Kenneth Y. Usuki, MD
Department of Radiation Oncology
University of Rochester School of Medicine
Rochester, New York

Balamurugan Vellayappan, MD
Department of Radiation Oncology
National University Cancer Institute Singapore
Singapore

Horia Vulpe, MD
Clinical Oncology Department
Sohag University Hospital
Sohag University
Sohag, Egypt

Tony J.C. Wang, MD
Clinical Oncology Department
Sohag University Hospital
Sohag University
Sohag, Egypt

George Weidlich, MD
Zap Surgical Inc.
San Carlos, California

Shun Wong, MD
Division of Radiation Oncology
Tokyo Metropolitan Cancer and Infectious
    Diseases Center
Komagome Hospital
Tokyo, Japan

Victor Yang, MD
Neurosurgery Division
Sunnybrook Health Science Centre
Toronto University
Toronto, Ontario, Canada

William T. Yuh, MD
Department of Radiology
University of Washington
Seattle, Washington

K. Liang Zeng, MD
Radiation Oncology Department
Sunnybrook Health Science Centre
Toronto University
Toronto, Ontario, Canada

1 Tumor Vascular Modulation: Role of Endothelial Cells, Ceramide, and Vascular Targeted Therapies

Deepa Sharma and Gregory J. Czarnota

Contents

1.1 INTRODUCTION

At present, treating extracranial tumors with stereotactic body radiation therapy (SBRT) delivered in a single high dose or a small number of fractions is considered a standard form of treatment. Initially, this type of radiation delivery was only feasible for treating cranial tumors using stereotactic radiosurgery (SRS). Initially, SRS was used to treat arteriovenous malformations (AVMs) which are an abnormality in the brain caused by poorly formed blood vessels. Brain AVMs are known to cause major dysfunction between arteries and veins and often require medical management. Studies suggest that SRS can obliterate 50% to 90% of AVMs depending on its volume, location, and the prescribed radiation dose [1–4]. After the successful implication of SRS for treating AVMs, this technique is now being used for the treatment of brain tumors and metastases.

Recent advancements in radiation therapy with image guidance and treatment planning have made treating cranial and extracranial tumors easier with SRS and SBRT, respectively. Several preclinical and clinical studies have had a high rate of success using these techniques to treat a variety of tumors [5–8]. Classical deoxyribonucleic acid (DNA)-damage response is recognized as one of the well-known effects of radiation therapy [9][10]. However, it has since been recognized that radiation delivered at a high dose in addition initiates a signaling cascade that generates a pro-apoptotic sphingolipid known as ceramide. The biosynthesis of ceramide starts with the hydrolysis of sphingomyelin by acidic sphingomyelinase (ASMase) on the outer leaflet of endothelial cell membranes. The clustering and aggregation of ceramide molecules on the cell membrane stimulate endothelial cell apoptosis. The addition of ASMase and/or ceramide inhibitors halts this entire process. These phenomena have been reported in numerous xenograft models including fibrosarcoma and melanoma transplanted in wild-type and ASMase knockout mice and are now established as predictive of a preclinical response [8]. The role of ceramide endothelial cell apoptosis has been elusive in clinical studies with few studies suggesting ceramide as a marker to distinguish between responding and non-responding patients. A study by Satiskumar *et al.* demonstrated that substantial increases in serum secretory sphingomyelinase (S-SMase) activity and ceramide levels in patients treated with a single high dose of 15 Gy were correlated with good clinical outcomes. Conversely, non-responding patients did not exhibit any increment in serum S-SMase and ceramide [11]. Similarly, patients with liver and lung oligometastases of colorectal cancer origin exhibited significant elevation in plasma ceramide levels subsequently resulting in reduced tumor volume. In contrast, the non-responding patients exhibited a drop in plasma ceramide and an increase in tumor volume [12].

The preclinical and clinical experiences with SRS/SBRT show remarkable outcomes. However, the biological mechanisms leading to these outcomes are not fully understood. High-dose radiotherapy can destroy tumor vasculature as a result of gross endothelial cell apoptosis that leads to additional indirect/secondary tumor cell death [8, 13]. Such indirect tumor cell death can further enhance an anticancer immune response, which is activated by the release of tumor antigens from dying tumor cells [14–17]. Thus, due to the enhanced cell kill and antitumor effects observed following SRS and SBRT, there has been a paradigm shift in standard radiobiological understanding.

1.2 CHALLENGES IN RADIOBIOLOGICAL MODELING FOLLOWING HIGH RADIATION DOSES

In 1975, Rodney Withers introduced four factors that determine the response to fractionated radiotherapy known as the 4R's of radiobiology: repair, repopulation, redistribution, and reoxygenation [18]. The 4R's of radiobiology can link to the success or failure of conventional fractionated radiation therapy. However, the model becomes ineffective when tumors are treated at high doses with SRS or SBRT. **Repair:** At higher doses, the repair of sublethal DNA damage rates reduces due to the saturation of repair mechanisms [19]. Also, a higher radiation dose delivered over a short duration can lead to increased DNA damage which might cause more complicated alterations making it difficult to repair [20]. **Repopulation:** SRS and SBRT treatments are typically given over the course of a week. The repopulation of tumor cells is almost impossible during this short time. **Redistribution:** When cells are exposed to extremely high doses (e.g., 20 Gy), cell cycle progression is interrupted resulting in an immediate cell cycle arrest. This causes the cells to die at the cell cycle phase they were in at the time of irradiation. This is different from when cells are treated with a low dose that causes the cells to preferentially die at G2/M-phase [21]. **Reoxygenation:** A radiation dose higher than 10 Gy per fraction causes severe vascular damage which can elevate the hypoxia level in the intratumoral microenvironment and halt the reoxygenation process for hypoxic cells due to the short overall treatment time of single-dose radiotherapy. With each fraction of conventional fractionated radiation therapy, the death of oxic tumor cells at the tumor periphery allows the reoxygenation of hypoxic cells deeper within the tumor, restoring radiosensitivity, unlike radiation delivered at high doses, which causes both oxic and hypoxic cells to undergo secondary cell death [22]. Thus, the 4R's of radiobiology are generally ineffective at modelling the response to the high doses of radiation given with SRS and SBRT. Another radiobiological model was introduced in 1989 by Fowler known as the linear-quadratic model (LQ model), which estimates the prediction of tumor survival in response to varying radiation doses [23]. It is reported that the LQ model can accurately predict cell kill resulting from DNA damage at conventionally fractionated doses. However, the model may overestimate cell kill at high doses due to the occurrence of both direct and indirect cell deaths [24]. The tumor cell survival curve based on the LQ model depicts a sharp bend in the curve in response to increasing radiation doses because of the additional tumor cell death that appears only at doses higher than 10 Gy. The LQ model is generated largely based on in vitro data that incorporates doses lower than what is used in SRS/SBRT making its utility inappropriate at high doses per fraction [25]. Therefore, while the 4R's and the LQ model of radiobiology can easily be implemented to estimate the response of conventional fractionated radiation therapy, the use of these principles for predicting the outcome of SRS and SBRT remains up for debate.

1.3 RADIOBIOLOGICAL DETERMINANTS OF HIGH-DOSE RADIOTHERAPY

The progression and metastases of cancer depend on the homeostasis of the tumor microenvironment, which comprises different cell populations [26, 27]. It is evident that cancer treatments, including chemotherapy and radiation therapy, target not only the tumor but also other cellular components like vascular endothelial cells and immune cells. The surge in interest in the contribution of endothelial cells to the tumor response began in 1971 when Judah Folkman first recognized that the growth and survival of tumors depend on angiogenesis [28, 29]. The process of angiogenesis relies on the proliferation, migration, and remodeling of endothelial cells [30, 31]. Endothelial cells are known to be the primary target for radiation-induced cell death because they are enriched (20-fold as compared to other cells) in secretory ASMase [32]. An increase in ASMase-induced ceramide generation is mandatory to achieve the endothelial apoptotic effect [8, 13, 33, 34].

A study by Garcia-Barros et al. demonstrated that endothelial cell apoptosis occurs as a primary event and is followed by secondary tumor cell death, which contributes to the overall radiation-induced tumor response. Experiments conducted with mice deficient in ASMase and Bcl-2-associated X protein (Bax), implanted with MCA/129 fibrosarcomas and B16F1 melanoma tumors, resulted in enhanced tumor growth by 200% to 400% compared to their wild-type counterparts. Wild-type (ASMase+/+) mice exhibited a significant increase in endothelial cell death within 1 to 6 hours following a dose of 15–20 Gy, while the tumor cells in the same mice remained intact for 2 to 3 days. The occurrence of tumor cell death days later, subsequently, led to increased tumor growth delay and overall tumor cure by 50% [8]. The mechanism regulating the endothelial cell apoptosis that contributes to the overall tumor response is known to be dependent on the activation of the ASMase–ceramide pathway. Within a few hours of irradiation, the accumulation of ceramide in the endothelial compartment causes its rapid destruction followed by an avalanche of tumor cell death. A study conducted by Santana et al. reported that lymphoblasts from Niemann–Pick patients who are ASMase-deficient abrogated the process of radiation-induced ceramide formation and apoptosis. A retroviral transfer of human ASMase cDNA reversed this phenomenon by inducing ceramide-dependent apoptosis. Furthermore, exposure of fibrosarcoma-bearing wild-type (ASMase+/+) mice to 20 Gy in a single dose demonstrated significant apoptotic cell death in the lung and thymic tissue. In (ASMase–/–) mice, the same radiation dose failed to induce ceramide generation and apoptosis in endothelial cells [34]. Pena and colleagues reached a similar conclusion demonstrating significant endothelial cell apoptosis in a dose- and time-dependent manner following irradiation of the central nervous system (CNS) of C57BL/6 mice with a dose of 5 to 100 Gy. It was found that endothelial cell death accounted for up to 20% of radiation-induced apoptosis in CNS specimens, peaking at 12 hours within a window of 4 to 24 hours after irradiation. Intravenous injection of fibroblast growth factor (FGF) and basic endothelial growth factor (bFGF), before and after giving the dose of 50 Gy, inhibited endothelial cell apoptosis [35]. Thus, these studies suggest that endothelial cell death happening after a high dose is primarily responsible for overall tumor response mediated by the ASMase–ceramide pathway. Some studies contradict these findings and emphasize that tumor cells are responsible for enhanced radiation response.

To determine the role of tumor cells in the radiation response, severe combined immunodeficiency (SCID) mice deficient in the DNA double-strand break repair gene DNA-dependent protein kinase (DNA-PKcs–/–) were inserted with a functional (DNA-PKcs+/+) gene. Exposure to a single dose of 30 Gy or 4 × 5 Gy fractions delivered over 2 days resulted in a substantial tumor growth delay of 1.5-fold in (DNA-PKcs–/–) mice compared to their (DNA-PKcs+/+) counterparts. Thus, the inoculation of the functional DNA repair gene into tumor cells restored radioresistance resulting in a reduced radiation response [36]. Furthermore, Moding and colleagues incorporated a dual recombinase technology to generate primary sarcomas in genetically engineered mouse models with targeted mutations in both endothelial cells and tumor cells. The study showed that primary sarcoma with Bax, a pro-apoptotic gene, and ataxia telangiectasia mutated (Atm), a DNA damage response gene, removed from mouse endothelial cells and tumor cells exhibited different outcomes. The removal of the Bax and Atm gene from mouse endothelial cells did not impact the primary sarcoma response to radiation therapy of 20 Gy. On the contrary, the same genes removed from mouse tumor cells resulted in a significant increase in tumor cell death and growth inhibition of primary sarcoma. Thus, the study revealed that tumor cells, but not endothelial cells, are the prime determinants of radiation response [37]. An interesting observation from Ogawa et al. indicated that tumor cells in nude mice are crucial for determining radiation response, whereas in SCID mice, damage to both tumor cells and endothelial cells governs the radiosensitivity [38].

After a controversial debate of whether or not endothelial cells determine tumor response to radiation therapy, Garcia-Barros *et al.* conducted experiments with SCID mice, a model known to carry a germline mutation in their DNA repair gene [39]. These mice are also 2.5- to 3.0-fold more radiosensitive compared to other mouse models [40–42]. To confirm the engagement of the endothelial component in radiation responses, MCA/129 fibrosarcomas and B16 melanomas grown in SCID (ASMase+/+) and C57BL/6 (ASMase+/+) mice were exposed to high-dose radiotherapy. A single dose of 20 Gy resulted in a significant endothelial cell death in both SCID (ASMase+/+) as well as C57BL/6 (ASMase+/+) mice. The tumor growth delay in SCID (ASMase+/+) mice occurred in a pattern similar to wild-type C57BL/6 (ASMase+/+) mice. Endothelial apoptosis and tumor growth were abrogated in tumors implanted in (ASMase–/–) mice. Thus, the study concluded that the endothelial compartment is solely responsible for enhanced radiation response and that the tumor cells do not impact the radiation-induced endothelial cell apoptosis and overall radiation response [5].

1.4 CELLULAR RESPONSE TO HIGH-DOSE RADIATION THERAPY

1.4.1 DIRECT AND INDIRECT CELL DEATH INDUCED BY HIGH-DOSE RADIATION

For many years, it was believed that cell death induced by ionizing radiation is mainly dependent on DNA damage [9, 10]. However, this belief was changed when an alternative mechanism was provided demonstrating single high doses of radiotherapy inducing plasma membrane alteration that can lead to the activation of the sphingomyelin pathway followed by ceramide generation [43, 44]. Ceramide, once formed, can serve as a second messenger, triggering various apoptotic signaling pathways. A study by Haimovitz-Friedman *et al.* confirmed the involvement of ceramide in the apoptotic response using bovine aortic endothelial cells (BAEC). The ceramide level reached its maximum within few minutes of radiation exposure (10 Gy in a single dose) in whole-cell lysates as well as in nuclei-free membranes prepared from BAEC. Thus, this study confirmed that radiation-induced apoptosis can be independent of DNA damage [33]. It is now evident that there is more than one pathway of radiation-induced cell death (Figure 1.1). Generally, a low dose of radiation induces cytotoxic effects on DNA eliciting DNA double-strand breaks, which cause direct tumor cell death [45–47]. High-dose radiation, on the other hand, can kill tumor cells directly, by causing DNA damage, or indirectly in p53-dependent manner or by causing massive tumor vasculature collapse through endothelial cell damage [8, 48, 49]. High-dose-induced vascular dysfunction can further cause tumor cell death by evoking tumor hypoxia and an immune response.

The occurrence of tumor cell death as a result of significant vascular endothelium damage following high-dose radiotherapy was first reported by Garcia-Barros and colleagues. Their results demonstrated that ASMase-deficient mice abrogated apoptosis of endothelial cells while the wild-type phenotype exhibited significant endothelial and tumor cell death. This confirmed that ASMase–ceramide activation is crucial for radiation-induced vascular endothelial damage [8]. Several other studies have also reported the involvement of ceramide-induced endothelial cell apoptosis in regulating the overall tumor response. A large body of work by Czarnota *et al.* has indicated that pre-treatment with ultrasound-stimulated microbubbles (USMB) before administering a radiation dose of 8 Gy can cause a significant elevation of ceramide leading to massive vascular endothelial cell death [50]. El Kaffas *et al.* investigated the dose-dependent effect of radiation in combination with USMB using MCA/129 fibrosarcoma-bearing wild-type (ASMase+/+) mice, knockout (ASMase–/–) mice, and wild-type mice treated with sphingosine-1-phosphate (S1P), a ceramide antagonist. In (ASMase+/+) mice, a combination of USMB and dose of 8 Gy resulted in the highest level of cell death of 8.7% at 3 hours, 53.2% at 24 hours, and 37.8% at 72 hours compared to radiation (8 Gy) only, which resulted in 10.0%, 17.3%, and 15.4% cell death at 3, 24, and 72 hours, respectively. Furthermore, USMB combined with radiation treatment resulted in a 40% attenuation of tumor blood flow within 24 hours, which persisted to 72 hours. The shutdown of the vasculature at 24 hours was reported to be accountable for endothelial cell death. The study further indicated that the ceramide level in (ASMase+/+) mice escalated within 24 hours of administering treatment with USMB and 8 Gy of radiation treatment, which confirmed the involvement of endothelial ASMase–ceramide activation leading to overall tumor vascular disruption [51].

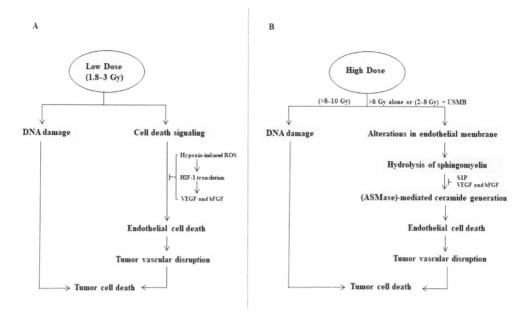

Figure 1.1 Endothelial and tumor cells' response to low- and high-dose radiotherapy.

Numerous other studies have also indicated a decrease in tumor perfusion concomitant with endothelial cell death. Irradiation of neuroblastoma xenografts following radiation (12 Gy) reduced the tumor blood volume by 63%, subsequently causing endothelial cell damage [52]. Similarly, rats bearing orthotopic human brain tumors exposed to a single dose of 20 Gy exhibited significant apoptosis and an 80% decrease in tumor blood flow within 2 hours of irradiation, suggesting that changes in vascularity correspond to endothelial damage [53].

Prior studies by Song and colleagues performed with FSaII fibrosarcoma tumors of mice demonstrated a severe decline in blood perfusion accompanied by elevated hypoxia within 1 to 5 days following a single dose of 20 Gy. The delayed secondary tumor cell death reported in this study was a ramification of extensive tumor vascular occlusion and increased intratumoral hypoxia [54]. Radiation-induced vascular damage is also known to cause hypoxic cell death. Upon exposure to a high dose, a fraction of hypoxic cells that survive the direct and indirect cell death later become devoid of nutrients, ultimately resulting in death. Also, vessels appearing nonfunctional due to reduced nutrients and oxygen resulting from massive radiation-induced tumor cell death might contribute to hypoxic cell death. It was reported that mice bearing FSaII fibrosarcoma exposed to a single dose of 20–30 Gy exhibited a decrease in cell survival by 3–4 logs, whereas a dose high up to 90 Gy was essential for the reduction of cell survival by 8 logs. The progressive cell survival loss that occurred after irradiation was caused by nutrient deprivation [55, 56]. Similar observations were reported by Hill and colleagues in a mouse KHT sarcoma model. Irradiation of tumors with a single dose of 20 Gy caused the death of hypoxic cells by a factor of 3 to 4 [57]. These data confirm the presence of hypoxic cell death in addition to tumor cell death as a result of vascular damage after high-dose radiotherapy.

In addition to the direct and indirect cell death effects, SRS and SBRT are known to trigger an antitumor immune response. Radiation-induced tumor cell death elicits the release of tumor antigens causing immunogenic cell death (ICD) [58]. Dying irradiated tumor cells release high mobility group box 1 (HMGB1) protein that interacts and activates toll-like receptor (TLR)-4 on the dendritic cells inducing an antitumor response [59]. Recently, a mathematical framework based on murine breast experimental data suggested that a radiation dose between 10 and 13 Gy per fraction is required to induce antitumor immunity [60]. Taken together, these studies suggest that SRS and SBRT are likely to kill more cancer cells as compared to conventional fractionated radiotherapy. The direct cell death in response to DNA damage and the indirect secondary tumor cell death mediated by vascular dysfunction account for the majority of cell killing. In addition, massive hypoxic cell death due to the deterioration of the intratumor microenvironment (assuming 20% of the tumor cells are hypoxic in solid tumors) combined with the tumor cell kill caused by an enhanced antitumor immune response further contribute to the cell death following SRS and SBRT.

1.4.2 RADIATION-INDUCED VASCULAR CHANGES

Our understanding of tumor vasculature has evolved over the past several years. Unlike normal vasculature, which is arranged hierarchically with evenly distributed arteries, veins, and capillaries, tumor vasculature remains highly disorganized, irregular, chaotic, dilated, leaky, and tortuous with no ability to differentiate between arterials and venules [61]. The endothelial layers in normal blood vessels are regularly shaped and are fully supported by organized pericytes that act as a basement membrane. On the contrary, the structure of tumor blood vessels is constructed with an inner single layer of endothelial cells that are poorly connected with uneven support of abnormal pericytes [62]. Due to these morphological abnormalities, tumor blood vessels are highly vulnerable to ionizing radiation. Several studies have reported drastic vascular changes following high-dose radiation therapy. Along with changes in structural integrity, fluctuations in tumor blood flow and tumor oxygenation are probably the most notable changes reported following a radiation dose higher than 8 to 10 Gy. Radiation-induced vascular changes in multiple tumor types have been reviewed in detail by Park *et al.* Collective data from human studies suggest a general trend of a slight increase in tumor blood flow, or no change in blood flow in some cases, observed at the beginning of a fractionated radiotherapy followed by a reduction toward the end of the course of treatment [63]. For preclinical animal models, exposure to conventionally fractionated radiotherapy of dose of 1.5 to 2.0 Gy causes no changes in vasculature in the early period of radiation with a slight vascular dysfunction reported at a later phase of irradiation [64, 65]. However, a single dose of 5 to 10 Gy causes moderate vascular damage and [64–66] increasing the dose to more than 10 Gy/fraction leads to extreme tumor vasculature deterioration [54].

Radiation-induced tumor vascular effects are known to be widely dependent on radiation dose, duration between the doses, tumor type, stages, and the site of the tumor. An extensive body of work has previously been reported by Song and colleagues regarding the vascular changes in irradiated tumors using Walker 256 carcinomas grown in the hind leg of rats. Tumors exposed to a single dose of 2, 5, 10, 30, and 60 Gy were monitored several days after irradiation (intravascular volume and extravasation rates of plasma protein/vascular permeability were measured). Radiation doses of 2 and 5 Gy did not induce much of a vascular response. However, a single dose of 10, 30, or 60 Gy resulted in a significant abolishment of vascular volume after the second, sixth, and twelfth days of irradiation [64]. Several other studies have also reported tumor blood flow reduction upon high-dose delivery. They suggest that a dose higher than 10 Gy in a single fraction is more effective in causing vascular damage than the same dose given in fractions [67]. Tumors treated with a fractionated dose initiate a transient increase followed by a rapid fall in vascular volume as the number of fractions increases. This is contrary to what is observed with a single high dose which results in a sharp decline in vascular volume throughout the tumor.

Radiation-induced vascular changes are known to greatly influence tumor oxygenation, but there are conflicting reports on whether the vascular damage contributes to changes in oxygen tension or not. Tumor oxygenation monitored in rats bearing rhabdomyosarcomas following a fractionated radiotherapy treatment of 60 Gy administered in 20 fractions over a period of 4 weeks showed no significant changes in partial pressure of oxygen (pO_2) measurements until week 3. However, at week 4, a significant decrease in tumor pO_2 was reported, which was attributed to be due to the obstruction of tumor capillaries [68, 69]. On the other hand, exposing A-07 human melanoma xenografts to a single dose of 10 Gy resulted in a 40% drop in blood perfusion and 25% increase in extracellular volume fraction within 72 hours. In this case, the tumor pO_2 content remained unchanged suggesting no correlation between vascular changes and oxygenation [70]. Heterogeneity in tumor vascular perfusion and oxygenation has frequently been reported. An observed phenomenon of reduced blood volume at the center of the tumor compared to the rim has been seen in many tumor types [71]. Patients with advanced non-small-cell lung cancer administered a dose of

27 Gy in 2, 4, and 6 fractions resulted in an increased vascular blood volume at the tumor rim by 31.6%, 49.3%, and 44.6%, respectively. The blood volume in the tumor center remained 16.4%, 19.9%, and 4.0% with the same fractions of radiotherapy [72]. An important observation by Mottram suggests that tumor cells at the periphery are more radiosensitive compared to the cells in the center of the tumor. Cells localized at the periphery are closer to nearby blood vessels and, therefore, have abundant oxygen supply [73]. In 1955, Thomlinson and Gray performed a detailed quantitative examination of carcinoma of the bronchus to study the radiosensitivity of marginal and central cells within a tumor. They found that the cells close to capillaries acquired sufficient amounts of nutrients/oxygen and remained proliferative, while the cells located at a distance greater than about 100 µm from the capillaries remained non-viable. The lower oxygen content in the central region during the time of irradiation made the cells more radioresistant compared to the peripheral cells that were well-oxygenated [74, 75].

1.5 COMBINING RADIOTHERAPY AND ANTI-ANGIOGENESIS STRATEGIES

In most solid tumors, the vasculature remains highly heterogeneous. Tumor angiogenesis keeps the tumor alive by supplying essential nutrients and oxygen [29]. The balance of several angiogenic regulators is required for the growth and metastases of tumors [76]. Targeting tumor vasculature for the treatment of solid tumors has shown exceptional success over several years. Currently, various vascular targeting agents, such as anti-angiogenic agents and vascular disrupting agents, are being extensively investigated in pre-clinical and clinical studies [77, 78]. Several clinical trials are currently being conducted to treat cancer in humans using angiogenesis inhibitors. Some of the agents approved for clinical trials are listed in Table 1.1. Anti-angiogenic agents are known to inhibit and, in some cases, completely stop the growth of new blood vessels [78], whereas vascular disrupting agents are designed to selectively decrease or shutdown the tumor blood flow [77, 79, 80].

Tumors promote the growth of new blood vessels by secreting numerous angiogenic growth factors such as basic fibroblast growth factor (bFGF) and vascular endothelial growth factor (VEGF). Endothelial cells in pre-existing vessels express several receptors to which this angiogenic growth factor binds and initiates various signaling pathways [81, 82]. The use of angiogenesis inhibitors/agents blocks the formation of new blood vessels, preventing the growth and progression of the tumor by alleviating ASMase-generated ceramide [83]. Work by Truman et al. indicated that radiation-induced ceramide acts as a rheostat for the survival and death of endothelial cells and that the timing of anti-angiogenic treatment is crucial for sensitizing tumors. The study showed bFGF and VEGF inhibited radiation-induced ASMase–ceramide activation, and apoptosis was reversed by the addition of endogenous C16 ceramide in MCA/129 fibrosarcoma. Pre-treatment of tumors with DC101, an angiogenesis inhibitor, 1 hour prior to radiotherapy with 13.5 Gy led to enhanced ASMase-generated ceramide, subsequently causing endothelial cell apoptosis. However, DC101 injected 1 hour after radiation remained ineffective [84]. Similar observations were reported by Rao et al. indicating that VEGF inhibitor axitinib administered 1 hour prior to radiation therapy increased tumor radiosensitivity. A dose of 27 to 40 Gy in a single exposure combined with axitinib caused endothelial cell death both in vitro in primary cultured cells and in vivo in mice bearing MCA/129 sarcoma or B16F1 melanoma. A growth delay of the tumor and complete response rate by 40% was also observed in these mice followed by a combination of radiation and axitinib [85]. Tumor response following treatment with a combination of angiogenic inhibitors/agents and radiation has also been reported in several studies; however, the exact mechanism of interaction between angiogenic inhibitors and radiotherapy is still unknown. An anti-angiogenic/vascular targeting agent combined with radiation is expected to enhance tumor response by reducing tumor blood perfusion and oxygenation. However, studies indicate the occurrence of increased tumor blood flow and oxygen concentration following a combination of both [86–89]. Breast (MDA-MB-231) xenografts, when exposed to a single dose of 8 and 16 Gy combined with sunitinib, a VEGF inhibitor, demonstrated significant cell death and a subsequent increase in tumor blood flow by 50% [89]. It was attributed that a single high dose of radiation can cause damage to abnormal blood vessels, while the addition of sunitinib may allow vessel normalization causing increased oxygenation.

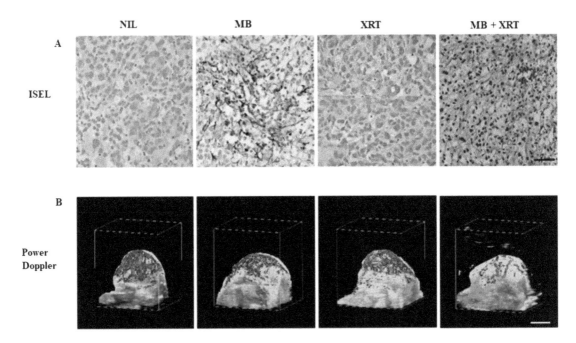

Figure 1.2 Histochemical staining of PC3 tumor xenograft endothelial cells with ISEL and power Doppler ultrasound images of PC3 tumor xenografts reveal response to treatment at 24 hours.

In recent years, USMB therapy has proven to be a novel form of targeted anti-angiogenic therapy. Microbubbles are small gas-filled bubbles ranging in size from 1 to 4 μm and are widely used as an ultrasound contrast agent due to their excellent acoustic response [90]. Upon contact with ultrasound waves, microbubbles can oscillate, expand, and collapse contributing to overall changes in the surrounding tissue environment. Disruption of the bubbles upon exposure to ultrasound acoustic pressure can cause a severe vascular disruption enhancing tumor response [91]. A large body of work by Czarnota and colleagues demonstrates that a combination of radiotherapy and USMB causes significant endothelial cell death followed by microvascular deterioration. A study performed on prostate tumor xenografts (PC3) treated with a radiation dose of either 2 or 8 Gy combined with a low or high dose of USMB treatment showed significant cell death of 44 ± 13% (mean ± standard error) with 2 Gy + USMB and 70 ± 8% with 8 Gy + USMB. Vascular disruption detected using power Doppler ultrasound indicated a decrease in tumor blood flow of 18 ± 22% (mean ± standard error) with radiation alone, 20 ± 37% with USMB alone, and 65 ± 8% with a combination of 8 Gy and USMB (Figure 1.2). Furthermore, the group receiving the combined treatment exhibited significant tumor growth delay and fewer proliferating cells [50]. Data from other mouse models bearing breast, bladder, and fibrosarcoma tumors have also revealed a similar effect using these combination therapies [51, 92–94]. The endothelial cell death-induced vascular dysfunction observed with a combination of radiation and USMB is found to be ceramide-dependent. USMB is known to cause a mechanical perturbation in the endothelial cell membrane leading to enhanced ceramide generation followed by vascular destruction. A study by Kim *et al.* reported 14-fold higher ceramide content in PC3 xenografts following treatment using a combination of USMB and radiation (8 Gy). The increased ceramide level was linked to enhanced tumor cell death and vascular damage [95]. Subsequently, Al-Mahrouki and colleagues extensively investigated the genetic pathway involved in the regulation of ceramide-mediated tumor vascular disruption following the administration of USMB and radiation therapy. In particular, they studied the role of UDP glycosyltransferase 8 (UGT8) in tumor response enhancement. UGT8 is a key enzyme that catalyzes the transfer of galactose to ceramide. Experiments were conducted with genetically modified PC3 cells and tumor xenografts generated from stably transfected PC3 cells with a downregulated UGT8 gene. A combination of USMB and dose of 8 Gy caused greater cell damage in the downregulated UGT8 tumor

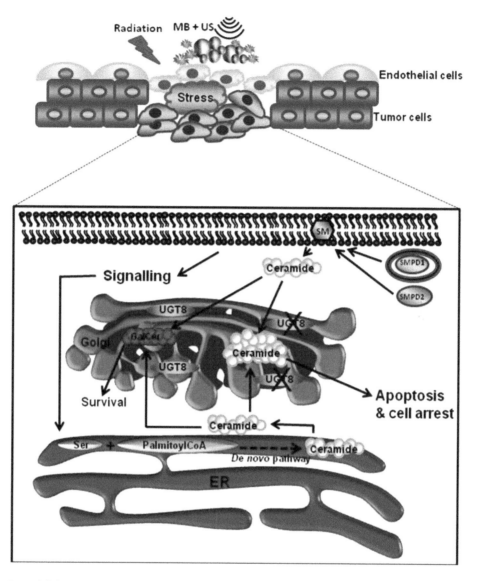

Figure 1.3 Model depicting UGT8 signaling and its role in ceramide biosynthesis.

model as compared to control tumors. In addition, they reported a significant decrease in the level of tumor blood flow and oxygen saturation in the UGT8 downregulated model. An increase in tumor response was found to be concomitant with a greater amount of ceramide accumulated due to the downregulation of the UGT8 gene (Figure 1.3) [96]. Thus, targeting UGT8 combined with vascular disrupting therapy might be a good starting point for the further exploration and optimization of this new treatment strategy for cancer.

1.6 CONCLUSION

SRS/SBRT is increasingly being recognized as one of the essential treatment options for cancer. A high radiation dose delivered in a single fraction or in a small number of fractions affects the tumor vasculature by causing ceramide-mediated endothelial apoptosis leading to indirect/secondary tumor cell death. The indirect tumor cell death further evokes an immune response resulting in an overall enhancement in radiation response. Ceramide generation by the activation of the ASMase pathway is a central determinant of

Table 1.1 **Angiogenesis inhibitors/agents undergoing clinical trials for treating human cancers**

ANGIOGENESIS INHIBITORS/AGENTS	BRAND NAME
Axitinib	Inlyta
Bevacizumab	Avastin
Cabozantinib	Cometriq
Everolimus	Afinitor
Lenalidomide	Revlimid
Lenvatinib mesylate	Lenvima
Pazopanib	Votrient
Ramucirumab	Cyramza
Regorafenib	Stivarga
Sorafenib	Nexavar
Sunitinib	Sutent
Thalidomide	Synovir, Thalomid
Vandetanib	Caprelsa
Ziv-aflibercept	Zaltrap

radiation-induced vascular endothelial cell damage. By upregulating ASMase-released ceramide using various angiogenesis inhibitors and/or anti-angiogenic therapy (USMB), tumor radiosensitivity can be restored.

Damage to endothelial cells and tumor cells appears to be instigated by both low dose (1.8–3 Gy) and single high dose (>8 Gy) alone or combined with USMB (2–8 Gy + USMB). (A) With each low-dose fraction, hypoxia-mediated ROS results in HIF-1 translation making the cells radioresistant. Inhibition of HIF-1 leads to massive endothelial cell death, microvascular damage, and increased tumor cell death. (B) High-dose-induced tumor cell death is mediated via rapid translocation of lysosomal ASMase to the extracellular leaflet of endothelial cell membranes resulting in ceramide generation. The accumulation of ceramide in endothelial cells causes its disruption followed by vascular collapse and tumor cell death. The addition of S1P, VEGF, and bFGF can halt this entire process. *Abbreviations*: ASMase, acid sphingomyelinase; bFGF, basic fibroblast growth factor; DNA, deoxyribonucleic acid; HIF-1, hypoxia-inducible factor 1; ROS, reactive oxygen species; S1P, sphingosine-1-phosphate; USMB, ultrasound-stimulated microbubbles; VEGF, vascular endothelial growth factor.

(A) Tumors treated with a combination of USMB and radiation (8 Gy dose) exhibited increased cell death confirmed with ISEL staining. ISEL-positive cells can be identified by a dark-stained nucleus. The scale bar represents 60 microns. (B) A significant drop in the blood flow signal was observed following a combination of USMB and radiation compared to control groups with no treatment or treatments including USMB only and radiation only. The scale bar represents 2 mm. Adapted from [50]. ISEL= *in situ* end-labeling; MB= microbubble; NIL= no microbubble; XRT= radiation.

The *de novo* biosynthesis of ceramide takes place in the endoplasmic reticulum. Elevated expression of UGT8 converts ceramide to galactosylceramide resulting in degradation of ceramide and inhibition in the apoptotic signaling pathway. Conversely, UGT8, when underexpressed, leads to elevated ceramide levels initiating a cell death signaling pathway [96]. ER, endoplasmic reticulum; GalCer, galactosylceramide; MB + US, microbubble + ultrasound; S1P, sphingosine-1-phosphate; Ser, serine; SM, sphingomyelin; SMPD1, sphingomyelin phosphodiesterase 1; SMPD2, sphingomyelin phosphodiesterase 2; UGT8, UDP glycosyltransferase 8.

REFERENCES

1. Ding D, Starke RM, Kano H, Mathieu D, Huang P, Kondziolka D, et al. Radiosurgery for erebral arteriovenous malformations in a randomized trial of unruptured brain arteriovenous malformations (ARUBA)-eligible patients: A multicenter study. *Stroke*. 2016;47.
2. Andrade-Souza YM, Ramani M, Scora D, Tsao MN, TerBrugge K, Schwartz ML. Radiosurgical treatment for rolandic arteriovenous malformations. *J. Neurosurg*. 2006.
3. Hanakita S, Koga T, Shin M, Igaki H, Saito N. Application of single-stage stereotactic radiosurgery for cerebral arteriovenous malformations >10 cm^3. *Stroke*. 2014.
4. Rubin BA, Brunswick A, Riina H, Kondziolka D. Advances in radiosurgery for arteriovenous malformations of the brain. *Neurosurgery*. 2014;74.
5. García-Barros M, Thin TH, Maj J, Cordon-Cardo C, Haimovitz-Friedman A, Fuks Z, et al. Impact of stromal sensitivity on radiation response of tumors implanted in SCID hosts revisited. *Cancer Res*. 2010.
6. Baumann R, Chan MKH, Pyschny F, Stera S, Malzkuhn B, Wurster S, et al. Clinical results of mean GTV dose optimized robotic-guided stereotactic body radiation therapy for lung tumors. *Front Oncol*. 2018;8.
7. Berkovic P, Gulyban A, Nguyen PV, Dechambre D, Martinive P, Jansen N, et al. Stereotactic robotic body radiotherapy for patients with unresectable hepatic oligorecurrence. *Clin Colorectal Cancer*. 2017;16.
8. Garcia-Barros M, Paris F, Cordon-Cardo C, Lyden D, Rafii S, Haimovitz-Friedman A, et al. Tumor response to radiotherapy regulated by endothelial cell apoptosis. *Science*. 2003;80.
9. Radford IR. The level of induced DNA double-strand breakage correlates with cell killing after x-irradiation. *Int J Radiat Biol*. 1985.
10. Ward JF. DNA damage produced by ionizing radiation in mammalian cells: Identities, mechanisms of formation, and reparability. *Prog Nucleic Acid Res Mol Biol*. 1988.
11. Sathishkumar S, Boyanovsky B, Karakashian AA, Rozenova K, Giltiay NV, Kudrimoti M, et al. Elevated sphingomyelinase activity and ceramide concentration in serum of patients undergoing high dose spatially fractionated radiation treatment. Implications for endothelial apoptosis. *Cancer Biol Ther*. 2005.
12. Dubois N, Rio E, Ripoche N, Ferchaud-Roucher V, Gaugler MH, Campion L, et al. Plasma ceramide, a real-time predictive marker of pulmonary and hepatic metastases response to stereotactic body radiation therapy combined with irinotecan. *Radiother Oncol*. 2016.
13. Paris F, Fuks Z, Kang A, Capodieci P, Juan G, Ehleiter D, et al. Endothelial apoptosis as the primary lesion initiating intestinal radiation damage in mice. *Science*. 2001;80.
14. Garnett CT, Palena C, Chakarborty M, Tsang KY, Schlom J, Hodge JW. Sublethal irradiation of human tumor cells modulates phenotype resulting in enhanced killing by cytotoxic T lymphocytes. *Cancer Res*. 2004;64.
15. Reits EA, Hodge JW, Herberts CA, Groothuis TA, Chakraborty M, Wansley EK, et al. Radiation modulates the peptide repertoire, enhances MHC class I expression, and induces successful antitumor immunotherapy. *J Exp Med*. 2006;203.
16. Verbrugge I, Gasparini A, Haynes NM, Hagekyriakou J, Galli M, Stewart TJ, et al. The curative outcome of radioimmunotherapy in a mouse breast cancer model relies on mTOR signaling. *Radiat Res*. 2014;182.
17. Sharabi AB, Nirschl CJ, Kochel CM, Nirschl TR, Francica BJ, Velarde E, et al. Stereotactic radiation therapy augments antigen-specific PD-1-mediated antitumor immune responses via cross-presentation of tumor antigen. *Cancer Immunol Res*. 2015;3.
18. Withers HR. The four R's of radiotherapy. 1975.
19. Dikomey E. The influence of the size of dose on the repair kinetics of X-ray-induced DNA strand breaks studied in CHO cells. 1993.
20. Moding EJ, Mowery YM, Kirsch DG. Opportunities for radiosensitization in the SBRT era. *Cancer J*. 2016;22.
21. Syljuåsen RG, McBride WH, Syljuasen RG. Radiation-induced apoptosis and cell cycle progression in jurkat T cells. *Radiat Res*. 1999;152.
22. Song CW, Lee YJ, Griffin RJ, Park I, Koonce NA, Hui S, et al. Indirect tumor cell death after high-dose hypofractionated irradiation: Implications for stereotactic body radiation therapy and stereotactic radiation surgery. *Int J Radiat Oncol Biol Phys*. 2015.
23. Fowler JF. The linear-quadratic formula and progress in fractionated radiotherapy. *Br. J. Radiol*. 1989.
24. McMahon SJ. The linear quadratic model: Usage, interpretation and challenges. *Phys Med Biol*. 2019;64.
25. Kirkpatrick JP, Meyer JJ, Marks LB. The linear-quadratic model is inappropriate to model high dose per fraction effects in radiosurgery. *Semin Radiat Oncol*. 2008;18.
26. Balkwill FR, Capasso M, Hagemann T. The tumor microenvironment at a glance. *J Cell Sci*. 2012;125.
27. Whiteside TL. The tumor microenvironment and its role in promoting tumor growth. *Oncogene*. 2008.
28. Sherwood LM, Parris EE, Folkman J. Tumor angiogenesis: Therapeutic implications. *N. Engl. J. Med*. 1971. pp. 1182–1186.
29. Folkman J. Tumor angiogenesis factor. *Cancer Res*. 1974.

30. Lamalice L, Le Boeuf F, Huot J. Endothelial cell migration during angiogenesis. *Circ. Res.* 2007.
31. Feng T, Yu H, Xia Q, Ma Y, Yin H, Shen Y, et al. Cross-talk mechanism between endothelial cells and hepatocellular carcinoma cells via growth factors and integrin pathway promotes tumor angiogenesis and cell migration. *Oncotarget.* 2017;8.
32. Tabas I. Secretory sphingomyelinase. *Chem Phys Lipids.* 1999.
33. Haimovitz-Friedman A, Kan CC, Ehleiter D, Persaud RS, Mc loughlin M, Fuks Z, et al. Ionizing radiation acts on cellular membranes to generate ceramide and initiate apoptosis. *J Exp Med.* 1994.
34. Santana P, Peña LA, Haimovitz-Friedman A, Martin S, Green D, McLoughlin M, et al. Acid sphingomyelinase-deficient human lymphoblasts and mice are defective in radiation-induced apoptosis. *Cell.* 1996.
35. Peña LA, Fuks Z, Kolesnick RN. Radiation-induced apoptosis of endothelial cells in the murine central nervous system: Protection by fibroblast growth factor and sphingomyelinase deficiency. *Cancer Res.* 2000.
36. Gerweck LE, Vijayappa S, Kurimasa A, Ogawa K, Chen DJ. Tumor cell radiosensitivity is a major determinant of tumor response to radiation. *Cancer Res.* 2006.
37. Moding EJ, Castle KD, Perez BA, Oh P, Min HD, Norris H, et al. Tumor cells, but not endothelial cells, mediate eradication of primary sarcomas by stereotactic body radiation therapy. *Sci Transl Med.* 2015.
38. Ogawa K, Boucher Y, Kashiwagi S, Fukumura D, Chen D, Gerweck LE. Influence of tumor cell and stroma sensitivity on tumor response to radiation. *Cancer Res.* 2007.
39. Fulop GM, Phillips RA. The SCID mutation in mice causes a general defect in DNA repair. *Nature.* 1990.
40. Biedermann KA, Sun J, Giaccia AJ, Tosto LM, Brown JM. SCID mutation in mice confers hypersensitivity to ionizing radiation and a deficiency in DNA double-strand break repair. *Proc Natl Acad Sci U S A.* 1991.
41. Budach W, Hartford A, Gioioso D, Freeman J, Taghian A, Suit HD, et al. Tumors arising in SCID mice share enhanced radiation sensitivity of SCID normal tissues. *Cancer Res.* 1992.
42. Chang C, Biedermann KA, Mezzina M, Brown JM. Characterization of the DNA double strand break repair defect in SCID mice. *Cancer Res.* 1993.
43. Peña LA, Fuks Z, Kolesnick R. Stress-induced apoptosis and the sphingomyelin pathway. *Biochem. Pharmacol.* 1997.
44. Fuks Z, Haimovitz-Friedman A, Kolesnick RN. The role of the sphingomyelin pathway and protein kinase C in radiation-induced cell kill. *Important Adv. Oncol.* 1995.
45. Lomax ME, Folkes LK, O'Neill P. Biological consequences of radiation-induced DNA damage: Relevance to radiotherapy. *Clin Oncol.* 2013;25.
46. Mladenov E, Magin S, Soni A, Iliakis G. DNA double-strand break repair as determinant of cellular radiosensitivity to killing and target in radiation therapy. *Front. Oncol.* 2013.
47. Cannan WJ, Pederson DS. Mechanisms and consequences of double-strand DNA break formation in chromatin. *J. Cell Physiol.* 2016.
48. Zhao H, Zhuang Y, Li R, Liu Y, Mei Z, He Z, et al. Effects of different doses of X-ray irradiation on cell apoptosis, cell cycle, DNA damage repair and glycolysis in HeLa cells. *Oncol Lett.* 2019;17.
49. Lowe SW, Schmitt EM, Smith SW, Osborne BA, Jacks T. P53 is required for radiation-induced apoptosis in mouse thymocytes. *Nature.* 1993.
50. Czarnota GJ, Karshafian R, Burns PN, Wong S, Al Mahrouki A, Lee JW, et al. Tumor radiation response enhancement by acoustical stimulation of the vasculature. *Proc Natl Acad Sci U S A.* 2012.
51. El Kaffas A, Al-Mahrouki A, Hashim A, Law N, Giles A, Czarnota GJ. Role of acid sphingomyelinase and ceramide in mechano-acoustic enhancement of tumor radiation responses. *J Natl Cancer Inst.* 2018.
52. Jani A, Shaikh F, Barton S, Willis C, Banerjee D, Mitchell J, et al. High-dose, single-fraction irradiation rapidly reduces tumor vasculature and perfusion in a xenograft model of neuroblastoma. *Int J Radiat Oncol.* 2016.
53. Brown SL, Nagaraja TN, Aryal MP, Panda S, Cabral G, Keenan KA, et al. MRI-tracked tumor vascular changes in the hours after single-fraction irradiation. *Radiat Res.* 2015.
54. Song CW, Lee YJ, Griffin RJ, Park I, Koonce NA, Hui S, et al. Indirect tumor cell death after high-dose hypofractionated irradiation: Implications for stereotactic body radiation therapy and stereotactic radiation surgery. *Int J Radiat Oncol Biol Phys.* 2015.
55. Clement JJ, Tanaka N, Song CW. Tumor reoxygenation and postirradiation vascular changes. *Radiology.* 1978;127.
56. Kim MS, Kim W, Park IH, Kim HJ, Lee E, Jung JH, et al. Radiobiological mechanisms of stereotactic body radiation therapy and stereotactic radiation surgery. *Radiat Oncol J.* 2015;33.
57. Hill RP. Radiation-induced changes in the in vivo growth rate of KHT sarcoma cells: Implications for the comparison of growth delay and cell survival. *Radiat Res.* 1980.
58. Golden EB, Apetoh L. Radiotherapy and immunogenic cell death. *Semin. Radiat. Oncol.* 2015.
59. Apetoh L, Ghiringhelli F, Tesniere A, Criollo A, Ortiz C, Lidereau R, et al. The interaction between HMGB1 and TLR4 dictates the outcome of anticancer chemotherapy and radiotherapy. *Immunol. Rev.* 2007.
60. Poleszczuk J, Enderling H. The optimal radiation dose to induce robust systemic anti-tumor immunity. *Int J Mol Sci.* 2018.

61. Siemann DW. The unique characteristics of tumor vasculature and preclinical evidence for its selective disruption by tumor-vascular disrupting agents. *Cancer Treat. Rev.* 2011.

62. Bergers G, Song S. The role of pericytes in blood-vessel formation and maintenance. *Neuro Oncol.* 2005;7.

63. Park HJ, Griffin RJ, Hui S, Levitt SH, Song CW. Radiation-induced vascular damage in tumors: Implications of vascular damage in ablative hypofractionated radiotherapy (SBRT and SRS). *Radiat Res.* 2012;177.

64. Song CW, Levitt SH. Vascular changes in Walker 256 carcinoma of rats following X irradiation. *Radiology.* 1971;100.

65. Wong HH, Song CW, Levitt SH. Early changes in the functional vasculature of Walker carcinoma 256 following irradiation. *Radiology.* 1973;108.

66. Song CW, Payne JT, Levitt SH. Vascularity and blood flow in x-irradiated Walker carcinoma 256 of rats. *Radiology.* 1972.

67. Clement JJ, Tanaka N, Song CW. Tumor reoxygenation and postirradiation vascular changes. *Radiology.* 1978.

68. Zywietz F, Hahn LS, Lierse W. Ultrastructural studies on tumor capillaries of a rat rhabdomyosarcoma during fractionated radiotherapy. *Acta Anat (Basel).* 1994;150.

69. Zywietz F, Reeker W, Kochs E. Tumor oxygenation in a transplanted rat rhabdomyosarcoma during fractionated irradiation. *Int J Radiat Oncol Biol Phys.* 1995.

70. Brurberg KG, Thuen M, Ruud E-BM, Rofstad EK. Fluctuations in pO_2 in irradiated human melanoma xenografts. *Radiat Res.* 2006.

71. Forster J, Harriss-Phillips W, Douglass M, Bezak E. A review of the development of tumor vasculature and its effects on the tumor microenvironment. *Hypoxia.* 2017;5.

72. Ng QS, Goh V, Milner J, Padhani AR, Saunders MI, Hoskin PJ. Acute tumor vascular effects following fractionated radiotherapy in human lung cancer: In vivo whole tumor assessment using volumetric perfusion computed tomography. *Int J Radiat Oncol Biol Phys.* 2007.

73. Mottram JC. A factor of importance in the radio sensitivity of tumours. *Br J Radiol.* 1936.

74. Gray LH, Conger AD, Ebert M, Hornsey S, Scott OC. The concentration of oxygen dissolved in tissues at the time of irradiation as a factor in radiotherapy. *Br J Radiol.* 1953;26.

75. Thomlinson RH, Gray LH. The histological structure of some human lung cancers and the possible implications for radiotherapy. *Br J Cancer.* 1955.

76. Nishida N, Yano H, Nishida T, Kamura T, Kojiro M. Angiogenesis in cancer. *Vasc. Health Risk Manag.* 2006.

77. Hinnen P, Eskens FALM. Vascular disrupting agents in clinical development. *Br. J. Cancer.* 2007.

78. Rajabi M, Mousa SA. The role of angiogenesis in cancer treatment. *Biomedicines.* 2017.

79. Wu XY, Ma W, Gurung K, Guo CH. Mechanisms of tumor resistance to small-molecule vascular disrupting agents: Treatment and rationale of combination therapy. *J. Formos. Med. Assoc.* 2013.

80. Mason RP, Zhao D, Liu L, Trawick ML, Pinney KG. A perspective on vascular disrupting agents that interact with tubulin: Preclinical tumor imaging and biological assessment. *Integr. Biol.* 2011.

81. Hofer E, Schweighofer B. Signal transduction induced in endothelial cells by growth factor receptors involved in angiogenesis. *Thromb Haemost.* 2007;97.

82. Fallah A, Sadeghinia A, Kahroba H, Samadi A, Heidari HR, Bradaran B, et al. Therapeutic targeting of angiogenesis molecular pathways in angiogenesis-dependent diseases. *Biomed. Pharmacother.* 2019.

83. Geel RMJM, Beijnen JH, Schellens JHM. Concise drug review: Pazopanib and axitinib. *Oncologist.* 2012;17.

84. Truman JP, García-Barros M, Kaag M, Hambardzumyan D, Stancevic B, Chan M, et al. Endothelial membrane remodeling is obligate for anti-angiogenic radiosensitization during tumor radiosurgery. *PLoS One.* 2010.

85. Rao SS, Thompson C, Cheng J, Haimovitz-Friedman A, Powell SN, Fuks Z, et al. Axitinib sensitization of high single dose radiotherapy. *Radiother Oncol.* 2014.

86. Teicher BA, Dupuis N, Kusomoto T, Robinson MF, Liu F, Menon K, et al. Antiangiogenic agents can increase tumor oxygenation and response to radiation therapy. *Radiat Oncol Investig.* 1994;2.

87. Teicher BA, Holden SA, Ara G, Dupuis NP, Liu F, Yuan J, et al. Influence of an anti-angiogenic treatment on 9L gliosarcoma: Oxygenation and response to cytotoxic therapy. *Int J Cancer.* 1995;61.

88. Griffin RJ, Williams BW, Wild R, Cherrington JM, Park H, Song CW. Simultaneous inhibition of the receptor kinase activity of vascular endothelial, fibroblast, and platelet-derived growth factors suppresses tumor growth and enhances tumor radiation response. *Cancer Res.* 2000;62.

89. El Kaffas A, Giles A, Czarnota GJ. Dose-dependent response of tumor vasculature to radiation therapy in combination with Sunitinib depicted by three-dimensional high-frequency power Doppler ultrasound. *Angiogenesis.* 2013.

90. Kaneko OF, Willmann JK. Ultrasound for molecular imaging and therapy in cancer. *Quant Imaging Med Surg.* 2012.

91. Liu Z, Gao S, Zhao Y, Li P, Liu J, Li P, et al. Disruption of tumor neovasculature by microbubble enhanced ultrasound: A potential new physical therapy of anti-angiogenesis. *Ultrasound Med Biol.* 2012;38.

92. Tran WT, Iradji S, Sofroni E, Giles A, Eddy D, Czarnota GJ. Microbubble and ultrasound radioenhancement of bladder cancer. *Br J Cancer.* 2012.

93. Al-Mahrouki AA, Iradji S, Tran WT, Czarnota GJ. Cellular characterization of ultrasound-stimulated micro-bubble radiation enhancement in a prostate cancer xenograft model. *DMM Dis Model Mech*. 2014.

94. Tarapacki C, Lai P, Tran WT, El Kaffas A, Lee J, Hupple C, et al. Breast tumor response to ultrasound mediated excitation of microbubbles and radiation therapy in vivo. *Oncoscience*. 2016.

95. Kim HC, Al-Mahrouki A, Gorjizadeh A, Karshafian R, Czarnota GJ. Effects of biophysical parameters in enhancing radiation responses of prostate tumors with ultrasound-stimulated microbubbles. *Ultrasound Med Biol*. 2013.

96. Al-Mahrouki A, Giles A, Hashim A, Kim HC, El-Falou A, Rowe-Magnus D, et al. Microbubble-based enhancement of radiation effect: Role of cell membrane ceramide metabolism. *PLoS One*. 2017.

2

Gamma Knife: From Single-Fraction SRS to IG-HSRT

Daniel M. Trifiletti, Jason Sheehan, and David Schlesinger

Contents

2.1 INTRODUCTION

Radiosurgery has traditionally been a high-dose, single-fraction treatment technique that has been found to be extremely effective for a large spectrum of malignant and benign neurosurgical conditions (Leksell, 1951). Delivery of high-dose radiotherapy in a single session leaves very little room for error, and as a result radiosurgery maintains a requirement for rigorous accuracy and precision management in treatment delivery. Gamma Knife (GK) radiosurgery (Elekta Instrument AB, Stockholm, Sweden) traditionally achieves this through the use of isocentric convergence of many small beamlets (201 or 192, depending on the model of the device) to create large-dose gradients and a rigid headframe that immobilizes the patient's head and defines a stereotactic coordinate system with a direct mechanical linkage between the patient's head and the isocenter of the Gamma Knife suitable for localization and targeting (Lunsford et al., 1988; Lindquist and Paddick, 2007). Image guidance has been based on up-front imaging of the patient using fiducial systems mounted to the patient's headframe to localize anatomy relative to the stereotactic frame of reference.

The development of radiosurgery did not end with the invention of the Gamma Knife, however. As experience accrued using linear accelerators as an alternative to the Gamma Knife for radiosurgery, it has become apparent that for certain clinical situations (for instance, tumors larger than what is typically indicated for radiosurgery or tumors directly adjacent to sensitive organs at risk [OARs]), delivery of the total dose over several fractions (hypofractionated stereotactic radiotherapy [HSRT]) creates some potential advantage, with similar tumoricidal effectiveness paired with further reduced normal tissue toxicity. The experience with linear accelerators also demonstrated the potential advantages to be gained by the use of in-room imaging techniques, making accurate and precise patient localization possible without the use of a rigid headframe and thereby making practical hypofractionated regimes.

Several techniques for both patient immobilization and image guidance have been developed that make hypofractionation on the Gamma Knife possible without compromising the historic precision characteristic of GKRS. This chapter explores the technology of frameless image-guided HSRT using the Gamma Knife Perfexion with the Extend™ system (Elekta Instrument AB, Stockholm, Sweden), as well as the Gamma Knife Icon, with a focus on the technology of each system, the benefits and limitations, and the flexibility in work-flow that each system supports. Additional potential HSRT platforms are explored in subsequent chapters.

2.2 TRADITIONAL GAMMA KNIFE RADIOSURGERY—SINGLE-FRACTION A PRIORI IMAGE-GUIDED RADIOSURGERY

2.2.1 IMMOBILIZATION

Traditional Gamma Knife radiosurgery is performed using a rigid stereotactic frame which is placed around the patient's head and fixed using four pins which are inserted to the outer table of the patient's skull. Frame placement is often performed in a small procedure room nearby the radiosurgery center, using local anesthetic to numb the pinsites as needed. Other centers prefer to administer light sedation in addition to the local anesthesia.

By design and definition, the stereotactic frame defines a coordinate system called the Leksell Coordinate System which has an origin superior, posterior, and right of the patient's head and increments towards the patient's left (+X), anterior (+Y), and inferior (+Z). The Leksell frame mounts mechanically to the treatment table on the Gamma Knife using an adapter in the case of the Perfexion (Figure 2.1), and therefore there is a mechanical correspondence between the stereotactic space defined by the frame and the coordinates of the Gamma Knife itself.

2.2.2 IMAGE GUIDANCE

Image guidance for traditional radiosurgery occurs a-priori of the procedure itself. Immediately following the frame placement, patients are generally sent for treatment planning imaging. The modalities involved may include MR, CT, and/or biplane angiography depending on the indication. Images are linked to the stereotactic coordinate system using modality-specific indicator boxes which are attached to the stereotactic frame during the imaging procedure. The indicator boxes result in fiducial markers in the resulting images, which, once they are registered with the treatment planning system, allow any point in the patient's brain anatomy to be referenced in stereotactic coordinate space (Figure 2.2 a, b). Non-stereotactic images (often

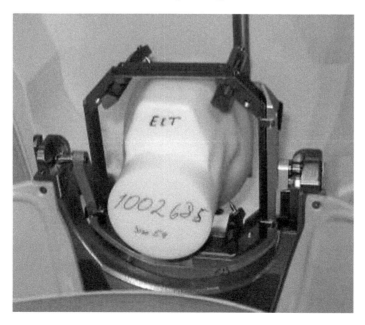

Figure 2.1 The Leksell stereotactic G-Frame attached to an anthropomorphic head phantom and docked to the Gamma Knife treatment bed. In addition to providing rigid immobilization, the stereotactic frame defines a coordinate system all over the patient's head that is mechanically linked to the machine coordinate system.

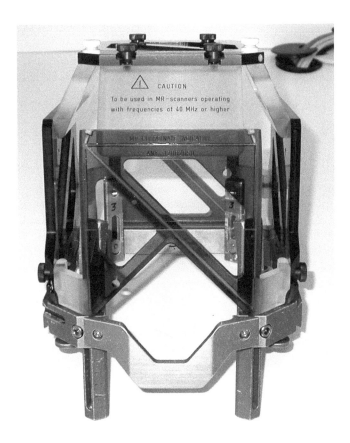

Figure 2.2a An MR indicator box mounted to a Leksell stereotactic frame. The indicator box channels are filled with copper sulfate solution that appears bright in MR images.

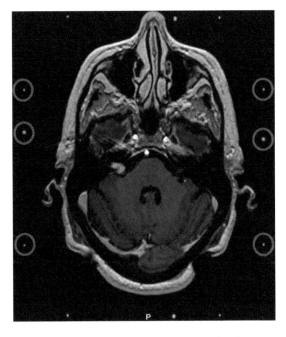

Figure 2.2b The fiducial marks (marked in red) on an MR image acquired with a stereotactic frame and MR indicator box.

including MR and/or PET) acquired prior to the frame placement are possible but must be co-registered to one of the stereotactic image studies to be useful.

2.3 LIMITATIONS OF TRADITIONAL GAMMA KNIFE SRS TECHNIQUES WHEN APPLIED TO IG-HSRT

There are several limitations to the traditional GK SRS frame, which limit its utility in HSRT. Most apparent is that the process of frame placement is invasive: pins are inserted into the outer table of the skull to create a rigid mechanical interface between the patient's head and the treatment machine. Moreover, this rigid association between the frame and the patient's skull is critical for the creation of the coordinate system used for localization and targeting, and any change in it invalidates the existing treatment plan.

A second limitation is that image guidance is a priori. This means that any change in the rigid association between frame and patient skull requires a new imaging study to re-establish the location of the patient anatomy relative to the coordinate system.

Beyond the inherent disadvantages of patient satisfaction, there are technical limitations to leaving a rigid head frame on for days as well. The presence alone of a rigid head frame does not ensure rigid fixation. Subtle shifts in the position of the head frame over the treatment course are possible, and daily measurements would be prudent to rule out these subtle systematic errors (either by digital probe or by imaging-based measurements).

2.4 REQUIREMENTS FOR GAMMA KNIFE IG-HSRT

2.4.1 ACCURACY AND PRECISION REQUIREMENTS

The most critical component of a hypofractionated immobilization system is that it can reliably and repeatedly localize the isocenter in three-dimensional space for each treatment fraction and that this localization must remain valid over the course of each treatment fraction. This tenant is the basis by which SRT is safe and feasible. A commonly acceptable tolerance for isocenter displacement is a non-systematic error of less than 1 mm. Although this tolerance is somewhat arbitrarily defined, there is evidence that adherence results in superior local control (Treuer et al., 2006). For the single-fraction case, the "gold standard" assumption has been that the rigid stereotactic frame provides superior immobilization performance over the relatively short time-frame required to deliver a radiosurgical treatment. Given the small geometric distances between tumor and sensitive OARs in the brain, any image-guided hypofractionated stereotactic radiation therapy (IG-HSRT) system cannot deviate far from the single-fraction standard.

2.4.2 REQUIREMENT FOR PATIENT ACCEPTANCE

A possibly overlooked component of rigid immobilization is that it is must be reasonably well tolerated by patients. Patient satisfaction has become a critical component of medical care, and other fractionated radiotherapy approaches (i.e., gynecologic brachytherapy) have led to psychosocial disorders in some patients, thought to be related to the discomfort of the applicator left in place between fractions (Kirchheiner et al., 2014). Any IG-HSRT system that limits patient discomfort would also lead to less inter- and intrafraction motion, require less intrafraction treatment breaks, and have faster daily patient set up.

2.4.3 REQUIREMENT FOR SOME POTENTIAL TO EXPAND INDICATIONS

Traditional single-fraction Gamma Knife radiosurgery has been a remarkably successful technique, with over 1.3 million patients treated worldwide between 1968 and 2019, with 95,000 treated in 2019 alone (Leksell Gamma Knife Society, 2019). A successful system for hypofractionated Gamma Knife treatments requires a rationale that creates an expansion of indications beyond those already effectively managed with the current system.

There is a great potential in a method of reliable GK immobilization and hypofractionation. As described in other chapters, HSRT would increase the scope of SRT by permitting radiosurgery in anatomic locations that have previously been treated with conventionally fractionated radiotherapy because of concern of adjacent normal tissue tolerance. Additionally, it is possible that some intracranial tumors have a biology that would demonstrate improved local control with multi-fraction radiosurgery (Jee et al., 2014; Minniti et al., 2014; Toma-Dasu et al., 2014; Casentini et al., 2015).

2.5 HISTORICAL ATTEMPTS TO HYPOFRACTIONATE RADIOSURGERY TREATMENTS

There have been a variety of historical attempts both within and outside the Gamma Knife subspecialty to create methods that could allow for hypofractionation. This section summarizes some historical attempts which form the basis for modern GK-HSRT.

2.5.1 PROTRACTED FRAME APPLICATION

The feasibility of this method was first reported by Simonová in the early 1990s as a method to achieve hypofractionated stereotactic radiotherapy while only utilizing available devices (Simonova et al., 1995). They reported on 48 patients who underwent head frame placement and then returned for once daily treatments for 2 to 6 days. The method was considered feasible, well tolerated, and relatively safe. However, patients were admitted for the duration of therapy, making this an expensive treatment option. Additionally, patient-reported outcomes were not included in this report. A similar "split-dose" approach was reported where the total SRS dose was divided into two equal fractions. Patients underwent frame placement, imaging for treatment planning, and then first treatment fraction in the evening of the first day, followed by a second fraction delivered approximately 14–15 hours later. The authors of the study reported that the treatment was well tolerated and showed a small survival benefit for patients receiving two-fraction SRS as compared to an earlier cohort receiving single-fraction SRS. However, the authors cautioned against the possibility of a frame becoming dislodged over the total time of the procedure (Davey et al., 2007).

2.5.2 REMOVABLE FRAME SYSTEMS

The TALON cranial fixation system (Nomos Corp., Sewickley, PA) is a removable frame system that permits rigid fixation of the skull to a head frame through attachment to base screws inserted into the patient's skull. These screws are attached to the TALON system and permit minute adjustments of the cranium after fixation. The screws are left in place between fractions (usually 2 to 5 days). Salter et al. reported on the TALON system's positional accuracy and estimated that 95% of true isocenter position between fractions would fall within 1.55 mm of the planned isocenter position. The TALON system was well tolerated by patients; however, three of nine patients included developed infections at the screw sites, and two patients had loosening of the screws between fractions requiring re-tightening (Salter et al., 2001). The TALON system was not attempted in a Gamma Knife SRT context.

2.5.3 RELOCATABLE FRAME SYSTEMS

Multiple relocatable head frame systems have been developed over the past 15 years. This includes rigid frames used for radiosurgery registration, which are not invasively attached to the patient (Reisberg et al., 1998; Alheit et al., 1999; Ryken et al., 2001; Baumert et al., 2005; Minniti et al., 2010; Ruschin et al., 2010). Examples include systems that have utilized bite blocks, head straps, thermoplastic masks, optical tracking, or some combination. In all cases, the important characteristics include relatively simple, noninvasive methods for placing the patient in a repeatable treatment position corresponding to the position at the time of treatment planning.

2.6 HISTORICAL DEVELOPMENT OF ONBOARD IMAGE GUIDANCE FOR RADIOSURGERY

The development of in-room image-guidance systems for radiotherapy was a significant development that enhanced the accuracy and precision by which a patient could be set in the correct treatment position. These systems, designed primarily for linear accelerator-based radiotherapy, were quickly adapted for use in radiosurgery contexts. Systems evolved from simple 2D MV portal imaging systems that used film (and later flat-panel detectors) that were exposed by the treatment beam to allow clinicians to verify whether the target was within the collimated field (Dong et al., 1997). The invention of amorphous-silicon flat-panel detectors motivated attempts to use the treatment machine itself as an megavoltage cone beam CT system (Pouliot et al., 2005). kV-CBCT systems were developed using x-ray tubes and detectors mounted orthogonally from the LINAC treatment beam (Jaffray, 2007). Dual ceiling/floor-mounted stereoscopic kV x-ray systems were developed specifically for radiosurgery applications.

The aforementioned developments for linear accelerators were motivated as much for extracranial stereotactic and non-stereotactic indications as they were for intracranial indications, as the stereotactic frame was a well-established and well-validated technique for intracranial radiosurgery. However, as noted earlier, an enhanced ability to hypofractionate is considered advantageous in certain clinical situations. To that end, David Jaffray's group at Princess Margaret Hospital developed a kV-CBCT system that they successfully integrated with a Gamma Knife Perfexion. The system uses a conventional 90 kVp rotating anode x-ray tube and an opposing detector. The system is supported by a set of vertical supports, which allows the system to translate from a parked position above the shield-doors of the Perfexion to an imaging position between the patient and the shield-doors. A rotational axis allows the system to rotate by 210° for imaging. Isotropic voxel resolutions (1mm or 0.5 mm) are achievable with a reconstruction field of view of 25.6 × 25.6 × 19.3 cm (Ruschin et al., 2013).

2.7 EXTEND SYSTEM FOR THE GAMMA KNIFE PERFEXION

While the previous section summarizes work that has been performed to explore options for GK IG-HSRT, the first clinically available commercial solution in practice to allow for hypofractionated Gamma Knife radiosurgery treatments was the Extend system (which at the time of publication of this edition is not actively marketed but is supported and clinically deployed). The Gamma Knife Extend System made reproducible, frameless stereotactic fixation of the head possible through a suctioned dental mold of the hard palate and maxillary teeth. The system removed the requirement for surgical intervention needed for frame placement, and no devices were left *in situ* between fractions, which could cause pain or serve as a nidus for infection (Ruschin et al., 2010).

2.7.1 MAIN COMPONENTS

The Extend frame system consists of a carbon-fiber front plate to which a dental impression/mouthpiece can be attached, a base plate to which the front-piece can be attached, and a vacuum cushion on which the patient's head sits. The Extend frame rigidly docks with the GK patient positioning system (PPS).

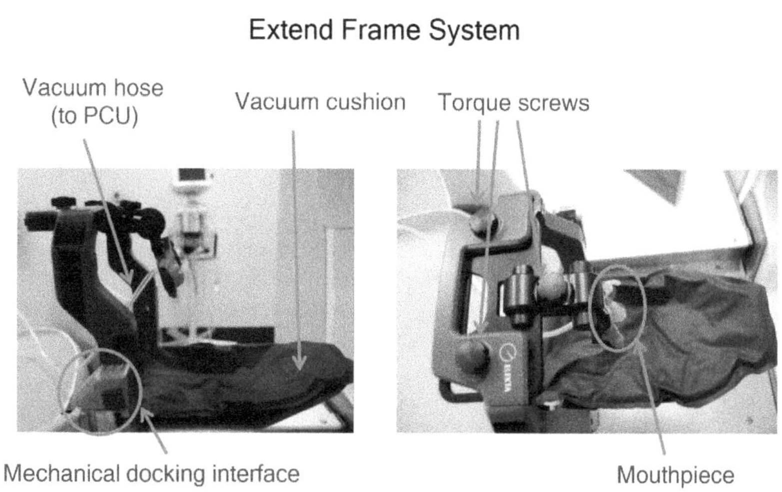

Extend Frame System

Figure 2.3 The Extend Frame system and its components.

Patient Control Unit (PCU)

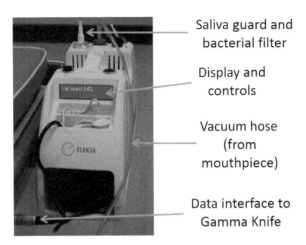

Saliva guard and bacterial filter

Display and controls

Vacuum hose (from mouthpiece)

Data interface to Gamma Knife

Figure 2.4 The Patient Control Unit (PCU) for the Gamma Knife Extend system. The PCU creates a vacuum that is used to monitor patient movement and sends data during treatment to the Gamma Knife control system to interrupt treatment if the vacuum level falls below a set threshold.

Reposition Check Tool (RCT)

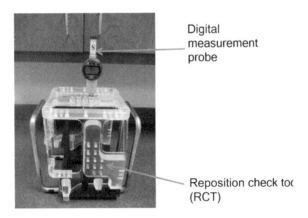

Digital measurement probe

Reposition check tool (RCT)

Figure 2.5 The reposition check tool (RCT) template and associated digital measurement probe. The red carrier doubles as a QA tool for the RCT.

The mouthpiece of the frame is attached via plastic tubing to the Patient Control Unit (PCU). The PCU consists of a vacuum pump and tubing that connects to the mouthpiece and interfaces with the patient and the treatment unit. The reposition check tool (RCT) consists of an acrylic measurement template and an associated set of digital measurements probes. The RCT fits into slots on the Extend frame. Measurement holes in the RCT template are used for the measurement of head position to confirm three-dimensional positioning between fractions.

2.7.2 DENTAL MOLD CREATION

The first step in the use of the Extend system is the selection of the mouthpiece and the creation of the dental mold. A dental impression is created using standard impression material (vinyl polysiloxane) using a mixing gun. A plastic spacer is placed between the mold and the hard palate before inserting the

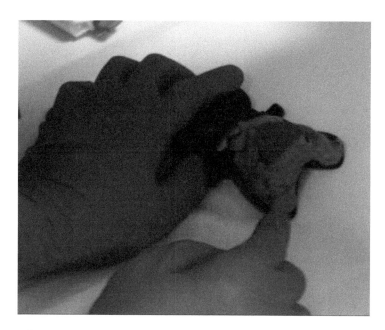

Figure 2.6 Creation of an Extend system dental impression. The impression material fills a mouthpiece, and the plastic spacer (in purple) creates a vacuum space within the dental material.

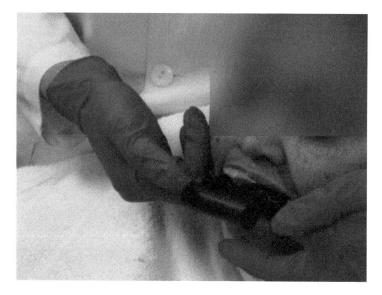

Figure 2.7 Placement of an Extend system mouthpiece in a patient's mouth. Even pressure must be applied for several minutes while the impression material cures. (Patient's face blurred for confidentiality.)

mouthpiece into the patient's mouth to create the impression. The spacer allows for an air space in which the vacuum can suction the mold to the palate, aligned by dental anatomy (Figure 2.6). Once the mouthpiece is placed in the patient's mouth, even pressure must be maintained along the palate to allow the impression material to cure (Figure 2.7). If there is insufficient material between the teeth and the mouthpiece or between the hard and soft palate, then reliable suction may be difficult. In addition, edentulous patients or patients without adequate dentition are contraindicated for Extend immobilization.

2.7.3 SETUP AT GAMMA KNIFE

Creation of the dental impression is followed by setup at the Gamma Knife and the construction of the Extend frame system using the completed mouthpiece.

2.7.3.1 Dental Mold Insertion/Frame Creation

The patient is placed in a comfortable position on the Gamma Knife treatment bed. The dental mold connection with the spacer and vacuum tubing is confirmed and is guided into the patient's mouth and abutted to the hard palate and maxillary dentition. The PCU vacuum is then tested with the mouthpiece in place to a vacuum level of 30% to 40% (as a percentage of atmospheric pressure). The PCU has a safety alarm that can detect a loss of suction (defined as a 10% change in suction from the set point).

With the dental mold in place and the vacuum activated, the head frame can be secured. This is first done by attaching the front piece to the mold and then by locking the front piece to the docking area (which is locked to the GK couch). When patient comfort is again confirmed, the mouthpiece and head frame are hand tightened and then secured with a torque wrench (Figure 2.8).

2.7.3.2 Vacuum Cushion Creation

In the supine position with the head frame attached, the vacuum cushion is molded to the scalp, and the PCU is used to evacuate air from the cushion. As the vacuum level in the cushion increases, the cushion becomes increasingly rigid and molded to the shape of the patient's head. When complete, the result is a rigid cushion containing a firm impression of the dorsal aspect of the scalp that will be maintained for each fraction.

2.7.3.3 Test Measurements/RCT Measurement Hole Selection

The completed patient-specific dental impression, frame, and vacuum cushion define the stereotactic alignment of the patient's head with the couch.

To confirm proper alignment of the head within the frame, daily reference measurements are taken using the reposition check tool (RCT) and are compared to measurements taken at the time of

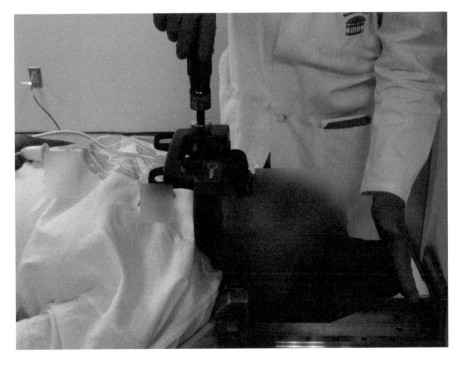

Figure 2.8 Creation of a patient-specific Extend frame by tightening the locking screws on the frame front plate with a torque wrench. (Patient's face blurred for confidentiality.)

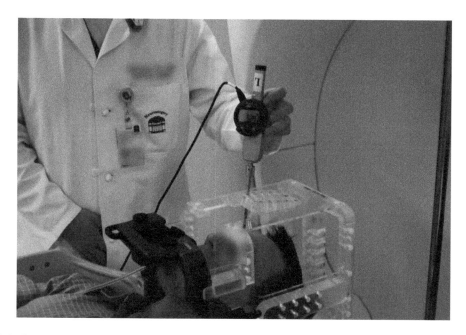

Figure 2.9 Physician acquiring reposition measurements using the RCT and the digital probe system.

image acquisition (computed tomography [CT] or magnetic resonance imaging [MRI]). During this initial setup step, RCT apertures are chosen for the measurements, and the distances to the head are recorded on a worksheet. The RCT consists of four plastic panels that surround the patient's head in the Extend frame (Figure 2.9). Measurements are taken with a pair of electronic linear measurement probes that are included in the Extend system (C150XB Digimatic Indicator, Mitutoyo Corp.). The probes measure the distance between preset holes in the RCT and the scalp. At least one aperture (and ideally more than one) must be chosen for each panel of the RCT. Apertures should be chosen to ideally allow normal incidence of the probe tips to the patient's head. Choosing apertures far apart from each other and avoiding areas of loose skin or fat can improve the precision and reproducibility of measurements.

It is important to note that any change in the vacuum pressure of the mouthpiece, of the vacuum cushion, or of the tension in the screws of the head frame can result in compromise of the rigidity and reproducibility of the Extend system. If these changes occur, the system should be reset from the beginning.

2.7.4 SIMULATION (CT) IMAGING

Following initial setup at the Gamma Knife, patients proceed to simulation imaging which will serve as the reference stereotactic images for treatment planning.

2.7.4.1 Simulation Imaging Setup and Reference RCT Measurements

The basic principle of the Extend system is that the patient position at the time of treatment must match (to within a small uncertainty threshold) the patient position at the time of simulation imaging. Therefore, at the time of stereotactic CT imaging, reference measurements are collected that will serve as the standard to compare future measurements to (prior to each treatment delivery). The process begins with the stereotactic immobilization of the patient as outlined earlier, but it is done on the CT couch as opposed to the GK couch. During any period that the head frame is assumed to be rigidly fixed, the PCU should be set to alarm for changes in vacuum, and the patient should be visually monitored to ensure comfort, as hand signals are preferred while the mouthpiece is in place. Measurements proceed as described earlier, using the measurement apertures chosen at the time of initial setup at the Gamma Knife. These measurements are read off of the display on the PCU and recorded on a worksheet for later use.

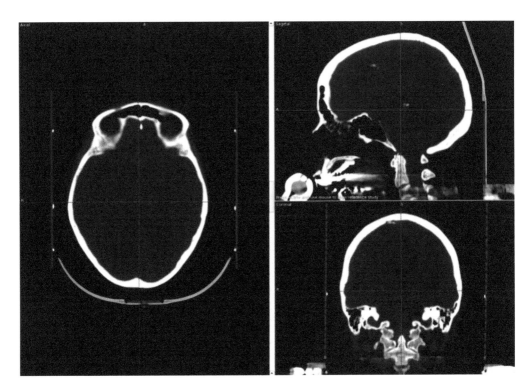

Figure 2.10 Stereotactic CT of an Extend patient. The CT field of view must cover the entire head and include the lateral fiducial markers.

2.7.4.2 Simulation Stereotactic CT Imaging

After proper immobilization is achieved and RCT measurements confirmed and recorded, the Extend CT indicator box is mounted to the frame. The CT indicator is a transparent box with implanted fiducial markers that can serve as rigid points in the GK treatment planning software (TPS, GammaPlan, Elekta AB, Stockholm, Sweden). CT images of the head are then obtained from vertex to mid-frame and with a field of view wide enough to include the entire CT indicator and corresponding fiducial markers (Figure 2.10). Intravenous contrast can be utilized as clinically indicated.

2.7.4.3 Post-CT Measurements

Immediately after the CT sequences are obtained and before releasing the vacuum suction, post-CT measurements with the RCT are important to verify that the patient did not shift during CT imaging. This is done by the same method as described earlier, through the same apertures as were used in the pre-CT measurements. Any difference of more than 0.5 mm from the pre-CT value should prompt the team to remove and reposition the Extend head frame, re-measure, re-obtain CT images, and then confirm measurements. After the repeat measurements are verified, the suction can be released and the head frame removed. Images are then transferred to the Gamma Knife treatment planning system.

2.7.5 INTEGRATION OF NON-STEREOTACTIC SCANS

The Extend system requires that a stereotactic CT be used as a stereotactic reference. CT images are less likely to suffer from localized geometric distortion, and the Extend frame does not fit within all MR head coils—two considerations which may be the source for this requirement. However, multi-modality images (especially MR) are critical for the visualization of most intracranial indications, so non-stereotactic images may be incorporated into treatment planning via image registration. The Gamma Knife treatment planning system includes cross-modality rigid co-registration algorithms for this purpose (Viola and Wells III, 1997).

2.7.6 TREATMENT PLANNING

After all stereotactic and non-stereotactic imaging has been imported into the Gamma Knife treatment planning system, the stereotactic CT images are registered to stereotactic space using the fiducial markers visible in the CT images. MR and other modality images are then co-registered to the reference CT images as described earlier. The details of target visualization and delineation vary by institution; however, the planning includes functionality to delineate target volume(s) and adjacent organs at risk (OARs) similar to traditional SRT planning. Isocenter-based "shots" are placed and customized based on target size, shape, and adjacent OARs (Figure 2.11). Total dose and the number of fractions are entered, and the plan is reviewed by the neurosurgeon, medical physicist, and radiation oncologist.

2.7.7 TREATMENT PROCEDURE

2.7.7.1 Entering Reference Measurements (1st Fraction)

When a patient treatment is started at the Gamma Knife console prior to delivery of the first treatment fraction, the system requires that the reference position measurements acquired at the time of CT imaging be entered into the system. Accuracy at this step is critical because it will create a reference point to which each fraction will be directly compared. As such the reference measurements should be carefully double checked (preferably by a separate member of the team) prior to moving forward.

2.7.7.2 Repositioning Measurements

When the patient is prepared to proceed with therapy, they should enter the GK vault and have the Extend treatment head attached in a manner similar to when it was attached prior to CT-simulation (supine, vacuum cushion in place, mouthpiece, vacuum, front piece, secure to couch). Then new measurements should be collected using the electronic probe though the same apertures that were used for the reference measurements. This process is guided by the GK Extend console, which automatically captures each probe measurement and compares to the previously entered reference measurements. Once all selected RCT apertures have been measured, the console software will calculate a three-dimensional translational vector of the difference in patient position as compared to the reference measurement. The system will warn the operator if the radial positional difference is greater than 1.0 mm and suggest the clinician consider repositioning to

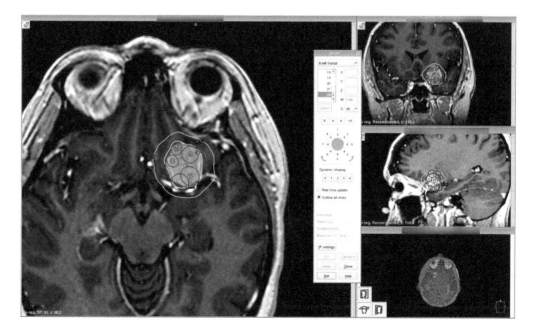

Figure 2.11 A treatment plan for a hypofractionated Gamma Knife radiosurgery case using the Extend system.

achieve a more favorable patient position. However, clinical judgment is ultimately involved in determining what level of positional uncertainty is acceptable.

2.7.7.3 Treatment

Following patient positioning, each individual SRT treatment is administered in the same way as a single-fraction Gamma Knife treatment. The Gamma Knife treatment couch translates the patient's head into the center of the radiation body of the unit, placing the patient's head at a location corresponding to the coordinates of each shot of the treatment plan in turn. Each position is maintained for a dwell duration as calculated by the treatment planning system in order to achieve the desired overall dose distribution. Treatments are monitored by the operator of the machine via video and audio surveillance. The patient is provided with a call button to alert the operator if they require assistance as the patient cannot speak with the Extend mouthpiece in place. It can be useful for the treatment team (usually a radiation therapist, medical physicist, neurosurgeon, and radiation oncologist) to develop a set of hand gestures that provide general communication.

2.7.7.4 Intrafraction Position Monitoring

Patient immobilization is monitored using the PCU vacuum surveillance system. Patient motion beyond a very small threshold will trigger a loss of suction in the Extend mouthpiece. Any loss of vacuum greater than 10% of vacuum level set at the time of patient positioning will trigger an interrupt which will shield the Gamma Knife 60Co sources and pause the treatment, automatically withdrawing the patient from the treatment position in the machine. This occurrence requires that the patient position be re-measured using the RCT and probe system and repositioned if required before the treatment can resume. If adequate repositioning is impossible, then a new stereotactic CT may be acquired and the treatment plan shifted to accommodate the new patient position.

2.7.8 ACCURACY COMPARED TO SIMILAR SYSTEMS

The mean setup uncertainty of the Gamma Knife Extend system has been shown to be reproducibility on the order of 0.4 to 1.3 mm (Ruschin et al., 2010; Sayer et al., 2011; Schlesinger et al., 2012; Ma et al., 2014). Table 2.1 reports the mean displacement of the patient after setup by a representative variety of proposed immobilization systems applicable to fractionated radiosurgery. As demonstrated, the setup uncertainty of the Extend system is comparable to other available relocatable frame systems.

Table 2.1 **Reports of the residual setup uncertainty of a variety of relocatable immobilization systems for fractionated radiation treatments**

AUTHOR	DEVICE	SETUP DISPLACEMENT, MM (SD)
(Sweeney et al., 1998)	Bite block + vacuum assist	<1.02*
(Rosenberg et al., 1999)	GTC frame	1.1 (0.6)#
(Ryken et al., 2001)	Mask + optically tracked bite block	0.16 (0.04)†
(Baumert et al., 2005)	Mask + bite block	2.2 (1.1)‡
(Kunieda et al., 2009)	Bite block + vacuum assist	0.93–1.09 (0.52–0.88)‡
(Minniti et al., 2010)	Relocatable frame + upper jaw support	0.5 (0.4)‡
(Ruschin et al., 2010)	Extend prototype	1.0^ / 1.3†
(Schlesinger et al., 2012)	Extend clinical system	0.64 (0.25)^

Note: Symbols indicate basis for setup displacement measurement (* fiducials versus surface landmarks; # orthogonal radiograph landmarks; † fiducials versus CBCT; ‡ simulation CT versus QA CT; ^ probe/depth measurements).

Source: Table adapted from Schlesinger et al., (2012).

2.7.9 LIMITATIONS OF EXTEND SYSTEM

The Extend system provides a reliable, noninvasive method for reproducible immobilization of patients for HSRT. However, as compared to HSRT systems for other treatment devices, the Extend system may have some potential limitations.

2.7.9.1 Complicated Workflow

One drawback to the Extend system is the complex logistic aspect of the mouthpiece creation, application, and RCT measurement system. The mouthpiece can be bulky and must fit snuggly and firmly for the frame to be reliably fixed. Generally, patients consider the first day of treatment arduous because dental mold creation, head frame fitting, CT, planning, first fraction, and numerous precise measurements can take hours to complete with the head frame in place. However, subsequent treatments are considered relatively convenient (Sayer et al., 2011).

2.7.9.2 Vacuum as a Proxy for Motion

The basis of the real-time monitoring of the Extend system is in the vacuum alert. The system relies on the assumption that any change in the vacuum pressure of greater than 10% equates to a displaced target. It does not detect possible patient motion in which the vacuum level changes by less than 10%. Conversely, the vacuum alert assumes that any change in vacuum of greater than 10% means that the patient moved systematically (as opposed to no movement or temporary movement). Additionally, after the alert is activated, it is impossible to tell if the change in pressure was indeed related to patient movement or potentially equipment failure in the vacuum or tubing, and the confirmation of functional equipment is warranted for unexpected vacuum alarm. In any case, the activation of the vacuum alarm results in a treatment pause, and the entire head frame should be removed, replaced, and re-measured to ensure proper placement and treatment accuracy.

2.7.9.3 Patient Contraindications (Dentition/Performance Status/Gag)

For a patient to be eligible for HST using the GK Extend system, they should be otherwise fit for radiosurgery (limited number of intracranial targets, adequate performance score, etc.). Although it should be noted that there are some aspects of the Extend system that may require additional patient cooperation compared to a standard head frame. Patients with a sensitive gag reflex may not be willing or able to tolerate a bulky mouthpiece for the duration of multiple treatments. Also as discussed previously, adequate dentition is critical to the reproducibility of the dental mold placement.

2.8 IG-HSRT WITH GAMMA KNIFE® ICON™

The Extend system proved to be a practical, if somewhat cumbersome, method for achieving a hypofractionated Gamma Knife technique. However, the system was limited in scope and functionality relative to comparable systems routinely used in Linac-based radiosurgery. Recognizing that the Extend system would not be an optimal solution on its own, Elekta Instrument, AB, the manufacturer of the Gamma Knife, began a development cycle intending to address some of the shortcomings of the Extend system for IG-HSRT. In particular, they redesigned and commercialized a prototype created at the University of Toronto (Ruschin et al., 2013) and created a new treatment solution with an integrated capability to verify and monitor patient position before and during treatment. This ultimately resulted in a new Gamma Knife platform, Gamma Knife Icon. Gamma Knife Icon replaces the older Extend system as a Gamma Knife IG-HSRT solution. The new system discards the cumbersome dental-impression-based frame in favor of thermoplastic mask immobilization; however, it also includes features that improve single-fraction G-frame treatments.

2.8.1 MAIN COMPONENTS

Gamma Knife Icon adds several new components to the Gamma Knife Perfexion platform. The most conspicuous addition is that of a CBCT mounted to the side of the unit. The system also includes a stereoscopic optical tracking system consisting of a folding infrared camera system mounted near the foot of the Gamma Knife PPS that is aimed toward the head of the unit. A new headrest is included that holds a

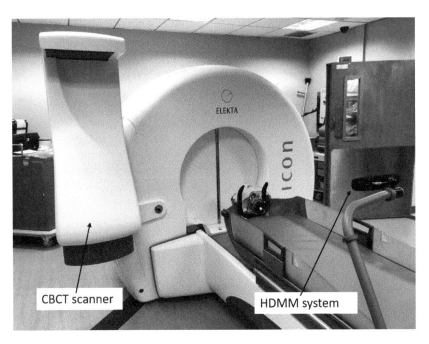

Figure 2.12 The Gamma Knife Icon system. Annotations show the onboard CBCT scanner and the High-Definition Motion Management (HDMM) infrared system.

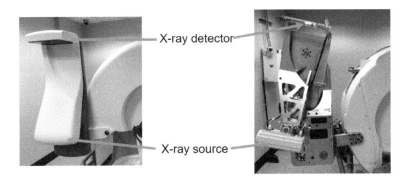

Figure 2.13 The Gamma Knife Icon CBCT system. Left: CBCT gantry with plastic covers intact. Right: CBCT gantry with plastic covers removed. Note: The flat-panel detector has been removed in the picture on the right.

patient-specific pillow, acts as a mount for a thermoplastic mask, and contains four infrared reflectors on rigid posts that act as reference markers for the patient position as described later (Figure 2.12).

The CBCT system is designed specifically for the task of determining the patient's stereotactic position both at time of initial setup and before each treatment session (Figure 2.13). The CBCT has 200 degree rotation, an imaging volume of 448 cm³, 0.5 mm voxel size, and a resolution of more than 7 line pairs/cm, using 332 projections at 90 kVp. The system has two imaging modes distinguished by their nominal computed tomography dose index; high-dose mode has a CTDI of approximately 6.5 mGy and has a slightly higher signal-to-noise ratio as compared to the low-dose mode with a CTDI of 2.3 mGy. Most importantly, the CBCT isocenter has a known geometrical relationship with the radiation isocenter of the unit, determined through a calibration and quality assurance procedure (AlDahlawi et al., 2017). CBCT scans acquired with the system are therefore in stereotactic coordinate space. Preliminary tests on localization uncertainty using a phantom suggest a mean positional uncertainty of less than 0.2 mm (Eriksson and Nordström, 2014; Eriksson et al., 2014).

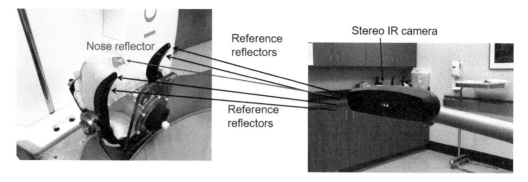

Figure 2.14 A close-up of the HDMM system. Right: A close-up view of the stereo infrared camera system. Left: A close-up view of the patient headrest/marker system with an anthropomorphic phantom setup. The system tracks the nose marker relative to the four reference markers found on the posts on each side of the headrest.

The optical imaging system, named the HDMM System, uses a stereoscopic infrared camera unit to track patient motion. The system is mounted near the foot of the PPS on a folding arm. When raised, the camera system has a view of a newly developed headrest that mounts to the head of the PPS in a manner similar to the stereotactic frame adapter for frame-based procedures. The headrest has lateral posts, each with two infrared-reflective markers that together serve as a static positional reference for the tracking system. During procedures, a fifth marker is placed on the patient's nose, and the system differentially tracks patient position relative to the four static markers (Figure 2.14). Phantom studies have reported a motion resolution of 0.06 mm (Wright et al., 2017).

Gamma Knife Icon is designed in a way that makes an in-field upgrade of the existing Gamma Knife Perfexion systems possible. The radiation body and collimator design of the two models are identical; the upgrade involves mounting the CBCT system to the radiation body, the HDMM system to the PPS bed, removal of the Extend system components (if applicable), and upgrading the treatment planning and control systems to support the new functionality (as well as some aesthetic updates).

2.8.2 GENERAL WORKFLOW

The general IG-HSRT workflow is analogous to the Extend system in principle. An immobilization solution is created for the patient, and the patient's reference position is determined using the CBCT. A treatment plan is created using this reference CBCT as a basis for stereotactic coordinates. Prior to each treatment session, the patient is placed in the immobilization system, and a CBCT image is acquired to determine the patient's current position. The system compares the current position to the reference position and automatically corrects the treatment plan to match the current patient position. During the treatment itself, the HDMM camera system tracks the patient's motion. Treatment is automatically gated off if the patient moves out of position beyond a clinical threshold and gates back on if the patient returns to the correct position. Additional CBCT scans and corrections can be obtained if the patient does not return to the correct position within a time threshold. There are several important advantages of the Icon system over the previous Extend system, including the ease in creating a thermoplastic mask relative to a dental impression, the ease and accuracy of acquiring CBCT images to determine patient position rather than cumbersome manual distance measurements, the capability of the system to automatically adjust the treatment plan to the patient's current position rather than attempting to correct the patient's position by moving the patient, and finally the accuracy in treatment delivery gained by tracking the patient throughout the procedure and gating the delivery as required.

2.8.2.1 Setup at Gamma Knife

Setup of a patient for an IG-HSRT treatment on the Gamma Knife Icon has three primary steps: creation of a custom head-cushion, creation of a thermoplastic mask, and acquiring a reference CBCT image defining the patient's stereotactic position for treatment planning.

2.8.3 PATIENT-SPECIFIC HEAD CUSHION

The Icon system headrest is designed to accept a patient-specific head cushion. The cushions are soft while in-package but begin to harden with exposure to room air. After the patient is placed in a comfortable position on the PPS, a cushion package is opened, the cushion placed behind the patient's head and molded to fit the patient's anatomy. The cushion begins to stiffen quickly and is fully cured in approximately 15 minutes.

2.8.4 THERMOPLASTIC MASK CREATION

Thermoplastic masks have long been used in linac radiotherapy, and the mask system for the Gamma Knife Icon is quite familiar in concept. Masks arrive in packaging stiff and flat. Upon heating to approximately 165° F, the mask becomes quite deformable within 10 to 15 minutes. The heated mask is placed over the patient's face, snapped onto the appropriate locations on the headrest, and molded to match the patient's anatomy. As the mask cools it stiffens and is primary cured after approximately 15 minutes.

2.8.5 REFERENCE CBCT IMAGING

Once the patient head cushion and thermoplastic mask are created, a reference CBCT is acquired that defines the patient's reference stereotactic position. The resulting images are transferred to the treatment planning system and are co-registered to previously acquired imaging (if any). Quality of the CBCT scans is optimized for patient positioning, not for anatomical visualization, so SRS-quality MR and/or CT scans are critical for IG-HSRT treatment planning using the Icon system.

2.8.6 TREATMENT PLANNING

Treatment planning proceeds in a manner similar to G-frame-based and Extend-frame treatments. For each target, the total dose may be distributed over one or multiple fractions. A single treatment plan may include different doses, but any given treatment plan may have only one fractionation schedule. (This does not preclude treating targets with different numbers of fractions; however, separate treatment plans are required). Once treatment planning is complete, the final plan is approved and exported to the Gamma Knife Icon console.

2.8.6.1 Pre-Treatment Workflow

Pre-treatment setup of a patient involves making the patient comfortable, placing the patient in the previously created immobilization system, starting patient tracking, acquiring the patient's current stereotactic position, correcting the treatment plan to match the patient's current stereotactic position, and commencing treatment.

2.8.7 MAKING THE PATIENT COMFORTABLE AND APPLYING IMMOBILIZATION SYSTEM

The patient is placed in a comfortable position on the PPS, ideally using therapist notes on the position the patient was in during the initial setup. The patient-specific head cushion is placed behind the patient's head, and the thermoplastic mask is applied and snapped onto the patient headrest.

2.8.8 PRE-TREATMENT SETUP CBCT IMAGING

Once the patient is comfortably set up on the table, a CBCT scan is acquired in a manner similar to the images acquired for the pretreatment setup. The purpose of this image is different, however; rather than to determine a reference position for the patient, the pre-treatment CBCT determines the patient's current position. This image is exported to the treatment planning system and is co-registered relative to the reference CBCT scan via a 3-D rigid registration. The registration matrix represents the difference in stereotactic coordinate systems between the patient's reference and current positions relative to the Icon system. The treatment planning system applies this difference as a correction to the treatment plan to move it to the patient's current position. The planning system allows the operator to view the correction as well as the predicted residual dose difference after applying the correction. Once reviewed, the corrections are sent to the Gamma Knife Icon control console.

2.8.9 MOTION TRACKING AND GATING

During the pre-treatment CBCT scan, the HDMM camera system tracks the position of the patient's nose reflector relative to the four static reflectors on the headrest. A time-averaged position of the patient during the scan is computed, and this serves as a reference position for the patient's nose marker.

2.8.10 RE-IMAGING TO BASELINE PATIENT

After the corrected treatment plan is sent to the console, treatment commences. During the treatment, the HDMM system continuously tracks the patient's nose marker and allows treatment to continue as long as the magnitude of nose motion is below a clinical threshold (default is 1.5 mm, although this is user-adjustable). In the event the patient moves beyond this threshold, the control system gates the beams off by moving the sources to a blocked position. If the patient returns to a position below the threshold within 30seconds, the system will gate back on (Figure 2.15). If the patient remains out of position beyond this time limit, or if the system gates the same shot more than five times, the system will automatically pause the treatment. In this instance, a new CBCT is acquired to determine a new baseline position for the patient, a new set of corrections are generated and sent to the treatment console, and the HDMM is re-referenced.

2.8.11 IMPLICATIONS FOR WORKFLOW FLEXIBILITY

The Icon system makes a highly flexible treatment workflow possible that can accommodate multiple-dose-fractionation schemes using both thermoplastic mask and G-frame-based immobilization systems. This is supported by advances in the treatment planning system, which allow delivered dose to be accumulated and used during treatment plan design and dose evaluation. Specifically, the system includes functionality to re-plan completed cases, taking into account dose delivered in prior treatments, amend multiple-fraction treatments in-between fractions, and amend partially delivered fractions, taking into account the partial dose delivered. In each case, targets can be added/subtracted, doses can be modified, and even the immobilization system can be changed from mask to frame or vice versa. This increased flexibility enables the creation of highly personalized treatments but also requires great care and organization to ensure the treatment team has the correct information for each treatment.

2.8.12 SYSTEM ACCURACY

HSRT provides a radiobiological safety margin for the treatment of indications that would be difficult to treat in a single fraction; however, it does not eliminate the requirement for extremely low-treatment delivery uncertainties. Several studies have investigated various sources of uncertainty in the Icon CBCT and HDMM systems.

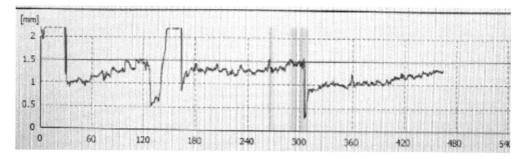

Figure 2.15 A screenshot of an HDMM trace showing the magnitude of the relative motion of the nose marker. Red line indicates the treatment gating threshold. Yellow highlights indicate times when system was gating beams because the patient was out of position beyond threshold.

2.8.12.1 Patient Motion and Gating

One of the most important sources of uncertainty for mask-based HSRT procedures on the Icon platform is the suitability of the relative motion of a reflective nose marker as a surrogate for motion of targeted anatomy. One study investigating this on both phantoms and clinical treatments found that on average, the reflective nose marker displaces about twice the magnitude of the corresponding intracranial target (Wright et al., 2019).

However, patients span a range of capacity in terms of being able to remain relatively still during treatment and tolerance of remaining in a desired position over the length of a treatment. A recent study of 462 mask patients used a neural network model to predict the probability of a treatment interruption requiring CBCT re-baselining using log-entries recorded by the HDMM system over the first 5 minutes of treatment as the model input. The analysis showed that the magnitude and frequency of nose marker motion events (relative to the most recent baseline nose position) recorded in the control system logfile could predict the occurrence of future treatment interruptions [AUC–ROC (area under the curve–receiver operating characteristics) = 0.84 for the test population]. This information could be useful during the mask creation process to determine if mask immobilization will be suitable for a given patient and could be used during the first few minutes of a treatment to measure the stability of the patient setup. The same study demonstrated that CBCT re-baselining could significantly reduce mean nose marker displacement over an entire treatment fraction (from a mean of 0.96 ± 0.96 mm to 0.62 ± 0.25 mm in the study population) (MacDonald et al., 2020).

2.8.12.2 Registration Uncertainty

Image-guided systems are critically dependent on co-registration for the calculation of the rotational and translational differences between a patient's current position with respect to the treatment machine and the position required for treatment (i.e., the treatment planning position). One study of the Icon registration system for thermoplastic mask-based immobilization cases found that the 3D image registration uncertainty as determined by anterior commissure/posterior commissure landmarks was on the order of 0.2 mm when co-registering CBCT to CBCT images, 0.5 mm when co-registering CT to CBCT images, and 0.8 mm when co-registering MR to CBCT images, with best results obtained when including the skull base in the registration region-of-interest (Chung et al., 2018).

2.8.12.3 Quality Assurance of Frame-Based Cases

While the Icon system was optimized for using thermoplastic mask immobilization, the Icon CBCT system offers important benefits for quality assurance for G-frame-based treatments. Because the CBCT images are natively acquired in stereotactic space, they can be used as an independent check for G-frame-based treatment plans where stereotactic coordinates are defined using an indicator box that mounts to the frame system itself. These two stereotactic coordinate systems should nominally be identical. In practice, there are uncertainties in each measurement, which can be estimated by acquiring a pre-treatment CBCT of a G-frame patient and co-registering the CBCT images to the frame/fiducial images used for treatment planning. The co-registration differences represent the difference between the independent stereotactic coordinate systems. Problems such as mis-applied indicator boxes at the time of imaging or more seriously a shift in the frame between treatment planning imaging and the CBCT will become apparent as an unusually large co-registration error. Studies using the CBCT system in this way have demonstrated that G-frames have an expectedly low setup uncertainty, with mean residual uncertainties after setup reported to be on the order of 0.3 mm in translation on each orthogonal axis and rotations below 0.5 degrees around each axis (Dutta et al., 2018). Rarely, a larger frame shift may occur due to errors in pin placement, pin length, and head size. The CBCT system makes these shifts simple to detect and correct before a procedure commences (Peach et al., 2018). CBCT verification of frame stability may be especially helpful in situations where a frame has been placed using only three of the four posts (Stieler et al., 2018).

2.8.13 LIMITATIONS OF THE GAMMA KNIFE ICON

The Icon system represents a significant improvement in support for IG-HSRT procedures on the Gamma Knife platform. However, it is not without some limitation.

2.8.13.1 Support for Multiple Single-Fraction Workflows

One common workflow used with mask-based treatments is to treat multiple small lesions in a patient using multiple single-fraction procedures. The use of thermoplastic masks makes this workflow practical to accomplish; however, at present the treatment planning system provides no way to create a comprehensive treatment plan and simply select which targets to treat on which treatment sessions. Instead, individual treatments must be replanned manually and the total dose evaluated by accumulating dose as treatments progress.

2.8.13.2 Radiobiological Effects

The dose accumulation functionality of the treatment planning system calculates total dose by simple dose addition, not through the use of any radiobiological model. This brings up the question of how to manage various multi-session and reirradiation scenarios (Sanders et al., 2019).

2.8.13.3 The Future of Gamma Knife IG-HSRT: Advances in Treatment Planning with Gamma Knife Lightning

Gamma Knife treatment planning has historically utilized a forward-planning technique. The individual operating the treatment planning software was responsible for manually placing isocenters, in the process determining the number and collimator sizes of isocenters, relative isocenter weighting, and prescription isodose lines. Practical treatment tradeoffs such as acceptable conformity and dose falloff versus treatment time were left up to the individual.

The development of the Icon platform and the ability for performing IG-HSRT procedures using a flexible workflow makes the idea of adaptive radiosurgery treatments practical. More frequent treatment planning in turn would benefit from a more automated and consistent treatment planning paradigm. While recent versions of the Gamma Knife treatment planning system have included functionality to automatically place isocenters and then optimize treatments against dose metrics such as conformity, selectivity, and beam-time (Schlesinger et al., 2010), the system has not included a complete inverse-planning solution based on dose–volume constraints and objectives.

There have been several historical attempts to create these kind of fully functional inverse treatment planning solutions for GKRS (Shepard et al., 2000; Wu et al., 2003; Ghobadi et al., 2012; Tian et al., 2020; Xu et al., 2020); however, these have not been commercially adopted (with few exceptions [Levivier et al., 2018]). Elekta Instrument, AB, recently (at the time of this publication) announced the availability of Gamma Plan Lightning, which includes this functionality. The new treatment planning algorithm proceeds through three steps. The first step is isocenter placement, which is dependent on contouring the intended target. Once isocenters are placed, they remain fixed. After isocenter placement, an optimization algorithm based on a linear programming model attempts to find a solution that best meets various dose/volume and other treatment planning objectives and constraints by optimizing individual sector durations. Finally, a shot-sequencing step recombines these individually optimized sector durations into deliverable shots. Importantly, the algorithm can include practical treatment considerations such as beam-on-time into the optimization as well as more traditional dose/volume objectives/constraints. Early tests of the algorithm report fast (median time 5.7 seconds) optimization times, equal or better treatment planning metrics, and a factor of 2–3 reduction in beam-on-time as compared to traditional forward plans on the current generation of computer hardware distributed by the manufacturer (Sjolund et al., 2019). The lightening system will also include improvements in contouring, an ability to determine the order in which multiple targets are treated, as well as back-end improvements for functions such as backup and recovery.

2.9 CONCLUSIONS

Radiosurgery has evolved remarkably as compared to previous decades, led by advances in existing surgical-delivery platforms including the Gamma Knife platform. Radiosurgery has become more flexible, expanding from a strictly single-fraction modality to encompassing a variety of multi-fraction treatment strategies. The Gamma Knife platform has evolved to take advantage of technologies that make this flexibility possible without sacrificing the precision and accuracy that have established Gamma Knife as the "gold standard" for radiosurgery. Future improvements may focus on radiobiological considerations that will allow further customization of Gamma Knife IG-HSRT and truly personalized radiosurgical care.

REFERENCES

AlDahlawi I, Prasad D, Podgorsak MB (2017) Evaluation of stability of stereotactic space defined by cone-beam CT for the Leksell Gamma Knife icon. *Journal of Applied Clinical Medical Physics* 18:67–72.

Alheit H et al. (1999) Stereotactically guided conformal radiotherapy for meningiomas. *Radiotherapy and Oncology: Journal of the European Society for Therapeutic Radiology and Oncology* 50:145–150.

Baumert BG, Egli P, Studer S, Dehing C, Davis JB (2005) Repositioning accuracy of fractionated stereotactic irradiation: Assessment of isocentre alignment for different dental fixations by using sequential CT scanning. *Radiotherapy and Oncology: Journal of the European Society for Therapeutic Radiology and Oncology* 74:61–66.

Casentini L, Fornezza U, Perini Z, Perissinotto E, Colombo F (2015) Multisession stereotactic radiosurgery for large vestibular schwannomas. *Journal of Neurosurgery*: 1–7.

Chung HT et al. (2018) Assessment of image co-registration accuracy for frameless Gamma Knife surgery. *PloS One* 13:e0193809.

Davey P, Schwartz M, Scora D, Gardner S, O'Brien PF (2007) Fractionated (split dose) radiosurgery in patients with recurrent brain metastases: Implications for survival. *British Journal of Neurosurgery* 21:491–495.

Dong L, Shiu A, Tung S, Boyer A (1997) Verification of radiosurgery target point alignment with an electronic portal imaging device (EPID). *Medical Physics* 24:263–267.

Dutta SW et al. (2018) Stereotactic shifts during frame-based image-guided stereotactic radiosurgery: Clinical measurements. *International Journal of Radiation Oncology, Biology, Physics* 102:895–902.

Eriksson M, Nordström H (2014) Design and performance characteristics of a cone beam CT system for the LGK perfexion. In: *17th International Leksell Gamma Knife Society Meeting*. New York, NY.

Eriksson M, Nutti B, Hennix M, Malmberg A, Nordström H (2014) Position accuracy analysis of the stereotactic reference defined by CBCT on LGK perfexion. In: *17th International Leksell Gamma Knife Society Meeting*. New York, NY.

Ghobadi K, Ghaffari H, Aleman D, Ruschin M, Jaffray D (2012) SU-D-211–03: An Automated inverse planning optimization approach for single-fraction and fractionated radiosurgery using Gamma Knife perfexion. *Medical Physics* 39:3610.

Jaffray DA (2007) Kilovoltage volumetric imaging in the treatment room. *Frontiers of Radiation Therapy and Oncology* 40:116–131.

Jee TK et al. (2014) Fractionated Gamma Knife radiosurgery for benign perioptic tumors: Outcomes of 38 patients in a single institute. *Brain Tumor Research and Treatment* 2:56–61.

Kirchheiner K et al. (2014) Posttraumatic stress disorder after high-dose-rate brachytherapy for cervical cancer with 2 fractions in 1 application under spinal/epidural anesthesia: Incidence and risk factors. *International Journal of Radiation Oncology, Biology, Physics* 89:260–267.

Kunieda E et al. (2009) The reproducibility of a HeadFix relocatable fixation system: Analysis using the stereotactic coordinates of bilateral incus and the top of the crista galli obtained from a serial CT scan. *Physics in Medicine and Biology* 54:N197–204.

Leksell Gamma Knife Society (2019) Leksell Gamma Knife: Indications treated 1968–2019. In: (Society LGK, ed).

Leksell L (1951) The stereotaxic method and radiosurgery of the brain. *Acta Chirurgica Scandinavica* 102:316–319.

Levivier M, Carrillo RE, Charrier R, Martin A, Thiran JP (2018) A real-time optimal inverse planning for Gamma Knife radiosurgery by convex optimization: Description of the system and first dosimetry data. *Journal of Neurosurgery* 129:111–117.

Lindquist C, Paddick I (2007) The Leksell Gamma Knife Perfexion and comparisons with its predecessors. *Neurosurgery* 61:130–140; discussion 140–131.

Lunsford LD, Flickinger JC, Steiner L (1988) The Gamma Knife. *JAMA* 259:2544.

Ma L, Pinnaduwage D, McDermott M, Sneed PK (2014) Whole-procedural radiological accuracy for delivering multi-session Gamma Knife radiosurgery with a relocatable frame system. *Technology in Cancer Research & Treatment* 13:403–408.

MacDonald RL, Lee Y, Schasfoort J, Soliman H, Sahgal A, Ruschin M (2020) Real-time infrared motion tracking analysis for patients treated with gated frameless image guided stereotactic radiosurgery. *International Journal of Radiation Oncology, Biology, Physics* 106:413–421.

Minniti G et al. (2010) Fractionated stereotactic radiotherapy for skull base tumors: Analysis of treatment accuracy using a stereotactic mask fixation system. *Radiation Oncology* (London, England) 5:1.

Minniti G et al. (2014) Fractionated stereotactic radiosurgery for patients with skull base metastases from systemic cancer involving the anterior visual pathway. *Radiation Oncology* (London, England) 9:110.

Peach MS, Trifiletti DM, Dutta SW, Larner JM, Schlesinger DJ, Sheehan JP (2018) Spatial shifts in frame-based Gamma Knife radiosurgery: A case for cone beam CT imaging as quality assurance using the Gamma Knife(R) Icon. *Journal of Radiosurgery and SBRT* 5:315–322.

Pouliot J et al. (2005) Low-dose megavoltage cone-beam CT for radiation therapy. *International Journal of Radiation Oncology, Biology, Physics* 61:552–560.

Reisberg DJ, Shaker KT, Hamilton RJ, Sweeney P (1998) An intraoral positioning appliance for stereotactic radiotherapy. *The Journal of Prosthetic Dentistry* 79:226–228.

Rosenberg I, Alheit H, Beardmore C, Lee KS, Warrington AP, Brada M (1999) Patient position reproducibility in fractionated stereotactic radiotherapy: An update after changing dental impression material. *Radiotherapy and Oncology: Journal of the European Society for Therapeutic Radiology and Oncology* 50:239–240.

Ruschin M et al. (2013) Cone beam computed tomography image guidance system for a dedicated intracranial radiosurgery treatment unit. *International Journal of Radiation Oncology, Biology, Physics* 85:243–250.

Ruschin M et al. (2010) Performance of a novel repositioning head frame for Gamma Knife perfexion and image-guided linac-based intracranial stereotactic radiotherapy. *International Journal of Radiation Oncology, Biology, Physics* 78:306–313.

Ryken TC et al. (2001) Initial clinical experience with frameless stereotactic radiosurgery: Analysis of accuracy and feasibility. *International Journal of Radiation Oncology, Biology, Physics* 51:1152–1158.

Salter BJ et al. (2001) The TALON removable head frame system for stereotactic radiosurgery/radiotherapy: Measurement of the repositioning accuracy. *International Journal of Radiation Oncology, Biology, Physics* 51:555–562.

Sanders J, Nordstrom H, Sheehan J, Schlesinger D (2019) Gamma Knife radiosurgery: Scenarios and support for re-irradiation. *Physica Medica: PM: An International Journal Devoted to the Applications of Physics to Medicine and Biology: Official Journal of the Italian Association of Biomedical Physics* 68:75–82.

Sayer FT, Sherman JH, Yen CP, Schlesinger DJ, Kersh R, Sheehan JP (2011) Initial experience with the eXtend System: A relocatable frame system for multiple-session Gamma Knife radiosurgery. *World Neurosurgery* 75:665–672.

Schlesinger D, Xu Z, Taylor F, Yen CP, Sheehan J (2012) Interfraction and intrafraction performance of the Gamma Knife Extend system for patient positioning and immobilization. *Journal of Neurosurgery* 117(Suppl):217–224.

Schlesinger DJ, Sayer FT, Yen CP, Sheehan JP (2010) Leksell GammaPlan version 10.0 preview: Performance of the new inverse treatment planning algorithm applied to Gamma Knife surgery for pituitary adenoma. *Journal of Neurosurgery* 113(Suppl):144–148.

Shepard DM, Ferris MC, Ove R, Ma L (2000) Inverse treatment planning for Gamma Knife radiosurgery. *Medical Physics* 27:2748–2756.

Simonova G, Novotny J, Novotny J, Jr., Vladyka V, Liscak R (1995) Fractionated stereotactic radiotherapy with the Leksell Gamma Knife: Feasibility study. *Radiotherapy and Oncology: Journal of the European Society for Therapeutic Radiology and Oncology* 37:108–116.

Sjolund J, Riad S, Hennix M, Nordstrom H (2019) A linear programming approach to inverse planning in Gamma Knife radiosurgery. *Medical Physics* 46:1533–1544.

Stieler F, Wenz F, Schweizer B, Polednik M, Giordano FA, Mai S (2018) Validation of frame-based positioning accuracy with cone-beam computed tomography in Gamma Knife Icon radiosurgery. *Physica Medica: PM: An International Journal Devoted to the Applications of Physics to Medicine and Biology: Official Journal of the Italian Association of Biomedical Physics* 52:93–97.

Sweeney R et al. (1998) Repositioning accuracy: Comparison of a noninvasive head holder with thermoplastic mask for fractionated radiotherapy and a case report. *International Journal of Radiation Oncology, Biology, Physics* 41:475–483.

Tian Z et al. (2020) A preliminary study on a multiresolution-level inverse planning approach for Gamma Knife radiosurgery. *Medical Physics* 47:1523–1532.

Toma-Dasu I, Sandstrom H, Barsoum P, Dasu A (2014) To fractionate or not to fractionate? That is the question for the radiosurgery of hypoxic tumors. *Journal of Neurosurgery* 121(Suppl):110–115.

Treuer H et al. (2006) Impact of target point deviations on control and complication probabilities in stereotactic radiosurgery of AVMs and metastases. *Radiotherapy and Oncology: Journal of the European Society for Therapeutic Radiology and Oncology* 81:25–32.

Viola P, Wells III WM (1997) Alignment by maximization of mutual information. *International Journal of Computer Vision* 24:137–154.

Wright G, Harrold N, Hatfield P, Bownes P (2017) Validity of the use of nose tip motion as a surrogate for intracranial motion in mask-fixated frameless Gamma Knife Icon therapy. *Journal of Radiosurgery and SBRT* 4:289–301.

Wright G, Schasfoort J, Harrold N, Hatfield P, Bownes P (2019) Intra-fraction motion gating during frameless Gamma Knife Icon therapy: The relationship between cone beam CT assessed intracranial anatomy displacement and infrared-tracked nose marker displacement. *Journal of Radiosurgery and SBRT* 6:67–76.

Wu QJ et al. (2003) Real-time inverse planning for Gamma Knife radiosurgery. *Medical Physics* 30:2988–2995.

Xu Q et al. (2020) Tuning-target-guided inverse planning of brain tumors with abutting organs at risk during Gamma Knife stereotactic radiosurgery. *Cureus* 12:e9585.

3 CyberKnife Image-Guided Hypofractionated Stereotactic Radiotherapy

Christopher McGuinness, Martina Descovich and Cynthia Chuang

Contents

3.1 HISTORICAL OVERVIEW OF THE CYBERKNIFE SYSTEM

The CyberKnife system is a fully integrated platform for stereotactic radiosurgery (SRS) and stereotactic body radiation therapy (SBRT) treatments. The delivery system consists of a linear accelerator (Linac) mounted on a robotic arm enabling the delivery of radiation from hundreds of noncoplanar, nonisocentric beams around the patient. Stereotactic targeting accuracy is achieved by combining real-time orthogonal x-ray images with advanced image recognition software for automatic tracking of bony landmarks, implanted fiducials, or clearly distinguishable tumors within the lung. This allows the delivery of highly conformal hypofractionated treatments in the entire body without the need for rigid fixation devices. Figure 3.1 shows a CyberKnife treatment vault with the important components labeled.

The CyberKnife system came to market in the late 1990s. The first prototype was installed at Stanford University. It was designed as a frameless alternative to the existing SRS systems for the treatment of brain and C spine lesions. The CyberKnife prototype was called Neutroton 1000. Since the initial design, Accuray Inc. (Sunnyvale, CA) released five CyberKnife models: the G3 system in 2002, the G4 system in 2005, the VSI system in 2009, the M6 system in 2012, and the S7 system in 2020.

Over the years, the development of new tracking methods (including fiducial-free spine and lung tracking) and the capability to track respiratory motion in real time allowed to expand the clinical applications of CyberKnife to several extracranial sites including the spine, lung, liver, pancreas, and prostate (Kilby et al., 2010). Further improvements in the beam collimator and delivery system resulted in a considerably faster treatment time and making it possible to treat larger lesions. Notably, the major change introduced by the M6™ system was the addition of the InCise™ micro-multileaf collimator (MLC) to the collimator system. The addition of the MLC has been shown to significantly reduce treatment time by 30% to 50% while maintaining or improving treatment quality (McGuinness et al., 2015; Kim et al., 2017).

Hardware improvements have been matched by advances in the treatment planning system (TPS). Multiplan® replaced the original On Target® in 2005, providing advanced dose optimization algorithms and beam/time reduction techniques (Schlaefer et al., 2008), Monte Carlo dose calculation (Ma et al., 2008), automatic segmentation, and deformable image registration. In 2017, the Precision® treatment planning system was released, which included the Volo™ optimizer. This new TPS and optimization algorithm significantly reduced the amount of time required for developing treatment plans while improving the quality and efficiency of the treatments delivered (Schüler et al., 2020).

3.1.1 SYSTEM SPECIFICATIONS

The CyberKnife system consists of an X-band cavity magnetron and a side-coupled standing wave LINAC mounted on a robotic manipulator (Kuka Roboter GmbH, Augsburg, Germany). The linac produces an unflattened 6 MV photon beam with a dose rate up to 1000 cGy/min. The beam is collimated using one of three collimator systems: (1) the fixed collimator assembly (FCA), (2) the Iris™ variable aperture collimator, and (3) the InCise™ micro-multileaf collimator (MLC). The FCA consists of 12 circular tungsten cones with diameters ranging from 5 to 60 mm. Field size is defined at a source-to-axis distance (SAD) of 800 mm. The Iris collimation system consists of two hexagonal banks of tungsten, producing a 12-sided aperture, with the same set of field sizes available as the FCA (Echner et al., 2009). The mechanical uncertainty of the Iris field sizes is 0.2 mm, which affects the output factor for the smallest field size (5, 7.5, and 10 mm). The uncertainty in output factor for the 5 mm aperture can be up to 10% and is approximately 1.4% for the 10 mm aperture. While the manufacturer restricts the use of the 5 mm aperture, we do not recommend using either 5 or 7.5 mm Iris aperture for clinical cases.

Plans generated with multiple apertures typically result in better quality (dose conformity and gradient) and require a lower number of monitor units (MUs) (Pöll et al., 2008). However, using multiple fixed cones is not practical because it requires multiple path traversals and results in excessively long treatment times.

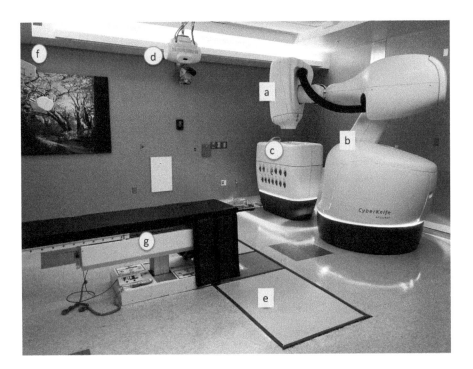

Figure 3.1 Image of a CyberKnife treatment suite. (a) Linear accelerator. (b) Robotic manipulator arm. (c) Exchange table with the fixed collimator assembly, the Iris™ variable aperture collimator, and the InCise™ micro-multileaf collimator. (d) X-ray imaging source. (e) Flat-panel detector. (f) Synchrony® camera array. (g) Patient positioning couch.

The Iris collimator allows using multiple apertures without these limitations. To further improve the delivery efficiency, the Incise MLC collimator was introduced in 2014, followed by an updated model in 2015. The first MLC model consists of 41 pairs of tungsten leaves with a width of 2.5 mm allowing a maximum field size of 120 mm (leaf motion direction) by 102.5 mm at 800 mm SAD. The second MLC model consists of 26 pairs of leaves with a width of 3.85 mm allowing a maximum field size of 115 mm by 100.1 mm. The leaves are interdigitated and can reach fully over-traveled positions. The leaf's height is 90 mm, and the maximum interleaf leakage is less than 0.5% (Accuray Inc., 2018). The addition of the micro-MLC has been shown to reduce MU and treatment time by 30% to 50% with equivalent or improved conformality, dose gradient, and critical organ sparing (Van De Water et al., 2011; McGuinness et al., 2015). Kim et. al. published a comparison of 144 cases of spine SBRT treatments where 78 were treated with fixed collimators, and 66 were treated with MLC collimator. They demonstrated a reduction in dose gradient and treatment times by 30% for the MLC cases while maintaining equivalent or improved dose coverage, conformity, and local recurrence rates between the two groups (Kim et al., 2017).

An automated exchange table system enables switching between the collimator housing. For early CyberKnife models, the exchange table contains receptacle storage spaces for the Iris housing, the FCA, and the 12-fixed tungsten cones and enables changing the cones automatically during treatment. For the M6 and S7 models, the exchange table contains the additional storage space for the MLC assembly. However, due to space limitation on the exchange table, the automatic exchange of the 12-fixed cones is no longer available.

Treatments are delivered from hundreds of beams arranged around the target. Each beam is defined by a source point, called a node, a direction, and a field size. Plans with the micro-MLCs may have several segments with different MLC leaf patterns for each beam. The complete set of nodes is called the path set and contains a different number of positions depending on the collimator type and treatment site. For the fixed and Iris collimator, the head path contains 179 nodes and the body path contains 117 nodes. For the MLC, the head path contains 171 nodes and the body path contains 102 nodes. The MLC path sets have fewer nodes to accommodate the slightly larger MLC housing.

The image-guided system consists of two diagnostic x-ray sources mounted in the ceiling and two amorphous silicon flat-panel detectors embedded in the floor, imaging the patient from two orthogonal oblique views at ±45°. Target localization during patient setup and treatment delivery is achieved by comparing the live x-ray images with a library of digitally reconstructed radiographs (DRRs) pre-generated from the planning CT at 45° angles through the imaging center. Based on this comparison, the tracking software calculates the differences in the three translational and three rotational directions between simulation and treatment positions as the couch correction parameters. Patients are positioned on a motorized treatment table with either five or six degrees of freedom, depending on the couch model. If the couch correction parameters are below the threshold set for treatment, the robot retargets the radiation beams, without the need to stop the treatment to move the patient couch.

The CyberKnife system includes an integrated treatment planning system allowing a fully autonomous environment for image fusion, contouring, DRR generation, treatment planning, plan evaluation, and patient-specific QA generation. In 2017, a new TPS, Precision, was released (Accuray, Inc.). It offers the same capabilities as the previous system (Multiplan) but with several improvements such as the new optimization algorithm, Volo, which reduces optimization time and improves plan quality (Schüler et al., 2020; Zeverino et al., 2019).

3.2 PATIENT SETUP AND TREATMENT SIMULATION

Proper patient setup and simulation is important for ensuring the full capabilities of the system during treatment planning and delivery. It is particularly important to ensure the patient is comfortable at the time of simulation so they can maintain the same position over the course of a 20- to 60-minute treatment. For brain lesions, a thermoplastic head mask with a headrest should be used. For cervical spine lesions, a head and shoulder mask should be used to minimize motion of the head and neck. For thoracic or lumbar spine lesions, a vacuum bag or foam cradle can be used to immobilize the thoracic, abdominal, or pelvic regions. Alternatively, patients can be positioned just on a foam pad to improve comfort, as patients positioned comfortably are less likely to move during treatment. For thoracic cases, the patient can be placed on a thick pad so their arms fall below the level of the body, thereby increasing the potential number of lateral beams that can be used without concern for beams passing through the arms. This is preferred instead of raising the arms overhead for two reasons: (1) to prevent the arms extending outside of the patient safety zone, which could potentially cause collision; (2) the position of arms overhead could be difficult and tiring for patients to maintain for the whole treatment duration. For lumbar and pelvic cases, the arms can rest on the patient's chest.

CT simulation is usually performed with the patient in the supine position. A CT scan with slice thickness between 1 and 1.5 mm is recommended. The slice thickness is important as finer slices result in higher-resolution DRRs and ultimately result in better tracking accuracy (Adler et al., 1999). The scan should be centered on the target extending 10 to 15 cm above and below the superior and inferior border of the target and/or encompassing all the organs at risk (OAR) such as lungs, bowels, stomach, or liver. This may be a longer scan than is typical for linac-based treatments because the noncoplanar beam arrangement in CyberKnife requires the scan to include any region along the patient anatomy where a potential beam will enter. The primary CT used for treatment planning must be a noncontrast CT as the contrast-enhancing agents might distort the quality of the DRR and impact tracking accuracy.

3.3 VOLUME DEFINITION AND TREATMENT PLANNING

A CT image is required for dose calculation during treatment planning and to generate DRRs used for patient setup and tracking during treatment delivery. Other imaging modalities such as MRI, PET, or additional CT scans can be incorporated directly in the TPS and registered to the primary CT image. Image registration and fusion can be performed manually or semiautomatically using fiducial marker positions or maximization of mutual information (Maurer and West, 2006). In the Precision TPS, a fast multi-modal method for deformable image registration (DIR) is also available (P Jordan et al., Accuray deformable image registration: description and evaluation, Accuray White Paper), enabling fast autosegmentation of cranial and head and neck anatomy. The autosegmentation tool is based on an atlas-based approach. Due to the variability of cranial anatomy, the system matches the patient image with multiple atlas images and chooses one optimal CT and three optimal MR atlas images. It uses a nonrigid registration algorithm to map the atlas image onto the patient's CT and T1-weighted MR image (Studholme et al., 1996). A set of warped contours is generated for the patient image from the set of atlas contours following the registration process.

The MR images are fused onto CT images for contouring, treatment planning, and evaluation. Brain lesions are contoured using gadolinium-enhanced T1-weighted Magnetization Prepared—RApid Gradient Echo (MPARAGE) or similar sequences and T2-weighted fluid-attenuated inversion recovery (FLAIR) MRI sequences. Primary brain tumor lesions, postoperative resection cavities, single and multiple brain metastasis, and benign diseases (such as trigeminal neuralgia or arteriovenous malformation) can be treated on the CyberKnife. Ma et al. evaluated the relationship of number of targets and radiosurgery platform with the dose to normal brain (Ma et al., 2011a) and developed an optimization technique to improve the planning quality of multiple metastasis treatments (Ma et al., 2011b). Small lesions are typically treated in a single fraction as in Gamma Knife radiosurgery. Larger lesions, or lesion located in critical areas (near the optic structures or the brainstem), are treated in 3–5 fractions. The high conformality and steep dose gradient for a case with multiple brain metastases can be seen in Figure 3.2. Brain metastases are usually treated with small planning target volume (PTV) margin (0–1 mm). For postoperative brain cases, the surgical resection cavity is usually expanded by 2 mm to create the clinical target volume (CTV)/PTV (Murphy, 2009).

Target volume definition for spinal SBRT is described in Radiation Therapy Oncology Group (RTOG) 0631 (Ryu, 2011) and in a consensus report (Cox et al., 2012). To summarize, MR and CT images are fused to help define the target volume and spinal cord. The CTV should encompass any abnormal marrow signal and adjacent normal bone. Single and multilevel spinal lesions can be treated with the CyberKnife. Figure 3.3 shows an example of dose distribution for a single thoracic spine lesion. A highly conformal dose distribution can be achieved with sharp dose falloff near the spinal cord. Notably, even the low-dose isodose line (i.e., 5 Gy) bends away from the spinal cord. Sahgal et al. developed a treatment planning approach to improve the dose distribution in multiple consecutive vertebral body metastases (Sahgal et al.,

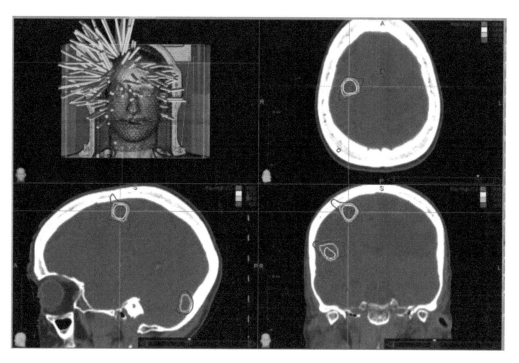

Figure 3.2 An example of a plan with three separate brain metastases. The prescription dose for this case was 19 Gy as shown in red. The 5 Gy isodose is shown in blue.

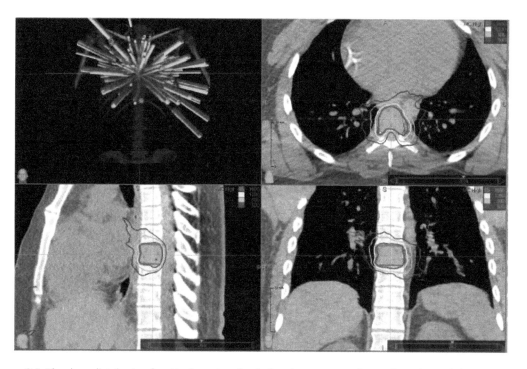

Figure 3.3 The dose distribution for this thoracic spine lesion demonstrates the conformality and sharp dose fall-off near the spinal cord that can be achieved with a large number of noncoplanar beams available on CyberKnife. The prescription isodose for this case was 16 Gy shown in red. The 8 and 5 Gy isodose levels are shown in green and blue, respectively.

2008). The spinal cord must be taken into special consideration for these treatments. Often a 2-mm expansion is included on the contoured spinal cord volume, and the expanded volume is subtracted from the PTV adding a safety margin to compensate for contouring and registration uncertainties and possible misalignment during treatment. Chuang et al. investigated the effects of residual target motion in CyberKnife radiosurgery and calculated patient-specific residual target motions on the order of 2 mm (Chuang et al., 2007). Fürweger et al. (2010) evaluated the targeting accuracy and residual motion in 260 patients treated with single-fraction CyberKnife radiosurgery and concluded that submillimeter targeting accuracy could be achieved despite patient motion. In a more recent study, Pantelis et al. (2018) reported the total geometric treatment uncertainty of a CyberKnife system using phantom-based and patient-based methods. The clinical targeting accuracy was estimated by analyzing treatment and follow-up data of a patient treated for a thalamic functional lesion. All their measurements demonstrated a total system uncertainty less than 1 mm for fiducial tracking, Xsight spine tracking, and 6D skull tracking methods.

3.4 PLAN OPTIMIZATION AND DOSE CALCULATION

CyberKnife can deliver both isocentric and non-isocentric plans. In isocentric plans, all the beams are directed to a single point in space, called treatment isocenter. Isocentric plans are adequate only for small spherical targets and have limited applications. The majority of treatments are delivered via non-isocentric beams directed to the periphery of the target. In Precision® TPS, non-isocentric plans can be generated using two optimization methods: Sequential and Volo.

The Sequential optimization algorithm proposed by Schlaefer and Schweikard (2008) was developed in early versions of Multiplan® TPS to mimic the decision-making process of a clinician. The optimization problem is framed, given thousands of possible beams defined by node position, beam angle, and collimator size (for fixed or Iris plans) or segment shape (for MLC plans). Once the beam parameters are chosen, the user can define dose–volume constraints and objectives, and the optimization algorithm finds the best subset of beams and beam weights to meet them. However, rather than setting weights to prioritize importance (as in simplex or iterative optimization algorithms), the objectives are defined in the order of decreasing importance. The optimizer manipulates beams and beam weights until the first objective is met and then proceeds to the next objectives sequentially. The solution for each prior step becomes a constraint with a user-defined relaxation factor as the optimizer moves to subsequent objectives. In this way, target coverage can be guaranteed before minimizing dose to OAR. While Sequential optimization works well for relatively simple cases, it has some limitations. In particular, for MLC plans the optimization process requires the generation of predefined segment shapes, resulting in long optimization times. In order to improve the optimization speed and to incorporate delivery efficiency in the optimization problem, the Volo optimizer was developed. In the Volo optimizer, the dose–volume goals, their importance (weighting), and the delivery efficiency objectives are all combined in a single cost function. For MLC plans, a fluence-based optimization step is followed by segmentation and aperture adaptation. It has been shown that plans optimized with Volo have superior dosimetric characteristics, are more efficient, and can be delivered in less time, compared to plan generated with the Sequential optimizer (Schüler et al, 2020).

Two dose calculation algorithms are available: ray tracing and Monte Carlo. The ray-tracing algorithm accounts for heterogeneity corrections along the primary path only. It computes an effective path length based on the electron density in the CT image but does not include effects of tissue inhomogeneity on scattered radiation. A contour correction is applied to the ray-tracing algorithm to estimate the effective depth of off-axis points. The beam for a given collimator size is divided into 12 equally spaced rays at 30° intervals around the perimeter of the cone, which are calculated using a trilinear interpolation with the nearest four rays. Contour correction improves the accuracy of dose calculation for oblique beam incidence and should be selected in the case of superficial targets.

The Monte Carlo algorithm includes the effect of tissue inhomogeneity on the scattered dose, which can be quite significant at air–tissue interfaces and somewhat significant at bone–tissue interfaces. Differences in dose calculations can be quite significant in lung cases when planning with ray tracing versus Monte Carlo algorithm (Wilcox et al., 2010). It is recommended to use Monte Carlo for final dose calculation in all thoracic cases and for targets near the sinuses or other air cavities.

3.5 TREATMENT DELIVERY AND IMAGE GUIDANCE

The CyberKnife system is capable of delivering highly conformal dose distributions with stereotactic imaging accuracy making it well suited for hypofractionated treatments. To ensure the conformal dose distribution is being delivered to the desired target volume while sparing adjacent OAR, highly accurate target localization and real-time tracking capabilities are implemented using sophisticated image guidance. A pair of orthogonal kilovoltage x-ray sources and detectors provides high contrast images of bony landmarks or fiducial markers which can be used for patient setup and accurate motion tracking in real time throughout the treatment. Images can be taken every 15 to 150 seconds (typical imaging frequency is 30 to 60 seconds, depending on treatment site).

3.5.1 FIDUCIAL TRACKING

Fiducial tracking uses radio-opaque markers for positioning. Ideally, three or more separate markers with adequate distance apart should be used to provide 6D corrections (three translations and three rotations). This is most commonly used for prostate and liver lesions where fiducial markers are implanted directly into the organ. It can also be used for lung cases though risk of pneumothorax due to fiducial implantation must be considered for this approach. There is a fiducial-free option for tracking lung lesions that can be clearly distinguished on orthogonal x-ray images. Screws or pins fixed to the vertebral body can also be used for fiducial tracking though this is rare as other fiducial-less tracking methods have been developed for spine lesions.

3.5.2 6D SKULL TRACKING

Skull tracking is used for intracranial cases or for any site that is considered fixed with respect to the skull. The patient's skull is imaged with 2D orthogonal images, and a transformation algorithm determines the best linear transformation between the image and the DRR. The transformations are combined and back projected to determine the 6D transformation that best aligns the current skull position to the original planning CT skull position. The algorithm is described by Fu and Kuduvalli (2008).

3.5.3 XSIGHT SPINE TRACKING

Xsight spine tracking is used for spine lesions—cervical, thoracic, lumbar, and sacral—or for any sites that are considered fixed with respect to the spine. Image registration is based on the differential contrast between bony features in the vertebral bodies. During planning, the user defines an imaging center that is just anterior to the spinal cord and midline relative to the vertebral body. A grid of 81 (9 × 9) nodes, shown in Figure 3.4, is displayed over each of the two orthogonal DRRs, usually encompassing several vertebral bodies. The user can adjust the overall size of the grid to maximize the number of nodes containing bony features. It is best to place the middle node (imaging isocenter) at a location with higher bony density in the DRR to ensure the algorithm's capability to calculate rotations consistently. A box matching algorithm computes local displacement vectors for each node point between the image taken of the patient during treatment and the original DRR and computes a final translation and rotation vector used to register the patient (Fu et al., 2006). The algorithm has been demonstrated to be very robust with a total system accuracy of 0.61 mm (Ho et al., 2007).

3.5.4 XSIGHT SPINE TRACKING IN THE PRONE POSITION

Spinal treatments delivered in the prone position can benefit from decreased dose given to anterior organs such as the heart and bowels (Descovich et al., 2012). This is due to the increased number of beams available from posterior directions that are unavailable when the patient is positioned supine due to physical limitation of the robot and couch. However, breathing motion becomes a significant problem for spine treatments when the patient is prone (Fürweger et al., 2011). Even if breathing motion is compensated, a 2-mm margin should be added to the CTV to account for the reduced accuracy of respiration-compensated tracking. This additional margin may reduce the potential dosimetric gain of prone treatments for spine lesions, and careful consideration criteria for patient selection should be applied (Fürweger et al., 2014).

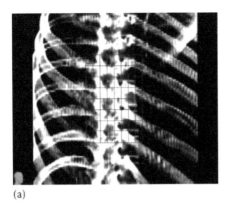 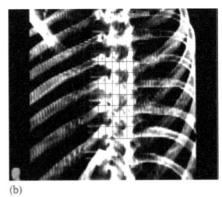

(a)   (b)

Figure 3.4 Digitally reconstructed radiographs (DRR) for the two orthogonal views are shown in panels (a) and (b). Xsight spine tracking compares features in orthogonal x-ray images taken during patient setup with DRR generated in the planning computed tomography to determine 6D corrections. The algorithm compares bony features within the blue grid shown in the figure. The user determines the grid size and location during the planning process.

3.5.5 SYNCHRONY RESPIRATORY TRACKING

Synchrony (Accuray, Inc.) is a motion management system that accounts for breathing motion. Synchrony can be used in combination with Fiducial, Xsight Lung, and Xsight Spine prone tracking. The robot position is continuously readjusted to follow a moving target based on the correlation model prediction of the target location. Prior to treatment, a series of x-ray images are used to develop a correlation model between the positions of a set of infrared light-emitting diodes (LEDs) on the patient's body surface and the target. The model is updated during treatment everytime x-ray images are taken, approximately every 60 seconds. As the patient breathes, the beams are adjusted to follow the motion. Overall tracking accuracy of less than 1.5 mm is possible using this tracking method (Sonja et al., 2011). Yang et. al. measured the 95% tracking confidence interval to be within 0.66 mm for sinusoidal respiratory motion of amplitude ≤20 mm (Yang et al., 2019).

CHECKLIST: KEY POINTS FOR CLINICAL PRACTICE

- CyberKnife enables accurate delivery of IG-HSRT treatment to patients with intracranial and spinal lesions
- The system consists of a compact linac attached to a robotic manipulator
- CyberKnife IG-HSRT is frameless as image guidance is performed constantly throughout the treatment
- Plans consist of hundreds of highly focused, noncoplanar radiation beams, which enable one to achieve highly conformal dose distribution with steep-dose gradient
- An understanding of the system operating principles is essential to plan simulation and delivery procedures appropriately

REFERENCES

Accuray Inc. (2018) CyberKnife treatment delivery system technical specifications.www.accuray.com/wp-content/uploads/cyberknife-treatment-delivery-system_-technical-specifications.pdf

Adler JR, Jr., Murphy MJ, Chang SD, Hancock SL (1999) Image-guided robotic radiosurgery. *Neurosurgery* 44(6):1299–1306.

Chuang C, Sahgal A, Lee L, Larson D, Huang K, Petti P, Verhey L, Ma L (2007) Effects of residual target motion for image-tracked spine radiosurgery. *Medical Physics* 34(11):4484.

Cox BW, Spratt DE, Lovelock M, Bilsky MH, Lis E, Ryu S, Sheehan J et al. (2012) International spine radiosurgery consortium consensus guidelines for target volume definition in spinal stereotactic radiosurgery. *International Journal of Radiation Oncology, Biology, Physics* 83(5):e597–605.

Descovich M, Ma L, Chuang CF, Larson DA, Barani IJ (2012) Comparison between prone and supine patient setup for spine stereotactic body radiosurgery. *Technol Cancer Research & Treatment* 11(3):229–236. www.ncbi.nlm.nih.gov/pubmed/22468994.

Echner GG, Kilby W, Lee M, Earnst E, Sayeh S, Schlaefer A, Rhein B et al. (2009) The design, physical properties and clinical utility of an iris collimator for robotic radiosurgery. *Physics Medical & Biology* 54(18):5359–5380.

Fu D, Kuduvalli G (2008) A fast, accurate, and automatic 2D–3D image registration for image-guided cranial radiosurgery. *Medical Physics* 35(5):2180.

Fu D, Kuduvalli G, Maurer CR, Adler JR (2006) 3D target localization using 2D local displacements of skeletal structures in orthogonal x-ray images for image-guided spinal radiosurgery. *International Journal of Computer Assist Radiological Surgery* 1:198–200.

Fürweger C, Drexler C, Kufeld M, Muacevic A, Wowra B, Schlaefer A (2010) Patient motion and targeting accuracy in robotic spinal radiosurgery: 260 single-fraction fiducial-free cases. *International Journal of Computer Assist Radiological Surgery* 78(3):937–945.

Füerweger C, Drexler C, Kufeld M, Wowra B (2011) Feasibility of fiducial-free prone-position treatments with CyberKnife for lower lumbosacral spine lesions. *Cureus* 3(1):e21.

Fürweger C, Drexler C, Muacevic A, Wowra B, de Klerck EC, Hoogeman MS (2014) CyberKnife robotic spinal radiosurgery in prone position: Dosimetric advantage due to posterior radiation access? *Journal of Applied Clinical Medical Physics/American College of Medical Physics* 15(4): 4427.

Ho AK, Fu D, Cotrutz C, Hancock SL, Chang ST, Gibbs IC, Maurer CR, Adler JR (2007) A study of the accuracy of CyberKnife spinal radiosurgery using skeletal structure tracking. *Neurosurgery* 60(2 Suppl 1):ONS147–ONS156.

Kilby W, Dooley JR, Kuduvalli G, Sayeh S, Maurer CR (2010) The CyberKnife robotic radiosurgery system in 2010. *Technology in Cancer Research & Treatment* 9(5):433–452.

Kim N, Lee H, Kim JS, Baek JG, Lee CG, Chang SK, Koom WS (2017). Clinical outcomes of multileaf collimator-based CyberKnife for spine stereotactic body radiation therapy. *The British Journal of Radiology*, 90(1079), 20170523.

Ma C-M, Li J S, Deng J, Fan J (2008) Implementation of Monte Carlo dose calculation for CyberKnife treatment planning. *Journal of Physics: Conference Series* 102:012016.

Ma L, Petti P, Wang B, Descovich M, Chuang C, Barani IJ, Kunwar S, Shrieve DC, Sahgal A, Larson DA (2011a) Apparatus dependence of normal brain tissue dose in stereotactic radiosurgery for multiple brain metastases. *Journal of Neurosurgery* 114(6):1580–1584.

Ma L, Sahfal A, Hwang A, Hu W, Descovich M, Chuang C, Barani I, Sneed PK, McDermott M, Larson DA (2011b) A two-step optimization method for improving multiple brain lesion treatments with robotic radiosurgery. *Technology in Cancer Research & Treatment* 10(4):331–338.

Maurer CR, Jr., West JB (2006) Medical image registration using mutual information. In: Heilbrun MP (ed.), *CyberKnife Radiosurgery: Practical Guide*, 2nd ed. Sunnyvale, CA: The CyberKnife Society.

McGuinness C, Gottschalk AR, Lessard E, Nakamura J, Pinnaduwage D, Pouliot J, Sims C, Descovich M (2015) Investigating the clinical advantages of a robotic linac equipped with a multi-leaf collimator in the treatment of brain and prostate cancer patients. *Journal of Applied Clinical Medical Physics* 16(5). doi:10.1120/jacmp.v16i5.5502.

Murphy MJ (2009) Intrafraction geometric uncertainties in frameless image-guided radiosurgery. *International Journal of Radiation Oncology, Biology, Physics* 73(5):1364–1368.

Pantelis E, Moutsatsos A, Antypas C, Zoros E, Pantelakos P, Lekas L, Romanelli P, Zourari K, Hourdakis CJ (2018) On the total system error of a robotic radiosurgery system: Phantom measurements, clinical evaluation and long-term analysis. *Physics Medical Biology* 20;63(16):165015.

Pöll JJ, Hoogeman MS, Prévost JB, Nuyttens JJ, Levendag PC, Heijmen BJ (2008) Reducing monitor units for robotic radiosurgery by optimized use of multiple collimators. Medical Physics 35(6):2294–2299.

Ryu S (2011) Radiation Oncology Group. RTOG 0631 protocol information. www.rtog.org/clinicaltrials/proto-coltable/studydetails.aspx?study=0631, Open to accrual date 2011.

Sahgal A, Chuang C, Larson D, Huang K, Petti P, Weinstein P, Ma L (2008) Split-volume treatment planning of multiple consecutive vertebral body metastases for CyberKnife image-guided robotic radiosurgery. *Medical Dosimetry* 33(3):175–179.

Schlaefer A, Schweikard A (2008) Stepwise multi-criteria optimization for robotic radiosurgery. *Medical Physics* 35(5):2094.

Schüler E, Lo A, Chuang CF, Soltys SG, Pollom EL, Wang L (2020). Clinical impact of the VOLO optimizer on treatment plan quality and clinical treatment efficiency for CyberKnife. *Journal of Applied Clinical Medical Physics* 21(5):38–47.

Sonja Dieterich Carlo Cavedon Cynthia F. Chuang Alan B. Cohen Jeffrey A. Garrett Charles L. Lee Jessica R. Lowenstein Maximian F. d'souza David D. Taylor Jr. Xiaodong Wu Cheng Yu (2011). Report of AAPM TG 135: Quality assurance for robotic radiosurgery. *Medical Physics* 38(6):2914–2936.

Studholme C, Hill DL, Hawkes DJ (1996) Automated 3-D registration of MR and CT images of the head. *Medical Image Analysis* 1(2):163–175.

Van De Water S, Hoogeman MS, Breedveld S, Nuyttens JJME, Schaart DR, Heijmen BJM (2011) Variable circular collimator in robotic radiosurgery: A time-efficient alternative to a mini-multileaf collimator? *International Journal of Radiation Oncology, Biology, Physics* 81(3):863–870.

Wilcox EE, Daskalov GM, Lincoln H, Shumway RC, Kaplan BM, Colasanto JM (2010) Comparison of planned dose distributions calculated by Monte Carlo and Ray-Trace algorithms for the treatment of lung tumors with CyberKnife: A preliminary study in 33 patients. *International Journal of Radiation Oncology, Biology, Physics* 77(1):277–284.

Yang B, Chiu TL, Law WK, Geng H, Lam WW, Leung TM, Yiu LH, Cheung KY, Yu SK (2019). Performance evaluation of the CyberKnife system in real-time target tracking during beam delivery using a moving phantom coupled with two-dimensional detector array. *Radiological Physics and Technology* 12(1):86–95.

Zeverino M, Marguet M, Zulliger C, Durham A, Jumeau R, Herrera F, Schiappacasse L, Bourhis J, Bochud FO, Moeckli R (2019) Novel inverse planning optimization algorithm for robotic radiosurgery: First clinical implementation and dosimetric evaluation. *Medical Physics* 64:230–237.

4 Linac-Based IG-HSRT Technology

Richard Popple

Contents

4.1 INTRODUCTION

Traditionally, high-precision, high-dose radiotherapy has required a stereotactic head-frame to transfer the target coordinates from the imaging frame of reference to the treatment delivery system frame of reference and to maintain the patient position during treatment. Because affixing a head frame is an invasive procedure, stereotactic radiotherapy was typically limited to single-fraction treatment of intracranial targets. Furthermore, imaging, planning, quality assurance, and treatment delivery had to be accomplished in a single day, making stereotactic procedures resource intensive. In the end, the planning and delivery techniques were limited to spherical dose distributions (shots), making planning and delivery challenging for large and complex target shapes. In 1994, Brada and Laing described the Royal Marsden Hospital experience using stereotactic radiotherapy (SRT) for brain tumors and, despite these limitations, concluded

> The technology of stereotactic radiotherapy is evolving, and it is likely that SRT will be integrated into conventional radiotherapy practice to become simply a high-precision technique of radiotherapy delivery in everyday use.
>
> **(Brada and Laing, 1994, 102)**

Their prediction is rapidly becoming a reality.

Brada and Laing identified four requirements for stereotactic radiotherapy: precise patient fixation, accurate target delineation, target localization, and means of delivery. Accurate target delineation has long been available in the form of diagnostic MR and CT imaging procedures. Precision patient fixation and target localization without a stereotactic frame have been enabled by the advent of image guidance. In-room megavoltage and kilovoltage imaging has made localization of a target volume possible without the need for a frame. Furthermore, image guidance in combination with appropriate patient immobilization has made stereotactic treatment possible for extracranial sites as well (Ryu et al., 2001). Six degree-of-freedom patient positioning systems accurately align the treatment planning and linear accelerator coordinate systems, further improving target localization. Intra-fraction motion monitoring systems have been developed to monitor the patient position during treatment, eliminating the need for a frame to maintain targeting accuracy. In the end, the linear accelerator as a means of delivery has significantly improved since the report of the Royal Marsden experience. Multileaf collimators (MLCs) coupled with computer optimized planning have made intensity modulated radiation therapy (IMRT) and modulated arc therapy (MAT) possible. IMRT and MAT can generate dose distributions that are highly conformal and have rapid dose fall-off, even for complex target shapes. IMRT and MAT are more efficient to deliver than shot-based techniques, reducing treatment times. Flattening filter free (FFF) beams, now entering widespread clinical use, have dose rates up to 2.4 times higher than conventional flattened beams. FFF beams further reduce the time required to deliver high-dose per fraction treatments.

4.2 LINEAR ACCELERATOR TECHNOLOGY

4.2.1 MULTILEAF COLLIMATOR

The multileaf collimator (MLC) provides beam shaping for modern C-arm linacs and is used to shape the aperture for three-dimensional conformal radiation therapy (3D-CRT) and dynamic conformal arc therapy (DCA) and to produce beam modulation for IMRT and modulated arc therapy. MLC technology has been extensively described elsewhere (see, for example, Van Dyk, 1999). Briefly, an MLC comprises variable position, high-density leaves arranged in opposing pairs such that the leaf edges are aligned parallel to the beam divergence. In the direction of leaf movement, there are two designs: double focused, for which the leaf end remains parallel to the beam divergence throughout the range of motion, and single-focused, for which the leaf moves perpendicular to the central axis. Single-focused MLCs have rounded leaf ends to maintain coincidence between the light and radiation fields and a nearly constant penumbra over the entire range of leaf motion. An MLC either replaces one set of secondary collimators or is a tertiary collimator located below the secondary collimators.

MLCs are the core technology enabling the C-arm linac delivery techniques used for hypofractionated radiation therapy. Three-dimensional conformal radiation therapy (3D-CRT) is a technique in which field apertures are designed to conform to the target shape for each of a number of beam directions, typically 3 to 9.3D-CRT has been in use for several decades and has been described extensively elsewhere (see, for example, Prado, Starkschall, and Mohan, 2012). Dynamic conformal arc (DCA) is an arc technique analogous to 3D-CRT in that the aperture shape varies to conform to the target shape as the gantry rotates. Intensity modulated radiation therapy (IMRT) uses multiple computer optimized apertures for each beam direction to generate non-uniform incident radiation fluence. The hallmark of IMRT is the ability to create concave dose distributions to better spare normal tissues. There are two types of MLC-based IMRT, static and dynamic. For static MLC (SMLC) IMRT, the MLC leaves are stationary while radiation is being delivered. For dynamic MLC (DMLC) IMRT, the MLC leaves move while radiation is being delivered. IMRT is in widespread use and has also been extensively described elsewhere (see, for example, Boyer, Ezell, and Yu, 2012) and volume 2 of Van Dyk, 1999). Modulated arc therapy (MAT) is a rotational technique in which the MLC aperture shape varies with gantry angle. In most modern applications, the dose rate and gantry speed also vary. MAT differs from conformal arc therapy in that the aperture shapes and weights are inverse planned to meet dosimetric objectives, rather than conforming to the target shape at all gantry angles.

The dosimetric properties of MLCs are similar between designs (Huq et al., 2002). The average of inter- and intra-leaf transmission is typically in the range of 1% to 2.5% and, for modulated techniques, has the effect of limiting the range of modulation achievable within the field defined by the secondary collimator. The effect of leaf leakage can be reduced by moving the secondary collimators to conform to the extrema of the leaf positions, a technique commonly referred to as jaw tracking or jaw following. For spine radiosurgery, jaw tracking has been shown to decrease the dose to the spinal cord for both dynamic MLC IMRT and MAT delivery techniques, although the effect for MAT was small and not of clinical significance (Snyder et al., 2014). The magnitude of the jaw tracking effect on DMLC relative to MAT was likely due to the different optimization techniques. The DMLC planning technique was fluence map optimization followed by leaf sequencing and resulted in leaf settings having small apertures, whereas the direct aperture optimization used for MAT resulted in larger apertures. Consequently, the modulation factor (a measure of the monitor units required to deliver the prescribed dose) was higher for DMLC relative to MAT, resulting in more dose due to leakage. Jaw tracking reduces the leakage dose and so had more effect on the DMLC plans than the MAT plans. The impact of jaw tracking on critical structure sparing is proportional to the degree of modulation.

MLC leaf widths, as projected to isocenter, typically range from 2.5 mm to 10 mm. The effect of leaf width on dosimetry is dependent on delivery technique, target volume, and target shape and is greatest for non-modulated techniques. For dynamic conformal arc (DCA), Dhabaan et al. found improvement in both conformity and normal tissue sparing for a 2.5 mm leaf width MLC compared to a 5 mm leaf width (Dhabaan et al., 2010). Jin et al. compared MLC leaf widths of 3 mm, 5 mm, and 10 mm for dynamic conformal arc and IMRT radiosurgery (Jin et al., 2005). For dynamic conformal arc, the conformity improved as the leaf width decreased with the largest improvement observed for the smallest targets. For IMRT, the conformity index was essentially the same for the 3 mm and 5 mm leaf widths; however, the narrower leaf width provided modestly better sparing of small critical structures for intracranial cases. Monk et al. compared MLCs having 3 mm and 5 mm leaf width for 3D conformal radiosurgery and found that although the conformity of the 5 mm MLC was not as good as that for the 3 mm MLC, the plans for the 5 mm leaf width met the RTOG clinical criteria (Monk et al., 2003). They concluded that the differences were small enough that the improvement in conformity index was not large enough to dictate equipment choice. Serna et al. found no improvement in conformity for a 2.5 mm width relative to a 5 mm leaf width but did find an improvement in dose falloff, particularly for targets smaller than 10 cm^3 (Serna et al., 2015). Wu et al. examined the effect of leaf width on the sparing of critical structures adjacent to the target volume, comparing a 2.5 mm and 5 mm leaf widths for SRS of targets abutting the brainstem or the spinal cord (Wu et al., 2009b). They found that the 2.5 mm leaf width had significantly improved sparing of adjacent critical structures, particularly for small targets and complex geometries. Chae et al. came to the same conclusion for spine targets, finding that target coverage was improved for both IMRT and MAT for a 2.5 mm leaf width, particularly for complex shape (Chae et al., 2014). The general conclusion of studies on the effect of leaf width is that a narrower leaf improves plan quality, but that the degree of improvement is dependent on the technique, the target size, and the target shape. The effect of leaf width on plan quality is greatest for dynamic conformal arc and least for modulated arc therapy. A smaller leaf width has the largest effect on plan quality for small targets and for targets with complex geometries.

4.2.2 MODULATED ARC THERAPY

Modulated arc therapy (MAT) is a rotational technique in which the multileaf collimator aperture shape varies with gantry angle. In most modern applications, the dose rate and gantry speed also vary. MAT differs from conformal arc therapy in that the aperture shapes and weights are inverse planned to meet dosimetric objectives, rather than conforming to the target shape at all gantry angles.

Modulated arc therapy (MAT) was first described by Yu in 1995 (Yu, 1995). Limited clinical implementation followed (Ma et al., 2001; Yu et al., 2002; Duthoy et al., 2004; Wong et al., 2005); however, due to lack of robust planning tools, it primarily remained a technique of academic interest. In 2008, the introduction of variable dose rate during gantry rotation by linac manufacturers in conjunction with clinically usable planning tools resulted in widespread adoption of MAT. There are two major types of modulated arc therapy, although the terminology for the two is often used interchangeably. Yu described intensity modulated arc therapy (IMAT), in which multiple arcs were delivered at a constant dose rate and gantry speed. The apertures of the arcs overlap to generate a nonuniform fluence from any given gantry angle. As originally conceived, the arcs were planned using a two-step process. The first step was fluence map optimization with no consideration of machine constraints, followed by a second step in which apertures were generated to approximate the optimal fluence map while considering the speed constraints of the

MLC leaves and the gantry. For IMAT, multiple arcs were necessary to deliver the optimized dose distribution. Otto described volumetric modulated arc therapy (VMAT) in which a single arc was combined with variable dose rate and gantry speed, and the aperture shapes and weights were directly optimized (Otto, 2008). In current use, the terms IMAT and VMAT are often used interchangeably to describe any type of inverse-planned therapy comprising one or more arcs during which the aperture shape changes during gantry rotation.

Conformal arc therapy has been widely used for linac intracranial radiosurgery (Shiu et al., 1997; Cardinale et al., 1998; Leavitt, 1998; Solberg et al., 2001); so it was a natural extension to use modulated arc therapy for stereotactic applications. The primary advantage of MAT relative to non-modulated techniques is the ability to improve conformity and/or sparing of critical structures. Multiple studies have compared MAT to dynamic conformal arc (DCA) (Wu et al., 2009a; Audet et al., 2011; Huang et al., 2014; Salkeld et al., 2014; Serna et al., 2015; Zhao et al., 2015). In general, these studies found that MAT produced more conformal dose distributions while maintaining similar low-dose spill, as measured by the volume of healthy brain receiving more than 50% of the prescription dose. Plan quality was found to be better for multiple noncoplanar arcs relative to a single arc. In addition to malignant disease, treatment of large intracranial arteriovenous malformations has been reported using MAT and hypofractionated RT with promising results (Subramanian et al., 2012). Other non-malignant lesions that are typically treated using non-modulated techniques are also amenable to MAT and can be treated with similar plan quality and higher efficiency (Abacioglu et al., 2014).

MAT has been applied to stereotactic spine radiotherapy, for which conformal techniques are not adequate (Wu et al., 2009a). Compared to conventional IMRT, MAT results in equivalent target dosimetry and reduced treatment time. The conformity of MAT is slightly improved relative to IMRT, because the limited number of entrance directions for IMRT results in more spillage of the prescription dose outside of the target volume; however, this difference is probably not of clinical significance. For the spinal cord, achieving equivalence of MAT to IMRT requires two or more arcs. Wu et al. found that a single arc was inferior to both two arcs and to IMRT. For example, for a prescription dose of 16 Gy, 1% of the spinal cord received more than 9 Gy for a single arc, whereas it received 8.5 Gy for two arcs and for conventional IMRT. The average dose–volume histograms for the PTV and spinal cord are shown in Figure 4.1. Interestingly, Wu et al. found that two arcs both improved plan quality and reduced treatment delivery time (8.56 minutes for one arc versus 7.88 minutes for two). This seemingly counter-intuitive result occurs because the increased freedom of two aperture shapes at each gantry position allows more efficient coverage of the target volume, resulting in fewer total monitor units from the two-arc plan.

Plan quality for MAT and IMRT is similar and so the primary advantage of MAT is reduced treatment time, although the degree of reduction depends on the specific techniques being compared. For example, for spine SBRT, Wu et al. found a treatment time reduction, whereas Kuijper found similar treatment times because Kuiper used more arcs for MAT and fewer beams for IMRT than Wu (Wu et al., 2009a; Kuijper et al., 2010). The largest component of the difference in treatment time between conventional IMRT and MAT is the time spent in-between beams waiting for data transfer and for gantry positioning. Increased automation of IMRT delivery, becoming available on modern linacs, will likely reduce the time difference between IMRT and MAT. Coupled with planning techniques that combine fixed and arc modulated fields (Matuszak et al., 2013), automated delivery systems will blur the distinction between fixed field IMRT and MAT (Popple et al., 2014).

4.2.3 FLATTENING FILTER FREE BEAMS

Linear accelerators generate megavoltage photon beams by bombarding a high-Z target with megavoltage electrons. The photon production is forward peaked and results in a non-uniform dose distribution. To ameliorate this issue, a nonuniform filter is introduced to differentially absorb photons such that the resulting dose distribution is uniform at a reference depth such as 10 cm. Introduction of the flattening filter comes at the penalty of reducing the dose rate throughout the field to that of the periphery. However, for small-field sizes such as those encountered in stereotactic irradiation, the field is reasonably uniform without the flattening filter. The use of fluence modulation techniques (IMRT and MAT) further reduces the utility of a flattening filter. Removing the flattening filter increases the dose rate near central axis by

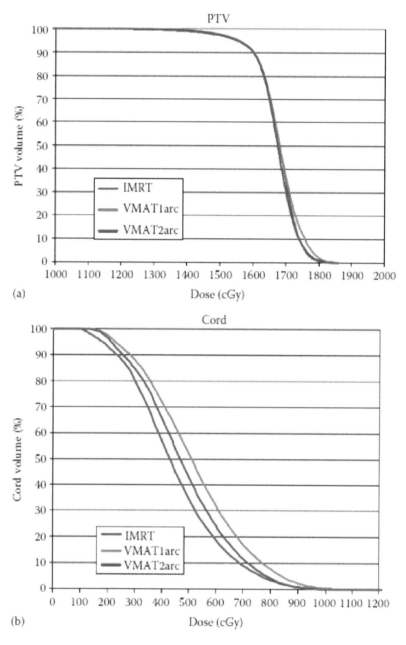

Figure 4.1 Average dose–volume histogram curves for the (a) planning target volume (PTV) and (b) spinal cord. (Reprinted from Wu et al., 2009a, copyright 2009, with permission from Elsevier.)

a factor of 2 to 4, depending on the beam energy, which can lead to significantly shorter delivery times for hypofractionated treatment regimens. In 1991, O'Brien reported removing the flattening filter from a 6 MV linac, achieving a dose rate increase of 2.75 and acceptable flatness (O'Brien et al., 1991). In 2007, Bayouth reported on image-guided radiosurgery, both intra- and extracranial, using a flattening filter free linear accelerator (Bayouth et al., 2007). Planning studies demonstrated feasibility for body radiosurgery as well (Vassiliev et al., 2009). C-arm linacs without a flattening filter became commercially available in 2010, and early clinical experience demonstrated treatment times for CNS radiosurgery approaching those of conventional fractionation, as shown in Figure 4.2.

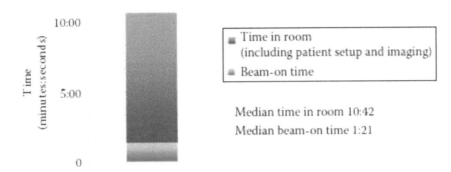

Figure 4.2 Graphical representation of treatment time for CNS SRS using FFF linac. The median radiation-beam on time was 1:21 while the median time the patient spent in the treatment room, including treatment setup and imaging, was 10:42. (Reprinted from Prendergast et al., 2011, copyright 2011, with permission from Old City Publishing.)

The flattening filter is a source of scattered photons that contribute significantly to leakage and peripheral dose. Flattening filter free beams reduce the peripheral dose by eliminating this source of scatter (Kry et al., 2010; Cashmore et al., 2011; Kragl et al., 2011). Furthermore, flattening filter free beams require less current at the target to deliver a given dose, thus reducing other sources of head leakage as well (Vassiliev et al., 2006). For 6 MV, removing the flattening filter reduces leakage radiation by 60% and the dose outside of the field edge by 11% (Cashmore, 2008). The impact of reduced leakage dose is illustrated in Figure 4.3, which compares the out-of-field dose for flattened and unflattened 6 MV beams due to intracranial irradiation in children (Cashmore et al., 2011). For SBRT, the use of flattening filter free beams has been shown to reduce the dose 20 cm from the field edge by 23% for 6 MV and 31% for 10 MV (Kragl et al., 2011).

A flattening filter free beam delivers a higher surface dose than a flattened beam of the same energy due to the lower energy components in the beam spectrum. However, the surface dose for a flattened beam has stronger field size dependence because the scatter from a flattening filter, which increases with increasing field size, is considerably lower in energy than the primary beam and increases the dose at shallow depths. The surface dose for large field sizes (>~30 cm) is similar between flattened and unflattened beams (Vassiliev et al., 2006; Cashmore, 2008; Kragl et al., 2009); however, at the smaller fields typical of hypofractionated radiation therapy, the surface dose is larger for filter free beams. For a 4×4 cm^2 field at 3 mm depth, the dose increase relative to a flattened beam of the same energy is approximately 20% for 6 MV (Vassiliev et al., 2006; Kragl et al., 2009) and 25% for 10 MV (Kragl et al., 2009). Surface dose decreases with increasing energy, and a 10 MV unflattened beam is about 25% less than a 6 MV unflattened beam and similar to that of a 6 MV flattened beam (Kragl et al., 2009). The surface dose is the same for fields collimated by MLCs or by the secondary collimators (Wang et al., 2012). Although the surface dose for flattening filter free beams is modestly higher than flattening filter, the increased dose is not likely of clinical significance, and utilizing a sufficient number of treatment fields or arcs, as is good SRS/SBRT practice, should be sufficient to mitigate the modestly higher relative surface dose of flattening filter free beams.

With respect to calibration, FFF beams have two characteristics different from conventional flattened beams. First, the beam spectrum has more low energy components compared to a flattened beam. However, the relationship between depth dose and stopping power ratio given in the American Association of Physicists in Medicine (AAPM) Task Group 51 (TG-51) protocol can be used (McEwen et al., 2014). The higher dose rate leads to concern about ion recombination; however, it has been demonstrated that the two-voltage technique remains valid for the dose rates produced by flattening filter free beams (Kry et al., 2012). If the reference chamber is sufficiently long, a small correction factor is needed to account for the change in dose rate over the length of the chamber; otherwise, the calibration of flattening filter free beams is straightforward and uses the same protocol as a flattened beam (McEwen et al., 2014).

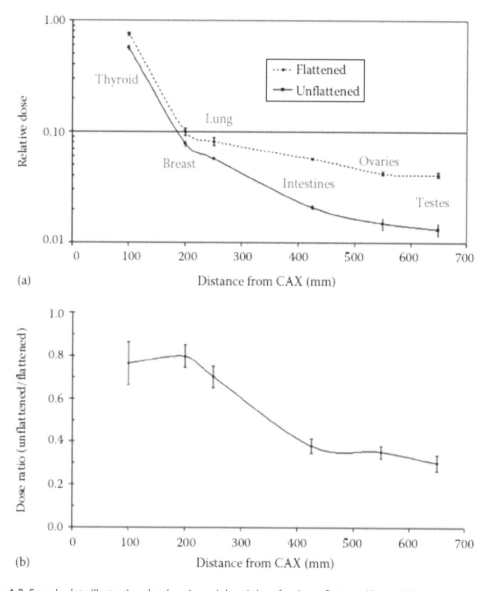

Figure 4.3 Sample data illustrating the drop in peripheral dose for the unflattened beam delivery in the clinical intensity modulated radiotherapy plans (a). (b) Plot of the average dose ratio (unflattened/flattened) for all of the plans delivered. CAX = central axis. (Reprinted from Cashmore et al., 2011, copyright 2011, with permission from Elsevier.)

The dose rate delivered by flattening filter free beams is approximately two to four times the dose rate of flattened beams, leading to concerns about differing radiobiological effect relative to conventional flattened beams. A number of in vitro studies have reported on cell survival in flattening filter beams. Most studies have found that there is no difference in cell survival over the dose rate range encompassed by flattened and unflattened beams (Sorensen et al., 2011; King et al., 2013; Verbakel et al., 2013); however, at least one study has reported that flattening filter free beams reduced clonogenic survival of cancer cells (Lohse et al., 2011). A review of the dose rate effects in external beam radiation therapy concluded that the dose rate effect is governed by the total fraction delivery time, not by the average linear accelerator dose rate or by the instantaneous dose per pulse (Ling et al., 2010). For the dose range typical of hypofractionated radiation therapy, the radiobiological effect is expected to increase when treatment time is reduced. As illustrated in Figure 4.4, late responding normal tissues have a larger increase rate than tumor and early responding

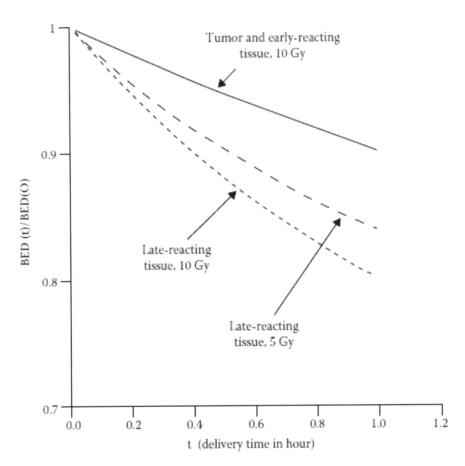

Figure 4.4 Normalized Biological Equivalent Dose (BED) as a function of overall beam on time for delivering 10 Gy to the tumor and early responding tissue and 5 or 10 Gy to late reacting tissue. (Reprinted from Ling et al., 2010, copyright 2010, with permission from Elsevier.)

normal tissues (Ling et al., 2010), suggesting caution when implementing any technology that significantly shortens overall treatment time. Early reports of experience with flattening filter free beams, typically in combination with modulated arc therapy, have shown low rates of acute toxicity (Alongi et al., 2012; Alongi et al., 2013; Prendergast et al., 2013; Scorsetti et al., 2011; Scorsetti et al., 2014; Scorsetti et al., 2015; Wang et al., 2014). There is a limited long-term experience with flattening filter free beams, and so the data on late effects is still maturing; however, preliminary results have not demonstrated unexpected toxicity.

4.2.4 IGRT

Image guidance has become an essential component of modern radiation therapy (De Los Santos et al., 2013). The goal of image-guided radiation therapy (IGRT) is the improvement of targeting accuracy to permit the use of smaller setup margins and consequently reduce irradiation of normal tissue. In the context of stereotactic treatment techniques, image guidance replaces the stereotactic frame for aligning the target with the linear accelerator coordinate system. To achieve stereotactic accuracy using relocatable fixation devices instead of a frame, image guidance is necessary before every treatment (Masi et al., 2008; Murphy, 2009). In-room image guidance began with megavoltage (MV) electronic portal imagers (EPIDs) that were developed as a direct replacement for film. In recent years, MV EPIDS have been largely supplanted by kilovoltage (kV) imaging. There are a variety of kV technologies available for image guidance (De Los

Santos et al., 2013), but the most common for stereotactic applications using C-arm linacs are gantry-mounted (on-board) kV imagers and room-fixed stereoscopic kV imagers.

An on-board kV imaging system is composed of an x-ray source mounted on the gantry opposite an amorphous silicon flat-panel detector. Both the source and detector are retractable. Presently available systems are mounted to the gantry orthogonal to the treatment beam axis (Varian Medical Systems, Palo Alto, CA; Elekta Oncology Systems, Crawley, UK), although one system that is no longer available was mounted parallel to the beam axis (Siemens, Concord, CA). The imager can be used to obtain radiographs from stationary gantry positions or can be rotated to acquire a large number of projection images to construct a cone-beam CT (CBCT). Planar radiographs can be used for stereoscopic image guidance, typically using orthogonal views. On-board imagers and CBCT have been extensively described elsewhere (De Los Santos et al., 2013).

The source and detector of an on-board imaging system undergo small motions as the gantry rotates, resulting in small misalignments with the MV radiation isocenter. Systematic motions are compensated for by measuring them as a function of gantry angle using one or more high-density balls having a known geometry with respect to the MV isocenter. The resulting relationship between image offset and gantry angle, referred to as a flexmap, is stored and used to realign each image to the MV isocenter. Realignment is accomplished either by physically shifting the detector position to compensate for the offset or by correcting the image coordinate system in software. Both methods have been shown to have sub-millimeter residual positioning error (Bissonnette et al., 2008), resulting in radiosurgery positioning accuracy comparable to that reported for stereotactic frames (Chang et al., 2007).

Stereoscopic x-ray imagers use a pair of x-ray sources and corresponding flat-panel detectors mounted in the treatment room. The central rays of the sources intersect at machine isocenter and are separated by an angle sufficient for stereoscopic visualization of patient anatomy (Verellen et al., 2003). Stereoscopic systems are capable of sub-millimeter accuracy (Verellen et al., 2003), comparable to a stereotactic frame (Gevaert et al., 2012b). Stereoscopic x-ray imaging serves as the primary image-guidance system for robotic radiosurgery systems (Adler et al., 1997), whereas it often complements the on-board imaging system for modern C-arm linacs. Presently, the BrainLAB ExacTrac system (BrainLAB, Germany) is the only stereoscopic x-ray system commercially available for use with C-arm linacs.

One advantage of room-fixed stereo-imaging systems relative to gantry-mounted systems is that images can be obtained over a wider range of gantry and couch positions. Gantry-mounted systems have significant limitations when the couch is not near zero (IEC coordinate system) because the imager or the source will collide with the couch. Although this is not a problem for pre-treatment imaging, it limits the utility of an on-board imaging system to verify correct patient positioning after a couch rotation.

Clinical use of both on-board and room-fixed image-guidance systems should include routine testing to evaluate the coincidence of the MV radiation isocenter with the isocenter of the imaging system, particularly for stereotactic applications. Daily end-to-end testing using a phantom-containing markers in known positions can be used to test localization accuracy. The AAPM Task Group 142 report recommends that IGRT systems used for stereotactic techniques demonstrate an accuracy of ±1 mm (Klein et al., 2009).

4.2.5 INTRAFRACTION POSITION MONITORING

For frameless image-guided stereotactic treatment techniques, pretreatment image guidance should be accompanied by intra-fraction motion monitoring. Hoogeman et al. studied intra-fraction motion for intracranial and both prone and supine extracranial patients (Hoogeman et al., 2008). The frequency distribution of the displacement vector magnitude at several intervals from the initial IGRT position correction is shown in Figure 4.5. For the intracranial targets, 95% of displacements were less than 1.6 mm over a 15-minute interval. For extracranial targets treated in a supine position, 95% of displacements were within 2.8 mm over the same interval. However, for prone treatments, the displacement was significantly larger, with 5% of displacements being larger than 3.1 mm after only 1 minute. The larger intra-fraction motion in the prone group was attributed in part to respiration. Hoogeman et al. recommended repeat imaging and patient setup correction at 5-minute intervals and suggested a 0.6 mm margin to account for residual motion for intracranial targets, and 1.0 and 2.0 mm for extracranial targets treated in the supine and prone position, respectively (Hoogeman et al., 2008). Similarly, Murphy at al. found that intra-fraction motion in

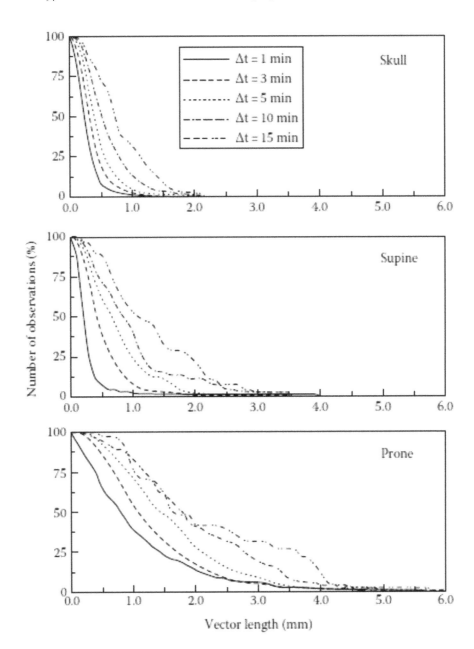

Figure 4.5 Cumulative vector length distributions of the three patient groups for different intervals. For all groups, the width of the cumulative distributions increased with time. For the skull and supine group, intervals of 1–3 minutes yielded distributions with a width in the 1 mm or even sub-millimeter range. The width of the distributions of the prone group was larger than of the other groups, which was partly caused by respiration-induced motion of the fiducial markers. The horizontal scale of the graphs was set to 6.0 mm to keep them readable. Note that the maximum value observed for the prone group was 12.3 mm, exceeding the scale of the graph. (Reprinted from Hoogeman et al., 2008, copyright 2008, with permission from Elsevier.)

the skull and spine was typically within 0.8 mm over 1–5 minute intervals and concluded that a tracking interval of 1 to 2 minutes is necessary for most intracranial and spinal radiosurgery applications (Murphy et al., 2003). If the target position is monitored at less-frequent intervals, the margin must be increased to assure that the target receives the prescribed dose (Murphy et al., 2003; Hoogeman et al., 2008; Murphy, 2009; Kang et al., 2013).

X-ray image-guidance systems can be used for intra-fraction motion monitoring, but on-board systems have limited applicability for noncoplanar geometries because of potential collision of the imaging system with the treatment table or the patient. Room-fixed (stereoscopic) systems do not have collision risks; however, there may be configurations in which the view of one of the detectors is obstructed by the gantry or the couch. Furthermore, x-ray systems are not real-time, and the time required for processing and decision-making limits the frequency for which such systems are practical. In addition, the small, but non-negligible, x-ray dose may be a consideration.

One solution to these concerns for cranial RT are optical systems that monitor the patient position in real time. There are two approaches to optical monitoring. The first is monitoring of infrared reflective or emissive targets affixed to the patient or a surrogate such as a bite block (Meeks et al., 2000; Wang et al., 2010). The second approach is an optical surface imaging (SI) system that uses a set of cameras to capture a 3D rendering of the patient surface.

Optically guided systems that use fiducial markers comprise infrared LEDs (Meeks et al., 2000), infrared markers attached to a bite block (Wang et al., 2010), or infrared markers attached to the patient's skin (Wang et al., 2001). Using a stereoscopic infrared camera, the targets are detected in images, and the 3D position is calculated and compared with the expected position. IR fiducial monitoring systems can achieve sub-millimeter accuracy (Gevaert et al., 2012b; Tagaste et al., 2012). While sufficient for position monitoring, the intra-fraction positioning accuracy is inadequate for pre-treatment positioning, and x-ray-based image guidance is necessary to achieve stereotactic accuracy (Wang et al., 2010).

Surface imaging systems use cameras oriented to view isocenter (Li et al., 2006). The cameras obtain a 3D point cloud that is compared with a reference surface to determine patient position and orientation. Multiple cameras are typically used to both improve accuracy and to ensure that sufficient cameras have an unobstructed line-of-sight over the entire range of gantry motion. When used for patient setup, the reference surface is derived from the treatment planning CT, whereas for intra-fraction motion monitoring the reference surface is obtained directly from the SI following patient setup. For intracranial SRS, optical surface imaging of the face can achieve sub-millimeter motion tracking accuracy at frame rates of approximately 5 per second (Peng et al., 2010; Li et al., 2011; Wiersma et al., 2013). The face must be exposed, so open-faced masks or other immobilization devices are needed when using SI. SI can maintain positioning accuracy similar to a stereotactic frame during treatment; however, to achieve that accuracy, CBCT or stereoscopic x-ray imaging is necessary for initial positioning (Peng et al., 2010; Li et al., 2011; Wiersma et al., 2013). The false positive rate for SI systems is small but non-negligible, particularly for non-zero table angles. Consequently, for stereotactic applications patients should be repositioned using CBCT or stereoscopic x-ray imaging rather than the SI system (Covington et al., 2019).

4.2.6 Six Degree-of-Freedom Positioning

Conventional treatment tables have four degrees of freedom (4DOF): three translations and one rotation about the vertical axis. Any rotational misalignment of the patient around the longitudinal axis (roll) or the lateral axis (pitch) cannot be corrected with a 4DOF table. Correcting rotational misalignment is not important for small, spherical targets that are easily localized; however, rotational misalignment can have significant dosimetric consequences for targets that are complex in shape, such as paraspinal lesions, do not have nearby localization anatomy, such as targets in the center of the skull, or are non-isocentric, such as multiple intracranial targets treated using a single isocenter. Robotic and manual systems have been developed that correct for pitch and roll. Robotic systems comprise a pitch and roll stage that is sandwiched between the couch translation stage and the couch top. These systems correct for all six degrees of freedom (6DOF) under computer control. Manual positioners generally consist of an extension to the treatment couch to support the head that can be tilted around the longitudinal and lateral axes and have inclinometers to display pitch and roll (Dhabaan et al., 2012). Phantom studies have demonstrated that robotic 6DOF systems have accuracies better than 0.5 mm and 0.5 degree (Meyer et al., 2007; Wilbert et al., 2010; Gevaert et al., 2012a). In vivo studies suggest similar accuracy as assessed by CBCT following 6DOF correction (Dhabaan et al., 2012; Lightstone et al., 2012), with one study reporting that 97% of patients were within 1 mm and 0.5 degree (Gevaert et al., 2012a). A number of investigations have reported on the improvement of target coverage for 6DOF relative to 4DOF or 3DOF setup. Gevaert et al. found that

4DOF correction resulted in 5% less target volume receiving the prescription dose relative to 6DOF correction, with an extreme case having 35% less target volume coverage relative to 6DOF (Gevaert et al., 2012a). Other groups (Schreibmann et al., 2011; Dhabaan et al., 2012) have found similar results. For spinal radiosurgery, ignoring rotational corrections (3DOF) can result in significant under-dosage of the tumor. The degree of underdose is dependent on the magnitude of rotational error and the size and shape of the tumor. Small tumors and irregular shape are more likely to suffer an underdose (Schreibmann et al., 2011).

4.3 TREATMENT PLANNING

4.3.1 TREATMENT PLAN EVALUATION

Treatment planning objectives can be classified along a spectrum ranging from conformity to organ avoidance. For a target volume embedded in isotropic normal tissue, conformity of the prescription isodose volume to the target and rapid dose fall-off in all directions are the treatment planning goals. A lesion located in brain or lung parenchyma and having no nearby critical structures, such as brainstem or proximal bronchial tree, is a typical example. Conversely, if a target volume is located in close proximity to critical structures that have dose tolerances significantly lower than the prescription dose and the tolerance of the surrounding soft tissue, the avoidance of the critical structures is more important than dose spill into the surrounding soft tissue. A typical example is a paraspinal lesion, for which the avoidance of the spinal cord is most important, and dose is allowed to spread into the adjacent soft tissue to preferentially spare the cord. Most treatment planning situations fall somewhere between the two extremes, although the majority of intracranial hypofractionated RT and SRS cases can be classified as having conformity as the primary objective and spine SBRT cases as having organ avoidance as the primary objective.

For intracranial targets, the primary metrics used to evaluate linac stereotactic plans are conformity index, gradient index, and healthy brain dose–volumes, including the volume of brain receiving more than 12 Gy (V12Gy) (Levegrun et al., 2004; Blonigen et al., 2010; Minniti et al., 2011). The simplest conformity index PITV, introduced by the Radiation Therapy Oncology Group (RTOG), is the planning isodose volume divided by the target volume (Shaw et al., 1993):

$$PITV = PIV/TV$$

where PIV is the volume enclosed by the prescription isodose surface and TV is the target volume. An ideal plan has PITV value of 1, whereas PITV less than 1 indicates under treatment and PITV greater than 1 overtreatment. This index is simple to calculate and is in widespread use. The flaw in the PITV index is that it does not take into account the location of the prescription isodose. A dose distribution having a PIV equal to the TV but located outside the target volume (a geometric miss) will nevertheless have an ideal PITV of 1. To remedy this problem, Paddick and Lippitz proposed an index, often referred to as the Paddick-CI, that captures both volume and position of the PIV relative to the TV. The Paddick-CI uses the target volume encompassed by the prescription isodose (TVPIV) and is defined as

$$Paddick\text{-}CI = TVPIV^2 / (TV \times PIV)$$

The conformity indices are designed to describe how well the prescription isodose volume conforms to the target. The gradient index is designed to quantify the dose fall off away from the target volume. Paddick and Lippitz defined the gradient index GI as the ratio of the volume of half the prescription isodose (V50) to the volume of the prescription isodose (the PIV) (Paddick and Lippitz, 2006).

$$GI = V50 / PIV$$

Caution must be exercised when using the gradient index to compare plans having different conformity indices and thus different PIV values in the denominator of the GI definition. Two plans having the same V50 but different PIVs will also have different gradient indices. When conformity is improved without change in V50, the GI increases and thus appears to be worse. This situation is illustrated in Figure 4.6.

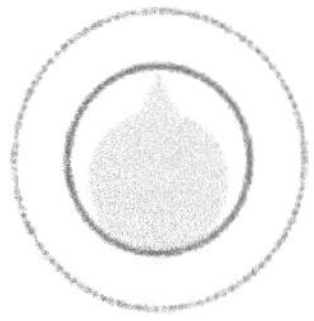

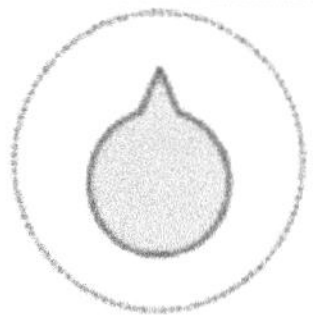

Parameters	TV = 1.64	TV = 1.64
	PIV = 3.05	PIV = 1.64
	TVRI = 1.64	TVRI = 1.64
	V50 = 12.21	V50 = 12.21
PITV	1.86	1
Paddick-CI	0.54	1
GI	4	7.44

Figure 4.6 Gradient index for dose distributions having the same V50 but different conformity indices. Solid red line represents the prescription dose and the dashed blue line half of the prescription dose.

Note that for two plans having the same V50, the ratio of the gradient indices is the inverse of the ratio of the PIVs.

Stereotactic treatment of the spine is an organ avoidance problem, and as such gradient and conformity indices have a less important role in evaluating plan quality. Although gradient index is rarely reported, conformity index is often reported as a means to evaluate target coverage relative to spill of the prescription dose outside of the target. However, conformity index is generally secondary to direct dose–volume measures of target coverage DV, the dose for which volume V of the target receives at least the dose D. The volume V can be expressed either as a percentage or an absolute value. If no units are given, the value is typically percentage volume. Evaluation of the spinal cord dose generally focuses on the high-dose portion of the DVH, and common metrics include the maximum point dose, D0.35cc, and D10% (Ryu et al., 2014). When evaluating spinal cord dose using percentage volume, rather than absolute volume, it is important to be aware of the contouring guidelines used for the spinal cord, because differing lengths of cord will yield different results for the same dose distribution. The RTOG-0631 protocol, for example, specified D10% in relation to the segment of spinal cord extending 5–6 mm beyond the target. Other organs at risk depend on the location of the target along the spine and include esophagus, lungs, and kidneys (Schipani et al., 2012). These structures are usually not in close proximity to the target, and limiting the dose received by these structures is generally not difficult.

4.3.2 TREATMENT PLANNING TECHNIQUES

4.3.2.1 Intracranial

Treatment planning for brain has conformity and rapid dose fall-off (gradient index) as the primary objectives. For 3D-CRT and DCA, the aperture size, number of fixed fields or arcs, and the field placement dictate conformity and gradient. However, for inverse-planned IMRT or MAT, conformity and gradient must be included in the objective function. Some planning systems have tools that incorporate dose fall-off into the objective function, such as the normal tissue objective (NTO) included in the Eclipse treatment planning system (Varian Medical Systems). Alternatively, conformity and dose fall-off can be included as objectives by assigning dose limits to a series of nested shells constructed around the target, as illustrated in Figure 4.7 (Audet et al., 2011; Clark et al., 2010; Clark et al., 2012). Each shell is created by expanding the target volume and then using Boolean operations to remove the target and the inner shells. Three shells are typically sufficient to provide good control over the dose distribution (Clark et al., 2010; Clark et al., 2012). The inner and middle shells are used to optimize the conformity index and the middle and outer shells to control the gradient index. The maximum dose limits for the inner and middle shells are the prescription dose and half the prescription dose, respectively. These limits force the dose to decrease from the prescription dose at the surface of the target volume to half of the prescription dose at the external surface of the inner shell, forcing the prescription dose to conform to the target. The outer shell is used to confine the 50% isodose volume to the interior of the middle shell, thus minimizing V50 and the gradient index. The upper limit for the outer shell must be less than half of the prescription dose but is otherwise somewhat arbitrary. One group has reported using 40% of the prescription dose (Clark et al., 2012). The dimensions of the shells are based on the desired dose falloff. The dimensions should be small to maximize falloff; however, if the dimensions are too small, the optimizer will not be able to achieve the desired falloff, resulting in a suboptimal target coverage. Fortunately, there is a body of literature describing the achievable dose falloff for linear accelerator radiosurgery that can provide guidance. The expansion of the target volume to obtain the inner shell should be R50, the distance from the prescription isodose line to the 50% isodose line. The expected value of R50 is dependent on technique and target volume, with R50 increasing as target volume increases. For a 10 mm diameter circular collimator, the R50 distance has been reported for a variety of arc techniques as ranging from approximately 3 to 5 mm (Pike et al., 1990). For MLC 3D-CRT and DCA, an average R50 of 4.1, 5.5, and 6.5 mm has been reported for tumors having diameters less than 20 mm, between 20 and 30 mm, and greater than 30 mm, respectively (Hong et al., 2011). The middle and outer shell thicknesses are somewhat arbitrary. The middle shell is designed to contain the 50% isodose surface, and the thickness determines the minimum dose gradient in the neighborhood of the 50% isodose surface. The purpose of the outer shell is to constrain the 50% isodose line to the interior of the inner shell. One group has reported using 10 mm and 30 mm expansions of the target, resulting in 5 mm and 20 mm thickness, for the middle and outer shells, respectively (Clark et al., 2012).

For 3D-CRT or DCA, the volume of normal tissue receiving less than V50 is generally not an explicit planning objective because it is determined by the field geometry. However, for computer-optimized techniques, low-dose spill will not be minimized if it is not included as an objective (Thomas et al., 2014). Thomas et al. showed that for treatment of multiple targets using MAT, the volume of normal brain receiving more than 25% of the prescription dose can be limited to between 25 cm^3 and 250 cm^3 and that a mean brain dose in the range 3% to 11% can be achieved (Thomas et al., 2014). Ultimately, the number, size, and locations of the targets will determine what is achievable.

Some groups report including target dose uniformity as a dosimetric objective for radiosurgical planning (Mayo et al., 2010; Audet et al., 2011; Subramanian et al., 2012). For single-fraction radiosurgery using the Gamma Knife, maximum doses as high as twice the prescription dose are common (Paddick and Lippitz, 2006). For Gamma Knife, plans having more dose uniformity had a worse gradient index (Paddick and Lippitz, 2006). This effect is also observed for cones (Meeks et al., 1998), 3DCRT (Hong et al., 2011), and DCA (Hong et al., 2011; Tanyi et al., 2012). Although this effect has not been reported for IMRT or MAT radiosurgery, it is reasonable to expect that forcing uniformity within the target will produce results because forcing uniformity within the target necessarily results in a reduced dose gradient at the surface of the target. There is a trade-off between target dose uniformity and dose gradient in the normal tissue, so the clinical benefits of each must be evaluated for each patient.

When conformity and gradient are the primary treatment planning objectives, field arrangement is not critical. Achieving a conformal dose distribution with rapid, uniform fall-off simply requires a large number of entrance angles, either using fixed beams or a sufficient number of noncoplanar arcs. A small number of fixed beam angles or arcs results in a non-uniform gradient having a faster falloff in some directions at the expense of slower falloff in other directions. For example, a single arc has a rapid dose falloff

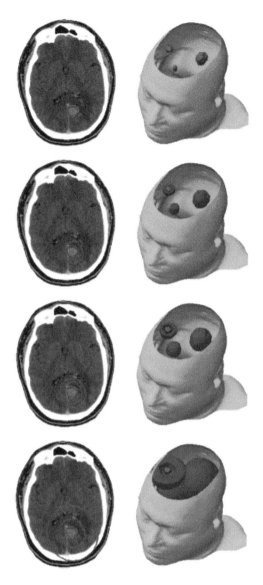

Figure 4.7 Dose control tuning structures utilized for dose optimization. Left side represents two-dimensional visualization for a single target patient. Right side represents three-dimensional visualization for a multitarget patient. From top to bottom (both sides): target(s) (red), inner control tuning structure (blue), middle control tuning structure (blue), and outer control tuning structure (blue). (Reprinted from Clark et al., 2012, copyright 2012, with permission from Elsevier.)

perpendicular to the arc plane, but a less rapid falloff in plane, whereas a four-arc geometry has less variation in dose gradient (Pike et al., 1990). Multiple, noncoplanar arcs are generally necessary for stereotactic applications (Podgorsak et al., 1989). For 3DCRT, seven to ten noncoplanar fields produces dose distributions similar to multiple noncoplanar arcs but with slightly higher peripheral dose. Increasing the number of fields reduces the peripheral dose (Bourland and McCollough, 1994). Hong et al. recommended one beam per 2 Gy of prescribed dose to limit the dose spill at 3 to 4 Gy (Hong et al., 2011). When organ-avoidance is an objective, field arrangement is more important. Bouquets of nine beams with optimized arrangements have been reported to achieve good results (Wagner et al., 2001). Similar field geometries are appropriate for IMRT and MAT (Benedict et al., 2001; Clark et al., 2010; Nath et al., 2010; Audet et al.,

2011) (9–11 Nath 2010; 4 arcs Audet). If body dose is a concern, geometries that are nearly parallel to the longitudinal body axis should be avoided.

4.3.2.2 Spine

In the spine, avoidance of the spinal cord is the primary goal. The relationship between the target volume and the spinal cord is complex, because the target is typically wrapped around the spinal cord. Consequently, DCA and 3DCRT techniques are not able to meet the treatment planning goals for spine SBRT and so intensity-modulated techniques are required (Yenice et al., 2003; Wu et al., 2009b). For fixed beam IMRT, Yenice et al. demonstrated that 7 beams directed posterior and obliquely having a spacing of 20 to 30 degrees results in acceptable dosimetry (Yenice et al., 2003). Kuiper et al. described similar beam arrangements, shown in Figure 4.8, that were designed based on the complexity of the target volume (vertebral body or entire vertebra) and the spinal level of the involved vertebral body (Kuijper et al., 2010). For MAT, two coplanar arcs having a full rotation are generally sufficient (Wu et al., 2009a; Kuijper et al., 2010); however, Kuiper et al. have reported using a third arc for cases in which the entire vertebral body is targeted (Kuijper et al., 2010). Using a single arc is not recommended because the achievable dosimetry in the spinal cord is inferior to static IMRT fields (Wu et al., 2009a).

The distance from the target volume to the spinal cord and the degree to which the spinal cord is enclosed by the target govern the achievable target coverage. The target often abuts the spinal cord, necessitating under-dosage at the periphery in order to limit the spinal cord dose to an acceptable level. Kuijper et al. found that when vertebral body is the target, 95% of the target received at least the prescription dose 16 Gy, whereas for the entire vertebral body only 85% received 16 Gy (Kuijper et al., 2010). Yenice et al. achieved typical target coverage of 95% for a 20-Gy prescribed dose (Yenice et al., 2003). A dose gradient of about 10% per millimeter should be achievable in the region between the target and the spinal cord (Yenice et al., 2003; Kuijper et al., 2010) and can be used to estimate the minimum distance between the cord and the prescription isodose surface prior to treatment planning. It is important to note that, similar to intracranial SRS, forcing dose uniformity within the target will likely reduce the gradient at the boundary of the target, reducing the achievable dose coverage. Maximum doses in the target between 115% and 140% have been reported (Yenice et al., 2003; Wu et al., 2009a; Kuijper et al., 2010).

Although conformity is not a primary objective for spine radiosurgery, the dose to the unspecified normal tissue outside of the target volume should be controlled. The stringent limitation of the spinal cord dose can cause unacceptable dose spillage outside the target, particularly for fixed IMRT techniques, which are particularly susceptible to streaks of high dose extending outside of the target. One approach to controlling dose spill is to create an annulus around the target volume similar to the inner control shell previously described for intracranial targets. Limiting the maximum dose inside the control structure to no more than the prescription dose will generally limit dose spill to acceptable levels. In this manner, a conformity index in the range of 0.9 to 1.1 can be achieved (Kuijper et al., 2010).

Depending on the location of the lesion within the spine, consideration of other critical structures, such as lungs, esophagus, kidneys, bowel, heart, and liver, may be necessary. Explicitly including dose limits

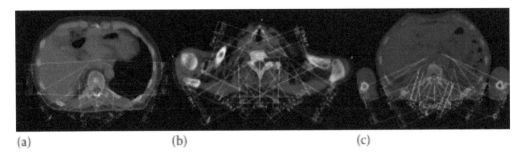

(a)   (b)   (c)

Figure 4.8 Three beam configurations used for conventional IMRT plans. Standard beam configuration of vertebral body (left), entire cervical vertebra case (middle), and entire lumbar vertebra case (right). (Reprinted from Kuijper et al., 2010, copyright 2010, with permission from Elsevier.)

during inverse planning for critical structures other than the cord is usually unnecessary; however, it is important to review all critical structure doses. Because of the complex modulation required to avoid the spinal cord, unexpectedly high dose can be delivered to structures that are distant from the target volume. In cases for which the tolerance dose is exceeded, incorporation of explicit dose constraints into the optimization system typically reduces the dose to an acceptable level without compromising target coverage.

4.3.3 AUTOMATED AND KNOWLEDGE-BASED PLANNING

Artificial intelligence is the study of systems that mimic human cognition. Artificial intelligence techniques are increasingly being applied to radiation therapy treatment planning and are moving from research to commercialization. These techniques encompass a broad spectrum of approaches, ranging from simple algorithms that automate planning tasks to machine learning models which utilize prior treatment plans to develop new treatment plans for similar cases.

Two commercial systems that utilize heuristic algorithms for linear-accelerator-based radiosurgery planning are HyperArc™ (Varian Medical Systems, Palo Alto, CA) and Multiple Metastasis Elements (BrainLAB AG, Munich, Germany). Both systems use a single isocenter to simultaneously treat multiple targets. HyperArc is based on a class solution for VMAT originally reported by Clark et al. (Clark et al., 2012) and refined by Thomas et al. in 2014 (Thomas et al., 2014). The solution relied on multiple nested tuning structures that were used to optimize the conformity index and gradient index, as well as an objective function that explicitly penalized the low-dose brain spill (Thomas et al., 2014; Yuan et al., 2018). HyperArc uses a pre-selected arc geometry and a radiosurgery specific normal tissue objective that eliminates the need for multiple nested tuning structures to achieve optimal conformity and gradient indices. HyperArc allows automated delivery without a requirement for room entry to change couch angles, decreasing the treatment delivery time. Multiple Metastases Elements (MME) uses DCAs selected from a templated arc geometry and sets the weight of each arc to optimize the overall target conformity. As shown

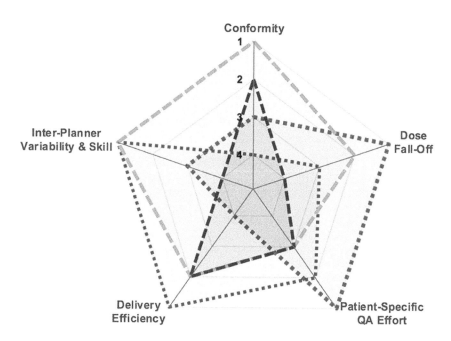

Figure 4.9 Spider plot graphically comparing the studied SRS techniques across the different categories of dosimetry and efficiency. SRS treatment modalities were ranked relative to each other per specified category on a scale of 1 through 4. (Reprinted from Vergalasova et al., 2019. https://doi.org/10.3389/fonc.2019.00483, distributed under the terms of the Creative Commons Attribution License (CC BY).)

in Figure 4.9, both systems produce clinically acceptable treatment plans and reduce the planning time and, importantly, the variability in plan quality among different planners (Vergalasova et al., 2019).

Machine learning is a type of artificial intelligence in which computer systems solve a problem without requiring a specific algorithm to do so. Instead, a machine learning system uses mathematical models to generalize from prior knowledge (Domingos, 2012). One machine learning approach for radiation therapy treatment planning uses prior treatment plans to estimate dose–volume histograms for a new case. This approach, referred to as knowledge-based planning (KBP), has been successfully applied to radiosurgery treatment planning (Shiraishi et al., 2015). For radiosurgery, KBP can produce plans equivalent or superior to manual plans for a majority of cases (Ziemer et al., 2017).

4.4 CONCLUSION

In the opinion of Brada and Laing, fractionated stereotactic radiotherapy was the way forward for the treatment of brain tumors and that C-arm linear accelerators are the preferred platform to accomplish that goal. Their predictions regarding the evolution of linear accelerator technology have been realized. With the combination of multileaf collimators, flattening filter free beams, IGRT, 6-DOF tables, and intra-fraction motion monitoring, frameless single or fractionated stereotactic techniques are possible in both the skull and spine. By combining these technologies, highly conformal dose distributions can be delivered with sub-millimeter accuracy.

The future promises further developments. The increasing sophistication of linear accelerator control systems and planning techniques will enable hybrid MAT and IMRT (Matuszak et al., 2013) and the efficient use of a large number of table orientations (Nguyen et al., 2014), further improving plan quality and decreasing treatment time. Knowledge-based planning systems (Shiraishi et al., 2015) have the potential to reduce the expertise required to deliver high-quality radiosurgery. By reducing the learning barriers, knowledge-based systems will facilitate the dissemination of hypofractionated stereotactic radiotherapy into more centers, fulfilling the prediction of Brada and Laing that stereotactic radiotherapy will become a routine tool in the radiotherapy armamentarium.

REFERENCES

Abacioglu U, Ozen Z, Yilmaz M, Arifoglu A, Gunhan B, Kayalilar N, Peker S, Sengoz M, Gurdalli S, Cozzi L (2014) Critical appraisal of RapidArc radiosurgery with flattening filter free photon beams for benign brain lesions in comparison to GammaKnife: A treatment planning study. *Radiation Oncology* 9:119.

Adler JR, Jr., Chang SD, Murphy MJ, Doty J, Geis P, Hancock SL (1997) The CyberKnife: A frameless robotic system for radiosurgery. *Stereotactic Functional Neurosurgery* 69:124–128.

Alongi F, Cozzi L, Arcangeli S, Iftode C, Comito T, Villa E, Lobefalo F, Navarria P, Reggiori G, Mancosu P, Clerici E, Fogliata A, Tomatis S, Taverna G, Graziotti P, Scorsetti M (2013) Linac based SBRT for prostate cancer in 5 fractions with VMAT and flattening filter free beams: Preliminary report of a phase II study. *Radiation Oncology* 8:171.

Alongi F, Fogliata A, Clerici E, Navarria P, Tozzi A, Comito T, Ascolese AM, Clivio A, Lobefalo F, Reggiori G, Cozzi L, Mancosu P, Tomatis S, Scorsetti M (2012) Volumetric modulated arc therapy with flattening filter free beams for isolated abdominal/pelvic lymph nodes: Report of dosimetric and early clinical results in oligometastatic patients. *Radiation Oncology* 7:204.

Audet C, Poffenbarger BA, Chang P, Jackson PS, Lundahl RE, Ryu SI, Ray GR (2011) Evaluation of volumetric modulated arc therapy for cranial radiosurgery using multiple noncoplanar arcs. *Medical Physics* 38:5863–5872.

Bayouth JE, Kaiser HS, Smith MC, Pennington EC, Anderson KM, Ryken TC, Buatti JM (2007) Image-guided stereotactic radiosurgery using a specially designed high-dose-rate linac. *Medical Dosimetry* 32:134–141.

Benedict SH, Cardinale RM, Wu Q, Zwicker RD, Broaddus WC, Mohan R (2001) Intensity-modulated stereotactic radiosurgery using dynamic micro-multileaf collimation. *International Journal of Radiation Oncology, Biology, Physics* 50:751–758.

Bissonnette JP, Moseley D, White E, Sharpe M, Purdie T, Jaffray DA (2008) Quality assurance for the geometric accuracy of cone-beam CT guidance in radiation therapy. *International Journal of Radiation Oncology, Biology, Physics* 71:S57–61.

Blonigen BJ, Steinmetz RD, Levin L, Lamba MA, Warnick RE, Breneman JC (2010) Irradiated volume as a predictor of brain radio necrosis after linear accelerator stereotactic radiosurgery. *International Journal of Radiation Oncology, Biology, Physics* 77:996–1001.

Bourland JD, McCollough KP (1994) Static field conformal stereotactic radiosurgery: Physical techniques. *International Journal of Radiation Oncology, Biology, Physics* 28:471–479.

Boyer AL, Ezell GA, and Yu CX (2012) Intensity-Modulated Radiation Therapy. In: *Treatment planning in radiation oncology* (Khan FM, Gerbi BJ, eds), pp. 201–228. Philadelphia: Lippincott Williams &Wilkins.

Brada M, Laing R (1994) Radiosurgery/stereotactic external beam radiotherapy for malignant brain tumours: The Royal Marsden Hospital experience. *Recent Results Cancer Research* 135:91–104.

Cardinale RM, Benedict SH, Wu Q, Zwicker RD, Gaballa HE, Mohan R (1998) A comparison of three stereotactic radiotherapy techniques; ARCS vs. noncoplanar fixed fields vs. intensity modulation. *International Journal of Radiation Oncology, Biology, Physics* 42:431–436.

Cashmore J (2008) The characterization of unflattened photon beams from a 6 MV linear accelerator. *Physics Medical Biology* 53:1933–1946.

Cashmore J, Ramtohul M, Ford D (2011) Lowering whole-body radiation doses in pediatric intensity-modulated radiotherapy through the use of unflattened photon beams. *International Journal of Radiation Oncology, Biology, Physics* 80:1220–1227.

Chae SM, Lee GW, Son SH (2014) The effect of multileaf collimator leaf width on the radiosurgery planning for spine lesion treatment in terms of the modulated techniques and target complexity. *Radiation Oncology* 9:72.

Chang J, Yenice KM, Narayana A, Gutin PH (2007) Accuracy and feasibility of cone-beam computed tomography for stereotactic radiosurgery setup. *Medical Physics* 34:2077–2084.

Clark GM, Popple RA, Prendergast BM, Spencer SA, Thomas EM, Stewart JG, Guthrie BL, Markert JM, Fiveash JB (2012) Plan quality and treatment planning technique for single isocenter cranial radiosurgery with volumetric modulated arc therapy. *Practice in Radiation Oncology* 2:306–313.

Clark GM, Popple RA, Young PE, Fiveash JB (2010) Feasibility of single-isocenter volumetric modulated arc radio-surgery for treatment of multiple brain metastases. *International Journal of Radiation Oncology, Biology, Physics* 76:296–302.

Covington EL, Fiveash JB, Wu X, Brezovich I, Willey CD, Riley K, Popple RA (2019) Optical surface guidance for submillimeter monitoring of patient position during frameless stereotactic radiotherapy. *Journal of Applied Clinical Medical Physics* 20:91–98.

De Los Santos J, Popple R, Agazaryan N, Bayouth JE, Bissonnette JP, Bucci MK, Dieterich S, Dong L, Forster KM, Indelicato D, Langen K, Lehmann J, Mayr N, Parsai I, Salter W, Tomblyn M, Yuh WT, Chetty IJ (2013) Image guided radiation therapy (IGRT) technologies for radiation therapy localization and delivery. *International Journal of Radiation Oncology, Biology, Physics* 87:33–45.

Dhabaan A, Elder E, Schreibmann E, Crocker I, Curran WJ, Oyesiku NM, Shu HK, Fox T (2010) Dosimetric performance of the new high-definition multileaf collimator for intracranial stereotactic radiosurgery. *Journal of Applied Clinical Medical Physics* 11:3040.

Dhabaan A, Schreibmann E, Siddiqi A, Elder E, Fox T, Ogunleye T, Esiashvili N, Curran W, Crocker I, Shu HK (2012) Six degrees of freedom CBCT-based positioning for intracranial targets treated with frameless stereotac-tic radiosurgery. *Journal of Applied Clinical Medical Physics* 13:3916.

Domingos P (2012) A few useful things to know about machine learning. *Communication ACM* 55:78–87.

Duthoy W, De Gersem W, Vergote K, Boterberg T, Derie C, Smeets P, De Wagter C, De Neve W (2004) Clinical implementation of intensity-modulated arc therapy (IMAT) for rectal cancer. *Journal of Applied Clinical Medical Physics* 60:794–806.

Gevaert T, Verellen D, Engels B, Depuydt T, Heuninckx K, Tournel K, Duchateau M, Reynders T, De Ridder M (2012a) Clinical evaluation of a robotic 6-degree of freedom treatment couch for frameless radiosurgery. *International Journal of Radiation Oncology, Biology, Physics* 83:467–474.

Gevaert T, Verellen D, Tournel K, Linthout N, Bral S, Engels B, Collen C, Depuydt T, Duchateau M, Reynders T, Storme G, De Ridder M (2012b) Setup accuracy of the Novalis ExacTrac 6DOF system for frameless radiosur-gery. *International Journal of Radiation Oncology, Biology, Physics* 82:1627–1635.

Hong LX, Garg M, Lasala P, Kim M, Mah D, Chen CC, Yaparpalvi R, Mynampati D, Kuo HC, Guha C, Kalnicki S (2011) Experience of micromultileaf collimator linear accelerator based single fraction stereotactic radiosurgery: Tumor dose inhomogeneity, conformity, and dose fall off. Med Phys 38:1239–1247.

Hoogeman MS, Nuyttens JJ, Levendag PC, Heijmen BJ (2008) Time dependence of intrafraction patient motion assessed by repeat stereoscopic imaging. *International Journal of Radiation Oncology, Biology, Physics* 70:609–618.

Huang Y, Chin K, Robbins JR, Kim J, Li H, Amro H, Chetty IJ, Gordon J, Ryu S (2014) Radiosurgery of multiple brain metastases with single-isocenter dynamic conformal arcs (SIDCA). *Radiotherapy & Oncology* 112:128–132.

Huq MS, Das IJ, Steinberg T, Galvin JM (2002) A dosimetric comparison of various multileaf collimators. *Physics Medical Biology* 47:N159–170.

Jin JY, Yin FF, Ryu S, Ajlouni M, Kim JH (2005) Dosimetric study using different leaf-width MLCs for treatment planning of dynamic conformal arcs and intensity-modulated radiosurgery. *Medical Physics* 32:405–411.

Kang KM, Chai GY, Jeong BK, Ha IB, Lee S, Park KB, Jung JM, Lim YK, Yoo SH, Jeong H (2013) Estimation of optimal margin for intrafraction movements during frameless brain radiosurgery. *Medical Physics* 40:051716.

King RB, Hyland WB, Cole AJ, Butterworth KT, McMahon SJ, Redmond KM, Trainer C, Prise KM, McGarry CK, Hounsell AR (2013) An in vitro study of the radiobiological effects of flattening filter free radiotherapy treatments. *Physics Medical Biology* 58:N83–94.

Klein EE, Hanley J, Bayouth J, Yin FF, Simon W, Dresser S, Serago C, Aguirre F, Ma L, Arjomandy B, Liu C, Sandin C, Holmes T, Task Group 142 AeAoPiM (2009) Task Group 142 report: Quality assurance of medical accelerators. *Medical Physics* 36:4197–4212.

Kragl G, af Wetterstedt S, Knausl B, Lind M, McCavana P, Knoos T, McClean B, Georg D (2009) Dosimetric characteristics of 6 and 10MV unflattened photon beams. *Radiotherapy & Oncology* 93:141–146.

Kragl G, Baier F, Lutz S, Albrich D, Dalaryd M, Kroupa B, Wiezorek T, Knoos T, Georg D (2011) Flattening filter free beams in SBRT and IMRT: Dosimetric assessment of peripheral doses. *Zeitschrift für Medizinische Physik* 21:91–101.

Kry SF, Popple R, Molineu A, Followill DS (2012) Ion recombination correction factors (P(ion)) for Varian TrueBeam high-dose-rate therapy beams. *Journal of Applied Clinical Medical Physics* 13:3803.

Kry SF, Vassiliev ON, Mohan R (2010) Out-of-field photon dose following removal of the flattening filter from a medical accelerator. *Physics Medical Biology* 55:2155–2166.

Kuijper IT, Dahele M, Senan S, Verbakel WF (2010) Volumetric modulated arc therapy versus conventional intensity modulated radiation therapy for stereotactic spine radiotherapy: A planning study and early clinical data. *Radiotherapy & Oncology* 94:224–228.

Leavitt DD (1998) Beam shaping for SRT/SRS. *Medical Dosimetry* 23:229–236.

Levegrun S, Hof H, Essig M, Schlegel W, Debus J (2004) Radiation-induced changes of brain tissue after radiosurgery in patients with arteriovenous malformations: Correlation with dose distribution parameters. *International Journal of Radiation Oncology, Biology, Physics* 59:796–808.

Li G, Ballangrud A, Kuo LC, Kang H, Kirov A, Lovelock M, Yamada Y, Mechalakos J, Amols H (2011) Motion monitoring for cranial frameless stereotactic radiosurgery using video-based three-dimensional optical surface imaging. *Medical Physics* 38:3981–3994.

Li S, Liu D, Yin G, Zhuang P, Geng J (2006) Real-time 3D-surface-guided head refixation useful for fractionated stereotactic radiotherapy. *Medical Physics* 33:492–503.

Lightstone AW, Tsao M, Baran PS, Chan G, Pang G, Ma L, Lochray F, Sahgal A (2012) Cone beam CT (CBCT) evaluation of inter- and intra-fraction motion for patients undergoing brain radiotherapy immobilized using a commercial thermoplastic mask on a robotic couch. *Technology in Cancer Research & Treatment* 11:203–209.

Ling CC, Gerweck LE, Zaider M, Yorke E (2010) Dose-rate effects in external beam radiotherapy redux. Radiother Oncol 95:261–268.

Lohse I, Lang S, Hrbacek J, Scheidegger S, Bodis S, Macedo NS, Feng J, Lutolf UM, Zaugg K (2011) Effect of high dose per pulse flattening filter-free beams on cancer cell survival. *Radiotherapy & Oncology* 101:226–232.

Ma L, Yu CX, Earl M, Holmes T, Sarfaraz M, Li XA, Shepard D, Amin P, DiBiase S, Suntharalingam M, Mansfield C (2001) Optimized intensity-modulated arc therapy for prostate cancer treatment. *International Journal of Cancer* 96:379–384.

Masi L, Casamassima F, Polli C, Menichelli C, Bonucci I, Cavedon C (2008) Cone beam CT image guidance for intracranial stereotactic treatments: Comparison with a frame guided set-up. *International Journal of Radiation Oncology, Biology, Physics* 71:926–933.

Matuszak MM, Steers JM, Long T, McShan DL, Fraass BA, Romeijn HE, Ten Haken RK (2013) FusionArc optimization: A hybrid volumetric modulated arc therapy (VMAT) and intensity modulated radiation therapy (IMRT) planning strategy. *Medical Physics* 40:071713.

Mayo CS, Ding L, Addesa A, Kadish S, Fitzgerald TJ, Moser R (2010) Initial experience with volumetric IMRT (RapidArc) for intracranial stereotactic radiosurgery. *International Journal of Radiation Oncology, Biology, Physics* 78:1457–1466.

McEwen M, DeWerd L, Ibbott G, Followill D, Rogers DW, Seltzer S, Seuntjens J (2014) Addendum to the AAPM's TG-51 protocol for clinical reference dosimetry of high-energy photon beams. *Medical Physics* 41:041501.

Meeks SL, Bova FJ, Wagner TH, Buatti JM, Friedman WA, Foote KD (2000) Image localization for frameless stereotactic radiotherapy. *International Journal of Radiation Oncology, Biology, Physics* 46:1291–1299.

Meeks SL, Buatti JM, Bova FJ, Friedman WA, Mendenhall WM (1998) Treatment planning optimization for linear accelerator radiosurgery. *International Journal of Radiation Oncology, Biology, Physics* 41:183–197.

Meyer J, Wilbert J, Baier K, Guckenberger M, Richter A, Sauer O, Flentje M (2007) Positioning accuracy of cone-beam computed tomography in combination with a HexaPOD robot treatment table. *International Journal of Radiation Oncology, Biology, Physics* 67:1220–1228.

Minniti G, Clarke E, Lanzetta G, Osti M, Trasimeni G, Bozzao A, Romano A, Enrici R (2011) Stereotactic radiosurgery for brain metastases: Analysis of outcome and risk of brain radionecrosis. *Radiation Oncology* 6:48.

Monk JE, Perks JR, Doughty D, Plowman PN (2003) Comparison of a micro-multileaf collimator with a 5-mm-leaf-width collimator for intracranial stereotactic radiotherapy. *International Journal of Radiation Oncology, Biology, Physics* 57:1443–1449.

Murphy MJ (2009) Intrafraction geometric uncertainties in frameless image-guided radiosurgery. *International Journal of Radiation Oncology, Biology, Physics* 73:1364–1368.

Murphy MJ, Chang SD, Gibbs IC, Le QT, Hai J, Kim D, Martin DP, Adler JR, Jr. (2003) Patterns of patient movement during frameless image-guided radiosurgery. *International Journal of Radiation Oncology, Biology, Physics* 55:1400–1408.

Nath SK, Lawson JD, Simpson DR, Vanderspek L, Wang JZ, Alksne JF, Ciacci J, Mundt AJ, Murphy KT (2010) Single-isocenter frameless intensity-modulated stereotactic radiosurgery for simultaneous treatment of multiple brain metastases: Clinical experience. *International Journal of Radiation Oncology, Biology, Physics* 78:91–97.

Nguyen D, Rwigema JC, Yu VY, Kaprealian T, Kupelian P, Selch M, Lee P, Low DA, Sheng K (2014) Feasibility of extreme dose escalation for glioblastoma multiforme using 4pi radiotherapy. *Radiation Oncology* 9:239.

O'Brien PF, Gillies BA, Schwartz M, Young C, Davey P (1991) Radiosurgery with unflattened 6-MV photon beams. *Medical Physics* 18:519–521.

Otto K (2008) Volumetric modulated arc therapy: IMRT in a single gantry arc. *Medical Physics* 35:310–317.

Paddick I, Lippitz B (2006) A simple dose gradient measurement tool to complement the conformity index. *Journal of Neurosurgery* 105(Suppl):194–201.

Peng JL, Kahler D, Li JG, Samant S, Yan G, Amdur R, Liu C (2010) Characterization of a real-time surface image-guided stereotactic positioning system. *Medical Physics* 37:5421–5433.

Pike GB, Podgorsak EB, Peters TM, Pla C, Olivier A, Souhami L (1990) Dose distributions in radiosurgery. *Medical Physics* 17:296–304.

Podgorsak EB, Pike GB, Olivier A, Pla M, Souhami L (1989) Radiosurgery with high energy photon beams: A comparison among techniques. *International Journal of Radiation Oncology, Biology, Physics* 16:857–865.

Popple RA, Balter PA, Orton CG (2014) Point/Counterpoint. Because of the advantages of rotational techniques, conventional IMRT will soon become obsolete. *Medical Physics* 41:100601.

Prado KL, Starkschall G, and Mohan R (2012) Three-Dimensional Conformal Radiation Therapy. In: *Treatment planning in radiation oncology* (Khan FM, Gerbi BJ, eds), pp. 169–200. Philadelphia: Lippincott Williams &Wilkins.

Prendergast BM, Dobelbower MC, Bonner JA, Popple RA, Baden CJ, Minnich DJ, Cerfolio RJ, Spencer SA, Fiveash JB (2013) Stereotactic body radiation therapy (SBRT) for lung malignancies: Preliminary toxicity results using a flattening filter-free linear accelerator operating at 2400 monitor units per minute. *Radiation Oncology* 8:273.

Prendergast BM, Popple RA, Clark GM, Guthrie BL, Markert JM, Spencer SA, Fiveash JB (2011) Improved clinical efficiency in CNS stereotactic radiosurgery using a flattening filter free linear accelerator. *Journal of Radiosurgery and BRT* 1:117–122.

Ryu SI, Chang SD, Kim DH, Murphy MJ, Le QT, Martin DP, Adler JR, Jr. (2001) Image-guided hypo-fractionated stereotactic radiosurgery to spinal lesions. *Neurosurgery* 49:838–846.

Ryu SI, Pugh SL, Gerszten PC, Yin FF, Timmerman RD, Hitchcock YJ, Movsas B, Kanner AA, Berk LB, Followill DS, Kachnic LA (2014) RTOG 0631 phase 2/3 study of image guided stereotactic radiosurgery for localized (1–3) spine metastases: Phase 2 results. *Practice in Radiation Oncology* 4:76–81.

Salkeld AL, Unicomb K, Hayden AJ, Van Tilburg K, Yau S, Tiver K (2014) Dosimetric comparison of volumetric modulated arc therapy and linear accelerator-based radiosurgery for the treatment of one to four brain metastases. *Journal of Medical Imaging Radiation Oncology* 58:722–728.

Schipani S, Wen W, Jin JY, Kim JK, Ryu S (2012) Spine radiosurgery: A dosimetric analysis in 124 patients who received 18 Gy. *International Journal of Radiation Oncology, Biology, Physics* 84:e571–576.

Schreibmann E, Fox T, Crocker I (2011) Dosimetric effects of manual cone-beam CT (CBCT) matching for spinal radiosurgery: Our experience. *Journal of Applied Clinical Medical Physics* 12:3467.

Scorsetti M, Alongi F, Castiglioni S, Clivio A, Fogliata A, Lobefalo F, Mancosu P, Navarria P, Palumbo V, Pellegrini C, Pentimalli S, Reggiori G, Ascolese AM, Roggio A, Arcangeli S, Tozzi A, Vanetti E, Cozzi L (2011) Feasibility and early clinical assessment of flattening filter free (FFF) based stereotactic body radiotherapy (SBRT) treatments. *Radiation Oncology* 6:113.

Scorsetti M, Alongi F, Clerici E, Comito T, Fogliata A, Iftode C, Mancosu P, Navarria P, Reggiori G, Tomatis S, Villa E, Cozzi L (2014) Stereotactic body radiotherapy with flattening filter-free beams for prostate cancer: Assessment of patient-reported quality of life. *Journal of Cancer Research in Clinical Oncology* 140:1795–1800.

Scorsetti M, Comito T, Cozzi L, Clerici E, Tozzi A, Franzese C, Navarria P, Fogliata A, Tomatis S, D'Agostino G, Iftode C, Mancosu P, Ceriani R, Torzilli G (2015) The challenge of inoperable hepatocellular carcinoma (HCC): Results of a single-institutional experience on stereotactic body radiation therapy (SBRT). *Journal of Cancer Research in Clinical Oncology* 141:1301–1309.

Serna A, Puchades V, Mata F, Ramos D, Alcaraz M (2015) Influence of multi-leaf collimator leaf width in radiosurgery via volumetric modulated arc therapy and 3D dynamic conformal arc therapy. *Physica Medica: European Journal of Medical Physics* 31:293–296.

Shaw E, Kline R, Gillin M, Souhami L, Hirschfeld A, Dinapoli R, Martin L (1993) Radiation Therapy Oncology Group: Radiosurgery quality assurance guidelines. *International Journal of Radiation Oncology, Biology, Physics* 27:1231–1239.

Shiraishi S, Tan J, Olsen LA, Moore KL (2015) Knowledge-based prediction of plan quality metrics in intracranial stereotactic radiosurgery. *Medical Physics* 42:908.

Shiu AS, Kooy HM, Ewton JR, Tung SS, Wong J, Antes K, Maor MH (1997) Comparison of miniature multileaf collimation (MMLC) with circular collimation for stereotactic treatment. *International Journal of Radiation Oncology, Biology, Physics* 37:679–688.

Snyder KC, Wen N, Huang Y, Kim J, Zhao B, Siddiqui S, Chetty IJ, Ryu S (2014) Use of jaw tracking in intensity modulated and volumetric modulated arc radiation therapy for spine stereotactic radiosurgery. *Practical Radiation Oncology* 5:e155–e162.

Solberg TD, Boedeker KL, Fogg R, Selch MT, DeSalles AA (2001) Dynamic arc radiosurgery field shaping: A comparison with static field conformal and noncoplanar circular arcs. *International Journal of Radiation Oncology, Biology, Physics* 49:1481–1491.

Sorensen BS, Vestergaard A, Overgaard J, Praestegaard LH (2011) Dependence of cell survival on instantaneous dose rate of a linear accelerator. *Radiotherapy & Oncology* 101:223–225.

Subramanian S, Srinivas C, Ramalingam K, Babaiah M, Swamy ST, Arun G, Kathirvel M, Ashok S, Clivio A, Fogliata A, Nicolini G, Rao KS, Reddy TP, Amit J, Vanetti E, Cozzi L (2012) Volumetric modulated arc-based hypofractionated stereotactic radiotherapy for the treatment of selected intracranial arteriovenous malformations: Dosimetric report and early clinical experience. *International Journal of Radiation Oncology, Biology, Physics* 82:1278–1284.

Tagaste B, Riboldi M, Spadea MF, Bellante S, Baroni G, Cambria R, Garibaldi C, Ciocca M, Catalano G, Alterio D, Orecchia R (2012) Comparison between infrared optical and stereoscopic X-ray technologies for patient setup in image guided stereotactic radiotherapy. *International Journal of Radiation Oncology, Biology, Physics* 82:1706–1714.

Tanyi JA, Doss EJ, Kato CM, Monaco DL, L ZM, Chen Y, Kubicky CD, Marquez CM, Fuss M (2012) Dynamic conformal arc cranial stereotactic radiosurgery: Implications of multileaf collimator margin on dose-volume metrics. *British Journal of Radiology* 85:e1058–1066.

Thomas EM, Popple RA, Wu X, Clark GM, Markert JM, Guthrie BL, Yuan Y, Dobelbower MC, Spencer SA, Fiveash JB (2014) Comparison of plan quality and delivery time between volumetric arc therapy (RapidArc) and Gamma Knife radiosurgery for multiple cranial metastases. *Neurosurgery* 75:409–417; discussion 417–408.

Van Dyk J (1999) *The modern technology of radiation oncology: A compendium for medical physicists and radiation oncologists*. Madison, WI: Medical Physics Pub.

Vassiliev ON, Kry SF, Chang JY, Balter PA, Titt U, Mohan R (2009) Stereotactic radiotherapy for lung cancer using a flattening filter free Clinac. *Journal of Applied Clinical Medical Physics* 10:14–21.

Vassiliev ON, Titt U, Ponisch F, Kry SF, Mohan R, Gillin MT (2006) Dosimetric properties of photon beams from a flattening filter free clinical accelerator. *Physics Medical Biology* 51:1907–1917.

Verbakel WF, van den Berg J, Slotman BJ, Sminia P (2013) Comparable cell survival between high dose rate flattening filter free and conventional dose rate irradiation. *Acta Oncology* 52:652–657.

Verellen D, Soete G, Linthout N, Van Acker S, De Roover P, Vinh-Hung V, Van de Steene J, Storme G (2003) Quality assurance of a system for improved target localization and patient set-up that combines real-time infrared tracking and stereoscopic X-ray imaging. *Radiotherapy & Oncology* 67:129–141.

Vergalasova I, Liu H, Alonso-Basanta M, Dong L, Li J, Nie K, Shi W, Teo BK, Yu Y, Yue NJ, Zou W, Li T (2019) Multi-institutional dosimetric evaluation of modern day Stereotactic Radiosurgery (SRS) treatment options for multiple brain metastases. *Frontiers in Oncology* 9:483.

Wagner TH, Meeks SL, Bova FJ, Friedman WA, Buatti JM, Bouchet LG (2001) Isotropic beam bouquets for shaped beam linear accelerator radiosurgery. *Physics Medical Biology* 46:2571–2586.

Wang JZ, Rice R, Pawlicki T, Mundt AJ, Sandhu A, Lawson J, Murphy KT (2010) Evaluation of patient setup uncertainty of optical guided frameless system for intracranial stereotactic radiosurgery. *Journal of Applied Clinical Medical Physics* 11:3181.

Wang LT, Solberg TD, Medin PM, Boone R (2001) Infrared patient positioning for stereotactic radiosurgery of extracranial tumors. Comput Biol Med 31:101–111.

Wang PM, Hsu WC, Chung NN, Chang FL, Jang CJ, Fogliata A, Scorsetti M, Cozzi L (2014) Feasibility of stereotactic body radiation therapy with volumetric modulated arc therapy and high intensity photon beams for hepatocellular carcinoma patients. *Radiation Oncology* 9:18.

Wang Y, Khan MK, Ting JY, Easterling SB (2012) Surface dose investigation of the flattening filter-free photon beams. *International Journal of Radiation Oncology, Biology, Physics* 83:e281–285.

Wiersma RD, Tomarken SL, Grelewicz Z, Belcher AH, Kang H (2013) Spatial and temporal performance of 3D optical surface imaging for real-time head position tracking. *Medical Physics* 40:111712.

Wilbert J, Guckenberger M, Polat B, Sauer O, Vogele M, Flentje M, Sweeney RA (2010) Semi-robotic 6 degree of freedom positioning for intracranial high precision radiotherapy; first phantom and clinical results. *Radiation Oncology* 5:42.

Wong E, D'Souza DP, Chen JZ, Lock M, Rodrigues G, Coad T, Trenka K, Mulligan M, Bauman GS (2005) Intensity-modulated arc therapy for treatment of high-risk endometrial malignancies. *International Journal of Radiation Oncology, Biology, Physics* 61:830–841.

Wu QJ, Wang Z, Kirkpatrick JP, Chang Z, Meyer JJ, Lu M, Huntzinger C, Yin FF (2009b) Impact of collimator leaf width and treatment technique on stereotactic radiosurgery and radiotherapy plans for intra- and extracranial lesions. *Radiation Oncology* 4:3.

Wu QJ, Yoo S, Kirkpatrick JP, Thongphiew D, Yin FF (2009a) Volumetric arc intensity-modulated therapy for spine body radiotherapy: Comparison with static intensity-modulated treatment. *International Journal of Radiation Oncology, Biology, Physics* 75:1596–1604.

Yenice KM, Lovelock DM, Hunt MA, Lutz WR, Fournier-Bidoz N, Hua CH, Yamada J, Bilsky M, Lee H, Pfaff K, Spirou SV, Amols HI (2003) CT image-guided intensity-modulated therapy for paraspinal tumors using stereotactic immobilization. *International Journal of Radiation Oncology, Biology, Physics* 55:583–593.

Yu CX (1995) Intensity-modulated arc therapy with dynamic multileaf collimation: An alternative to tomotherapy. *Physics Medical Biology* 40:1435–1449.

Yu CX, Li XA, Ma L, Chen D, Naqvi S, Shepard D, Sarfaraz M, Holmes TW, Suntharalingam M, Mansfield CM (2002) Clinical implementation of intensity-modulated arc therapy. *International Journal of Radiation Oncology, Biology, Physics* 53:453–463.

Yuan Y, Thomas EM, Clark GA, Markert JM, Fiveash JB, Popple RA (2018) Evaluation of multiple factors affecting normal brain dose in single-isocenter multiple target radiosurgery. *Journal of Radiosurgery & SBRT* 5:131–144.

Zhao B, Yang Y, Li X, Li T, Heron DE, Saiful Huq M (2015) Is high-dose rate RapidArc-based radiosurgery dosimetrically advantageous for the treatment of intracranial tumors? *Medical Dosimetry* 40:3–8.

Ziemer BP, Shiraishi S, Hattangadi-Gluth JA, Sanghvi P, Moore KL (2017) Fully automated, comprehensive knowledge-based planning for stereotactic radiosurgery: Preclinical validation through blinded physician review. *Practice in Radiation Oncology* 7:e569–578.

5

Advanced MRI for Brain Metastases

Hatef Mehrabian, Michael Chan, Paula Alcaide Leon, Sten Myrehaug, Hany Soliman, and Chris Heyn

Contents

5.1 INTRODUCTION

Brain metastases are a common manifestation of metastatic tumors and occur in 20% to 40% of all cancer patients with significant implications regarding morbidity and mortality. Imaging plays a crucial role in diagnosis and detection of disease as well as semiquantitative assessment of its response to therapy. In the age of personalized medicine with increasing technological advancements and the development of targeted cancer therapies, there is not only an increasing need for rigorous evaluation of the anatomic details but also the metabolic, functional, and micro-structural details such as tumor cellularity, metabolism, and pH, oxygenation, and perfusion. These factors are often the target of novel therapeutic agents or strategies and have been shown to have prognostic implications. In addition, there has been a growing need for accurate and precise quantitative imaging tools, especially for predicting patient outcomes and assessing early response to therapy, and managing treatment-related late-effects from tumor recurrence.

The focus of this chapter will be on magnetic resonance imaging (MRI) techniques, which are the current gold standard for imaging brain metastases and have been shown to be superior to other modalities, such as computed tomography (CT), for this purpose. While basic MRI techniques provide unparalleled tissue contrast that is essential in the brain, advanced MR techniques—including MR spectroscopy (MRS), chemical exchange saturation transfer (CEST), magnetization transfer (MT), perfusion imaging, diffusion-weighted imaging (DWI), hyperpolarized ^{13}C, and blood-oxygen-level-dependent (BOLD) MRI—provide the opportunity to explore important metabolic, functional, and micro-structural aspects of tumors. By manipulating various parameters, MR represents an extremely adaptable technique with essentially limitless opportunities for generating information on tumor biomarkers that can be used in the future to guide and direct management.

5.2 MR SPECTROSCOPY

MRS is a technique used to quantify the chemical composition of tissues. Using this methodology, the presence of molecules containing MR-detectable nuclei—such as protons (^1H), phosphorus (^{31}P), sodium (^{23}Na), carbon (^{13}C), or fluorine (^{19}F)—can be measured and mapped. MRS is most commonly performed on ^1H owing to the large natural abundance of this isotope, the large biological abundance of protons in organic tissues, and the favorable gyromagnetic ratio of protons, which results in better signal-to-noise ratio (SNR) compared to other nuclear species. To detect different nuclear species, a radiofrequency (RF) coil tuned to the specific frequency of that nucleus is needed. Proton MRS can be performed using the RF coils used for clinical MRI, whereas MRS of other species requires a specially built and tuned RF coil.

5.2.1 BASIC PHYSICS AND TECHNIQUE

The physics of MRS is based on the fact that nuclei within a molecule will resonate at a frequency that is proportional to the gyromagnetic ratio (a constant for a given nuclear species) and the local magnetic field experienced by the nucleus. Changes in the local magnetic field result in changes in the resonance frequency of the nucleus, a phenomenon known as chemical shift. This is dependent on the magnetic shielding of nuclei by electrons around the nucleus, which in turn is related to the chemical bonding of the nucleus within the molecule. Depending on the chemical bonding, nuclei within a molecule will resonate at specific frequencies that are characteristic for a given molecule and result in a unique signature that can be detected using MRI. In MRS, the resonance frequencies for a given molecule are expressed in terms of parts per million (ppm), which is a unit of frequency that is the same regardless of the field strength at which the measurement took place. Preferably, MRS should be performed at as high a magnetic field strength as possible because of the SNR advantage gained at higher fields.

In the brain, the most abundant proton species is from water, which resonates at a frequency of 4.7 ppm. To improve the detection of brain metabolites, which have concentrations that are orders of magnitude smaller than water, different methodologies are used to selectively suppress the signal from water. There are many ways to accomplish this, but a commonly employed strategy is through the application of RF pulses centered on the resonance frequency of water protons with the aim of saturating and decreasing the signal from water.

The echo time (TE) that is used in MRS is selected based on the specific application and the metabolites of interest. Generally, this can be done with short TE (e.g., 20–40 ms) or long TE (e.g., >135 ms). At lower TE, higher SNR can be achieved allowing the detection of many more metabolites. Visually, the spectra are more complex as many more peaks are visible. Longer TE spectra have lower SNR but are visually simpler to interpret.

Two major MRS techniques are used clinically: single-voxel spectroscopy (SVS) and magnetic resonance spectroscopic imaging (MRSI). SVS allows the characterization and quantification of metabolites in a defined region of interest (ROI) or volume of interest (VOI). This can be achieved by using magnetic field gradients and sequential slice-selective RF excitation to interrogate protons in the VOI (typically a cube with dimensions on the order of centimeters). The two most commonly employed methodologies are point-resolved spectroscopy (PRESS) and stimulated echo acquisition mode (STEAM) (Drost et al., 2002). Of these, PRESS is used more often. The sequence consists of RF pulses to initially suppress water followed by one 90° and two sequential 180° RF pulses. Each RF pulse is applied with a sequential orthogonal magnetic field gradient to selectively excite and refocus signal within the ROI. The sequence is a spin echo sequence that maximizes signal-to-noise ratio through spin echo refocusing. Compared to STEAM, which does not employ spin echo refocusing, PRESS can achieve higher signal-to-noise ratio but suffers from longer TE and less precise localization of signal from the VOI. SVS is faster to acquire than MRSI, which allows separate acquisitions of short and long TE data. Additionally, quantification of tissue metabolites, which can be performed with various models such as LC model, is more robust with SVS (Provencher, 1993).

MRSI allows for the spatial localization of metabolites. The basic design of MRSI sequences is similar to SVS with format based on PRESS or STEAM, for instance. The addition of phase encode gradients to these sequences is used to encode spatial frequency information, which are used to fill a spectroscopy grid. The spatial frequency information can then be used to reconstruct the spatial distribution of spectra via a Fourier transform. One of the main differences between MRSI and MRI is that frequency encoding gradients are not used with MRSI, as the frequency dimension holds the information related to chemical shift. The need for a separate phase encode step to encode each point within the spectroscopy grid is time consuming, which is one of the limitations of the technique. The major advantage over SVS is the ability to map the spatial distribution of metabolites, which can vary from region to region within a heterogeneous tumor.

5.2.2 INTERPRETATION

The three most abundant species in the brain are N-acetylaspartate (NAA), creatine (Cr), and choline (Cho) (Soares and Law, 2009). The most abundant brain metabolite is NAA located upfield from the water at approximately 2.0 ppm. It is synthesized in the mitochondria of neurons and is therefore a neuronal cell marker and a marker of neuronal cell viability, although it can also be found in glial cells. Pathologies that lead to neuronal cell death or replacement will result in a decrease in NAA. Cr is the second most abundant brain metabolite and is found at approximately 3.0 ppm. The Cr peak is derived from protons found on Cr and phosphocreatine, which are products of energy metabolism. Generally, the level of Cr in the brain is less affected by pathology and can be used as an internal reference. For example, the ratio of a brain metabolite such as NAA can be crudely normalized measuring the NAA to Cr ratio, thereby allowing the NAA levels in a pathological portion of the brain to be compared to normal brain, for example. Cho is found at approximately 3.2 ppm and is the third most abundant metabolite after Cr. The signal is derived from protons on Cho and metabolites of Cho, which are generally found in cell membranes. Cho is therefore a marker of cell membrane turnover and cellular proliferation. Consequently, Cho is typically elevated in brain neoplasms including metastases and generally decreased in areas of tumor necrosis. Elevated Cho is also demonstrated in brain inflammation and other pathologies and is therefore not specific. Figure 5.1 demonstrates a typical SVS proton spectrum for normal brain demonstrating these three major brain metabolites.

Other important metabolites which can be detected include lipids (Lip, 0.9–1.2 ppm), myoinositol (Myo, 3.56 ppm), and amino acid peaks such as glutamine/glutamate (~2.0–2.5 ppm) and alanine (~1.48 ppm). Lactate (Lac) is a doublet centered at 1.3 ppm, which undergoes inversion (seen below the baseline) at TE = 135 ms, thereby allowing it to be seen separately from Lip peaks, which occur at this location at short TE. While Lac can be found in healthy adult brains, it is generally not detectable at physiologic concentrations using most commonly employed proton MRS technique. Increases in brain Lac and detection with MRS indicate increasing underlying anaerobic metabolism, which can be seen with brain neoplasms, as well as a wide range of pathologies including hypoxia, acute inflammation, infection, and metabolic brain disease. Lip can be seen as a result of contamination by fat from adjacent structures (e.g., subcutaneous fat). It is also present in pathologies resulting in cell membrane degradation/necrosis. Myo is a degradation product of myelin and found in glial cells. It can be increased with glial cell proliferation as is seen with inflammation, gliosis, and gliomas. Ala can be seen in meningiomas, and Glx peaks can be seen in certain metabolic brain diseases such as hepatic encephalopathy.

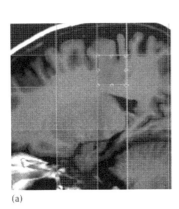
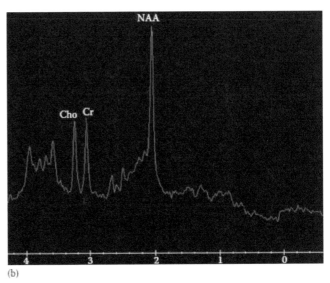

(a)   (b)

Figure 5.1 T_1-weighted images of normal white matter with overlay of single-voxel spectroscopy region of interest (box delimited with diamonds) (a) and corresponding MR spectroscopy (point-resolved spectroscopy, echo time 35 ms) (b) showing the typical appearance of the three major metabolic peaks (choline, creatine, and N-acetylaspartate).

5.2.3 CLINICAL APPLICATION

Brain tumors generally show decreased NAA and elevated Cho compared to normal brain (Soares and Law, 2009). Metastases can show an absence of NAA as they are not neuronal derived, although the small size of most metastases and low resolution of SVS or MRSI result in partial volume averaging of adjacent brain containing NAA, so this observation is not commonly made. With tumor necrosis, Lip peaks can be demonstrated. Furthermore, Lac may be elevated in metastases as a result of derangements of tumor cell energy metabolism (e.g., Warburg effect). Generally, there are no specific metabolic differences on MRS for metastases arising from different primary tumors, although one small study did report elevated Lip in untreated colorectal metastases (Chernov et al., 2006).

Although MRS has been used extensively in differentiating brain metastases from primary brain tumors (Caivano et al., 2013; Ishimaru et al., 2001; Ortiz-Ramón, Ruiz-España, Mollá-Olmos, and Moratal, 2020), the limited number of studies examining the use of proton MRS for predicting response of brain metastases to stereotactic radiosurgery (SRS) has been disappointing. A study involving a small cohort of 26 patients with intracranial metastases (predominantly lung, breast, colorectal) treated with SRS with follow-up at least 3 months after treatment examined the relationship between baseline proton MRS and treatment response (Chernov et al., 2007). In this cohort, approximately 50% of patients had treatment response (defined as 50% reduction in tumor volume), and 46% had local progression (defined as a 25% increase in tumor volume). The authors found no correlation between the metabolic profile of tumors at baseline and treatment response. However, in another study including 21 brain metastases, pretreatment MRS spectra were correlated with 5-month survival of the patients (Sjøbakk et al., 2007).

Shortly after SRS, metastases undergo metabolic changes that can be observed by proton MRS even before changes in tumor size are observed. In a small cohort of 81 patients with 85 brain metastases, proton MRS performed within 16–18 hours after SRS showed significant reductions in Cho/Cr ratios compared to baseline measurements despite a lack of morphologic change in size of metastases at this early time point (Chernov et al., 2004). This decrease was greater for tumors with initially high Cho/Cr ratios. The authors attributed this metabolic change to a reduction in cell proliferation and cell death.

A more longitudinal study of SRS-treated brain metastases demonstrated increase in NAA/Cr ratio and decreases in Cho and Lip content within the first month after treatment in responders (Chernov et al.,

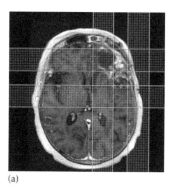
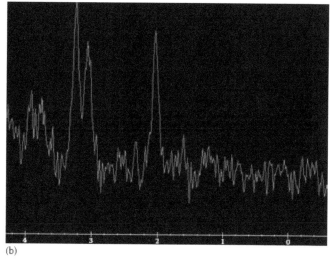

(a)   (b)

Figure 5.2 A 73-year-old woman with a history of breast cancer metastasis to the left frontal lobe treated with surgery and stereotactic radiotherapy. This patient subsequently developed an enlarging enhancing component along the posterior treatment margin. T_1-weighted postcontrast images with overlay of single-voxel spectroscopy region of interest (box delimited with diamonds) (a) and corresponding MR spectroscopy (point-resolved spectroscopy, echo time 144 ms) (b) demonstrate slightly elevated choline-to-creatine ratio and reduction in N-acetylaspartate, which were suspicious for tumor recurrence.

2009). No significant alteration was observed in patients with stable tumors, and decreases in NAA/Cr and increases in Lip and Cho were observed in progressors.

MRS has been used extensively in differentiating between radiation necrosis and tumor recurrence, where most studies show high specificity and suboptimal sensitivity in distinguishing these two entities. After radiation therapy, regions of necrosis within a tumor undergo progressive increases in Lac and Lip peaks, as well as a transient increase in a Cho peak leading to an eventual decrease in Cho with ongoing necrosis. In contrast, tumor recurrence is characterized by a persistently increasing Cho peak (Chernov et al., 2005). Multiple metabolite peak ratios have been proposed in the literature for differentiating between tumor recurrence and radiation-induced necrosis, particularly combinations of the Cho, NAA, and Cr peaks. For example, in a study of 33 intra-axial metastases treated with SRS, or fractionated radiation therapy, the best MRS parameter for predicting tumor progression was the Cho/nCho (Cho in tumor to Cho in normal contralateral brain) ratio, which yielded a sensitivity of 33% and specificity of 100% for Cho/nCho of more than 1.2 (Huang et al., 2011). The same study also compared MR perfusion which showed superior performance with an area under curve (AUC) of 0.802 versus 0.612 for MRS. In 29 brain metastases, Weybright reported significantly higher Cho/Cr and Cho/NAA ratios for tumor recurrence compared to radiation necrosis (Weybright et al., 2005). A meta-analysis of 13 studies involving a total of 397 lesions also concluded that recurrent tumors show higher Cho/Cr and Cho/NAA ratios (Chuang, Liu, Tsai, Chen, and Wang, 2016). Figure 5.2 shows SVS findings suspicious for disease recurrence after SRS, and Figure 5.3 shows SVS findings in a patient with radiation necrosis. Metabolic imaging with MRS has shown some interesting trends, but in everyday practice, the performance of MRS in predicting treatment response and differentiating radiation injury in treated brain metastases has been mediocre at best.

5.3 SATURATION TRANSFER MRI

Saturation transfer MRI involves chemical exchange saturation transfer (CEST) that probes tissue metabolism and magnetization transfer (MT) that characterizes macromolecular content of the tissue (Henkelman et al., 1993; Ward et al., 2000). The CEST contrast mechanisms can be used to detect intrinsic molecular species and chemical environments such as amino acids, peptides, proteins, and tissue pH (Jin, Wang,

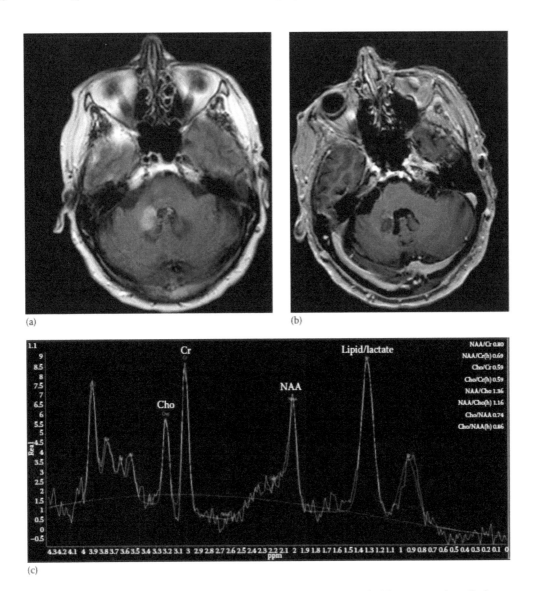

(a)   (b)

(c)

Figure 5.3 A 68-year-old woman with history of lung cancer metastasis treated with stereotactic radiotherapy. Axial (a) FLAIR and (b) T_1-weighted postcontrast images demonstrate a minimally enhancing mass in the right cerebellum and brachium pontis. Single-voxel MR spectroscopy analysis (echo time 144 ms) (c) demonstrates a large lipid/lactate peak, normal choline peak, and decreased *N*-acetylaspartate peak. While these findings are not specific for any entity in particular, a large lipid/lactate peak can be seen in the setting of radiation-induced necrosis. This mass was found to decrease in size on subsequent follow-up imaging without any evidence of recurrence.

Zong, and Kim, 2012; Ward and Balaban, 2000) and MT probes lipids associated with myelin and cell membranes and has been shown to be sensitive to inflammation and demyelination due to multiple sclerosis (Horsfield, 2005).

Imaging for both CEST and MT is similar and involves applying an off-resonance RF saturation pulse and then imaging signal from water protons. By comparing pre-excitation and post-excitation images, the distribution of the molecular target can be mapped. Repeating this process for a range of offset frequencies provides a CEST or MT spectrum. The difference between CEST and MT is the metabolite or molecule that is being targeted by the saturation pulses. MT also generally uses higher saturation power. CEST selectively excites mobile proton pools (typically hydroxyl, amine, or amide

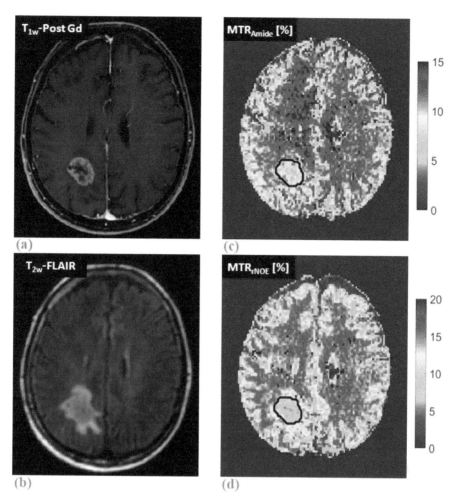

Figure 5.4 A patient with brain metastasis. Axial (a) T_1-weighted postgadolinium and (b) T_2-weighted FLAIR images demonstrate a rim-enhancing mass. Axial chemical exchange saturation transfer (CEST)-weighted maps corresponding to (c) magnetization transfer ratio (MTR) of amide (MTR_{Amide}) peak at 3.5 ppm and (d) MTR of rNOE (MTR_{rNOE}) peak at −3.5 ppm. The tumor is shown with black contour on the CEST maps. The tumor and also edema regions have lower MTR values compared to normal white matter.

protons) on the molecular target, which chemically exchanges with a larger free water pool and have resonance frequencies close to water protons (approximately ±6 ppm). MT however applies the saturation pulse to offsets larger than 10 ppm or 20 ppm and excites semi-solid macromolecules that exchange their magnetization with water protons through cross-relaxation. The end result for both CEST and MT is saturation or reduction in signal intensity of the free water pool. Because of the exchange of protons or magnetization with the larger free water pool, an amplification of the signal response occurs, and the sensitivity of CEST and MT are therefore far greater than what would be possible via the direct detection of proton species on the molecular target. The most commonly used CEST metrics in cancer are the CEST signal corresponding to amides (3.5 ppm), amines (2 ppm), and relayed nuclear Overhauser effect, rNOE (−3.5 ppm). MT is commonly used as the magnetization transfer ratio (MTR) at approximately 20 to 30 ppm. Quantification of the MT effect (qMT) using a two-pool model has also been performed (Levesque et al., 2010; Mehrabian, Myrehaug, Soliman, Sahgal, and Stanisz, 2018). Figure 5.4 shows a representative CEST maps of a brain metastasis, which was produced by computing the MTR at the amide and rNOE CEST peaks.

CEST can also be used to detect exogenously administered molecular species, which act as contrast agents or molecular tracers. This can be done with naturally occurring molecules or molecular constructs which incorporate paramagnetic (Aime, Delli Castelli, Fedeli, and Terreno, 2002), diamagnetic (McMahon et al., 2008), or hyperpolarized species (Schroder, Lowery, Hilty, Wemmer, and Pines, 2006) (PARACEST, DIACEST, HYPERCEST, respectively). Recently, exogenously administered glucose and glucose analogs (e.g., ^{19}F-fluorodeoxyglucose [^{19}FDG] and 2-deoxyglucose) have been mapped using CEST (Rivlin, Horev, Tsarfaty, and Navon, 2013; Walker-Samuel et al., 2013). These studies indicate the potential of MR for metabolic imaging with capability similar to that of ^{18}FDG-PET.

Several studies have used CEST in differentiating brain tumors and also grading primary brain tumors (Jing et al., 2014; Jones et al., 2006; Paech et al., 2018; Zhao et al., 2013). For instance, Wen showed the distribution of peptides in primary brain tumors correlate with tumor grade (Wen et al., 2010). The high sensitivity of CEST to treatment-induced metabolic changes in the tumor has been used in assessing response of brain metastases to SRS 1 week after treatment showing significant reduction in CST metric for responders and increase for non-responders (Desmond et al., 2017). Also, certain CEST metrics at baseline (before treatment) could predict response of brain metastases to SRS (Desmond et al., 2017). In differentiating radiation necrosis from tumor progression, CEST showed promising results in preclinical studies and in 17 patients with brain metastases where tumor progression had higher CEST than the radiation necrosis, and a perfect separation of these two cohorts was achieved (Mehrabian, Desmond, Soliman, Sahgal, and Stanisz, 2017; Zhou et al., 2011).

Significantly reduced MT effect has been reported in tumors compared to normal white matter which can be used in tumor detection and also differentiating brain metastases from other brain tumors (Ainsworth et al., 2017; Garcia et al., 2015; Mehrabian, Lam, Myrehaug, Sahgal, and Stanisz, 2018). The MT metric in differentiating radiation necrosis from tumor progression provided statistically significant separation, but its performance was not as good as the CEST metrics (Mehrabian, Desmond, Soliman, et al., 2017).

5.4 PERFUSION IMAGING

Perfusion MRI provides information about the tumor microcirculation. The importance of measuring the hemodynamic characteristics in tumors arises from the notion that more aggressive tumors are characterized by endothelial hyperplasia and neovascularization. This technique can be performed using a variety of methods relying on the use of intravascular tracers. In MR, this is most commonly accomplished with intravenous administration of gadolinium-based contrast agents. The two dominant contrast-enhanced MR perfusion techniques are dynamic susceptibility contrast (DSC)-enhanced and dynamic contrast-enhanced (DCE) perfusion imaging. In addition to contrast-enhanced techniques, there are emerging non-contrast-enhanced methodologies for evaluating tissue perfusion. The dominant technique is called arterial spin labeling (ASL).

MR contrast agents work in a fundamentally different way from CT contrast agents, and understanding this difference is important for appreciating how various contrast-enhanced MR perfusion techniques work with their advantages and limitations. CT contrast agents work by increasing the amount of x-rays absorbed (attenuated) by tissues containing the contrast agent. For CT perfusion, as a bolus of contrast agent enters the microcirculation of a tissue, the attenuation of x-rays is increased proportional to the effective concentration of the contrast agent within the tissue of interest. MRI works through the detection of signal generated by magnetization of protons within the tissues being imaged. Gadolinium contrast agents alter the MR signal via two major mechanisms. There is the effect of gadolinium contrast agents on the T_1 of water interacting with the contrast agent. For this to occur, the water molecules must directly interact with the gadolinium ion within the chelate (e.g., this occurs over molecular distances). This interaction results in the shortening of T_1 and enhancement of MR signal on T_1-weighted images, which is not a simple relationship that scales linearly with dose. Competing with the signal enhancement effects caused by T_1 shortening is the effect of the gadolinium contrast agent causing shortening of tissue T_2 and T_2^*. These effects occur over larger distances compared to the T_1-shortening effects of gadolinium. The T_2 and T_2^* effects are caused by the effects of gadolinium bolus as it passes through the microcirculation. The

Table 5.1 **Standard kinetic parameters derived from the Tofts model**

PARAMETER	PHYSIOLOGICAL MEANING
v_p (plasma volume fraction)	The blood plasma volume fraction of the whole tissue.
v_e (fraction of extravascular extracellular space)	Fraction on tissue volume corresponding to extravascular extracellular space.
K^{trans} (transfer constant of contrast from the plasma to the tissue extravascular extracellular space)	Related to the balance between capillary permeability and blood flow. When capillary permeability is high, the amount of contrast that leaks out of the vessels depends on the amount of contrast that gets to the capillaries per unit of time. In this case, K^{trans} reflects the blood plasma flow. In cases of low permeability, K^{trans} equals permeability surface area product.
k_{ep} (rate constant)	Rate constant between the interstitial space and blood plasma ($k_{ep} = K^{trans}/v_e$).

Source: Tofts, P.S. et al., J. Magn. Reson. Imaging (US), 10, 223, 1999.

gadolinium within the vasculature results in a differential magnetic susceptibility of blood and surrounding tissue, leading to micro- and macroscopic magnetic field inhomogeneities causing signal loss via altered T_2 and T_2^* effects. The effect on T_2 or T_2^* is complex and will depend on a number of parameters including the concentration of gadolinium in the intravascular space, extravascular extravasation of contrast, the configuration and size of the vasculature, diffusion of water through the bulk magnetic susceptibility field of the vasculature, and specific imaging parameters. The different flavors of contrast-enhanced MR perfusion exploit either the T_1-shortening effects (e.g., DCE) or T_2/T_2^*-shortening effects (DSC) of gadolinium contrast agents.

5.4.1 DYNAMIC CONTRAST-ENHANCED PERFUSION

In the healthy brain, the blood–brain barrier precludes contrast leakage into the interstitial space. In brain metastases, vessels are abnormal and leaky, allowing for contrast agent extravasation resulting in MRI signal enhancement. DCE perfusion uses the T_1-shortening signal enhancement effect of extravasated gadolinium on the interstitial tumor water pool. The technique is therefore based on the acquisition of sequential T_1-weighted images with temporal resolution of a few seconds. The analysis of DCE perfusion data is performed through pharmacokinetic modeling of tumor gadolinium contrast concentration estimated by measuring dynamic signal intensity changes on T_1-weighted images. Other authors use a simpler approach assessing changes in signal intensity directly and obtaining semiquantitative parameters (Kuhl et al., 1999; Padhani et al., 2000; Engelbrecht et al., 2003; Arasu et al., 2011). Commonly used semiquantitative parameters include initial area under the gadolinium concentration curve and wash-in and wash-out rates (Leach et al., 2005; Lankester et al., 2007). These parameters are easier to obtain than model-derived parameters but have lower reproducibility and no direct physiologic meaning. For these reasons, the model-derived approach is preferred (Leach et al., 2005). It is currently unclear which model is most suitable for the various tumor types, tumor sites, and treatment methods. The most widespread model is the modified Tofts-Kety model (Tofts et al., 1999). Parameters obtained with this model are described in Table 5.1.

DCE-MRI has been shown to be useful in radiated brain metastases for monitoring treatment response and radiation side effects. The use of pretreatment DCE-MRI as a tool for predicting response to radiotherapy has also been widely investigated in other regions outside the brain.

5.4.1.1 Predicting Tumor Response

Many clinical studies have correlated DCE-MRI-derived parameters with important histopathological features that are related to tumor radiosensitivity (Zahra et al., 2007). An important determinant of tumor response to radiotherapy is tumor oxygenation, with hypoxic tumors being more resistant. A preclinical study in tumor xenografts of eight human melanoma lines showed an inverse relationship between perfusion parameters (K^{trans} and v_e) and tumor hypoxia (Egeland et al., 2012). A similar study in gliomas also showed promising results (Jensen et al., 2014). Applying more advanced models to DCE-MRI that allow for quantification of water exchange rate between intracellular and extracellular spaces (k_{IE}) has shown early changes in k_{IE} after SRS correlates with tumor volume change at later time points, allowing for early identification of non-responders to treatment (Mehrabian, Desmond, Chavez, et al., 2017).

5.4.1.2 Monitoring Response to Treatment

DCE-MRI has been shown to be useful in monitoring response after radiation in brain metastases. A study in 20 patients with brain metastases treated with whole-brain radiotherapy found that an early change in the specific subvolume of a brain metastatic tumor showing pretreatment high relative cerebral blood volume (rCBV) and high K^{trans} is a better predictor for postradiation therapy response than a tumor volumetric change (Farjam et al., 2013). A later study on 26 patients with cerebral metastases treated with SRS showed an overall reduction in K^{trans} values of the cerebral metastases in the early posttreatment period. Furthermore, an increase in K^{trans} values was predictive of tumor progression (Almeida-Freitas et al., 2014). Non-model-derived parameters have also shown value in predicting tumor response after whole-brain radiotherapy in brain metastases (Farjam et al., 2014b). Another study of 20 brain metastases treated with SRS showed a significant decrease in plasma volume fraction (v_p) 1 month after treatment (Kapadia et al., 2017). In 123 lung cancer brain metastases changes in K^{trans}, v_e and v_p were significantly associated with response to SRS (Kuchcinski et al., 2017).

5.4.1.3 Distinguishing Radionecrosis from Recurrence

Differentiation between radionecrosis and recurrent metastasis can be very challenging. Radionecrosis consists of fibrinoid necrosis of the blood vessel walls followed by necrosis of the surrounding parenchyma. Capillaries of brain metastases are different from those of the brain as they usually resemble the organ from which the cancer arose and do not have a blood–brain barrier (Long, 1979). On conventional T_1-weighted images, both radiation necrosis and recurrent metastases demonstrate enhancement and nonenhancing areas of necrosis. However, several studies have shown a difference in the dynamic nature of the enhancement reflecting different permeability, vessel morphology, and vessel density. Hazle, Jackson, Schomer, and Leeds (1997) were able to distinguish between recurrence, radiation necrosis, and a combination of both as they found that radiation necrosis and tumors enhance at different rates. A study of 44 melanoma brain metastases patients undergoing immunotherapy showed significantly lower relative vascular volume fraction (rv_p) in pseudoprogression compared to tumor progression (Umemura et al., 2020). A study of 61 patients with brain metastases treated with SRS also showed a significantly lower K^{trans} in radionecrosis compared to disease progression (Knitter et al., 2018).

A study in patients with gliomas and brain metastases describes the use of delayed enhancement maps after 75 minutes of contrast administration to distinguish between active tumor and nontumoral tissues. The former is characterized by faster clearance than accumulation, and the latter is characterized by slower clearance than accumulation (Zach et al., 2012).

5.4.2 DYNAMIC SUSCEPTIBILITY CONTRAST-ENHANCED PERFUSION

DSC relies on the T_2/T_2^*-shortening effect of a bolus of intravascular gadolinium as it passes through the microcirculation of the tumor. Usually a gradient echo T_2-weighted echo planar sequence is used, which allows for whole-brain coverage with temporal resolution less than 2 seconds (Petrella and Provenzale, 2000). The drop in MR signal, which occurs as the bolus passes through the tumor microcirculation, is used to calculate the CBV, which reflects microvascular density and has proven to be very useful for the differentiation of lesions on the basis of their microvasculature. Figure 5.5 illustrates CBV map for a typical clinical case. Another parameter derived from the T_2 signal–time curves is the percentage of signal recovery (Barajas et al., 2009). However, this parameter has been shown to be highly dependent on acquisition parameters with very low reproducibility between different centers (Boxerman et al., 2013). There are some limitations of DSC-MRI, the main one being the presence of susceptibility artifact in the ROI. This is most commonly due to hemorrhage but also caused by air and bone in the inferior temporal and frontal regions. This is a very common source of false negatives when evaluating CBV maps from

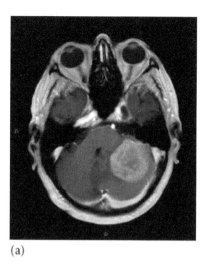
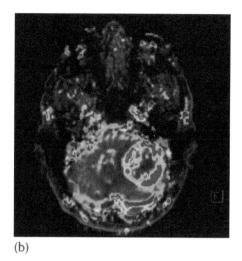

(a)   (b)

Figure 5.5 A 72-year-old man with esophageal cancer and a 2-week history of headaches, dizziness, nausea, and vomiting. Axial (a) T$_1$-weighted postgadolinium images demonstrate a large enhancing mass in the left cerebellar hemisphere. Axial (b) dynamic susceptibility contrast perfusion-derived CBV map overlaid on T$_1$-weighted postgadolinium image shows a ring of high CBV within the lesion. Resection of the lesion revealed metastatic adenocarcinoma.

DSC-MR perfusion. The presence of fast contrast leakage during the first pass of the gadolinium bolus is another source of artifact as extravascular gadolinium tends to increase the signal of the tissue, competing with the T$_2$ shortening caused by intravascular gadolinium and results in artificially low CBV values. The effect of contrast extravasation can be solved by contrast preloading or by postprocessing correcting algorithms.

With regard to radiated brain metastases, DSC-MR perfusion has proven to be useful in the early posttreatment period predicting patient outcomes as well as in the long-term follow-up distinguishing between recurrent metastases and radiation necrosis.

5.4.2.1 Early Posttreatment Phase

Essig studied 18 patients with cerebral metastases by using DSC-MRI to assess if preradiation and early time point CBV measurements of metastases could predict outcome in patients treated with SRS (Essig et al., 2003). Measurements of regional CBV changes in metastases and normal brain after treatment were also performed. The authors found that pretreatment CBV was not able to predict outcome. Measurements of CBV at 6 weeks posttreatment were a more sensitive and specific biomarker of treatment outcome. In particular, a reduction in regional CBV was predictive of treatment outcome with sensitivity in excess of 90% compared to sensitivity of 64% for a change in tumor volume alone. CBV values of normal brain were in line with expected physiologic range and were unchanged with therapy (Essig et al., 2003). Similar results were obtained by Weber with a slightly larger sample (25 patients) (Weber et al., 2004).

5.4.2.2 Late Posttreatment Phase

Hoefnagels investigated the ability of perfusion MRI to differentiate between tumor recurrence and radiation necrosis in patients showing radiological progression of disease. They concluded that when lesions display a CBV greater than 1.85 times the normal gray matter, necrosis can be excluded (specificity of 100% and sensitivity of 70%) (Hoefnagels et al., 2009). A similar study in patients treated with fractionated radiation therapy revealed 56% sensitivity and 100% specificity for CBV as a predictor of tumor progression (Huang et al., 2011). Knitter also showed a significantly lower rCBV in radionecrosis compared to disease progression in 61 patients with brain metastases treated with SRS (Knitter et al., 2018). A sensitivity of 91% and a specificity of 72% for a relative CBV to white matter threshold of 1.54 for identifying recurrent tumor versus radionecrosis have been described (Barajas et al., 2009). In another study, a relative CBV to white matter greater than 2.1 provided 100% sensitivity and 95.2% specificity (Mitsuya et al., 2010). Muto also recently reconfirmed that the 2.1 cut off of rCBV is the most accurate and reliable perfusion metric for differentiating radiation necrosis from tumor progression (Muto et al., 2018).

In contrast to the aforementioned studies, which demonstrated that an increased CBV after SRS serves as a marker for tumor recurrence several months after treatment, perfusion shows a different trend early after treatment. A recent study found an association between early 1-month posttreatment CBV reductions and tumor progression (Jakubovic et al., 2014). The authors postulated that these apparently contradicting findings may be explained by the time-dependent evolution of vascular changes after radiotherapy. In fact, a study exploring postradiation vascular changes in a tumor model demonstrated a transient switch between two different types of angiogenesis—sprouting angiogenesis to a nonsprouting (intussusceptive) angiogenesis (Hlushchuk et al., 2008, 2011). This "angiogenic switch" was hypothesized to correspond with early reductions in CBV. As time elapses, tumor recurrence is accompanied by a switch back to sprouting angiogenesis with increasing tumor vasculature and increasing CBV, whereas a favorable treatment response is associated with further CBV reduction.

5.4.3 ARTERIAL SPIN LABELING

ASL is an emerging clinical tool that allows the characterization of blood flow without the administration of exogenous contrast agents. In ASL, intravascular water is magnetically labeled using RF pulses. By comparing the MR signal of tissue imaged with and without magnetic intravascular water labeling, the blood flow to the tissue can be quantified. There are a number of specific methodologies for accomplishing this including continuous ASL (CASL), pulsed ASL (PASL), and a hybrid of the two methods termed pseudocontinuous ASL (pCASL). In CASL of the brain, a slice positioned at the level of the extracranial internal carotid arteries is excited using an inversion RF pulse, and a separate acquisition of the brain is acquired. The inverted magnetization from in-flowing blood from the neck will reduce the signal in the brain image. By comparing the signal intensity of the brain image to an acquisition without intravascular water labeling, the amount of signal resulting from blood flow can be quantified. Generally, CASL has had limited clinical utility because of the significant hardware demands required to achieve the proper conditions for blood water labeling and other effects (e.g., magnetization transfer), which complicate the quantification of blood flow. One type of PASL methodology called flow-sensitive alternating inversion recovery is based on an acquisition scheme, which employs two acquisitions of the target tissue: one acquisition with slice-selective inversion RF pulse and one with a nonselective inversion pulse. By comparing the signal from the tissue using these two acquisitions, the amount of signal resulting from flowing blood in the slice-selective acquisition can be calculated. The advantage of PASL is that the labeling of blood water is more easily implemented; however, the signal-to-noise ratio of the methodology is inferior to CASL. pCASL overcomes many of the shortcomings of CASL and PASL and is now a commonly used ASL methodology.

ASL has been used in the study of brain metastases treated with SRS. Soni reported a strong correlation between blood flow measurement with DSC perfusion and ASL perfusion in brain metastases, suggesting ASL is a reliable non-contrast replacement for DSC-MRI (Soni, Dhanota, Kumar, Jaiswal, and Srivastava, 2017). In a small cohort of 25 patients (28 total brain metastases), a decrease in relative Cerebral Blood Flow (CBF) for metastases at 6 weeks posttherapy measured using a pCASL methodology was predictive of treatment response (Weber et al., 2004). Cohen showed pCASL is useful in detecting post-operative residual lesion in hyper-perfused brain metastases (Cohen et al., 2020). Several studies have also used pCASL in differentiating brain metastases from primary brain neoplasms (Abdel Razek, Talaat, El-Serougy, Abdelsalam, and Gaballa, 2019; Ganbold et al., 2017; Geerts et al., 2019; Soni et al., 2019).

As ASL becomes more widely available on clinical systems, the application of this methodology in neuro-oncology will increase. A number of challenges remain, however, including the accuracy of CBF measurements made with ASL in tissues with longer blood transit times, for example, where the effects of T_1 relaxation on the blood water pool become significant. Furthermore, the reproducibility of the methodology across vendors as well as repeatability within subjects requires further study. Nonetheless, the ability to evaluate tumor physiology without the administration of IV contrast will certainly be a tremendous advantage in evaluating patients with brain metastases particularly in studies that require multiple examinations over a short time interval, where the administration of repeated gadolinium contrast is undesirable and impractical.

5.5 DIFFUSION-WEIGHTED IMAGING

Diffusion-Weighted Imaging (DWI) is a promising technique for evaluating treatment response. Using this methodology, the Brownian motion of free water molecules within a tissue can be estimated, and a map of the apparent diffusion coefficient (ADC) of the tissue can be produced. By measuring the microscopic motion of water within a tissue, the structure of the tissue at the cellular and subcellular level can be inferred. For example, water that is primarily compartmentalized within the cells of a tissue will have movement through space that is more restricted compared to water in the extracellular compartment. Thus, areas of high cellularity will demonstrate more restricted water motion and lower ADC values than areas of low cellularity or necrosis.

However, there are limitations to this simplistic interpretation of ADC values. Immediately after surgery or treatment, a low ADC within a tumor could indicate cytotoxic injury, which is characterized by cell swelling and water shifting from the extracellular compartment that allows for free movement to the motion-restricted intracellular compartment. Furthermore, a low ADC does not always imply cellularity or cellular water, as it can also be seen in the context of interactions between water and proteins or other substances that hinder its motion. For example, blood clot, highly proteinaceous material, or pus can restrict water motion and result in low ADC.

5.5.1 CLINICAL APPLICATION

A number of studies have explored how ADC changes after radiation therapy and how these changes correlate with response to treatment. In comparing results from the literature, there are important differences between studies that have to be considered. First, the timing of when diffusion measurements are made is an important consideration. Generally, most studies using DWI to predict treatment response have DWI measurements taken days to a few weeks after treatment. Furthermore, there are a number of other variables that may limit comparison between studies and a universal interpretation of results, including what imaging parameters were used, how quantification was performed, whether it was done using an ROI tool or quantified on a voxel-by-voxel basis, as well as how nonsolid, necrotic, or cystic components of the tumor were addressed.

One of the first studies examining the utility of DWI was performed on a small cohort of eight patients (six with brain metastases) (Mardor et al., 2003). These tumors were mainly treated with single-fraction SRS (16 Gy), with MRI performed at 0.5 T. Measurements were taken at baseline, 1, 7, and 14 days posttreatment. ROI analysis was performed on the solid portion of the metastasis (no necrosis was seen in any of the metastases studied). Outcome was tumor volume measured as volume of enhancing tumor approximately 48 days after treatment. The authors found a statistically significant moderate correlation between tumor volume change and change in ADC calculated from baseline and 7 days after therapy. Specifically, tumors demonstrating greater volume reduction and positive response to treatment showed greater increases in ADC at 1 week after treatment compared with tumors that did not respond.

A trend of increasing ADC in treated metastases has been confirmed in other larger studies. For example, in a prospective observational study on 86 patients (including 38 brain metastases) treated with SRS and MRI performed at 1.5 T, ROI analysis performed on solid portion demonstrated increases in ADC measured at 1 month after treatment (Huang et al., 2010). An analysis of the relationship between ADC changes and treatment response was not carried out. A larger study performed on 107 patients with brain metastases treated with SRS demonstrated a statistically significant decrease in ADC for progressors compared to patients with stable or responsive disease (Lee et al., 2014). A major limitation of ROI-based analysis is the question of how best to deal with metastases that have a heterogeneous imaging appearance pretreatment and/or develop imaging heterogeneity after treatment. Imaging heterogeneity is presumably related to underlying histological heterogeneity in a tumor. Decisions on where to place the ROI on the pretreatment scan or posttreatment scans can lead to significant measurement differences. Furthermore, an ROI-based analysis is statistically weakened when multiple voxels within the ROI are grouped together during the analysis.

A number of voxel-based analyses have emerged, which can potentially overcome these problems and provide a very powerful picture of imaging changes that occur after radiation therapy. Using voxel-based analysis, similar trends (to the aforementioned ROI-based analysis) were shown for ADC of viable tumor volume in 38 brain metastases where responders had increase in ADC while it decreased for non-responders (Mahmood, Hjorth Johannesen, Geertsen, and Hansen, 2020). In a prospective multi-center study in 223 brain metastases, Zakaria demonstrated the added benefit of considering ADC in clinical scoring systems may improve individualization of the treatment (Zakaria et al., 2020). A more representative voxel-based technique is functional diffusion mapping in which a voxel-by-voxel calculation of ADC changes between pre- and posttreatment scans is performed by spatially coregistering the two imaging datasets (Moffat et al., 2005). The power in this type of analysis is that it does not make an assumption that all regions of the tumor are initially the same or respond the same after treatment. It also effectively deals with tumor heterogeneity issues by analyzing and treating each voxel within the analysis volume independently. By examining the distribution of ADC changes across the entire tumor, functional diffusion mapping has been shown to be a promising biomarker for assessing treatment response in primary brain tumors. One of the limitations of this technique is related to problems associated with coregistering imaging data after significant changes in tumor volume or tissue distortion. Other voxel-based quantitative diffusion techniques without reliance on image coregistration—such as the diffusion abnormality index—have recently emerged and have shown utility in predicting treatment response in brain metastases (Farjam et al., 2014a). Advanced analysis techniques such as texture analysis and machine learning are also gaining traction in assessing ADC of brain metastases (Payabvash, Aboian, Tihan, and Cha, 2020; Zhang et al., 2019).

5.6 INTRAVOXEL INCOHERENT MOTION

Intravoxel incoherent motion (IVIM) is a diffusion MR technique that allows approximation of tumor blood flow and blood volume without the administration of IV contrast. Essentially, the motion of water within the microcirculation of a tumor can be modeled as a random (incoherent) process, which is similar to molecular water diffusion but occurs at a macroscopic scale (Le Bihan et al., 1986, 1988). By performing a diffusion MR experiment using measurements with several weak magnetic gradients (low b values) and stronger gradients (high b values), the contribution of water motion from blood flow and molecular water diffusion can be separated. Instead of reporting an "apparent" diffusion coefficient which has contributions from water motion at both scales, the ADC can be unpacked into a value which approaches the true molecular water diffusion (D), a parameter called the perfusion fraction (f), which is the fraction of water within the intravascular compartment (similar to blood volume), and a parameter called the pseudodiffusion coefficient (D*), which has been compared to blood flow (Le Bihan and Turner, 1992). Figure 5.6 illustrates a treated metastasis analyzed using IVIM methodology.

While the initial description of the IVIM methodology is quite old, only recently has the methodology found an increasing use in the CNS and in studying brain tumors (Federau et al., 2014; Kim et al., 2014b). A recent study examining IVIM in treated brain metastases showed improved accuracy of IVIM combined with DSC perfusion over DWI in combination with DSC, IVIM, and DSC alone in distinguishing between treatment response and tumor recurrence for brain metastases (Kim et al., 2014a). Conklin proposed a simplified IVIM that uses the high b value images ($b > 200$ s/mm^{-2}) and provides only molecular water diffusion (D) and perfusion fraction (f) (Conklin et al., 2016). Kapadia compared the perfusion fraction from the simplified IVIM model and the plasma fraction from DCE-MRI and reported significant change in these perfusion markers 1 month after SRS (Kapadia et al., 2017). Detsky also used the simplified IVIM model using six b values in differentiating radiation necrosis from tumor progression in patients with brain metastases treated with SRS showing higher f in tumor progression cohort (Detsky et al., 2017).

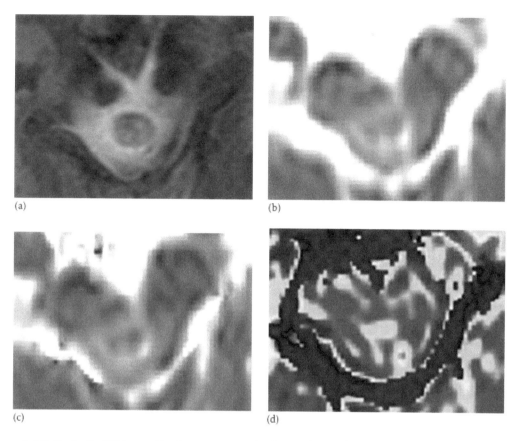

Figure 5.6 Patient with SRS-treated brainstem metastasis. Cropped magnified images of the brainstem showing round T_2 FLAIR hypointense midbrain lesion with surrounding vasogenic edema (a). Diffusion MRI conventional apparent diffusion coefficient (ADC) map (b) shows tumor to have diffusion characteristics similar to the brain but less water diffusion than surrounding edema. The diffusion coefficient map (c) and perfusion fraction, f map (d) calculated from intravoxel incoherent motion model. The diffusion coefficient map is similar to ADC as expected. The f map shows elevated perfusion fraction in the treated metastasis (cyan) compared to surrounding edema. The cerebral aqueduct is displaced posteriorly and laterally to the left and shows high CSF flow (red), which is captured by this technique.

5.7 BLOOD OXYGEN LEVEL-DEPENDENT MRI

Hypoxia is a key driver of an array of malignant processes including angiogenesis, alterations in energy metabolism, and cancer cell metastasis and invasion. Tumor hypoxia is also correlated with a higher risk of metastatic disease, radioresistance, and mortality (Harada, 2011). Thus, an ability to noninvasively quantify tumor hypoxia could provide valuable insight into the significance of this factor and may also be important in tailoring therapy, particularly to tumors demonstrating higher levels of hypoxia. While there are PET imaging techniques for evaluating tissue hypoxia such as with direct oxygen extraction measurements with $^{15}O_2$ or with hypoxia markers such as ^{18}F-FAZA, the availability of these radiotracers and inherent technical limitations of PET (e.g., low spatial resolution) limit their widespread use for evaluating metastatic disease. MRI techniques based on the principle of blood oxygen level-dependent (BOLD) imaging may offer a methodology for evaluating tumor hypoxia using technology that is widely accessible.

The BOLD effect is most commonly used in functional MRI (fMRI) to map areas of brain activation. It is predicated on the concept of neurovascular coupling in which blood flow to regions of activated brain is transiently increased resulting in a local decrease in deoxyhemoglobin within the microcirculation. Whereas oxyhemoglobin is diamagnetic with a negligible effect on the local magnetic field, deoxyhemoglobin is paramagnetic, and its presence within blood vessels results in magnetic susceptibility effects on the intravascular compartment as well as local tissues in the extravascular compartment. These effects on the MR signal are best detected using T_2^*-weighted sequences. In the case of brain activation, the lower deoxyhemoglobin results in a very small increase in MR signal that can be characterized using statistical methods. By measuring these small signal changes and correlating them with a specific task (e.g., finger tapping), areas of the brain activation can be mapped.

Magnetic susceptibility effects in any given tissue are complex and are affected by a number of factors that can confound BOLD measurements. Recently, there has been progress in modeling and estimating tissue oxygenation based on measurements of tissue T_2 and T_2^*. The work by He et al. has shown a relationship between tissue oxygenation, tissue transverse relaxation, hematocrit, and blood volume fraction. Using this methodology, tissue oxygenation in normal brain tissue has been measured and validated (He et al., 2008; Christen et al., 2011). The application of this theory in neoplasia has been accomplished in a rat glioma model (Christen et al., 2012). It has also been applied in Glioblastoma (GBM) patients to detect hypoxic regions in and around the tumor (Maralani et al., 2018; Tóth et al., 2013) but has yet to be applied in brain metastases.

Overall, quantitative BOLD techniques are promising, but much work has yet to be done to validate the methodology and the assumptions underlying the theory, which may be valid in normal tissue but violated in disease states such as tumors. Additionally, the methodology in its present state is complex and time consuming, which will limit its implementation as a tool in clinical practice.

5.8 HYPERPOLARIZED ^{13}C

Hyperpolarized ^{13}C MRS is a functional imaging technique that uses exogenous ^{13}C-labeled molecules, such as pyruvate or fumarate, and MRS methodology to investigate metabolic processes in normal and pathologic tissues. In MRI, SNR depends on a number of factors, which include the polarization of the nuclei generating the signal. Polarization refers to the difference in the number of nuclei with spins that are parallel with the magnetic field and the number of nuclei with spins that are antiparallel. Whereas with nonhyperpolarized ^{13}C nuclei in normal thermal equilibrium, there is only a slightly higher proportion of nuclei with spins aligned in parallel compared to antiparallel, the process of hyperpolarizing ^{13}C nuclei— a process that leads to the number of nuclear spins aligned in parallel being several orders of magnitude higher than antiparallel nuclear spins—results in a significantly higher signal. In fact, the recent development of a novel process using dynamic nuclear polarization for polarizing nuclear spins in solution has allowed for a 10,000-fold signal increase over other conventional MRS methods (Ardenkjaer-Larsen et al., 2003; Golman et al., 2003).

To date, hyperpolarized [1–^{13}C] pyruvate is the most widely studied and used hyperpolarized substrate. Following intravenous injection and uptake of [1–^{13}C] pyruvate by cells, the hyperpolarized [1–^{13}C] pyruvate molecules may undergo metabolism in one of three pathways: conversion to Lac catalyzed by lactate dehydrogenase (LDH), transamination to form alanine catalyzed by alanine aminotransferase, or decarboxylation to form carbon dioxide catalyzed by pyruvate dehydrogenase (Kurhanewicz et al., 2011). The kinetics of these metabolic reactions and relative concentrations of the metabolites—which differ depending on the state of the tissue—can be measured and monitored in real time, thus providing important metabolic information about the tissue of interest. In tumors, the metabolism of hyperpolarized [1–^{13}C] pyruvate molecules within cancer cells depends on a number of factors, including how much of the agent is delivered to the tumor, how much is taken up by the cancer cells, as well as the LDH concentration and activity (Brindle, 2012). Figure 5.7 shows Lac signal within a breast cancer xenograft in a rat model following intravenous injection with hyperpolarized [1–^{13}C] pyruvate. Time-resolved metabolic data demonstrate changes in the Lac and pyruvate signal following the injection.

The use of hyperpolarized [1–^{13}C] pyruvate MRS has been shown in multiple preclinical studies to have a potential for predicting treatment outcomes and assessing early response to therapy (Brindle et al., 2011; Kurhanewicz et al., 2011; Yen et al., 2011; Brindle, 2012). High reproducibility in differentiating malignant glioma tumors from normal brain tissues in a mouse and nonhuman primate tumor model by measuring pyruvate and Lac signals using hyperpolarized ^{13}C MRS imaging has been demonstrated (Park et al., 2010, 2014). Changes in pyruvate metabolism following radiotherapy in rat glioma models using a similar

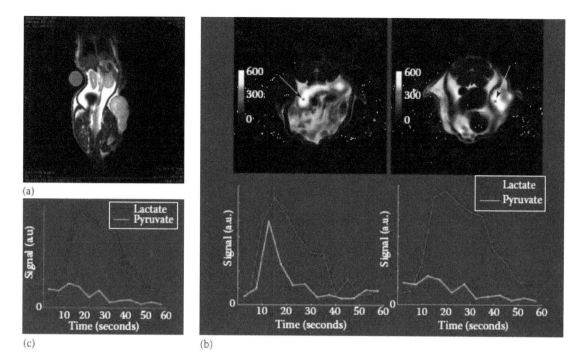

Figure 5.7 A rat model with an MDA-MB-231 breast cancer xenograft tumor treated with 16 Gy radiation. (a) A coronal T_2-weighted image demonstrates the tumor in the left flank. (b) An axial image with lactate signal overlayed on a T_2-weighted anatomic image following intravenous injection with hyperpolarized [1–^{13}C] pyruvate shows increased signal within the tumor. (c) Time-resolved metabolic data demonstrate changes in the lactate and pyruvate signal within the voxel indicated in (b) following the injection.

technique have also been shown (Day et al., 2011). Hyperpolarized [1–^{13}C] pyruvate has also been shown to have a role in detecting response to the PI3K inhibitor in both glioblastoma and breast cancer mouse models. In these models, reductions in the hyperpolarized Lac signal correlated with reductions in LDH activity due to PI3K pathway inhibition (Ward et al., 2010; Chaumeil et al., 2012).

5.9 MULTIPARAMETRIC APPROACHES

An emerging technique in oncology imaging is the application of multiparametric approaches for predicting treatment response or distinguishing between tumor recurrence and radionecrosis. A summary of the MR techniques discussed in this chapter with respect to differentiating tumor recurrence from radiation necrosis is provided in Table 5.2. While the preceding sections illustrate the progress in advanced MR techniques to answer these questions, it is unlikely that any one single imaging technique will provide a definitive answer for these complex clinical questions. Multiparametric approaches, which use multiple imaging measures to predict specific outcomes, may be one possible solution. In this methodology, a specific clinical outcome is correlated with a set of imaging measurements in a training patient dataset. The combination of imaging parameters determined from the training set is then used to predict outcomes prospectively in a test patient dataset. For example, a multiparametric analysis based on measurements of ADC, normalized CBV, and initial AUC from tumor perfusion was used to predict early tumor progression from pseudoprogression in glioblastoma (Park et al., 2015). The study confirmed a superior diagnostic accuracy of the multiparametric approach compared to the performance of the individual imaging parameters alone. Such multiparametric approaches have also found utility in predicting glioma grade for instance. The application of multiparametric techniques in treated brain metastases is an emerging area of research.

Table 5.2 **Summary of magnetic resonance techniques for differentiating tumor recurrence from radiation necrosis**

MR TECHNIQUE	TUMOR RECURRENCE	RADIATION NECROSIS
Conventional MRI	Persistent increase in the size of the enhancing lesion and degree of vasogenic edema	Transient increase in lesion size after SRS with eventual decrease in the size of the enhancing lesion and degree of vasogenic edema
	May demonstrate nonenhancing areas of necrosis	Peripheral rim enhancement with central decreased enhancement
MR spectroscopy	Persistently increasing Cho peak on serial imaging	Transient increased Cho peak with eventual decreasing Cho peak on serial imaging
	Nonspecific decreased NAA peak with or without increased lipid/lactate peaks	Decreased NAA peak and increased lipid/lactate peaks
DCE-MRI	Increased K^{trans} compared to necrosis	Reduced K^{trans} compared to recurrence
DSC-MRI	Time-dependent changes, but overall higher CBV-compared necrosis	Time-dependent changes, but overall lower CBV-compared recurrence
ASL	Increase in relative CBF	Decrease in relative CBF
DWI	Decreased ADC values compared to radiation necrosis	Increased ADC values following SRS compared to recurrence
IVIM	Increased f value and decreased D value	Decreased f value and increased D value
CEST	High CEST and MT values for amide and rNOE peaks	Very low CEST and MT signals at amide and rNOE peaks
BOLD MRI, MT, and hyperpolarized ^{13}C MRS	More research is required	More research is required

5.10 CONCLUSION

As cancer treatment strategies improve and patients live longer with systemic disease, the burden of brain metastases will continue to grow, as will the need for accurate and precise imaging tools for evaluating this disease. MRI has emerged as a powerful tool not only for diagnosing and detecting brain metastases but also for predicting patient outcome and evaluating early response to therapy. The techniques described earlier provide only a brief introduction into the vast possibilities offered by MRI for generating information regarding tumor biomarkers that can be used to guide and direct therapy.

ACKNOWLEDGMENTS

The authors would like to thank Dr. Charles Cunningham (Associate Professor at the University of Toronto and Senior Scientist at the Sunnybrook Research Institute) for contributing the hyperpolarized ^{13}C MRS images from his work.

REFERENCES

Abdel Razek AAK, Talaat M, El-Serougy L, Abdelsalam M, Gaballa G (2019) Differentiating glioblastomas from solitary brain metastases using arterial spin labeling perfusion and diffusion tensor imaging-derived metrics. *World Neurosurgery, 127,* e593–598.

Aime S, Delli Castelli D, Fedeli F, Terreno E (2002) A paramagnetic MRI-CEST agent responsive to lactate concentration. *Journal of American Chemical Society, 124*, 9364–9365.

Ainsworth NL, McLean MA, McIntyre DJO, Honess DJ, Brown AM, Harden SV, Griffiths JR (2017) Quantitative and textural analysis of magnetization transfer and diffusion images in the early detection of brain metastases. *Magnetic Resonance in Medicine, 77*(5), 1987–1995.

Almeida-Freitas DB, Pinho MC, Otaduy MC, Braga HF, Meira-Freitas D, da Costa Leite C (2014) Assessment of irradiated brain metastases using dynamic contrast-enhanced magnetic resonance imaging. *Neuroradiology, 56*, 437–443.

Arasu VA, Chen RC, Newitt DN, Chang CB, Tso H, Hylton NM, Joe BN (2011) Can signal enhancement ratio (SER) reduce the number of recommended biopsies without affecting cancer yield in occult MRI-detected lesions? *Academy Radiology, 18*, 716–721.

Ardenkjaer-Larsen JH, Fridlund B, Gram A, Hansson G, Hansson L, Lerche MH, Servin R, Thaning M, Golman K (2003) Increase in signal-to-noise ratio of >10,000 times in liquid-state NMR. *Proceedings of National Academy of Science USA, 100*, 10158–10163.

Barajas RF, Chang JS, Sneed PK, Segal MR, McDermott MW, Cha S (2009) Distinguishing recurrent intra-axial metastatic tumor from radiation necrosis following Gamma Knife radiosurgery using dynamic susceptibility-weighted contrast-enhanced perfusion MR imaging. *American Journal of Neuroradiology, 30*, 367–372.

Boxerman JL, Paulson ES, Prah MA, Schmainda KM (2013) The effect of pulse sequence parameters and contrast agent dose on percentage signal recovery in DSC-MRI: Implications for clinical applications. *American Journal of Neuroradiology, 34*, 1364–1369.

Brindle KM (2012) Watching tumours gasp and die with MRI: The promise of hyperpolarised 13C MR spectroscopic imaging. *British Journal of Radiology, 85*, 697–708.

Brindle KM, Bohndiek SE, Gallagher FA, Kettunen MI (2011) Tumor imaging using hyperpolarized 13C magnetic resonance spectroscopy. *Magnetic Resonance in Medicine, 66*, 505–519.

Caivano R, Lotumolo A, Rabasco P, Zandolino A, D'Antuono F, Villonio A, . . . Cammarota A (2013) 3 Tesla magnetic resonance spectroscopy: Cerebral gliomas vs. metastatic brain tumors. Our experience and review of the literature. *International Journal of Neuroscience, 123*(8), 537–543.

Chaumeil MM, Ozawa T, Park I, Scott K, James CD, Nelson SJ, Ronen SM (2012) Hyperpolarized 13C MR spectroscopic imaging can be used to monitor everolimus treatment in vivo in an orthotopic rodent model of glioblastoma. *Neuroimage, 59*, 193–201.

Chernov MF, Hayashi M, Izawa M, Abe K, Usukura M, Ono Y, Kubo O, Hori T (2004) Early metabolic changes in metastatic brain tumors after Gamma Knife radiosurgery: 1H-MRS study. *Brain Tumor Pathology, 21*, 63–67.

Chernov MF, Hayashi M, Izawa M, Nakaya K, Ono Y, Usukura M, Yoshida S et al. (2007) Metabolic characteristics of intracranial metastases, detected by single-voxel proton magnetic resonance spectroscopy, are seemingly not predictive for tumor response to Gamma Knife radiosurgery. *Minim Invasive Neurosurgery, 50*, 233–238.

Chernov MF, Hayashi M, Izawa M, Ochiai T, Usukura M, Abe K, Ono Y et al. (2005) Differentiation of the radiation-induced necrosis and tumor recurrence after Gamma Knife radiosurgery for brain metastases: Importance of multi-voxel proton MRS. *Minim Invasive Neurosurgery, 48*, 228–234.

Chernov MF, Hayashi M, Izawa M, Nakaya K, Tamura N, Ono Y, Abe K et al. (2009) Dynamics of metabolic changes in intracranial metastases and distant normal-appearing brain tissue after stereotactic radiosurgery: A serial proton magnetic resonance spectroscopy study. *Neuroradiology Journal, 22*, 58–71.

Chernov MF, Ono Y, Kubo O, Hori T (2006) Comparison of 1H-MRS-detected metabolic characteristics in single metastatic brain tumors of different origin. *Brain Tumor Pathology, 23*, 35–40.

Christen T, Lemasson B, Pannetier N, Farion R, Remy C, Zaharchuk G, Barbier EL (2012) Is T2* enough to assess oxygenation? Quantitative blood oxygen level-dependent analysis in brain tumor. *Radiology, 262*, 495–502.

Christen T, Lemasson B, Pannetier N, Farion R, Segebarth C, Remy C, Barbier EL (2011) Evaluation of a quantitative blood oxygenation level-dependent (qBOLD) approach to map local blood oxygen saturation. *NMR Biomed, 24*, 393–403.

Chuang MT, Liu YS, Tsai YS, Chen YC, Wang CK (2016) Differentiating radiation-induced necrosis from recurrent brain tumor using MR perfusion and spectroscopy: A meta-analysis. *PLoS ONE, 11*(1), 1–13.

Cohen C, Law-Ye B, Dormont D, Leclercq D, Capelle L, Sanson M, . . . Pyatigorskaya N (2020) Pseudo-continuous arterial spin labelling shows high diagnostic performance in the detection of postoperative residual lesion in hyper-vascularised adult brain tumours. *European Radiology, 30*(5), 2809–2820.

Conklin J, Heyn C, Roux M, Cerny M, Wintermark M, Federau C (2016) A Simplified Model for Intravoxel Incoherent Motion Perfusion Imaging of the Brain. *American Journal of Neuroradiology, 37*(12), LP2251–2257.

Day SE, Kettunen MI, Cherukuri MK, Mitchell JB, Lizak MJ, Morris HD, Matsumoto S, Koretsky AP, Brindle KM (2011) Detecting response of rat C6 glioma tumors to radiotherapy using hyperpolarized [1–^{13}C] pyruvate and 13C magnetic resonance spectroscopic imaging. *Magnetic Resonance Medicine, 65*, 557–563.

Desmond KL, Mehrabian H, Chavez S, Sahgal A, Soliman H, Rola R, Stanisz GJ (2017) Chemical exchange saturation transfer for Predicting response to stereotactic radiosurgery in human brain metastasis. *Magnetic Resonance in Medicine, 78*(3), 1110–1120.

Detsky JS, Keith J, Conklin J, Symons S, Myrehaug S, Sahgal A, . . . Soliman H (2017) Differentiating radiation necrosis from tumor progression in brain metastases treated with stereotactic radiotherapy: Utility of intravoxel incoherent motion perfusion MRI and correlation with histopathology. *Journal of Neuro-Oncology, 134*(2), 433–441.

Drost DJ, Riddle WR, Clarke GD, AAPM MR Task Group #9 (2002) Proton magnetic resonance spectroscopy in the brain: Report of AAPM MR task group #9. *Medical Physics, 29*, 2177–2197.

Egeland TA, Gulliksrud K, Gaustad JV, Mathiesen B, Rofstad EK (2012) Dynamic contrast-enhanced-MRI of tumor hypoxia. *Magnetic Resonance in Medicine, 67*, 519–530.

Engelbrecht MR, Huisman HJ, Laheij RJ, Jager GJ, van Leenders GJ, Hulsbergen-Van De Kaa CA, de la Rosette JJ, Blickman JG, Barentsz JO (2003) Discrimination of prostate cancer from normal peripheral zone and central gland tissue by using dynamic contrast-enhanced MR imaging. *Radiology, 229*, 248–254.

Essig M, Waschkies M, Wenz F, Debus J, Hentrich HR, Knopp MV (2003) Assessment of brain metastases with dynamic susceptibility-weighted contrast-enhanced MR imaging: Initial results. *Radiology, 228*, 193–199.

Farjam R, Tsien CI, Feng FY, Gomez-Hassan D, Hayman JA, Lawrence TS, Cao Y (2013) Physiological imaging-defined, response-driven subvolumes of a tumor. *International Journal of Radiation Oncology, Biology, Physics, 85*, 1383–1390.

Farjam R, Tsien CI, Feng FY, Gomez-Hassan D, Hayman JA, Lawrence TS, Cao Y (2014) Investigation of the diffusion abnormality index as a new imaging biomarker for early assessment of brain tumor response to radiation therapy. *Neuro Oncology, 16*, 131–139.

Farjam R, Tsien CI, Lawrence TS, Cao Y (2014) DCE-MRI defined subvolumes of a brain metastatic lesion by principle component analysis and fuzzy-c-means clustering for response assessment of radiation therapy. *Medical Physics, 41*, 011708.

Federau C, Meuli R, O'Brien K, Maeder P, Hagmann P (2014) Perfusion measurement in brain gliomas with intravoxel incoherent motion MRI. *American Journal of Neuroradiology, 35*, 256–262.

Ganbold M, Harada M, Khashbat D, Abe T, Kageji T, Nagahiro S (2017) Differences in high-intensity signal volume between arterial spin labeling and contrast-enhanced T1-weighted imaging may be useful for differentiating glioblastoma from brain metastasis. *Journal of Medical Investigation, 64*(1–2), 58–63.

Garcia M, Gloor M, Bieri O, Radue EW, Lieb JM, Cordier D, Stippich C (2015) Imaging of primary brain tumors and metastases with fast quantitative 3-dimensional magnetization transfer. *Journal of Neuroimaging, 25*(6), 1007–1014.

Geerts B, Leclercq D, Tezenas du Montcel S, Law-ye B, Gerber S, Bernardeschi D, . . . Pyatigorskaya N (2019) Characterization of skull base lesions using pseudo-continuous arterial spin labeling. *Clinical Neuroradiology, 29*(1), 75–86.

Golman K, Olsson LE, Axelsson O, Mansson S, Karlsson M, Petersson JS (2003) Molecular imaging using hyperpolarized 13C. *British Journal of Radiology, 76* Spec No 2, S118–127.

Harada H (2011) How can we overcome tumor hypoxia in radiation therapy? *Journal of Radiation Research, 52*, 545–556.

Hazle JD, Jackson EF, Schomer DF, Leeds NE (1997) Dynamic imaging of intracranial lesions using fast spin-echo imaging: Differentiation of brain tumors and treatment effects. *Journal of Magnetic Resonance Imaging, 7*, 1084–1093.

He X, Zhu M, Yablonskiy DA (2008) Validation of oxygen extraction fraction measurement by qBOLD technique. *Magnetic Resonance Imaging, 60*, 882–888.

Henkelman RM, Huang X, Xiang QS, Stanisz GJ, Swanson SD, Bronskill MJ (1993) Quantitative interpretation of magnetization transfer. *Magnetic Resonance in Medicine: Official Journal of the Society of Magnetic Resonance in Medicine/Society of Magnetic Resonance in Medicine, 29*, 759–766.

Hlushchuk R, Makanya AN, Djonov V (2011) Escape mechanisms after antiangiogenic treatment, or why are the tumors growing again? *International Journal of Developmental Biology, 55*, 563–567.

Hlushchuk R, Riesterer O, Baum O, Wood J, Gruber G, Pruschy M, Djonov V (2008) Tumor recovery by angiogenic switch from sprouting to intussusceptive angiogenesis after treatment with PTK787/ZK222584 or ionizing radiation. *American Journal of Pathology, 173*, 1173–1185.

Hoefnagels FW, Lagerwaard FJ, Sanchez E, Haasbeek CJ, Knol DL, Slotman BJ, Vandertop WP (2009) Radiological progression of cerebral metastases after radiosurgery: Assessment of perfusion MRI for differentiating between necrosis and recurrence. *Journal of Neurology, 256*, 878–887.

Horsfield MA (2005) Magnetization transfer imaging in multiple sclerosis. *Journal of Neuroimaging, 15*, 58S–67S.

Huang CF, Chiou SY, Wu MF, Tu HT, Liu WS, Chuang JC (2010) Apparent diffusion coefficients for evaluation of the response of brain tumors treated by Gamma Knife surgery. *Journal of Neurosurgery, 113*(Suppl), 97–104.

Huang J, Wang AM, Shetty A, Maitz AH, Yan D, Doyle D, Richey K et al. (2011) Differentiation between intra-axial metastatic tumor progression and radiation injury following fractionated radiation therapy or stereotactic radiosurgery using MR spectroscopy, perfusion MR imaging or volume progression modeling. *Magnetic Resonance Imaging, 29*, 993–1001.

Ishimaru H, Morikawa M, Iwanaga S, Kaminogo M, Ochi M, Hayashi K (2001) Differentiation between high-grade glioma and metastatic brain tumor using single-voxel proton MR spectroscopy. *European Radiology, 11*(9), 1784–1791.

Jakubovic R, Sahgal A, Soliman H, Milwid R, Zhang L, Eilaghi A, Aviv RI (2014) Magnetic resonance imaging-based tumour perfusion parameters are biomarkers predicting response after radiation to brain metastases. *Clinical Oncology (RColl Radiol), 26*, 704–712.

Jensen RL, Mumert ML, Gillespie DL, Kinney AY, Schabel MC, Salzman KL (2014) Preoperative dynamic contrast-enhanced MRI correlates with molecular markers of hypoxia and vascularity in specific areas of intratumoral microenvironment and is predictive of patient outcome. *Neuro Oncology, 16*, 280–291.

Jin T, Wang P, Zong X, Kim S-G (2012) Magnetic resonance imaging of the Amine-Proton EXchange (APEX) dependent contrast. *NeuroImage, 59*(2), 1218–1227.

Jing Y, Shuzhong CD, Jinyuan ZS, Qinwei Z, . . . Yi-Xiang W (2014) Amide proton transfer-weighted imaging of the head and neck at 3 T: A feasibility study on healthy human subjects and patients with head and neck cancer. *NMR in Biomedicine, 27*(10), 1239–1247.

Jones CK, Schlosser MJ, van Zijl PCM, Pomper MG, Golay X, Zhou J (2006) Amide proton transfer imaging of human brain tumors at 3T. *Magnetic Resonance in Medicine, 56*(3), 585–592.

Kapadia A, Mehrabian H, Conklin J, Symons SP, Maralani PJ, Stanisz GJ, . . . Heyn C (2017) Temporal evolution of perfusion parameters in brain metastases treated with stereotactic radiosurgery: Comparison of intravoxel incoherent motion and dynamic contrast enhanced MRI. *Journal of Neuro-Oncology, 135*(1), 119–127.

Kim DY, Kim HS, Goh MJ, Choi CG, Kim SJ (2014a) Utility of intravoxel incoherent motion MR imaging for distinguishing recurrent metastatic tumor from treatment effect following Gamma Knife radiosurgery: Initial experience. *American Journal of Neuroradiology, 35*, 2082–2090.

Kim HS, Suh CH, Kim N, Choi CG, Kim SJ (2014b) Histogram analysis of intravoxel incoherent motion for differentiating recurrent tumor from treatment effect in patients with glioblastoma: Initial clinical experience. *American Journal of Neuroradiology, 35*, 490–497.

Knitter JR, Erly WK, Stea BD, Lemole GM, Germano IM, Doshi AH, Nael K (2018) Interval change in diffusion and perfusion MRI parameters for the assessment of pseudoprogression in cerebral metastases treated with stereotactic radiation. *American Journal of Roentgenology, 211*(1), 168–175.

Kuchcinski G, Le Rhun E, Cortot AB, Drumez E, Duhal R, Lalisse M, . . . Delmaire C (2017) Dynamic contrast-enhanced MR imaging pharmacokinetic parameters as predictors of treatment response of brain metastases in patients with lung cancer. *European Radiology, 27*(9), 3733–3743.

Kuhl CK, Mielcareck P, Klaschik S, Leutner C, Wardelmann E, Gieseke J, Schild HH (1999) Dynamic breast MR imaging: Are signal intensity time course data useful for differential diagnosis of enhancing lesions? *Radiology, 211*, 101–110.

Kurhanewicz J, Vigneron DB, Brindle K, Chekmenev EY, Comment A, Cunningham CH, Deberardinis RJ et al. (2011) Analysis of cancer metabolism by imaging hyperpolarized nuclei: Prospects for translation to clinical research. *Neoplasia, 13*, 81–97.

Lankester KJ, Taylor JN, Stirling JJ, Boxall J, d'Arcy JA, Collins DJ, Walker-Samuel S, Leach MO, Rustin GJ, Padhani AR (2007) Dynamic MRI for imaging tumor microvasculature: Comparison of susceptibility and relaxivity techniques in pelvic tumors. *Journal of Magnetic Resonance Imaging, 25*, 796–805.

Leach MO, Brindle KM, Evelhoch JL, Griffiths JR, Horsman MR, Jackson A, Jayson GC et al. (2005) The assessment of antiangiogenic and antivascular therapies in early-stage clinical trials using magnetic resonance imaging: Issues and recommendations. *British Journal of Cancer, 92*, 1599–1610.

Le Bihan D, Breton E, Lallemand D, Aubin ML, Vignaud J, Laval-Jeanet M (1988) Separation of diffusion and perfusion in intravoxel incoherent motion MR imaging. *Radiology, 168*, 497–505.

Le Bihan D, Breton E, Lallemand D, Grenier P, Cabanis E, Laval-Jeanet M (1986) MR imaging of intravoxel incoherent motions: Application to diffusion and perfusion in neurologic disorders. *Radiology, 161*, 401–407.

Le Bihan D, Turner R (1992) The capillary network: A link between IVIM and classical perfusion. *Magnetic Resonance Medicine, 27*, 171–178.

Lee CC, Wintermark M, Xu Z, Yen CP, Schlesinger D, Sheehan JP (2014) Application of diffusion-weighted magnetic resonance imaging to predict the intracranial metastatic tumor response to Gamma Knife radiosurgery. *Journal of Neuro-Oncology, 118*, 351–361.

Levesque IR, Giacomini PS, Narayanan S, Ribeiro LT, Sled JG, Arnold DL, Pike GB (2010) Quantitative magnetization transfer and myelin water imaging of the evolution of acute multiple sclerosis lesions. *Magnetic Resonance in Medicine, 63*(3), 633–640.

Long DM (1979) Capillary ultrastructure in human metastatic brain tumors. *Journal of Neurosurgery, 51*, 53–58.

Mahmood F, Hjorth Johannesen H, Geertsen P, Hansen RH (2020) Diffusion MRI outlined viable tumour volume beats GTV in intra-treatment stratification of outcome. *Radiotherapy and Oncology, 144*, 121–126.

Maralani PJ, Das S, Mainprize T, Phan N, Bharatha A, Keith J, . . . Mikulis D (2018) Hypoxia detection in infiltrative astrocytoma: Ferumoxytol-based quantitative BOLD MRI with intraoperative and histologic validation. *Radiology, 288*(10), 172601.

Mardor Y, Pfeffer R, Spiegelmann R, Roth Y, Maier SE, Nissim O, Berger R et al. (2003) Early detection of response to radiation therapy in patients with brain malignancies using conventional and high b-value diffusion-weighted magnetic resonance imaging. *Journal of Clinical Oncology, 21*, 1094–1100.

McMahon MT, Gilad AA, DeLiso MA, Berman SM, Bulte JW, van Zijl PC (2008) New "multicolor" polypeptide diamagnetic chemical exchange saturation transfer (DIACEST) contrast agents for MRI. *Magnetic Resonance Medicine, 60*, 803–812.

Mehrabian H, Desmond, KL, Chavez S, Bailey C, Sahgal A, Czarnota JG, . . . Stanisz GJ (2017) Water exchange rate constant as a biomarker of treatment efficacy in patients with brain metastases undergoing stereotactic radiosurgery. *International Journal of Radiation Oncology, Biology, Physics, 98*(1), 47–55.

Mehrabian H, Desmond KL, Soliman H, Sahgal A, Stanisz GJ (2017) Differentiation between radiation necrosis and tumor progression using chemical exchange saturation transfer. *Clinical Cancer Research, 23*(14), 3667–3675.

Mehrabian H, Lam WW, Myrehaug S, Sahgal A, Stanisz GJ (2018) Glioblastoma (GBM) effects on quantitative MRI of contralateral normal appearing white matter. *Journal of Neuro-Oncology, 139*(1), 97–106.

Mehrabian H, Myrehaug S, Soliman H, Sahgal A, Stanisz GJ (2018) Quantitative magnetization transfer in monitoring glioblastoma (GBM) response to therapy. *Scientific Reports, 8*(1), 2475.

Mitsuya K, Nakasu Y, Horiguchi S, Harada H, Nishimura T, Bando E, Okawa H, Furukawa Y, Hirai T, Endo M (2010) Perfusion weighted magnetic resonance imaging to distinguish the recurrence of metastatic brain tumors from radiation necrosis after stereotactic radiosurgery. *Journal of Neurooncology, 99*, 81–88.

Moffat BA, Chenevert TL, Lawrence TS, Meyer CR, Johnson TD, Dong Q, Tsien C et al. (2005) Functional diffusion map: A noninvasive MRI biomarker for early stratification of clinical brain tumor response. *Proceedings of National Academy of Science USA, 102*, 5524–5529.

Muto M, Frauenfelder G, Senese R, Zeccolini F, Schena E, Giurazza F, Jäger HR (2018) Dynamic susceptibility contrast (DSC) perfusion MRI in differential diagnosis between radionecrosis and neoangiogenesis in cerebral metastases using rCBV, rCBF and K2. *Radiologia Medica, 123*(7), 545–552.

Ortiz-Ramón R, Ruiz-España S, Mollá-Olmos E, Moratal D (2020) Glioblastomas and brain metastases differentiation following an MRI texture analysis-based radiomics approach. *Physica Medica, 76*, 44–54.

Padhani AR, Gapinski CJ, Macvicar DA, Parker GJ, Suckling J, Revell PB, Leach MO, Dearnaley DP, Husband JE (2000) Dynamic contrast enhanced MRI of prostate cancer: Correlation with morphology and tumour stage, histological grade and PSA. *Clinical Radiology, 55*, 99–109.

Paech D, Windschuh J, Oberhollenzer J, Dreher C, Sahm F, Meissner JE, . . . Radbruch A (2018) Assessing the predictability of IDH mutation and MGMT methylation status in glioma patients using relaxation-compensated multipool CEST MRI at 7.0 T. *Neuro-Oncology, 20*(12), 1661–1671.

Park I, Larson PE, Tropp JL, Carvajal L, Reed G, Bok R, Robb F et al. (2014) Dynamic hyperpolarized carbon-13 MR metabolic imaging of nonhuman primate brain. *Magnetic Resonance Medicine, 71*, 19–25.

Park I, Larson PE, Zierhut ML, Hu S, Bok R, Ozawa T, Kurhanewicz J et al. (2010) Hyperpolarized 13C magnetic resonance metabolic imaging: Application to brain tumors. *Neuro Oncology, 12*, 133–144.

Park JE, Kim HS, Goh MJ, Kim SJ, Kim JH (2015) Pseudoprogression in patients with glioblastoma: Assessment by using volume-weighted voxel-based multiparametric clustering of MR imaging data in an independent test set. *Radiology, 275*, 792–802.

Payabvash S, Aboian M, Tihan T, Cha S (2020) Machine Learning Decision Tree Models for Differentiation of Posterior Fossa Tumors Using Diffusion Histogram Analysis and Structural MRI Findings. *Frontiers in Oncology, 10*.

Petrella JR, Provenzale JM (2000) MR perfusion imaging of the brain: Techniques and applications. *American Journal of Roentgenology, 175*, 207–219.

Provencher SW (1993) Estimation of metabolite concentrations from localized in vivo proton NMR spectra. *Magnetic Resonance Medicine, 30*, 672–679.

Rivlin M, Horev J, Tsarfaty I, Navon G (2013) Molecular imaging of tumors and metastases using chemical exchange saturation transfer (CEST) MRI. *Science Representation, 3*, 3045.

Schroder L, Lowery TJ, Hilty C, Wemmer DE, Pines A (2006) Molecular imaging using a targeted magnetic resonance hyperpolarized biosensor. *Science, 314*, 446–449.

Sjøbakk TE, Johansen R, Bathen TF, Sonnewald U, Kvistad KA, Lundgren S, Gribbestad IS (2007) Metabolic profiling of human brain metastases using in vivo proton MR spectroscopy at 3T. *BMC Cancer, 7*, 141.

Soares DP, Law M (2009) Magnetic resonance spectroscopy of the brain: Review of metabolites and clinical applications. *Clinical Radiology, 64*, 12–21.

Soni N, Dhanota DPS, Kumar S, Jaiswal AK, Srivastava AK (2017) Perfusion MR imaging of enhancing brain tumors: Comparison of arterial spin labeling technique with dynamic susceptibility contrast technique. *Neurology India, 65*(5), 1046–1052.

Soni N, Kumar S, Srindharan K, Mishra P, Gupta N, Bathla G, . . . Behari S (2019) Comparative Evaluation of Brain Tuberculosis and Metastases Using Combined Analysis of Arterial Spin Labeling Perfusion and Diffusion Tensor Imaging. *Current Problems in Diagnostic Radiology, 48*(6), 547–553.

Tofts PS, Brix G, Buckley DL, Evelhoch JL, Henderson E, Knopp MV, Larsson HB et al. (1999) Estimating kinetic parameters from dynamic contrast-enhanced T(1)-weighted MRI of a diffusable tracer: Standardized quantities and symbols. *Journal of Magnetic Resonance Imaging, 10*, 223–232.

Tóth V, Förschler A, Hirsch NM, Den Hollander J, Kooijman H, Gempt J, . . . Preibisch C (2013) MR-based hypoxia measures in human glioma. *Journal of Neuro-Oncology, 115*(2), 197–207.

Umemura Y, Wang D, Peck KK, Flynn J, Zhang Z, Fatovic R, . . . Young RJ (2020) DCE-MRI perfusion predicts pseudoprogression in metastatic melanoma treated with immunotherapy. *Journal of Neuro-Oncology, 146*(2), 339–346.

Walker-Samuel S, Ramasawmy R, Torrealdea F, Rega M, Rajkumar V, Johnson SP, Richardson S et al. (2013) In vivo imaging of glucose uptake and metabolism in tumors. *National Medicine, 19*, 1067–1072.

Ward CS, Venkatesh HS, Chaumeil MM, Brandes AH, Vancriekinge M, Dafni H, Sukumar S et al. (2010) Noninvasive detection of target modulation following phosphatidylinositol 3-kinase inhibition using hyperpolarized 13C magnetic resonance spectroscopy. *Cancer Research, 70*, 1296–1305.

Ward KM, Aletras AH, Balaban RS (2000) A new class of contrast agents for MRI based on proton chemical exchange dependent saturation transfer (CEST). *Journal of Magnetic Resonance Imaging, 143*, 79–87.

Ward KM, Balaban RS (2000) Determination of pH using water protons and chemical exchange dependent saturation transfer (CEST). *Magnetic Resonance in Medicine, 44*(5), 799–802.

Weber MA, Thilmann C, Lichy MP, Gunther M, Delorme S, Zuna I, Bongers A et al. (2004) Assessment of irradiated brain metastases by means of arterial spin-labeling and dynamic susceptibility-weighted contrast-enhanced perfusion MRI: Initial results. *Invest Radiology, 39*, 277–287.

Wen Z, Hu S, Huang F, Wang X, Guo L, Quan X, Wang S, Zhou J (2010) MR imaging of high-grade brain tumors using endogenous protein and peptide-based contrast. *Neuroimage, 51*, 616–622.

Weybright P, Sundgren PC, Maly P, Hassan, DG, Nan B, Rohrer S, Junck L (2005) Differentiation between brain tumor recurrence and radiation injury using MR spectroscopy. *AJR. American Journal of Roentgenology, 185*(6), 1471–1476.

Yen YF, Nagasawa K, Nakada T (2011) Promising application of dynamic nuclear polarization for in vivo (13)C MR imaging. *Magnetic Resonance Medicine Science, 10*, 211–217.

Zach L, Guez D, Last D, Daniels D, Grober Y, Nissim O, Hoffmann C et al. (2012) Delayed contrast extravasation MRI for depicting tumor and non-tumoral tissues in primary and metastatic brain tumors. *PLoS One, 7*, e52008.

Zahra MA, Hollingsworth KG, Sala E, Lomas DJ, Tan LT (2007) Dynamic contrast-enhanced MRI as a predictor of tumour response to radiotherapy. *Lancet Oncology, 8*, 63–74.

Zakaria R, Chen YJ, Hughes DM, Wang S, Chawla S, Poptani H, . . . Mohan S (2020) Does the application of diffusion weighted imaging improve the prediction of survival in patients with resected brain metastases? A retrospective multicenter study. *Cancer Imaging, 20*(1).

Zhang G, Chen X, Zhang S, Ruan X, Gao C, Liu Z, Wei X (2019) Discrimination between solitary brain metastasis and glioblastoma multiforme by using ADC-based texture analysis: A comparison of two different ROI placements. *Academic Radiology, 26*(11), 1466–1472.

Zhao X, Wen Z, Zhang G, Huang F, Lu S, Wang X, . . . Zhou J (2013) Three-dimensional turbo-spin-echo amide proton transfer MR imaging at 3-Tesla and its application to high-grade human brain tumors. *Molecular Imaging and Biology, 15*(1), 114–122.

Zhou J, Tryggestad E, Wen Z, Lal B, Zhou T, Grossman R, . . . Van Zijl PCM (2011) Differentiation between glioma and radiation necrosis using molecular magnetic resonance imaging of endogenous proteins and peptides. *Nature Medicine, 17*(1), 130–134.

6 From Frame to Frameless: Brain Radiosurgery

Young Lee and Steven Babic

Contents

6.1 INTRODUCTION

The word *stereotactic* originating from the Greek word στερεός [*stereos*], which means "solid," and "-taxis", which means "order", defines the 3D localization of a point in space by a unique set of coordinates that correspond to a fixed external reference frame. Historically, such a frame has acted as a support for hollow probes that guide electrodes or biopsy needles to precise locations within an animal or human brain.

Since the 1950s, the stereotactic principle has been adopted by neurosurgeons, and the first dedicated unit for stereotactic radiosurgery (SRS) known as the Gamma Knife (Elekta AB, Stockholm, Sweden) was developed by a Swedish neurosurgeon. The patient was immobilized using a stereotactic head frame, and narrow beams of cobalt-60 gamma rays (Gamma Knife, GK) were focused to a small target within the brain. Since the mid-1980s, adaptations of medical linear accelerators (linacs) to produce similarly precise megavoltage x-ray beams have made this technique accessible to many hospitals. With the development of relocatable head frames used with CT-based fiducial marker systems, dedicated treatment planning systems, and hardware modifications to the linac apparatus for high-precision delivery, single-fraction SRS and fractionated stereotactic radiation therapy (SRT) for the brain are readily available in many oncology centers.

Lesions suitable for single fraction SRS are generally less than 3 to 4 cm in diameter and range from small malignant brain metastases, to benign-aggressive lesions such as acoustic neuromas, and to truly benign conditions such as arteriovenous malformations (AVMs). To reduce toxicity to sensitive structures such as the brainstem and optic nerves, some lesions may be treated with SRT, that is, with a fractionated prescription dose.

In recent years, the dominant indication for SRS is brain metastases. The utilization has increased dramatically as we have learned that although whole-brain treatments for multiple brain metastases can be an effective option, there is a considerable expense associated with neurocognitive deficits in long-term survivors. Moreover, when other metastases appear post whole-brain radiation therapy, treatment options are limited (Sahgal et al., 2015). As a result, at present, withholding whole-brain radiation and treating patients with SRS/SRT, who present with a limited number of brain metastases (up to 5), is the standard of care and increasingly for patients who present with even a greater number of multiple metastases, that is, greater than 5 (Nguyen et al., 2019). There is a technical challenge in treating multiple targets with high radiation doses since accurate treatment positioning is crucial in avoiding geometrical misses and maintaining delivery of low doses to normal tissues. Treatment times may be long; consequently, the invasive frame can be difficult to maintain for a patient, and SRT is almost impossible to deliver. Figure 6.1 summarizes different aspects of the SRS system and workflow, and how each element relates to the accuracy needed.

SRS has traditionally been based on utilizing an invasive nonmoveable head frame to immobilize and position the patient's head to afford the highest degree of accuracy and precision. With frame-based systems, the tumor's stereotactic location within the rigid frame is defined, and it can be reproduced at the time of treatment using a stereotactic coordinate localization device. When treating with a linear accelerator, the tumor's stereotactic location within the rigid frame can be reproduced using a stereotactic coordinate localization device and the lasers within the treatment room.

It is imperative that the spatial relationship of the frame relative to the skull is upheld since any slippage of the frame prior to treatment will result in a positioning error and if not corrected for, it may lead to geographic miss of the target and/or normal critical structures being overdosed. Although stereotactic frame slippage is not common, it can be one significant drawback of utilizing rigid nonmoveable head frames. In addition to this, a number of other disadvantages include the need for a neurosurgeon for the purpose of frame placement; once the frame is attached, simulation, imaging, planning, quality assurance (QA) checks, and treatment need to be completed all within the same day; fractionated treatment is not practical; there is pain and uneasiness for the patient; and there exists a risk of bleeding and infection at the site of frame attachment. As a result of these limitations and with the advancement of image-guided radiation therapy (IGRT), where setup verification images can be utilized to precisely check the isocenter location using bony anatomy, there has been a move toward noninvasive moveable frames and frameless systems for SRS and SRT.

The transition of "frame" to "frameless" brain radiosurgery is discussed in this chapter with respect to the influence of evolving technology including the changes in immobilization systems.

6.2 FRAME-BASED SRS: USING LOCALIZER BOXES

Frame-based SRS can be used with GK and linacs equipped with external cones and/or micro-multileaf collimators (micro-MLCs). Compared to the CyberKnife robotic radiosurgery system, real-time imaging is not normally utilized to correct for intrafractional motion with GK-based SRS nor is it commonly used during linac delivery (Wowra et al., 2012). Furthermore, it can be argued that real-time imaging is not necessary with an invasive frame-based system as the intrafractional motion is negligible. As discussed later in this chapter, it should be noted that real-time tracking systems are becoming more readily available in both GK and linac-based SRS systems.

Central to the stereotactic method is the requirement for imageable fiducial reference markers that are attached to the stereotactic system immobilizing the patient. These markers are vital in providing accurate geometrical information on the coordinates of the planned isocenter. They are commonly in the form of

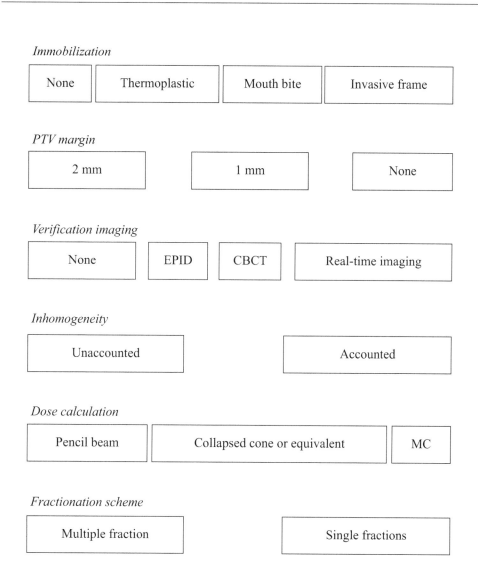

Figure 6.1 The dependence of different aspects of SRS delivery on accuracy. The accuracy increases from left to right, and the effect on planning target volume (PTV) margin is illustrated. Other factors can influence accuracy such as imaging and fusion. However, imaging and fusion are not mutually exclusive in the influence on accuracy. *Abbreviations*: EPID, electronic portal imaging device; CBCT, cone-beam computed tomography; MC, Monte Carlo.

"crowns" containing various rod configurations, etchings on the sides of plastic fiducial boxes, and/or wires stretched between rigid spacers. The fundamental and ideal requirements of a fiducial system are (1) no significant scan artifacts are generated that obscure the images used for target and organs-at-risk (OARs) delineation, (2) precise and rapid docking onto the patient's immobilization system, (3) an unambiguous and preferably simple marker arrangement which enables manual checking of computed target coordinates, and (4) capability of correcting for the effect of imaging slices being nonperpendicular to the scanner couch which has a tendency to sag from the weight of the patient as it passes through the scan plane. Although the calculation software that gives general solutions to the equations of fiducial markers is readily available, it is often preferable to adjust the frame tilt such that the patient CT scans are orthogonal to the couch and parallel to the scan plane. This enables the ease of checking and keeps an intuitive feel for the precise isocenter setup geometry.

Some of the commercial systems and the requirements and advantages of each system in the context of SRS are discussed in the following sections.

6.2.1 NEUROSURGICAL INVASIVE FRAMES

Some examples of neurosurgical frames that remain the most reliable and stable platforms include the Cosman–Roberts–Wells (CRW) frame presented in Figure 6.2 and Leksell frame shown in Figure 6.3. Accurate fixation of the stereotactic frame to the patient's head is achieved by means of three to four steel pins that are inserted into tiny holes drilled into the patient's skull. Frame fitting with this procedure consequently requires a local anesthetic and a neurosurgeon to both place and remove the frame. All of the processes required for the treatment procedure need to be carried out in a single day. Although such frames can be removed and refitted for limited fractionation regimes, this is not ideal as placing the frame in the same position is difficult and can cause significant patient discomfort. The following sections briefly describe the two most widely used commercially available invasive frames.

6.2.1.1 Brown–Roberts–Wells/Cosman–Roberts–Wells

The original Brown–Roberts–Wells (BRW) system, which consisted of a skull base ring with carbon epoxy head posts that offers minimal CT interference, was created at the University of Utah in 1977. The frame ring is attached to the patient with screws that are tightened into the skull. The localizer unit is secured to the ring with three ball-and-socket interlocks and consists of six vertical posts and three diagonal posts, creating an N-shaped appearance (Figure 6.2b). This "N" construct establishes the axial CT plane relative to the skull base by calculating the relative distance of the oblique to the vertical rods. Target coordinates are established by identifying the axial slice that best features the lesion. The x and y coordinates for each of the nine fiducial rods are identified on the CT or MRI, as are the x and y coordinates for the target. All coordinates are then converted to coordinates in stereotactic space. In the 1980s, Wells and Cosman simplified and improved the BRW by designing an arc guidance frame similar to the Leksell frame. The arc system directs a stereotactic probe isocentrically around the designated target, thus avoiding a fixed entry point.

The CRW system included some of the same design elements as the BRW system, including a phantom frame, the same CT localizer, and the same probe depth fixed at 16 cm. New innovations included the introduction of MRI-compatible frames and localizers and versatility in arc-to-frame applications that enabled inferior trajectories into the posterior fossa or lateral routes into the temporal lobe. One of the compatible SRS treatment planning systems with the CRW frame is the Radionics radiosurgery treatment planning software (Integra NeuroSciences, Burlington, MA). Printouts of stereotactic coordinate templates (Figure 6.2a) and how they fit onto the treatment localizer box are crucial in setting up the patient in the correct position.

(a)   (b)

Figure 6.2 (a) Radionics box with plastic templates placed on the box with printed templates and (b) the Cosman–Roberts–Wells system placed on the patient on a CT scanner couch.

When not corrected, caution should be taken using geometrically distorted MRI for SRS with metal frames. Using images of phantoms with CRW and Leksell frames scanned on a GE Signa 1.5 T machine, Burchiel et al. (1996) showed that the CRW frame caused larger MRI distortions compared to the Leksell frame. The study concluded that properties of frame systems used for stereotactic neurosurgery may greatly influence the accuracy of frame-based stereotactic neurosurgery and that the accuracy of these frame systems is testable.

Li et al. (2011) had compared the BRW to the frameless PinPoint system using the VisionRT surface-matching system. On average, for 11 patients (19 lesions), the translational and rotational magnitudes (1 standard deviation) were observed to be 0.3 (0.2) mm and 0.2 (0.2)°, respectively.

6.2.1.2 Leksell

The Leksell Coordinate Frame G, made of titanium, is fixed to the patient's head using four self-tapping screws, which keep the frame firmly and accurately in place. It is lighter than the CRW frame, and it is fully MR compatible. Its small frame size fits in most MR head coils and minimizes distortion. MR, CT, and angiography localizer boxes (Figure 6.3) ensure parallel and equidistant images, and there are table

(a)

(b)

(c)

(d)

Figure 6.3 Images of the Leksell frame system. (a) Phantom placed in the frame with the fiducial box for CT imaging, (b) treatment planning system rendered bony anatomical image with the CT fiducial box, (c) fiducial box for angiography imaging, and (d) phantom in the frame with the fiducial box used in treatment.

adapters for CT, angiography, and treatment units, including linac and GK. These integrated tabletop adapters ensure patient fixation is consistent at all stages of the process, assuring accuracy of target localization and patient setup to submillimeter mechanical accuracy.

The original neurosurgical frame consisted of a semicircular arc with a moveable probe carrier. The arc is fixed to the patient's head in such a manner that its center corresponds with a selected cerebral target. The electrodes are always directed toward the center and, hence, to the target. Rotation of the arc around the axis rods in association with lateral adjustment of the electrode carrier enables any convenient point of entrance of the electrodes to be chosen, independent of the site of the target (Leksell, 1971; Lundsford et al., 1988).

The model G base frame is rectangular and has dimensions of 190 mm by 210 mm. A straight or curved front piece can be used, as it allows airway access in emergencies. The x, y, and z axes on the frame recapitulate the CT and MRI axes.

The frame center coordinates are 100, 100, and 100, whereas a hypothetical frame origin (x, y, and $z = 0$) resides in the superior posterior right side of the frame. In the neurosurgical frame, the semicircular arc attached to the base frame has a radius of 190 mm (Louw, 2003). Leksell frame is supported by many treatment planning systems, but it is the only frame that is currently in clinical use with the Leksell GK system.

6.2.2 NONINVASIVE MOVEABLE FRAMES

Unlike rigid stereotactic head frames, these immobilization devices can be easily removed since they are not permanently attached to the patient, thereby significantly reducing patient anxiety and discomfort. There are several different types of noninvasive moveable frames which offer accurate and reproducible cranial immobilization comparable to the previously described nonmoveable frame-based systems. As with the invasive frames, for treatment planning, a stereotactic fiducial-based localizer system can be attached to the moveable head frame at the time of CT simulation and at the time of treatment delivery. A stereotactic localizer can be utilized for setting the patient up to the treatment isocenter. A few varieties of these noninvasive relocatable frames are described later, and they include the Gill–Thomas–Cosman (GTC) frame, the BrainLAB mask system, and the Laitinen Stereoadapter 5000 frame.

6.2.2.1 Gill–Thomas–Cosman Frame

The GTC frame consists of an aluminum alloy base ring that is attached to the treatment couch for rigid immobilization, a dental plate/oral appliance, an occipital headrest pad, and Velcro straps. The dental plate is made by taking a dental impression of the patient's teeth. The headrest pad is shaped to the posterior skull encompassing the occipital protuberance. Both of these patient-specific devices are securely mounted to the head ring. The Velcro strap is posteriorly fixed to the headrest and is also connected anteriorly to the base ring. Together, the three different components uphold the frame's position on the patient. To enable daily setup reproducibility, the Velcro strap lengths can be marked at each side, and a clear plastic hemispherical dome with fixed holes or portals, called a depth confirmation helmet, can be used. The depth confirmation helmet is placed over the head ring, and a rod with a millimeter scale is inserted into each hole to measure the distance to the cranial surface. The distance readings from each hole can be compared to the readings obtained at the time of CT simulation to ensure that the frame has been accurately placed. The GTC frame is a commercially available (Integra, Plainsboro, NJ) relocatable head frame that was adapted to be compatible with the BRW stereotactic coordinate system. Its design is originally based on the Gill–Thomas frame (Gill et al., 1991; Graham et al., 1991).

Using the depth helmet before each treatment fraction, Das et al. (2011) measured the daily relocation error of the GTC frame. Based on 10 patients, they found a mean vector displacement or radial error of 1.03 ± 0.34 mm. In the mediolateral, anteroposterior, and craniocaudal directions, the mean errors were 0.38, 0.15, and 0.17 mm, respectively, and all errors were within ± 2 mm, 97% to 99% of all cases. Using the depth confirmation helmet as well, Burton et al. (2002) evaluated the setup reproducibility on 31 patients, and they reported mean errors of 0.1, 0.1, and 0.4 mm in the mediolateral, anteroposterior, and craniocaudal directions, respectively. They also determined a mean displacement vector of 1.2 mm, with 92% of the displacement vectors less than 2 mm and 97% less than 2.5 mm. Utilizing 126 anterior and

123 lateral daily pretreatment portal images coregistered to the digitally reconstructed radiographs (DRRs) from planning CTs (the reference images) of 15 patients, Kumar et al. (2005) determined a total 3D mean displacement vector of 1.8 ± 0.8 mm with a range of 0.3–3.9 mm. In an earlier study from measurements based upon 20 patients, the GTC frame was found to have a superior relocalization accuracy on the order of ±0.4 mm.

6.2.2.2 BrainLAB Mask System

Based upon the GTC frame, the BrainLAB mask system (BrainLAB, Munich, Germany) consists of a patient-specific thermoplastic mask, a U-shaped frame, vertical posts, and an optional bite block that attaches to an upper jaw device for additional support. Vertical posts fasten the thermoplastic mask to the head ring, and an adaptor is utilized to attach the head ring to the imaging or treatment couch. At the time of CT simulation, the thermoplastic mask is custom shaped to the patient's head both anteriorly and posteriorly, and a CT localizer box is attached to the head ring for image localization. For the purpose of patient setup at the time of treatment, a stereotactic localizer box or target positioner is fixed to the head ring.

Many authors have investigated the setup errors and intrafraction motion of patients immobilized with the BrainLAB mask (Alheit et al., 2001; Minniti et al., 2011; Theelen et al., 2012). Alheit et al. (2001) utilized simulator films and electronic portal imaging device (EPID) to show that patient position reproducibility using the BrainLAB mask is less than 2 mm. Both Minniti et al. (2011) and Theelen et al. (2012) used serial CT scans for BrainLAB mask-positioning verification and reported a mean 3D displacement of 0.5 ± 0.7 mm (maximum of 2.9 mm) and 1.16 ± 0.68 mm (maximum of 2.25 mm), respectively. Both observed the largest translational deviation to be in the superior–inferior direction.

Using pretreatment kV imaging, Bednarz et al. (2009) investigated the setup accuracy of the BrainLAB mask and found a mean 3D displacement of 3.17 ± 1.95 mm from the isocenter. This result was consistent with those of Willner et al. (1997), who reported a mean 3D vector deviation of 2.4 ± 1.3 mm from the isocenter. Ali et al. (2010) utilized kV onboard imaging and reported mean shifts of 0.1 ± 2.2, 0.7 ± 2.0, and 1.6 ± 2.6 mm in anterior–posterior, medial–lateral, and superior–inferior directions, respectively. Using posttreatment kV imaging, Ramakrishna et al. (2010) investigated the intrafraction motion with the BrainLAB mask and found a mean intrafraction shift of 0.7 ± 0.5 mm.

6.2.2.3 Laitinen Stereoadapter 5000 Frame

The Laitinen Stereoadapter 5000 (Sandstrom Trade and Technology Inc., Welland, Ontario, Canada) is a noninvasive relocatable stereotactic frame that is attached to the patient's head using (1) two earplugs that are bilaterally inserted into the external auditory canals and (2) a nasal support assembly that rests against the bridge of the nose (Laitinen et al., 1985; Kalapurakal et al., 2001). The earplugs are, respectively, attached to two lateral plates. The lateral plates are joined together at the vertex using a connector plate and are also attached to the nasal support assembly through two side arms. The nasal support assembly is equipped with an adjustable thumbscrew, which is used to secure the earplugs against the lateral plates and to push them into the external auditory canals. For further frame stabilization, a strap is attached to the lateral plates and wraps posteriorly around the head. A couch adaptor device secures the frame's connector plate at the vertex to the imaging or treatment couch. Since the frame is made of an aluminum alloy and plastic, it is both MRI and CT compatible. To accommodate patients with varying sizes of external auditory canals, different sizes of earplugs are available. This frame has been found to be well tolerated by both children and adults, although a mild pressure sensation at the ear canals does yield some patient discomfort (Golden et al., 1998; Kalapurakal et al., 2001).

For reproducible setups and repositioning accuracy, a number of components (i.e., connector plate, arms of the nasal support assembly) of the Stereoadaptor 5000 frame have graduated scales in millimeters that together with target plates (these attach to the lateral side plates) can be utilized to establish the frame's reference coordinates and to set up the patient to the treatment isocenter. A number of authors have evaluated the repositioning accuracy of this frame. Using orthogonal portal images, Golden et al. (1998) found the reproducibility to be generally about 2 mm. Utilizing portal images coregistered to CT scout images, Kalapurakal et al. (2001) reported a mean isocenter shift of 1.0 ± 0.7, 0.8 ± 0.8, and 1.7 ± 1.0 mm in the lateral (x), anterior/posterior (y), and superior/inferior (z) directions, respectively. Testing for accuracy and reproducibility of repeated mountings, Delannes et al. (1991) determined a mean distance error of 0.9, 0.6, and 0.9 mm in the x, y, and z coordinates, respectively.

6.2.3 SUMMARY

Frame-based SRS using localizer or fiducial boxes enable accurate setup of patients without the need for patient setup verification images. With invasive frames, patients can be immobilized accurately to less than 1 mm setup errors. Though rare, slippage is possible with invasive frames, and without imaging, these errors may not be easily caught. This highlights the importance of SRS and SRT.

A small decrease in accuracy is inevitable with moveable frame systems. However, the moveable frames can allow similar accuracy in patient immobilization to invasive frames in fractionated treatments without the discomfort to patients. To account for the small decrease in accuracy, a larger margin should be used to create a planning target volume (PTV, Figure 6.1).

The loss in accuracy from invasive to moveable frames, however, maybe lessened with the advancement of IGRT, as it is widely available for linac treatments and being incorporated into SRS and SRT. The vital role that IGRT plays in SRS and SRT will be discussed within the subsequent sections.

6.3 NONINVASIVE FRAMELESS SYSTEMS

In order to compensate for the loss of rigid immobilization associated with invasive head frames, a high-precision IGRT must be added to all frameless systems. Imaging at the time of treatment is used to directly determine the position of the target and rectify for any patient movement and or positioning errors. As a result, the accurate correlation between the patient anatomy and immobilization device, which is key to the frame-based stereotactic approach is no longer essential.

In general, the daily repositional accuracy of a relocatable frame should vary by less than 1 mm. Relocation checks are best carried out by fusing the bony anatomy of repeat sets of cone-beam CT (CBCT), but if not possible, anterior–posterior and lateral EPID can be taken in order to quantify the displacements of reference markers attached to the frame. Alternatively, physical depth measurements to the surface of the head from a reference "depth helmet," mounted onto the stereotactic frame, may be made. Optical video methods can be used based on fixed geometry wall-mounted cameras in the treatment room and reflective markers on the patient. For frameless positioning, treatment room image-guided systems must include onboard CBCT, onboard MV EPID, or onboard kV images and/or kV x-ray systems mounted on the ceiling and floor. Some commercially available noninvasive frameless systems that will be discussed in the following sections include the GK eXtend frame and Icon systems, PinPoint frame, optically guided bite block, and BrainLAB frameless system.

6.3.1 eXTEND

The eXtend frame system (Elekta, Stockholm, Sweden) is a noninvasive vacuum bite block repositioning head frame for cranial immobilization. The main components of this system include a carbon fiber frame body that is attached to the treatment couch, a headrest, and a mouthpiece that is affixed to the frame body using a frontpiece (Ruschin et al., 2010). Prior to treatment, a patient-specific dental impression of the upper mouth is obtained together with a cushion impression of the back of the skull. A vacuum device is used to suction the custom bite block to the patient's upper hard palate. To verify that the patient's head is accurately positioned within the eXtend frame, a spring-loaded digital dial gauge is inserted through slotted holes in a repositioning check tool (RCT) that is attached to the frame. The distance between the frame and the patient's head is measured and compared to reference values measured on the initial day of treatment. The patient can then be repositioned at the time of setup if the difference exceeds a predefined tolerance (e.g., 1 mm). In one study that evaluated four patients immobilized with the eXtend frame and treated on a GK machine (Perfexion, Elekta, Stockholm, Sweden), the mean radial positioning error was found to be between 0.33 and 0.84 mm (Sayer et al., 2011). Using the RCT and CBCT image guidance, Ruschin et al. (2010) reported on the setup accuracy and intrafraction motion of the eXtend frame system on 12 patients. Specifically with CBCT, the mean 3D intrafraction motion was found to be 0.4 ± 0.3 mm, and with the RCT it was 0.7 ± 0.5 mm. For patients treated on a linac and on a GK machine (Perfexion,

Elekta, Stockholm, Sweden), the mean 3D setup error was 0.8 and 1.3 mm, respectively, thus confirming eXtend frame's excellent immobilization performance.

6.3.2 GK ICON

The GK Icon system with a stereotactic CBCT enables increased workflow flexibility and additional treatment options to the Perfexion system, which only allowed invasive frame treatments. Patient can be immobilized using a mask, and the stereotactic coordinate of the LGK can now be determined using the integrated CBCT system, paralleling the coordinate system defined by the frame and localizer boxes. As such, unlike previous generations, the Icon allows for CBCT imaging of the patient anatomy, frameless treatments, a more cohesive MR-only planning, and accurate high-definition motion management (HDMM) based on infrared (IR) detector. The HDMM system monitors the IR reflector placed on the patient's nose such that its position is calculated against the fixed reflectors on the mask frame (see Figure 6.4) to monitor the patient's location throughout the treatment fraction. Icon allows for a fractionated stereotactic radiotherapy (FSRT) and is now clinically operating at many institutions (Sarfehnia et al., 2018; Knutson et al., 2019; Zeverino et al., 2017).

In the image-guided workflow for the Icon, the stereotactic reference for the plan is defined by the CBCT images taken prior to start of treatment, usually referred to in the Icon workflow as "planning images", which are usually taken using "Preset 2" (see Table 6.1) [ref: White paper: Design and performance characteristics of a Cone Beam CT system for Leksell Gamma Knife Icon]. Co-registration of the "planning images" with reconstructed CBCT images acquired at the time of treatment (usually taken with

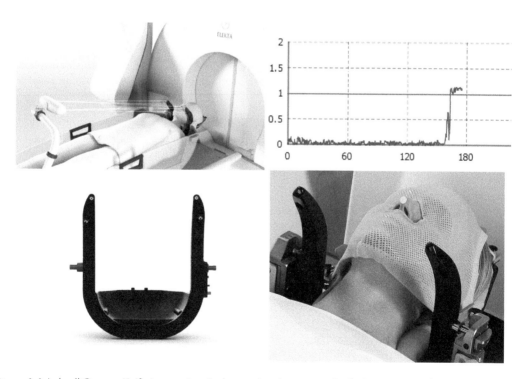

Figure 6.4 Leksell Gamma Knife Icon system is the previously accurate Perfexion system with the CBCT imaging and HDMM capabilities (top-left image). Intrafractional motion monitoring (in mm) can be seen by the trace shown from baseline (top-right image). A user-defined maximum allowed movement is indicated by the red line. The mask adaptor with integrated markers is shown in the bottom-left image. A thermoplastic mask is used to immobilize the patient and an IR marker is attached to the patient's nose (bottom-right image). Both the patient marker and integrated markers on the mask adaptor are monitored by the IR camera mounted at the middle of the patient positioning system. (From: White Paper: High Definition Motion Management—Enabling stereotactic Gamma Knife® radiosurgery with non-rigid patient fixations.)

Table 6.1 **Two-imaging presets defined on the GK Icon system. Preset 2 is chosen to define the stereotactic space (i.e., used for planning image) and Preset 1 is chosen for daily setup imaging**

	PRESET 1	**PRESET 2**
mAs/projection	0.4	1.0
kVp	90	90
Number of projections	332	332
Image volume (voxels)	448^3	448^3
Voxel size	0.5 mm	0.5 mm
Resolution	7 lp/cm	8 lp/cm
CTDI	2.5 mGy	6.3 mGy
CNR	1	1.5

Preset 1—see Table 6.1) gives a geometric transformation that is used to correct the delivery of the plan according to the current patient position [ref: white paper: Accuracy of co-registration of planning images with Cone Beam CT images, White Paper, 1509393.02]. GK's very precise couch uses translational movements to move the patient into the radiation focus with high accuracy by using a mask frame fixation that is related to a stereotactic coordinate system. The remaining rotational corrections are accounted for by the dedicated treatment planning system, and a final dose distribution is displayed against the planning distribution and must be accepted for the treatment to proceed.

At installation, the Leksell coordinate system is calibrated in relation to the radiation focus. In order to ensure that the calibration remains constant over time, a focus precision quality assurance method is designed to verify the radiological focus position in relation to the Leksell coordinates. The localization of the CBCT images is calibrated to the Leksell coordinate system for a given CBCT scanning position. (From white paper Geometric Quality Assurance for Leksell Gamma Knife® Icon™.)

6.3.3 PINPOINT (AKTINA)

Similar in design to the eXtend frame, the PinPoint (Aktina Medical, Congers, NY) is a commercially available noninvasive frameless system equipped with a vacuum fixation bite block device for patient localization and fixation (see Figure 6.5). It consists of an internal and external component. The internal component contains a custom-made patient-specific dental mouthpiece with a continuous mild vacuum suction to the upper hard palate. A gentle vacuum suction is applied between the dental mouthpiece and the upper hard palate to assure tight contact. The external component consists of the dental mouthpiece secured to a metal arch frame that is in turn locked into a carbon fiber couch board equipped with a thermoplastic head support. The head support is patient specific and is formed by creating an impression of the back of the skull. It limits head motion in the left, right, superior, and posterior directions. An adjustable, rigid connector defines the inferior and anterior limits and confines head rotations particularly to nodding (pitch) and shaking (roll). Once the patient is immobilized in the PinPoint frame, the patient's head cannot move without losing suction. To set up the patient to treatment isocenter, a localizer box with three imbedded spherical CT visible markers is attached over the bridge of the PinPoint system.

Li et al. (2011) evaluated the PinPoint frameless system and its ability to immobilize patients and restrict head motion. A video-based 3D optical surface imaging system with three ceiling-mounted camera pods (AlignRT, Vision RT Ltd, London, UK) was utilized to verify treatment setup as well as to monitor and quantify head motion near real time during treatment. Two hypofractionated stereotactic radiotherapy (HF-SRT) and two single-fraction SRS patients (10 treatment fractions in total) were immobilized with PinPoint, and the magnitude of motion was compared against that attained with 11 SRS patients immobilized with the invasive BRW head frame. In terms of setup verification, the mean 3D translational difference using PinPoint was 0.9 ± 0.3 mm. The mean translations and rotations were 0.3 ± 0.2 mm and 0.2° ± 0.1°, respectively. These values were found to be consistent with the magnitude of motion quantified with

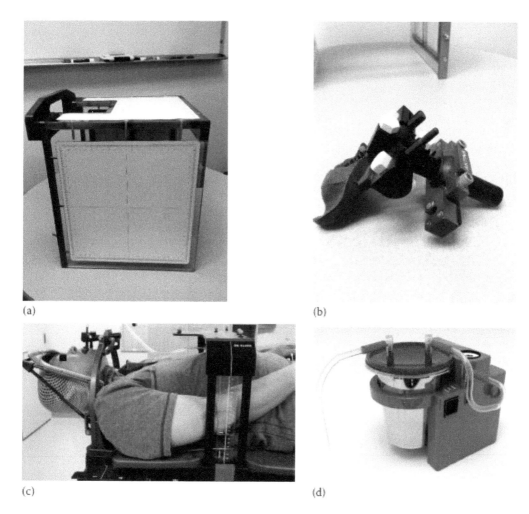

Figure 6.5 PinPoint frameless system. (a) The cranial localizer box that allows for three-point localization, (b) vacuum fixation mouthpiece with patient-specific dental impression, (c) patient immobilized in the thermoplastic support frame together with mouthpiece attached to an external arch block, and (d) portable suction unit that provides vacuum suction.

the BRW head frame. Although a slow head-drifting motion was observed with the PinPoint frame, for 98% of the time the magnitude of head motion was within 1.1 mm and 1.0° thus showing that this frameless system is adequate to tightly limit head motion during stereotactic radiation treatments.

Babic et al. (2018) utilized the noninvasive PinPoint system on 15 HF-SRT and 21 SRS patients treated using intensity-modulated radiation therapy and gantry-mounted stereotactic cones, respectively, on a linear accelerator. The residual error determined from the post-treatment kV CBCT verification image was used as a surrogate for intrafractional head motion during treatment. The mean intrafractional motion over all fractions with the PinPoint system was 0.62 ± 0.33 mm and 0.45 ± 0.33 mm for the HF-SRT and SRS cohort of patients (P value = 0.266), respectively. Of all fractions treated, no intrafractional motion exceeded 1.5 mm. These results were compared against 15 SRS patients immobilized with the CRW frame. For the CRW frame, the mean intrafractional motion was 0.30 ± 0.21 mm, and no intrafractional motion greater than 0.75 mm was observed. While both PinPoint and CRW frame were determined to be stringent immobilization systems, CRW frame provided superior head immobilization compared to the PinPoint system. Based on these results, the authors concluded that the invasive CRW frame is to be used in cases where minimal exposure to the surrounding neural tissue is critical such as lesions in the brainstem. For all other cases, the noninvasive PinPoint is the head immbolization system of choice for SRS.

6.3.4 OPTICALLY GUIDED FRAMELESS SYSTEM

The optically guided frameless system consists of several components: a cushion to hold the patient's head posteriorly, a thermoplastic mask that immobilizes the patient anteriorly from the forehead to the upper lip, a custom bite block molded to the patient's upper jaw, an optical array of fiducial markers that attaches to the bite block, and an in-room infrared camera system mounted on the ceiling. This system was developed at the University of Florida (Bova et al., 1997) and is commercially available through Varian Medical Systems (Palo Alto, CA). The unique feature of this system is that immobilization is separated from localization since the bite block is not fixed to the thermoplastic mask, thus making this system also compatible with rigid stereotactic frames (Meeks et al., 2000). The localization is achieved through the optical reference array and the infrared camera system. The infrared camera is equipped with illuminators that emit infrared light. The light is subsequently reflected off of infrared fiducials on the optical array and detected by the camera's charge-coupled device optics to accurately locate the position of the fiducials. The position of the fiducials relative to the treatment isocenter is predefined in a stereotactic coordinate system, and by combining this with a calibration matrix that relates the position of the camera to the treatment machine isocenter, target localization and real-time tracking relative to the isocenter are enabled (Kamath et al., 2005).

Relative to the stereotactic rigid head frame, Meeks et al. (2000) showed that the optic-guided bite-plate system provides a mean patient localization accuracy of 1.1 ± 0.3 mm. Phillips et al. (2000) found that with this system, positioning a target point in the radiation field had an accuracy of 1.0 ± 0.2 mm. Using orthogonal kV planar imaging verification, Wang et al. (2010) conducted a retrospective analysis of the setup accuracy of 56 patients using the optically guided frameless system and reported an average 3D isocenter localization error of 0.37 mm with a maximum error of 2 mm. Peng et al. (2010) reported a mean setup error of 1.2 ± 0.7 mm using a combined optical tracking and 3D ultrasound imaging system (SonArray system, Varian, Palo Alto, CA) and CBCT. Based on 15 patients, Ryken et al. (2001) determined the average localization accuracy at isocenter to be submillimeter at 0.82 ± 0.41 mm, and they concluded that in terms of target localization and accuracy, the optically guided frameless system is comparable to frame-based systems.

6.3.5 BRAINLAB FRAMELESS MASK

The BrainLAB frameless thermoplastic mask (BrainLAB AG, Feldkirchen, Germany) is a commercially available immobilization system that is strengthened with a custom-made mouthpiece and three reinforcing straps attached under the mask and covering the forehead, chin, and the area below the nose (Gevaert et al., 2012a). For localization and real-time tracking, six infrared markers are placed on top of the thermoplastic mask and are detected by an infrared camera system (ExacTrac) mounted to the ceiling. In conjunction with this optical guidance system, stereoscopic, planar kV x-ray images (Novalis Body) are taken and registered in six degrees of freedom (6DOF) with the DRRs from the planning CT images. Once the registration is accepted, any departures from the treatment isocenter are determined. Translational and rotational positioning errors in 6DOF can be corrected for with the treatment couch and a robotic tilt module underneath the tabletop (Verellen et al., 2003; Gevaert et al., 2012b).

Gevaert et al. (2012a) investigated the setup errors and intrafraction motion of 40 patients immobilized with the BrainLAB frameless mask system. Prior to 6DOF correction, the setup errors were found to be significantly larger in the lateral and longitudinal directions, and the mean 3D setup error was 1.91 ± 1.25 mm. Intrafractional errors were found to be significantly larger in the longitudinal direction (mean shift of 0.11 ± 0.55 mm), and the mean 3D intrafractional motion was 0.58 ± 0.42 mm. The mean intrafractional rotations were comparable for the vertical, longitudinal, and lateral directions and all within ± 0.03°. These results were found to be comparable to the mean 3D intrafraction motion of the BRW invasive head frame (considered the gold standard of immobilization) and reported by Ramakrishna et al. (2010) to be 0.40 ± 0.30 mm. Verbakel et al. (2010) investigated the positional accuracy of the BrainLAB frameless system using a hidden target test with a head phantom and found the accuracy to be approximately 0.3 mm in each direction (1 standard deviation). Utilizing posttreatment x-ray verification on 43 patients, they also determined the intrafraction motion and reported it to be 0.35 ± 0.21 mm (maximum of 1.15 mm). Based on their findings, Verbakel et al. (2010) concluded that patient setup with the BrainLAB frameless mask together with ExacTrac/Novalis Body is accurate and stable, and intrafraction motion is very small.

6.3.6 IMAGE-GUIDED FRAMELESS SRS

For frameless SRS, image guidance plays a crucial role in (1) localizing/identifying the target that may or may not be within a defined coordinate system and (2) ensuring cranial immobilization comparable to that attained with rigid frames (viewed as the gold standard of immobilization). Image guidance is typically utilized in the initial patient setup, pretreatment verification, intrafraction monitoring, and posttreatment verification. Prior to treatment, the patient is initially aligned to fiducials on the immobilization mask or localization device that is placed during simulation. After this is done, planar and/or volumetric imaging is completed to look for large setup variations. Planar imaging may include kV or EPID, as well as stereoscopic kV imaging (e.g., BrainLAB Novalis Body). Volumetric image acquisition (e.g., kV CBCT) followed by 3D–3D matching with the treatment planning CT offers the greatest amount of information for 6DOF patient setup. The geometric accuracy of planar and volumetric radiation-based imaging systems has been extensively studied and reported to be 1–2 and ≤1 mm, respectively (De Los et al., 2013). To correct for initial patient setup variations that exceed a certain threshold (e.g., 1 mm translation and 1° rotational), robotic 6DOF treatment couches (e.g., Hexapod [Elekta, Stockholm, Sweden], Robotic Tilt Module [BrainLAB AG]) can be utilized to reposition the patient. Pretreatment verification images (again with planar or volumetric imaging) may then be taken to confirm that the patient repositioning was accurate. These may also be used as a reference to evaluate intrafraction motion.

Following treatment, posttreatment images should be taken and compared against the reference pretreatment images to reveal and quantify any unexpected intrafraction motion. Intrafraction motion can also be done in real time so that patient movement is continuously monitored during treatment. Some commercial systems that have been developed for this purpose include a radiation-based system called ExacTrac (BrainLAB AG, Feldkirchen, Germany) that combines infrared and 2D orthogonal kV images for "snapshot" images during treatment and a non-radiation-based system called AlignRT (VisionRT, London, UK), which uses two or more cameras to perform rapid patient surface imaging. Both systems have been shown to have a geometric accuracy of 1–2 mm.

6.3.7 SUMMARY

For frameless immobilization systems, IGRT is crucial for the system to work with accuracies required for SRS and SRT. The system relies on the accuracy and the precision of the IGRT system without the external localization box. It is still crucial that the immobilization is rigid and stable as most IGRT does not account for patient motion intrafractionally, and the treatments can be long especially with increasing multiple metastases treatments. When a linac is used for SRS, extra-QA system must be put into place to maintain the accuracy required. A thorough, specific SRS QA program that includes the regular testing of the imaging system is crucial in delivering a successful SRS.

6.4 STEREOTACTIC TREATMENT PLANNING AND DELIVERY

To successfully treat with SRS and SRT, immobilization is only one part of a complicated planning and treatment delivery process as outlined in Figure 6.6. The following sections briefly describe the procedures throughout the SRS and SRT processes and how the selection of the immobilization system can affect each stage.

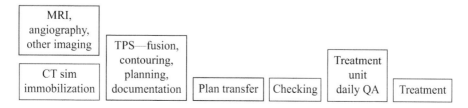

Figure 6.6 A typical SRS and stereotactic radiation therapy planning to treatment process.

6.4.1 SIMULATION

CT/MRI or MRI alone is usually required to localize the target volumes as well as the OARs. CT can be used to account for inhomogeneity in the tissue for dose calculation but, in the head, using only MRI with electron density overrides has been deemed clinically acceptable.

In general terms, attempts should be made to minimize distortion in MRI. Impulse generators for deep brain stimulators should have their amplitude reduced to zero to prevent side effects. A protocol should be in place with the radiology department to ensure frequent calibration to minimize field heterogeneities. The Leksell frame is engineered to very high standards, and Burchiel et al. (1996) have indicated that its metallic purity is such that it induces little distortion relative to other commercially available frames. Specific MR pulse sequences are available in many articles and are beyond the scope of this topic. However, it should be noted that T1-weighted sequences with thin cuts reduce spatial distortion, and inversion recovery images sharply demarcate anatomical structures.

For MRI and PET imaging, sealed tubes containing the appropriate liquid and in similar configurations to CT localizers can be used but have been superseded by image fusion techniques. Any imaging that will be used for localization require well thought-out QA procedures. For angiography, imaging modality that can aid in localizing AVMs for radiosurgery and small markers or rulers built into the three faces of fiducial plates (one anterior to the patient, and two left and right to the patient) are usually used. In this application, it is particularly useful to have the flexibility to image the lesion without restrictions to the source–patient–image plane geometry. However, more recently, systems to register 2D-3D imaging have been introduced commercially by BrainLAB in the radiation therapy workflow, though this has not been readily adopted.

For all images, the frame (and metal screws for invasive frames) can cause image distortions so caution should be taken when imaging the patient to try and minimize artifacts from occurring near the target and OAR regions.

6.4.2 TREATMENT PLANNING

Commercially available stereotactic planning systems are often associated with complete stereotactic, surgical, brachytherapy, and radiosurgery packages. The ability of a treatment planning system (TPS) to register or fuse multimodality images is essential as MRI is an important imaging tool for the brain, but CT scans are generally accepted as the most geometrically accurate, distortion-free, tomographic imaging modality.

Manipulation of the virtual patient showing the target volume and sensitive structures has become essential for forward planning, and beam optimization software is important as the complexity of sensitive structure avoidance makes intuitive planning more difficult. With multiple treatment targets becoming more of the norm and also patients returning for more treatments, there is a need for inverse planning modules to help the planning process. Plan analysis tools for the calculation of dose–volume data and assessment of optimized "cost functions" are also required. Automated and optimized inverse planning, which requires specification of dose–volume constraints, is emerging in many radiation therapy clinics.

Stereotactic treatment planning requires the planning of multiple noncoplanar arcs focused to single or several isocenters or multiple, fixed, and noncoplanar conformal beams. The planner must be aware of the limitations of the given treatment machine geometry regarding potential collimator–couch collisions as most of the collision software available on the treatment planning systems are limited. Adjustable beams-eye-views, which show the proximity of sensitive structures such as the brainstem and optic chiasm, are essential features. Overlapping of beam entrance and exits should ideally be avoided, and in general, arc planes should be separated according to the expression 180/n degrees, where n is the number of arcs to be used. The planner needs to be aware and avoid creating beams or arcs that enter through any high-density regions of the immobilization device as this can cause errors in the dose calculation. Avoidance of sensitive, often previously treated structures mean that optimization software and biological tumor control probability and normal tissue complication probability modeling are increasingly important developments. In general, the axial, sagittal, coronal, and arc planes are displayed. Dose–volume histogram analyses are used to assist the clinical team in deciding the optimum beam arrangement. The planning system must also have the ability to import and export images and plan information. For frame-based systems, there must be a transfer of the stereotactic coordinates from the planning system to the unit in addition to the normal transfer of information.

Three to five arcs and 7 to 13 static beams produce sufficient normal tissue sparing for SRT. However, when planning SRT or radiosurgery for benign conditions, particularly in young patients, it is relevant to consider increasing the number of arcs to reduce the exit dose to the brain from any sagittally orientated arcs, although these may be interrupted to avoid possibly sensitive exiting segments. For larger and/or irregular target volumes (35–70 cc), four to six fixed, noncoplanar fields are likely to be a more appropriate technique. In this case, either conformal blocking or MLC of each portal beam's eye view of the target can be used. The added efficiency of the MLC device always has to be balanced against the gold standard of target conformation given by customized lead alloy blocks (corresponding crudely to a micro-MLC having infinitely small leaves). One should note that increasing the number of noncoplanar arcs or beams (in order to improve dose conformality) adds to the total treatment time; hence, the trade-off between these two needs to be carefully considered.

6.4.3 TREATMENT SETUP AND QUALITY ASSURANCE

Prior to treatment, it is important to be able to confirm the planned setup using the CT scanner or treatment machine itself, depending on the type of lesion to be treated. Automatic or assisted setups are invaluable in limiting potential errors, and imaging can provide the desirable confirmation of accuracy for this type of treatment using exposures displaying the whole-head anatomy.

The linac requires additional QA checks at differing intervals. These checks depend on the precise technique used. The following brief summary of quality control procedures is representative of those undertaken with typical frequency of SRS checks, with suggested tolerances given in brackets. Note these quality and safety tests (refer toSolberg et al., 2012) are in addition to the standard machine and on-board imaging QA tests described in AAPM Task Group Report 142 (American Association of Physicists in Medicine [AAPM] Report No. 54, 142).

Daily
1. Frame fixation to patient (<1 mm)
2. Comparison of the left lateral, right lateral, and ceiling lasers with the corresponding light field centers at the cardinal gantry angles (<0.5 mm)
3a. Movement of the light field center cross for 180° collimator rotation (<0.5 mm)
3b. Movement of the radiation field center cross for 180° collimator rotation using onboard EPID (<0.5 mm)
4. CBCT imaging and treatment coordinate coincidence (<1 mm)
5. Light field symmetry of the selected tertiary collimator about the field center cross (<0.5 mm)

Monthly (frequency may depend on how often treatments are done on the unit)
1. A single transverse arc is executed over a water phantom with a calibrated chamber at the isocenter. The chamber dose is compared to a computer plan of the same treatment for representative monitor units used on actual treatments (±2%).

Commissioning and annually
1. EPID (or film) of a radio-opaque object set at the laser defined isocenter taken at regular gantry, collimator, and couch intervals using a tertiary collimator/micro-MLC attached to the treatment head (Lutz et al., 1988; Podgorsak et al., 1989) (<1 mm). NOTE: This test should be repeated more frequently if deemed necessary.
2. *End-to-end test*: With a Lucy head phantom mounted on the stereotactic frame (Figure 6.7), CT planning and treatment of a "lesion" in the phantom are given. Multiple inserts (Figure 6.8) consisting of film, diamond detectors, scintillators, and optically stimulated luminescence dosimeters, respectively, may be placed inside the Lucy phantom to measure the point dose or 2D planar dose and compare against the corresponding multiple arc plan (less than 5% dose and 1-mm profile displacement). Gel dosimetry shows promise for the verification of the 3D isodose envelopes treated with SRS; however, it requires expertise in reading out the dose using optical CT or MRI.

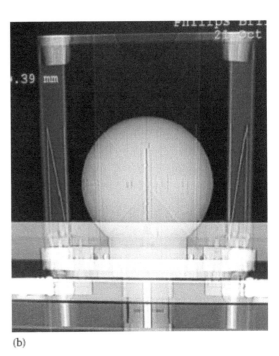

(a)   (b)

Figure 6.7 A typical SRS phantom localized for end-to-end test. (a) Lucy 3D phantom set up in a Leksell stereo-tactic frame for CT simulation and (b) CT scout view of a Lucy phantom setup.

Figure 6.8 Lucy phantom with multiple inserts. Lucy phantom hemispheres (a), film insert with pinpricks for positioning accuracy testing (b), insert accommodating detector (PTW microDiamond detector or the Exradin W1 scintillator) for dosimetry testing (c), positional guidance insert with markers at effective point of measurements of the detectors (only used during imaging) (d), insert with contrast-fillable organ contours (e), insert filled with mineral oil for MRI (f). (Sarfehnia et al., 2018.)

Other checks, such as checking that the tertiary collimators are not damaged and any interlocks (such as the barcode reader on the Elekta [Stockholm, Sweden] tertiary cone system) are in working order, should be done. The QA just discussed encompasses general tests recommended for SRS. These listed tests are not exhaustive and may follow a different frequency. Note that within these tests, it is important to establish position and dose at different gantry/couch positions. For SRT, different MLC tests should be added to check position and motion of the MLC in volumetric modulated arc therapy and intensity-modulated radiation therapy.

In general, the daily repositional accuracy of a relocatable frame should vary by less than 1 mm. Relocation checks are best carried out by overlaying the bony anatomy of repeat sets of CBCT images but if not possible, anterior–posterior and lateral EPIDs can be taken in order to quantify the displacement of reference markers attached to the frame. Alternatively, physical depth measurements to the surface of the head from a reference depth helmet mounted onto the stereotactic frame may be made. Optical video methods can be used on the basis of fixed geometry, wall-mounted cameras in the treatment room, and reflective markers placed on the patient.

6.4.4 OVERALL ACCURACY AND MARGINS

The overall accuracy achievable with a linac-based system should be less than 2 mm under routine conditions, provided sufficient emphasis is placed on the importance of QA. However, localization of the target volume remains the greatest uncertainty in SRS/SRT, and it is important to view the accuracy in its full clinical perspective. Assessing the accuracy of a given system involves rationalizing the cumulative effect of small, typically 0.5–1 mm errors, which generally translate in practice to a 1–3 mm safety margin around the lesion.

With SRS systems, smaller margins may be justified depending on the immobilization system used. The cumulative effect of all the errors in treatment should not be ignored, but each clinic needs to assess and evaluate this error and its effect on the OAR dose. With increasing number of cases of brain metastases presenting to radiation oncology clinics, minimizing dose to the normal brain is essential, especially within regions where radiation dose is linked to a decrease in cognitive function. Keeping cumulative dose to the normal brain low is a major concern as more and more patients are receiving focal irradiation and also returning for further irradiation to other sites.

Often with SRT, a variable PTV margin may be required where some compromise is needed close to sensitive organs. This PTV is then covered by the prescribed isodose surface, commonly 90% or 95%. If multiple isocenters are used with arcing circular cross-sectional beam configurations, dose uniformity is sacrificed for a more conformal target coverage. In this case, the prescribed isodose may be as low as 50% of the maximum "hot spots" in the overlapping regions of two spherical dose distributions.

6.5 CONCLUSION

Frame to frameless SRS and SRT requires the assessment and understanding of the whole workflow from patient demographics, general treatment philosophy, the availability of the imaging techniques, QA processes, and what PTV margin is needed for safe treatment. With a minimal reduction in patient setup accuracy that includes stringent QA and IGRT processes, "frameless" procedures can allow for more patient comfort and flexibility in how the radiation can be delivered to multiple targets even in patients who have undergone previous courses of radiation therapy. See Table 6.2 for a summary of all the frame and frameless systems discussed in this chapter. In order to account for a reduction in accuracy when using

Table 6.2 **Summary of all the frames discussed with achievable accuracy of the immobilization systems**

IMMOBILIZATION DESIGN	DEVICE NAME	IMAGE GUIDANCE REQUIRED (PLANAR AND/OR VOLUMETRIC)	ACCURACY (INCLUDING SETUP AND VECTOR DISPLACEMENT) (MM)
Invasive frame	BRW/CRW	No	<1
	Leksell	No	<1
Noninvasive moveable frame	GTC	No	1–3
	BrainLAB mask system	No	1–3
	Laitinen Stereoadapter 5000	No	1–2
Noninvasive frameless	eXtend	Yes	<1.5
	GK Icon Mask	Yes	<1.0
	PinPoint	Yes	<1.5
	Optically guided bite block	Yes	<2
	BrainLAB frameless mask	Yes	<2

Note: Image guidance is not required for "frame" immobilization designs but may add to positioning accuracy if added.

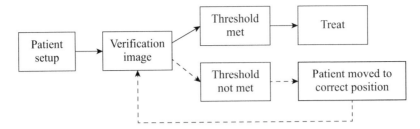

Figure 6.9 An example of image-guided radiation therapy procedure for linear accelerator delivery. Dashed lines indicate a loop which would be repeated until the threshold is met.

a frameless device, margins must be used to create a PTV from the target volumes. This process should follow a study by each clinic that takes into account all the approaches that have been highlighted in this chapter such as the immobilization system, imaging, fusion, planning, delivery system, and IGRT. With the IGRT system in place, there should be a treatment verification process in place such that patient position is corrected when a threshold position (translational and rotational) is observed (Figure 6.9). This can increase patient treatment time, and, therefore, care should be taken with initial patient setup in order to minimize reimaging of the patient.

With all the correct procedures in place, frameless immobilization promises to deliver similar accuracy in patient setup to both SRS and SRT, and its refined methods of patient fixation, treatment planning, delivery, and verification make it an important component in the evolution of high-precision radiation.

REFERENCES

AAPMReport No. 54 (June 1995). Stereotactic radiosurgery. Report of Task Group 42, Radiation Therapy Committee. American Institute of Physics, Inc. Woodbury, NY.

AAPMReport No. 142 (September 2009) Task Group 142 report: Quality assurance of medical accelerators. Report of Task Group 142, Radiation Therapy Committee.

Alheit H, Dornfeld S, Dawel M, Alheit M, Henzel B, Steckler K, Blank H, Geyer P (2001) Patient position reproducibility in fractionated stereotactically guided conformal radiotherapy using the BrainLAB mask system. *Strahlenther Onkol* 177:264–268.

Ali I, Tubbs J, Hibbitts K, Algan O, Thompson S, Herman T, Ahmad S (2010) Evaluation of the setup accuracy of a stereotactic radiotherapy head immobilization mask system using kV on-board imaging. *Journal of Applied Clinical Medical Physics* 11:3192.

Babic S, Lee Y, Ruschin M, Lochray F, Lightstone A, Atenafu E, Phan N, Mainprize T, Tsao M, Soliman H, Sahgal (2018) To frame or not to frame? Cone-beam CT-based analysis of head immobilization devices specific to linac-based stereotactic radiosurgery and radiotherapy. *Journal of Applied Clinical Medical Physics* 19:111.

Bednarz G, Machtay M, Werner-Wasik M, Downes B, Bogner J, Hyslop T, Galvin J, Evans J, Curran W, Jr., Andrews D (2009) Report on a randomized trial comparing two forms of immobilization of the head for fractionated stereotactic radiotherapy. *Medical Physics* 36:12–17.

Bova FJ, Buatti JM, Friedman WA, Mendenhall WM, Yang CC, Liu C (1997) The University of Florida frameless high-precision stereotactic radiotherapy system. *International Journal of Radiation Oncology, Biology, Physics* 38:875–882.

Burchiel KJ1, Nguyen TT, Coombs BD, Szumoski J (1996) MRI distortion and stereotactic neurosurgery using the Cosman-Roberts-Wells and Leksell frames. *Stereotact Functional Neurosurgery* 66(1–3):123–136.

Burton KE, Thomas SJ, Whitney D, Routsis DS, Benson RJ, Burnet NG (2002) Accuracy of a relocatable stereotactic radiotherapy head frame evaluated by use of a depth helmet. *Clinical Oncology (R Coll Radiol)* 14:31–39.

Das S, Isiah R, Rajesh B, Ravindran BP, Singh RR, Backianathan S, Subhashini J (2011) Accuracy of relocation, evaluation of geometric uncertainties and clinical target volume (CTV) to planning target volume (PTV) margin in fractionated stereotactic radiotherapy for intracranial tumors using relocatable Gill–Thomas–Cosman (GTC) frame. *Journal of Applied Clinical Medicine Physics* 12:3260.

Delannes M, Daly NJ, Bonnet J, Sabatier J, Tremoulet M (1991) Fractionated radiotherapy of small inoperable lesions of the brain using a non-invasive stereotactic frame. *International Journal of Radiation Oncology, Biology, Physics* 21:749–755.

De Los SJ, Popple R, Agazaryan N, Bayouth JE, Bissonnette JP, Bucci MK, Dieterich S et al. (2013) Image guided radiation therapy (IGRT) technologies for radiation therapy localization and delivery. *International Journal of Radiation Oncology, Biology, Physics* 87:33–45.

Gevaert T, Verellen D, Engels B, Depuydt T, Heuninckx K, Tournel K, Duchateau M, Reynders T, De RM (2012a) Clinical evaluation of a robotic 6-degree of freedom treatment couch for frameless radiosurgery. *International Journal of Radiation Oncology, Biology, Physics* 83:467–474.

Gevaert T, Verellen D, Tournel K, Linthout N, Bral S, Engels B, Collen C et al. (2012b) Setup accuracy of the Novalis ExacTrac 6DOF system for frameless radiosurgery. *Int International Journal of Radiation Oncology, Biology, Physics* 82:1627–1635.

Gill SS, Thomas DG, Warrington AP, Brada M (1991) Relocatable frame for stereotactic external beam radiotherapy. *International Journal of Radiation Oncology, Biology, Physics* 20:599–603.

Golden NM, Tomita T, Kepka AG, Bista T, Marymont MH (1998) The use of the Laitinen stereoadapter for three-dimensional conformal stereotactic radiotherapy. *Journal of Radiosurgery* 1(3):191–200.

Graham JD, Warrington AP, Gill SS, Brada M (1991) A non-invasive, relocatable stereotactic frame for fractionated radiotherapy and multiple imaging. *Radiotherapy & Oncology* 21:60–62.

Kalapurakal JA, Ilahi Z, Kepka AG, Bista T, Goldman S, Tomita T, Marymont MH (2001) Repositioning accuracy with the Laitinen frame for fractionated stereotactic radiation therapy in adult and pediatric brain tumors: Preliminary report. *Radiology* 218:157–161.

Kamath R, Ryken TC, Meeks SL, Pennington EC, Ritchie J, Buatti JM (2005) Initial clinical experience with frameless radiosurgery for patients with intracranial metastases. *International Journal of Radiation Oncology, Biology, Physics* 61:1467–1472.

Knutson NC, Hawkins BJ, Bollinger D, Goddu SM, Kavanaugh JA, Santanam L, Mitchell TJ, Zoberi JE, Tsien C, Huang J, Robinson CG, Perkins SM, Dowling JL, Chicoine MR, Rich KM, Dunn GP, Mutic S (2019) Characterization and validation of an intra-fraction motion management system for mask-based radiosurgery. *Journal of Applied Clinical Medical Physics* 20:21–26.

Kumar S, Burke K, Nalder C, Jarrett P, Mubata C, A'hern R, Humphreys M, Bidmead M, Brada M (2005) Treatment accuracy of fractionated stereotactic radiotherapy. *Radiotherapy & Oncology* 74:53–59.

Laitinen LV, Liliequist B, Fagerlund M, Eriksson AT (1985) An adapter for computed tomography-guided stereotaxis. *Surgical Neurology* 23:559–566.

Leksell L (1971) *Stereotaxis and Radiosurgery: An Operative System.* Springfield, IL: Thomas.

Li G, Ballangrud A, Kuo LC, Kang H, Kirov A, Lovelock M, Yamada Y, Mechalakos J, Amols H (2011) Motion monitoring for cranial frameless stereotactic radiosurgery using video-based three-dimensional optical surface imaging. *Medical Physics* 38:3981–3994.

Louw (2003) Stereotactic surgery with the Leksell frame. In: Schulder M, Gandhi CD (eds.), *Handbook of Stereotactic and Functional Neurosurgery*, pp. 27–33. New York: Dekker.

Lundsford LD, Leksell D (1988) The Leksell system. In: *Modern Stereotactic Neurosurgery*, pp. 27–46. Boston, MA: Martinus Nijhoff Publishing.

Lutz W, Winston KR, Maleki N (1988) A system for stereotactic radiosurgery with a linear accelerator. *International Journal of Radiation Oncology, Biology, Physics* 14:373–381.

Meeks SL, Bova FJ, Wagner TH, Buatti JM, Friedman WA, Foote KD (2000) Image localization for frameless stereotactic radiotherapy. *International Journal of Radiation Oncology, Biology, Physics* 46:1291–1299.

Minniti G, Scaringi C, Clarke E, Valeriani M, Osti M, Enrici RM (2011) Frameless linac-based stereotactic radiosurgery (SRS) for brain metastases: Analysis of patient repositioning using a mask fixation system and clinical outcomes. *Radiation Oncology* 6:158.

Nguyen TK, Sahgal A, Detsky J, Soliman H, Myrehaug S, Tseng, CL, Husain ZA, Carty A, Das S, Yang V, Lee Y, Sarfehnia A, Chugh B, Yeboah C, Ruschin M (2019) Single-fraction stereotactic radiosurgery versus hippocampal-avoidance whole brain radiation therapy for patients with 10 to 30 brain metastases: A dosimetric analysis. *International Journal of Radiation Oncology, Biology, Physics* 105:349–399.

Peng LC, Kahler D, Samant S, Li J, Amdur R, Palta JR, Liu C (2010) Quality assessment of frameless fractionated stereotactic radiotherapy using cone beam computed tomography. *International Journal of Radiation Oncology, Biology, Physics* 78:1586–1593.

Phillips MH, Singer K, Miller E, Stelzer K (2000) Commissioning an image-guided localization system for radiotherapy. *International Journal of Radiation Oncology, Biology, Physics* 48:267–276.

Podgorsak EB, Pike GB, Olivier A, Pla M, Souhami L (1989) Radiosurgery with high energy photon beams: A comparison among techniques. *International Journal of Radiation Oncology, Biology, Physics* 16:857–865.

Ramakrishna N, Rosca F, Friesen S, Tezcanli E, Zygmanszki P, Hacker F (2010) A clinical comparison of patient setup and intra-fraction motion using frame-based radiosurgery versus a frameless image-guided radiosurgery system for intracranial lesions. *Radiotherapy & Oncology* 95:109–115.

Ruschin M, Nayebi N, Carlsson P, Brown K, Tamerou M, Li W, Laperriere N et al. (2010) Performance of a novel repositioning head frame for Gamma Knife perfexion and image-guided linac-based intracranial stereotactic radiotherapy. *International Journal of Radiation Oncology, Biology, Physics* 78:306–313.

Ryken TC, Meeks SL, Pennington EC, Hitchon P, Traynelis V, Mayr NA, Bova FJ, Friedman WA, Buatti JM (2001) Initial clinical experience with frameless stereotactic radiosurgery: Analysis of accuracy and feasibility. *International Journal of Radiation Oncology, Biology, Physics* 51:1152–1158.

Sahgal A, Aoyama H, Kocher M, Neupane B, Collette S, Tago M, Shaw P, Beyene J, Chang EL (2015) Phase 3 trials of stereotactic radiosurgery with or without whole-brain radiation therapy for 1 to 4 brain metastases: Individual patient data meta-analysis. *International Journal of Radiation Oncology, Biology, Physics* 91:710–717.

Sarfehnia A, Ruschin M, Chugh B, Yeboah C, Becker N, Cho YB, Lee Y (2018) Performance characterization of an integrated cone-beam CT system for dedicated gamma radiosurgery. *Medical Physics* 45:4179–4190.

Sayer FT, Sherman JH, Yen CP, Schlesinger DJ, Kersh R, Sheehan JP (2011) Initial experience with the eXtend system: A relocatable frame system for multiple-session Gamma Knife radiosurgery. *World Neurosurgery* 75:665–672.

Solberg TD, Balter JM, Benedict SH, Fraass BA, Kavanagh B, Miyamoto C, Pawlicki T, Potters L, Yamada Y (2012) Quality and safety considerations in stereotactic radiosurgery and stereotactic body radiation therapy: Executive summary. *Practice in Radiation Oncology* 2(1):2–9.

Theelen A, Martens J, Bosmans G, Houben R, Jager JJ, Rutten I, Lambin P, Minken AW, Baumert BG (2012) Relocatable fixation systems in intracranial stereotactic radiotherapy. Accuracy of serial CT scans and patient acceptance in a randomized design. *Strahlenther Onkol* 188:84–90.

Verbakel WF, Lagerwaard FJ, Verduin AJ, Heukelom S, Slotman BJ, Cuijpers JP (2010) The accuracy of frameless stereotactic intracranial radiosurgery. *Radiotherapy & Oncology* 97:390–394.

Verellen D, Soete G, Linthout N, Van AS, De RP, Vinh-Hung V, Van de Steene J, Storme G (2003) Quality assurance of a system for improved target localization and patient set-up that combines real-time infrared tracking and stereoscopic x-ray imaging. *Radiotherapy & Oncology* 67:129–141.

Wang JZ, Rice R, Pawlicki T, Mundt AJ, Sandhu A, Lawson J, Murphy KT (2010) Evaluation of patient setup uncertainty of optical guided frameless system for intracranial stereotactic radiosurgery. *Journal of Applied Clinical Medical Physics* 11:92–100.

Willner J, Flentje M, Bratengeier K (1997) CT simulation in stereotactic brain radiotherapy-analysis of isocenter reproducibility with mask fixation. *Radiotherapy & Oncology* 45:83–88.

Wowra B, Muacevic A, Tonn JC (2012) CyberKnife radiosurgery for brain metastases. *Progress in Neurological Surgery* 25:201–209.

Zeverino M, Jaccard M, Patin D, Ryckx N, Marguet M, Tuleasca C, Schiappacasse L, Bourhis J, Levivier M, Bochud FO, Moeckli R (2017) Commissioning of the Leksell Gamma Knife® Icon™. *Medical Physics* 44:355–363.

Elekta White papers:
Position accuracy analysis of the stereotactic reference defined by the CBCT on Leksell Gamma Knife® Icon™
Geometric Quality Assurance for Leksell Gamma Knife® Icon™
Design and performance characteristics of a Cone Beam CT system for Leksell Gamma Knife® Icon™
High Definition Motion Management—enabling stereotactic Gamma Knife® radio surgery with non-rigid patient fixations

7 Principles of Image-Guided Hypofractionated Stereotactic Radiosurgery for Brain Tumors

Sean S. Mahase, Susan C. Pannullo, John T. McKenna,
and Jonathan P.S. Knisely

Contents

7.1 RATIONALE, ADVANTAGES, AND DISADVANTAGES OF HYPOFRACTIONATED STEREOTACTIC RADIOSURGERY

7.1.1 TUMOR RADIOBIOLOGY

A hypofractionated approach has several theoretical advantages in comparison with single-fraction radiosurgery. Fractionation exploits fundamental radiobiological principles such as reassortment and reoxygenation. Hypofractionating treatment promotes the transition of cells from radioresistant cell cycle phases (S and G2) to more sensitive phases (G1 and M). Additionally, resection cavities may be relatively hypoxic, reducing production of oxygen-based free radicals during irradiation that induce the DNA breaks that ultimately cause cell death. This premise, known as the oxygen enhancement ratio (Ward, 1994), was a fundamental supporting premise of preoperative stereotactic radiosurgery (SRS) trials reducing definitive dosing by 20% compared to those used for post-operative cases (Shaw *et al.*, 2000). With hypofractionation, each treatment will differentially eliminate the most oxygenated cells and improve oxygen diffusion to previously hypoxic regions that will be again differentially affected by subsequent fractions. Taking advantage of these radio-biological principles may shift the therapeutic ratio, leading to improved tumor control while decreasing late normal tissue effects through enhancing tumor kill through the aforementioned mechanisms and decreasing late normal tissue effects through permitting inter-fraction repair of normal tissue sublethal DNA damage (Hall and Brenner, 1993). It is important that the physical dosimetric advantages and steep dose falloff that are a hallmark of single-fraction SRS are maintained with a fractionated approach.

Additionally, recent advances in understanding immunological responses to cancer have permitted experiments that reveal some potentially very important data on the use of checkpoint inhibitors and radiation. In the presence of checkpoint inhibitors, the use of hypofractionated treatment regimens increases systemic immunological responses relative to high single-dose irradiation and can even induce abscopal responses (Vanpouille-Box *et al.*, 2017; Galluzzi *et al.*, 2018).

Radiosurgery-induced release of tumor antigens and double-strand DNA fragments stimulates the secretion of Type I interferon through the cGAS-STING pathway. Higher dose single fraction radiosurgery leads to higher levels of double-strand DNA fragments in the cytoplasm, which stimulates the induction of Trex1, a DNA exonuclease. Trex1 dampens the immune response by degrading those DNA fragments that were helping stimulate Type I interferon secretion. Hypofractionation of radiosurgery likely provides additional advantages in eradication of tumors that are still being elucidated, even in the absence of pharmacological immune checkpoint inhibition (Golden *et al.*, 2020).

7.1.2 BALANCE BETWEEN TUMOR SIZE AND TOXICITY TO NORMAL BRAIN

Treating larger lesions inevitably encompasses larger volumes of normal tissue in the radiation field. To address this issue, lower prescription doses are frequently employed to stay within dose constraints when treating tumors larger than appoximately10 cc (or >2.5 cm in diameter) with a single fraction.

However, lower SRS doses of less than 18 Gy are often selected for tumor with diameters more than 3 cm, based on toxicity data from the Radiation Therapy Oncology Group (RTOG) 90–05 (Shaw *et al.*, 2000) dose-finding trial for single-fraction SRS in recurrent brain tumors. These lower doses are associated with lower local control rates. Ebner and colleagues treated 754 brain metastases with single fraction SRS, of which 93 were classified as large tumors (3 cm or greater). They reported improved 1-year local control rates in smaller tumors, 86% versus 68% ($p < 0.001$) (Ebner *et al.*, 2015). Higher local recurrence rates have also been observed with cystic or irregularly enhancing lesions treated with single-fraction SRS (Shiau *et al.*, 1997; Mori *et al.*, 1998). These suboptimal outcomes must also be weighed against the known risk of treatment-related toxicity when higher-dose single-fraction SRS treatments are employed for larger intracranial tumors (Shaw *et al.*, 2000; Blonigen *et al.*, 2010; Soltys *et al.*, 2015).

Hypofractionated SRS (HSRS), delivered over 2–5 fractions, employs the steep dose gradients and tight planning margins used in single-fraction SRS while possessing the radiobiological advantages of fractionation with the goal of shifting the therapeutic ratio. However, even with this approach, the risk of developing prolonged reactive edema leading to microischemic and necrotic parenchymal changes is increased when normal brain tissue volumes of more than 23 cc receive >4 Gy per fraction in a treatment prescribed to a total of ≥20 Gy (Ernst-Stecken *et al.*, 2006). Also, early complications after SRS, such as seizure or the worsening of neurologic symptoms, have been recognized for larger volumes and are proportional to the radiation dose delivered to normal brain (Ernst-Stecken *et al.*, 2006). A significant increase in side effects occurs typically if more than 10 cc of normal brain is treated with more than 10 Gy in a single fraction (Levegrün *et al.*, 2004; Lawrence *et al.*, 2010; Sneed *et al.*, 2015). A retrospective review of 173 lesions prescribed to a median dose of 18 Gy using linear accelerator-based single-fraction SRS in patients who predominantly previously received fractionated radiotherapy reported that the volume of normal brain receiving a dose between 8 and 16 Gy was the best predictor for radionecrosis. They proposed that patients with V10 Gy more than 10.5 cc or V12 Gy more than 7.9 cc be considered for hypofractionated treatment to minimize the risk of toxicity (Blonigen *et al.*, 2010).

Hypofractionation is an effective alternative to single-fraction SRS in several scenarios. Ernst-Stecken and colleagues concluded that HSRS with 5 fractions of 6–7 Gy per fraction is a safe and effective treatment for patients with cerebral metastatic disease not amenable to single-dose radiosurgery due to gross tumor volume (GTV) more than 3 cc, location involving critical anatomical sites, or if normal brain volume constraints would otherwise be exceeded (Ernst-Stecken *et al.*, 2006). A study comparing brain metastases of more than2 cm in diameter treated with either single fraction SRS to 15–18 Gy or HSRS delivered in 27 Gy over 3 fractions reported that HSRS resulted in better local control (91% versus 77%; *p* = 0.01) and a reduction in the risk of radiation necrosis (9% versus 18%; *p* = 0.01) (Minniti *et al.*, 2016). Kim *et al.* treated brain metastases more than 3 cm in diameter with HSRS doses of 24 Gy, 27 Gy, or 30 Gy in three fractions. They demonstrated that 27 Gy was the optimal dose, providing a better progression-free survival rate than the 24 Gy cohort (80% versus 65%) while having a lower incidence of radionecrosis than the 30 Gy cohort (9% versus 37%) (Kim *et al.*, 2019). Solimon and colleagues treated 137 resection cavities to a median total dose of 30 Gy over 5 fractions, resulting in an 84% 1-year local control rate with radionecrosis in seven (6%) patients (Solimon *et al.*, 2019). Several additional studies substantiate HSRS as an alternative to single-fraction SRS in providing improved local control with relatively low toxicity (Higuchi *et al.*, 2009; Yomo *et al.*, 2012; Angelov *et al.*, 2018).

7.1.3 ELOQUENT AREAS

The clinical manifestations of any radiation-induced toxicity will be contingent on the location of the treated volume within the brain. For example, the frontal and occipital lobes appear to tolerate significantly larger radiosurgical treatment volumes in comparison to the temporal lobe, basal ganglia, and brainstem. It has been shown that delivering single-fraction SRS doses to more than 3 cc of normal brain tissue in the brainstem, basal ganglia, mesencephalon, or internal capsule will result in significantly more toxicity, warranting consideration of HSRS when targeting disease within these eloquent areas of the brain (Flickinger *et al.*, 1997).

7.1.4 DISADVANTAGES OF HYPOFRACTIONATION

There are caveats to employing HSRS. Patients must be present for multiple sessions, which may be difficult for patients and their families. This is particularly true for patients with poor performance status or who live at a distance from the treatment center. Additionally, multiple treatments also entail greater utilization of departmental resources and personnel. Studies on HSRS have predominantly been retrospective in nature, thus the ability to achieve durable local control for radioresistant tumors (melanoma, sarcoma, and renal cell carcinoma) treated with 2–5 fractions compared to more radiosensitive tumors (breast, small-cell lung cancer [SCLC], and non-SCLC) is as yet still under investigation. Single-fraction approaches for these diverse histologies did not result in local control differences, and it does not appear that any substantial data exists regarding HSRS dose-escalation strategies to try to further improve local control probabilities (Oermann *et al.*, 2013; Hamel-Perreault *et al.*, 2019; Pessina *et al.*, 2016). In the end, the efficacy of HSRS is contingent on being able to reproducibly target the tumor volume with each treatment.

7.2 HYPOFRACTIONATED RADIOSURGICAL TREATMENT PLANNING FOR BRAIN TUMORS

Prior to the initiation of any patient's HSRT treatment, a careful commissioning of the treatment process should be completed. This commissioning process is resource intensive and requires significant expertise, specialized tools, and careful attention to detail (Halvorsen *et al.*, 2017; Benedict *et al.*, 2010; Schell *et al.*, 1995; Brezovich *et al.*, 2019; IAEA Technical Report Series No. 483). Because of this, a medical physicist should be consulted immediately once a hospital organization decides that it wishes to perform HSRT treatments. Due to the complex nature of delivery caused by generally small treatment fields, tight setup margins and high doses per fraction, extreme care and attention to detail must be followed. The commissioning physicist must be provided with numerous specialized tools as a prerequisite for initiating an HSRT or SRS program and he or she should be given ample time to perform the specialized measurements necessary to ensure that the treatment planning systems, patient setup systems, and processes are in place to ensure safe delivery of the treatment with high quality standards.

Aside from the extreme technical requirements, there are also many processes that should be setup and made clear from the beginning of initiation of the treatment program. Ideally, a committee should be formed that will finally produce a policy and procedure manual that will outline every step of the treatment process. This committee should be interdisciplinary since these treatments require significant coordination across multiple departments. Also, an interdisciplinary approach will likely result in better outcomes for the patient and better employee engagement throughout the organization. In our practice, the multidisciplinary team consisting of radiation oncologists, neurosurgeons, neuroradiologists, nurses, and advanced practice providers, together with radiation therapists, dosimetrists, and medical physicists meets weekly to discuss upcoming and current cases and review previously treated cases for possible complications and need for further treatment.

It is important to note that this is beyond the usual purview of the various clinical tumor boards. By including the entire range of people involved with providing the specialized services, a degree of standardization across the organizational components can be achieved. Examples include the neurosurgeons informing the radiation oncologists of surgical complications necessitating delay in receipt of postoperative radiotherapy. Another important consideration is the sequencing of radiotherapy with systemic therapy as some combinations may enhance radiosensitivity. Thus, the time between the receipt of last or next systemic therapy and the patient's first or last fraction of radiation, respectively, must be coordinated. Principles of combining SRS with immunotherapy will be discussed in Chapter 29.

7.2.1 PREPLANNING FRAMEWORK

Delivering high radiation doses per fraction in the proximity of various OARs requires accurate patient positioning to ensure treatment efficacy and safety. A frameless treatment technique using a thermoplastic immobilization mask that enables a reproducible setup in combination with image guidance will provide the high precision that is ideal for an HSRS approach (Thomas et al., 2018).

A detailed understanding of the relevant anatomy (vasculature, sinuses, adjacent OARs) especially as it relates to patterns of tumor spread and recurrence is absolutely necessary when delineating radiosurgical target volumes (TV). Preoperative and timely post-operative contrast-enhanced MR-based diagnostic imaging is essential to evaluate the size and shape of the lesion or cavity for TV determination. The pathology and radiobiology of the tumor type may also be taken into account when selecting hypofractionated dose schedules. The culmination of these aforementioned consideration will guide the creation of the CTV. End-to-end testing of the imaging, planning, and precision and reliability of delivering several treatments to the same delineated location will influence the PTV size of a hypofractionated course of therapy (Lightstone et al., 2005). These characteristics will aid in predicting the expected clinical trajectory and appropriateness of treatment technique, dose, and fractionation.

The following questions should be answered when starting the planning process for any radiosurgical case:

1. Did the patient receive prior whole-brain radiotherapy and/or SRS?
2. Are there any patient-related factors (e.g., performance status, anxiety or claustrophobia, transportation difficulties) that may potentially impair their ability to comply with multiple treatments?
3. What is the size of lesion or cavity?
4. If this is a post-operative case, what is the anticipated time between resection and start of radiotherapy?
5. What was the duration between the acquisition of the planning MRI and the start of treatment?
6. What is the appropriate dose required to obtain durable tumor control?
7. What are clinical target volume (CTV) margins needed to avoid a marginal recurrence?
8. What are the PTV margins required to account for setup error?
9. Is hypofractionation required to decrease the risk of treatment-related toxicity to an acceptable level?
10. Is the patient receiving systemic therapies concurrently with radiotherapy? Alternatively, what will be the interval between systemic therapy and radiotherapy?

7.2.2 DIAGNOSTIC IMAGING

A contrast-enhanced MRI should be obtained using a standardized SRS MRI protocol. We perform frameless MRI radiosurgery imaging for inpatients on a wide range of 1.5 and 3.0 T scanners, using an institutional imaging protocol designed to fit within a 15–30-minute time slot. It includes a post-contrast 3D axial T2 FLAIR, axial T1, and sagittal 3D T1 SPACE or CUBE sequences without and with gadolinium contrast. The axial T2 FLAIR and axial T1 are both acquired with 3 mm slice thickness, while the 3D T1 is acquired in the sagittal plane with 1 mm isotropic voxels to facilitate reconstruction in axial and coronal planes. We stipulate a 10–20-minute delay between contrast injection and scanning, which increases the detection of small volume brain metastases (Kushnirsky *et al.*, 2016). The volumetric T1 series used for planning employs an image matrix of 256 × 256. At times, we ask for additional standard brain MRI sequences, including precontrast 3D T1, axial T2, axial diffusion weighted image, and axial susceptibility weighted imaging to help delineate critical normal structures and tumor volumes, Close collaboration with a dedicated neuroradiologist will help ensure studies of quality adequate for planning SRS.

Because MRI may introduce subtle geometric distortions that are inapparent to even practiced observers, routine quality assurance calibrations of the imaging process with specialized phantoms to identify, reduce, and correct these distortions to acceptably low levels are important to assure that the fundamental studies used to plan treatment are as spatially accurate (Sun *et al.*, 2014). Higher-field strength magnets are more prone to geometric distortion than lower-field strength magnets, which should be taken into consideration when choosing a platform for performing imaging for planning SRS.

If the patient is having surgery prior to SRS, the timing of postoperative imaging acquisition for planning SRS is a variable that merits attention. Based on data garnered from gliomas, a postoperative MRI should be performed within 72 hours of the resection to optimize differentiation between residual tumor and blood (Albert *et al.*, 1994; Forsyth *et al.*, 1997). Depending on the size of the cavity and the tumor histology, it may be prudent to allow time for postoperative cavity involution in order to decrease the volume of normal brain treated. Approximately one-third of cavities transiently experience increases in volume (Patel *et al.*, 2011). Most shrinkage will occur within a few days of surgery, with gradual shrinkage occurring over a protracted period of time, which may be advantageous for planning stereotactic irradiation, particularly for indolent tumors where there is no need to hasten radiation treatment delivery (Atalar *et al.*, 2013). We try to complete postoperative irradiation within 4 weeks of the date of surgery. Because the timing of postoperative SRS may be dependent on interactions with other caregivers such as rehabilitation medicine, medical oncology, we almost always obtain a specialized 'quick' MRI scan for planning SRS for these patients. This is important because the resection bed geometry needs to be accurately (and contemporaneously) determined, as do the sizes and locations of any other metastatic foci that require radiosurgical management. If more than a week passes between the treatment planning MRI and the date of starting radiosurgery, there are rising probabilities that one or more target will not have treatment coverage as per the original plan (Salkeld *et al.*, 2018).

7.2.3 SIMULATION

The patient treatment planning process commences with the setup and immobilization of the patient. There are multiple immobilization systems available each with their own properties. When selecting an immobilization system, one must consider not only the cost but also the rigidity of the system and the image-guidance system that will be used during treatment. The ideal system will support rigidly and reproducibly both the anterior and posterior aspect of the head. It will be also beneficial to select a system that further reproducibly supports the neck of the patient. For HSRT, we typically use either a three-piece (one posterior and two anterior components) thermoplastic immobilization masks or a thermoplastic mask coupled with an Accuform that is large enough to support the whole cranium and neck. Regardless of the chosen system, it is very important to follow very carefully the vendor recommendations in not only placement of the mask, but also the drying time. With many systems, if the mask is removed too early, then it may shrink or deform prior to the actual treatment day and may result in the need to resimulate the patient. Further, during mask creation it is important for the mask-maker to manually keep the patient from moving until the mask is at least partially dry. If the patient is left too early in the drying process, he or she may move resulting in a mask that is too loose to properly immobilize the patient during the treatment.

The CT protocol should be standardized. In our practice, we use a standard protocol for all HSRT and SRS patients. The CT imaging required for planning HSRS can also help with assessing MRI scans' geometric accuracy, if enough landmarks in the region of interest can be identified in both studies to assure an accurate rigid registration has been achieved (Poetker *et al.*, 2005). We use CT settings at the time of simulation of 120 kV and a field of view that includes the entire cranium and any external localizing devices—from the vertex to the bottom of the C3 vertebral body. Helical scanning will permit reformatting to any desired slice thickness—we use reconstructed axial slices of 0.625 mm thickness. As a supplemental aid to achieving a reproducible setup, BBs (radio-opaque markers) are placed on the isocentric "origin" marks on the thermoplastic mask. We do not administer contrast at the time of CT simulation for planning brain metastasis radiosurgery.

During the initial setup of the protocol, we included a diagnostic medical physicist to set the parameters of the machine. These parameters were chosen to prioritize high pixel resolution. It is important to ensure a high resolution since this CT scan will be used later as a reference in subsequent patient setup by the image-guidance system. It is also important to remember to include all setup devices that might intersect with the treatment fields in the field of view.

Once the planning MRI and CT have been acquired, they should both be imported into an appropriate contouring system. These systems have advanced significantly over the years and will provide appropriate results when done correctly. During the fusion process, two image sets are registered to one another. This can be done rigidly or deformably. It is important that the fused image sets be of sufficient resolution for this process to be accurate. Ideally, if the CT scan will be your final reference image (i.e., contours will be copied from MRI to CT for treatment), it will have a resolution superior to the MRI resolution. If the CT resolution is lower than the MRI then there is a danger of distorting the contours when resampling to the lower resolution dataset. Recently, certain commercial algorithms capable of correcting MRI distortions to

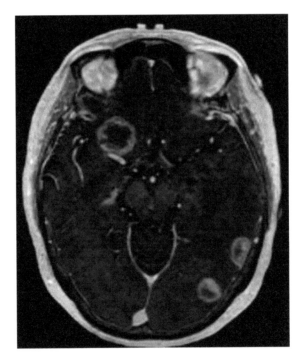

Figure 7.1 A 79-year-old never-smoker with a history of stage IV, T1N2M1 EGFR-mutated adenocarcinoma of the lung managed with osimertinib developed metastatic disease involving the CNS that resulted in her having head-aches and experiencing falls. An MRI of the brain revealed multiple intracranial foci consistent with metastatic disease. A dominant lesion which had significant peritumoral edema was identified in the inferior right frontal lobe measuring 3.6 × 2.9 × 2.5 cm. She was taken to the operating room where a right frontotemporal crani-otomy with a 95% resection of the metastasis was performed (tumor adherent to the M1 segment of the middle cerebral artery could not be safely excised). Three weeks later, she was treated with HSRS to the resection bed and five additional unresected metastases. A dose of 25 Gy in 5 fractions was used for the right frontal metastasis resection bed and 30 Gy in 5 fractions for the unresected metastases. The proximity of the right optic nerve to the resection bed precluded higher doses being delivered to this site. This reconstructed axial image from her preoperative MRI shows the right frontal tumor in contact with the middle cerebral artery with mass effect on the optic apparatus as well as two additional metastases in the left occipital and posterior left temporal lobes.

deformably match the CT scan are being sold. Great care must be used in applying these corrections; with careful quality assurance procedures and high-resolution matching, they may not be significant.

During CT simulation, the patient is positioned supine on the simulator couch with arm position dictated by patient comfort. A knee pillow is provided, and the patient is straightened and aligned immediately prior to frameless stereotactic mask fabrication. Sufficient time to permit the mask to cool and become rigid will help assure patient comfort when the patient returns for treatment. A thermoplastic mask specifically designed for linear accelerator-based cranial radiosurgery is the contemporary standard of care—masks that lack the required structural rigidity will not permit submillimetric targeting accuracy. Investigations into the use of surface-guided radiation therapy technology to guide intracranial radiosurgery may provide benefit for selected patients, but this is still a developing technological approach (Covington et al., 2019; Swinnen et al., 2020).

7.2.4 3D RIGID IMAGE FUSION

Carrying out accurate rigid image registration is a critical next step. High-resolution scans are a prerequisite, of course. Algorithms in commercial software packages provide acceptable registrations between MRI and CT for intracranial radiosurgery. The goal of the process is to carry out appropriate translations and rotations of the MRI study so that the voxels of the MRI scan can then be reformatted and displayed with the geometry of the treatment planning CT scan (Bond et al., 2003). The results should be verified by a physicist or physician well-versed in neuroanatomy.

As a quality assurance measure, it is also essential to first check the dates of all scans to be certain that the correct CT/MRI combinations are being fused (postoperative vs preoperative MRI or correct CT simulation date in the case of multiple past treatments at the same institution). Fusion of the T1-weighted, gadolinium-enhanced volumetric acquisition may be initiated manually with an emphasis on alignment at the site of disease (anterior, middle, or posterior cranial fossa). This is not always necessary; tools that delineate a cuboidal region of interest "box" can be placed to encompass all relevant structures including tumor and adjacent bony or parenchymal anatomy and aligned using an automated algorithm such as normalized mutual information to rigidly align the two 3D studies. Fine or coarse adjustments to guide and refine the fusion process can be made accordingly until an accurate 3D alignment has been achieved. For example, for a patient with a vestibular schwannoma, the intracanalicular portion of the tumor should fit perfectly into the bony internal auditory canal (Poetker et al., 2005).

The fusion should be verified in its entirety using one or more tools such as a "spyglass" or contrasting color phases to ensure accurately matching tumor volume and adjacent bony anatomy (Figure 7.2). No matter who performs this process, all aspects of a 3D rigid fusion should be independently verified by the radiation oncologist prior to segmentation of the target volume and normal structures. Failure to do so may potentially squander valuable time if errors are subsequently identified before treatment and risks poorer tumor control and an increased risk of injury to the patient if errors inherent in this process are not identified prior to treatment.

7.2.4.1 Fusion Summary

1. SRS protocol contrast-enhanced MRI
2. T1-weighted postgadolinium volumetric MRI fused to dedicated treatment planning CT obtained using a customized immobilization mask
3. Set localization box for guiding fusion to include tumor and/or postoperative cavity as well as adjacent fixed skull base or calvarial anatomy
4. Manually align to roughly approximate scans in all three planes
5. Autoalign
6. Verify fusion at multiple regions surrounding target area with spy glass window and with contrasting color phases before saving under name to include MRI and CT-simulation dates and initials of the operator
7. Physician (time-out) to independently verify dates of all MRIs and CT simulation are again correct and belong to the correct patient and that the fusion is geometrically precise in the volume of greatest interest before contouring

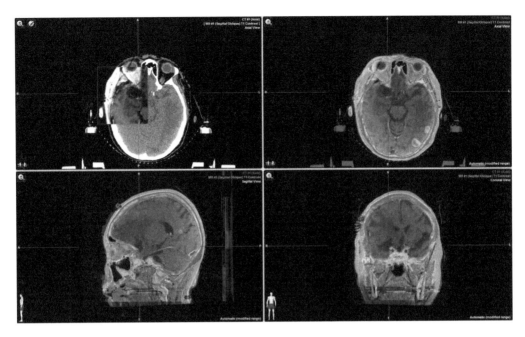

Figure 7.2 Fusion of the T1 post-spoiled gradient recalled (SPGR) diagnostic contrast-enhancing MRI sequence with the CT simulation scan is shown here using iPlan® treatment planning software (BrainLAB AG, Munich, Germany). The "image fusion mode" is utilized to verify the image fusion between the stereotactic MRI and the treatment planning CT scan. With tools such as blue/amber phasing and the rectangular spyglass (top left panel), special attention is paid to the degree of bony anatomy alignment between CT (blue) and MRI (amber), which can be faded in and out to verify close agreement especially with respect to the left temporal region where the tumor resection bed is located in this case.

7.2.5 TARGET VOLUME DELINEATION

During the target delineation phase, the GTV is outlined by the radiation oncologist and all critical structures (i.e., brain parenchyma and vessels, cochlea, optic apparatus, brainstem, and spinal cord), whether autosegmented or manually segmented by residents or physics staff, are also adjusted and verified. It is important, rather than relying on written reports alone, to review the proposed GTV with a neurosurgeon and a neuroradiologist before finalizing target delineation especially if there is any uncertainty as to the extent of postoperative change and residual disease. In this situation, postoperative pre- and postcontrast scans can be used to define the GTV more accurately. Additionally, when there is difficulty in identifying the exact location of a postoperative bed, including a preoperative contrast-enhanced MRI scan can help confirm that the correct volume is targeted. In post-operative cases, the involved neurosurgeon can add useful information regarding areas of particular concern for recurrence. The GTV is defined to include all areas of suspicious enhancement plus the postoperative cavity in the case of prior resection.

A GTV expansion of 1 mm has been shown to be adequate in achieving an effective PTV when using six-degree-of-freedom registration with a frameless radiosurgical technique (Figures 7.3 and 7.4) (Dhabaan *et al.*, 2012; Prabhu *et al.*, 2013). Furthermore, data support the notion that small expansions of 1–2 mm to arrive at a PTV does not result in significantly increased marginal recurrence rates compared to infield recurrences after hypofractionated SRS to treat brain metastases (Eaton *et al.*, 2013). A phase III trial evaluating 1 and 3 mm PTV expansions for single-fraction SRS identified higher rates of radionecrosis with 3 mm PTV expansions (Kirkpatrick *et al.*, 2015). Notably, volume increases proportionally as the cube of the radius ($V = 4/3 \pi r_1 r_2 r_3$), and isometrically increasing the radius of a TV by a few millimeters increases the volume of the final PTV at a startling rate.

The implementation of PTV margins to the GTV was founded on the notion that highly conformal plans increase the risk of marginal miss secondary to difficulty defining the postoperative cavity (Soltys

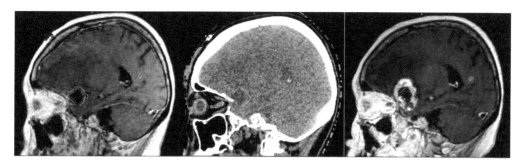

Figure 7.3 The fused postoperative MRI scan (left) with the resection bed (orange contour) and a 2.0 mm PTV margin (fluorescent green) shown in a sagittal image. Because of uncertainty about the accuracy of delineation of the resection bed that could not be resolved by referring to the CT scan (center), an additional fusion was requested of the preoperative MRI scan (right) to help confirm the position of the postoperative resection bed by showing the preoperative tumor extent (red). More posteriorly, a smaller, unresected metastasis is shown with a purple GTV contour and a blueish green PTV contour (1.0 mm PTV margin). Edema can be seen within the left frontal lobe on both the pre- and postoperative MRI scans.

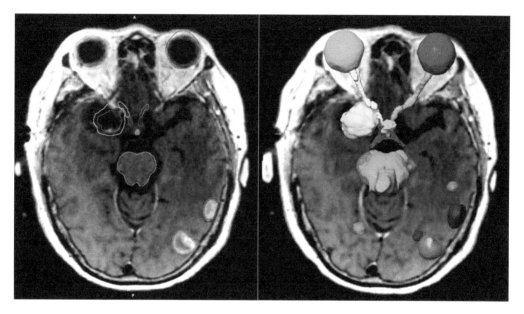

Figure 7.4 Every individual tumor volume and critical normal tissues should be contoured on the co-registered postoperative MRI scan (left), with assistance, when needed, from the CT scan and other co-registered volumetric imaging studies. In these images, the right and left optic nerves are shown in yellow and orange, the anterior-most portion of the chiasm in pink, the pituitary stalk in sky blue, the brainstem in green, and different colors were selected for each individual metastasis and its corresponding PTV. The axial postoperative tumor bed shows the middle cerebral artery at its most posterior edge. The operative note indicated that this artery had tumor adherent to it that could not be resected, and it is appropriately contained within the GTV, along with all contrast-enhancing postoperative parenchymal tissue and the corresponding fluid-filled cavity. The differences in PTV margins (2.0 mm margin for the resection bed and 1.0 mm for unresected metastases) are clearly seen on the left image. The MR image on the right shows the contoured PTV and critical normal structures from a caudad perspective.

et al., 2008) and the recognition that there are some postoperative shifts that may affect targeting between when imaging studies are performed and when treatment is delivered. Based on studies from Stanford University, when treating a postoperative resection bed, we expand the margin by 2.0 mm as this was shown to improve 1-year local control rates from 84% to 97% without a significant increase in toxicity (3% with a 2-mm margin versus 8% with no margin) (Choi *et al.*, 2012). We always confirm the GTV and

PTV with the neurosurgeon involved with the case both as a courtesy and to confirm that we have accurately defined these volumes and relevant critical normal tissues.

7.2.5.1 Target Delineation Summary

1. GTV = Contrast-enhancing gross (or residual disease to include operative cavity if postoperative) based on T1-weighted postcontrast volumetric MRI sequence
2. PTV = GTV + 1 mm (or 2 mm for postoperative resection beds)

7.2.6 PRESCRIPTION AND DOSE-FRACTIONATION SELECTION

Radiosurgical dose selection is primary dictated by tumor (target) volume. Larger GTVs are associated with higher with local recurrence rates (Eaton *et al.*, 2013). Additionally, the volume receiving 10 Gy or 12 Gy (V10 or V12) is important validated predictor of symptomatic radionecrosis (Sahgal *et al.*, 2009; Pinkham *et al.*, 2015). For roughly spherical lesions, the diameter derived from initial diagnostic imaging can be used in selecting an appropriate dose and fractionation. With increasing spatial complexity from spherical to irregularly shaped lesions, the volume of normal brain treated tends to increase, in part due to the difficulty with generating dose plans that adequately protect normal brain in concave areas of the TV. The volume of adjacent normal parenchymal tissue irradiated can be minimized by relying on highly conformal techniques employing noncoplanar arc-based intensity-modulated treatment delivery (Thomas *et al.*, 2018). HSRS should be strongly considered once the GTV/cavity surpasses a volume of approximately 14 cc, which is roughly equivalent to a GTV sphere with a 1.5 cm radius. The corresponding volume for a postoperative PTV (assuming a 2 mm isometric expansion of the GTV) for which HSRS should be used is approximately 20–21 cc.

7.2.7 BED CALCULATIONS

The most clinically applied radiobiological concept involves the biologically equivalent dose (BED), which can be thought of as a common 'currency' through which different dose–fractionation schedules are exchanged and compared (Table 7.1). A linear quadratic model is used to estimate the BED of the neoplastic or normal tissue cells by determining the alpha/beta (α/β) ratio, which represents the dose at which the linear and quadratic components of cell kill are equal (Hall and Giaccia, 2012). Due to their rapid growth, the α/β ratio for tumors is higher (α/β = 10) than that for senescent brain cells (α/β = 2–3), enabling hypofractionation to preferentially spare normal tissues while maintaining its therapeutic effect on tumor cells. Slower growing, benign neoplasms also possess lower α/β ratios. As an alternative to performing manual BED calculations, numerous online calculators and applications are available.

Table 7.1 Recommended dose constraints for relevant central nervous system organs at risk

ORGAN	CONSTRAINT	1 FRACTION	3 FRACTIONS	5 FRACTIONS	ENDPOINT (> GRADE 3 TOXICITY)
Optic nerve/ optic chiasm	Volume	<0.2 cc	<0.2 cc	<0.2 cc	Neuritis
	Volume max	8 Gy	15.3 Gy (5.1 Gy/fraction)	23 Gy (4.6 Gy/fx)	
	Max point dose*	10 Gy	17.4 Gy (5.8 Gy/fx)	25 Gy (5 Gy/fx)	
Cochlea	Max point dose	9 Gy	17.1 Gy (5.7 Gy/fx)	25 Gy (5 Gy/fx)	Hearing loss
Brainstem	Volume	<0.5 cc	<0.5 cc	<0.5 cc	Cranial neuropathy
	Volume max	10 Gy	18 Gy (6 Gy/fx)	23 Gy (4.6 Gy/fx)	
	Max point dose	15 Gy	23.1 Gy (7.7 Gy/fx)	31 Gy (6.2 Gy/fx)	

*"Point" defined as 0.035 cc or less.
Source: Timmerman, 2008.

7.3 DOSIMETRY

7.3.1 BEAM GEOMETRY AND PLANNING

The ultimate goal in radiosurgical treatment planning is simply to cover the PTV with as close to 100% of the prescription dose as possible while, at the same time, maintaining a very sharp dose falloff at the PTV edge to preserve adjacent normal brain and critical structures. Many SRS delivery techniques described here and in other chapters have been devised to further achieving this goal. These include traditional approaches such as step-and-shoot fixed gantry methods and arc-based delivery approaches that have both recently been adapted to include intensity modulation.

For smaller, roughly spherical lesions, circular collimators can be used to generate noncoplanar, arc-based plans that have very steep falloff of dose outside the TV. Dynamic arc therapy with high-resolution multileaf collimators can also generate highly conformal plans, but the dose falloff is not quite as steep because of the finite size of collimator leaves in adapting to lesion geometry. Novel techniques such as volumetric intensity-modulated arc-based therapy (VMAT) have been devised to improve conformality while not sacrificing the speed of treatment delivery (Figure 7.5). This approach achieves conformal dose

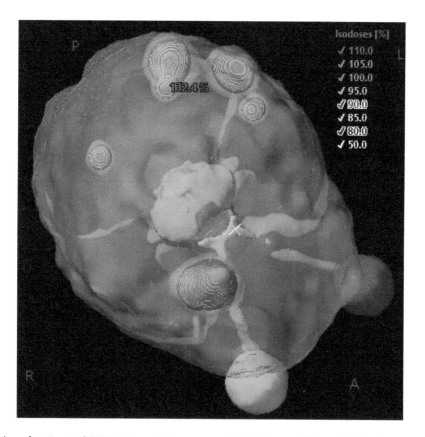

Figure 7.5 A hypofractionated SRS treatment plan generated with Elements™ cranial SRS planning software (BrainLAB AG, Munich, Germany) using 20 arcs (10 noncoplanar arcs treated in a back-and-forth approach) and 6 MV flattening filter-free photons with a single isocenter is visualized from a top–down (cranial) perspective. The display of the plan was taken from the Eclipse™ treatment planning system from Varian Medical Industries (Palo Alto, CA) This approach was chosen to expeditiously deliver conformal doses to all of the planning target volumes (PTV) with minimal incidental dose delivery to normal tissues. A total dose of 30 Gy in 5 fractions was delivered to the 100% isodose surface of the unresected metastases PTVs and 25 Gy in 5 fractions to the 100% isodose surface of the left inferior frontal PTV using a Novalis TX® stereotactic linear accelerator (Varian Medical Industries, Palo Alto, CA; BrainLAB AG, Munich, Germany). Isodose surfaces of 50% (15 Gy) and up are depicted in this image. The hot spot of 132.4% (39.7 Gy) was located within one of the two closely spaced left occipital metastases. 18.4 cc of the brain (including the PTVs) got a dose of 4 Gy per fraction.

distributions through simultaneously modulating the photon dose rate, gantry rotation speed, and beam shape/aperture through multileaf collimation (Wang *et al.*, 2012; Zhao *et al.*, 2015). This technique has been shown to offer much higher degrees of target conformality and lower dose to normal brain, particularly when noncoplanar VMAT arcs are used to deliver treatment (Zhao *et al.*, 2015).

Field/arc placement should be determined with respect to the PTV location and its spatial relationship to critical structures. As an example, it may be prudent when feasible to avoid placing beams that would enter or exit through the eye/optic apparatus, cochlea, or brainstem during treatment, but inverse treatment planning approaches can minimize dose delivery to the critical normal tissues so delineated, or the degrees of arc through which treatment is delivered can be increased so as to render the incidental dose delivered to these structures trivial.

There are several advantages and disadvantages to using coplanar versus noncoplanar arcs. Advantages of noncoplanar arcs include the ability to improve plan conformality at both high and lower isodose levels and, if desirable, dose homogeneity. One potential drawback for these types of plans is the inability to image when the table is at non-zero positions. If the treating center does not have the capability to monitor the patient after table kicks, then table inaccuracies or patient motion can affect the quality of the delivery. For this reason, it is ideal to have a system that can verify patient position at nonzero table positions. Surface guidance may provide the ability to confirm patient position accuracy (Covington *et al.*, 2019; Swinnen *et al.*, 2020).

If the center does not have this capability, then more rigid immobilization may be necessary and more lengthy quality control measures (e.g., Winston Lutz or couch angle) should be undertaken. For these reasons, we use stereotactic image guidance. The kV pairs we employ are not gantry-mounted imaging systems and can acquire stereotactic 2D x-rays without any concern for collision with the patient or couch regardless of the table angle. This is in contrast to gantry-mounted kV/kV pairs, kV/MV pairs or CBCT, which can only be acquired at a couch angle of zero. Ideally, if CBCT will be used for patient setups using noncoplanar treatments, then some secondary system will be used for patient motion monitoring (such as optical guidance systems). Additionally, CBCT utilizes significantly more radiation than a pair of 2D x-rays, which should be taken into account as that higher dose of radiation delivered with the CBCT is largely being delivered to normal tissues.

7.3.2 PLAN ASSESSMENT AND QUALITY ASSURANCE

When reviewing a plan, it is the role of the physician to scrutinize the following:

1. Heterogeneity of coverage: Heterogeneity ('hot spots') is favored if the target volume encompasses little normal tissue. Homogeneity is preferred if there are critical normal tissues that are included in the PTV—this should minimize the risk of consequential damage to these normal tissues
2. What isodose surface provides 100% coverage?
3. Size and location of "hot" and "cold" spots
4. Dose–volume histogram (DVH) constraints (may need to consider dosing from prior treatments, particularly to critical normal tissues)
5. Assess the degree of agreement between PTV and prescription isodose (PI) dose conformity indices

Heterogeneity of coverage (prescribing to a lower isodose surface) may be advantageous in terms of dose falloff outside the PTV and for intensifying dose delivery to tumor cells within the PTV. Similarly, having a small number of target voxels relatively underdosed may be appropriate, depending on the locations of intra-PTV hot spots, the treatment volume that would be irradiated to the prescription dose if these few voxels were included within the PI surface, and the risk of recurrence that these cold spots may introduce. Dose conformity indices can help with selection of PI surfaces. Different dose conformity indices exist, and each has advantages under different scenarios. The RTOG recommends that the PITV, defined as the ratio of the prescription isodose (PI) volume over the target volume (TV), be used; many others have been proposed to address inherent shortcomings of this approach (Knöös *et al.*, 1998; Paddick, 2000; Wagner *et al.*, 2003; Wu *et al.*, 2003; Feuvret *et al.*, 2006; Paddick and Lippitz, 2006).

7.3.3 DOSE–VOLUME HISTOGRAM CONSTRAINTS

The "volume receiving 12 Gy" or V_{12} has been widely adopted as the standard method of reporting the dose to normal brain parenchyma in single-fraction SRS procedures (Sahgal *et al.*, 2009; Pinkham *et al.*, 2015;

Lawrence *et al.*, 2010). This approach has also been adapted for reporting outcomes using HSRS, providing a common endpoint to compare single and multi-fraction SRS. As an example, in a case prescribed to a nominal dose of 35 Gy delivered over 5 fractions (7 Gy delivered per fraction), the normal brain volume (defined as the normal brain minus the GTV) receiving more than 4 Gy/fraction ideally will not exceed 20 cc (Ernst-Stecken *et al.*, 2006). As mentioned earlier, however, different locations may possess different tolerances, and therefore the treated location should be defined and reported (e.g., midbrain versus frontal cortex) for subsequent evaluation and refinement of treatment prescription philosophy. There are, as yet, no published and generally accepted constraint doses for the hippocampi when radiosurgery is employed for intracranial targets. The steep fall-off of the high-dose volume outside the PTV will help protect the hippocampi, but what the impact of partial hippocampal irradiation to moderate doses is on neurocognition is still unknown. It is likely there are different impacts for left versus right and for anterior versus posterior locations in the hippocampi.

7.3.4 OPTIC APPARATUS CONSTRAINTS

The conglomerate imaging data garnered from CT-MRI fusion (already performed for TV definition) is the preferred approach to define and contour the optic structures (Table 7.1). The optic nerves can be clearly delineated on a CT scan, given their conspicuity within the fatty tissues of the orbit before they pass through the optic canals. Using the bony window will help delineate the optic nerve as it traverses the bony orbital apex. Intracranially, the volumetric MRI pulse sequences can clearly show the chiasm as a structure separate from the cerebrospinal fluid in the perichiasmatic cistern, intracranial arteries, and contiguous normal frontal cortex. The chiasm is sometimes easiest to appreciate on coronal imaging and can be used to locate the distal aspects of both optic nerves. It is advisable to contour the entire optic apparatus in continuity through the proximal optic tracts.

There is a relatively low risk of radiation-induced optic neuritis (RION) for a single-fraction D_{max} of less than 10 Gy, and one major study has indicated an acceptably low risk of RION with a single-fraction D_{max} of less than 12 Gy (Mayo *et al.*, 2010). In the case of hypofractionation, a maximum dose limit to the chiasm of 19.5 Gy delivered in 3 fractions or 25 Gy delivered in 5 fractions has been quoted in the literature based on retrospective case series (Grimm *et al.*, 2011). In the end, it is important to consider other clinical factors that may potentiate the risk of optic nerve damage from irradiation, such as age (Parsons *et al.*, 1994; Bhandare *et al.*, 2005) and a history of demyelinating disease (Daniels *et al.*, 2009) or comorbid conditions that may cause microvascular damage such as diabetes mellitus or uncontrolled hypertension. In such cases, choosing a milder dose fractionation regimen may be preferable to keep the dose to the anterior visual pathways below constraint tolerances.

7.3.5 COCHLEAR CONSTRAINTS

Cochlear toxicity risk is largely based on the proximity of the tumor volume to the petrous bone. Factors that warrant careful consideration during planning include the conformality of the treatment and protection of normal structures, including the brainstem and cochlea (Linskey, 2008). Clinical factors to consider when planning include concurrent use of ototoxic systemic therapies. It may be advisable to coordinate treatment between systemic therapy cycles (with time given for drug clearance). Notably, ototoxocity from systemic agents will likely present bilaterally, while radiotherapy toxicity will be limited to the treated side.

In the case of treating vestibular tumors, several factors will determine the risk and thus the potential for ipsilateral cochlear sparing (Table 7.1). Prior to treatment, a baseline audiometric evaluation should be obtained. For patients with useful hearing who will receive single-fraction SRS, the prescription dose to the margin of the vestibular schwannoma should be limited to ≤12 Gy to maximize the probability of hearing preservation.

A few studies on single-fraction SRS data show that limiting the mean cochlea dose to approximately 4 Gy will improve the probability of hearing preservation. From limited data, hypofractionation (especially in the case of acoustic neuroma) provides durable tumor control and effectively preserves hearing (Bhandare *et al.*, 2010; Hayden Gephart *et al.*, 2013), but there is a paucity of data demonstrating the optimal dose to minimize cochlea toxicity. After treatment, hearing should be tested with routine audiometry starting 6 months after RT and continuing twice yearly thereafter until any posttreatment changes become manifest and stabilize.

7.3.6 BRAINSTEM CONSTRAINTS

Constraining dose delivery to the brainstem is essential for posterior fossa lesions (Table 7.1). The dose delivered to the brainstem and the volume of brainstem treated relate importantly to complication probabilities. This location limits surgical interventions for radiation-induced toxicities, underscoring the need to minimize complications. For single-fraction SRS, a maximum brainstem dose of 12.5 Gy is associated with low (<5%) risk of complications. Higher doses (15.2 Gy) to small volumes have been used with a low reported incidence of complications in patient groups with a poor prognosis for long-term survival (e.g., brainstem metastases) (Kased *et al.*, 2008; Lorenzoni *et al.*, 2009; Lehrer *et al.*, 2020).

REFERENCES

Albert FK, Forsting M, Sartor K, Adams HP, Kunze S (1994) Early postoperative magnetic resonance imaging after resection of malignant glioma: Objective evaluation of residual tumor and its influence on regrowth and prognosis. *Neurosurgery* 34:45–60.

Angelov L. Mohammadi AM, Bennett EE, Abbassy M, Elson P, Chao ST, Montgomery JS, *et al.* (2018) Impact of 2-staged stereotactic radiosurgery for treatment of brain metastases ≥2 cm. *Journal of Neurosurgery* 129:366–382.

Atalar B, Choi CY, Harsh GR, Chang SD, Gibbs IC, Adler JR, Soltys SG (2013) Cavity volume dynamics after resection of brain metastases and timing of postresection cavity stereotactic radiosurgery. *Neurosurgery* 72:180–185.

Benedict SH, Yenice KM, Followill D, Galvin JM, Hinson W, Kavanagh B, Keall P, et al. (2010) Stereotactic body radiation therapy: The report of AAPM task group 101. *Medical Physics* 37(8):4078–4101.

Bhandare N, Jackson A, Eisbruch A, Pan CC, Flickinger JC, Antonelli P, Mendenhall WM (2010) Radiation therapy and hearing loss. *International Journal of Radiation Oncology, Biology, Physics* 76:S50–57.

Bhandare N, Monroe AT, Morris CG, Bhatti MT, Mendenhall WM (2005) Does Altered Fractionation Influence the Risk of Radiation-Induced Optic Neuropathy? *International Journal of Radiation Oncology, Biology, Physics* 15;62(4):1070–1077.

Blonigen BJ, Steinmetz RD, Levin L, Lamba MA, Warnick RE, Breneman JC (2010) Irradiated volume as a predictor of brain radionecrosis after linear accelerator stereotactic radiosurgery. *International Journal of Radiation Oncology, Biology, Physics* 77: 996–1001.

Bond JE, Smith V, Yue NJ, Knisely JPS (2003) Comparison of an image registration technique based on normalized mutual information with a standard method utilizing implanted markers in the staged radiosurgical treatment of large arteriovenous malformations. *International Journal of Radiation Oncology, Biology, Physics* 57:1150–1158.

Brezovich IA, Wu X, Popple RA, *et al.* (2019) Stereotactic radiosurgery with MLC-defined arcs: Verification of dosimetry, spatial accuracy, and end-to-end tests. *Journal of Applied Clinical Medical Physics* 20(5):84–98.

Choi CYH, Chang SD, Gibbs IC, Adler JR, Harsh GR, Lieberson RE, Soltys SG (2012) Stereotactic radiosurgery of the postoperative resection cavity for brain metastases: Prospective evaluation of target margin on tumor control. *International Journal of Radiation Oncology, Biology, Physics* 84:336–342.

Covington EL, Fiveash JB, Wu X, Brezovich I, Willey CD, Riley K, Popple RA (2019) Optical surface guidance for submillimeter monitoring of patient position during frameless stereotactic radiotherapy. *Journal of Applied Clinical Medical Physics* 20(6):91–98.

Daniels TB, Pollock BE, Miller RC, Lucchinetti CF, Leavitt JA, Brown PD (2009) Radiation-induced optic neuritis after pituitary adenoma radiosurgery in a patient with multiple sclerosis. *Journal of Neuro-Oncology* 93:263–267.

Dhabaan A, Schreibmann E, Siddiqi A, Elder E, Fox T, Ogunleye T, Esiashvili N, Curran W, Crocker I, Shu HK (2012) Six degrees of freedom CBCT-based positioning for intracranial targets treated with frameless stereotactic radiosurgery. *Journal of Applied Clinical Medicine Physics* 13:3916.

Eaton BR, Gebhardt B, Prabhu R, Shu H-K, Curran WJ, Crocker I (2013) Hypofractionated radiosurgery for intact or resected brain metastases: Defining the optimal dose and fractionation. *Radiation Oncology* 8:135.

Ebner D, Rava P, Gorovets D, Cielo D, Hepel JT (2015) Stereotactic radiosurgery for large brain metastases. *Journal of Clinical Neuroscience* 22:1650–1654.

Ernst-Stecken A, Ganslandt O, Lambrecht U, Sauer R, Grabenbauer G (2006) Phase II trial of hypofractionated stereotactic radiotherapy for brain metastases: Results and toxicity. *Radiotherapy & Oncology* 81:18–24.

Feuvret L, Noël G, Mazeron JJ, Bey P (2006) Conformity index: A review. *International Journal of Radiation Oncology, Biology, Physics* 64:333–342.

Flickinger JC, Kondziolka D, Pollock BE, Maitz AH, Lunsford LD (1997) Complications from arteriovenous malformation radiosurgery: Multivariate analysis and risk modeling. *International Journal of Radiation Oncology, Biology, Physics* 38:485–490.

Forsyth PA, Petrov E, Mahallati H, Cairncross JG, Brasher P, MacRae ME, Hagen NA, Barnes P, et al. (1997) Prospective study of postoperative magnetic resonance imaging in patients with malignant gliomas. *Journal of Clinical Oncology*15:2076–2081.

Galluzzi L, Vanpouille-Box C, Bakhoum SF, Demaria S (2018) SnapShot: CGAS-STING Signaling. *Cell*173(1):276–276.e1.

Golden EB, Marciscano AE, Formenti SC (2020) Radiation and *in-situ* tumor vaccination. *International Journal of Radiation Oncology, Biology, Physics* 108:3–5.

Grimm J, LaCouture T, Croce R, Yeo I, Zhu Y, Xue J (2011) Dose tolerance limits and dose volume histogram evaluation for stereotactic body radiotherapy. *Journal of Applied Clinical Medicine Physics* 12:3368.

Hall EJ, Brenner DJ (1993) The radiobiology of radiosurgery: Rationale for different treatment regimes for AVMs and malignancies. *International Journal of Radiation Oncology, Biology, Physics* 25:381–385.

Hall EJ, Giaccia AJ (2012) *Radiobiology for the Radiologist*. 7th ed. Philadelphia: Lippincott Williams & Wilkins.

Halvorsen PH, Cirino E, Das IJ, Garrett JA, Yang J, Yin F-F, Fairobent LA (2017) AAPM -RSS medical physics practice guideline 9.a. for SRS -SBRT. *Journal of Applied Clinical Medicine Physics* 18:10–21.

Hamel-Perreault E, Mathieu D, Masson-Cote L. (2019) Factors influencing the outcome of stereotactic radiosurgery in patients with five or more brain metastases. *Current Oncology* 26(1):e64–69.

Hayden Gephart MG, Hansasuta A, Balise RR, Choi C, Sakamoto GT, Venteicher AS, Soltys SG et al. (2013) Cochlea radiation dose correlates with hearing loss after stereotactic radiosurgery of vestibular schwannoma. *World Neurosurgery* 80:359–363.

Higuchi Y, Serizawa T, Nagano O, Matsuda S, Ono J, Sato M, Iwadate Y, Saeki N (2009) Three-staged stereotactic radiotherapy without whole brain irradiation for large metastatic brain tumors. *International Journal of Radiation Oncology, Biology, Physics* 74:1543–1548.

International Atomic Energy Agency, Dosimetry of Small Static Fields Used in External Beam Radiotherapy, Technical Reports Series No. 483, IAEA, Vienna (2017).

Kased N, Huang K, Nakamura JL, Sahgal A, Larson DA, McDermott MW, Sneed PK (2008) Gamma Knife radiosurgery for brainstem metastases: The UCSF experience. *Journal of Neuro Oncology* 86:195–205.

Kim KH, Kong D, Cho KR, Lee MH, Choi J, Seol HJ, Kim ST, et al. (2019) Outcome evaluation of patients treated with fractionated Gamma Knife radiosurgery for large (>3 cm) brain metastases: A dose-escalation study. *Journal of Neurosurgery* 16;1–10.

Kirkpatrick JP, Wang Z, Sampson JH, McSherry F, Herndon JE, Allen KJ, DuffyE et al. (2015) Defining the optimal planning target volume in image-guided stereotactic radiosurgery of brain metastases: Results of a randomized trial. *International Journal of Radiation Oncology* 91:100–108.

Knöös T, Kristensen I, Nilsson P (1998) Volumetric and dosimetric evaluation of radiation treatment plans: Radiation conformity index. *International Journal of Radiation Oncology, Biology, Physics* 42:1169–1176.

Kushnirsky M, Nguyen V, Katz JS, Steinklein J, Rosen L, Warshall C, Schulder M, Knisely JP (2016) Time delayed contrast enhanced MRI improves detection of brain metastases and apparent treatment volumes. *Journal of Neurosurgery* 11:1–7.124:489–495.

Lawrence YR, Li XA, El NI, Hahn CA, Marks LB, Merchant TE, Dicker AP (2010) Radiation dose-volume effects in the brain. *International Journal of Radiation Oncology, Biology, Physics* 76:S20–27.

Lehrer EJ, Snyder MH, Desai BD, i CE, Narayan A, Trifiletti, DM, Schlesinger D, et al.(2020) Clinical and radiographic adverse events after Gamma Knife radiosurgery for brainstem lesions: A dosimetric analysis. *Radiotherapy & Oncology* 147:200–209.

Levegrün S, Hof H, Essig M, Schlegel W, Debus J (2004) Radiation-induced changes of brain tissue after radiosurgery in patients with arteriovenous malformations: Correlation with dose distribution parameters. *International Journal of Radiation Oncology, Biology, Physics* 59:796–808.

Lightstone AW, Benedict SH, Bova FJ, Solberg TD, Stern RL (2005) Intracranial stereotactic positioning systems: Report of the American Association of Physicists in Medicine Radiation Therapy Committee Task Group no. 68. *Medical Physics* 32:2380–2398.

Linskey ME (2008) Hearing preservation in vestibular schwannoma stereotactic radiosurgery: What really matters? *Journal of Neurosurgery* 109(Suppl):129–136.

Lorenzoni JG, Devriendt D, Massager N, Desmedt F, Simon S, Van Houtte P, Brotchi J, Levivier M (2009) Brainstem metastases treated with radiosurgery: Prognostic factors of survival and life expectancy estimation. *Surgical Neurology* 71:188–196.

Mayo C, Martel MK, Marks LB, Flickinger J, Nam J, Kirkpatrick J (2010) Radiation dose-volume effects of optic nerves and chiasm. *International Journal of Radiation Oncology, Biology, Physics* 76:S28–35.

Minniti G, Scaringi C, Paolini S, Lanzetta G, Romano A, Cicone F, Mattia Osti M, *et al.*(2016) Single-fraction versus multifraction (3 × 9 Gy) stereotactic radiosurgery for large (>2 cm) brain metastases: A comparative analysis of local control and risk of radiation-induced brain necrosis. *International Journal of Radiation Oncology, Biology, Physics* 95:1142–1148.

Mori Y, Kondziolka D, Flickinger JC, Kirkwood JM, Agarwala S, Lunsford LD (1998) Stereotactic radiosurgery for cerebral metastatic melanoma: Factors affecting local disease control and survival. *International Journal of Radiation Oncology, Biology, Physics* 42:581–589.

Oermann EK, Kress MA, Todd JV, Collins BT, Hoffman R, Chaudhry H, Collins SP, Morris D, Ewend MG (2013) The impact of radiosurgery fractionation and tumor radiobiology on the local control of brain metastases. *Journal of Neurosurgery* 119:1131–1138.

Paddick I (2000) A simple scoring ratio to index the conformity of radiosurgical treatment plans. Technical note. *Journal of Neurosurgery* 93(Suppl 3):219–222.

Paddick I, Lippitz B (2006) A simple dose gradient measurement tool to complement the conformity index. *Journal of Neurosurgery* 105(Suppl):194–201.

Parsons JT, Bovis FJ, Fitzgerald CR, Mendenhall WM, Million RR (1994). Radiation optic neuropathy and megavoltage external-beam radiation: Analysis of time-dose factors. *International Journal of Radiation Oncology, Biology, Physics* 30(4):755–763.

Patel TR, McHugh BJ, Bi WL, Minja FJ, Knisely JPS, Chiang VL (2011) A comprehensive review of MR imaging changes following radiosurgery to 500 brain metastases. *American Journal of Neuroradiology* 32:1885–1892.

Pessina F, Navarria P, Cozzi L, Ascolese AM, Maggi G, Riva M, Masci G, D'Agostino G, Finocchiaro G, Santoro A, Bello L, Scorsetti M. (2016) Outcome evaluation of oligometastatic patients treated with surgical resection followed by hypofractionated stereotactic radiosurgery (HSRS) on the tumor bed, for single, large brain metastases. *PLoS One* 27;11(6):e0157869.

Pinkham MB, Whitfield GA, Brada M (2015) New developments in intracranial stereotactic radiotherapy for metastases. *Clinical Oncology (R Coll Radiol)* 27(5):316–323.

Poetker DM, Jursinic PA, Runge-Samuelson CL, Wackym PA (2005) Distortion of magnetic resonance images used in Gamma Knife radiosurgery treatment planning: Implications for acoustic neuroma outcomes. *Otology & Neurotology* 26:1220–1228.

Prabhu RS, Dhabaan A, Hall WA, Ogunleye T, Crocker I, Curran WJ, Shu H-JS (2013) Clinical outcomes for a novel 6 degrees of freedom image guided localization method for frameless radiosurgery for intracranial brain metastases. *Journal of Neuro-Oncology* 113:93–99.

Sahgal A, Ma L, Chang E Shiu A, Larson DA, Laperriere N, Yin F, *et al.* (2009) Advances in technology for intracranial stereotactic radiosurgery. *Technology in Cancer Research Treatment* 8(4):271–280.

Salkeld AL, Hau EKC, Nahar N, Sykes JR, Wang W, Thwaites DI. (2018) Changes in brain metastasis during radiosurgical planning. *International Journal of Radiation Oncology, Biology, Physics* 102:727–733.

Schell MC, Bova FJ, Larson DA Leavitt DD, Lutz WR, Podgorsak EB, Wu A. (1995) "Stereotactic Radiosurgery Report of Task Group 42 Radiation Therapy Committee" AAPM Report No. 54. Published for the American Association of Physicists in Medicine by the American Institute of Physics, Woodbury, NY.

Shaw E, Scott C, Souhami L, Dinapoli R, Kline R, Loeffler J, Farnan N (2000) Single dose radiosurgical treatment of recurrent previously irradiated primary brain tumors and brain metastases: Final report of RTOG protocol 90–05. *International Journal of Radiation Oncology, Biology, Physics* 47(2):291–298.

Shiau CY, Sneed PK, Shu HK, Lamborn KR, McDermott MW, Chang S, Nowak P *et al.*(1997) Radiosurgery for brain metastases: Relationship of dose and pattern of enhancement to local control. *International Journal of Radiation Oncology, Biology, Physics* 37:375–383.

Sneed PK, Mendez J, Vemer-van den Hoek JG, *et al.* (2015) Adverse radiation effect after stereotactic radiosurgery for brain metastases: Incidence, time course, and risk factors. *Journal of Neurosurgery* 123(2):373–386.

Soliman H, Myrehaug S, Tseng C, Ruschin M, Hashmi A, Mainprize T, Spears J, et al. (2019) Image-guided, linac-based, surgical cavity-hypofractionated stereotactic radiotherapy in 5 daily fractions for brain metastases. *Neurosurgery* 1;85(5):E860–869.

Soltys SG, Adler JR, Lipani JD, Jackson PS, Choi CYH, Puataweepong P, White S, *et al.* (2008) Stereotactic radiosurgery of the postoperative resection cavity for brain metastases. *International Journal of Radiation Oncology, Biology, Physics* 70(1):187–193.

Soltys SG, Seiger K, Modlin LA, Gibbs IC, Hara W, Kidd EA, Hancock SL, *et al.*(2015) A phase I/II dose-escalation trial of 3-fraction stereotactic radiosurgery (SRS) for large resection cavities of brain metastases. *International Journal of Radiation Oncology, Biology, Physics* 93:S38.

Sun J, Barnes M, Dowling J, Menk F, Stanwell P, Greer PB (2014) An open source automatic quality assurance (OSAQA) tool for the ACR MRI phantom. *Australasian Physical and Engineering Sciences in Medicine* 38:39–46. Available at: http://link.springer.com/10.1007/s13246-014-0311-8.

Swinnen ACC, Öllers MC, Loon Ong C, Verhaegen F (2020) The potential of an optical surface tracking system in non-coplanar single isocenter treatments of multiple brain metastases. *Journal of Applied Clinical Medical Physics* 21(6):63–72.

Thomas EM, Popple RA, Bredel M, Fiveash JB (2018) *Linac-Based Stereotactic Radiosurgery and Hypofractionated Stereotactic Radiotherapy*. Adult CNS Radiation Oncology. Eds. Cham, Switzerland: Springer International Publishing AG, 639–663.

Timmerman RD (2008) An overview of hypofractionation and introduction to this issue of seminars in radiation oncology. *Seminar in Radiation Oncology* 18:215–222.

Vanpouille-Box C, Alard A, Aryankalayil MJ, Sarfraz Y, Diamond JM, Schneider RJ, Inghirami G *et al.* (2017) DNA exonuclease Trex1 regulates radiotherapy-induced tumour immunogenicity. *Nature Communications* 8:15618.

Wagner TH, Bova FJ, Friedman WA, Buatti JM, Bouchet LG, Meeks SL (2003) A simple and reliable index for scoring rival stereotactic radiosurgery plans. *International Journal of Radiation Oncology, Biology, Physics* 57:1141–1149.

Wang JZ, Pawlicki T, Rice R, Mundt AJ, Sandhu A, Lawson J, Murphy KT (2012) Intensity-modulated radiosurgery with RapidArc for multiple brain metastases and comparison with static approach. *Medical Dosimetry* 37:31–36.

Ward JF (1994) The complexity of DNA damage: Relevance to biological consequences. *International Journal of Radiation Biology* 66(5):427–432.

Wu Q-RJ, Wessels BW, Einstein DB, Maciunas RJ, Kim EY, Kinsella TJ (2003) Quality of coverage: Conformity measures for stereotactic radiosurgery. *Journal of Applied Clinical Medical Physics* 4:374–381.

Yomo S., Hayashi M. Nicholson C (2012) A prospective pilot study of two-session Gamma Knife surgery for large metastatic brain tumors. *Journal of Neuro-Oncology* 109:159–165.

Zhao B, Yang Y, Li X, Li T, Heron DE, Huq MS (2015) Is high-dose rate RapidArc-based radiosurgery dosimetrically advantageous for the treatment of intracranial tumors? *Medical Dosimetry* 40:3–8.

Principles of Image-Guided Hypofractionated Radiotherapy of Spine Metastases

Johannes Roesch, Stefan Glatz, and Matthias Guckenberger

Contents

8.1 INTRODUCTION

With its increasing use beginning with cranial tumors, stereotactic radiotherapy (SRT) has successively opened up a broad field of indications. In 2011, a survey among radiation oncologists in the United States showed that 39% of the physicians were using stereotactic body radiotherapy (SBRT) for the treatment of spinal tumors. Both SBRT users and nonusers were planning to increase the number of SBRT treatments (Pan et al., 2011). Likewise, a survey in six European countries reported that one-third of the centers were practicing SBRT for vertebral tumors and half of the centers considered the currently available evidence as sufficient for routine treatment outside of clinical trials (Dahele et al., 2015). New techniques and technology have opened the way to improve old and achieve new objectives of spinal irradiation. Pain relief and prevention/reduction of morbidities from bone metastases, like neurological deficits or vertebral instability, are still the main goals (Lutz et al., 2011). Furthermore, disease control in selected oligometastatic patients and delay of systemic disease progression are emerging indications. Recent developments and growing experience in this field have led to recommendations particularly pertaining to delineation and image-guided treatment delivery.

8.2 PATIENT SELECTION

There is no controversy that SBRT of spine metastasis is associated with a higher workload and increased cost compared to conventional RT. More rigid patient immobilization and eventually longer treatment time might cause pain and more stress to the patient. It is therefore of utmost importance to preselect patients who do benefit from the added advantages of SBRT. There are, in particular, high local control rates of up to 80% to 90%, especially in long-term survivors (Husain et al., 2017; Moussazadeh et al., 2015). Furthermore, pain response appears being earlier and more durable compared to conventional RT (Sprave et al., 2018).

After checking for feasibility of SBRT, the basis hereby is to identify and become aware of the treatment aim within the individual situation. In general, three main goals can be prioritized: 1. Prevention/treatment of neurological deficits 2. Prolonged local control/pain control 3. Prolonged overall survival in non-metastasized or oligometastatic cases.

First, for patients with spinal instability, spinal cord/thecal sac/nerve root compression, and/or severe acute possibly reversible neurological deficits, a multidisciplinary approach is necessary. The same applies to patients without neurologic symptoms but imminent spinal cord compression due to possible fracture after therapy. This might be the case for lytic lesion, as these are known to be associated with compression fractures following SBRT (Sahgal et al., 2013b; Tseng et al., 2018). Today, surgery still provides the fastest way of central nervous tissue decompression or spinal stabilization, and all patients should be evaluated for their operability. Surgery after SBRT, although feasible in the hands of an experienced neurosurgeon, might be associated with more complicated surgical conditions (Roesch et al., 2017; Wise et al., 1999). A preoperative neurological deficit due to close proximity of the tumor to neural structures is also a factor increasing the risk for surgical complications, indicating that these are high-risk patients in general (Finkelstein et al., 2003; Pascal-Moussellard et al., 1998). One promising strategy is separation surgery which involves circumferential decompression of central nervous tissue by epidural disease which effectively creates "space" between the disease and spinal cord before SBRT: this small surgical intervention might be associated with a favorable toxicity profile and then allow more effective SBRT for the residual disease (Bate et al., 2015; Rothrock et al., 2020; Turel et al., 2017).

Regarding SBRT alone in patients with metastatic spinal cord compression, these patients are usually excluded from clinical trials, as are patients with high-grade epidural disease (Guckenberger et al., 2012; Ryu et al., 2019). Nonetheless, there is retrospective data that SBRT of metastases with epidural disease extensions and even symptomatic spinal cord compression is feasible and beneficial (Guckenberger et al., 2014; Lee et al., 2014). Ryu et al. (2010)(Ryu et al., 2010) reported a series where 85 lesions in 62 patients with epidural compression, but with a motor strength of at least four out of five, were treated with single-fraction SBRT. With reported neurological function improvement of 81%, SBRT appears to have an advantage compared to conventional irradiation in these cases. It should be noted that the evaluation of these patients is a challenge as they often are in palliative care and, upon relapse, are less likely to be aggressively evaluated given the severity of the neurologic deficits. Taken together, it is of utmost important to discuss these patients within a multidisciplinary team of experienced neurosurgeons and radiation oncologists, all with sufficient experience in management of spinal metastases (Glicksman et al., 2020).

Second, the patient should have a sufficiently long-life expectancy to potentially benefit from the proposed higher and prolonged local control and pain control rates of SBRT compared to conventional RT. This patient selection criterion is based on the absence of short-term benefits regarding pain relief of single-fraction SBRT in the recently reported NRG Oncology/RTOG 0631 trial (Ryu et al., 2019). The question whether patients will benefit from SBRT within his/her expected lifetime is important and simultaneously challenging as the large uncertainties of estimating metastatic patient's life expectancy are very well known. Indeed, OS is short with a median OS of 6 to 7 months in unselected patients treated with radiotherapy for painful vertebral metastases (Mizumoto et al., 2008; van der Linden et al., 2005). Several different scoring systems have been evaluated sharing similar prognostic factors. Not surprising, the Karnofsky/

ECOG Performance Status is a very strong predictor of OS. Furthermore, primary tumor type is an important factor influencing OS, with long life expectancy especially observed for breast cancer, prostate cancer, and lymphoma/myeloma patients (Mizumoto et al., 2008). As a matter of fact, the histopathological influence regarding local control is more under discussion. Whereas some studies reported lower LC for supposedly more radio-resistant tumor types, for example, NSCLC, renal cell carcinoma and melanoma (Guckenberger et al., 2014; Heron et al., 2012; Tseng et al., 2018), no such correlation was found in other studies (Garg et al., 2012; Laufer et al., 2013). Furthermore, the extent of metastatic spread and the time from primary diagnosis as indicators for disease activity appear to correlate with OS (Chao et al., 2012). Absence of visceral metastases and oligometastatic status by the time of treatment is associated with a better prognosis regarding OS (Barzilai et al., 2019). Importantly, most historical data for estimating OS of patients with spinal metastases were generated before wide-spread availability of systemic targeted therapy or immunotherapy options. Since the introduction of these systemic therapies, OS has improved substantially for, for example, malignant melanoma patients or NSCLC patients with high PD-L1 status or EGFR mutation, making such patients potential candidates for spinal SBRT.

Third, patients with oligometastatic disease might profit by local treatment of all visible lesions in terms of a prolonged OS. Recently, first results of a randomized controlled phase II trial comparing palliative standard of care systemic therapy with or without local stereotactic ablative radiotherapy to all visible lesions in patients with oligometastatic disease (SABR-COMET) were presented (Palma et al., 2019). This multicenter trial included 99 patients with a maximum of five metastatic lesions. With 28 months in the SOC arm compared to 41 months in the SBRT arm, median OS was significantly improved. Additionally, PFS was superior in the SBRT arm (6 months versus 12 months). About one-third of the patients in this study were treated for bone metastases, and more detailed information about the location of bone metastases is not available.

8.3 TARGET VOLUME AND ORGANS-AT-RISK DEFINITION

8.3.1 CLINICAL INFORMATION AND IMAGING MODALITIES

Similar to evaluation for other treatment modalities, the patient's general condition, symptomatology, and the physician's assessment provide useful information with regard to the patient's suitability for SBRT. Notably, localization and characteristics of the pain and neurological deficits like paresis, dysesthesia, autonomic dysfunction, or back pain resulting from compression of spinal nerves or the spinal cord itself can add important information on the location and extent of involvement of the vertebral metastases.

Diagnostic imaging is necessary to determine the precise local tumor extension and the overall disease burden of the spine and the entire body. Furthermore, it can help in patient selection, identification of possible contraindications, and treatment planning. In an international survey of five SBRT-experienced centers, spinal instability or neurologic deficit resulting from bony compression of neural structures was regarded as an exclusion criterion for SBRT (Guckenberger et al., 2011). Symptomatic spinal cord compression was unanimously regarded to be a contraindication for SBRT by all five centers. Epidural involvement alone influenced the decision for or against SBRT only in two centers. Compression fractures and vertebral instabilities were discussed with surgeons in all five institutions, and there was consensus that surgical intervention would be offered prior to SBRT, especially among the four North American centers. The RTOG 0631 study protocol excluded patients with compression fractures causing spinal instability, more than 50% loss of vertebral body height, as well as spinal cord compression or displacement or epidural compression within 3 mm of the spinal cord due to tumor or bony fragments (Ryu et al., 2014).

The whole spine is frequently assessed by T1- and T2-weighted sagittal magnetic resonance imaging (MRI) with additional pulse sequences and imaging planes, for example, short tau inversion recovery and axial T1/T2 sequences of suspicious vertebrae (Dahele et al., 2011). In cases where spinal cord compression is suspected, MRI is absolutely the preferred imaging technique (Loblaw et al., 2005). Pathological lesions present usually hypointense to fatty bone marrow on native T1-weighted images and present hyperintense when gadolinium-enhanced or T2-weighted imaging is used (Vanel, 2004). Bone scintigraphy, being another instrument to assess the disease burden of the spine, was previously described to be inferior to MRI (Algra et al., 1991).

^{18}F-FDG PET/CT will add valuable information especially regarding staging of the metastatic disease burden. Additionally, ^{18}F-FDG PET/CT may also provide valuable information about the local disease extent; however, its limited resolution needs to be considered. Similar benefits are expected for prostate cancer and PET/CT imaging using Choline and especially PSMA traces.

8.3.2 IMAGING IN TREATMENT PLANNING

The importance of high-quality imaging is critical, given the requirement for accurate target and normal tissue delineation and the steep dose gradients between organs at risk (OARs) such as spinal cord and cauda equina and the target volume.

The basic imaging modalities for treatment planning are widely agreed upon. The planning CT scan should be obtained with maximum 2 mm slice thickness in axial direction (Cox et al., 2012; Guckenberger et al., 2011). Contrast enhancement is recommended especially in cases of difficult tumor and OAR discrimination (Dahele et al., 2011). Although delineation of target volume and OAR without registration of additional diagnostic imaging such as MRI has been reported (Gibbs et al., 2007; Ryu et al., 2003), the full extent of the tumor may not be easily visible in the presence of bone destruction or paraspinal and epidural tumor spread. Furthermore, to visualize the spinal cord, CT alone is suboptimal, and spinal cord volumes can be better delineated by using MRI (Geets et al., 2005). Therefore, a dedicated planning MRI is common practice in many centers. T1 sequence with contrast enhancement and a slice thickness of 1–2 mm is reported to be the most commonly used (Guckenberger et al., 2011). Furthermore, T1 sequence without contrast, T2 sequence, and least commonly fluid-attenuated inversion recovery (FLAIR) sequences can help to define the exact tumor extent especially in cases where there is tumor infiltration of the surrounding soft tissue or the epidural space (Loughrey et al., 2000). Moreover, fat-suppressed or heavily T2-weighed sequences like fast spin echo (FSE) or constructive interference in steady state are used as myelographic MRI sequences and are used to define the spinal cord.

Standard diagnostic MRI studies are usually suitable for treatment planning, but reimaging is indicated when there are patient position discrepancies impacting the CT–MRI registration accuracy, surgery in the time interval between imaging and radiotherapy, and the development of new symptoms.

CT myelography can be used as an alternative to MRI but is more invasive and less available. Nonetheless, it was shown to provide excellent differentiation of the spinal cord within submillimeter accuracy (Thariat et al., 2009). Therefore, it should be considered in cases where MRI is not feasible due to claustrophobia, or presence of cardiac pacer, or when there is presence of metal implants. Potential side effects include headache, infection, neurological deficit, and allergic reaction. Another problem of CT myelography is that it might be impeded by contrast blockage due to the tumor or postoperative adhesions.

References to 4D CTs to account for breathing-related organ motion are rare (Nelson et al., 2009). Nelson et al. (2009) reported stable axial position of both the tumor and the spinal cord in respiration-correlated 4D CT and, therefore, did not recommend its use in standard practice. Along the same lines, these results were confirmed in a later analysis (Wang et al., 2016). Dieleman et al. (2007) described significant mobility especially of the distal esophagus within 8–9 mm. Similarly, Wang et al. (2019) described intrafraction motion increasing from 1 mm for the upper esophagus to 6.5 mm for the lower. This movement was found to cause a dose-blurring effect within the organ at risk itself, lowering the D_{2ccm} by 6.4%. On the other hand, the effect on D_{5ccm} was found to be also lower or higher, the latter accounting for movement in direction of the target volume. Fortunately, this did not lead to a violation of the maximum dose. The same applies for the interfraction motion.

In the postoperative setting, reduction of metal-related artifacts in CT and MRI can be achieved through various ways (Stradiotti et al., 2009). Orthopedic metal implants tend to absorb a high proportion of radiation from the CT scanner, causing incomplete or faulty projections. The artifacts are formed during image reconstruction and may be reduced by filter algorithms, adjustment of tube current, and others. These artifacts are related to the materials of x-ray beam attenuation coefficient, favoring titanium before stainless steel or cobalt chrome implants (Stradiotti et al., 2009). Artifacts in MRI occur due to magnetic field inhomogeneities, causing geometrical distortion on the resulting image. Avoidance strategies include optimal patient positioning in the center of the field, the usage of small voxel volumes, and metal artifact reduction sequences, such as FSE. Furthermore, the use of radiolucent carbon-fiber composite implants leads to a significant reduction of imaging artifacts, resulting in easier image registration, target delineation, and more accurate physical treatment planning (Tedesco et al., 2017).

PET imaging is currently not routinely used for tumor delineation in spinal SBRT. However, its use in retreatment of postoperative spine cases has been reported (Gwak et al., 2006). In three cases, a reduction of the tumor volume by a factor of 2.2 based on the PET-based versus the CT-based planning was described resulting in local control at 6 months in two of three cases. Besides tumor delineation, PET imaging might be a suitable imaging modality to detect recurrence.

8.3.3 IMAGE REGISTRATION

Although not done at every institution, image registration of MRI to the planning CT is a common practice and highly recommended if possible, especially in cases of vertebral metastases with a soft tissue component and epidural involvement (Dahele et al., 2011; Guckenberger et al., 2012). Image registration errors are reported to be small in most studies. However, the accuracy of rigid image registration of MRI and CT datasets can be affected by different factors. Intrascanning motion, especially when patients suffer from pain, can affect image quality of the CT and the MRI and therefore impede the registration. Motion between different MRI sequences and between acquisitions of different image modalities is another important factor. General technical limitations are reported to be less than 1 mm in the cranial area (Nakazawa et al., 2014). Ideally, the MRI would be acquired in treatment position by using compatible immobilization devices. Rigid image registration instead of deformable registration is recommended despite the flexibility of the vertebral column. Nonrigid image registration, an intense focus of research, might play a role in the future (Crum et al., 2004). Furthermore, radiation oncologists should be aware of possible pitfalls in MRI-imaging (Putz et al., 2020). Since diagnostic imaging is more focused on qualitative assessment regarding the lesion's extent, the representation of the exact anatomical situation plays a secondary role. For example, this may result in missing 3D distorsion correction due to resource saving workflows in diagnostic imaging, which may result in anatomical displacement of 1–2 mm at the image-edges. With MRI-techniques as synthetic CT or MRI-linacs urging into the field of radiation oncology, this becomes more and more important.

8.3.4 DOSE AND FRACTIONATION

In general, widely accepted doses in SBRT for vertebral metastases are 16–24 Gy delivered in 1 fraction and 12 Gy × 2, 7–10 Gy × 3, or 6–8 Gy × 5 for the multiple-fraction schemes (Glicksman et al., 2020; Husain et al., 2017; Pan et al., 2011; Zeng et al., 2019). Doses are often prescribed to an isodose line of 80% for single-fractionation or 90% to 100% for multiple-fractionation schemes covering the planning target volume.

Whereas a dose–response relation seems to be logical and is established within the one fraction setting (Yamada et al., 2008), there are no consensus optimal fractionated SBRT regimens and whether about the selection of single-fraction versus multiple-fraction SBRT. Several studies reported better local control rates of more aggressive and less fractionated dose regimens compared to more fractionated ones in the postoperative setting (Al-Omair et al., 2013; Laufer et al., 2013). On the other hand, none or not even a contrary correlation was found in other series (Guckenberger et al., 2014; Heron et al., 2012). As mentioned earlier, high local control rates of up to 80% to 90% are reported by the majority of studies in the context of spinal SBRT, irrespective of the chosen SBRT dose and fractionation regimen (Husain et al., 2017).

Regarding early pain response 3 months after treatment, single fraction SBRT seems to be equally effective compared to conventional external beam radiotherapy (EBRT) in randomized controlled trials (Ryu et al., 2019; Sprave et al., 2018). Within the RTOG 0631 trial, single-fraction SBRT with 16–18 Gy achieved an improvement in pain score at the index site of -3, which was not significantly different to conventional radiotherapy (1 × 8 Gy) with -3.88. Pain response was seen in 40.3% of patients. Similarly, the smaller German trial with only 55 patients reported a response rate (RR) of 69.6% after single fraction SBRT. Notably, the latter study suggests a faster and more durable pain response of SBRT patients, as RR and complete response rate (CR) were 73.7% and 57% in the SBRT-group compared to 35% and 10% in EBRT-group 6 months after treatment. These findings correspond well to a 54% complete response rate half a year after treatment in another single arm phase 1–2 trial with 149 patients treated with

Table 8.1 **Spinal cord tolerance doses and corresponding biologically equivalent doses to the epidural tumor**

	1 FRACTION (GY)	2 FRACTIONS (GY)	3 FRACTIONS (GY)	5 FRACTIONS (GY)	10 FRACTIONS (GY)	20 FRACTIONS (GY)
Spinal cord tolerance (total physical dose) Equivalent to minimum dose to the epidural tumor component	10	17	20	25	35	45
Biologically equivalent dose to the epidural tumor component in 2 Gy fractions (EQD2/10)	17	26	27	31	39	46

three-fraction SBRT (Wang et al., 2012). Although not directly comparable, pain response appears favorable in the DOSIS trial: 57 patients with 63 spinal metastases were either treated with 35 Gy in 5 fractions or 48.5 Gy in 10 fractions dependent on life expectancy (Guckenberger et al., 2012). A pain score reduction of -4, combined with a response rate of 82% after 3 months, seems to be at least equivalent to the results described earlier. As expected, quality of life was associated with long-term pain-response, which was 84% after 5 years.

Radiobiological modeling may suggest a further advantage of moderate hypofractionation compared to single-fraction SBRT especially in the situation of epidural disease spread. Table 8.1 shows the modeled relationship between number of treatment fractions, accepted spinal cord tolerance doses, and the effective dose to the epidural tumor component, converted in 2 Gy equivalent doses (alpha/beta ratio of 10 Gy). Given the problem of covering the CTV with the prescription dose while keeping the spinal cord to safe levels, potential underdosage may occur (Figure 8.1). Consequently, relative underdosing of epidural disease may occur with single-fraction SBRT in particular, leading to worse tumor control (Bishop et al., 2015). In fact, a lower equivalent dose to the epidural disease may be observed in single-fraction SBRT as compared to higher-dose conventional palliative radiation approaches like 30 Gy in 10 fractions.

8.3.5 ORGANS-AT-RISK DEFINITION

There is wide acceptance that a homogeneous dose of 50 Gy to the spinal cord using conventional fractionation (1.8–2.0 Gy/day) holds a low risk of myelopathy (Emami et al., 1991; Kirkpatrick et al., 2010; Sahgal et al., 2008), whereas the 5% incidence level is reached between 57 and 61 Gy (Schultheiss et al., 1995). In SBRT of the spine, isodose distributions to the spinal cord differ substantially from those of conventional radiotherapy. Smaller volumes are irradiated, and the dose gradient is frequently within the spinal cord itself. In addition to the well-described dose–response effect, a volume effect for spinal cord toxicity is clearly established (Bijl et al., 2002; Hopewell et al., 1987). A nonlocal repair mechanism, acting from nonirradiated to irradiated tissue covering a distance of 2–3 mm, was postulated as part of the explanation (Bijl et al., 2002; van Luijk et al., 2005). Based on mouse model, Bijl et al. (2003) reported an increase in paralysis if adjacent spinal cord suffers from a low-dose bath, or treated lesions are too close together.

Despite the common usage of SBRT, the incidence of myelopathy after SBRT is low in the literature. To avoid spinal cord injury, Ryu et al. (2007) proposed that 10 Gy should not exceed 10% of the actual spinal cord volume defined as 6 mm above and below the radiosurgery target. This statement was based on their data of one case of myelopathy out of 86 patients treated with spinal SBRT and a minimum follow-up of at least 1 year. Combined data from Stanford University Medical Center and University of Pittsburgh Medical Center reported a radiation myelopathy incidence of 5 out of 1075 patients, and no specific dosimetric factors contributing to this complication could be identified. The authors concluded that biological equivalent dose estimates were not useful for defining spinal cord tolerance to hypofractionated dose schedules, and limiting the volume of spinal cord treated above 8 Gy in a single fraction to 1 cm^3 at maximum was recommended (Gibbs et al., 2009). For stereotactic treatment, incidence of radiation myelopathy is estimated to be less than 1% for a maximum dose in the spinal cord of 13 Gy within 1 fraction or 20 Gy within 3 fractions (Kirkpatrick et al., 2010). Along the same lines, there was no case of radiation-induced myelopathy in a large multi-institutional study of more than 300 cases treated with heterogeneous dose and fractionation regimens. The authors recommended (1) the use of a planning risk volume, containing the spinal cord plus a small safety margin of 1–2 mm, (2) limiting the dose maximum of the spinal cord to 60 Gy$_{2/2}$, (3) daily image guidance and online correction of setup errors for accurate treatment delivery, and (4) customized patient immobilization yields a high level of safety (Guckenberger et al., 2014). Likewise, the maximum dose received by 0.1 cm^3 of the planning spinal cord was restricted to 23.75 Gy in 5 fractions or 35 Gy in 10 fractions in the prospective DOSIS-II trial (Guckenberger et al., 2012).

Sahgal et al. analyzed a group of nine irradiation naïve patients who underwent spine SBRT in multiple centers and developed radiation-induced myelopathy. In one case, SBRT was given as a boost to conventional hypofractionated radiotherapy of 30 Gy given in 10 fractions. This group was compared to a control cohort of 66 patients, who did not suffer radiation myelopathy after SBRT. A significant difference in small volume analysis and in maximum point volume dose was found between the two groups. Based on these data, a logistic regression model was used to generate a probability profile for human radiation myelopathy (RM) after SBRT treatment of the spine. The authors concluded that a maximum point dose inside the thecal sac of 12.4 Gy in 1 fraction, 17.0 Gy in 2 fractions, 20.3 Gy in 3 fractions, 23.0 Gy in 4 fractions, and 25.3 Gy in 5 fractions poses a low risk for radiation myelopathy (<5%) (Sahgal et al., 2013a). Furthermore, it was shown that small isodose volume doses, for example, D$_{max}$ correlate with larger isodose volume doses of the spinal cord (Ma et al., 2018). Therefore, this group shares the opinion that dose hot spots in particular affect spinal cord tolerance following SBRT (Sahgal et al., 2012a).

In a similar study, Sahgal et al. compared a group of 5 patients suffering from radiation-induced myelopathy after conventional radiotherapy and SBRT reirradiation of the spine to 14 patients without radiation-induced myelopathy. Again, a significant dosimetric difference was found for the normalized biologically effective point maximum volume dose (P$_{max}$ nBED). Recommendations were given to keep the P$_{max}$ nBED dose within the thecal sac to no more than 20–25 Gy$_{2/2}$, not to exceed a cumulative P$_{max}$ nBED above 70 Gy$_{2/2}$ considering conventional and stereotactic treatment and to have at least a time interval of 5 months between both treatments (Sahgal et al., 2012b). A comprehensive review is provided in Chapter 20.

With regard to delineation, the spinal cord should be defined as seen in the coregistered T2-weighted MRI including at least one vertebral level superior and inferior (SI) to the planning target volume (PTV) (Guckenberger et al., 2011, 2012). The same extension applies for the thecal sac, which serves as the surrogate for the cauda equina. If MRI is not available, the thecal sac can be defined using the treatment planning CT.

Spinal cord motion has been evaluated in a few studies. A cardiac pulse-triggered wavelike motion pattern has been described (Figley and Stroman, 2007). Cai et al. (2007) and Figley and Stroman (2007) have described that the mean spinal cord motion is less than 0.5–0.6 mm with the largest axial amplitude in the anterior–posterior direction. Oztek et al. also described the phenomenon of a "dancing" spinal cord within the thecal sack using dynamic MRI (Oztek et al., 2020). A safety margin or planning-at-risk volume (PRV) expansion of 1–2 mm is often used for treatment planning.

Depending on the dose constraints, the spinal canal as seen on the CT might be used as a surrogate for the spinal cord (Dahele et al., 2011). This accounts for a small safety margin of the organ at risk, and at the same time is much better to define on the planning CT than the actual structure (Gerszten et al., 2007). However, this may result in suboptimal coverage of the clinical target volume (CTV), leading to a higher risk of recurrence.

Apart from the spinal cord, other OARs such as esophagus, nerve plexuses, nerve roots, ureters, lung, and liver should also be contoured (Dahele et al., 2011). Compared to SBRT for other body sites, contouring of those structures is not different for spine SBRT and should be done according to existing guidelines (e.g., RTOG contouring atlases).

8.3.6 PATTERNS OF FAILURE

Local and/or clinical tumor recurrence occurs in 6% to 23% of cases after spinal SBRT (Chang et al., 2007; Sahgal et al., 2008; Zeng et al., 2019). Tumor recurrence can be found within, at the edge of, or outside the target volume. Whereas infield recurrence might be more dependent on tumor phenotype (Gerszten et al., 2007), failures occurring contiguous to the target volume may correlate with the target delineation.

Patterns of failure after SBRT for spinal metastases have been reported. Within the immediate superior or inferior vertebrae, neither Gerszten et al. nor Ryu et al. observed any case of tumor progression (Gerszten et al., 2007; Ryu et al., 2004). In a series of 74 cases, Chang et al. (2007) reported tumor progression occurring in the adjacent vertebra in one case. In the same study, the authors identified 17 image-documented cases of progression, and the main sites of recurrence were the epidural space (8/17) (due in part to underdosing as a result of spinal cord constraints) and untreated pedicles, posterior elements (3/17), or posterior parts of the vertebral body (2/17). In this study, the vertebral body alone without the posterior elements was included into the target volume. Likewise, Bishop et al. (2015) described an association between marginal tumor recurrences and worse target dose coverage, mostly due to sparing of spinal cord. A minimum dose of 34 Gy BED (equivalent to 14 Gy in 1 fraction or 21 Gy in three fractions) was associated with a significant better 1-year local control rate of 94% versus 80% when less than 34 Gy BED. This corresponds well to another analysis finding minimum PTV doses more than 15 Gy in a single fraction to significantly improve local control (Lovelock et al., 2010).

Similarly, Nelson et al. (2009) reported on 4 out of 33 recurrences, which were located either in direct relation to the epidural space or the pedicles. In these cases, the PTV included the gross tumor volume (GTV) plus a margin of 3–6 mm. Gerszten et al. (2005) reported six cases of recurrence (6/60) based on a series of renal cell carcinoma spine metastases and SBRT treatments targeting GTV only. Since the recurrences occurred often at the edge of the target volume, the group recommended further inclusion of the adjacent normal-appearing vertebral body. Sahgal et al. (2009) reported of 60 cases treated in a similar way, and the majority of the 8 cases of recurrence were situated within 1 mm around the spinal cord. Other sites of recurrence such as within the paraspinal region are rare or directly related to infield recurrence (Chang et al., 2007). Patel et al. (2012) correlated local control rates of 154 extradural spinal lesions to the extent of the vertebral body irradiation. Although not statistically significant, a trend for lower 2-year local tumor progression was found for the treatment of the whole vertebral body (21.1%) compared to partial vertebral body irradiation (34.9%).

The question of GTV-to-CTV margins remains open. Neither Sahgal et al. nor Dahele et al. could establish an association between outcome and different delineation strategies in their reviews (Dahele et al., 2011; Sahgal et al., 2008), and differences in opinions about delineation concepts, dose fractionations, and treatment techniques most likely impeded the analysis. However, based on all available data, the main risk for recurrence is at the sites of underdosing or geographic miss of the tumor, either because of proximity to critical OAR or exclusion of imaging-defined uninvolved vertebral segments like the posterior vertebral elements (Lovelock et al., 2010).

8.3.7 TARGET VOLUME DEFINITION

Sahgal et al. (2008) described two fundamentally different concepts of target volume definition. One is to contour only the radiographically visible tumor without the application of anatomic margins for potential sites of microscopic disease, as derived from the classical intracranial stereotactic radiosurgery teachings. This GTV may then be directly transformed into the PTV by adding geometrical margins ranging from 0 mm (GTV = PTV) (Gerszten et al., 2007; Sahgal et al., 2007) to 2 mm (Gibbs et al., 2007). The second is to apply a CTV to include anatomic areas at risk within the spinal segment and then apply the PTV (Guckenberger et al., 2012) (Figure 8.1).

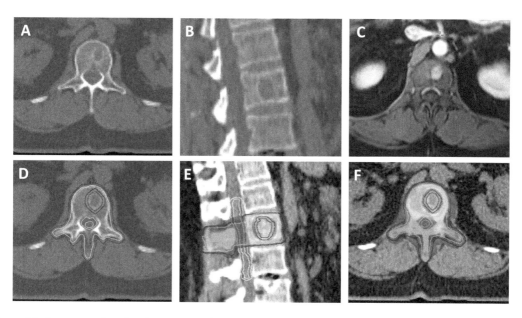

Figure 8.1 Case example using the two-step dose concept according to DOSIS study (Guckenberger et al., 2012): Osteolytic metastasis in CT (A and B) and MRI planning imaging (C); high-dose and low-dose CTV and PTV (D); and VMAT plan using a simultaneous-integrated boost concept (E and F).

The increasing usage of spinal SBRT and the necessity of outcome comparison across different institutions, treatment platforms, and dose fractionation schedules have led to the formation of consensus definitions to standardize nomenclature and delivery specific to spinal SBRT. In this context, recommendations for target volume definition based on the principal concept of a GTV, CTV, and PTV were established. The International Spine Radiosurgery Consortium published recommendations for target delineation based on a series of cases with the CTV delineated by an expert group of radiation oncologists and neurosurgeons. Cox et al. (2012) suggested the implementation of a modified Weinstein–Boriani–Biagini (WBB) system for evaluation of tumor extent (Figure 8.2). In that study, the target volumes of 10 spinal SBRT cases were independently contoured by 10 physicians and analyzed for consistency by a mathematical model. The GTV represented the complete extent of gross metastatic disease identified by clinical information and all available imaging, including paraspinal and epidural tumor. The contoured CTV was consistently shown to account for tissues suspicious for subclinical microscopic invasion, like regions with abnormal bone marrow signal, by all participating institutions. As a rule, it entirely includes all affected and, if in relation to, adjacent sectors according to the suggested WBB concept. In most cases, the whole involved vertebral body was contoured, and with regard to the GTV extension, the ipsilateral and/or contralateral pedicle was included into the CTV. Direct infiltration of pedicles leads to further enclosure of the lamina. If both pedicles and lamina are involved, the entire posterior element is required to be included. An extraosseous CTV expansion is not necessary if the GTV is strictly confined to the bone.

There is variation in practice as to treating the CTV to a single prescribed dose or to adapt the treatment using a simultaneous integrated boost technique. For example, the target volume concept as published in the dose-intensified image-guided fractionated radiosurgery for spinal metastases (DOSIS) study is a two-dose-level approach with high-risk and low-risk target volumes (Guckenberger et al., 2012). The low-risk PTV includes the entire vertebra of the involved levels with appropriate margin to account for treatment setup errors. The high-risk PTV is defined accordingly to the WBB system described earlier. The survey published by Guckenberger et al. (2011) discussed additional important aspects. No more than three vertebral levels should be treated as one single target volume to avoid uncertainties due to deformation.

Margin expansion of CTV to PTV depends on inter- and intrafraction motion management, treatment platforms, immobilization systems, fractionation, method of prescription, and magnitude of dose. In general, a 3D expansion around the CTV of 3 mm or less wherever applicable is recommended (Guckenberger

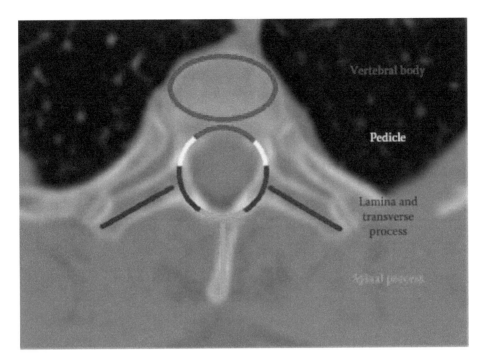

Figure 8.2 Modified Weinstein–Boriani–Biagini model as proposed by the International Spine Radiosurgery Consortium for consensus target volumes for spine radiosurgery.

et al., 2011; Cox et al., 2012). Some investigators, especially those treating spinal metastases with CyberKnife (Accuray, Sunnyvale, CA), reported that a PTV expansion from GTV was not performed. OAR, particularly the spinal cord and the cauda equina, should always be excluded from PTV. Further, PTV cropping may be performed at the discretion of the treating physician to spare the adjacent normal tissues while still ensuring adequate CTV coverage.

8.3.8 TARGET VOLUME DEFINITION IN POSTOPERATIVE STEREOTACTIC BODY RADIOTHERAPY

In 2017, a consensus contouring guideline regarding post-operative SBRT for metastatic solid tumor malignancies to the spine was established (Redmond et al., 2017). Ten specialists in SBRT of the spine contoured GTV, CTV, and PTV independently in each of 10 representative post-operative cases. Agreement between contours was calculated, and optimized consensus contours were derived. The GTV is stated as the postoperative residual disease with particular attention to tumor adjacent to the spinal cord. Recommendations for CTV include treatment of the entire preoperative extent of bony and epidural disease, the postoperative region plus immediately adjacent bony anatomic compartments at risk of microscopic disease extension (see Figure 9.2). In particular, a "donut-shaped" CTV was consistently applied in cases of preoperative circumferential epidural extension. In case of epidural extension, an anatomic expansion up to 5 mm should be considered. Surgical instrumentation and incision is excluded from CTV unless preoperatively involved. The PTV includes the CTV plus a margin of up to 2.5 mm. Further modification to the discretion of the physician is the exclusion of the spinal cord avoidance structure as well as other critical organs at risk.

8.4 SETUP AND IMAGE-GUIDANCE PROCEDURE

The proximity of the spinal tumor to the spinal cord, combined with very steep dose gradients in between, requires highly accurate patient setup and immobilization, most likely the most accurate treatment in extracranial SRT. A prerequisite to a stable patient position, as well as an accurate and suitable treatment, is the patient's comfort in the immobilization device. A comfortable position improves the accuracy of the treatment and prevents involuntary movement, especially with back pain being the most frequent indication for radiotherapy. No specific premedication should be applied, routinely. However, a mild sedative or anxiolytic drug such as lorazepam for agitated patients and medication according to the WHO analgesic ladder for patients in pain may be considered.

8.4.1 IMMOBILIZATION

Traditional stereotactic frame-based immobilization systems fixated the patient invasively to the treatment reference system (Verhey, 1995). Such systems became irrelevant in times of image guidance.

Today, most centers prefer frameless immobilization devices. Treatment delivery without any patient immobilization system is not encouraged, unless thorough active patient monitoring or tracking is applied. Dahele et al. reported excellent spine stability even without rigid patient fixation: posttreatment errors were 1 mm or less and 1° or less in 97% of the treatment fractions. They combined pre- and posttreatment imaging, fast treatment delivery techniques using volumetric modulated arc therapy (VMAT), and support devices like a thin mattress as well as knee supports to improve patient comfort (Dahele et al., 2012).

The immobilization system needs to be suitable for the spinal level to be treated. For cervical spinal tumors, thermoplastic head shoulder masks are frequently used, sometimes combined with vacuum cushions for the caudal part of the body (Lohr et al., 1999). In the thoracic and lumbar region, immobilization systems in spine SBRT are similar to other systems used in high-precision radiotherapy. Most importantly, the patient should be placed in a stable and typically supine position in the immobilization device.

Sahgal and Li et al. analyzed intrafraction patient motion using cone-beam CT (CBCT) imaging and compared three different immobilization devices. They compared the results of evacuated cushions in immobilizing the thoracic and lumbar spine in SBRT to the results of a vacuum device with polyethylene sheet combined with a full-body evacuated cushion in SBRT, and to the results of an S-frame thermoplastic mask immobilizing the cervical spine (Li et al., 2012). They found similar setup errors across the three immobilization systems. Intrafractional patient motion was smallest using the vacuum fixation device, which allowed a small PTV margin of only 2 mm.

8.4.2 PRETREATMENT IMAGE GUIDANCE

Guckenberger et al. (2007) recommended evaluating setup errors in six degrees of freedom by CBCT. To keep the dose distribution to the spinal cord within ±5% of the prescribed dose, deviations exceeding 1 mm in the transverse plane, 4 mm in SI direction, or rotations of 3.5° or more should not be tolerated. It is therefore of highest relevance to correct the translational error component online prior to every treatment fraction; the action level should not exceed 1 mm (Kim et al., 2009). Nonrigid deformation might become relevant especially in multilevel SBRT but such errors cannot be corrected by current image-guidance procedures (Guckenberger et al., 2006; Knybel et al., 2019; Langer et al., 2005; Wang et al., 2008).

Kim et al. (2009) confirmed that the steep dose gradients in spine SBRT, combined with close proximity of the target volume and the spinal cord, require very high treatment accuracy. The implementation of daily image guidance combined with online correction of setup errors in all six degrees of freedom provides precise target localization. Wang et al. (2008) recommended a second image-guided radiation therapy (IGRT) verification immediately after correction to ensure that residual positioning errors are ≤1 mm and ≤1° in any direction or axis.

The IGRT procedure can be performed using either CBCT or stereoscopic x-ray systems (e.g., ExacTrac). Chang et al. (2010) performed a comparison of both guidance modalities reporting modest setup discrepancies between CBCT and planar x-ray in spinal SBRT. In their comparison, CBCT achieved more accurate patient positioning information than the stereoscopic x-ray images. In recent years, radiotherapy systems with online MRI were implemented in daily practice (Ma et al., 2017). Hereby, the primary benefits are superior soft tissue contrast, improved direct monitoring of anatomical changes in real time per fraction and throughout the entire treatment course, and avoidance of additional dose exposure to the patient associated with CBCT imaging. These will allow a shift in workflow with respect to online or potentially even real time adaptive radiotherapy.

8.4.3 INTRAFRACTION IMAGING

As for intrafraction motion due to breathing seems negligible (Wang et al., 2016) there seems to be a dependency on overall treatment time. Ma and Sahgal et al. recorded intrafraction target motion every 50–100 seconds using kV x-ray imaging and reported high rates of nonrandom target movements. Cervical spine targets showed the largest motion, whereas thoracic and lumbar locations were less prone to nonrandom motion. Intrafraction motion increased with time, and average treatment durations needed to maintain submillimeter and subdegree accuracy were 5.5, 5.9, and 7.1 minutes for cervical, thoracic, and lumbar locations, respectively. Intrafractional imaging combined with periodic interventions to overcome patient-specific target motion was recommended (Ma et al., 2009). Tseng et al. assessed the motion of the spinal cord and cauda equina in patients with spinal metastases using dynamic MRI. They differentiated between physiologic oscillatory motion due to cardiorespiratory activity and random shifts due to patient motion. Physiologic oscillatory motion was found to be small, whereas significant bulk motion was present in all observed 65 patients. Displacements were largest in SI direction (median 0.66 mm/maximum 3.90 mm) compared to the lateral (median 0.59 mm/maximum 2.87 mm) or anterior–posterior axis (median 0.51 mm/maximum 2.21 mm) (Tseng et al., 2015).

Digital tomosynthesis (DTS) enables spine position tracking and may also increase the treatment accuracy and precision. Verbakel et al. (2015) used DTS and triangulation for spine detection. This technique generates volumetric datasets from projection images acquired over a limited gantry angle. CBCT projection images over different gantry rotation angles were used to generate single-slice DTS images. These slices then could be registered to digitally reconstruct DTS images derived from the planning CT. With multiple DTS slices, the 3D position of the spine could be determined through triangulation. The interest in DTS in radiotherapy has grown over the past years. It may offer an interesting addition to the established imaging techniques for intrafractional imaging (Gurney-Champion et al., 2013). These results emphasize the need for sufficient margins, short treatment durations, and/or intrafraction imaging in spine SBRT.

Volumetric modulated arc therapy (VMAT) can potentially shorten treatment times, but its utility depends on the capabilities of the available treatment device. Many studies describe the higher efficiency of VMAT over static intensity-modulated radiotherapy (IMRT) with comparable PTV coverage (Kuijper et al., 2010; Matuszak et al., 2010). Wu et al. evaluated the feasibility of VMAT for spine SBRT with respect to three major goals, namely, to achieve a highly conformal dose distribution, to reduce the dose to the spinal cord, and to shorten the overall treatment time. The plan comparison between IMRT and VMAT using one or two arcs revealed an improved conformity index for VMAT (1.06 vs 1.15). Furthermore, two-arc VMAT and IMRT showed comparable OAR sparing. The mean treatment time of IMRT was 15.86 minutes for IMRT compared to 7.88 minutes for VMAT with two arcs, confirming a substantially improved treatment efficiency (Wu et al., 2009).

Flattening filter-free (FFF) beams with high-dose rates reduce beam-on time and subsequently treatment duration. This technology is increasingly used for SBRT. Stieb et al. analyzed outcome and toxicity after SBRT to different body sites with FFF beams in 84 patients. Median follow-up was 11 months, and no case of severe acute toxicity was observed. Only one patient developed grade 3 late toxicity. Hence, the use of FFF beams in SBRT was concluded to be safe (Stieb et al., 2015). Ong and Verbakel et al. confirmed that SBRT FFF plans provide comparable PTV coverage and OAR sparing. Moreover, the average beam-on time was reduced by a factor of up to 2.5 compared to treating with non-FFF plans (Wu et al., 2009).

With implementation of the first MR-Linacs, there is another possibility to further minimize intrafraction motion. First planning studies support similar plan quality compared to VMAT plans delivered by a "conventional" LINAC, as clinical outcomes are still to be awaited (Yadav et al., 2019).

8.4.4 FIDUCIAL MARKERS

The use of implanted fiducial markers for improved image guidance, which has been well established for lung and prostate radiotherapy, has not been adopted for spine (Gerszten et al., 2007). Pan et al. (2011) reported the proportion of physicians having used fiducial markers for stereotactic treatment of the spine to be 12.4%. Considering the natural immobility of these tumors in relationship to the osseous structures of the spine and excellent visibility of the bony structures with all available image-guidance technologies, the rare usage of fiducial markers is understandable. Nevertheless, there might be an advantage in intrafraction motion compensation or in CT–MRI registration prior to treatment (Mani and Rivazhagan, 2013).

8.5 CONCLUSION

Target delineation in SBRT of the spine varies among different institutions, and consensus guidelines are emerging. Patterns of recurrence studies showed that special attention should be brought to delineation of the posterior vertebral elements, even more so in cases of epidural disease. The use of high conformal doses and steep dose gradients requires highly accurate setup and treatment delivery. Therefore, noninvasive immobilization systems and daily CBCT-IG are current treatment standard as MR-Linacs are urging into the field.

CHECKLIST: KEY POINTS FOR CLINICAL PRACTICE

Planning CT	• Axial slice thickness 1–1.5 mm • IV contrast-enhanced CT may be considered for mass-like lesions
Dedicated planning MRI in treatment position	• Axial slice thickness 1–2 mm • Sequences: T1 native, T1 gadolinium-enhanced, T2 FLAIR • Planning CT and MRI acquired ideally on the same day, maximum time interval 2 weeks • Planning CT and MRI acquired ideally in treatment position
Rigid registration of CT and MRI	
Target volume delineation primarily according to contrast-enhanced CT and gadolinium-enhanced T1 MRI	• GTV: morphologic visible tumor • CTV: affected and directly adjacent segments of the vertebral level • PTV: geometrical CTV expansion of 2 mm and exclusion of the spinal cord
Spinal cord definition: various concepts	• Spinal cord as delineated in T2 MRI images with 1–2 mm PRV margin • Thecal sac • Spinal canal
Immobilization	• Patient placed in comfortable position • Sufficient pain medication • Robust patient immobilization and/or continuous active patient monitoring
Image guidance	• Daily image guidance with online correction of setup errors • Action level of maximum 1 mm • Correction of setup errors preferably in six degrees of freedom • Verification imaging after the couch shift immediately prior to treatment especially in single fraction radiosurgery • Repeated intrafraction imaging or active patient monitoring
Treatment delivery	• Minimization of treatment delivery time by preferred use of VMAT and FFF beam delivery

REFERENCES

Algra, P.R., Bloem, J.L., Tissing, H., Falke, T.H., Arndt, J.W., Verboom, L.J., 1991. Detection of vertebral metastases: Comparison between MR imaging and bone scintigraphy. *Radiogr. Rev. Publ. Radiol. Soc. N. Am. Inc* 11, 219–232. https://doi.org/10.1148/radiographics.11.2.2028061

Al-Omair, A., Masucci, L., Masson-Cote, L., Campbell, M., Atenafu, E.G., Parent, A., Letourneau, D., Yu, E., Rampersaud, R., Massicotte, E., Lewis, S., Yee, A., Thibault, I., Fehlings, M.G., Sahgal, A., 2013. Surgical resection of epidural disease improves local control following postoperative spine stereotactic body radiotherapy. *Neuro Oncology* 15, 1413–1419. https://doi.org/10.1093/neuonc/not101

Barzilai, O., Versteeg, A.L., Sahgal, A., Rhines, L.D., Bilsky, M.H., Sciubba, D.M., Schuster, J.M., Weber, M.H., Pal Varga, P., Boriani, S., Bettegowda, C., Fehlings, M.G., Yamada, Y., Clarke, M.J., Arnold, P.M., Gokaslan, Z.L., Fisher, C.G., Laufer, I., the AO Spine Knowledge Forum Tumor, 2019. Survival, local control, and health-related quality of life in patients with oligometastatic and polymetastatic spinal tumors: A multicenter, international study. *Cancer* 125, 770–778. https://doi.org/10.1002/cncr.31870

Bate, B.G., Khan, N.R., Kimball, B.Y., Gabrick, K., Weaver, J., 2015. Stereotactic radiosurgery for spinal metastases with or without separation surgery. *Journal of Neurosurgical Spine* 22, 409–415. https://doi.org/10.3171/2014.10.SPINE14252

Bijl, H.P., van Luijk, P., Coppes, R.P., Schippers, J.M., Konings, A.W., van der Kogel, A.J., 2002. Dose-volume effects in the rat cervical spinal cord after proton irradiation. *International Journal of Radiation Oncology, Biology, Physics* 52, 205–211.

Bijl, H.P., van Luijk, P., Coppes, R.P., Schippers, J.M., Konings, A.W., van der Kogel, A.J., 2003. Unexpected changes of rat cervical spinal cord tolerance caused by inhomogeneous dose distributions. *International Journal of Radiation Oncology, Biology, Physics* 57, 274–281.

Bishop, A.J., Tao, R., Rebueno, N.C., Christensen, E.N., Allen, P.K., Wang, X.A., Amini, B., Tannir, N.M., Tatsui, C.E., Rhines, L.D., Li, J., Chang, E.L., Brown, P.D., Ghia, A.J., 2015. Outcomes for spine stereotactic body radiation therapy and an analysis of predictors of local recurrence. *International Journal of Radiation Oncology* 92, 1016–1026. https://doi.org/10.1016/j.ijrobp.2015.03.037

Cai, J., Sheng, K., Sheehan, J.P., Benedict, S.H., Larner, J.M., Read, P.W., 2007. Evaluation of thoracic spinal cord motion using dynamic MRI. *Radiotherapy & Oncology* 84, 279–282. https://doi.org/10.1016/j.radonc.2007.06.008

Chang, E.L., Shiu, A.S., Mendel, E., Mathews, L.A., Mahajan, A., Allen, P.K., Weinberg, J.S., Brown, B.W., Wang, X.S., Woo, S.Y., Cleeland, C., Maor, M.H., Rhines, L.D., 2007. Phase I/II study of stereotactic body radiotherapy for spinal metastasis and its pattern of failure. *Journal of Neurosurgical Spine* 7, 151–160.

Chang, Z., Wang, Z., Ma, J., O'Daniel, J.C., Kirkpatrick, J., Yin, F.-F., 2010. 6D image guidance for spinal non-invasive stereotactic body radiation therapy: Comparison between ExacTrac X-ray 6D with kilo-voltage cone-beam CT. *Radiother. Oncol. J. Eur. Soc. Ther. Radiol. Oncol.* 95, 116–121. https://doi.org/10.1016/j.radonc.2009.12.036

Chao, S.T., Koyfman, S.A., Woody, N., Angelov, L., Soeder, S.L., Reddy, C.A., Rybicki, L.A., Djemil, T., Suh, J.H., 2012. Recursive partitioning analysis index is predictive for overall survival in patients undergoing spine stereotactic body radiation therapy for spinal metastases. *International Journal of Radiation Oncology, Biology, Physics* 82, 1738–1743. https://doi.org/10.1016/j.ijrobp.2011.02.019

Cox, B.W., Spratt, D.E., Lovelock, M., Bilsky, M.H., Lis, E., Ryu, S., Sheehan, J., Gerszten, P.C., Chang, E., Gibbs, I., Soltys, S., Sahgal, A., Deasy, J., Flickinger, J., Quader, M., Mindea, S., Yamada, Y., 2012. International spine radiosurgery consortium consensus guidelines for target volume definition in spinal stereotactic radiosurgery. *International Journal of Radiation Oncology, Biology, Physics* 83, e597–605. https://doi.org/10.1016/j.ijrobp.2012.03.009

Crum, W.R., Hartkens, T., Hill, D.L.G., 2004. Non-rigid image registration: Theory and practice. *British Journal of Radiology* 77, S140–153. https://doi.org/10.1259/bjr/25329214

Dahele, M., Hatton, M., Slotman, B., Guckenberger, M., 2015. Stereotactic body radiotherapy: A survey of contemporary practice in six selected European countries. *Acta Oncol. Stockh. Swed.* 54, 1237–1241. https://doi.org/10.3109/0284186X.2014.1003961

Dahele, M., Verbakel, W., Cuijpers, J., Slotman, B., Senan, S., 2012. An analysis of patient positioning during stereotactic lung radiotherapy performed without rigid external immobilization. *Radiother. Oncol. J. Eur. Soc. Ther. Radiol. Oncol.* 104, 28–32. https://doi.org/10.1016/j.radonc.2012.03.020

Dahele, M., Zindler, J.D., Sanchez, E., Verbakel, W.F., Kuijer, J.P.A., Slotman, B.J., Senan, S., 2011. Imaging for stereotactic spine radiotherapy: Clinical considerations. *International Journal of Radiation Oncology* 81, 321–330. https://doi.org/10.1016/j.ijrobp.2011.04.039

Dieleman, E.M.T., Senan, S., Vincent, A., Lagerwaard, F.J., Slotman, B.J., van Sörnsen de Koste, J.R., 2007. Four-dimensional computed tomographic analysis of esophageal mobility during normal respiration. *International Journal of Radiation Oncology, Biology, Physics* 67, 775–780. https://doi.org/10.1016/j.ijrobp.2006.09.054

Emami, B., Lyman, J., Brown, A., Coia, L., Goitein, M., Munzenrider, J.E., Shank, B., Solin, L.J., Wesson, M., 1991. Tolerance of normal tissue to therapeutic irradiation. *International Journal of Radiation Oncology, Biology, Physics* 21, 109–122.

Figley, C.R., Stroman, P.W., 2007. Investigation of human cervical and upper thoracic spinal cord motion: Implications for imaging spinal cord structure and function. *Magn. Reson. Med. Off. J. Soc. Magn. Reson. Med. Soc. Magn. Reson. Med.* 58, 185–189. https://doi.org/10.1002/mrm.21260

Finkelstein, J.A., Zaveri, G., Wai, E., Vidmar, M., Kreder, H., Chow, E., 2003. A population-based study of surgery for spinal metastases. Survival rates and complications. *J. Bone Joint Surg. Br.* 85, 1045–1050.

Garg, A.K., Shiu, A.S., Yang, J., Wang, X.S., Allen, P., Brown, B.W., Grossman, P., Frija, E.K., McAleer, M.F., Azeem, S., Brown, P.D., Rhines, L.D., Chang, E.L., 2012. Phase 1/2 trial of single-session stereotactic body radiotherapy for previously unirradiated spinal metastases. *Cancer* 118, 5069–5077. https://doi.org/10.1002/cncr.27530

Geets, X., Daisne, J.-F., Arcangeli, S., Coche, E., De Poel, M., Duprez, T., Nardella, G., Grégoire, V., 2005. Inter-observer variability in the delineation of pharyngo-laryngeal tumor, parotid glands and cervical spinal cord: Comparison between CT-scan and MRI. *Radiother. Oncol. J. Eur. Soc. Ther. Radiol. Oncol.* 77, 25–31. https://doi.org/10.1016/j.radonc.2005.04.010

Gerszten, P.C., Burton, S.A., Ozhasoglu, C., Vogel, W.J., Welch, W.C., Baar, J., Friedland, D.M., 2005. Stereotactic radiosurgery for spinal metastases from renal cell carcinoma. *Journal of Neurosurgical Spine* 3, 288–295. https://doi.org/10.3171/spi.2005.3.4.0288

Gerszten, P.C., Burton, S.A., Ozhasoglu, C., Welch, W.C., 2007. Radiosurgery for spinal metastases: Clinical experience in 500 cases from a single institution. *Spine* 32, 193–199.

Gibbs, I.C., Kamnerdsupaphon, P., Ryu, M.R., Dodd, R., Kiernan, M., Chang, S.D., Adler, J.R., Jr., 2007. Image-guided robotic radiosurgery for spinal metastases. *Radiotherapy & Oncology* 82, 185–190.

Gibbs, I.C., Patil, C., Gerszten, P.C., Adler, J.R., Jr., Burton, S.A., 2009. Delayed radiation-induced myelopathy after spinal radiosurgery. *Neurosurgery* 64, A67–72.

Glicksman, R.M., Tjong, M.C., Neves-Junior, W.F.P., Spratt, D.E., Chua, K.L.M., Mansouri, A., Chua, M.L.K., Berlin, A., Winter, J.D., Dahele, M., Slotman, B.J., Bilsky, M., Shultz, D.B., Maldaun, M., Szerlip, N., Lo, S.S., Yamada, Y., vera-badillo, f.e., marta, g.n., moraes, f.y., 2020. Stereotactic ablative radiotherapy for the management of spinal metastases: A review. *JAMA Oncology* 6, 567. https://doi.org/10.1001/jamaoncol.2019.5351

Guckenberger, M., Hawkins, M., Flentje, M., Sweeney, R.A., 2012. Fractionated radiosurgery for painful spinal metastases: DOSIS—a phase II trial. *BMC Cancer* 12, 530. https://doi.org/10.1186/1471-2407-12-530

Guckenberger, M., Mantel, F., Gerszten, P.C., Flickinger, J.C., Sahgal, A., Létourneau, D., Grills, I.S., Jawad, M., Fahim, D.K., Shin, J.H., Winey, B., Sheehan, J., Kersh, R., 2014. Safety and efficacy of stereotactic body radiotherapy as primary treatment for vertebral metastases: A multi-institutional analysis. *Radiation Oncology* 9, 226. https://doi.org/10.1186/s13014-014-0226-2

Guckenberger, M., Meyer, J., Vordermark, D., Baier, K., Wilbert, J., Flentje, M., 2006. Magnitude and clinical relevance of translational and rotational patient setup errors: A cone-beam CT study. *International Journal of Radiation Oncology, Biology, Physics* 65, 934–942.

Guckenberger, M., Meyer, J., Wilbert, J., Baier, K., Bratengeier, K., Vordermark, D., Flentje, M., 2007. Precision required for dose-escalated treatment of spinal metastases and implications for image-guided radiation therapy (IGRT). *Radiotherapy & Oncology* 84, 56–63.

Guckenberger, M., Sweeney, R.A., Flickinger, J.C., Gerszten, P.C., Kersh, R., Sheehan, J., Sahgal, A., 2011. Clinical practice of image-guided spine radiosurgery-results from an international research consortium. *Radiotherapy & Oncology* 6, 172. https://doi.org/10.1186/1748-717X-6-172

Gurney-Champion, O.J., Dahele, M., Mostafavi, H., Slotman, B.J., Verbakel, W.F.A.R., 2013. Digital tomosynthesis for verifying spine position during radiotherapy: A phantom study. *Physics in Medical & Biology* 58, 5717–5733. https://doi.org/10.1088/0031-9155/58/16/5717

Gwak, H.-S., Youn, S.-M., Chang, U., Lee, D.H., Cheon, G.J., Rhee, C.H., Kim, K., Kim, H.-J., 2006. Usefulness of (18)F-fluorodeoxyglucose PET for radiosurgery planning and response monitoring in patients with recurrent spinal metastasis. *Minim. Invasive Neurosurgery MIN* 49, 127–134. https://doi.org/10.1055/s-2006-932181

Heron, D.E., Rajagopalan, M.S., Stone, B., Burton, S., Gerszten, P.C., Dong, X., Gagnon, G.J., Quinn, A., Henderson, F., 2012. Single-session and multisession CyberKnife radiosurgery for spine metastases-University of Pittsburgh and Georgetown University experience. *Journal of Neurosurgical Spine* 17, 11–18. https://doi.org/10.3171/2012.4.SPINE11902

Hopewell, J.W., Morris, A.D., Dixon-Brown, A., 1987. The influence of field size on the late tolerance of the rat spinal cord to single doses of X rays. *British Journal of Radiology* 60, 1099–1108. https://doi.org/10.1259/0007-1285-60-719-1099

Husain, Z.A., Sahgal, A., De Salles, A., Funaro, M., Glover, J., Hayashi, M., Hiraoka, M., Levivier, M., Ma, L., Martínez-Alvarez, R., Paddick, J.I., Régis, J., Slotman, B.J., Ryu, S., 2017. Stereotactic body radiotherapy for *de novo* spinal metastases: Systematic review: International Stereotactic Radiosurgery Society practice guidelines. *Journal of Neurosurgical Spine* 27, 295–302. https://doi.org/10.3171/2017.1.SPINE16684

Kim, S., Jin, H., Yang, H., Amdur, R.J., 2009. A study on target positioning error and its impact on dose variation in image-guided stereotactic body radiotherapy for the spine. *International Journal of Radiation Oncology, Biology, Physics* 73, 1574–1579. https://doi.org/10.1016/j.ijrobp.2008.12.023

Kirkpatrick, J.P., van der Kogel, A.J., Schultheiss, T.E., 2010. Radiation dose-volume effects in the spinal cord. *International Journal of Radiation Oncology, Biology, Physics* 76, S42–49. https://doi.org/10.1016/j.ijrobp.2009.04.095

Knybel, L., Cvek, J., Cermakova, Z., Havelka, J., Pomaki, M., Resova, K., 2019. Evaluation of spine structure stability at different locations during SBRT. *Biomedical Papers*. https://doi.org/10.5507/bp.2019.027

Kuijper, I.T., Dahele, M., Senan, S., Verbakel, W.F., 2010. Volumetric modulated arc therapy versus conventional intensity modulated radiation therapy for stereotactic spine radiotherapy: A planning study and early clinical data. *Radiotherapy & Oncology* 94, 224–228. https://doi.org/10.1016/j.radonc.2009.12.027

Langer, M.P., Papiez, L., Spirydovich, S., Thai, V., 2005. The need for rotational margins in intensity-modulated radiotherapy and a new method for planning target volume design. *International Journal of Radiation Oncology, Biology, Physics* 63, 1592–603.

Laufer, I., Iorgulescu, J.B., Chapman, T., Lis, E., Shi, W., Zhang, Z., Cox, B.W., Yamada, Y., Bilsky, M.H., 2013. Local disease control for spinal metastases following "separation surgery" and adjuvant hypofractionated or high-dose single-fraction stereotactic radiosurgery: Outcome analysis in 186 patients. *Journal of Neurosurgical Spine* 18, 207–214. https://doi.org/10.3171/2012.11.SPINE12111

Lee, I., Omodon, M., Rock, J., Shultz, L., Ryu, S., 2014. Stereotactic radiosurgery for high-grade metastatic epidural cord compression. *Journal of Radiosurgery & SBRT* 3, 51–58.

Loblaw, D.A., Perry, J., Chambers, A., Laperriere, N.J., 2005. Systematic review of the diagnosis and management of malignant extradural spinal cord compression: The cancer care Ontario practice guidelines initiative's neuro-oncologydisease site group. *Journal of Clinical Oncology* 23, 2028–2037. https://doi.org/10.1200/JCO.2005.00.067

Lohr, F., Debus, J., Frank, C., Herfarth, K., Pastyr, O., Rhein, B., Bahner, M.L., Schlegel, W., Wannenmacher, M., 1999. Noninvasive patient fixation for extracranial stereotactic radiotherapy. *International Journal of Radiation Oncology, Biology, Physics* 45, 521–527.

Loughrey, G.J., Collins, C.D., Todd, S.M., Brown, N.M., Johnson, R.J., 2000. Magnetic resonance imaging in the management of suspected spinal canal disease in patients with known malignancy. *Clinical Radiology* 55, 849–855. https://doi.org/10.1053/crad.2000.0547

Lovelock, D.M., Zhang, Z., Jackson, A., Keam, J., Bekelman, J., Bilsky, M., Lis, E., Yamada, Y., 2010. Correlation of local failure with measures of dose insufficiency in the high-dose single-fraction treatment of bony metastases. *International Journal of Radiation Oncology, Biology, Physics* 77, 1282–1287. https://doi.org/10.1016/j.ijrobp.2009.10.003

Lutz, S., Berk, L., Chang, E., Chow, E., Hahn, C., Hoskin, P., Howell, D., Konski, A., Kachnic, L., Lo, S., Sahgal, A., Silverman, L., von Gunten, C., Mendel, E., Vassil, A., Bruner, D.W., Hartsell, W., 2011. Palliative radiotherapy for bone metastases: An ASTRO evidence-based guideline. *International Journal of Radiation Oncology, Biology, Physics* 79, 965–976. https://doi.org/10.1016/j.ijrobp.2010.11.026

Ma, L., Sahgal, A., Hossain, S., Chuang, C., Descovich, M., Huang, K., Gottschalk, A., Larson, D.A., 2009. Nonrandom intrafraction target motions and general strategy for correction of spine stereotactic body radiotherapy. *International Journal of Radiation Oncology, Biology, Physics* 75, 1261–1265. https://doi.org/10.1016/j.ijrobp.2009.04.027

Ma, L., Wang, L., Lee, Y., Tseng, C.-L., Soltys, S., Braunstein, S., Sahgal, A., 2018. Correlation between small-volume spinal cord doses for spine stereotactic body radiotherapy (SBRT). *J. Radiosurgery SBRT* 5, 229–236.

Ma, L., Wang, L., Tseng, C.-L., Sahgal, A., 2017. Emerging technologies in stereotactic body radiotherapy. *Chinese Clinical Oncology* 6, S12–12. https://doi.org/10.21037/cco.2017.06.19

Matuszak, M.M., Yan, D., Grills, I., Martinez, A., 2010. Clinical applications of volumetric modulated arc therapy. *International Journal of Radiation Oncology, Biology, Physics* 77, 608–616. https://doi.org/10.1016/j.ijrobp.2009.08.032

Mizumoto, M., Harada, H., Asakura, H., Hashimoto, T., Furutani, K., Hashii, H., Takagi, T., Katagiri, H., Takahashi, M., Nishimura, T., 2008. Prognostic factors and a scoring system for survival after radiotherapy for metastases to the spinal column: A review of 544 patients at Shizuoka Cancer Center Hospital. *Cancer* 113, 2816–2822. https://doi.org/10.1002/cncr.23888

Moussazadeh, N., Lis, E., Katsoulakis, E., Kahn, S., Svoboda, M., DiStefano, N.M., McLaughlin, L., Bilsky, M.H., Yamada, Y., Laufer, I., 2015. Five-year outcomes of high-dose single-fraction spinal stereotactic radiosurgery. *International Journal of Radiation Oncology* 93, 361–367. https://doi.org/10.1016/j.ijrobp.2015.05.035

Nelson, J.W., Yoo, D.S., Sampson, J.H., Isaacs, R.E., Larrier, N.A., Marks, L.B., Yin, F.F., Wu, Q.J., Wang, Z., Kirkpatrick, J.P., 2009. Stereotactic body radiotherapy for lesions of the spine and paraspinal regions. *International Journal of Radiation Oncology, Biology, Physics* 73, 1369–1375. https://doi.org/10.1016/j.ijrobp.2008.06.1949

Oztek, M.A., Mayr, N.A., Mossa-Basha, M., Nyflot, M., Sponseller, P.A., Wu, W., Hofstetter, C.P., Saigal, R., Bowen, S.R., Hippe, D.S., Yuh, W.T.C., Stewart, R.D., Lo, S.S., 2020. The Dancing cord: Inherent spinal cord motion and its effect on cord dose in spine stereotactic body radiation therapy. *Neurosurgery*. https://doi.org/10.1093/neuros/nyaa202

Palma, D.A., Olson, R., Harrow, S., Gaede, S., Louie, A.V., Haasbeek, C., Mulroy, L., Lock, M., Rodrigues, G.B., Yaremko, B.P., Schellenberg, D., Ahmad, B., Griffioen, G., Senthi, S., Swaminath, A., Kopek, N., Liu, M., Moore, K., Currie, S., Bauman, G.S., Warner, A., Senan, S., 2019. Stereotactic ablative radiotherapy versus standard of care palliative treatment in patients with oligometastatic cancers (SABR-COMET): A randomised, phase 2, open-label trial. *The Lancet* 393, 2051–2058. https://doi.org/10.1016/S0140-6736(18)32487-5

Pan, H., Simpson, D.R., Mell, L.K., Mundt, A.J., Lawson, J.D., 2011. A survey of stereotactic body radiotherapy use in the United States. *Cancer* 117, 4566–4572. https://doi.org/10.1002/cncr.26067

Pascal-Moussellard, H., Broc, G., Pointillart, V., Siméon, F., Vital, J.M., Sénégas, J., 1998. Complications of vertebral metastasis surgery. *Eur. Spine J. Off. Publ. Eur. Spine Soc. Eur. Spinal Deform. Soc. Eur. Sect. Cerv. Spine Res. Soc.* 7, 438–444.

Patel, V.B., Wegner, R.E., Heron, D.E., Flickinger, J.C., Gerszten, P., Burton, S.A., 2012. Comparison of whole versus partial vertebral body stereotactic body radiation therapy for spinal metastases. *Technol. Cancer Res. Treat.* 11, 105–115.

Putz, F., Mengling, V., Perrin, R., Masitho, S., Weissmann, T., Rösch, J., Bäuerle, T., Janka, R., Cavallaro, A., Uder, M., Amarteifio, P., Doussin, S., Schmidt, M.A., Dörfler, A., Semrau, S., Lettmaier, S., Fietkau, R., Bert, C., 2020. Magnetic resonance imaging for brain stereotactic radiotherapy: A review of requirements and pitfalls. *Strahlenther. Onkol.* 196, 444–456. https://doi.org/10.1007/s00066-020-01604-0

Redmond, K.J., Robertson, S., Lo, S.S., Soltys, S.G., Ryu, S., McNutt, T., Chao, S.T., Yamada, Y., Ghia, A., Chang, E.L., Sheehan, J., Sahgal, A., 2017. Consensus Contouring Guidelines for Postoperative Stereotactic Body Radiation Therapy for Metastatic Solid Tumor Malignancies to the Spine. *International Journal of Radiation Oncology* 97, 64–74. https://doi.org/10.1016/j.ijrobp.2016.09.014

Roesch, J., Cho, J.B.C., Fahim, D.K., Gerszten, P.C., Flickinger, J.C., Grills, I.S., Jawad, M., Kersh, R., Letourneau, D., Mantel, F., Sahgal, A., Shin, J.H., Winey, B., Guckenberger, M., 2017. Risk for surgical complications after previous stereotactic body radiotherapy of the spine. *Radiation Oncology* 12, 153. https://doi.org/10.1186/s13014-017-0887-8

Rothrock, R., Pennington, Z., Ehresman, J., Bilsky, M.H., Barzilai, O., Szerlip, N.J., Sciubba, D.M., 2020. Hybrid Therapy for Spinal Metastases. *Neurosurgery Clinics of North America* 31, 191–200. https://doi.org/10.1016/j.nec.2019.11.001

Ryu, S., Deshmukh, S., Timmerman, R.D., Movsas, B., Gerszten, P.C., Yin, F.F., Dicker, A.P., Shiao, S.L., Desai, A.B., Mell, L.K., Iyengar, P., Hitchcock, Y.J., Allen, A.M., Burton, S.A., Brown, D.R., Sharp, H.J., Chesney, J., Siddiqui, S., Chen, T.H., Kachnic, L.A., 2019. Radiosurgery compared to external beam radiotherapy for localized spine metastasis: Phase III results of NRG Oncology/RTOG 0631. *International Journal of Radiation Oncology* 105, S2–3. https://doi.org/10.1016/j.ijrobp.2019.06.382

Ryu, S., Fang Yin, F., Rock, J., Zhu, J., Chu, A., Kagan, E., Rogers, L., Ajlouni, M., Rosenblum, M., Kim, J.H., 2003. Image-guided and intensity-modulated radiosurgery for patients with spinal metastasis. *Cancer* 97, 2013–2018.

Ryu, S., Jin, J.Y., Jin, R., Rock, J., Ajlouni, M., Movsas, B., Rosenblum, M., Kim, J.H., 2007. Partial volume tolerance of the spinal cord and complications of single-dose radiosurgery. *Cancer* 109, 628–636.

Ryu, S., Pugh, S.L., Gerszten, P.C., Yin, F.-F., Timmerman, R.D., Hitchcock, Y.J., Movsas, B., Kanner, A.A., Berk, L.B., Followill, D.S., Kachnic, L.A., 2014. RTOG 0631 phase 2/3 study of image guided stereotactic radiosurgery for localized (1–3) spine metastases: Phase 2 results. *Practical Radiation Oncology* 4, 76–81. https://doi.org/10.1016/j.prro.2013.05.001

Ryu, S., Rock, J., Jain, R., Lu, M., Anderson, J., Jin, J.-Y., Rosenblum, M., Movsas, B., Kim, J.H., 2010. Radiosurgical decompression of metastatic epidural compression. *Cancer* 116, 2250–2257. https://doi.org/10.1002/cncr.24993

Ryu, S., Rock, J., Rosenblum, M., Kim, J.H., 2004. Patterns of failure after single-dose radiosurgery for spinal metastasis. *Journal of Neurosurgery* 101(Suppl 3), 402–405.

Sahgal, A., Ames, C., Chou, D., Ma, L., Huang, K., Xu, W., Chin, C., Weinberg, V., Chuang, C., Weinstein, P., Larson, D.A., 2009. Stereotactic body radiotherapy is effective salvage therapy for patients with prior radiation of spinal metastases. *International Journal of Radiation Oncology, Biology, Physics* 74, 723–731. https://doi.org/10.1016/j.ijrobp.2008.09.020

Sahgal, A., Chou, D., Ames, C., Ma, L., Chuang, C., Lambom, K., Huang, K., Chin, C.T., Weinstein, P., Larson, D., 2007. Proximity of spinous/paraspinous radiosurgery metastatic targets to the spinal cord versus risk of local failure. *International Journal of Radiation Oncology, Biology, Physics* 69, S243.

Sahgal, A., Larson, D.A., Chang, E.L., 2008. Stereotactic body radiosurgery for spinal metastases: A critical review. *International Journal of Radiation Oncology, Biology, Physics* 71, 652–665.

Sahgal, A., Ma, L., Fowler, J., Weinberg, V., Gibbs, I., Gerszten, P.C., Ryu, S., Soltys, S., Chang, E., Wong, C.S., Larson, D.A., 2012a. Impact of dose hot spots on spinal cord tolerance following stereotactic body radiotherapy: A generalized biological effective dose analysis. *Technology in Cancer Research and Treatment* 11, 35–40.

Sahgal, A., Ma, L., Weinberg, V., Gibbs, I.C., Chao, S., Chang, U.K., Werner-Wasik, M., Angelov, L., Chang, E.L., Sohn, M.J., Soltys, S.G., Letourneau, D., Ryu, S., Gerszten, P.C., Fowler, J., Wong, C.S., Larson, D.A., 2012b. Reirradiation human spinal cord tolerance for stereotactic body radiotherapy. *International Journal of Radiation Oncology, Biology, Physics* 82, 107–116. https://doi.org/10.1016/j.ijrobp.2010.08.021

Sahgal, A., Weinberg, V., Ma, L., Chang, E., Chao, S., Muacevic, A., Gorgulho, A., Soltys, S., Gerszten, P.C., Ryu, S., Angelov, L., Gibbs, I., Wong, C.S., Larson, D.A., 2013a. Probabilities of radiation myelopathy specific to stereotactic body radiation therapy to guide safe practice. *International Journal of Radiation Oncology, Biology, Physics* 85, 341–347. https://doi.org/10.1016/j.ijrobp.2012.05.007

Sahgal, A., Whyne, C.M., Ma, L., Larson, D.A., Fehlings, M.G., 2013b. Vertebral compression fracture after stereotactic body radiotherapy for spinal metastases. *Lancet Oncol* 14, e310–320. https://doi.org/10.1016/S1470-2045(13)70101-3

Schultheiss, T.E., Kun, L.E., Ang, K.K., Stephens, L.C., 1995. Radiation response of the central nervous system. *International Journal of Radiation Oncology, Biology, Physics* 31, 1093–1112.

Sprave, T., Verma, V., Förster, R., Schlampp, I., Bruckner, T., Bostel, T., Welte, S.E., Tonndorf-Martini, E., Nicolay, N.H., Debus, J., Rief, H., 2018. Randomized phase II trial evaluating pain response in patients with spinal metastases following stereotactic body radiotherapy versus three-dimensional conformal radiotherapy. *Radiotherapy & Oncology* 128, 274–282. https://doi.org/10.1016/j.radonc.2018.04.030

Stradiotti, P., Curti, A., Castellazzi, G., Zerbi, A., 2009. Metal-related artifacts in instrumented spine. Techniques for reducing artifacts in CT and MRI: State of the art. *European Spine Journal* 18, 102–108. https://doi.org/10.1007/s00586-009-0998-5

Tedesco, G., Gasbarrini, A., Bandiera, S., Ghermandi, R., Boriani, S., 2017. Composite PEEK/Carbon fiber implants can increase the effectiveness of radiotherapy in the management of spine tumors. *Journal of Spine Surgery* 3, 323–329. https://doi.org/10.21037/jss.2017.06.20

Thariat, J., Castelli, J., Chanalet, S., Marcie, S., Mammar, H., Bondiau, P.-Y., 2009. CyberKnife stereotactic radiotherapy for spinal tumors: Value of computed tomographic myelography in spinal cord delineation. *Neurosurgery* 64, A60–66. https://doi.org/10.1227/01.NEU.0000339129.51926.D6

Tseng, C.-L., Soliman, H., Myrehaug, S., Lee, Y.K., Ruschin, M., Atenafu, E.G., Campbell, M., Maralani, P., Yang, V., Yee, A., Sahgal, A., 2018. Imaging-based outcomes for 24 Gy in 2 daily fractions for patients with *de novo* spinal metastases treated with spine stereotactic body radiation therapy (SBRT). *International Journal of Radiation Oncology* 102, 499–507. https://doi.org/10.1016/j.ijrobp.2018.06.047

Tseng, C.-L., Sussman, M.S., Atenafu, E.G., Letourneau, D., Ma, L., Soliman, H., Thibault, I., Cho, B.C.J., Simeonov, A., Yu, E., Fehlings, M.G., Sahgal, A., 2015. Magnetic resonance imaging assessment of spinal cord and cauda equina motion in supine patients with spinal metastases planned for spine stereotactic body radiation therapy. *International Journal of Radiation Oncology, Biology, Physics* 91, 995–1002. https://doi.org/10.1016/j.ijrobp.2014.12.037

Turel, M., Kerolus, M., O'Toole, J., 2017. Minimally invasive "separation surgery" plus adjuvsant stereotactic radiotherapy in the management of spinal epidural metastases. *Journal of Craniovertebral Junction Spine* 8, 119. https://doi.org/10.4103/jcvjs.JCVJS_13_17

van der Linden, Y.M., Dijkstra, S.P., Vonk, E.J., Marijnen, C.A., Leer, J.W., 2005. Prediction of survival in patients with metastases in the spinal column: Results based on a randomized trial of radiotherapy. *Cancer* 103, 320–328. https://doi.org/10.1002/cncr.20756

van Luijk, P., Bijl, H.P., Konings, A.W., van der Kogel, A.J., Schippers, J.M., 2005. Data on dose-volume effects in the rat spinal cord do not support existing NTCP models. *International Journal of Radiation Oncology, Biology, Physics* 61, 892–900. https://doi.org/10.1016/j.ijrobp.2004.10.035

Verbakel, W.F.A.R., Gurney-Champion, O.J., Slotman, B.J., Dahele, M., 2015. Sub-millimeter spine position monitoring for stereotactic body radiotherapy using offline digital tomosynthesis. *Radiother. Oncol. J. Eur. Soc. Ther. Radiol. Oncol.* https://doi.org/10.1016/j.radonc.2015.04.004

Verhey, L.J., 1995. Immobilizing and positioning patients for radiotherapy. *Seminar in Radiation Oncology* 5, 100–114. https://doi.org/10.1054/SRAO00500100

Wang, H., Shiu, A., Wang, C., O'Daniel, J., Mahajan, A., Woo, S., Liengsawangwong, P., Mohan, R., Chang, E.L., 2008. Dosimetric effect of translational and rotational errors for patients undergoing image-guided stereotactic body radiotherapy for spinal metastases. *International Journal of Radiation Oncology, Biology, Physics* 71, 1261–1271. https://doi.org/10.1016/j.ijrobp.2008.02.074

Wang, X., Ghia, A.J., Zhao, Z., Yang, J., Luo, D., Briere, T.M., Pino, R., Li, J., McAleer, M.F., Weksberg, D.C., Chang, E.L., Brown, P.D., Yang, J.N., 2016. Prospective evaluation of target and spinal cord motion and dosimetric changes with respiration in spinal stereotactic body radiation therapy utilizing 4-D CT. *Journal of Radiosurgery & SBRT* 4, 191–201.

Wang, X., Yang, J., Zhao, Z., Luo, D., Court, L., Zhang, Y., Weksberg, D., Brown, P.D., Li, J., Ghia, A.J., 2019. Dosimetric impact of esophagus motion in single fraction spine stereotactic body radiotherapy. *Physics in Medicine & Biology* 64, 115010. https://doi.org/10.1088/1361-6560/ab1c2b

Wang, X, Rhines, L.D., Shiu, A.S., Yang, J.N., Selek, U., Gning, I., Liu, P., Allen, P.K., Azeem, S.S., Brown, P.D., Sharp, H.J., Weksberg, D.C., Cleeland, C.S., Chang, E.L., 2012. Stereotactic body radiation therapy for management of spinal metastases in patients without spinal cord compression: A phase 1–2 trial. *Lancet Oncology* 13, 395–402. https://doi.org/10.1016/S1470-2045(11)70384-9

Wise, J.J., Fischgrund, J.S., Herkowitz, H.N., Montgomery, D., Kurz, L.T., 1999. Complication, survival rates, and risk factors of surgery for metastatic disease of the spine. *Spine* 24, 1943–1951.

Wu, Q.J., Yoo, S., Kirkpatrick, J.P., Thongphiew, D., Yin, F.F., 2009. Volumetric arc intensity-modulated therapy for spine body radiotherapy: Comparison with static intensity-modulated treatment. *International Journal of Radiation Oncology, Biology, Physics* 75, 1596–604. https://doi.org/10.1016/j.ijrobp.2009.05.005

Yadav, P., Musunuru, H.B., Witt, J.S., Bassetti, M., Bayouth, J., Baschnagel, A.M., 2019. Dosimetric study for spine stereotactic body radiation therapy: Magnetic resonance guided linear accelerator versus volumetric modulated arc therapy. *Radiology & Oncology* 53, 362–368. https://doi.org/10.2478/raon-2019-0042

Yamada, Y., Bilsky, M.H., Lovelock, D.M., Venkatraman, E.S., Toner, S., Johnson, J., Zatcky, J., Zelefsky, M.J., Fuks, Z., 2008. High-dose, single-fraction image-guided intensity-modulated radiotherapy for metastatic spinal lesions. *International Journal of Radiation Oncology, Biology, Physics* 71, 484–490. https://doi.org/10.1016/j.ijrobp.2007.11.046

Zeng, K.L., Tseng, C.-L., Soliman, H., Weiss, Y., Sahgal, A., Myrehaug, S., 2019. Stereotactic body radiotherapy (SBRT) for oligometastatic spine metastases: An overview. *Frontiers in Oncology* 9, 337. https://doi.org/10.3389/fonc.2019.00337

9 Spine Stereotactic Radiosurgery for the Treatment of *De Novo* Spine Metastasis

Ehsan H. Balagamwala, Mihir Naik, Lilyana Angelov, John H. Suh, Simon S. Lo, Arjun Sahgal, Eric Chang, and Samuel T. Chao

Contents

9.1 INTRODUCTION

Skeleton is the third most common site of metastasis, and up to 70% of all cancer patients may develop spine metastases during the natural course of their disease. Spine metastases most frequently present with back pain. As spine metastases progress, focal neurologic symptoms may develop due to epidural disease and/or cord compression. Approximately 10% to 20% of patients with spine metastases will progress to develop symptomatic spinal cord compression (Fornasier and Horne, 1975; Grant et al., 1991; Finn et al., 2007). Typically, spine metastases are treated with a combination of conventionally fractionated external beam radiotherapy (EBRT) and/or surgery, with surgery often reserved for patients with mechanical spine instability or cord compression (Hartsell et al., 2005; Patchell et al., 2005; Lutz et al., 2011; Kaloostian et al., 2014). However, the biggest disadvantage of EBRT is the relative inability to spare nearby organs-at-risk (OAR), most importantly the spinal cord, thereby limiting dose escalation. With recent improvements in systemic treatments which are likely to improve control of micrometastatic disease, developing innovative ways to improve control rates of spinal metastasis will be important to maximize disease control and minimize toxicity of treatment. Recently, the radiation oncology community has witnessed the development of and rapid adoption of spine stereotactic radiosurgery (SRS), which allows the delivery of high doses of radiation in a single or few fractions with the promise of improved palliation and local control. Radiosurgery was first described by Lars Leksell in 1951 for the treatment of intracranial lesions, and since then intracranial radiosurgery has become a cornerstone for the treatment of both benign and malignant intracranial diseases (Leksell, 1951; Guo et al., 2008; Suh, 2010; Murphy and Suh, 2011). With advancements in immobilization, image guidance, intensity modulation, and computerized treatment planning, extracranial radiosurgery has become both a reality and routine at multiple centers across the country (Videtic and Stephans, 2010; Shin et al., 2011; Zaorsky et al., 2013). With SRS, it is possible to deliver high radiotherapy doses to the tumor and its vasculature to overcome inherent radioresistance to fractionated radiotherapy, thereby achieving superior local control and pain relief (Balagamwala et al., 2012b). The current understanding of the biology of hypofractionated radiotherapy was discussed in Chapter 1.

In this chapter, we will discuss the indications, techniques, treatment planning, outcomes as well as toxicities of spine SRS.

9.2 INDICATIONS

Spine SRS is typically performed in a single or limited number of fractions. Indications tend to be institution-specific; however, patients with a long predicted life expectancy, good Karnofsky Performance Score (KPS), radioresistant histology, limited spine metastases, oligometastases, and well-controlled systemic disease are generally considered good candidates for spine SRS. Spine SRS is also a good treatment option in reirradiation of spine metastases, and this is discussed in more detail in Chapter 10. Relative contraindications include spinal cord compression, mechanical instability of the spine, and active connective tissue disorder. In the past, prior SRS to the same level was considered a relative contraindication; however, recent data by Thibault et al. showed excellent safety of fractionated SRS with no grade 3 or higher toxicity (Thibault et al., 2015). Some centers exclude radiosensitive histologies; however, in our institutional experience, select patients with traditionally radiosensitive histologies have done well with spine SRS. A summary of indications and relative contraindications of spine SRS is presented in Table 9.1. The American Society of Radiation Oncology (ASTRO) and American College of Radiology (ACR) have published guidelines for the treatment of spine metastases, with a section dedicated to the role of spine SRS (Lutz et al., 2011, 2017; Lo et al., 2012, 2015).

9.3 TECHNICAL AND TREATMENT PLANNING CONSIDERATIONS

Spine SRS is a resource-intensive treatment modality requiring extensive experience and expertise as well as multidisciplinary involvement of radiation oncologists, neurosurgeons, medical physicists, and radiation therapists. Given the high dose per fraction delivered in spine SRS and the proximity of the spinal cord, it is imperative to achieve accuracy of 1–2 mm (Sahgal et al., 2008). In order to perform spine SRS safely and effectively, the following components are essential: a linear accelerator equipped with a high-definition multileaf collimator and onboard image guidance with cone-beam CT, a body immobilization system,

Table 9.1 **Spine SBRT indications and contraindications**

INDICATIONS	RELATIVE CONTRAINDICATIONS
• Long predicted life expectancy	• Poor expected survival
• KPS ≥70	• KPS 40–60
• Radioresistant histology	• Radiosensitive histology
• ≥3–5 mm separation from spinal cord	• Spinal cord compression
• Prior conventional EBRT	• Multilevel or diffuse spine metastasis
• Limited spine metastases	• Mechanical spine instability
• Oligometastatic disease	• Poorly controlled systemic disease
• Well-controlled systemic disease	• Active connective tissue disease
• Post-separation surgery	

and a sophisticated treatment planning system. Other systems have also been used effectively including CyberKnife (Accuray Inc., Sunnyvale, CA) which utilizes a linear accelerator mounted on a robotic arm. More recently, integrated MRI linear accelerators and image-guided Gamma Knife Icon system with abilities to treat base of skull and upper cervical spine lesions are also showing promise (Ma et al., 2017).

9.3.1 IMMOBILIZATION

The utility of spine SRS lies in maximizing dose distribution within the target volume with a steep dose gradient outside the target volume, thereby sparing the spinal cord. Therefore, spine SRS requires a translational accuracy of less than 2 mm and a rotational accuracy of less than 2° (Chang et al., 2004; Lo et al., 2010). Although respiration does not significantly impact motion of spinal tumors, reproducibility of the spine setup especially in multifraction regimens poses a significant clinical challenge. Furthermore, recent studies have shown that respiration does impact spinal cord motion and may affect spinal cord during SRS (Oztek et al., 2020). Although Hamilton et al. have described an invasive rigid spine fixation device, similar to the head frame utilized in Gamma Knife radiosurgery, it is not practical for routine utilization (Hamilton et al., 1995). Therefore, most centers have utilized near rigid immobilization systems. Many centers utilize commercially available solutions whereas other centers such as Memorial Sloan-Kettering Cancer Center and the University of Heidelberg, Germany, have developed noninvasive in-house systems for near rigid immobilization.

At the Cleveland Clinic, we perform CT-based simulation in the supine position and utilize a five-point head and neck mask for cervical and upper thoracic spinal lesions or the Elekta BodyFIX stereotactic body frame (Medical Intelligence, Schwabmunchen, Germany) for mid-to-low thoracic and lumbosacral spinal lesions. The BodyFIX system consists of a carbon fiber base plate, a whole-body vacuum cushion, a vacuum system, and a plastic fixation sheet (Figure 9.1). Hyde et al. reported their experience with the BodyFIX

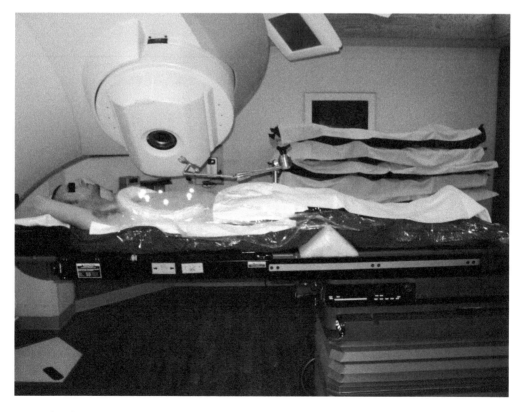

Figure 9.1 The Elekta BodyFIX stereotactic body frame (Medical Intelligence, Schwabmünchen, Germany) which consists of a carbon fiber base plate, whole-body vacuum cushion, vacuum system, and plastic fixation sheet for thoracic and lumbar lesions. (Adapted from Balagamwala et al., 2012c.)

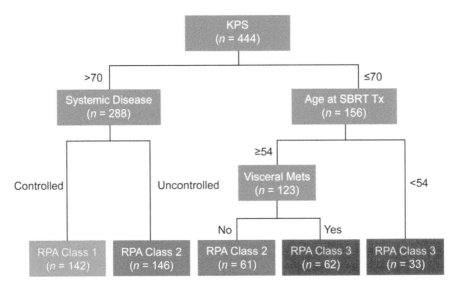

Figure 9.2 Recursive partitioning analysis allows optimal selection of patients for spine SRS. RPA class 1 and class 2 patients may benefit from upfront spine SRS whereas RPA class 3 patients are likely best served with conventional palliative radiotherapy. (Adapted from Balagamwala et al., 2018.)

system in 42 consecutive patients. They found that using this system allowed for excellent precision: 90% of treatments had less than 1 mm translational error and 97% treatments had less than 1° rotational error. They also found that when they utilized a stricter threshold for repositioning the patient (1 mm versus 1.5 mm), the intrafraction translational motion was improved suggesting that accurate patient position prior to treatment delivery improves the overall precision of spine SRS (Hyde et al., 2012).

9.3.2 TREATMENT PLANNING

9.3.2.1 Target Volume Delineation

The gross tumor volume (GTV) is defined as the radiographically visible tumor based on contrast-enhanced MRI. The clinical target volume (CTV) is defined as the margin applied to the GTV to account for microscopic disease in the vicinity of the GTV. The planning target volume (PTV) is formed by adding a margin to the CTV to account for daily patient setup errors. There are considerable differences between institutions regarding target volume definition as well as dose prescription (Guckenberger et al., 2011). Furthermore, there are differences between institutions on the imaging modalities used to delineate the target volume: some institutions rely on CT only, others rely on CT/MRI fusion while others obtain CT or invasive myelogram.

At the Cleveland Clinic, our practice involves obtaining a simulation CT as well as a high-definition (HD) MRI (1.5 mm slice thickness) of the region of interest. If there is an instrumentation that significantly distorts the MRI images, a CT myelogram is performed to define the spinal cord and epidural disease. We subsequently fuse the simulation CT and the HD MRI (or CT myelogram if obtained) in MIM® (MIM Software Inc., Cleveland, Ohio) or iPlan (BrainLAB, AG). For a lesion within the vertebral body, the CTV is defined as the entire vertebral body +/– right or left pedicle depending on involvement. For a lesion involving only the posterior elements, the CTV may only include the spinous process and the lamina. Given the accuracy of immobilization as discussed in section 9.3.1 as well as image guidance (section 9.3.2.4), we have not routinely added a PTV. Our target delineation methodology is similar to the one utilized in the Radiation Therapy Oncology Group (RTOG) 0631 trial, which was a Phase II/III study of image-guided SRS for localized spine metastasis (Ryu et al., 2014). Furthermore, in order to minimize inter-observer variability in target volume delineation consensus, contouring recommendations for spine SRS have been published for spinal metastasis in the definitive and postoperative setting as well as in the management of sacral metastasis (Cox et al., 2012b; Redmond et al., 2017; Dunne et al., 2020).

9.3.2.2 Delineation of Organs at Risk (OARs) and Dose Limits

The principal OAR in spine SRS is the spinal cord and/or the cauda equina because radiation myelopathy can be a devastating complication. This is especially a concern in patients treated for diseases with a long natural history such as ependymoma. There is little consensus in the literature on the imaging modality used to delineate the neural structures or the definition of spinal cord and cauda equina (Balagamwala et al., 2012c). Furthermore, different institutions utilize different spinal cord and the cauda equina dose constraints.

At the Cleveland Clinic, we define the spinal cord and the cauda equina based on the HD MRI. The spinal cord is defined as the spinal cord at the vertebral level(s) being treated with a 4.5 mm cranial and caudal margin. Since the cauda equina is composed of nerve roots floating in the thecal sac, we define the cauda equina as the entire thecal sac at the vertebral level(s) being treated with a 4.5 mm cranial and cauda margin. We limit ≤10% of the contoured spinal cord to ≥10 Gy and limit the maximum point dose (0.03 cc) to less than 14 Gy. Since the cauda equina is composed of nerve roots with a potentially higher dose tolerance, we limit ≤10% of the contoured cauda equina to ≥12 Gy and limit the maximum point dose (0.03 cc) to less than 16 Gy.

Other OARs to consider during treatment planning include esophagus, kidneys, and bowel in select patients. Esophagus is an important OAR to consider especially for cervical and thoracic spine SRS. Although no firm dose constraints for esophagus have been published, there have been reports of grade 3–4 esophageal toxicities (Yamada et al., 2008; Moulding et al., 2010; Cox et al., 2012a). At the Cleveland Clinic, we attempt to limit the maximum dose to the esophagus to 16 Gy when possible. During treatment planning, it is also important to be mindful about avoiding entrance and exit dose through the kidney. This is especially important in patients with renal cell carcinoma who may already have undergone a nephrectomy and have limited renal reserve. Similarly, minimizing dose to the bowel can prevent acute nausea, vomiting, and diarrhea in patients undergoing lumbar spine SRS. While there is institutional variability in normal tissue constraints, dose constraints for spine SRS treatments from the UK are in line with those utilized in various international protocols (Hanna et al., 2018).

9.3.2.3 Treatment Dose and Fractionation

Similar to the institutional differences described earlier, there is considerable institutional variation in terms of the optimal dose fractionation. Institutions utilize either a single fraction regimen or a multifraction regimen. Single fraction doses tend to range between 16 and 24 Gy with a recent trend toward delivering higher doses. Hypofractionated regimens include 24 Gy in 2–3 fractions, 27–30 Gy in 3 fractions, and 30–40 Gy in 5 fractions. As discussed below, there is evidence suggesting that ≥20 Gy per fraction leads to a higher risk for development of vertebral compression fractures (Rose et al., 2009; Sahgal et al., 2013a). Therefore, institutions that support a multifraction regimen argue that this approach allows the delivery of an equivalent biologically effective dose (BED) compared to a high-dose single fraction regimen while minimizing the risk for vertebral compression fractures. Tseng et al. reported a 1-year local control rate of 90.3% while a 1-year risk for vertebral compression fracture of 8.5% (Tseng et al., 2018). Retrospective data has suggested potential higher pain control and local control with single-fraction SRS vs multifraction; however, prospective data has been limited (Heron et al., 2012; Folkert et al., 2014; Ghia et al., 2016). A small prospective trial of 50 patients receiving spine SRS to 75 lesions from Taiwan has shown some suggestion that single fx may be associated with improved local control, but further studies are needed (Lai et al., 2019).

At the Cleveland Clinic, we utilize single fraction spine SRS, and our favored approach is 18 Gy in a single fraction based on our institutional data suggesting a dose–response relationship (unpublished). For certain radioresistant histologies (such as sarcomas and metastatic chordomas), we use 30 Gy in 3 fractions to enhance local control. There certainly remains a need for high-quality evidence comparing the efficacy and side effect profile of single fraction and multifraction regimens to guide future practice. We have recently embarked upon a prospective phase II trial comparing single fraction with two fraction spine SRS as discussed in more detail in subsequent sections (Chao and Balagamwala, 2020).

9.3.2.4 Image Guidance for Treatment Delivery

Safe and accurate delivery of high doses of radiation treatment is hinged upon accurate verification of patient position prior to and during the treatment. Incorporation of image guidance in radiotherapy has

Table 9.2 **Treatment planning summary at the Cleveland Clinic**

TREATMENT PLANNING SUMMARY	
Immobilization	Cervical and upper thoracic spine: 5 point head and neck mask Lower thoracic and lumbosacral spine: BodyFIX
Target Volume Delineation	GTV ± vertebral body ± right and left pedicles ± spinous process and lamina (depending on GTV location)
OAR Delineation	Spinal cord: spinal cord + 4.5 mm cranial and caudal extensions Cauda equina: thecal sac + 4.5 mm cranial and caudal extensions
Prescription Dose	16–18 Gy in 1 fraction
Image Guidance	Cone-beam CT (Varian Edge, NovalisTx) ± orthogonal x-rays (NovalisTx)

allowed for the meteoric rise of SRS not only in spine radiotherapy but also multiple other extracranial applications. Ma et al. identified a high incidence of translational variation of more than 1 mm and rotational variation of more than 1° with treatment times greater than 5 minutes (Ma et al., 2009). This study showed an important aspect of spine SRS that is often underappreciated: fast treatment delivery is imperative. The majority of patients treated for spine metastases have significant pain associated with their disease, and therefore positional changes are prevalent with long treatment durations. If long treatment duration is anticipated, an intrafraction break for image-verification could be utilized, and any positional changes can be readjusted (Rossi et al., 2020).

Image-guidance technologies include 2D techniques such as orthogonal x-rays as well as 3D techniques such as cone-beam CT to minimize setup errors (Degen et al., 2005). The major benefit of 3D techniques is the ability to visualize target volumes, OARs as well as adjusting patient position based on soft tissue in addition to bone. For selected patients, a MV-CBCT may offer mitigation of metal artifact, however, at the expense of soft tissue resolution (Dahele et al., 2011). At the Cleveland Clinic, we utilize CBCT for every spine SRS case and adjust patient position using a "6D" robotic couch. A summary of our treatment methodology is demonstrated in Table 9.2.

9.4 SPINE SRS OUTCOMES

Initial experience with spine SRS was published when Hamilton et al. reported their experience with five patients using a rigid stereotactic frame and linear accelerator in 1995 (Hamilton and Lulu, 1995). However, due to rigid immobilization, spine SRS was not rapidly adopted. Introduction of near-rigid, noninvasive immobilization (as discussed in section 9.3.1) has led to a significant increase in the utilization of spine SRS.

In one of the earliest reports of spine SRS, Benzil et al. described a series of 26 metastases in 22 patients (Benzil et al., 2004). They reported that 94% of these patients experienced significant pain relief within 72 hours, which was durable for up to 3 months. Moreover, 63% of patients experienced improvement in neurologic deficits. This study set the foundation for the adoption of spine SRS.

Since the early reports, many retrospective studies have been reported (select series are summarized in Table 9.3 [Garg et al., 2012]). Overall, spine SRS has resulted in excellent radiographic and clinical control (>85%). In 2004, Gerszten et al. published their initial results of 125 spinal segments (115 patients) treated with CyberKnife (Gerszten et al., 2004). At 1-month follow-up, 94% of patients experienced pain relief, while 89% experienced control of neurologic deficits. Radiographic control was likewise excellent (96%). Gerszten and colleagues updated their initial results and reported their outcomes in 393 patients with 500 spinal segments treated with single fraction spine SRS. Median dose was 20 Gy (range, 12.5–25 Gy) in a single fraction (Gerszten et al., 2007). The 1-year radiographic control was 90%, pain control was 86%, and 84% of patients experienced improvement of their neurologic deficits. Due to its large numbers, this study established spine SRS as a safe and efficacious modality for the treatment of spine metastasis.

Ryu et al. evaluated their experience at Henry Ford in 49 patients with 61 spinal segments. They reported pain control of 85% and reported a relapse of pain at the treated spinal segment in 7%. Importantly, they

Table 9.3 **Select Outcomes for single fraction spine SBRT**

SERIES (YEAR)	STUDY DESIGN	# PATIENTS (# SEGMENTS)	MEDIAN DOSE (RANGE) (GY)	RADIOGRAPHIC CONTROL	PAIN CONTROL
Ryu et al. (2004)	Retrospective	49 (61)	14 (10–16)	96%	93%
Gerszten et al. (2007)	Retrospective	393 (500)	20 (12.5–25)	89%	86%
Yamada et al. (2008)	Retrospective	93 (103)	24 (18–24)	90%	NR
Garg et al. (2012)	Prospective	61 (63)	18 (16–24)	88%	NR
Ryu et al. (RTOG 0631) (2014)	Prospective (Phase II)	44 (55)	16	NR	NR

Abbreviation: NR, not reported.

Table 9.4 **Outcome of spine SBRT in renal cell carcinoma (RCC)**

SERIES (YEAR)	STUDY DESIGN	# PATIENTS (# SEGMENTS)	MEDIAN DOSE (RANGE) (GY)	MEDIAN # OF FRACTIONS	RADIOGRAPHIC CONTROL	PAIN CONTROL
Nguyen et al. (2010)	Retrospective	48 (55)	27 (24–30)	3	78%	75%
Balagamwala et al. (2012a)	Retrospective	57 (88)	15 (8–16)	1	71%	68%
Thibault et al. (2014)	Prospective	37 (71)	24 (18–30)	2	83%	NR
Sohn et al. (2014)	Retrospective	13 (31)	38 (mean)	4	86%	100%

Abbreviation: NR, not reported.

reported an adjacent radiographic failure rate of 5% (Ryu et al., 2004). At the Cleveland Clinic, we also evaluated our risk for marginal failure, that is, tumor recurrence in one vertebral level above and below the treated spinal level (Koyfman et al., 2012). A total of 149 patients with 208 spinal segments were included in this study, and the rate of marginal failure was 12.5% and occurred a median time of 7.7 months. Patients who had paraspinal disease and those receiving less than 16 Gy were found to be at a higher risk for marginal failure. In 2011, Klish et al. published a prospective series of 65 spinal segments (58 patients) which were routinely irradiated in conjunction with adjacent segments. Eleven percent experienced failure in adjacent segments and at multiple other spinal segments, while only 3% of patients failed only in the adjacent segments. Given the high rate of failure outside of the irradiated field, these results suggested that routinely treating uninvolved adjacent segments with spine SRS was unnecessary (Klish et al., 2011).

Chang et al. evaluated the patterns of failure in a prospective single institution series (Chang et al., 2007). This study included 63 patients with 74 segments, and the 1-year local control was 84% after a median follow-up of 21.3 months. Pattern of failure analysis demonstrated two primary mechanisms of failure: recurrence in the bone adjacent to the site of previous treatment and recurrence in the epidural space adjacent to the spinal cord. It is thought that epidural space failure is often due to the relative underdosing of the epidural disease in order to meet spinal cord constraints. Anand et al. retrospectively evaluated their experience in patients with epidural compression and found that in patients with limited epidural disease, adequate dose was delivered to the epidural component with no difference in local control or pain relief (Anand et al., 2015). Similarly, Patel et al. evaluated whole or partial vertebral body spine SRS in a retrospective series of 154 segments (117 patients). Patients treated with the whole vertebral body approach achieved a superior radiographic local control (89% versus 71%, p = 0.029) and a lower retreatment rate (11% versus 19%, p = 0.285), albeit not statistically significant, compared to patients treated with the partial vertebral body approach (Patel et al., 2012).

Although many retrospective institutional studies have shown excellent radiographic and pain control rates, few studies have specifically evaluated radiation dose–response effects. Yamada and colleagues reported their initial experience with single-fraction high-dose spine SRS in 93 patients with 103 spinal segments. They reported that although tumor histology was not a significant predictor of local control, dose more than 23 Gy was associated with better local control compared to dose less than 23 Gy (95% versus 80%, p = 0.03) (Yamada et al., 2008). They also evaluated this dose–response relationship in the post-spinal cord decompression and spinal instrumentation setting. In this small study of 21 patients, the authors found that 3 out of 5 patients (60%) who underwent low-dose radiosurgery experienced local failure compared to 1 out of 16 patients (6%) who underwent high-dose radiosurgery. They estimated that the 1-year cumulative incidence of local failure was 6.3% for the high-dose (24 Gy) group and 20% for the low-dose (<24 Gy) group (p = 0.0175) (Moulding et al., 2010). The same group also evaluated whether a dose–response relationship exists for the treatment of paraspinal disease in the recurrent setting. Among 97 treatments, those patients treated with 30 Gy total dose had a significantly lower risk of developing local failure compared to those treated with 20 Gy (HR 0.51, p = 0.04). Other factors such as tumor size or histology were not associated with local failure (Damast et al., 2011).

At the Cleveland Clinic, we have also analyzed our experience in single-fraction spine SRS to evaluate dosimetric factors correlated with local control. A total of 189 patients with 256 spinal segments were included. Median prescription dose was 15 Gy (range, 8–16 Gy). We found that the presence of epidural disease, multilevel spinal disease as well as lung cancer histology were associated with radiographic failure (Chao et al., 2012a). When we restricted the analysis to traditionally radiosensitive histologies (i.e., non-renal cell, non-melanoma), we found that lung cancer histology was associated with an increased risk of radiographic failure. We also found that higher total dose and maximum dose to the target volume were associated with improved pain control (Balagamwala et al., 2012d). These data show that a dose–response relationship exists for radiosensitive histologies and suggests that there may be a role for dose escalation, especially for radioresistant histologies. Therefore, our institutional practice is to treat our patients with 18 Gy in 1 fraction as long as OAR constraints are met. It is important to note that although dose escalation leads to superior local control, the risk for development of vertebral compression fractures also increases, as discussed in detail below. In order to achieve dose escalation while minimizing risk for vertebral compression fracture, some have suggested utilization of fractionated spine SRS. We are currently evaluating single-fraction versus two-fraction spine SRS in a prospective phase 2 trial with primary end point of vertebral compression fracture and secondary end points of local control and pain control (Chao and Balagamwala, 2020).

There is also an increased interest in the utilization of SRS for the treatment of oligometastatic disease. Ho et al. conducted a retrospective analysis of patients with oligometastases enrolled on institutional phase I/II prospective protocols. Of the 209 patients on trial, 38 patients were treated for solitary spine metastases. The median OS was 75.7 months and the 2- and 5-year OS were 84% and 60%, respectively. Patients with prior spine surgery or higher KPS had improved survival, whereas patients with a prior history of radiotherapy had worse OS. Median time to change in systemic therapy was 41 months. This study suggested that spine SRS can be utilized safely with good OS outcomes for patients with oligometastatic disease (Ho et al., 2016). With accumulating data from trials such as SABR-COMET showing improved OS in patients with oligometastatic disease, we anticipate that spine SRS will be increasingly utilized in the management of cancer patients (Palma et al., 2019, 2020).

Due to the ability to deliver a high BED with spine SRS, it is most often utilized for the treatment of metastases from radioresistant tumors such as renal cell carcinoma. Nguyen et al. reported one of the first such series of 48 patients with 55 spinal metastases with a median follow-up time of 13.1 months. The actuarial 1-year progression free survival was 82.1%. The complete pain response rate at 1 and 12 months post-SRS was an impressive 44% and 52% (compared to baseline rate of 23%) (Nguyen et al., 2010). We also evaluated our series of 57 patients with 88 treated spinal segments treated with single fraction spine SRS. Median time to radiographic failure and pain progression was 26.5 months and 26.0 months, respectively. The median time to pain relief was 0.9 months. In our series, the rate of vertebral compression fractures was 14% (Balagamwala et al., 2012a). In a more recently analysis, Thibault and colleagues evaluated 37 patients with 71 spinal segments and noted a vertebral compression risk of 16% after spine SRS for renal cell carcinoma metastasis (Thibault et al., 2014). These results are summarized in Table 9.4.

There is also a significant interest in potentiating an enhanced response with SRS. Miller et al. evaluated the potentiating effects of concurrent tyrosine kinase inhibitors (TKIs) with spine SRS in patients with metastatic RCC. They performed a retrospective cohort analysis of 151 patients and divided the patients into those receiving concurrent first-line TKI therapy (A), systemic therapy-naïve patients (B), and patients who were undergoing SRS with (C) or without (D) concurrent TKI treatment after failure of first-line therapy. Multivariate competing risk modeling revealed that the risk of failure was lowest in cohort A (4%) versus those patients in cohorts B–D (19% to 27%) and highest in cohort E (57%), which was the negative control cohort receiving TKI-only. Incidence of vertebral compression fracture (VCF) or pain flare was similar across cohorts. This study suggested that local control can be significantly enhanced with concurrent first-line TKIs and spine SRS (Miller et al., 2016).

Sohn and colleagues performed a matched-pair analysis comparing spine SRS versus conventionally fractionated radiotherapy for spinal metastases from renal cell carcinoma. They found that more SRS patients had complete or partial pain responses (however, the difference was not statistically different). Furthermore, they found that the progression-free survival was significantly higher for the SRS patients ($p = 0.01$). There were no differences in toxicity (Sohn et al., 2014). Similarly, Hunter et al. compared conventional radiotherapy to spine SRS for renal cell carcinoma metastases and found that although the overall pain relief rates were not different between the two groups, spine SRS led to higher incidence of complete pain relief (33% versus 12%, $p = 0.01$) (Hunter et al., 2012).

Chao and colleagues performed a recursive partitioning analysis (RPA) for patients undergoing spine SRS at their institution. They evaluated 174 patients who underwent single fraction spine SRS with a median follow-up of 8.9 months. RPA analysis resulted in three classes ($p < 0.01$). Class 1 was defined as time from primary disease (TPD) of more than 30 months and KPS more than 70. Class 2 was defined as TPD more than 30 months and KPS ≤70 or TPD ≤30 months and age less than 70 years. Class 3 was defined as TPD ≤30 months and age ≥70 years. Median overall survival was 21.1 months for Class 1, 8.7 months for Class 2, and 2.4 months for Class 3 (Chao et al., 2012b). This RPA was updated with a total of 444 patients who underwent single fraction spine SRS. Median follow-up was 11.7 months, and median OS was 12.9 months. RPA identified three distinct classes. Class 1 was defined as KPS greater than 70 with controlled systemic disease ($n = 142$); class 3 was defined as KPS less than or equal to 70 and age less than 54 years or KPS ≤70, age 54 years and presence of visceral metastases ($n = 95$); all remaining patients comprise class 2 ($n = 207$). Median overall survival was 26.7 months for class 1, 13.4 months for class 2, and 4.5 months for class 3 ($p < 0.01$) (Balagamwala et al., 2018).

An important consideration when evaluating treatment options for palliation is quality of life (QOL). Although clinical trials in recent year are incorporating QOL measures, much of the retrospective evidence does not include QOL measures. Degen et al. reported the earliest quality of life (QOL) outcomes after spine SRS using the Short Form Health Survey (SF-12) at regular treatment intervals. No significant differences in QOL were observed up to 24 months after SRS, suggesting that SRS preserves patient QOL (Degen et al., 2005). RTOG 0631 was a phase II/III trial comparing conventional external beam radiotherapy (8 Gy in 1 fraction) to spine SRS (16 Gy or 18 Gy in 1 fraction), which sought not only to establish the safety and compare the efficacy of SRS to conventional radiotherapy, but also aimed to establish the impact of radiotherapy on the QOL (Ryu et al., 2014, p. 0). The phase II results of 44 patients were recently published which showed a feasibility success rate of 74% in a rigorous quality-controlled setting (Ryu et al., 2014). The results from the phase III component of the trial were presented at ASTRO 2019 and showed no difference in adverse effects with the utilization of spine SRS; however, no clear benefit in terms of pain control was identified with spine SRS. This was likely due to the lower than anticipated pain control rates in the SRS arm (Ryu et al., 2019). Full publication of the manuscript is eagerly awaited. Sahgal et al. presented the outcomes of the SC24 trial at the ASTRO 2020 Annual Meeting comparing 24 Gy in 2 fractions vs 20 Gy in 5 fractions for spine metastases (Sahgal, 2015). They enrolled 229 patients on the trial, and median follow-up was 6.7 months. They found that complete pain response rate at 3 months was significantly higher with spine SRS compared to conventional radiotherapy, 36% versus 14%, respectively

($p < 0.001$). This CR benefit was retained at 6 months as well (33% versus 16%, $p = 0.004$). This trial showed a significant benefit in complete pain response rate with spine SRS, validating the importance of spine SRS in clinical practice (Sahgal, late breaking abstract ASTRO 2020).

9.5 TOXICITIES

9.5.1 PAIN FLARE

The goal of palliative radiotherapy for spine metastases has traditionally been short-term pain relief. With the advent of spine SRS and the delivery of ablative radiotherapy doses, we are not only able to achieve adequate palliation, but also hope to achieve excellent local control. With hypofractionated, high-dose radiotherapy, there is concern for the development of pain flare, which is a temporary worsening of bone pain at the treated site. Pain flare usually occurs within 1–2 weeks following radiotherapy and responds readily to steroids. With conventional radiotherapy, the incidence of pain flare has ranged between 16% and 41% (Chow et al., 2005; Loblaw et al., 2007; Hird et al., 2009). Recently, we have gained a better understanding of the incidence of pain flare in patients treated with spine SRS.

Chiang et al. performed a prospective observational study in 41 patients undergoing multifraction spine SRS to total doses of 24–35 Gy in 2–5 fractions (Chiang et al., 2013). They reported an incidence of 68.3%, and pain flare was most commonly observed on day 1 after SRS (29%). The majority of patients were successfully treated with dexamethasone. Higher KPS and cervical or lumbar spine locations were associated with higher incidence of pain flare. Given the high incidence of pain flare, they have initiated a prophylactic dexamethasone protocol in all their patients undergoing spine SRS at their institution (Khan et al., 2015). Pan et al. retrospectively evaluated patients enrolled on institutional spine SRS phase I/II clinical trials at M.D. Anderson Cancer Center and found an incidence of 23%. Median time to pain flare was 5 days, and multifraction SRS was associated with higher incidence of pain flare (Pan et al., 2014). In our institutional experience of single-fraction SRS (14–16 Gy), the incidence of pain flare has been approximately 15%, and we do not routinely premedicate with dexamethasone (Jung et al., 2013). Schaub and colleagues have nicely summarized the modern literature on spine SRS toxicities and the strategies to mitigate them (Schaub et al., 2019).

9.5.2 VERTEBRAL COMPRESSION FRACTURE

With the delivery of high doses per fraction in spine SRS, a major concern for late toxicity is the occurrence of vertebral compression fractures. Rose et al. evaluated their experience of high dose (18–24 Gy in 1 fx) spine SRS and reported a vertebral fracture incidence of 39%. They found that lytic lesions involving more than 40% of the vertebral body as well as lesions caudal to T10 were 6.8 and 4.6 times, respectively, more likely to develop vertebral compression fractures (Rose et al., 2009). Similarly, Boehling et al. retrospectively evaluated patients treated on phase I/II trials with spine SRS to doses of 18–30 Gy in 1–5 fractions. They reported a fracture incidence of 20%. In their series, age more than 55, a pre-existing fracture and baseline pain were associated with increased risk for developing vertebral compression fractures (Boehling et al., 2012). Cunha et al. utilized the Spinal Instability Neoplastic Scoring (SINS) system to perform a more rigorous analysis of the risk factors predisposing for the development of vertebral compression fractures in patients treated with spine SRS. They retrospectively evaluated 90 patients with 167 spinal segments and found that the incidence of compression fractures was 11%. Of the fractures, 63% were *de novo*, whereas 37% were fracture progression. Their analysis demonstrated that alignment, lytic lesions, lung and hepatocellular primary histologies, and dose more than 20 Gy per fraction were significant predictors of vertebral fractures (Cunha et al., 2012).

Sahgal et al. performed a multi-institutional retrospective study of 252 patients with 410 treated spinal segments and evaluated the risk of VCF. This study also utilized the SINS criteria. After a median follow-up of 11.5 months, the incidence of VCF was 14%, and the median time to fracture was 2.46 months. Of the fractures, 47% were new fractures, and 53% were progression of preexisting fractures. Multivariate analysis demonstrated dose per fraction (greatest risk for ≥ 24 Gy versus 20 to 23 Gy versus ≤19 Gy), baseline VCF, lytic tumor, and spine deformity were predictive of VCF (Sahgal et al., 2013a). We evaluated our experience in 348 patients with 507 treatments and found a VCF incidence of 15%. Multivariate analysis showed pre-existing VCF and baseline pain were significant predictors for the development of VCF post-spine SRS (Balagamwala et al., 2013). Faruqi et al. performed a systematic literature review and identified that the most frequently identified risk factors for the development of VCF include lytic disease, baseline VCF, higher dose per fraction, spine deformity, older age and more than 40% to 50% involvement of the vertebral body (Faruqi et al., 2018).

At the Cleveland Clinic, we are currently conducting a prospective randomized phase II trial comparing 24 Gy in 2 fraction versus 18 Gy in 1 fraction with the primary end point of incidence of vertebral compression fractures. Secondary end points include local control, pain control, and incidence of pain flare. Target accrual is 130 patients (Chao and Balagamwala, 2020).

9.5.3 NEUROLOGIC TOXICITY

Radiation myelopathy is the most feared complication of spine radiotherapy. Radiation myelopathy is a late toxicity of spine SRS and rarely occurs less than 6 months after treatment and almost always presents within 3 years after treatment (Abbatucci et al., 1978). The risk of radiation myelopathy from spine SRS has been estimated to be less than 1% (Kirkpatrick et al., 2010).

Sahgal et al. performed a multi-institutional retrospective analysis of 5 patients who developed myelopathy after spine SRS and compared it to 19 patients without myelopathy (Sahgal et al., 2010a). Patients who developed myelopathy received the following maximum point doses to the thecal sac: 10.6 Gy/13.1 Gy/14.8 Gy in 1 fx, 25.6 Gy in 2 fx, and 30.9 Gy in 3 fx. When compared to those patients who did not develop myelopathy, the analysis showed that maximum dose up to 10 Gy in 1 fx to the thecal sac is safe.

In those patients who have undergone previous conventionally fractionated radiotherapy, the dose tolerance of the spinal cord maybe different. This was evaluated by Sahgal et al. in five patients with radiation myelopathy (Sahgal et al., 2010b). He reported that the risk of radiation myelopathy is very low when spine SRS is delivered more than 5 months after conventional radiotherapy; normalized biologically effective dose (nBED) to the thecal sac is 20–25 Gy provided the total nBED does not exceed 70 Gy and the SRS thecal sac nBED is not more than 50% of the total nBED. More recently, Sahgal et al. reported the first logistic regression model yielding estimates of radiation myelopathy from spine SRS. They estimated that the risk of myelopathy is less than 5% when limiting maximum thecal sac dose to 12.4 Gy in 1 fx, 17.0 Gy in 2 fx, 20.3 Gy in 3 fx, 23.0 Gy in 4 fx, and 25.3 Gy in 5 fx (Sahgal et al., 2013b). A dedicated chapter on spinal cord tolerance will provide in-depth analyses of this important topic.

9.6 CONTROVERSIES

Over the past decade, we have gained a great deal of understanding regarding the efficacy of and the toxicities associated with spine SRS in the upfront as well as reirradiation setting. Early results from RTOG 0631, a prospective, randomized phase III trial comparing single fraction spine SRS with single fraction conventional radiotherapy, showed no benefit compared to conventional radiotherapy; however, we await final publication. We also await the final results from a Canadian randomized trial evaluating 24 Gy in 2 fractions compared to 20 Gy in 5 fractions. Data from multiple institutions shows that spine SRS is very efficacious, with radiographic and pain control rates of more than 85%. However, the major differences among institutions pertain to the technical nuances of spine SRS including delineation of target volume and OARs, dose prescription, and most importantly single versus multifraction SRS. Comparative data regarding the safety and efficacy of single versus multifraction SRS do not exist, and current practice is based upon institutional preference.

9.7 FUTURE DIRECTIONS

Spine SRS has established itself as the standard of care in the reirradiation setting, and the groundwork for its role in the upfront setting has been laid based upon the numerous studies showing its safety and efficacy. Although early experience shows that spine SRS can be utilized in the setting of cord compression with good results, further work needs to be done to establish its role in the medically inoperable setting (Ryu et al., 2010; Suh et al., 2012). Furthermore, the role of spine SRS after vertebral augmentation has not been clearly defined, and the jury is out on single versus multiple fraction spine SRS. Multiple institutions have small prospective trials currently openly evaluating the role of spine SRS with concurrent immunotherapy. As systemic chemotherapy continues to improve and patients with metastatic disease continue to have longer life expectancy, the role of repeat spine SRS will have to be established.

CHECKLIST: KEY POINTS FOR CLINICAL PRACTICE

✓	ACTIVITY	SOME CONSIDERATIONS
	Patient selection	**Is the patient appropriate for SRS?** • Patient has limited burden of metastatic disease • Other sites of metastases well controlled on systemic agents • At least 2 mm between epidural disease and the spinal cord • KPS ≥60 • Radioresistant histology
	SRS versus EBRT	**Is the lesion amenable to SRS?** • Non-lymphoma/non-myeloma histology • Expected survival >6 months • ≤3 contiguous vertebral bodies involved
	Simulation	**Immobilization** • Supine • Cervical and upper thoracic lesions: 5-point aquaplastic mask • Thoracic and lumbosacral lesions: stereotactic body immobilization system (Linac) or Alpha Cradle (CyberKnife) **Imaging** • CT simulation • High-definition planning MRI with T1 and STIR sequences • CT myelogram if patient unable to undergo MRI • MRI to simulation CT fusion of the region of interest

	Treatment planning	**Contours**
		• CTV per published guidelines, dependent on extent of spine involvement
		• Spinal cord/cauda equina contour includes 2 slices above and below the vertebral level(s) of interest
		Treatment planning
		• Inverse treatment planning
		• CTV = PTV
		• Dose
		  • 16–18 Gy in 1 fraction, 24 Gy in 2 fractions, 27–30 Gy in 3 fractions, or 30–40 Gy in 5 fractions to PTV
		• Coverage
		  • Achieve PTV coverage 90%
		  • May accept lower coverage in epidural space to achieve cord tolerance
		• Dose constraints
		  • Dependent upon dose fractionation

	Treatment delivery	**Imaging** • Patients are first set up conventionally using room lasers and manual shifts • Orthogonal kV matching to relevant bony anatomy • Cone-beam CT to match to tumor (CTV or PTV) and spinal cord • Treatment couch with six degrees of freedom to correct translational and rotational shifts if available • Repeat CBCT for patient movement or long treatment time interval (10–15 minutes) • Utilize volumetric modulated arc therapy and flattening filter free mode whenever feasible to reduce treatment duration

(Continued)

✓	ACTIVITY	SOME CONSIDERATIONS
	Outcomes	**Local control** • Excellent (>85%) radiographic local control as well as pain control rates
	Toxicity	**Acute** • Risk of pain flare ranges from 15% to 50% and is adequately treated with short course of steroids • Esophagitis may also occur and is treated with supportive management **Chronic** • >20 Gy per fraction leads to a significant increase in risk for the development of vertebral compression fractures • Risk for spinal cord myelopathy is very low (<1%) so long as cord tolerance is not exceeded

REFERENCES

Abbatucci JS, Delozier T, Quint R, Roussel A, Brune D (1978) Radiation myelopathy of the cervical spinal cord: Time, dose and volume factors. *International Journal of Radiation Oncology, Biology, Physics* 4:239–248.

Anand AK, Venkadamanickam G, Punnakal AU, Walia BS, Kumar A, Bansal AK, Singh HM (2015) Hypofractionated stereotactic body radiotherapy in spinal metastasis—with or without epidural extension. *Clinical Oncology (R Coll Radiol G B)* 27:345–352.

Balagamwala EH, Angelov L, Koyfman SA, Suh JH, Reddy CA, Djemil T, Hunter GK, Xia P, Chao ST (2012a) Single-fraction stereotactic body radiotherapy for spinal metastases from renal cell carcinoma. *Journal of Neurosurgical Spine* 17:556–564.

Balagamwala EH, Chao ST, Suh JH (2012b) Principles of radiobiology of stereotactic radiosurgery and clinical applications in the central nervous system. *Technology in Cancer Research and Treatment* 11:3–13.

Balagamwala EH, Cherian S, Angelov L, Suh JH, Djemil T, Lo SS, Sahgal A, Chang E, Teh BS, Chao ST (2012c) Stereotactic body radiotherapy for the treatment of spinal metastases. *International Journal of Radiation Oncology* 1:255–265.

Balagamwala EH, Jung DL, Angelov L, Suh JH, Reddy CA, Djemil T, Magnelli A, Soeder S, Chao ST (2013) Incidence and risk factors for vertebral compression fractures from spine stereotactic body radiation therapy: Results of a large institutional series. *International Journal of Radiation Oncology, Biology, Physics* 87:S89.

Balagamwala EH, Miller JA, Reddy CA, Angelov L, Suh JH, Tariq MB, Murphy ES, Yang K, Djemil T, Magnelli A, Mohammadi AM, Soeder S, Chao ST (2018) Recursive partitioning analysis is predictive of overall survival for patients undergoing spine stereotactic radiosurgery. *Journal of Neuro-Oncology* 137:289–293.

Balagamwala EH, Suh JH, Reddy CA, Angelov L, Djemil T, Magnelli A, Soeder S, Chao ST (2012d) Higher dose spine stereotactic body radiation therapy is associated with improved pain control in radiosensitive histologies. *International Journal of Radiation Oncology, Biology, Physics* 84:S632–633.

Benzil DL, Saboori M, Mogilner AY, Rocchio R, Moorthy CR (2004) Safety and efficacy of stereotactic radiosurgery for tumors of the spine. *Journal of Neurosurgery* 101(Suppl 3):413–418.

Boehling NS, Grosshans DR, Allen PK, McAleer MF, Burton AW, Azeem S, Rhines LD, Chang EL (2012) Vertebral compression fracture risk after stereotactic body radiotherapy for spinal metastases. *Journal of Neurosurgical Spine* 16:379–386.

Chang EL, Shiu AS, Lii M-F, Rhines LD, Mendel E, Mahajan A, Weinberg JS, Mathews LA, Brown BW, Maor MH, Cox JD (2004) Phase I clinical evaluation of near-simultaneous computed tomographic image-guided stereotactic body radiotherapy for spinal metastases. *International Journal of Radiation Oncology, Biology, Physics* 59:1288–1294.

Chang EL, Shiu AS, Mendel E, Mathews LA, Mahajan A, Allen PK, Weinberg JS, Brown BW, Wang XS, Woo SY, Cleeland C, Maor MH, Rhines LD (2007) Phase I/II study of stereotactic body radiotherapy for spinal metastasis and its pattern of failure. *Journal of Neurosurgical Spine* 7:151–160.

Chao ST, Balagamwala EH (2020) Single- vs. two-fraction spine stereotactic radiosurgery for the treatment of vertebral metastases. *clinicaltrials.gov*. Available at: https://clinicaltrials.gov/ct2/show/NCT04218617 [Accessed October 12, 2020].

Chao ST, Balagamwala EH, Reddy CA, Angelov L, Djemil T, Magnelli A, Soeder S, Suh JH (2012a) Spine stereotactic body radiation therapy outcomes correlated to dosimetric factors. *International Journal of Radiation Oncology, Biology, Physics* 84:S212.

Chao ST, Koyfman SA, Woody N, Angelov L, Soeder SL, Reddy CA, Rybicki LA, Djemil T, Suh JH (2012b) Recursive partitioning analysis index is predictive for overall survival in patients undergoing spine stereotactic body radiation therapy for spinal metastases. *International Journal of Radiation Oncology, Biology, Physics* 82:1738–1743.

Chiang A, Zeng L, Zhang L, Lochray F, Korol R, Loblaw A, Chow E, Sahgal A (2013) Pain flare is a common adverse event in steroid-naïve patients after spine stereotactic body radiation therapy: A prospective clinical trial. *International Journal of Radiation Oncology, Biology, Physics* 86:638–642.

Chow E, Ling A, Davis L, Panzarella T, Danjoux C (2005) Pain flare following external beam radiotherapy and meaningful change in pain scores in the treatment of bone metastases. *Radiotherapy & Oncology* 75:64–69.

Cox BW, Jackson A, Hunt M, Bilsky M, Yamada Y (2012a) Esophageal toxicity from high-dose, single-fraction paraspinal stereotactic radiosurgery. *International Journal of Radiation Oncology, Biology, Physics* 83:e661–667.

Cox BW, Spratt DE, Lovelock M, Bilsky MH, Lis E, Ryu S, Sheehan J, Gerszten PC, Chang E, Gibbs I, Soltys S, Sahgal A, Deasy J, Flickinger J, Quader M, Mindea S, Yamada Y (2012b) International Spine Radiosurgery Consortium consensus guidelines for target volume definition in spinal stereotactic radiosurgery. *International Journal of Radiation Oncology, Biology, Physics* 83:e597–605.

Cunha MVR, Al-Omair A, Atenafu EG, Masucci GL, Letourneau D, Korol R, Yu E, Howard P, Lochray F, da Costa LB, Fehlings MG, Sahgal A (2012) Vertebral compression fracture (VCF) after spine stereotactic body radiation therapy (SBRT): Analysis of predictive factors. *International Journal of Radiation Oncology, Biology, Physics* 84:e343–349.

Dahele M, Zindler JD, Sanchez E, Verbakel WF, Kuijer JPA, Slotman BJ, Senan S (2011) Imaging for stereotactic spine radiotherapy: Clinical considerations. *International Journal of Radiation Oncology, Biology, Physics* 81:321–330.

Damast S, Wright J, Bilsky M, Hsu M, Zhang Z, Lovelock M, Cox B, Zatcky J, Yamada Y (2011) Impact of dose on local failure rates after image-guided reirradiation of recurrent paraspinal metastases. *International Journal of Radiation Oncology, Biology, Physics* 81:819–826.

Degen JW, Gagnon GJ, Voyadzis J-M, McRae DA, Lunsden M, Dieterich S, Molzahn I, Henderson FC (2005) CyberKnife stereotactic radiosurgical treatment of spinal tumors for pain control and quality of life. *Journal of Neurosurgical Spine* 2:540–549.

Dunne EM, Sahgal A, Lo SS, Bergman A, Kosztyla R, Dea N, Chang EL, Chang U-K, Chao ST, Faruqi S, Ghia AJ, Redmond KJ, Soltys SG, Liu MC (2020) International consensus recommendations for target volume delineation specific to sacral metastases and spinal stereotactic body radiation therapy (SBRT). *Radiother Oncol J Eur Soc Ther Radiol Oncol* 145:21–29.

Faruqi S, Tseng C-L, Whyne C, Alghamdi M, Wilson J, Myrehaug S, Soliman H, Lee Y, Maralani P, Yang V, Fisher C, Sahgal A (2018) Vertebral compression fracture after spine stereotactic body radiation therapy: A review of the pathophysiology and risk factors. *Neurosurgery* 83:314–322.

Finn MA, Vrionis FD, Schmidt MH (2007) Spinal radiosurgery for metastatic disease of the spine. *Cancer Control J Moffitt Cancer Cent* 14:405–411.

Folkert MR, Bilsky MH, Tom AK, Oh JH, Alektiar KM, Laufer I, Tap WD, Yamada Y (2014) Outcomes and toxicity for hypofractionated and single-fraction image-guided stereotactic radiosurgery for sarcomas metastasizing to the spine. *International Journal of Radiation Oncology, Biology, Physics* 88:1085–1091.

Fornasier VL, Horne JG (1975) Metastases to the vertebral column. *Cancer* 36:590–594.

Garg AK, Shiu AS, Yang J, Wang X-S, Allen P, Brown BW, Grossman P, Frija EK, McAleer MF, Azeem S, Brown PD, Rhines LD, Chang EL (2012) Phase 1/2 trial of single-session stereotactic body radiotherapy for previously unirradiated spinal metastases. *Cancer* 118:5069–5077.

Gerszten PC, Burton SA, Ozhasoglu C, Welch WC (2007) Radiosurgery for spinal metastases: Clinical experience in 500 cases from a single institution. *Spine* 32:193–199.

Gerszten PC, Ozhasoglu C, Burton SA, Vogel WJ, Atkins BA, Kalnicki S, Welch WC (2004) CyberKnife frameless stereotactic radiosurgery for spinal lesions: Clinical experience in 125 cases. *Neurosurgery* 55:89–98; discussion 98–99.

Ghia AJ, Chang EL, Bishop AJ, Pan HY, Boehling NS, Amini B, Allen PK, Li J, Rhines LD, Tannir NM, Tatsui CE, Brown PD, Yang JN (2016) Single-fraction versus multifraction spinal stereotactic radiosurgery for spinal metastases from renal cell carcinoma: Secondary analysis of Phase I/II trials. *Journal of Neurosurgical Spine* 24:829–836.

Grant R, Papadopoulos SM, Greenberg HS (1991) Metastatic epidural spinal cord compression. *Clinical Neurology* 9:825–841.

Guckenberger M, Sweeney RA, Flickinger JC, Gerszten PC, Kersh R, Sheehan J, Sahgal A (2011) Clinical practice of image-guided spine radiosurgery—results from an international research consortium. *Radiat Oncol Lond Engl* 6:172.

Guo S, Chao ST, Reuther AM, Barnett GH, Suh JH (2008) Review of the treatment of trigeminal neuralgia with Gamma Knife radiosurgery. *Stereotact Functional Neurosurgery* 86:135–146.

Hamilton AJ, Lulu BA (1995) A prototype device for linear accelerator-based extracranial radiosurgery. *Acta Neurochir* (Suppl 63):40–43.

Hamilton AJ, Lulu BA, Fosmire H, Stea B, Cassady JR (1995) Preliminary clinical experience with linear accelerator-based spinal stereotactic radiosurgery. *Neurosurgery* 36:311–319.

Hanna GG, Murray L, Patel R, Jain S, Aitken KL, Franks KN, van As N, Tree A, Hatfield P, Harrow S, McDonald F, Ahmed M, Saran FH, Webster GJ, Khoo V, Landau D, Eaton DJ, Hawkins MA (2018) UK consensus on normal tissue dose constraints for stereotactic radiotherapy. *Clinical Oncology (R Coll Radiol G B)* 30:5–14.

Hartsell WF, Scott CB, Bruner DW, Scarantino CW, Ivker RA, Roach M, Suh JH, Demas WF, Movsas B, Petersen IA, Konski AA, Cleeland CS, Janjan NA, DeSilvio M (2005) Randomized trial of short- versus long-course radiotherapy for palliation of painful bone metastases. *Journal of the National Cancer Institute* 97:798–804.

Heron DE, Rajagopalan MS, Stone B, Burton S, Gerszten PC, Dong X, Gagnon GJ, Quinn A, Henderson F (2012) Single-session and multisession CyberKnife radiosurgery for spine metastases-University of Pittsburgh and Georgetown University experience. *Journal of Neurosurgical Spine* 17:11–18.

Hird A, Chow E, Zhang L, Wong R, Wu J, Sinclair E, Danjoux C, Tsao M, Barnes E, Loblaw A (2009) Determining the Incidence of Pain Flare Following Palliative Radiotherapy for Symptomatic Bone Metastases: Results From Three Canadian Cancer Centers. *International Journal of Radiation Oncology, Biology, Physics* 75:193–197.

Ho JC, Tang C, Deegan BJ, Allen PK, Jonasch E, Amini B, Wang XA, Li J, Tatsui CE, Rhines LD, Brown PD, Ghia AJ (2016) The use of spine stereotactic radiosurgery for oligometastatic disease. *Journal of Neurosurgical Spine* 25:239–247.

Hunter GK, Balagamwala EH, Koyfman SA, Bledsoe T, Sheplan LJ, Reddy CA, Chao ST, Djemil T, Angelov L, Videtic GMM (2012) The efficacy of external beam radiotherapy and stereotactic body radiotherapy for painful spinal metastases from renal cell carcinoma. *Practice in Radiation Oncology* 2:e95–100.

Hyde D, Lochray F, Korol R, Davidson M, Wong CS, Ma L, Sahgal A (2012) Spine stereotactic body radiotherapy utilizing cone-beam CT image-guidance with a robotic couch: Intrafraction motion analysis accounting for all six degrees of freedom. *International Journal of Radiation Oncology, Biology, Physics* 82:e555–562.

Jung DL, Balagamwala EH, Angelov L, Suh JH, Reddy CA, Djemil T, Magnelli A, Soeder S, Chao ST (2013) Incidence and risk factors for pain flare following spine radiosurgery. *International Journal of Radiation Oncology, Biology, Physics* 87:S568–569.

Kaloostian PE, Yurter A, Zadnik PL, Sciubba DM, Gokaslan ZL (2014) Current paradigms for metastatic spinal disease: An evidence-based review. *Annals of Surgical Oncology* 21:248–262.

Khan L, Chiang A, Zhang L, Thibault I, Bedard G, Wong E, Loblaw A, Soliman H, Fehlings MG, Chow E, Sahgal A (2015) Prophylactic dexamethasone effectively reduces the incidence of pain flare following spine stereotactic body radiotherapy (SBRT): A prospective observational study. *Support Care in Cancer Official Journal of Multinational Association of Supportive Care in Cancer* 23(10):2937–2943.

Kirkpatrick JP, van der Kogel AJ, Schultheiss TE (2010) Radiation dose-volume effects in the spinal cord. *International Journal of Radiation Oncology, Biology, Physics* 76:S42–49.

Klish DS, Grossman P, Allen PK, Rhines LD, Chang EL (2011) Irradiation of spinal metastases: Should we continue to include one uninvolved vertebral body above and below in the radiation field? *International Journal of Radiation Oncology, Biology, Physics* 81:1495–1499.

Koyfman SA, Djemil T, Burdick MJ, Woody N, Balagamwala EH, Reddy CA, Angelov L, Suh JH, Chao ST (2012) Marginal recurrence requiring salvage radiotherapy after stereotactic body radiotherapy for spinal metastases. *International Journal of Radiation Oncology, Biology, Physics* 83:297–302.

Lai SF, Chen YF, Xiao FR, Hsu FM (2019) A prospective randomized phase II trial of single-fraction versus multi-fraction stereotactic spine radiosurgery for spinal metastases: An initial analysis. *International Journal of Radiation Oncology, Biology, Physics* 105:S48.

Leksell L (1951) The stereotaxic method and radiosurgery of the brain. *Acta Chir Scand* 102:316–319.

Lo SS, Lutz S, Chang EL, Galanopoulos N, Howell DD, Kim EY, Konski AA, Pandit-Taskar ND, Rose PS, Ryu S, Silverman LN, Sloan AE, Van Poznak C (2012) ACR Appropriateness Criteria® spinal bone metastases. American College of Radiology Available at: https://acsearch.acr.org/docs/71097/Narrative/ [Accessed March 21, 2015].

Lo SS, Ryu S, Chang E, Galanopoulos G, Jones J, Kim E, Kubicky C, Lee C, Rose P, Sahgal A, Sloan A, Teh B, Traughber B, Van Poznak C, Vassil A (2015) ACR Appropriateness Criteria® metastatic epidural spinal cord compression and recurrent spinal metastasis. *Journal of Palliative Medicine* 18 Available at: https://pubmed.ncbi.nlm.nih.gov/25974663/ [Accessed October 13, 2020].

LoSS, Sahgal A, Wang JZ, Mayr NA, Sloan A, Mendel E, Chang EL (2010) Stereotactic body radiation therapy for spinal metastases. *Discovery Medicine* 9:289–296.

Loblaw DA, Wu JSY, Kirkbride P, Panzarella T, Smith K, Aslanidis J, Warde P (2007) Pain flare in patients with bone metastases after palliative radiotherapy—A nested randomized control trial. *Support Care Cancer* 15:451–455.

Lutz S, Balboni T, Jones J, Lo S, Petit J, Rich SE, Wong R, Hahn C (2017) Palliative radiation therapy for bone metastases: Update of an ASTRO evidence-based guideline. *Practice in Radiation Oncology* 7:4–12.

Lutz S, Berk L, Chang E, Chow E, Hahn C, Hoskin P, Howell D, Konski A, Kachnic L, Lo S, Sahgal A, Silverman L, von Gunten C, Mendel E, Vassil A, Bruner DW, Hartsell W (2011) Palliative radiotherapy for bone metastases: An ASTRO evidence-based guideline. *International Journal of Radiation Oncology, Biology, Physics* 79:965–976.

Ma L, Sahgal A, Hossain S, Chuang C, Descovich M, Huang K, Gottschalk A, Larson DA (2009) Nonrandom intrafraction target motions and general strategy for correction of spine stereotactic body radiotherapy. *International Journal of Radiation Oncology* 75:1261–1265.

Ma L, Wang L, Tseng C-L, Sahgal A (2017) Emerging technologies in stereotactic body radiotherapy. *Chinese Clinical Oncology* 6:S12.

Miller JA, Balagamwala EH, Angelov L, Suh JH, Rini B, Garcia JA, Ahluwalia M, Chao ST (2016) Spine stereotactic radiosurgery with concurrent tyrosine kinase inhibitors for metastatic renal cell carcinoma. *Journal of Neurosurgical Spine* 25:766–774.

Moulding HD, Elder JB, Lis E, Lovelock DM, Zhang Z, Yamada Y, Bilsky MH (2010) Local disease control after decompressive surgery and adjuvant high-dose single-fraction radiosurgery for spine metastases. *Journal of Neurosurgical Spine* 13:87–93.

Murphy ES, Suh JH (2011) Radiotherapy for vestibular schwannomas: A critical review. *International Journal of Radiation Oncology, Biology, Physics* 79:985–997.

Nguyen Q-N, Shiu AS, Rhines LD, Wang H, Allen PK, Wang XS, Chang EL (2010) Management of spinal metastases from renal cell carcinoma using stereotactic body radiotherapy. *International Journal of Radiation Oncology, Biology, Physics* 76:1185–1192.

Oztek MA, Mayr NA, Mossa-Basha M, Nyflot M, Sponseller PA, Wu W, Hofstetter CP, Saigal R, Bowen SR, Hippe DS, Yuh WTC, Stewart RD, Lo SS (2020) The dancing cord: Inherent spinal cord motion and its effect on cord dose in spine stereotactic body radiation therapy. *Neurosurgery* 87(6):1157–1166.

Palma DA et al. (2019) Stereotactic ablative radiotherapy versus standard of care palliative treatment in patients with oligometastatic cancers (SABR-COMET): A randomised, phase 2, open-label trial. *The Lancet* 393:2051–2058.

Palma DA et al. (2020) Stereotactic ablative radiotherapy for the comprehensive treatment of oligometastatic cancers: Long-term results of the SABR-COMET phase II randomized trial. *Journal of Clinical Oncology* 38:2830–2838.

Pan HY, Allen PK, Wang XS, Chang EL, Rhines LD, Tatsui CE, Amini B, Wang XA, Tannir NM, Brown PD, Ghia AJ (2014) Incidence and predictive factors of pain flare after spine stereotactic body radiation therapy: Secondary analysis of phase 1/2 trials. *International Journal of Radiation Oncology, Biology, Physics* 90:870–876.

Patchell RA, Tibbs PA, Regine WF, Payne R, Saris S, Kryscio RJ, Mohiuddin M, Young B (2005) Direct decompressive surgical resection in the treatment of spinal cord compression caused by metastatic cancer: A randomised trial. *Lancet* 366:643–648.

Patel VB, Wegner RE, Heron DE, Flickinger JC, Gerszten P, Burton SA (2012) Comparison of whole versus partial vertebral body stereotactic body radiation therapy for spinal metastases. *Technology in Cancer Research and Treatment* 11:105–115.

Redmond KJ, Robertson S, Lo SS, Soltys SG, Ryu S, McNutt T, Chao ST, Yamada Y, Ghia A, Chang EL, Sheehan J, Sahgal A (2017) Consensus contouring guidelines for postoperative stereotactic body radiation therapy for metastatic solid tumor malignancies to the spine. *International Journal of Radiation Oncology, Biology, Physics* 97:64–74.

Rose PS, Laufer I, Boland PJ, Hanover A, Bilsky MH, Yamada J, Lis E (2009) Risk of fracture after single fraction image-guided intensity-modulated radiation therapy to spinal metastases. *Journal Clinical Oncology Off Journal American Society Clinical Oncology* 27:5075–5079.

Rossi E, Fiorino C, Fodor A, Deantoni C, Mangili P, Di Muzio NG, Del Vecchio A, Broggi S (2020) Residual intra-fraction error in robotic spinal stereotactic body radiotherapy without immobilization devices. *Physics and Imaging in Radiation Oncology* 16:20–25.

Ryu S et al. (2019) Radiosurgery compared to external beam radiotherapy for localized spine Metastasis: Phase III results of NRG oncology/RTOG 0631. *International Journal of Radiation Oncology* 105:S2–3.

Ryu S, Pugh SL, Gerszten PC, Yin F-F, Timmerman RD, Hitchcock YJ, Movsas B, Kanner AA, Berk LB, Followill DS, Kachnic LA (2014) RTOG 0631 phase 2/3 study of image guided stereotactic radiosurgery for localized (1–3) spine metastases: Phase 2 results. *Practice in Radiation Oncology* 4:76–81.

Ryu S, Rock J, Jain R, Lu M, Anderson J, Jin J-Y, Rosenblum M, Movsas B, Kim JH (2010) Radiosurgical decompression of metastatic epidural compression. *Cancer* 116:2250–2257.

Ryu S, Rock J, Rosenblum M, Kim JH (2004) Patterns of failure after single-dose radiosurgery for spinal metastasis. *Journal of Neurosurgery* 101(Suppl 3):402–405.

Sahgal A (2015) Study comparing stereotactic body radiotherapy vs conventional palliative radiotherapy (CRT) for spinal metastases—full text view. *ClinicalTrials.gov*. Available at: https://clinicaltrials.gov/ct2/show/NCT02512965 [Accessed October 13, 2020].

Sahgal A, Atenafu EG, Chao S, Al-Omair A, Boehling N, Balagamwala EH, Cunha M, Thibault I, Angelov L, Brown P, Suh J, Rhines LD, Fehlings MG, Chang E (2013a) Vertebral compression fracture after spine stereotactic body radiotherapy: A multi-institutional analysis with a focus on radiation dose and the spinal instability neoplastic score. *Journal Clinical Oncology Off Journal American Society Clinical Oncology* 31:3426–3431.

Sahgal A, Larson DA, Chang EL (2008) Stereotactic body radiosurgery for spinal metastases: A critical review. *International Journal of Radiation Oncology, Biology, Physics* 71:652–665.

Sahgal A, Ma L, Gibbs I, Gerszten PC, Ryu S, Soltys S, Weinberg V, Wong S, Chang E, Fowler J, Larson DA (2010a) Spinal cord tolerance for stereotactic body radiotherapy. *International Journal of Radiation Oncology, Biology, Physics* 77:548–553.

Sahgal A, Ma L, Weinberg V, Gibbs IC, Chao S, Chang U-K, Werner-Wasik M, Angelov L, Chang EL, Sohn M-J, Soltys SG, Létourneau D, Ryu S, Gerszten PC, Fowler J, Wong CS, Larson DA (2010b) reirradiation HUMAN Spinal Cord Tolerance for Stereotactic Body Radiotherapy. *International Journal of Radiation Oncology, Biology, Physics* 82:107–116.

Sahgal A, Weinberg V, Ma L, Chang E, Chao S, Muacevic A, Gorgulho A, Soltys S, Gerszten PC, Ryu S, Angelov L, Gibbs I, Wong CS, Larson DA (2013b) Probabilities of radiation myelopathy specific to stereotactic body radiation therapy to guide safe practice. *International Journal of Radiation Oncology, Biology, Physics* 85:341–347.

Schaub SK, Tseng YD, Chang EL, Sahgal A, Saigal R, Hofstetter CP, Foote M, Ko AL, Yuh WTC, Mossa-Basha M, Mayr NA, Lo SS (2019) Strategies to mitigate toxicities from stereotactic body radiation therapy for spine metastases. *Neurosurgery* 85:729–740.

Shin JH, Chao ST, Angelov L (2011) Stereotactic radiosurgery for spinal metastases: Update on treatment strategies. *Journal of Neurosurgical Science* 55:197–209.

Sohn S, Chung CK, Sohn MJ, Chang U-K, Kim SH, Kim J, Park E (2014) Stereotactic radiosurgery compared with external radiation therapy as a primary treatment in spine metastasis from renal cell carcinoma: A multicenter, matched-pair study. *Journal of Neuro-Oncology* 119:121–128.

Suh JH (2010) Stereotactic radiosurgery for the management of brain metastases. *The New England Journal of Medicine* 362:1119–1127.

Suh JH, Balagamwala EH, Reddy CA, Angelov L, Djemil T, Magnelli A, Soeder S, Chao ST (2012) The use of spine stereotactic body radiation therapy for the treatment of spinal cord compression. *International Journal of Radiation Oncology, Biology, Physics* 84:S631.

Thibault I, Al-Omair A, Masucci GL, Masson-Côté L, Lochray F, Korol R, Cheng L, Xu W, Yee A, Fehlings MG, Bjarnason GA, Sahgal A (2014) Spine stereotactic body radiotherapy for renal cell cancer spinal metastases: Analysis of outcomes and risk of vertebral compression fracture. *Journal of Neurosurgical Spine* 21:711–718.

Thibault I, Campbell M, Tseng C-L, Atenafu EG, Letourneau D, Yu E, Cho BCJ, Lee YK, Fehlings MG, Sahgal A (2015) Salvage stereotactic body radiotherapy (SBRT) following in-field failure of initial SBRT for spinal metastases. *International Journal of Radiation Oncology, Biology, Physics* 93:353–360.

Tseng C-L, Soliman H, Myrehaug S, Lee YK, Ruschin M, Atenafu EG, Campbell M, Maralani P, Yang V, Yee A, Sahgal A (2018) Imaging-based outcomes for 24 Gy in 2 daily fractions for patients with *de Novo* spinal metastases treated with spine stereotactic body radiation therapy (SBRT). *International Journal of Radiation Oncology, Biology, Physics* 102:499–507.

Videtic GMM, Stephans KL (2010) The role of stereotactic body radiotherapy in the management of non-small cell lung cancer: An emerging standard for the medically inoperable patient? *Current Oncology Reports* 12:235–241.

Yamada Y, Bilsky MH, Lovelock DM, Venkatraman ES, Toner S, Johnson J, Zatcky J, Zelefsky MJ, Fuks Z (2008) High-dose, single-fraction image-guided intensity-modulated radiotherapy for metastatic spinal lesions. *International Journal of Radiation Oncology, Biology, Physics* 71:484–490.

Zaorsky NG, Harrison AS, Trabulsi EJ, Gomella LG, Showalter TN, Hurwitz MD, Dicker AP, Den RB (2013) Evolution of advanced technologies in prostate cancer radiotherapy. *National Review Urology* 10:565–579.

10 Image-Guided Hypofractionated Stereotactic Radiotherapy for Reirradiation of Spinal Metastases

Kevin Diao, Amol J. Ghia, and Eric L. Chang

Contents

10.1 INTRODUCTION

The spine is the most common site for bony metastases and may be involved in up to 40% of all patients with cancer (Ecker et al., 2005; Klimo and Schmidt, 2004). Spinal metastases are responsible for significant morbidity to patients and can cause severe pain, reduced mobility, spinal instability, pathologic fractures, and paralysis secondary to malignant epidural spinal cord compression (MESCC). As a result, palliative intervention is often indicated to improve patient quality of life and preserve neurological function.

Patients with bone metastases, including those with spinal metastases, have typically been treated with longer 5–10 fraction courses of conventionally fractionated external beam radiotherapy (cEBRT). However, shorter fractionation schemes for palliation of bone metastases have been the subject of multiple large randomized trials in the late 1990s and early 2000s, which included patients with spinal metastases. The results of these studies overall demonstrated comparable pain response rates, toxicity, pathologic fracture risk, and risk for development of MESCC with single fraction radiotherapy (Chow et al., 2007). However, the length of follow-up after completing treatment was short, and there were significantly higher rates of retreatment observed in the single fraction arms. Furthermore, the majority of patients enrolled in these trials had radiosensitive histologies (Hartsell et al., 2005; Steenland et al., 1999; Kaasa et al., 2006). As a result, there are concerns about both the durability of pain and local disease control and generalizability to other patient populations.

With advancements in systemic therapy and other cancer treatment modalities, patients are surviving longer in the metastatic state making use of local treatment options that offer extended pain relief and disease control even more importantly (Spratt et al., 2017). Novel scoring systems for predicting patient survival with bone metastases have been developed and are useful for dispositioning patients to individualized courses of radiotherapy that may obviate their need for reirradiation if appropriately selected (Rades et al., 2019). In this modern cohort of bone metastasis patients, a favorable risk group of patients was identified with a 1-year overall survival of 72% and median overall survival of 24 months, demonstrating the changing objectives of managing patients with metastatic disease.

Due to the aforementioned trends, the reirradiation of previously treated areas of spinal disease will become a more common scenario that clinicians will face. The treatment of progressive disease within a previously irradiated spine is a significant challenge due in large part to the proximity of spinal disease to the spinal cord. In this setting, stereotactic body radiotherapy (SBRT) is useful because of its ability to deliver highly conformal radiation therapy to effectively control tumor while sparing the spinal cord. In fact, spine SBRT was initially conceived with reirradiation as its primary application. The first reported outcomes of a spine SBRT series was published in 1995 (Hamilton et al., 1995). According to a modern survey, the spine is now the second most commonly treated site using SBRT (Pan et al., 2011). The past decade has seen major advances in the application of SBRT, and spine SBRT is now considered a standard of care for reirradiation of spinal metastases (Sahgal et al., 2019). In this chapter, we summarize the most up-to-date literature on retreatment of spinal metastases with SBRT and describe our experience with important spine SBRT techniques.

10.2 INDICATIONS

Historically, invasive surgical procedures such as en bloc resection, cEBRT or a combination of the two, have been the cornerstones of the management of spinal metastases (Spratt et al., 2017). However, rapid advances in radiation, surgery, minimally invasive and systemic therapy options have made multidisciplinary management of patients essential for optimizing outcomes and have been the subject of decision algorithms such as the neurologic, oncologic, mechanical, and systemic (NOMS) framework (Laufer et al., 2013). In our practice, all patients considered for spine SBRT are presented at a multidisciplinary conference consisting of radiation oncologists, neurosurgeons, and diagnostic radiologists with input from interventional radiologists, medical oncologists, and pathologists as needed.

Despite the lack of high-level randomized data, spine reirradiation is now a standard indication for the use of spine SBRT, in both the *de novo* and post-operative settings, because of its ability to deliver high ablative doses of radiation while sparing critical organs-at-risk (OARs) such as the spinal cord using conformal techniques (Detsky et al., 2020). In the setting of reirradiation, high doses for retreatment are important as patients whose disease has progressed through initial radiotherapy may be relatively more radioresistant, and being able to conform to radiation dose around the previously irradiated spinal cord is important to mitigate risk of serious toxicity, without which an effective dose of radiation could not be delivered safely. For example, a review of data from patients who were retreated for bone metastases using cEBRT, a significant proportion of whom had spinal disease, demonstrated unfavorable results with an overall response rate of only 58% and complete response rates of 16% to 28%—likely in part due to attempts to maintain doses below the cumulative tolerance of the spinal cord and because of more radioresistant disease (Huisman et al., 2012). While most cases of reirradiation follow the use of initial cEBRT, there is also prospective data to support the use of spine SBRT for reirradiation of sites initially treated with *de novo* spine SBRT (Thibault et al., 2015).

Other widely accepted indications for spine SBRT include treatment of patients with oligometastases and radioresistant histologies (Sahgal et al., 2019). However, there are certain situations in which spine SBRT alone may not be the best option for treatment. Patients with significant neurologic symptoms as a result of MESCC, or due to mechanical instability, may require surgical intervention when possible. While not an absolute contraindication, there should be careful consideration prior to using spine SBRT in patients with extensive spinal or extraspinal disease given the likelihood of progression at sites outside the treatment field, the labor intensiveness and technical feasibility of a spine SBRT plan, and limited patient prognosis (Ryu et al., 2015).

Patients with a short time interval since the prior course of radiation are at significantly higher risk of toxicity from reirradiation and may also indicate patients with radioresistant disease or who are rapidly progressing elsewhere. We do not recommend reirradiation with spine SBRT within a time interval of less than 3 months and prefer at least a 6-month interval between courses. The recently published Hypofractionated Effects in the Clinic (HyTEC) report on spine SBRT recommended a minimum time interval to reirradiation of 5 months (Sahgal et al., 2019).

The epidural space is one of the most common sites for failure following radiotherapy due to intentional underdosing of the area in order to meet spinal cord dose constraints (Chang et al., 2007). While in the past, patients with MESCC were not considered candidates for spine SBRT due to proximity of the spinal cord to the tumor and inability to adequately dose the tumor while maintaining an acceptable spinal cord dose, recent phase I data has been published that demonstrates excellent local control outcomes with low toxicity using *de novo* spine SBRT (Ghia et al., 2018). In a cohort of 32 high-risk and inoperable patients with MESCC, relaxation of the spinal cord maximum dose to 16 Gy in a single fraction resulted in excellent rates of local control and no cases of radiation myelitis (RM). In light of these data, patients with spinal metastases presenting with MESCC who are not deemed operative candidates should not be excluded from the consideration of spine SBRT. However, such patients require multidisciplinary evaluation, careful consideration of treatment parameters, and discussion and documentation of risks and benefits prior to proceeding with treatment.

10.3 TREATMENT

10.3.1 IMMOBILIZATION AND IMAGING

Patient immobilization and maximizing patient comfort are of critical importance in all SBRT treatments, and this is even more true when reirradiating the spine. High-dose gradients around the spinal cord could lead to disastrous consequences if there is a significant misalignment or intra-fraction motion. The key principles in the delivery of spine SBRT are to (1) adequately immobilize the patient, (2) determine the position of the target prior to and during radiation delivery, and (3) ensure that the target is in the same position as it was during the time of CT simulation.

In our practice, patient immobilization is achieved using a semirigid vacuum-assisted body cushion such as the BodyFIX system (Elekta AB, Stockholm, Sweden) for patients being treated to the mid-thoracic spine or lower and a thermoplastic head and shoulder mask in combination with a long stereotactic cradle for patients being treated to the cervical or upper thoracic spine. Alignment is performed with a stereotactic localizer and target positioning frame (Integra-Radionics, Burlington, MA). An on-board imaging system such as a cone-beam CT (CBCT), CT-on-rails, or stereoscopic x-ray (i.e., ExacTrac, BrainLAB AG, Feldkirchen, Germany) is required. These image-guidance systems can be used to monitor and manage both inter-fraction and intra-fraction motions. Correction of translation and rotational shifts are achieved with either robotic couch movements or via the robotic treatment arm in the case of CyberKnife (Accuray, Sunnyvale, CA). It is important to note that even with the use of systems that provide real-time tracking such as CyberKnife, some degree of immobilization is important for both patient comfort and to minimize body movements that may occur during the SBRT treatment. For systems that utilize frequent intrafractional imaging every 15 seconds, for example, a simpler Vac-Lok or Alpha Cradle system may suffice, but it is our preference to use whole-body immobilization whenever possible (Lo et al., 2010).

A treatment planning CT, with or without contrast, is obtained with 1–2 mm slices through the area of interest. A contrast-enhanced MRI of the spine area of interest is also obtained for treatment planning and fused to the treatment planning CT when feasible. The MRI should be performed as close to the time of simulation as possible and preferably in a similar position, although patient positioning may be limited by MRI bore size. It has been noted that the spinal cord itself can shift up to 2 mm based on position alone (Van Mourik et al., 2014). Accurate delineation of the extent of disease on spine MRI with a robust CT/MRI fusion is required as significant contouring information is contained within the MRI.

After surgery, artifact from metallic hardware frequently limits the visibility of the spinal cord and epidural tumor on both MRI and CT. In such instances, a CT myelogram is highly recommended

(Redmond et al., 2017; Sahgal et al., 2011). However, CT myelography does not assist with artifact masking of residual tumor or "at-risk" areas in the post-surgical spine. Verified fusion of the preoperative MRI with the planning CT and communication with the operating neurosurgeon is important to accurately delineate the target volume in these cases.

10.3.2 TARGET VOLUME AND ORGAN AT RISK CONTOURING

Target volumes are typically defined on the treatment planning CT with the aid of fused MRI sequences obtained as described earlier. Gross tumor volume (GTV) is defined as gross disease seen on T1- and T2-weighted MRI sequences. Historically, there has been a great deal of variability regarding the delineation of target volumes in the spine from institution to institution. Currently, we define clinical treatment volume (CTV) according to the standards outlined in the Radiation Therapy Oncology Group (RTOG) 0631 protocol and the international standard, as published by the International Spine Radiosurgery Consortium (Cox et al., 2012; Redmond et al., 2017). A further expansion of the CTV to generate a planning treatment volume (PTV) may be considered up to 2 mm depending on the institution, type of immobilization, and pretreatment imaging used. However, we do not routinely employ a PTV margin.

Accurate contouring of the spinal cord is essential, as it is the primary dose-limiting structure. This is important in all spine SBRT cases, but even more so in the reirradiation setting, where commonly used dose constraints for the spinal cord are often exceeded and the margin for error is presumed to be small. Fusion of the diagnostic MRI to the treatment planning CT can aid delineation of the tumor and spinal cord itself. However, one must account for spatial distortion inherent in MRI as well as uncertainties related to fusion of patient anatomy from different imaging modalities where patient position may vary. In the postoperative setting, hardware artifact often makes accurate delineation of the spinal cord impossible, and it highly advantageous to use a CT myelogram to delineate the spinal cord in this setting.

We do not create a PRV expansion around the spinal cord, although it is common practice at other institutions to contour either the thecal sac (which is roughly equivalent to a 1.5 mm expansion around the cord) or generate a uniform 1.5–2 mm expansion around the spinal cord to create a spinal cord planning risk volume (PRV) in order to account for uncertainties in positioning and organ movement. Regardless of the method, it is important to understand and consider imaging, fusion, setup margin, treatment delivery uncertainties, and spinal cord constraints reported in literature based on different definitions of the spinal cord volume so as not to underdose the GTV or exceed recommended spinal cord safety margins, which will be discussed later on in this chapter. Relevant at-risk organs near the spine level to be treated, typically 1 cm superior and inferior to the PTV, are also contoured which include, but not limited to, the kidney, esophagus, heart, great vessels, bowel, and liver.

10.3.3 TREATMENT PLANNING—DOSE, FRACTIONATION, AND CONSTRAINTS

In delivering reirradiation, it is necessary to know the prior dose given to the spinal cord. In instances where the patient was treated at an outside institution, efforts must be made to obtain copies of the treatment records, including treatment planning images and dose–volume histograms (DVH) of OARs. It cannot simply be assumed that the spinal cord uniformly received the prescribed dose, even when using opposed fields. In many cases, the cord may have received doses above the prescribed dose, depending on the beam arrangement, possible wedges, and beam weightings. If the patient was treated emergently using a hand-calculated plan, the relevant treatment parameters should be obtained and the fields reconstructed using a computerized treatment planning system in order to accurately model prior dose to the spinal cord. As a general rule, uncertainty regarding prior dose should lead one to err on the side of caution, as the results of an inadvertent cord overdose could be significant.

Once accurate models of prior radiation dose are obtained and reirradiation dose constraints for OARs determined, SBRT planning for retreatment is much the same as would occur if no prior radiation had been done. Inverse planning is performed with inverse-treatment software such as Pinnacle (Philips Medical Systems, Andover, MA). A radiation plan can be generated using intensity-modulated radiotherapy and multiple step-and-shoot static fields typically ranging from 7 to 9 based on spine class solutions, or volumetric modulated arc therapy (VMAT), which has the advantage of decreased treatment delivery times and therefore also reducing the amount of intrafraction patient motion.

Commonly prescribed fractionation schemes in the reirradiation spine SBRT setting include 30 Gy in 5 fractions, 30 Gy in 4 fractions, 27 Gy in 3 fractions, 24 Gy in 3 fractions, 20 Gy in 5 fractions, and 20.6 Gy in 1 fraction (Myrehaug et al., 2017). Unfortunately, there is a paucity of high-quality data comparing fractionation schemes and little consensus as to the optimal dose and fraction number. For *de novo* spine metastases, single fraction SBRT may be associated with improved local control compared to multifraction regimens (Ghia et al., 2016). However, we generally favor multifraction regimens in the reirradiation setting off trial due to the potential for increased late toxicities with single fraction regimens. Early studies have demonstrated the superiority of 30 Gy in 5 fractions over 20 Gy in 5 fractions, although the latter is generally not considered to be within the ablative range (Damast et al., 2011). In our practice of reirradiation, we prescribe 27 Gy in 3 fractions to the GTV using a simultaneous integrated boost (SIB) technique and either 24 Gy or 21 Gy to the CTV for radioresistant or radiosensitive histologies, respectively. Comparing single versus multifraction treatment in the salvage reirradiation setting is the topic of an ongoing phase II trial (NCT03028337).

While dosimetric coverage of the target volumes to 95% or more of the prescribed dose is routine for other anatomic sites, this level of coverage may be difficult to achieve given the stricter cord constraints used in the retreatment setting. In our experience, coverage as low as 70% may be acceptable, as dictated by spinal cord limits, other OARs, and the clinical scenario. However, the competing goals of tumor coverage to prevent disability from local tumor progression and spinal cord avoidance to prevent treatment-related toxicities must be carefully balanced. As stated earlier in Chapter 10.2, certain clinical situations may warrant relaxation of commonly used spinal dose constraints, or in other cases consideration of limited procedures such as separation surgery or laser interstitial thermotherapy, which will be discussed in more detail later on.

Constraints to the spinal cord are dictated by the prior dose received without accounting for time elapsed between treatments, as there has been no reliable model to date for spinal cord recovery following radiotherapy. Sahgal et al. published a report comparing reirradiation spinal cord doses in 5 patients who developed radiation myelopathy compared to 14 control cases. The authors found that the risk of myelopathy was low provided that (1) the cumulative maximum dose to the thecal sac normalized to 2 Gy fractions assuming an alpha/beta ratio of 2 (i.e., $EQD2_2$) was ≤ 70 Gy, (2) the SBRT maximum thecal sac $EQD2_2$ was ≤ 25 Gy, and (3) the SBRT component comprised less than 50% of the cumulative normalized dose (Sahgal et al., 2012). The recently published HyTEC report continues to support these constraints due to lack of additional data (Sahgal et al., 2019). Although certain retrospective series have reported treating to mean cumulative spinal cord doses as high as $EQD2_2$ of 83.4 Gy without cases of RM, the data are generally not felt to be of high enough quality to guide treatment recommendations (Chang et al., 2012). A table summarizing the constraints on the spinal cord for various prior dose and fraction schemes is contained within the original Sahgal et al. 2012 publication.

While the linear quadratic model is frequently used to model biologically effective dose (BED) for conventionally fractionated radiotherapy, multiple studies have found it to be inappropriate for application to SBRT and fractional doses of 5–6 Gy or more (Lee et al., 2015). In addition, the aforementioned dose tolerance studies did not include patients treated with single fraction cEBRT as initial therapy. Therefore, limited radiobiological and clinical data in regard to risk for the development of RM in this specific scenario should engender caution when planning retreatment for patients who, for example, were initially treated with 8 Gy in a single fraction. Other OAR constraints within the treatment plan should be used according to published American College of Radiology (ACR) or institutional guidelines (Potters et al., 2010).

10.3.4 TREATMENT DELIVERY

At our institution, we apply a rigorous sequence of image guidance for multiple layers of confirmation of inter- and intrafractional patient positioning. We utilize a linear accelerator with micro multileaf collimators (MLCs) in combination with a 6D robotic treatment couch. Each day, patients are first aligned with ExacTrac, followed by CBCT, kilovoltage (kV) and megavoltage (mV) orthogonal films, and an additional ExacTrac verification image after each beam sequence. This method has been prospectively studied and shown to be as accurate as the use of implantable fiducial markers (Weksberg et al., 2016). In other practices, treatment delivery is performed using pretreatment alignment via kV imaging to match to bony anatomy followed by CBCT or CT-on-rails. The CyberKnife system will adjust based on the alignment to bony anatomy using stereoscopic kV imaging in real time while radiation dose is being delivered. Adjustments are made with the goal of having the target and spinal cord volumes aligned within a preset tolerance (preferably <1 mm) in all directions, including rotation (<1°), which may be adjusted automatically or manually, depending on the capabilities of the treatment table and delivery system. Once treatment is initiated, realignment should be performed as needed, either after a preset elapsed time or if there is suspicion that there has been a movement of the target.

10.4 CASE STUDY

A 58-year-old woman with follicular thyroid carcinoma had spinal metastases discovered on routine staging evaluation at the T3 and T6 vertebral bodies. She went on to receive 30 Gy in 10 fractions palliative radiotherapy from T1-T7 (Figure 10.1). She later had progression of disease at both T3 and

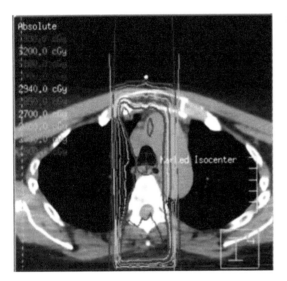
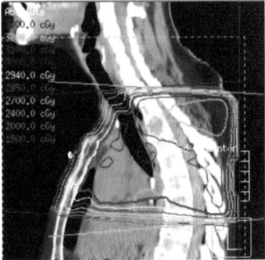
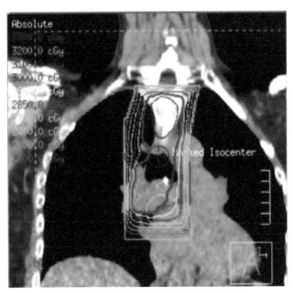

Figure 10.1 Treatment plan images in the axial (a), sagittal (b), and coronal (c) planes from a conventional plan delivering 30 Gy in 10 fractions to the T1–T7 vertebral levels with anterior–posterior and posterior–anterior beams. Isodose lines correspond to 32 Gy or 107% (green), 30 Gy or 100% (red), and 27 Gy or 90% (teal).

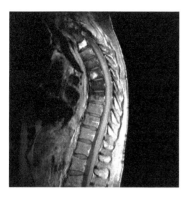
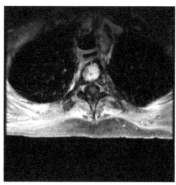
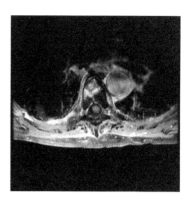

Figure 10.2 T1-weighted post-contrast magnetic resonance images in the sagittal plane (a) depicting the T3 and T6 lesions and axial plane images depicting the progressive T3 lesion (b) and T6 lesion (c).

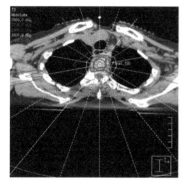
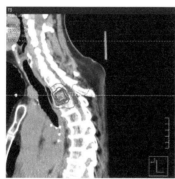
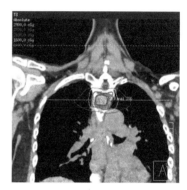

Figure 10.3 Treatment plan images for the *T3* vertebral level SBRT reirradiation plan delivering 27 Gy to the GTV and 21 Gy to the CTV in 3 fractions with a nine-field IMRT setup. Isodose lines correspond to 27 Gy (blue), 21 Gy (red), 16 Gy (yellow), and 10 Gy (pink). Note the avoidance of the spinal cord with the 10 Gy isodose line.

T6 without significant epidural disease approximately 3 years after initial treatment (Figure 10.2). After multidisciplinary evaluation, she was dispositioned to reirradiation spine SBRT. The CTV was delineated according to the International Spine Radiosurgery Consortium guidelines and in this case, as the GTV only involved the vertebral bodies of T3 and T6, included the entire vertebral bodies and bilateral pedicles. The GTV was prescribed 27 Gy and the CTV prescribed 21 Gy in 3 fractions for both the T3 (Figure 10.3) and the T6 (Figure 10.4) lesions, which were treated simultaneously on 3 consecutive days.

Based on a maximum spinal cord dose of 32 Gy in 10 fractions from the initial treatment, which corresponds to an $EQD2_2$ of 41.6 Gy, an effort was made to keep the 10 Gy isodose line out of the proper spinal cord, and her SBRT maximum spinal cord dose was ultimately limited to ≤12 Gy in 3 fractions, which corresponds to an $EQD2_2$ of 18 Gy. She tolerated the treatment well without any toxicity and had stable thoracic spine disease on surveillance imaging more than 7 years after spine SBRT retreatment.

10.5 OUTCOMES

The published literature on reirradiation of spinal metastases with SBRT reports results that are promising. Outcomes from selected studies have been summarized in Table 10.1. The interpretation of tumor control rates vary from study to study, as it is uncommon for suspected failures to be confirmed by biopsy. With

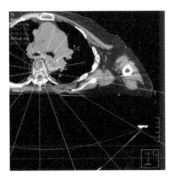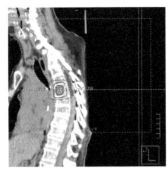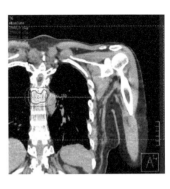

Figure 10.4 Treatment plan images for the *T6* vertebral level SBRT reirradiation plan delivering 27 Gy to the GTV and 21 Gy to the CTV in 3 fractions with a nine-field IMRT setup. Isodose lines correspond to 27 Gy (blue), 21 Gy (red), 16 Gy (yellow), and 10 Gy (pink). Note the avoidance of the spinal cord with the 10 Gy isodose line.

this caveat, 1-year local control rates range from 73% to 92%, with the majority of studies reporting 1-year local control rates over 80%. In a series from the University of California, San Francisco (UCSF), no significant difference was noted in failure rates between SBRT targets in a previously irradiated field and those that had not received prior radiation (Sahgal et al., 2009).

The most common site of failure in patients retreated with SBRT is within the epidural space. This has been noted in a number of published reports (Chang et al., 2007; Sahgal et al., 2009; Choi et al., 2010; Garg et al., 2011). Failure was correlated with increasing proximity of the tumor to the thecal sac. This is expected, given the need to constrain the dose to the spinal cord, which is even more stringent in the reirradiation setting. It may also be caused by inadequate coverage of the "at-risk" epidural space in order to reduce the dose given circumferentially around the spinal cord. Better understanding of which areas of the epidural space are at highest risk for microscopic extension combined with utilization of standardized contouring guidelines will increase our understanding of epidural patterns of failures. Other common sites of failure include paraspinal tissue and segments of the spinal column not included within the treatment volume, such as the pedicles and lamina of the posterior elements (Detsky et al., 2020; Masucci et al., 2011).

The use of separation surgery may help offset the increased risk of epidural failure in patients with MESCC undergoing spine SBRT. Separation surgery is performed with decompression, spinal stabilization, and resection of epidural disease without significant vertebral body resection (Moussazadeh et al., 2014). Large retrospective series have demonstrated favorable outcomes, making it particularly useful in cases where a greater distance between disease and spinal cord is needed (Bate et al., 2015; Laufer et al., 2013). Advocates of combining separation surgery with post-operative spine SBRT have suggested that the paradigm has radically changed the goals of surgical management of spine metastases (Laufer et al., 2013). However, high quality comparative studies are still needed to validate this approach.

A recent randomized, phase II trial evaluated pain response in patients with spinal metastases dispositioned to either 24 Gy in a single fraction or 30 Gy in 10 fractions and found a quicker and more marked pain response with the spine SBRT arm. There was an impressive 74% overall pain response rate compared to 35% for the cEBRT arm at 6 months (Sprave et al., 2018). Another randomized, phase II non-inferiority study of patients with painful non-spine bone metastases reported superior pain response rates and durability using single fraction SBRT compared to cEBRT (Nguyen et al., 2019). Pain control in retreatment spine SBRT series is in the range of 65% to 85%, which appears to be slightly lower than that reported for *de novo* spine SBRT (Masucci et al., 2011). Despite inherent differences in patient selection and heterogeneity, pain control following retreatment seems to be slightly better for spine SBRT compared to cEBRT, where pain response rates are reported to be around 58% (Huisman et al., 2012). It may be the case that pain control is more difficult to achieve in patients undergoing reirradiation spine SBRT, but patients with recurrent

Table 10.1 Select published outcomes for reirradiation of spinal tumors using stereotactic body radiation therapy

STUDY	# OF PATIENTS /#TREATMENTS	MEDIAN DOSE FIRST RT (GY)	MEDIAN INTERVAL (MO)	REIRRADIATION TD/ FRACTION NUMBER	MEDIAN CUMULATIVE DOSE TO SPINAL CORD (EQD2)	PLAN-NING	SETUP/ IMAGING	FOLLOW-UP (MO)	RM	LOCAL CONTROL	PAIN CONTROL
Detsky et al. (2020)	3/83	20	13.6	20–30 Gy/1–5	61.1 Gy	IMRT	CBCT	12.4	0%	86% at 1 year	N/A
Kawashiro et al. (2016)	23/23	30	13	24.5 Gy/5	N/A	IMRT	CBCT or CT-on-rails	10	0%	88% at 1 year	78.9%
Thibault et al. (2015)	40/56	24	12.9	30 Gy/4	54.6 Gy	IMRT	CBCT	6.8	0%	81% at 1 year	N/A
Thibault et al. (2015)	37/71	30	16	24 Gy/2	37.76 Gy	IMRT	CBCT	12.3	0%	73% at 1 year	N/A
Chang et al. (2012)	49/54	39.2	25	27 Gy/3	83.4 Gy	CK	kV tracking	17.3	0%	80.8% at 1 year	80.8% at 1 year
Damast et al. (2011)	95/97	30	N/A	20–30 Gy/5	54.3 Gy	IMRT	CBCT or daily ports	12.1	0%	74% at 1 year (30 Gy/5)	77%
Garg et al. (2011)	59/63	30	N/A	27–30 Gy/3–5	N/A	IMRT	CBCT or daily CT-on-rails	13	n = 2, G3 peripheral nerve injury	76% at 1 year	N/A
Mahadevan et al. (2011)	60/81	30	20	24–30 Gy/3–5	N/A	CK	kV tracking	12	n = 3, radicular pain; n = 1, lowerextremity weakness	93% at f/u	65% at 1 month
Choi et al. (2010)	42/51	40	19	20 Gy/2	76 Gy	CK	kV tracking	7	n = 1, G4	73% at 1 year	65%
Sterzing et al. (2010)	36/36	36.3	17.5	34.8 Gy/11	46.5 Gy	Tomo	Daily MVCT	7.5	0%	76% at 1 year	N/A
Sahgal et al. (2009)	25/37	36	11	24 Gy/3	N/A	CK	kV tracking	7	0%	90%	N/A
Mahan et al. (2005)	8/8	30	N/A	30 Gy/15	48 Gy	Tomo	Daily MVCT	15.2	0%	100% at f/u	100% at f/u
Milker-Zabel et al. (2003)	18/19	38	17.7	39.6 Gy	N/A	IMRT	Stereotactic	12.3	0%	94.7% at 1 year	81.3%

Abbreviations: RT, radiotherapy; TD, total dose; N/A, not available; G, grade; CBCT, cone-beam computed tomography; kV, kilovoltage; IMRT, intensity-modulated radiotherapy; MVCT, megavoltage computed tomography; sIMRT, intensity-modulated radiotherapy; f/u, follow-up; CK, CyberKnife; mo, months; Tomo, tomotherapy; RM, radiation myelopathy.

pain should be still be considered for retreatment based on the data above demonstrating that the majority of patients will achieve some pain relief.

10.6 TOXICITY

The most feared late toxicity of spine SBRT is radiation myelitis (RM). RM presents as a late complication involving progressive myelopathy that is irreversible. The clinical presentation can range from minor sensory deficits to complete paralysis. Other side effects associated with spine SBRT include a self-limited pain flare, Lhermitte's syndrome, vertebral compression fracture, and radiculopathy. However, data analyzing these complications are sparse in the *de novo* spine SBRT literature and completely absent from the reirradiation literature. Therefore, we focus our discussion of toxicity in this setting to RM.

Fortunately, RM secondary to spine SBRT is a rare event. Following *de novo* spine SBRT, the largest series to date reported a rate of less than 1% (Katsoulakis et al., 2017). However, its rare incidence also makes it difficult to accurately model toxicity. In the reirradiation setting, this is further complicated by the heterogeneity in initial and retreatment dose fractionation schedules as well as time interval between radiation treatments. Animal models have historically been used to investigate the tolerance of the spinal cord to radiation. One group of investigators reirradiated monkey spinal cords after administering an initial treatment of 44 Gy. Using two different conventionally fractionated retreatment doses, it was found that in a conservative model, 61% recovery of the cord could be assumed after 3 years and in more optimistic models even higher rates of recovery could occur (Ang et al., 2001). Another study used pigs whose cords had been previously irradiated to 30 Gy in 10 fractions. After 1 year, an inhomogeneous ablative dose of radiation was delivered in a single fraction lateral to the spinal cord. No increase in myelopathy was seen in pigs that had been previously irradiated when compared to a control group, suggesting that significant, perhaps even complete, cord recovery had occurred at 1 year (Medin et al., 2012).

Data from clinical studies are limited by small patient numbers but have informed practice. An early clinical study examining data from patients treated with multiple courses of EBRT found that significant spinal cord recovery did occur and that cumulative doses of $BED_2 \leq 135.5$ Gy, which translates to approximately $EQD2_2 \leq 70$ Gy, were associated with low risk of RM (Nieder et al., 2006). An analysis of the reirradiation spinal cord tolerance for SBRT contained five patients who developed RM, an indication of the rarity of the complication. The results, summarized earlier in the Treatment Planning section, suggest that cumulative doses of $EQD2_2 \leq 70$ Gy could be delivered safely (Sahgal et al., 2012). A reanalysis of the same cohort using the generalized linear–quadratic model found a similar result, with no cases of RM observed under an $EQD2_2$ of 70 Gy (Huang et al., 2013).

Given the agreement in the prior studies, a cumulative dose threshold of $EQD2_2 \leq 70$ Gy can be used as a rough guideline with the caveat that the evidence remains weak. Clinical judgment is required in all cases. Based on institutional experience, our practice is to limit patients who previously received the dose of 30 Gy in 10 fractions to the spinal cord to a maximum SBRT spinal cord dose of 10 Gy in 3 fractions, which is a more conservative approach. Ongoing clinical trials are investigating safe reirradiation dose escalation (NCT02278744).

It is also worth noting that the median latency time from treatment to RM is shorter with spine SBRT and even shorter with reirradiation. In a recent literature review, median time to RM following reirradiation spine SBRT was 6 months, a time half the 12-month latency time following *de novo* spine SBRT and one-third the 18-month latency time following *de novo* cEBRT (Sahgal et al., 2019).

Other OARs exposed to radiation during a prior course of radiation can also receive significant radiation dose during reirradiation spine SBRT. However, it is unknown whether this fact has resulted in increased rates of toxicity. The esophagus is another important organ that can be injured by SBRT, and a report examining esophageal toxicity when treating spine and lung tumors urged caution, especially after prior radiotherapy and/or chemotherapy (Abelson et al., 2012). A study examining patients treated with single-fraction SBRT for paraspinal metastases abutting the esophagus found a grade ≥ 3 toxicity rate of 6.8%, while another study examining patients treated with spine SBRT in the thoracic region reported a grade ≥3 toxicity rate of 1.8% (Huo et al., 2017). Overall, toxicity of non-spine OARs is a rare event in spine SBRT, with most reported cases being of mild severity (Chawla et al., 2013).

10.7 CONTROVERSIES

Reirradiation of any tumor is challenging since radioresistant biology might lead to additional radiation treatments that would be both risky and futile. However, the radiobiology of doses in the ablative range, such as those used with SBRT, is now understood to be distinct from that of cEBRT (Brown et al., 2014). Although these differences are still poorly understood, local control rates associated with ablative doses are comparable whether or not prior radiation has been used. Other concerns regarding the use of SBRT primarily involve the use of a high resource treatment in patients whose prognosis may be limited.

The use of SBRT in cases of MESCC is an emerging indication that requires significant institutional resources and expertise to perform. A retrospective review of 33 patients with MESCC treated with spine SBRT, published in 2014, demonstrated a 67% ambulatory rate and epidural tumor response rate of 74% with over a year of follow-up; however, one patient developed RM in this cohort (Lee et al., 2014). Thirty-two patients with inoperable MESCC, including 10 patients with high-grade (Bilsky 2 or 3) disease, were treated with initial spine SBRT with a 1-year local control of 89% and no cases of RM in a phase I trial (Ghia et al., 2018). These data support the feasibility of spine SBRT for patients with MESCC, although additional data on safety and efficacy are still needed to support this expanding indication.

10.8 FUTURE DIRECTIONS

Emerging techniques combining minimally invasive surgical techniques with spine SBRT have shown favorable outcomes and are potentially valuable options in the setting of a locally recurrent spine tumor, especially in those patients with high-grade epidural disease. Laser interstitial thermotherapy (LITT) is a novel surgical approach to spinal tumors invading the epidural space. LITT involves image-guided placement of catheters that provide thermal ablation of tumor at the interface of dura mater. A recent matched-group study comparing LITT with post-operative spine SBRT in 80 patients with at least Bilsky grade 1c epidural disease showed similar overall survival and progression-free survival between the groups with significantly lower estimated blood loss, hospital length of stay, overall complication rate, days until radiotherapy, and days until systemic treatment in the LITT group (De Almeida Bastos et al., 2020). Notably, one-third of the patients in the LITT group had received prior radiotherapy, showing the concept is feasible in cases of reirradiation spine SBRT.

The combination of immunotherapy and SBRT has been an area of active interest in multiple disease sites due to the proposed immunogenicity of SBRT and potential for a synergistic immune-modulated response toward cancer cells (Brooks et al., 2016). Retrospective studies have found a survival benefit when combining immunotherapy and SRS for brain metastases (Diao et al., 2018). However, little is known about the potential for synergy and role of combination therapy for spinal metastases. In addition, more information is needed regarding the safety of immunotherapy and targeted therapy when delivering spine SBRT to guide practice. On the opposite side of the spectrum, work is being undertaken to better understand the mechanism of RM as there are therapeutic agents that may prevent or mitigate RM risk (Wong et al., 2015).

In the end, high-quality studies are needed to compare and optimize dose and fractionation schedules used in SBRT. As the number of patients reirradiated with spine SBRT increases, a new problem may arise as to how to manage multiple spine retreatment failures. Multi-institutional collaboration and improved reporting standards are necessary to better define the maximal human spinal cord tolerance to SBRT in both the *de novo* and the reirradiation setting, while additional radiobiological studies are needed to accurately model the effects of ultra-hypofractionated radiation on normal tissue (Hrycushko et al., 2019).

CHECKLIST: KEY POINTS FOR CLINICAL PRACTICE

✓	ACTIVITY	CONSIDERATIONS
	Patient selection	• Does not have high-grade spinal cord compression, spinal instability, or is not a surgical candidate • Expected survival >6 months • KPS >50 • Interval from initial radiation >3 months (5–6 months preferred) • Oligometastatic or oligoprogressive disease or radioresistant histology • Patient evaluated by multidisciplinary team • Able to lay flat comfortably • No more than 2–3 contiguous vertebral bodies involved
	Simulation	*Immobilization* • Supine with semirigid stereotactic body immobilization system (Linac) or Alpha Cradle (CyberKnife) *Imaging* • Simulation CT scan with 1–2 mm slice thickness • Recent diagnostic MRI with gadolinium contrast • Verified CT to MRI fusion over area of tumor involvement • Consider CT myelogram if spine instrumentation present
	Target delineation	*Contours* • GTV as defined on MR and CT imaging • CTV per International Spine Radiosurgery Consortium guidelines, dependent on the extent of spine involvement • PTV expansion of 0–2 mm, based on institutional standards • Normal OARs contoured 1 cm superior and inferior to PTV

	Treatment planning	• Inverse treatment planning with IMRT or VMAT
		• Dose
		  • Institution dependent but consider 24–30 Gy in 2–5 fractions
		  • SIB may be considered
		• Coverage
		  • PTV prescription coverage 70% to 95% (highly dependent on clinical scenario)
		• Reirradiation dose constraints
		  • Spinal cord: thecal sac cumulative $EQD2_2 \leq 70$ Gy, SBRT maximum $EQD2_2 \leq 25$ Gy

	Treatment delivery	• Patients are first set up using room lasers and manual shifts or with stereoscopic X-ray imaging • CBCT to match to tumor (CTV or PTV) and spinal cord • Orthogonal kV or mV matching to relevant bony anatomy • Treatment couch with six degrees of freedom to correct translational and rotational shifts if available • Repeat CBCT or stereoscopic x-ray imaging at regular treatment time intervals or for suspected patient movement

REFERENCES

Abelson JA, Murphy JD, Loo BW, Chang DT, Daly ME, Wiegner EA, Hancock S et al. Esophageal tolerance to high-dose stereotactic ablative radiotherapy. *Diseases of the Esophagus*. 2012;25(7):623–629.

Ang KK, Jiang GL, Feng Y, Stephens LC, Tucker SL, Price RE. Extent and kinetics of recovery of occult spinal cord injury. *International Journal of Radiation Oncology, Biology, Physics*. 2001;50(4):1013–1020.

Bate BG, Khan NR, Kimball BY, Gabrick K, Weaver J. Stereotactic radiosurgery for spinal metastases with or without separation surgery. *Journal of Neurosurgical Spine*. 2015;22(4):409–415.

Brooks ED, Schoenhals JE, Tang C, et al. Stereotactic ablative radiation therapy combined with immunotherapy for solid tumors. *Cancer Journal*. 2016;22(4):257–266.

Brown JM, Carlson DJ, Brenner DJ. The tumor radiobiology of SRS and SBRT: Are more than the 5 Rs involved? *International Journal of Radiation Oncology, Biology, Physics*. 2014;88(2):254–262.

Chang EL, Shiu AS, Mendel E, et al. Phase I/II study of stereotactic body radiotherapy for spinal metastasis and its pattern of failure. *Journal of Neurosurgical Spine*. 2007;7(2):151–160.

Chang U-K, Cho W-I, Kim M-S, Cho CK, Lee DH, Rhee CH. Local tumor control after retreatment of spinal metastasis using stereotactic body radiotherapy; comparison with initial treatment group. *Acta Oncologica (Stockh Swed)*. 2012;51(5):589–595.

Chawla S, Schell MC, Milano MT. Stereotactic body radiation for the spine: A review. *American Journal of Clinical Oncology*. 2013;36(6):630–636.

Choi CYH, Adler JR, Gibbs IC, Chang SD, Jackson PS, Minn AY, Lieberson RE, Soltys SG. Stereotactic radiosurgery for treatment of spinal metastases recurring in close proximity to previously irradiated spinal cord. *International Journal of Radiation Oncology, Biology, Physics*. 2010;78(2):499–506.

Chow E, Harris K, Fan G, Tsao M, Sze WM. Palliative radiotherapy trials for bone metastases: A systematic review. *Journal of Clinical Oncology*. 2007;25(11):1423–1436.

Cox BW, Spratt DE, Lovelock M, Bilsky MH, Lis E, Ryu S, Sheehan J et al. International spine radiosurgery consortium consensus guidelines for target volume definition in spinal stereotactic radiosurgery. *International Journal of Radiation Oncology, Biology, Physics*. 2012;83(5):e597–605.

Damast S, Wright J, Bilsky M, Hsu M, Zhang Z, Lovelock M, Cox B, Zatcky J, Yamada Y. Impact of dose on local failure rates after image-guided reirradiation of recurrent paraspinal metastases. *International Journal of Radiation Oncology, Biology, Physics*.2011;81(3):819–826.

De Almeida Bastos DC, Everson RG, De oliveira santos BF, et al. A comparison of spinal laser interstitial thermotherapy with open surgery for metastatic thoracic epidural spinal cord compression. *Journal of Neurosurgical Spine*. 2020;1–9.

Detsky JS, Nguyen TK, Lee Y, et al. Mature imaging-based outcomes supporting local control for complex reirradiation salvage spine stereotactic body radiotherapy [in-press, 2020]. *Neurosurgery*. doi:10.1093/neuros/nyaa109.

Diao K, Bian SX, Routman DM, et al. Stereotactic radiosurgery and ipilimumab for patients with melanoma brain metastases: Clinical outcomes and toxicity. *Journal of Neurooncology*. 2018;139(2):421–429.

Ecker RD, Endo T, Wetjen NM, Krauss WE. Diagnosis and treatment of vertebral column metastases. *Mayo Clinic Proceedings*. 2005;80(9):1177–1186.

Garg AK, Wang X-S, Shiu AS, Allen P, Yang J, McAleer MF, Azeem S, Rhines LD, Chang EL. Prospective evaluation of spinal reirradiation by using stereotactic body radiation therapy: The University of Texas MD Anderson Cancer Center experience. *Cancer* 2011;117(15):3509–3516.

Ghia AJ, Chang EL, Bishop AJ, et al. Single-fraction versus multifraction spinal stereotactic radiosurgery for spinal metastases from renal cell carcinoma: Secondary analysis of Phase I/II trials. *Journal of Neurosurgical Spine*. 2016;24(5):829–836.

Ghia AJ, Guha-thakurta N, Hess K, et al. Phase 1 study of spinal cord constraint relaxation with single session spine stereotactic radiosurgery in the primary management of patients with inoperable, previously unirradiated metastatic epidural spinal cord compression. *International Journal of Radiation Oncology, Biology, Physics*. 2018;102(5):1481–1488.

Hamilton AJ, Lulu BA, Fosmire H, Stea B, Cassady JR. Preliminary clinical experience with linear accelerator-based spinal stereotactic radiosurgery. *Neurosurgery*. 1995;36(2):311–319.

Hartsell WF, Scott CB, Bruner DW, et al. Randomized trial of short- versus long-course radiotherapy for palliation of painful bone metastases. *Journal of the National Cancer Institute*. 2005;97(11):798–804.

Hrycushko B, Van der kogel AJ, Phillips L, et al. Spinal nerve tolerance to single-session stereotactic ablative radiation therapy. *International Journal of Radiation Oncology, Biology, Physics*. 2019;104(4):845–851.

Huang Z, Mayr NA, Yuh WT, Wang JZ, Lo SS. Reirradiation with stereotactic body radiotherapy: Analysis of human spinal cord tolerance using the generalized linear-quadratic model. *Future Oncology (Lond Engl)*. 2013;9(6):879–887.

Huisman M, Van den bosch MA, Wijlemans JW, Van vulpen M, Van der linden YM, Verkooijen HM. Effectiveness of reirradiation for painful bone metastases: A systematic review and meta-analysis. *International Journal of Radiation Oncology, Biology, Physics.* 2012;84(1):8–14.

Huo M, Sahgal A, Pryor D, Redmond K, Lo S, Foote M. Stereotactic spine radiosurgery: Review of safety and efficacy with respect to dose and fractionation. *Surgical Neurology International.* 2017;8:30.

Kaasa S, Brenne E, Lund JA, et al. Prospective randomised multicenter trial on single fraction radiotherapy (8 Gy x 1) versus multiple fractions (3 Gy x 10) in the treatment of painful bone metastases. *Radiotherapy & Oncology.* 2006;79(3):278–284.

Katsoulakis E, Jackson A, Cox B, Lovelock M, Yamada Y. A detailed dosimetric analysis of spinal cord tolerance in high-dose spine radiosurgery. *International Journal of Radiation Oncology, Biology, Physics.* 2017;99(3):598–607.

Kawashiro S, Harada H, Katagiri H, et al. Reirradiation of spinal metastases with intensity-modulated radiation therapy: An analysis of 23 patients. *Journal of Radiation Research.* 2016;57(2):150–156.

Klimo P, Schmidt MH. Surgical management of spinal metastases. *Oncologist.* 2004;9(2):188–196.

Laufer I, Iorgulescu JB, Chapman T, et al. Local disease control for spinal metastases following "separation surgery" and adjuvant hypofractionated or high-dose single-fraction stereotactic radiosurgery: Outcome analysis in 186 patients. *Journal of Neurosurgical Spine.* 2013;18(3):207–214.

Laufer I, Rubin DG, Lis E, Cox BW, Stubblefield MD, Yamada Y, Bilsky MH. The NOMS framework: Approach to the treatment of spinal metastatic tumors. *Oncologist.* 2013;18(6):744–751.

Lee I, Omodon M, Rock J, Shultz L, Ryu S. Stereotactic radiosurgery for high-grade metastatic epidural cord compression. *Journal of Radiosurgery & SBRT.* 2014;3(1):51–58.

Lee SH, Lee KC, Choi J, et al. Clinical applicability of biologically effective dose calculation for spinal cord in fractionated spine stereotactic body radiation therapy. *Radiology & Oncology.* 2015;49(2):185–191.

Lo SS, Sahgal A, Wang JZ, Mayr NA, Sloan A, Mendel E, Chang EL. Stereotactic body radiation therapy for spinal metastases. *Discovery Medicine.* 2010;9(47):289–296.

Mahadevan A, Floyd S, Wong E, Jeyapalan S, Groff M, Kasper E. Stereotactic body radiotherapy reirradiation for recurrent epidural spinal metastases. *International Journal of Radiation Oncology, Biology, Physics.* 2011;81(5):1500–1505.

Mahan SL, Ramsey CR, Scaperoth DD, Chase DJ, Byrne TE. Evaluation of image-guided helical tomotherapy for the retreatment of spinal metastasis. *International Journal of Radiation Oncology, Biology, Physics.* 2005;63(5):1576–1583.

Masucci GL, Yu E, Ma L, Chang EL, Letourneau D, Lo S, Leung E et al. Stereotactic body radiotherapy is an effective treatment in reirradiating spinal metastases: Current status and practical considerations for safe practice. *Expert Review of Anticancer Therapy* 2011;11(12):1923–1933.

Medin PM, Foster RD, van der Kogel AJ, Sayre JW, McBride WH, Solberg TD. Spinal cord tolerance to reirradiation with single-fraction radiosurgery: A swine model. *International Journal of Radiation Oncology, Biology, Physics.* 2012;83(3):1031–1037.

Milker-Zabel S, Zabel A, Thilmann C, Schlegel W, Wannenmacher M, Debus J. Clinical results of retreatment of vertebral bone metastases by stereotactic conformal radiotherapy and intensity-modulated radiotherapy. *International Journal of Radiation Oncology, Biology, Physics.* 2003;55(1):162–167.

Moussazadeh N, Laufer I, Yamada Y, Bilsky MH. Separation surgery for spinal metastases: Effect of spinal radiosurgery on surgical treatment goals. *Cancer Control.* 2014;21(2):168–174.

Myrehaug S, Sahgal A, Hayashi M, et al. Reirradiation spine stereotactic body radiation therapy for spinal metastases: Systematic review. *Journal of Neurosurgical Spine.* 2017;27(4):428–435.

Nieder C, Grosu AL, Andratschke NH, Molls M. Update of human spinal cord reirradiation tolerance based on additional data from 38 patients. *International Journal of Radiation Oncology, Biology, Physics.* 2006;66(5):1446–1449.

Nguyen QN, Chun SG, Chow E, et al. Single-fraction stereotactic vs conventional multifraction radiotherapy for pain relief in patients with predominantly nonspine bone metastases: A randomized phase 2 trial. *JAMA Oncology.* 2019;5(6):872–878.

Pan H, Simpson DR, Mell LK, Mundt AJ, Lawson JD. A survey of stereotactic body radiotherapy use in the United States. *Cancer.* 2011;117(19):4566–4572.

Potters L, Kavanagh B, Galvin JM, Hevezi JM, Janjan NA, Larson DA, Mehta MP et al. American Society for Therapeutic Radiology and Oncology (ASTRO) and American College of Radiology (ACR) practice guideline for the performance of stereotactic body radiation therapy. *International Journal of Radiation Oncology, Biology, Physics.* 2010;76(2):326–332.

Rades D, Haus R, Schild SE, Janssen S. Prognostic factors and a new scoring system for survival of patients irradiated for bone metastases. *BMC Cancer.* 2019;19(1):1156.

Redmond KJ, Lo SS, Soltys SG, et al. Consensus guidelines for postoperative stereotactic body radiation therapy for spinal metastases: Results of an international survey. *Journal of Neurosurgical Spine.* 2017;26(3):299–306.

Ryu S, Yoon H, Stessin A, Gutman F, Rosiello A, Davis R. Contemporary treatment with radiosurgery for spine metastasis and spinal cord compression in 2015. *Radiation & Oncology Journal*. 2015;33(1):1–11.

Sahgal A, Ames C, Chou D, Ma L, Huang K, Xu W, Chin C et al. Stereotactic body radiotherapy is effective salvage therapy for patients with prior radiation of spinal metastases. *International Journal of Radiation Oncology, Biology, Physics*. 2009;74(3):723–731.

Sahgal A, Bilsky M, Chang EL, Ma L, Yamada Y, Rhines LD, Létourneau D et al. Stereotactic body radiotherapy for spinal metastases: Current status, with a focus on its application in the postoperative patient. *Journal of Neurosurgical Spine*. 2011;14(2):151–166.

Sahgal A, Chang JH, Ma L, et al. Spinal cord dose tolerance to stereotactic body radiotherapy. *International Journal of Radiation Oncology, Biology, Physics*. 2019; in-press.

Sahgal A, Ma L, Weinberg V, Gibbs IC, Chao S, Chang U-K, Werner-Wasik M et al. Reirradiation human spinal cord tolerance for stereotactic body radiotherapy. *International Journal of Radiation Oncology, Biology, Physics*. 2012;82(1):107–116.

Spratt DE, Beeler WH, De moraes FY, et al. An integrated multidisciplinary algorithm for the management of spinal metastases: An international spine oncology consortium report. *Lancet Oncology*. 2017;18(12):e720–730.

Sprave T, Verma V, Förster R, et al. Randomized phase II trial evaluating pain response in patients with spinal metastases following stereotactic body radiotherapy versus three-dimensional conformal radiotherapy. *Radiotherapy & Oncology*. 2018;128(2):274–282.

Steenland E, Leer JW, Van houwelingen H, et al. The effect of a single fraction compared to multiple fractions on painful bone metastases: A global analysis of the Dutch Bone Metastasis Study. *Radiotherapy & Oncology*. 1999;52(2):101–109.

Sterzing F, Hauswald H, Uhl M, et al. Spinal cord sparing reirradiation with helical tomotherapy. *Cancer*. 2010;116(16):3961–3968.

Thibault I, Campbell M, Tseng CL, et al. Salvage stereotactic body radiotherapy (SBRT) following in-field failure of initial SBRT for spinal metastases. *International Journal of Radiation Oncology, Biology, Physics*. 2015;93(2):353–360.

Van Mourik AM, Sonke JJ, Vijlbrief T, et al. Reproducibility of the MRI-defined spinal cord position in stereotactic radiotherapy for spinal oligometastases. *Radiotherapy & Oncology*. 2014;113(2):230–234.

Weksberg DC, Yang JN, Tam AL, et al. Prospective validation of treatment accuracy using implanted fiducial markers for spinal stereotactic body radiation therapy. *Journal of Radiosurgery & SBRT*. 2016;4(1):7–14.

Wong CS, Fehlings MG, Sahgal A. Pathobiology of radiation myelopathy and strategies to mitigate injury. *Spinal Cord*. 2015;53, 574–580.

11 IG-HSRT for Benign Tumors of the Spine

Phillip M. Pifer, Hima Bindu Musunuru, John C. Flickinger, and Peter C. Gerszten

Contents

11.1 INTRODUCTION

Benign tumors of the spine represent a wide variety of histologies that occur within the intradural space, as well as epidural, paraspinal, and vertebral body locations. The primary treatment option for most benign spinal neoplasms is open surgical resection. The safety and effectiveness of such surgery has been clearly documented (Seppala et al., 1995a,b; Klekamp and Samii, 1998; Gezen et al., 2000; Conti et al., 2004; Parsa et al., 2004; Dodd et al., 2006). The majority of spinal meningiomas, schwannomas, and neurofibromas are noninfiltrative and can be completely and safely resected using microsurgical techniques (McCormick, 1996a; Kondziolka et al., 1998; Asazuma et al., 2004; Parsa et al., 2004). When complete tumor removal is achieved, recurrence is unlikely (Roux et al., 1996; Cohen-Gadol et al., 2003; Conti et al., 2004; Parsa et al., 2004).

In certain circumstances, however, some patients are less than ideal candidates for open standard surgical resection because of age, medical comorbidities, the recurrent nature of their tumor, or anatomical location of the lesion (Dodd et al., 2006). Tumors that have recurred after open surgical resection may make safe surgical resection challenging or not possible. Multiple benign spinal tumors, as are common in familial neurocutaneous disorders, may be a pattern of spinal pathology better suited for a less invasive radiosurgical option. It is in such clinical circumstances that radiosurgery might serve as an important treatment option for these patients.

Radiation therapy has been used for the treatment of numerous benign diseases since the discovery of the therapeutic potential of ionizing radiation (Lukacs et al., 1978; Solan and Kramer, 1985; Klumpar et al., 1994; Seegenschmiedt et al., 1994; Keilholz et al., 1996). Although similar benign tumor histologies can be found in and around the spine, the initial radiosurgical instruments were frame based and unable to treat such extracranial targets (Dodd et al., 2006). The emergence of frameless image-guided radiosurgery allows for the treatment of such tumors throughout the body (Adler et al., 1997; Ryu et al., 2001; Gerszten and Bilsky, 2006). Stereotactic radiosurgery (SRS) for the treatment of a variety of benign intracranial lesions has become widely accepted with excellent long-term outcomes and minimal toxicity (Steiner et al., 1974; Lunsford et al., 1991; Flickinger et al., 1996; Kondziolka et al., 2003; Sachdev et al., 2011). Kondziolka et al. presented a long-term control rate of 93% after the use of the Gamma Knife (Elekta AB, Stockholm, Sweden) to treat 1045 intracranial benign meningiomas (2008). In a subset of 488 patients for whom serial imaging was obtained, 215 tumors regressed and 256 were unchanged, resulting in a control rate of 97% (Murovic and Chang, 2014).

The development of frameless image-guided radiosurgery allows for the ability to treat such benign tumors throughout the body (Adler et al., 1997; Ryu et al., 2001; Dodd et al., 2006; Gerszten and Bilsky, 2006; Gerszten, Quader et al., 2012). The current evidence supporting the use of extracranial radiosurgery for malignant spine tumors is considerable (Hitchcock et al., 1989; Colombo et al., 1994; Hamilton et al., 1995; Pirzkall et al., 2000; Chang and Adler, 2001; Ryu et al., 2001, 2003, 2004; Medin et al., 2002; Sperduto et al., 2002; Milker-Zabel et al., 2003; Benzil et al., 2004; Bilsky et al., 2004; Chang et al., 2004; De Salles et al., 2004; Gerszten and Welch, 2004, 2007; Rock et al., 2004; Yin et al., 2004; Degen et al., 2005). There is much less experience regarding the use of radiosurgery for the treatment of benign tumors of the spine (Adler et al., 1997; Ryu et al., 2001; Gerszten and Bilsky, 2006; Murovic and Charles Cho, 2010; Gerszten, Quader et al., 2012; Chin et al., 2018). Radiosurgery has only recently become a part of the multimodality management of benign spinal tumors.

11.2 INDICATIONS AND SPECIAL CONSIDERATIONS FOR BENIGN TUMORS

Open surgical extirpation remains the primary treatment option for benign spinal tumors if feasible. The majority of spinal meningiomas, schwannomas, and neurofibromas are noninfiltrative and can be completely and safely resected using microsurgical techniques (McCormick, 1996a, 1996b; Asazuma et al., 2004; Parsa et al., 2004). However, the limitations of surgical options for some patients with benign tumors of the spine make radiosurgery an attractive alternative. Such limitations include medical comorbidities that preclude open surgery, recurrent tumors, patients with familial phakomatoses, or anatomical constraints such as tumor location relative to the spinal cord and/or motor nerve roots which increases the chances of incomplete resection or postoperative functional impairment.

The current clinical indications for benign spine tumor radiosurgery include tumors located in surgically difficult regions of the spine, recurrent benign spinal tumors after prior surgical resection, and benign spine tumors in patients who have significant medical comorbidities that preclude open surgery. Relative contraindications for radiosurgery for benign spine tumors include tumors without well-defined margins, tumors with significant spinal cord compression resulting in acute neurologic symptoms, prior radiation treatment to the spinal cord dose tolerance, and tumors that can easily be resected with conventional surgical techniques (Gibbs et al., 2015).

In the most recently published series from our institution, 55% of cases demonstrated recurrence after prior open surgical resection (Gerszten, Quader et al., 2012). In our previously published experience, only 23% of patients had undergone prior open surgical resection (Gerszten et al., 2008). This may reflect an increase in awareness among clinicians of radiosurgery as a noninvasive yet effective alternative to repeat open surgery in the event of tumor recurrence. However, as evidence of successful clinical outcomes of radiosurgery continue to increase, radiosurgery is more commonly being used as a primary treatment modality.

In addition to treating poor surgical candidates and recurrent or residual tumor after surgery, radiosurgery is most appropriate for well-circumscribed lesions associated with minimal spinal cord or nerve root compromise and no biomechanical instability (Gerszten, 2007). A theoretical advantage of utilizing spinal radiosurgery as the frontline management for spinal tumors is the possibility that such treatment may possibly act as prophylaxis against future spinal instability or neural element compression, obviating the need for extensive spinal surgery and instrumentation. Moreover, early conformal radiosurgery may obviate the need for large-field external beam radiation, which is known to suppress bone marrow function. Tumor shrinkage and complete obliteration aside, the minimally invasive technique of spinal radiosurgery may develop into an effective palliative strategy solely through local tumor control. In the end, the ability to perform spinal radiosurgery in the outpatient setting is an advantage that may spare patients with spinal tumors both time and the morbidities associated with hospitalization.

11.3 TECHNICAL AND DOSE CONSIDERATIONS

11.3.1 TARGET DELINEATION

The accurate geometric visualization of the tumor target on imaging is essential for radiosurgery contouring and treatment planning. Proper target delineation and contouring is indispensable for a safe and effective radiosurgery treatment. Although benign intradural, extramedullary tumors of the spine are often conspicuous by their homogeneous contrast enhancement, the large and often irregular shape of spinal neoplasms makes contouring a major challenge. The diagnostic imaging modality of choice for benign spinal tumors is magnetic resonance imaging (MRI). However, optimum radiosurgery treatment planning today rests on the modality of computed tomography (CT). Modern commercial radiosurgery systems use CT-based imaging for planning and delivery, and benign extramedullary spinal tumors are typically well visualized with postcontrast CT imaging. Most systems today also permit image fusion between MRI and CT images. Such image fusion may improve the target definition for spinal tumors, especially when such neoplasms exhibit heterogeneous contrast enhancement.

MRI–CT image fusion in the spine is usually more challenging than that required for intracranial radiosurgery, because it is more dependent upon the technical aspects of image acquisition and patient positioning during image acquisition. The quality of MR spinal image fusion often requires that the patient's imaging position closely matches the intended treatment position. Issues of spatial distortion need to be considered if MRI is used directly for planning. Because signal intensities of MR images do not reflect a direct relationship with electron densities, unless attenuation coefficients are manually assigned to the region of interest, spatial distortion limits the accuracy of using MRI directly in radiosurgery planning (Gibbs et al., 2015). The ability to identify tumors on CT, together with the probability of generating an adequate image fusion with MRI, is key to defining the radiosurgery target. Most benign tumors of the spine enhance brightly and have well-defined margins, making target delineation rather straightforward. Because virtually all extramedullary spinal tumors show some degree of contrast enhancement, postcontrast CT is sometimes used directly to define the target. CT myelography is an alternative imaging strategy which can provide superior tumor and spinal cord visualization in some intradural cases.

At our institution, patients are immobilized with the BodyFIX (Elekta AB, Stockholm, Sweden) when treatment sites are below T6; otherwise, a head and shoulder mask with S-board (CIVCO, Kalona, IA) is used. The target volume concept employed in all cases is the gross tumor volume (GTV) as seen on enhanced imaging. A clinical target volume (CTV) is not employed for these benign tumors. Our experience has shown that accurate contouring of benign spine tumors is nearly impossible without the use of MRI fusion. High-resolution sequence MRI often improves resolution degraded by instrumentation. In cases of instrumentation, we have attempted to resolve questions of tumor definition with CT myelography. Titanium implants are preferred over stainless steel in order to decrease imaging artifact. In some cases when radiosurgery is already anticipated prior to open surgery, such as the case for large "dumbbell" foraminal tumors, spinal instrumentation is only placed on the *contralateral* side of the tumor in order to allow for maximum tumor definition for radiosurgery.

The International Commission on Radiation Units and Measurements has formalized the contouring process by defining three different volumes: The GTV represents the unambiguous radiologic confines of the target neoplasm. Clinical target volume (CTV) includes nearby anatomic regions such in vertebral body metastases from malignant tumors where microscopic tumor extension is anticipated, but since benign spinal tumors are usually well circumscribed and do not exhibit metastasis, the CTV will more closely approximate the GTV. The planned target volume (PTV) is an augmented form of the CTV that corrects for movement and accuracy of treatment delivery to account for errors in target delineation including MR–CT co-registration and potential MR distortion and errors in delivery such as inaccurate cone-beam registration and patient movement during treatment.

11.3.2 DOSE SELECTION AND DELIVERY

The goal of spinal radiosurgery for benign spinal tumors is to deliver a clinically significant radiation dose to the tumor via a plan that respects the radiation dose tolerance limits of the nearby spinal cord, cauda equina, and surrounding organs such as the intestines, esophagus, kidneys, larynx, and liver. By virtue of their origin along the dura and spinal nerve roots, extramedullary spinal tumors can significantly impinge upon the spinal cord or cauda equina, making successful dose delivery more challenging (Gibbs et al., 2015). The degree of impingement of the tumor on the spinal cord may prevent the generation of a suitable radiosurgery treatment plan. Similar to radiosurgery doses prescribed for intracranial tumors, spine radiosurgery doses generally range from 12 to 20 Gy in a single fraction; doses as high as 30 Gy have been delivered when the treatment was hypofractionated in up to five sessions (Table 11.1). While spinal radiosurgery often is delivered using a single radiation fraction technique, the formal definition includes hypofractionated dosing plans with a maximum of five treatment sessions. Doses as high as 30 Gy have been reported in the setting of hypofractionation, and even higher total doses have been administered when one considers those patients who received conventional radiotherapy prior to spinal radiosurgery. The linear–quadratic equation remains the most accepted mathematical model of cell kill secondary to ionizing radiation (Yamada et al., 2007). This equation has been used to compare various dose fractionation schedules.

The spinal cord is one of the most radiosensitive structures and the dose-limiting organ for spine radiosurgery. Radiation-induced spinal cord injury, also known as radiation myelopathy (RM) though rare, can result in significant morbidity including pain, paranesthesia, sensory disturbances, weakness/paralysis, and bladder/bowel dysfunction. In the setting of stereotactic body radiation therapy (SBRT), radiation myelopathy has a median latent time of 12 months and 6 months in the *de novo* and reirradiation settings, respectively (Sahgal et al. 2019). Rarity of RM has complicated the ability to identify the spinal cord tolerance to radiosurgery/SBRT in long-term follow-up studies. Borrowing again from the intracranial radiosurgery experience, some physicians have fashioned a cord avoidance strategy based on dose tolerance data for the optic nerves and chiasm. Nevertheless, some data point to a less than 5% probability of myelopathy at 5 years when the cord receives a 60 Gy dose using standard fractionation (Yamada et al., 2007). Most centers have tailored plans to ensure that the spinal cord is exposed to no more than 20 Gy during fractionated SBRT therapy (Gerszten et al., 2006). A recent publication from Sahgal et al. examined reported cases of RM from spinal SBRT in malignant and benign tumors to determine the spinal cord dose tolerance. In radiation naïve patients, a thecal sac D_{max} of 44.6Gy, which corresponds to 12.4, 20.3, and 25.3 Gy in 1, 3, and 5 fractions, respectively, predicted for 5% or less risk of radiation myelopathy. In patients undergoing reirradiation SBRT, thecal cumulative dose $D_{max} \leq 70$ Gy, thecal SBRT EQD2 ≤ 25 Gy, SBRT dose to cumulative dose ratio ≤ 0.5, and an interval greater than 5 months since prior irradiation were associated with lower risk of radiation myelopathy (Sahgal et al., 2019). For a more in-depth discussion, this topic is extensively reviewed in the chapter "Spinal Cord Tolerance".

In contrast to patients with metastatic cancer, patients receiving radiosurgery for benign paraspinal neoplasms are expected to survive longer with higher functional status. Consequently, it is prudent that planning for benign tumors err on underestimating spinal cord tolerance in case an unrecognized form of RM may manifest over decades. Therefore, choosing spinal cord dose limits for 1% to 2% risks of myelopathy is usually more appropriate than 5%. Whether or not prior radiosurgery will sensitize the spinal cord and cauda equina to degenerative insults in aging patients is currently not known.

In our institutional experience with benign spine tumors, the mean maximum dose received by the GTV is 16 Gy (range, 12–24 Gy) delivered in a single fraction in the majority of cases. In a few cases in which the tumor was found to be intimately associated with the spinal cord with distortion of the spinal cord itself, the prescribed dose to the GTV was delivered in 3 fractions. The mean lowest dose received by

Table 11.1 Series of radiosurgery for benign spine tumors

SERIES	MENINGIOMA	SCHWANNOMA	NEUROFIBROMA	MEAN AGE (YEARS)	N	INDICATION	DOSE PER FRACTIONS	LENGTH OF F/U (MONTHS)	OUTCOME
Dodd et al. (2006)	16	30	9	46	51	51% postsurgical recurrent/residual	16–30 Gy/1–5 fractions	25	96% stable/decreased 3 repeat surgery 1 progression 1 new myelopathy
Chopra et al. (2005)	0	0	1	12	1	Residual	12.5 Gy/1 fraction	20	Stable
De Salles et al. (2004)	1	1	1	62	3	Not reported	12–15 Gy/1 fraction	6	Stable
Benzil et al. (2004)	1	2	0	61	3	Not reported	5–50.4 Gy/variable	Not reported	Rapid pain relief
Sahgal et al. (2007)	2	0	11	58	13	10 postsurgical 2 primary therapy	21 Gy/3 fractions	25	2 with progression (both NF1)
Gerszten et al. (2008)	13	35	25	44	73	30 pain 18 postsurgical 14 primary therapy 9 neuro deficit	12–20 Gy/1 fraction	37	73% pain improvement 100% stable/decreased 3 new myelopathy

(*Continued*)

Table 11.1 (Continued) Series of radiosurgery for benign spine tumors

SERIES	MENINGIOMA	SCHWANNOMA	NEUROFIBROMA	MEAN AGE (YEARS)	N	INDICATION	DOSE PER FRACTIONS	LENGTH OF F/U (MONTHS)	OUTCOME
Selch et al. (2009)	NA	25 nerve sheath tumors	NA	61	25	12 pain 22 neurologic deficit	12–15 Gy/1 fraction	6	100% stable
Sachdev et al. (2011)	32	47	24	53	87	Surgery contraindicated	14–30 Gy/1–5 fractions	33	99% stable or decreased, 1 progressed with new myelopathy, 6 persistent symptoms
Gerszten, Chen et al. (2012)	10	16	14	52	45	24 primary 21 postsurgical	12–24 Gy	32	100% control
Kalash et al. (2018)	18	18	NA	58	47	22 primary 25 postsurgical	9–21 Gy/ 1–3 fractions	54	5-year local control rate was 76%
Chin et al. (2019)	39	84	26	54	149	106 primary 43 postsurgical	12–30 Gy/1–5 fractions	49	Local failure rate of 2%, 5%, and 12% at 3, 5, 10 years

the GTV was 12 Gy (range, 8–16 Gy). The GTV volume ranged from 0.37 to 94.5 cm^3 (mean 13.7 cm^3; median 5.9 cm^3). Recent publication from our institution demonstrated similar 5-year local control rate for low-dose group (BED$_{10gy}$ < 30 Gy) compared to the high-dose group (BED$_{10gy}$ > 30 Gy) raising the possibility of dose de-escalation in this group of spine tumors (Kalash et al., 2018).

A planning target volume (PTV) expansion is sometimes employed at our institution in order to account for targeting inaccuracies. In the majority of our cases (85%), a PTV expansion of 2 mm was employed. The remaining PTV expansion ranged from 0 to 3 mm. As a rule, the PTV prescription dose was 2 Gy less than the prescription dose to the GTV. The mean number of beams used to deliver the radio-surgery treatment was 10 (median 9, range 7–14 beams).

A planning target expansion of usually 2 mm is employed for all tumors of the neural foramen at the level of the spinal cord, as well as tumors of the cauda equina and paraspinal locations. However, in instances in which the tumor is intimately associated with the spinal cord itself, no such planned target expansion is used. For cases in which the spinal cord itself is deformed by the tumor within the spinal canal, the radiosurgery is delivered in three separate sessions.

At our institution, cone-beam CT (CBCT) is used to ensure accurate patient setup prior to treatment delivery. In a series of 30 patients, CBCT was obtained before, midway, and upon completion of SBRT treatment. At the halfway point, translational variations were less than 0.5 mm in the lateral, longitudinal, and anteroposterior directions with very similar numbers at the completion of treatment (Monserrate et al. 2017). IGRT using CBCT is vital in treatment delivery as it facilitates treatment delivery with smaller PTV margins thereby reducing the dose to adjacent organs at risk.

11.4 OUTCOMES DATA

One of the first reports of the treatment of a benign spinal tumor, a hemangioblastoma, was reported by Chang et al. (1998). In 2001, the feasibility of image-guided spine radiosurgery for benign tumors was established when researchers at Stanford University reported the first clinical experience, which included two spinal schwannomas and one spinal meningioma (Ryu et al., 2001; Gibbs et al., 2015). Despite this relatively early use of stereotactic radiotherapy for benign spinal tumors, there has been a relative paucity of reports detailing clinical outcomes compared to spine malignancies. One issue that limits outcomes report-ing for benign spine radiosurgery is that the evaluation of radiosurgery for these lesions requires longer follow-up to confirm durable safety and efficacy than for malignant tumors of the spine.

Nevertheless, there is a growing amount of experience to date with clinical outcomes after spine radio-surgery for benign tumors (Gerszten et al., 2003; Benzil et al., 2004; De Salles et al., 2004; Chopra et al., 2005; Dodd et al., 2006; Gibbs et al., 2015; Kalash et al. 2018; Chin et al. 2019). Similar to the application of radiosurgery for intracranial pathology, spine radiosurgery has been primarily embraced as an adjunc-tive treatment technique for tumor recurrence or residual tumor after open surgery; tumor progression after failure of conventional radiation treatment in patients who have significant medical comorbidities that preclude open surgery; or as "salvage" therapy when further conventional irradiation or surgery is not appropriate.

Given their pathological similarities, it has been speculated that benign spinal lesions would be equally responsive to radiosurgery as their intracranial counterparts (Dodd et al., 2006). It is for this reason that centers experienced with radiosurgery have explored this developing technology to treat benign spine tumors. The extramedullary intradural spinal neoplasms treated with radiosurgery have primarily included meningiomas, schwannomas, and neurofibromas. Ryu et al. (2001) published the first clinical cohort description of radiosurgery being used to treat such lesions. In the initial experience from Stanford using the CyberKnife for 15 benign spinal tumors, no tumor progression was reported with a short follow-up period of 12 months.

The largest published series to date for benign intradural extramedullary spinal tumors from Chin et al. (2019) reported the results after radiosurgery treatment of 149 such tumors (84 schwannomas, 26 neurofi-bromas, and 39 meningiomas). Total treatment doses ranged from 12 to 30 Gy delivered in 1–5 fractions to tumor volumes that varied from 0.0 to 118.4 cm^3. Most common fractionations were 16 Gy in 1 fraction and 20 Gy in 2 fractions. In this cohort, there were nine tumors that progressed after SRS, and four of these tumors were in the same patient. Median imaging follow-up was 5 years, and local control rate was 98%, 95%, and 88% at 3, 5, and 10 years, respectively. The following sections, subdivided by histopathology, summarize available clinical outcome data for benign spinal tumor radiosurgery.

11.5 SPINAL MENINGIOMAS

Spinal meningiomas are arachnoid cap cell-derived tumors most common in the fifth to seventh decade that have a female predominance and occur mainly in the thoracic region. They arise from cells of the meningeal coverings of the central nervous system and occur more frequently within the brain than the spine, with a 5:1 ratio. Gross total surgical resection optimizes outcome (Peker et al., 2005). Spinal meningiomas, in general, have a more favorable prognosis relative to their intracranial counterparts. In a study of histological and microarray data of meningiomas, Sayaguès et al. (2006) determined that spinal meningiomas were most commonly associated with lower proliferative rates and more indolent histologies (psammomatous, transitional variants) and showed characteristic genetic and genomic differences compared with intracranial meningiomas. Radiosurgery has become a well-accepted treatment option for the management of intracranial meningioma and schwannoma (Chang and Adler, 2001; Kondziolka et al., 2003). A more recent publication of over 1,000 meningiomas revealed a 97% tumor control rate with a follow-up of up to 18 years (Kondziolka et al., 2008). Other studies have shown similar efficacy and safety of radiosurgery for meningiomas (Chang and Adler, 1997; Lee et al., 2002; Kondziolka et al., 2003).

In the initial report, Dodd et al. (2006) reported 16 treated spinal meningiomas (mean dose 20 Gy, mean tumor volume 2.4 cm^3, mean follow-up 27 months) and demonstrated radiologic stabilization in 67% and radiologic tumor decrease in 33% of the 15 who had radiographic follow-up. Only one patient required open surgery, and another sustained the only complication reported in the series. Seventy percent of the meningiomas treated in this series were symptomatically stable or improved. Most patients experienced an improvement in pain and strength with radiosurgery (Dodd et al., 2006). Subsequently, in the follow-up study by Sachdev et al. (2011) from Stanford University, all 32 meningiomas with radiographic follow-up were controlled at a mean follow-up of 33 months (range, 6–87) for all benign tumors. At last follow-up, 47% of meningiomas were stable, while 53% had decreased in volume.

From our institution's published series, 13 spinal meningiomas were treated using a single-fraction technique (mean dose 21 Gy, mean tumor volume 4.9 cm^3). Eleven of 13 patients had radiosurgery as an adjunctive treatment for residual or recurrent tumor following open surgical resection. Radiographic tumor control was demonstrated in all cases with a median follow-up of 17 months (Gerszten et al., 2008). Of the 11 patients who had undergone previous open surgical resection with residual or recurrent tumors, none demonstrated radiographic tumor progression on subsequent serial imaging after the radiosurgery treatment. Radiographic tumor control was also demonstrated for those two patients in whom radiosurgery was used as a primary treatment modality with median follow-up imaging at 14 months. A representative case is presented in Figure 11.1.

In a smaller series, Sahgal et al. (2007) reported treating two spine meningiomas with radiosurgery (mean dose 23 Gy delivered in 2 fractions, mean tumor volume 1.6 cm^3) with no evidence of radiographic tumor progression. Benzil et al. (2004) and De Salles et al. (2004) have also reported their experience with treating spine meningiomas with good long-term radiographic follow-up (Chang and Adler, 1997).

11.6 SPINAL SCHWANNOMAS

Nerve sheath tumors comprise schwannomas and neurofibromas. Schwannoma is the most common spinal tumor and has no proclivity for spinal region or gender (Seppala et al., 1995b). Nerve sheath tumors comprise schwannomas and neurofibromas. Schwannomas account for nearly one-third of primary spinal tumors, whereas neurofibromas constitute only 3.5% (Klekamp and Samii, 1998; Gezen et al., 2000; Conti et al., 2004; Parsa et al., 2004). Spinal schwannomas typically arise from the posterolaterally placed dorsal nerve root. Given their posterior position relative to the spinal cord or cauda equina, their removal via a laminectomy approach is usually straightforward. Patients with nerve sheath tumors typically present with local pain, radiating pain, and/or paraparesis and a relatively long duration of symptoms varying from 6 weeks to over 5 years.

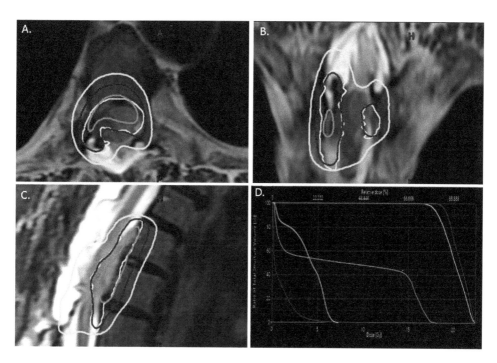

Figure 11.1 A 77-year-old female with a recurrent T2–T5 spinal meningioma who presented with rapidly progressive myelopathy. Patient underwent redo T1–T5 laminectomies with debulking of the intradural tumor. Post-operative gadolinium-enhanced axial MRI demonstrated persistent enhancing tumor present along the ventral thecal sac from T2 to T5. Two months after radiosurgery, patient was neurologically stable. (A, B, C) Axial, coronal, sagittal projections of PTV (yellow) and the isodose lines (15 Gy represented by green, 18 Gy represented by blue, and 21 Gy represented by orange) of the treatment plan. (D) Dose–volume histogram for treatment plan. PTV was treated with a prescribed dose of 18 Gy in 3 fractions (volume of 11.82 cm^3), with a spinal cord D$_{max}$ of 18.3 Gy, using VMAT. GTV represented by red, PTV represented by yellow, spinal cord represented by pink, carina represented by green, and lungs represented by blue.

Although many reports in the literature describe these tumors collectively, there are sufficient differences between the two tumor types to warrant a separate discussion for each. Schwannomas arise most commonly in the dorsal nerve root, are more commonly completely intradural (>80%), and are generally amenable to complete resection (Conti et al., 2004; Gibbs et al., 2015). On the contrary, neurofibromas arise more commonly in the ventral nerve root, are predisposed to multiple tumors by a strong association with NF1, and present with both intradural and extradural components in 66% of cases. These tumors are also uniquely different with respect to predisposing genetic defects. The merlin/schwannomin gene on chromosome 22 is associated with schwannomas in NF2, whereas the neurofibromin gene on chromosome 17 is associated with NF1 and neurofibromas. NF2 is an autosomal dominant genetic disorder that predisposes patients to developing multiple central and peripheral nervous system tumors. Schwannomas are the most common spinal tumor that occurs in these patients. Though sporadic occurrences of schwannomas unrelated to NF2 are not uncommon, those tumors associated with NF2 are more aggressive and recur more often after treatment. In a retrospective review of 87 patients with spinal nerve sheath tumors removed by surgery, 17 of whom had NF2-associated schwannomas, all NF2-related tumors recurred by 9 years compared with a 10-year recurrence rate of 28% in tumors not associated with NF2 (Klekamp and Samii, 1998). Factors strongly predictive of recurrence after surgery were partial resection, prior recurrence, NF2, and advanced age (Klekamp and Samii, 1998).

There is vast experience with the use of radiosurgery for the treatment of intracranial schwannomas. The long-term rate of growth control after radiosurgery for vestibular schwannoma has been documented as 95% to 98% (Kondziolka et al., 1998; Prasad et al., 2000). These highly encouraging tumor control rates should be applicable for spinal schwannomas as well. Chin et al. from Stanford reported on 84 tumors (median dose 18 Gy, median tumor volume 3.2 cm³,) and all but four tumors had radiographic tumor control after radiosurgery. However, 6 of the 84 patients required post-SRS resection. In two patients, post-SRS swelling caused neurological deficits, and in the remaining four patients, tumors were stable or smaller tumors, but had persistent or progressing symptoms. In a previous Stanford report, one-third of patients reported improvement in pain, weakness, or sensation, but 18% had a clinical decline after treatment (Dodd et al., 2006). Forty percent of patients in this series had spinal schwannomas in the setting of NF2. Forty-one percent of these patients were treated for recurrent or residual tumor after surgery.

From our institution, 35 schwannomas were treated with spine radiosurgery (mean dose 22 Gy, mean tumor volume 11.0 cm³). In patients whom pain was primary indication for radiosurgery (*n* = 17), 14 patients (82%) described significant pain improvement. Radiographic tumor control was demonstrated in six of seven patients (86%) for which radiosurgery was used as the primary treatment. A total of three patients underwent surgical resection for new or persistent neurological deficits (Gerszten et al., 2008). Benzil et al. (2004) and De Salles et al. (2004) have also reported favorable long-term outcomes following SRS for spine schwannomas. A representative case is presented in Figure 11.2.

Selch et al. (2009) retrospectively reviewed 20 patients with 25 nerve sheath tumors. Four patients each had NF1 and had NF2, respectively. Histopathology was available in seven patients after subtotal tumor removal 2–36 months prior to radiosurgery (four schwannomas, three neurofibromas). These patients underwent radiosurgery because of clinical and imaging evidence of tumor regrowth or persistent symptoms. Of the remaining 18 lesions, presumptive histopathology in nine was established after the removal of peripheral nerve sheath tumors elsewhere in the patient (five neurofibromas, four schwannomas). Nine tumors without histopathologic confirmation were treated based on symptoms and imaging consistent with nerve sheath tumor. At a median follow-up of 12 months, there were no local failures. Tumor size remained stable in 18 cases, and 28% demonstrated more than 2 mm reduction in tumor size.

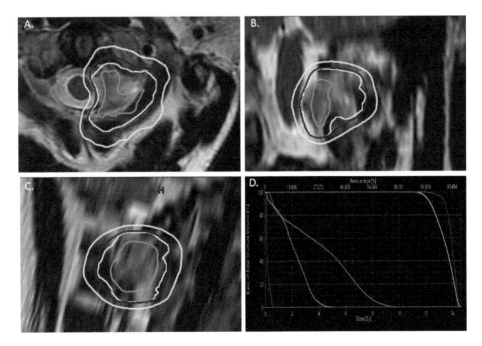

Figure 11.2 A 50-year-old male with a left C2 nerve root peripheral sheath tumor who presented with left-sided suboccipital neck pain. Gadolinium-enhanced axial MRI demonstrated heterogeneous enhancing extradural mass centered within the left C1–C2 foramen. Patient's symptoms and imaging are stable 2 years after SRS treatment. (A, B, C) Axial, coronal, and sagittal projections of the PTV (yellow) and isodose lines (8 Gy represented by green, 11 Gy represented by blue, and 14 Gy represented by orange) of the treatment plan. (D) Dose–volume histogram for treatment plan. PTV was treated with a prescribed dose of 11 Gy in 1 fraction (volume of 9.90 cm³), with a spinal cord D_{max} of 9.8 Gy, using VMAT. GTV represented by red, PTV represented by yellow, spinal cord represented by pink, left parotid represented by green, and right parotid gland represented by blue.

11.7 SPINAL NEUROFIBROMAS

Neurofibromas are benign nerve sheath tumors that may arise from either peripheral or spinal nerve roots. Neurofibromas of the spine are often multiple, predominate in the cervical region, and are commonly associated with NF1. Neurofibromas are less common than schwannomas, constituting only 3.5% of primary spinal tumors. Approximately 2% of patients with NFl will develop symptomatic spinal tumors. Multiple spinal tumors are not uncommon (Seppala et al., 1995a, 1995b). As with other nerve sheath tumors, patients present with pain and paraparesis. Two-thirds of neurofibromas occur in the cervical spine. These tumors grow both in the intradural as well as extradural spaces. Surgical extirpation of these tumors usually requires sectioning of the originating nerve root in order to completely resect the lesion.

Sahgal et al. (2007) reported a series of 11 treated neurofibromas (mean dose 21 Gy delivered in 3 fractions, mean tumor volume 6.0 cm^3). Radiographic control was documented in nine patients. Three patients had NF1, and two of these suffered progression. In the Dodd et al. series from Stanford, nine neurofibromas in seven patients with NF1 (mean dose 10.6 Gy, mean tumor volume 4.3 cm^3) were treated with radiosurgery, and tumor stabilization on imaging was documented in six of seven (86%) patients. After a mean follow-up of 20 months, half of the patients described an improvement in symptoms after radiosurgery while the remaining half documented worsening pain, weakness, or numbness at their last follow-up (Dodd et al., 2006). However, the tumors were radiographically stable in all patients. The authors therefore caution that the role of radiosurgery for neurofibromas remains unclear, particularly considering that a significant number of the NF1 patients were myelopathic at presentation. They further state that the most realistic and attainable goal for neurofibroma treatment in myelopathic patients is tumor control without significant expectations for symptomatic improvement (Dodd et al., 2006).

Our institution has published an initial experience with 25 neurofibroma cases together with 35 schwannoma cases and 13 meningioma cases (mean dose 21.3 Gy, mean tumor volume 12.6 cm^3) (Gerszten et al., 2008). Similar to the Stanford experience, no patient has had evidence of radiographic tumor progression on follow-up. Twenty-one of these patients had NF1, and nine had NF2. Radiosurgery ameliorated discomfort in 8 of 13 patients (61.5%) treated for pain. All patients with no improvement in pain had NF1. These findings echo outcomes from the Stanford series that found suboptimal pain control in NF1-associated spinal neurofibromas following radiosurgery (Dodd et al., 2006). Similarly, results following microsurgery for neurofibromas in NF 1 patients have also been observed in patients with NF1 (Seppala et al., 1995b). The multiplicity of neurofibromas in NF1 may be partially to blame as this factor makes identifying the symptomatic neurofibroma in need of treatment more difficult. Furthermore, given that many of the patients with neurofibromas have multiple lesions along their spine, it can often be difficult to determine whether symptom progression is due to the treated lesion or from any of the other neurofibroma lesions within the spine (Roux et al., 1996). Moreover, the infiltrating nature of neurofibromas, in contrast to the other benign extramedullary intradural spinal tumors, may engender more irreversible neural damage and increase the susceptibility of the native nerve root to injury both from microsurgical and radiosurgical treatments. In the end, future genomic investigations may reveal that intrinsic genetic differences in NF1-associated neurofibromas predispose to a weaker radiobiologic response.

In the Sachdev et al. (2011) series, at a mean follow-up of 33 months (range, 6–87) for all benign tumors, 82% of the neurofibromas were stable, while 18% had decreased in volume. Considering pain as a separate component, 17% reported improvement, 50% reported minimal change, and 33% reported worsening of pain. Given these results, the role of radiosurgery for neurofibromas remains unclear, particularly considering that a significant number of the NF1 patients were myelopathic at presentation. The poor clinical responses seen in this study appear to mimic the finding by Seppala et al. (1995a) in that only 1 of 15 patients who were alive at long-term follow-up after surgery reported complete freedom from symptoms. It is likely that the most realistic and attainable goal of neurofibroma treatment in myelopathic patients is tumor control without significant expectations for symptomatic improvement. Furthermore, given that many of the patients with neurofibromas had multiple lesions along their spine, it can often be difficult to determine whether symptom progression was due to the treated lesion or from any of the other neurofibroma spinal lesions (Murovic and Charles Cho, 2010; Gibbs et al., 2015).

11.8 SPINAL HEMANGIOBLASTOMAS

Hemangioblastomas are vascular tumors that are intramedullary in location and situated near the pial surface. Hemangioblastomas are the most common tumors associated with von Hippel–Lindau (VHL) disease. However, these tumors also can occur sporadically. Chang et al. (2011) evaluated 30 benign spinal tumors in 20 patients of which 8 tumors were hemangioblastomas. Three tumors were associated with VHL disease. All tumors were treated with 25.8 Gy as the mean SRS dosage, which was given in 1 fraction. The mean follow-up period was 50 months with a range between 23 and 72 months. Pain was not evaluated in this series. Of six patients who were asymptomatic prior to radiosurgery treatment, all (100%) remained asymptomatic after radiosurgery. In one patient who had gait difficulties, the gait improved. In another patient with unspecified neurological deficits, the neurological examination remained stable. Using imaging evaluation, six tumors regressed while one tumor stabilized and one tumor progressed. No complications reported in this publication.

Moss et al. (2009) evaluated 92 cranial and spinal hemangioblastomas in 31 patients who were treated with radiosurgery. Of these patients, 17 tumors were located in the spinal cord. The median follow-up time for the spinal hemangioblastoma patients was 37 months with a mean of 37.2 months. The tumor dose ranged from 20 to 25 Gy and was delivered in 1–3 fractions. Target volumes ranged from 0.06 to 2.65 cm^3. Radiographically, 15 of the 16 hemangioblastomas either remained stable (9) or improved (6). Only one patient's tumor increased in size during the follow-up period. Importantly, in this series of 16 hemangioblastomas managed with radiosurgery, no patients developed radiation myelopathy. Radiosurgery seems to be a safe and effective management strategy for spinal hemangioblastomas, especially in the setting of VHL. A representative case is presented in Figure 11.3.

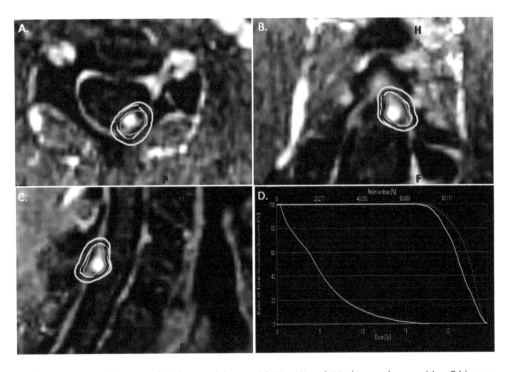

Figure 11.3 A representative case of a 40-year-old man with Von Hippel–Lindau syndrome with a C4 hemangioblastoma causing neck stiffness and upper extremities paresthesia. Gadolinium-enhanced axial MRI demonstrated a 5-mm enhancing C4 hemangioblastoma along the dorsal aspect of the spinal cord. (A, B, C) Axial, coronal, and sagittal projections of PTV (Yellow) and isodose lines (15 Gy represented by green, 18 Gy represented by blue, and 21 Gy represented by orange) of the treatment plan. (D) Dose–volume histogram for treatment plan. PTV was treated with a prescribed dose of 18 Gy in 3 fractions (volume of 0.47 cm^3), with a spinal cord D$_{max}$ of 17.8 Gy, using VMAT. GTV is represented by red, PTV is represented by yellow, and spinal cord is represented by pink.

11.9 TOXICITIES

In spine radiosurgery, the spinal cord and cauda equina are the organs at risk that most frequently limit the prescribed target dose. Spinal cord injury is arguably the most feared complication in radiotherapy and has historically limited the aggressiveness of spinal tumor treatment, whether benign or malignant (Gibbs et al., 2009). There has been considerable attention given to attempted determination of the human and animal radiation tolerance of spinal cord and cauda equina to stereotactic body radiotherapy (Ryu et al., 2003; Bijl et al., 2005; Sahgal et al., 2007, 2012, 2019). Complications from the radiosurgical treatment of benign spinal tumors are fortunately rare. Previous studies have suggested that the risk of RM is a function of total dose, fraction size, length of spinal cord exposed, and duration of treatment (Rampling and Symonds, 1998; Isaacson, 2000).

Dodd et al. (2006) reported the first published case of RM after radiosurgery for benign spinal tumors. This case involved a 29-year-old woman with a cervicothoracic spinal meningioma who developed myelopathic symptoms 8 months after radiosurgery taking a dose of 24 Gy in three sessions. It was felt that the relatively large volume of spinal cord (1.7 cm^3) irradiated above 18 Gy (3 fractions of 6 Gy) may have been the contributing factor. In their dose–volume analysis, the irradiated volume of spinal cord in this patient represented an outlier compared with other patients in the series. The volumes of spinal cord that received 8 Gy and 27 Gy were 4.7 cm^3 and 0.1 cm^3, respectively. The maximum spinal cord dose was 29.9 Gy. This patient developed posterior column dysfunction during the course of the myelitis but became neurologically stable after intervention with corticosteroids. The tumor volume had decreased at the time of intervention, and the tumor remained radiographically controlled at last follow-up. Edema seen on initial imaging resolved over time and was replaced by evidence of myelomalacia in the region. In the update of the Stanford experience by Sachdev et al. with 103 tumors in 87 patients with a mean follow-up of 33 months, the previously described patient remained the only patient in the entire series with RM. In a series of 19 benign spine tumors from the University of California San Francisco, no late toxicity, including myelopathy, occurred (Sahgal et al., 2007).

The benign spinal neoplasms that have thus far been treated with radiosurgery represent a heterogeneous group of neoplasms, some of which have had no prior treatment and some which have been treated with surgery and/or external beam radiation. Although already small, the risk of RM with radiosurgical treatment of benign spinal tumors may be further minimized once physicians understand how spinal cord radiation tolerance is influenced by previous treatments such a microsurgery.

In our initial series, comprising 73 benign tumors, two schwannoma patients and one meningioma patient developed radiosurgery-related cord injury manifesting as a Brown-Séquard syndrome, 5–13 months posttreatment (Gerszten et al., 2008). These three patients were treated with a combination of steroids, vitamin E, and gabapentin; one patient underwent hyperbaric oxygen therapy. A common factor in all three cases was prior open surgical resection of the tumors, which may have caused a predisposition of the spinal cord to radiation injury.

In our most recent series using cone-beam CT image guidance for radiosurgery delivery, no subacute or long-term spinal cord or cauda equina toxicity was noted at a median follow-up of 32 months (Gerszten, Chen et al., 2012). Forty-five consecutive benign spine tumors were treated using the Elekta Synergy S 6-MV linear with a beam modulator and cone-beam CT image guidance for target localization. The mean maximum dose received by the gross tumor volume (GTV) was 16 Gy (range, 12–24 Gy) delivered in a single fraction in 39 cases. The mean lowest dose received by the GTV was 12 Gy (range, 8–16 Gy). The GTV ranged from 0.37 to 94.5 cm^3 (mean 13.7 cm^3). In the majority of cases, a planning target volume expansion of 2 mm^3 was used. The technique used at our institution may serve as an important reference for treating benign spine tumors, with a high safety profile. Our prescribed dose to the target volume for benign tumors has decreased over time due to promising long-term radiographic control at lower SRS doses, and this has been further reinforced by the absence of radiation myelopathy.

11.10 CONTROVERSIES

Patients with benign lesions of the spine have prolonged life expectancy compared to their malignant counterparts. Therefore, the potential for delayed radiation myelopathy is of special concern when radiosurgery is considered. In addition, benign spine tumors have unique presentation, relationship to the spinal cord, and radiobiologic response to radiosurgery, any of which could represent unique challenges to the safe and effective application of radiosurgical ablation (Dodd et al., 2006). Because the life expectancy of most of these patients is considered to be normal and because radiation injury can take years to manifest (Dodd et al., 2006), there is more controversy regarding radiosurgery for the management of benign tumors of the spine. Especially in patients with greater survival and longer life span due to benign spinal tumor histology, the potential for delayed RM must be entertained. The low radiation tolerance of the spinal cord is the primary limiting factor when dosing therapeutic ionizing radiation for spinal tumors. In fact, the radiosensitivity of the spinal cord has often required that the treatment dose is far below the optimal therapeutic dose (Faul and Flickinger, 1995; Loblaw and Laperriere, 1998; Ryu et al., 2001). However, as discussed previously, our experience with dose de-escalation may provide a framework to retain high levels of local control with decreased risk of radiation myelopathy (Kalash et al. 2018).

Spinal radiosurgery aims to deliver a highly conformal radiation dose to the treatment volume, which should increase the likelihood of successful tumor control and minimize the risk of spinal cord injury (Ryu et al., 2001, 2003; Benzil et al., 2004; Bilsky et al., 2004). The intradural nature of benign spinal tumors engenders a close proximity between tumor and spinal cord or cauda equina, an anatomical relationship that may influence the risk of neural toxicity. Moreover, since benign spinal tumors are prone to late recurrence and the late toxicity of radiation to the spinal cord may take years to develop, the evaluation of radiosurgical treatment in terms of efficacy, safety, and durability will necessitate longer follow-up than that which has been granted to metastatic spine tumors (Gerszten et al., 2002). In the end, since benign spinal tumors are rarely life threatening or destabilizing to the spinal column, treatment with high-dose ionizing radiation or by any other means creates more controversy than that seen in the context of malignant spinal tumors; morbidity from any intervention, whether immediate or delayed, becomes less acceptable in this usually younger, healthier patient population (Gerszten, 2011).

11.11 FUTURE DIRECTIONS

Radiosurgery for benign spinal tumors is feasible and safe. Delivering highly conformal tumoricidal doses of radiation to spinal tumors is an enormous technical hurdle that has recently been surmounted. Spinal radiosurgery is arguably more complex than intracranial radiosurgery and poses unique challenges. Preliminary studies have demonstrated that spine radiosurgery can afford patients with benign spinal tumors a high probability of symptomatic relief and excellent long-term radiographic tumor control. However, longer-term follow-up data are necessary to assess the durability of radiosurgical treatment. The optimum dose–fraction regimens required for lesion ablation and the dose tolerance limit of the normal spinal cord and cauda equina remain to be determined. As the availability of magnetic resonance-guided radiotherapy increases, there will be an opportunity for real-time image guidance during treatment delivery thus potentially minimizing the risk of radiation myelopathy (Yadav et al, 2019). The initial benign spine radiosurgery experience is very positive and should encourage future studies to help poise radiosurgery as a less invasive treatment option for benign spinal tumors, similar to the current role of intracranial radiosurgery. Its role in patients with neurofibromatosis will also need to be further defined.

Further clinical experience with radiosurgery for benign tumors of the spine also opens the opportunity for the investigation of the use of radiosurgery for nonneoplastic extracranial indications. Given our current knowledge base regarding the accuracy and conformality of current delivery technologies, combined with our understanding of dose tolerance for the spinal cord and cauda equina and neuroanatomical targets, the next logical step is for the development of "functional" spine radiosurgery. Such benign disease processes might include pain syndromes such as failed back surgery syndrome, reflex sympathetic dystrophy, facet-mediated pain, as well as other conditions such as hyperhydrosis and spinal cord injury. While perhaps there has been a trend to lower doses for benign spine tumors with an emphasis on safety, there is no reason why extremely high and highly conformal doses similar to those used for trigeminal neuralgia could not be employed for the functional ablation of extracranial targets.

CHECKLIST: KEY POINTS FOR CLINICAL PRACTICE

✓	ACTIVITY	SOME CONSIDERATIONS
	Patient selection	*Is the patient appropriate for SBRT?* • The lesion is not amenable to open surgical resection • The lesion has recurred after prior open surgical resection • The patient has significant medical comorbidities that preclude open surgery • Absence of significant spinal cord compression or other acute neurologic symptoms
	SBRT vs external beam radiation therapy (EBRT)	*Is the lesion amenable to SBRT?* • The lesion has well-defined margins suitable for contouring and target delineation • Expected survival >6 months
	Simulation	*Immobilization* • Supine with appropriate stereotactic body immobilization system *Imaging* • MRI with gadolinium enhancement. T1/T2 sequences with 1–2 mm slide thickness • Possible use of CT scan ± intrathecal contrast for spinal cord delineation • Verified CT to MRI fusion over area of tumor involvement
	Treatment planning	*Contours* • GTV should include all portions of the tumor without a margin • CTV is not employed for these benign tumors • PTV expansion of 1–2 mm if necessary • No PTV employed if tumor causes direct compression of the spinal cord *Treatment planning* • Inverse treatment planning • Dose • 12–18 Gy in a single fraction if possible • Fractionation employed only in cases of possible spinal cord overdose • Coverage • GTV coverage of 80% to 95% • Dose constraints • Spinal cord, 10–12 Gy maximum point dose
	Treatment delivery	*Imaging* • Patients are first set up conventionally using room lasers and manual shifts • Orthogonal kV matching to relevant bony anatomy • Cone-beam CT to match to tumor (GTV or PTV) • Treatment couch with six degrees of freedom to correct translational and rotational shifts if available

REFERENCES

Adler J Jr, Chang S, Murphy M, Doty J, Geis P, Hancock S (1997) The CyberKnife: A frameless robotic system for radiosurgery. *Stereotact Funct Neurosurg* 69:124–128.

Asazuma T, Toyama Y, Maruiwa H, Fujimura Y, Hirabayashi K (2004) Surgical strategy for cervical dumbbell tumors based on a three-dimensional classification. *Spine* 29:E10–14.

Benzil DL, Saboori M, Mogilner AY, Rochio R, Moorthy CR (2004) Safety and efficacy of stereotactic radiosurgery for tumors of the spine. *J Neurosurg* 101:413–418.

Bijl H, van Luijk P, Coppes R, Schippers J, Konings A, van Der Kogel A (2005) Regional differences in radiosensitivity across the rat cervical spinal cord. *International Journal of Radiation Oncology, Biology, Physics* 61:543–551.

Bilsky M, Yamada Y, Yenice K, et al. (2004) Intensity-modulated stereotactic radiotherapy of paraspinal tumors: A preliminary report. *Neurosurgery* 54:823.

Chang E, Shiu A, Lii M-F, et al. (2004) Phase I clinical evaluation of near-simultaneous computed tomographic image-guided stereotactic body radiotherapy for spinal metastases. *International Journal of Radiation Oncology, Biology, Physics* 59:1288–1294.

Chang S, Adler J Jr (1997) Treatment of cranial base meningiomas with linear accelerator radiosurgery. *Neurosurgery* 41:1019–1025.

Chang S, Adler J (2001) Current status and optimal use of radiosurgery. *Oncology* 15:209–221.

Chang S, Murphy M, Geis P, Martin D, Hancock S, Doty J, Adler J Jr (1998) Clinical experience with image-guided robotic radiosurgery (the CyberKnife) in the treatment of brain and spinal cord tumors. *Neurol Med Chir* 38:780–783.

Chang U, Rhee C, Youn S, Lee D, Park S (2011) Radiosurgery using the CyberKnife for benign spinal tumors: Korea cancer center hospital experience. *J Neurooncol* 101:91–99.

Chin A, Fujimoto D, Kumar K et al. (2018) Long-term update of stereotactic radiosurgery for benign spinal tumors. *Neurosurgery* 85:5;708–716.

Chopra R, Morris C, Friedman W, Mendenhall W (2005) Radiotherapy and radiosurgery for benign neurofibromas. *Am J Clin Oncol* 28:317–320.

Cohen-Gadol A, Zikel O, Koch C, et al. (2003) Spinal meningiomas in patients younger than 50 years of age: A 21-year experience. *J Neurosurg Spine* 98:258.

Colombo F, Pozza F, Chierego G, Casentini L, De Luca G, Francescon P (1994) Linear accelerator radiosurgery of cerebral arteriovenous malformations: An update. *Neurosurgery* 34:14–20.

Conti P, Pansini G, Mouchaty H, Capuano C, Conti R (2004) Spinal neurinomas: Retrospective analysis and long-term outcome of 179 consecutively operated cases and review of the literature. *Surg Neurol* 61:34–44.

Degen J, Gagnon G, Voyadzis J, McRae D, Lunsden M, Dieterich S, Molzahn I, Henderson F (2005) CyberKnife stereotactic radiosurgical treatment of spinal tumors for pain control and quality of life. *J Neurosurg Spine* 2:540–549.

De Salles AA, Pedroso A, Medin P, Agazaryan N, Solberg T, Cabatan-Awang C, Epinosa DM, Ford J, Selch MT (2004) Spinal lesions treated with novalis shaped beam intensity modulated radiosurgery and stereotactic radiotherapy. *J Neurosurg* 101:435–440.

Dodd RL, Ryu MR, Kammerdsupaphon P, Gibbs IC, Chang J, Steven D, Adler J, John R (2006) CyberKnife radiosurgery for benign intradural extramedullary spinal tumors. *Neurosurgery* 58:674–685.

Faul CM, Flickinger JC (1995) The use of radiation in the management of spinal metastases. *J Neurooncol* 23:149–161.

Flickinger JC, Pollock BE, Kondziolka D (1996) A dose-response analysis of arteriovenous malformation obliteration after radiosurgery. *International Journal of Radiation Oncology, Biology, Physics* 36:873–879.

Gerszten P (2007) The role of minimally invasive techniques in the management of spine tumors: Percutaneous bone cement augmentation, radiosurgery and microendoscopic approaches. *Orthop Clin North Am* 38:441–450.

Gerszten P (2011) Radiosurgery for benign spine tumors and vascular malformations. In: Winn HR (ed.), *Youmans Neurological Surgery*, Vol. 3, pp. 2686–2692. Philadelphia, PA: Elsevier Saunders.

Gerszten P, Burton S, Ozhasoglu C, McCue K, Quinn A (2008) Radiosurgery for benign intradural spinal tumors. *Neurosurgery* 62:887–895.

Gerszten P, Chen S, Quadar M, Xu Y, Novotny J Jr, Flickinger J (2012) Radiosurgery for benign tumors of the spine using the synergy S with cone beam CT image guidance. *J Neurosurg* 117:197–202.

Gerszten P, Ozhasoglu C, Burton S, Kalnicki S, Welch WC (2002) Feasibility of frameless single-fraction stereotactic radiosurgery for spinal lesions. *Neurosurg Focus* 13:1–6.

Gerszten P, Quader M, Novotny J Jr, Flickinger J (2012) Radiosurgery for benign tumors of the spine: Clinical experience and current trends. *Technol Cancer Res Treat* 11:133–139.

Gerszten P, Welch W (2004) CyberKnife radiosurgery for metastatic spine tumors. *Neurosurg Clin North Am* 15:491.

Gerszten P, Welch W (2007) Combined percutaneous transpedicular tumor debulking and kyphoplasty for pathological compression fractures. *J Neurosurg Spine* 6:92–95.

Gerszten PC, Bilsky MH (2006) Spine radiosurgery. *Contemp Neurosurg* 28:1–8.

Gerszten PC, Burton SA, Ozhasoglu C, Vogel WJ, Quinn AE, Welch WC (2006). Radiosurgery for the management of spinal metastases. In: Kondziolka D (ed.), *Radiosurgery*, Vol. 6, pp. 199–210. Basel, Switzerland: Karger.

Gerszten PC, Ozhasoglu C, Burton S, Vogel WJ, Atkins BA, Kalnicki S (2003) CyberKnife frameless single-fraction stereotactic radiosurgery for benign tumors of the spine. *Neurosurg Focus* 14:1–5.

Gezen F, Kahraman S, Canakci Z, Beduk A (2000) Review of 36 cases of spinal cord meningioma. *Spine* 27:727–731.

Gibbs I, Chang S, Dodd R, Adler J (2015) Radiosurgery for benign extramedullary tumors of the spine. In: Gerszten P and Ryu S (eds.), *Spine Radiosurgery*, pp. 164–169. New York: Thieme.

Gibbs I, Patil C, Gerszten P, Adler J Jr, Burton S (2009) Delayed radiation-induced myelopathy after spinal radiosurgery. *Neurosurgery* 64:A67–72.

Hamilton A, Lulu B, Fosmire H, Stea B, Cassady J (1995) Preliminary clinical experience with linear accelerator-based spinal stereotactic radiosurgery. *Neurosurgery* 36:311–319.

Hitchcock E, Kitchen G, Dalton E, Pope B (1989) Stereotactic LINAC radiosurgery. *Br J Neurosurg* 3:305–312.

Isaacson S (2000). Radiation therapy and the management of intramedullary spinal cord tumors. *J Neurooncol* 47:231–238.

Kalash R, Glaser S, Flickinger J, et al. (2018) Stereotactic body radiation therapy for benign spine tumors: Is dose de-escalation appropriate? *J Neurosurg Spine*29:220–225

Keilholz L, Seegenschmiedt M, Sauer R (1996) Radiotherapy for prevention of disease progression in early-stage Dupuytren's contracture: Initial and long-term results. *International Journal of Radiation Oncology, Biology, Physics* 36:891–897.

Klekamp J, Samii M (1998) Surgery of spinal nerve sheath tumors with special reference to neurofibromatosis. *Neurosurgery* 42:279–290.

Klumpar D, Murray J, Ancher M (1994) Keloids treated with excision followed by radiation therapy. *J Am Acad Dermatol* 31:225–231.

Kondziolka D, Lunsford L, McLaughlin M (1998) Long-term outcomes after radiosurgery for acoustic neuromas. *N Engl J Med* 339:1426–1433.

Kondziolka D, Mathieu D, Lunsford L, Martin J, Madhok R, Niranjan A, Flickinger J (2008) Radiosurgery as definitive management of intracranial meningiomas. *Neurosurgery* 62:53–58.

Kondziolka D, Nathoo N, Flickinger JC (2003) Long term results after radiosurgery for benign intracranial tumors. *Neurosurgery* 53:815–821; discussion 821–822.

Lee J, Niranjan A, McInerney J, Kondziolka D, Flickinger JC, Lunsford LD (2002) Stereotactic radiosurgery providing long-term tumor control of cavernous sinus meningiomas. *J Neurosurg* 97:65–72.

Loblaw DA, Laperriere NJ (1998) Emergency treatment of malignant extradural spinal cord compression: An evidence-based guideline. *J Clin Oncol*16:1613–1624.

Lukacs S, Braun-Falco O, Goldschmidt H (1978) Radiotherapy of benign dermatoses: Indications, practice, and results. *J Dermatol Surg Oncol* 4:620–625.

Lunsford LD, Kondziolka D, Flickinger JC (1991) Stereotactic radiosurgery for arteriovenous malformations of the brain. *J Neurosurg* 75:512–524.

McCormick P (1996a) Surgical management of dumbbell and paraspinal tumors of the thoracic and lumbar spine. *Neurosurgery* 38:67–74.

McCormick P (1996b) Surgical management of dumbbell tumors of the cervical spine. *Neurosurgery* 38:294–300.

Medin P, Solberg T, DeSalles A (2002) Investigations of a minimally invasive method for treatment of spinal malignancies with LINAC stereotactic radiation therapy: Accuracy and animal studies. *International Journal of Radiation Oncology, Biology, Physics* 52:1111–1122.

Milker-Zabel S, Zabel A, Thilmann C, Schlegel W, Wannemacher M, Debus J (2003) Clinical results of retreatment of vertebral bone metastases by stereotactic conformal radiotherapy and intensity-modulated radiotherapy. *International Journal of Radiation Oncology, Biology, Physics* 55:162–167.

Monserrate A, Zussman B, Ozpinar A, et al. (2017) Sterotactic radiosurgery for intradural spine tumors using cone-beam CT image guidance. *Neurosurg Focus* 42(1):E11.

Moss J, Choi C, Adler J Jr, Soltys SG, Gibbs IC, Chang S (2009) Stereotactic radiosurgical treatment of cranial and spinal hemangioblastomas. *Neurosurgery* 65:79–85.

Murovic J, Charles S (2010) Surgical strategies for managing foraminal nerve sheath tumors: The emerging role of CyberKnife ablation. *Eur Spine J*19:242–256.

Murovic J, Chang S (2014) Treatment of benign spinal tumors with radiosurgery. In: Sheehan J and Gerszten P (eds.), *Controversies in Stereotactic Radiosurgery Best Evidence Recommendations*, pp. 240–246. New York: Thieme.

Parsa A, Lee J, Parney I, Weinstein P, McCormick M, Ames C (2004) Spinal cord and intraductal-extraparenchymal spinal tumors: Current best care practices and strategies. *J Neurooncol*69:219–318.

Peker S, Cerci A, Ozgen S, Isik N, Kalelioglu M, Pamir M (2005) Spinal meningiomas: Evaluation of 41 patients. *J Neurosurg Sci*49:7–11.

Pirzkall A, Lohr F, Rhein B, Höss A, Schlegel W, Wannenmacher M, Debus J (2000) Conformal radiotherapy of challenging paraspinal tumors using a multiple arc segment technique. *International Journal of Radiation Oncology, Biology, Physics* 48:1197–1204.

Prasad D, Steiner M, Steiner L (2000) Gamma surgery for vestibular schwannoma. *J Neurosurg* 92:745–759.

Rampling R, Symonds P (1998) Radiation myelopathy. *Curr Opin Neurol* 11:627–632.

Rock J, Ryu S, Yin F (2004) Novalis radiosurgery for metastatic spine tumors. *Neurosurg Clin North Am*15:503.

Roux F, Nataf F, Pinaudeau M, Borne G, Devaux B, Meder J (1996) Intraspinal meningiomas: Review of 54 cases with discussion of poor prognosis factors and modern therapeutic management. *Surg Neurol* 46:458–463.

Ryu S, Chang S, Kim D, Murphy M, Quynh-Thu L, Martin D, Adler J (2001) Image-guided hypo-fractionated stereotactic radiosurgery to spinal lesions. *Neurosurgery* 49:838–846.

Ryu S, Fang Yin F, Rock J, Zhu J, Chu A, Kagan E, Rogers L, Ajlouni M, Rosenblum M, Kim J (2003) Image-guided and intensity-modulated radiosurgery for patients with spinal metastasis. *Cancer* 97:2013–2018.

Ryu S, Rock J, Rosenblum M, Kim J (2004) Patterns of failure after single-dose radiosurgery for spinal metastasis. *J Neurosurg* 101(Suppl 3):402–405.

Sachdev S, Dodd R, Chang S, Soltys S, Adler J, Luxton G, Choi C, Tupper L, Gibbs I (2011) Stereotactic radiosurgery yields long-term control for benign intradural, extramedullary spinal tumors. *Neurosurgery* 69:533–539.

Sahgal A, Chou D, Ames C et al. (2007) Image-guided robotic stereotactic body radiotherapy for benign spinal tumors: The University of California San Francisco preliminary experience. *Technol Cancer Res Treat* 6:595–604.

Sahgal A, Chang J, Ma L et al. (2019) Spinal cord dose tolerance to stereotactic body radiation therapy. *International Journal of Radiation Oncology, Biology, Physics.*

Sahgal A, Ma L, Weinberg V et al. (2012) Reirradiation human spinal cord tolerance for stereotactic body radiotherapy. *International Journal of Radiation Oncology, Biology, Physics* 82:107–116.

Sayagués J, Tabernero M, Maíllo A et al. (2006) Microarray-based analysis of spinal versus intracranial meningiomas: Different clinical, biological, and genetic characteristics associated with distinct patterns of gene expression. *J Neuropathol Exp Neurol* 65:445–454.

Seegenschmiedt M, Martus P, Goldmann A, Wölfel R, Keilholz L, Sauer R (1994) Preoperative versus postoperative radiotherapy for prevention of heterotopic ossification (HO): First results of a randomized trial in high-risk patients. *International Journal of Radiation Oncology, Biology, Physics* 30:63–73.

Selch M, Lin K, Agazaryan N, Tenn S, Gorgulho A, DeMarco J, DeSalles A (2009) Initial clinical experience with image-guided linear accelerator-based spinal radiosurgery for treatment of benign. *Surg Neurol* 72:668–674.

Seppala M, Haltia M, Sankila R, Jaaskelainen J, Heiskanen O (1995a) Long-term outcome after removal of spinal neurofibroma. *J Neurosurg* 82:572–577.

Seppala M, Haltia M, Sankila R, Jaaskelainen J, Heiskanen O (1995b) Long term outcome after removal of spinal schwannoma: A clinicopathological study of 187 cases. *J Neurosurg* 83:621–626.

Solan M, Kramer S (1985) The role of radiation therapy in the management of intracranial meningiomas. *International Journal of Radiation Oncology, Biology, Physics* 11:675–677.

Sperduto P, Scott C, Andrews D (2002) Stereotactic radiosurgery with whole brain radiation therapy improves survival in patients with brain metastases: Report of radiation therapy oncology group phase III study. *International Journal of Radiation Oncology, Biology, Physics* 54:3a.

Steiner L, Leksell L, Forster D (1974) Stereotactic radiosurgery in intracranial arterio-venous malformations. *Acta Neurochir* (Suppl 21):195–209.

Yadav P, Musunuru HB, Witt J et al. (2019) Dosimetric study for spine stereotactic body radiation therapy: Magnetic resonance guided linear accelerator versus volumetric modulated arc therapy. *Radiol Oncol* 53(3):362–368.

Yamada Y, Lovelock D, Bilsky M (2007) A review of image-guided intensity-modulated radiotherapy for spinal tumors. *Neurosurgery* 61:226–235.

Yin F, Ryu S, Ajlouni M, Yan H, Jin J, Lee S, Kim J, Rock J, Rosenblum M, Kim J (2004) Image-guided procedures for intensity-modulated spinal radiosurgery. Technical note. *J Neurosurg* 101(Suppl 3):419–424.

Contents

12.1 INTRODUCTION

Spinal metastatic disease is common among patients with cancer and has traditionally been associated with a poor prognosis. Approximately 40% of cancer patients are affected by spinal metastases during their disease course (Klimo and Schmidt, 2004), and autopsy series have estimated that 30% to 70% of patients with known malignancy have evidence of spinal metastatic disease on postmortem examination (Fornasier and Horne, 1975; Wong et al., 1990; Harrington, 1993).

The incidence of spinal metastases is anticipated to increase as systemic therapies continue to improve and prolong survival of patients with metastatic disease. As such, a shift toward definitive rather than palliative management of spinal metastases has become increasingly important. In general, the management of spinal metastatic disease consists of radiation therapy, surgery, and systemic therapy, as well as combined modality approaches such as surgery followed by postoperative radiation therapy. The focus of this chapter is on postoperative spine image-guided hypofractionated stereotactic radiation therapy (IG-HSRT) with an emphasis upon treatment considerations and technique, clinical outcomes, and future directions.

12.1.1 HISTORICAL PERSPECTIVE

Surgery and radiation therapy are the mainstays of treatment for spinal metastases. Traditionally, conventionally fractionated low-dose radiotherapy alone has been used for patients with oncologic/nonmechanical pain with minimal epidural disease. Conventionally fractionated palliative radiotherapy is the primary treatment option for patients with widely metastatic spinal disease that cannot be addressed with surgical resection, patients who are medically unfit for surgery, and patients with limited life expectancy. Patients with spinal instability, malignant epidural spinal cord compression (MESCC), and/or mechanical pain resulting from pathologic fracture are generally best served by up-front surgery.

A landmark trial by Patchell et al. (2005) established the standard of care in the management of MESCC by demonstrating the superiority of surgical decompression followed by postoperative low-dose radiotherapy over low-dose radiotherapy alone. This multi-institutional prospective study randomized 101 patients to immediate direct circumferential decompressive surgery and postoperative radiotherapy or up-front radiation therapy alone (Patchell et al., 2005). Radiation dose, fractionation, and delivery were identical in both arms with 30 Gy in 10 fractions delivered with anterior–posterior/posterior–anterior technique to a treatment field encompassing a vertebral body above and below the level of involvement (enrollment restricted MESCC to a single area of involvement). A significantly higher proportion of patients in the surgical arm were ambulatory following treatment, and those treated in the surgical arm retained the ability to walk for a significantly longer period of time than those treated with radiation therapy alone. Among patients who were nonambulatory prior to treatment, those treated with surgery and postoperative radiation therapy were more likely to regain the ability to walk. Additionally, patients treated with surgery and postoperative radiation therapy had a significant benefit with regard to maintenance of continence, motor strength as assessed by the American Spinal Injury Association (ASIA) score, functional ability as assessed by Frankel score, and survival. Although criticized for small sample size and lack of standardization of surgical technique, this study was influential in establishing the role of surgery in the management of MESCC as other series had previously failed to demonstrate the benefit of surgery alone or in combination with radiotherapy over radiotherapy alone (Gilbert et al., 1978; Yong et al., 1980; Findlay, 1984). Furthermore, it gave credence to the concept that aggressive intervention combining surgery and postoperative radiotherapy is able to improve patient outcomes in the appropriate clinical scenario.

12.1.2 RECENT ADVANCES

Improvements in surgical technique and spinal instrumentation over the past two decades have expanded the armamentarium against spinal metastatic disease and MESCC (Sciubba and Gokaslan, 2006). An emphasis upon minimally invasive approaches for decompression and tumor resection, focusing on the removal of only the compressive epidural tumor or bone fragment, has served to lessen perioperative morbidity, improve surgical outcomes, and quicken recovery, thus reducing the interval to administration of adjuvant therapies. Indeed, innovative new approaches such as minimal access spinal surgery (MASS), minimally invasive separation surgery, and spinal laser interstitial thermal therapy (LITT) represent significant advances in this growing field where fewer options such as posterior laminectomy and major invasive decompressive surgeries were traditionally available (Massicotte et al., 2012; Sharan et al., 2014; Tatsui et al., 2015). Management of spinal instability has also matured with percutaneous vertebroplasty and kyphoplasty allowing for minimally invasive stabilization. Anterior and posterior stabilization approaches have accordingly evolved when spinal stabilization and fixation are warranted after invasive decompressive surgeries (Sciubba et al., 2010).

Radiotherapy technique and administration have similarly evolved at a rapid pace. Historically, the concerns and limitations of radiation therapy largely revolved around the relatively low tolerance of the spinal cord and potential for spinal cord injury (SCI) and/or radiation-induced myelopathy. As such, low-dose conventionally fractionated radiotherapy has been utilized as a palliative treatment for temporary control of oncologic pain and neurological symptoms. This is a particularly salient issue among patients with favorable histologies and longer life expectancy as well as patients with radioresistant histologies where low-dose conventional radiotherapy has limited success at achieving durable local tumor control (Greenberg et al., 1980; Maranzano and Latini, 1995). Klekamp and Samii reported on their experience with surgery followed by postoperative low-dose conventionally fractionated radiotherapy for spinal metastases. The overall local recurrence rate was approximately 70% at 1 year. On multivariate analysis, preoperative ambulatory status, favorable histology, complete tumor resection, and low number of involved vertebral levels were identified as independent predictors of a lower rate of recurrence (Klekamp and Samii, 1998). Although limited by the single-institution and retrospective nature of this work, this report alludes to the poor long-term local control rates achieved with low-dose conventional radiotherapy and the significant potential for disease progression and recurrence even after treatment with a combined modality approach.

Long-term, high-quality evidence exists supporting the safe and efficacious use of single-fraction stereotactic radiosurgery (SRS) and HSRT in the management of intracranial benign and malignant tumors. This technology has since been applied to numerous extracranial sites, including the spine. Efficacy data for spine IG-HSRT continue to emerge, but studies to date have demonstrated excellent local control with acceptable toxicity. Advances in image guidance, precise spine and body immobilization, multileaf collimation, robotic technologies, and intensity-modulated treatment planning have afforded radiation oncologists the ability to deliver highly conformal ablative doses of radiation with steep dose gradients, thereby sparing the spinal cord and other nearby critical structures. With regard to spinal metastases, IG-HSRT has several advantages compared to conventionally fractionated radiotherapy: (1) highly precise and accurate beam delivery allows for a higher biologically effective dose (BED) to be delivered and (2) smaller treatment volumes and sharp dose fall-off allow for sparing of critical organs at risk (OAR) such as the spinal cord, esophagus, bowel, and kidneys. Given the potential for spine IG-HSRT to overcome shortcomings associated with low-dose conventional radiotherapy, there have been significant efforts to evaluate spine IG-HSRT in the definitive setting and, more recently, in the postoperative setting with encouraging results.

With the complementary evolution of both surgical and IG-HSRT technologies, there has been a clear paradigm shift from the historic goal of short-term palliation toward offering definitive management, with long-term local control as the primary goal of therapy. With respect to postoperative spine management, IG-HSRT can serve as a potent adjuvant treatment to enhance local control. Fundamentally, if a patient undergoes a major spinal procedure, it is intuitive that the adjuvant therapy be equally as aggressive, hence the role of IG-HSRT. Surgery is also now being reconceptualized as neoadjuvant to definitive postoperative IG-HSRT. By focusing the surgical goals on epidural tumor resection and spinal cord decompression using minimally invasive techniques, with or without stabilization, the morbidity of surgery can be further minimized with delivery of tumoricidal doses with IG-HSRT as the primary therapy. This approach will also increase the utilization of spine surgery as currently it is thought of as a highly selective treatment for highly selected patients with single-level MESCC.

12.2 INDICATIONS

As the available treatment options continue to expand, it is of critical importance to better understand which patients will benefit from up-front surgical management versus those who should be addressed with radiotherapy alone. Ultimately, optimal treatment selection requires a multidisciplinary approach with consideration of (1) patient-specific factors such as neurologic symptoms, etiology of pain, medical comorbidities, surgical contraindications, systemic disease burden, and prognosis, as well as (2) disease-specific factors, including spinal instability, degree of epidural spinal cord compression (ESCC), relative tumor radiosensitivity, and prior radiation treatment.

Several groups have sought to develop patient stratification schema to guide management. The Spine Oncology Study Group developed a novel comprehensive classification system to diagnose neoplastic spinal instability based upon patient symptoms and radiographic criteria (Fisher et al., 2010). The objective of the spinal instability neoplastic score (SINS) criteria is to help identify patients who may benefit from surgical intervention and stabilization. The SINS is a qualitative score assigned based on the importance of several factors including pain, location, spinal alignment, presence of vertebral compression fracture (VCF), type of lesion, and posterolateral involvement of spinal elements (e.g., as shown in Table 12.1). By SINS, patients with a total score ranging 0–6 are considered stable, those scoring 7–12 are considered potentially unstable, and those with a total score ≥13 are considered unstable, and surgical management is highly recommended. The SINS has been validated as a reliable, reproducible tool among radiologists, radiation

Table 12.1 **Spinal instability neoplastic score (SINS)**

PATIENT-SPECIFIC	SPINE-SPECIFIC	TUMOR-SPECIFIC	TOTAL SINS SCORE
Pain (max score = 3)	**Location (max score = 3)**	**Type of lesion (max score = 2)**	**Stable = 0–6**
Mechanical pain (3)	Junctional: occiput—C2, C7–T2, T11–L1, L5–S1 (3)	Osteolytic (3)	**Potentially unstable = 7–12**
Occasional pain, nonmechanical (1)	Mobile: C3–C6, L2–L4 (2)	Mixed (1)	**Unstable ≥13**
No pain (0)	Semi-rigid: T3–T10 (1)	Osteosclerotic (0)	
	Rigid: S2–S5 (0)	**Posterolateral involvement of spinal elements** (max score =3)	
	Spinal alignment (max score = 4)	Bilateral (3)	
	Subluxation/translation (4)	Unilateral (1)	
	Kyphosis/scoliosis (2)	None (0)	
Normal (0)			
	Presence of vertebral compression fracture (max score = 3)		
	≥50% collapse (3)		
	<50% collapse (2)		
	No collapse with ≥50% involvement (1)		
	None (0)		

Source: Reproduced from Fisher, C.G. et al., *Spine* (1976), 35(22), E1221, 2010. With permission.

oncologists, and surgeons (Fourney et al., 2011; Fisher, Versteeg et al., 2014; Fisher, Schouten et al., 2014). The ASIA has utilized the ASIA impairment scale (AIS) to classify the degree of SCI (Kirshblum et al., 2011; Kirshblum and Waring, 2014). In the context of MESCC, surgeons and oncologists have used this scale to help stratify patients and the need for surgical intervention. In general, surgical resection should be considered in patients presenting with AIS grades A–D without improvement after high-dose corticosteroid administration, especially if the neurologic impairment developed over the preceding 48 hours (e.g., as shown in Table 12.2).

Bilsky and colleagues developed and validated a six-point grading system to define ESCC (Bilsky et al., 2010). Prior iterations have used myelography and MRI to define the degree of CSF space impingement and spinal cord compression to assist surgical decision making (Bilsky et al., 2001). The advent of spine IG-HSRT necessitated a more detailed system to specify the degree of thecal sac impingement as this information is important in assessing the safety and feasibility of a definitive IG-HSRT approach. In general, patients with low-grade ESCC (ESCC scale grades 0–1b) have sufficient distance at the spinal cord–tumor interface to permit safe IG-HSRT administration (e.g., as shown in Figure 12.1). On the ESCC scale, grade 0 indicates bone-only disease; grade 1a indicates epidural impingement without deformation of the thecal sac; and grade 1b indicates deformation of the thecal sac without spinal cord abutment. Grade 1c disease (deformation of the thecal sac with spinal cord abutment but without cord compression) is an intermediate between consideration as low-grade versus high-grade ESCC, and management is dependent upon multidisciplinary communication between the surgeon and radiation oncologist as well as other clinical factors. High-grade ESCC consists of grade 2 and grade 3 disease, which often warrants separation surgery with

Table 12.2 **ASIA impairment scale (AIS)**

A = Complete	No sensory or motor function is preserved in the sacral segments S4–S5.
B = Sensory incomplete	Sensory but no motor function is preserved below the neurological level and includes the sacral segments S4–S5 (light touch, pin prick at S4–S5, or deep anal pressure) and no motor function is preserved more than three levels below the motor level on either side of the body.
C = Motor incomplete	Motor function is preserved below the neurological level[a] and more than half of the key muscle functions below the single neurological level of injury have a muscle grade <3.
D = Motor incomplete	Motor function is preserved below the neurological level[a] and at least half of the key muscle functions below the neurological level of injury have a muscle grade of ≥3.
E = Normal	If sensation and motor function are tested and graded as normal in all segments and the patient had prior deficits, then the AIS grade is E. Someone without an initial spinal cord injury does not receive an AIS grade.

Source: Reproduced from Kirshblum, S.C. et al., *J. Spinal Cord Med.*, 34(6), 535, 2011. With permission.

Note: When assessing the extent of motor sparing below the level for distinguishing between AIS grades B and C, the motor level on each side is used, whereas to differentiate between AIS grades C and D the single neurological level is used.

[a] For an individual to receive grade C or D (motor incomplete), they must have either (1) voluntary anal sphincter contraction or (2) sacral sensory sparing with sparing of motor function more than three levels below the motor level for that side of the body.

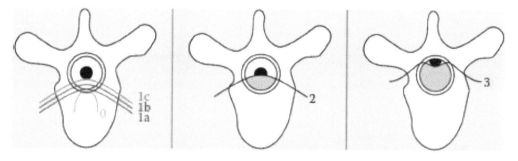

Figure 12.1 Schematic representation of the six-point epidural spinal cord compression scale. A grade of 0 indicates bone-only disease; 1a, epidural impingement, without deformation of the thecal sac; 1b, deformation of the thecal sac, without spinal cord abutment; 1c, deformation of the thecal sac with spinal cord abutment, but without cord compression; 2, spinal cord compression, but with CSF visible around the cord; and 3, spinal cord compression, no CSF visible around the cord. (Adapted from Bilsky, M.H. et al., *J. Neurosurg. Spine*, 13(3), 324, 2010. With permission.)

epidural tumor resection in order to achieve spinal cord decompression and sufficient separation between the spinal cord–tumor interface to permit safe postoperative IG-HSRT (Moussazadeh et al., 2014). On the ESCC scale, grade 2 disease denotes evidence of spinal cord compression with CSF visible around the spinal cord, and grade 3 disease denotes spinal cord compression without CSF visible around the cord.

Memorial Sloan Kettering Cancer Center (MSKCC) has adopted the NOMS framework as a decision-making algorithm for patients with metastatic spinal cancers (Laufer et al., 2013). The acronym NOMS depicts the neurologic, oncologic, mechanical, and systemic considerations that guide patient stratification and treatment selection (e.g., as shown in Figure 12.2). In general, patients with low-grade ESCC and without myelopathy do not require decompressive surgery. Radiosensitive histologies may be addressed with conventionally fractionated low-dose radiotherapy, and radioresistant histologies are treated with IG-HSRT. Patients presenting with high-grade ESCC and/or myelopathy should be considered for

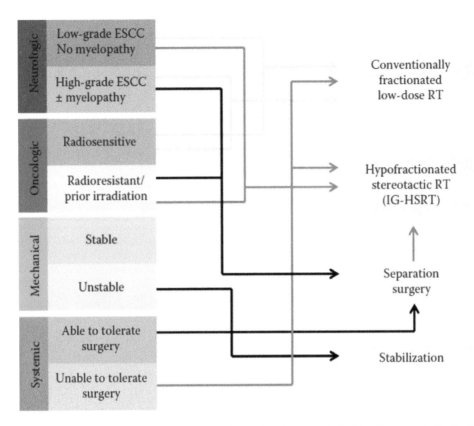

Figure 12.2 The NOMS (neurologic, oncologic, mechanical, and systemic) decision framework. By the NOMS algorithm, patients with radiosensitive histologies are treated with conventionally fractionated low-dose RT, regardless of myelopathy or degree of epidural spinal cord compression (ESCC). Patients with radioresistant histologies can be treated with up-front separation surgery followed by image-guided hypofractionated stereotactic radiation therapy (IG-HSRT) or definitive IG-HSRT alone, depending on the presence of myelopathy or degree of ESCC. Patients with mechanical spinal instability should be considered for spinal stabilization procedures. Patients unfit for surgery should be treated with conventionally fractionated low-dose RT or IG-HSRT depending on the degree of ESCC, presence of myelopathy, safety, and relative radiosensitivity. (Adapted from Laufer, I. et al., *Oncologist*, 18(6), 744, 2013a. With permission. Copyright Clearance Center, Inc.)

surgical intervention with stabilization and/or decompression. Radioresistant histologies may benefit from decompressive or separation surgery prior to delivery of IG-HSRT; however, it is also reasonable to manage patients with radiosensitive histologies with conventionally fractionated low-dose radiotherapy alone. Patients who are unable to tolerate surgery with high-grade ESCC and/or myelopathy should be managed in a palliative fashion with administration of conventionally fractionated low-dose radiotherapy, as the spinal cord–tumor interface is often insufficient to safely deliver IG-HSRT. Of note, any patient with evidence of mechanical spinal instability should be considered for up-front stabilization via a percutaneous or open approach. While the NOMS framework is a useful tool for patient stratification, there are alternative management strategies available for patients with spinal metastatic disease. For example, the application of spine IG-HSRT may be considered for radiosensitive histologies to enhance local control in the definitive setting. Additionally, patients presenting with high-grade ESCC or myelopathy are often considered for stabilization/decompression followed by IG-HSRT regardless of histologic radiosensitivity.

Additional considerations for selecting between surgical and nonsurgical approaches are impending spinal instability (SINS 7–12) with mechanical pain, prior irradiation to the involved region, need for pathologic diagnosis, older age, and oligometastatic disease favoring definitive intervention, particularly with radioresistant tumor histologies (Chi et al., 2009; Redmond et al., 2016). Table 12.3 summarizes the absolute and relative indications for surgical intervention in the patient with metastatic spinal disease.

Table 12.3 **Absolute and relative indications for surgical intervention**

ABSOLUTE INDICATIONS	RELATIVE INDICATIONS
SINS ≥ 13 (unstable)	ESCC grade 2 or 3
ASIA grade A–D, due to ESCC, especially if not responsive to high-dose corticosteroids and onset < 48 hours	Radioresistant histology in patient with anticipated overall survival longer than a year
	Local recurrence following prior radiotherapy
	Need for pathologic diagnosis
	Intractable pain from mechanical instability
	Oligometastatic disease

12.3 TREATMENT PLANNING

12.3.1 SIMULATION

Safe and accurate delivery of postoperative spine IG-HSRT requires the construction of a highly reproducible setup with rigid body immobilization at the time of CT simulation. Proper immobilization reduces the potential for intrafraction and interfraction shifts and position changes, which is particularly relevant for treatment systems that do not utilize real-time imaging (Sahgal et al., 2011). The location of the spinal lesion dictates the type of immobilization device, as outlined in Table 12.4.

Once the patient is immobilized with a reproducible setup, a thin-slice (≤2 mm) CT simulation scan with and without intravenous iodinated contrast is performed. Patients with contraindications to CT contrast should forego contrast administration. Subsequently, a thin-slice MRI is acquired with a region of interest that encompasses the target area and at least one vertebral body above and below the target. If there are no contraindications to MRI and/or gadolinium contrast, an MRI simulation scan with and without contrast administration should be performed. At a minimum, the following MRI sequences should be acquired: (1) T1-weighted axial images with and without gadolinium for tumor delineation and (2) T2-weighted or short tau inversion recovery (STIR) axial images for visualization of spinal cord and further tumor delineation. Sagittal reconstruction images and other MRI sequences may assist fusion and treatment planning and, therefore, should be considered. Additionally, incorporation of preoperative MRI axial images is critical to understanding the initial extent of disease and therefore clinical target volume (CTV) delineation. In situations where metallic artifact from spinal instrumentation obscures the delineation of critical neural structures (spinal cord, thecal sac) on MRI, a CT myelogram should be obtained. Following completion of simulation scans, the following datasets are fused within the treatment planning software: (1) CT simulation, (2) MRI simulation, (3) preoperative MRI, and (4) CT myelogram (if performed).

12.3.2 TARGET DELINEATION

Accurate target delineation is essential to successful IG-HSRT. Redmond, Lo et al. (2017) reported consensus guidelines for postoperative spinal IG-HSRT based on an international panel of 20 radiation

Table 12.4 **Reasonable spine IG-HSRT immobilization strategies for treatment planning**

SPINAL LOCATION	IMMOBILIZATION
Cervical and upper-thoracic (C1–T3)	Long thermoplastic cranial mask with mold cushioning for support
Mid-thoracic and lumbar (T4–L3)	Alpha Cradle® and wing board set up with arms positioned above the head with appropriate headrest support and memory foam
Inferior (L4–sacrum)	Hevezi pad or Vac Lok® immobilization with the patient's arms positioned at their side

oncologists and neurosurgeons. The consensus gross tumor volume (GTV) was residual disease on post-operative MRI. The CTV is based on the *preoperative* extent of disease. In general, the CTV should encompass all areas of gross disease on preoperative MRI, any postoperative residual disease, the involved anatomic compartments, and sites of potential microscopic disease spread including the adjacent ana-tomic compartment. Consensus guidelines for bony CTV delineation are shown in Table 12.5 (Redmond, Robertson et al., 2017). The surgical incision and scar are generally not included within the CTV unless considered to be at high risk of subclinical involvement or recurrence. Likewise, spinal hardware and instrumentation is generally not included within the CTV unless there is an increased concern for sub-clinical involvement. This necessitates a firm understanding of the surgical approach as documented in the operative report and through discussions with the surgeon. Although sometimes unavoidable owing to the initial or postoperative extent of disease, efforts should be taken to limit circumferential or "donut" treatment fields when possible, as this can impact subsequent treatment planning and the achievement of OAR constraints and target coverage. The planning target volume (PTV) includes the CTV with a geometric expansion of 1.0–2.0 mm. In situations where the PTV margin extends into the spinal cord or nearby OARs, a planning organ-at-risk volume (PRV) is generated to account for residual setup error and intrafractional organ motion for patients and is subtracted from the PTV. PRV delineation is discussed in Section 12.3.3. A representative postoperative spine IG-HSRT treatment plan is shown in Figure 12.3.

Table 12.5 **Guidelines for CTV delineation in postoperative spine IG-HSRT based on preoperative epidural and bony involvement**

PREOPERATIVE EPIDURAL INVOLVEMENT	PREOPERATIVE ISRC BONY INVOLVEMENT	POSTOPERATIVE CTV DESCRIPTION AND ISRC ANATOMIC REGIONS
Circumferential epidural disease	1–3, 5, 6, ±4	Circumferential treatment including the preoperative body, bilateral pedicles, bilateral transverse processes, bilateral laminae, and spinous process (1–6)
Anterior epidural involvement in region of central body	1	Preoperative body (1)
Anterior epidural involvement in lateral region of body	1	Preoperative body plus ipsilateral pedicle ± lamina (1, 2)
Epidural involvement anteriorly in the region of the body and unilaterally in the region of pedicle	1, 2	Preoperative body plus ipsilateral pedicle, ipsilateral transverse process and ipsilateral lamina (1–3)
Epidural involvement anteriorly in the region of the body, unilaterally in the region of pedicle, and posteriorly in the region of the spinous process	1, 4–6	Preoperative body plus ipsilateral pedicle, bilateral transverse process, bilateral laminae, and spinous process (1, 3–6, ±2)
Posterior epidural involvement in region of spinous process	4	Preoperative spinous process, bilateral laminae, and bilateral transverse processes (3–5)
Any of the above plus extensive paraspinal extension	As above	As above plus coverage of the entire preoperative extent of paraspinal extension

Source: Adapted from Redmond K.J. et al. *Int J Radiat Oncol Biol Phys.* 2017 Jan 1; 97(1): 64–74. With permission.
Abbreviations: IG-HSRT, image-guided hypofractionated stereotactic radiotherapy; CTV: clinical target volume; ISRC: International Spine Radiosurgery Consortium.

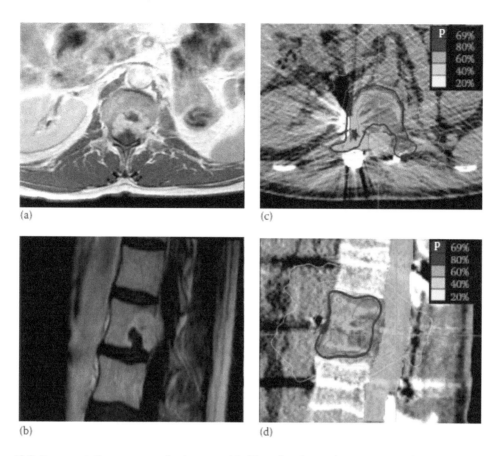

Figure 12.3 Representative postoperative image-guided hypofractionated stereotactic radiation therapy (IG-HSRT) treatment plan. This patient with metastatic clear cell renal cell carcinoma developed lower back pain. MRI demonstrated an L1 vertebral body lesion with bilateral pedicle involvement and extension into the anterior epidural space with likely impingement of the L2 nerve root and mild left neuroforaminal narrowing. (a) Preoperative axial T1-weighted MRI. (b) Preoperative sagittal T1-weighted MRI. The patient underwent L1 corpectomy, left L1 laminectomy, left T12–L1, and L1–L2 facetectomy. Posterior stabilization with instrumented allograft arthrodesis from T12 to L2 was performed. (c, d) Noncircumferential postoperative IG-HSRT plan taking into account initial bony involvement and epidural disease. The PTV is denoted in red, and thecal sac organ at risk is denoted in green. Isodose line color scheme: prescription (blue), 80% (violet), 60% (orange), 40% (maize), and 20% (yellow).

12.3.3 ORGANS AT RISK AND NORMAL TISSUE CONSTRAINTS

The location of the postoperative target volume dictates which nearby OARs should be contoured. At minimum, all OARs within one vertebral body superior and inferior to the PTV must be contoured, and the superior–inferior extent should be increased for nonisocentric planning systems where hot spots may be found remotely from the target. The critical neural structures, including the spinal cord and/or cauda equina, are delineated using T2-variant MRI or CT myelogram in cases of significant metal artifact from spinal instrumentation, A 0–2 mm expansion is used as the spinal cord PRV for planning purposes. The cauda equina OAR structure is defined as the thecal sac without an expansion. To further characterize spinal cord tolerance in the context of IG-HRST, Sahgal and colleagues analyzed the risk of radiation myelopathy (RM) among patients previously treated with conventionally fractionated radiotherapy and reirradiated with IG-HSRT (Sahgal, Ma et al., 2012). Example spinal cord constraints for various fractionation schemes are shown in Table 12.6. More recently, Ghia et al. (2018) performed a phase I trial of relaxation of the single-fraction point max dose constraint to the spinal cord in inoperable patients with MESCC and no history of prior radiotherapy and found no cases of RM even using a single-fraction point max constraint of 16 Gy, after a median follow-up of 17 months. Further studies have helped to define safe practice guidelines by developing a percentage probability of RM based upon point max doses to the thecal sac as referenced in Table 12.7 (Sahgal et al., 2010; Sahgal, Weinberg et al., 2012; Masucci et al., 2011). In addition to the use of point max as a spinal cord dose constraint parameter, the maximum dose per small volume (0.1–0.2 cc) is also utilized. Reasonable spine IG-HSRT dose constraints for other critical structures based on American Association of Physicist in Medicine Task Group 101 recommendations are outlined in Table 12.8 (Benedict et al., 2010).

12.3.4 DOSE SELECTION

The optimal dose and fractionation schema for postoperative spine IG-HSRT remains undefined. Dose prescription is often a complex decision that is contingent upon multiple considerations including nearby OARs, target volume, extent of epidural disease, number of involved vertebral levels, and prior irradiation

Table 12.6 Reasonable reirradiation IG-HSRT doses to the thecal sac P_{max} following common initial conventional radiotherapy regimens

PRIOR CONVENTIONAL RADIOTHERAPY DOSE nBED	IG-HSRT DOSE TO THECAL SAC P_{max}				
	1 FRACTION	2 FRACTIONS (GY)	3 FRACTIONS (GY)	4 FRACTIONS (GY)	5 FRACTIONS
0 Gy	10 Gy	14.5	17.5	20	22 Gy 25 Gy to <0.1 cc
20 Gy in 5 fractions (30 Gy$_{2/2}$)	9 Gy	12.2	14.5	16.2	18 Gy
30 Gy in 10 fractions (30 Gy$_{2/2}$)	9 Gy	12.2	14.5	16.2	18 Gy
37.5 Gy in 15 fractions (42 Gy$_{2/2}$)	9 Gy	12.2	14.5	16.2	18 Gy
40 Gy in 20 fractions (40 Gy$_{2/2}$)	n/a	12.2	14.5	16.2	18 Gy
45 Gy in 25 fractions (43 Gy$_{2/2}$)	n/a	12.2	14.5	16.2	18 Gy
50 Gy in 25 fractions (50 Gy$_{2/2}$)	n/a	11	12.5	14	15.5 Gy

Source: Reproduced from Sahgal, A. et al., Int. J. Radiat. Oncol. Biol. Phys., 82(1), 107, 2012a. With permission.
Abbreviations: nBED, normalized biologically effective dose; P_{max}, point max.

Table 12.7 Predicted % probability of radiation myelopathy (RM) by P_{max} volume absolute doses in Gy for IG-HSRT

% RISK OF RM	P_{max} LIMIT (GY)				
	1 FRACTION	2 FRACTIONS	3 FRACTIONS	4 FRACTIONS	5 FRACTIONS
1% probability	9.2	12.5	14.8	16.7	18.2
2% probability	10.7	14.6	17.4	19.6	21.5
3% probability	11.5	15.7	18.8	21.2	23.1
4% probability	12.0	16.4	19.6	22.2	24.4
5% probability	12.4	17.0	20.3	23.0	25.3

Source: Reproduced from Sahgal, A. et al., Int. J. Radiat. Oncol. Biol. Phys., 85(2), 341, 2012b. With permission.
Abbreviations: P_{max}: point max to the thecal sac.

(Sahgal et al., 2011). The existing literature for postoperative spine IG-HSRT is highly variable with regard to dose selection and fractionation making it difficult to offer generalized recommendations. Current data suggest that higher dose per fraction delivered over less fractions may enhance local control, particularly in radioresistant histologies such as sarcoma (Bishop et al., 2017), while other series have contradictory findings. Moreover, lower BED and hypofractionated regimens have been associated with a lower risk of radiation-associated toxicities (Cunha et al., 2012; Al-Omair, Masucci et al., 2013; Sahgal, Atenafu et al., 2013). There is likely a delicate balance of dose escalation and hypofractionation needed to achieve optimal local control and to minimize toxicities. The aforementioned controversies and uncertainties outline the need for future randomized prospective studies. In general, 85% to 95% target coverage by 90% to 100% of the prescription dose is acceptable; however, dose and fractionation selection is ultimately contingent upon the clinical context and physician preference. Given the potentially disastrous consequences, PTV coverage should always be compromised in order to meet OAR constraints. Reasonable fractionation regiments include 18–20 Gy in 1 fraction, 24 Gy in 2 fractions, 27–30 Gy in 3 fractions, and 30–40 Gy in 5 fractions.

12.3.5 TREATMENT DELIVERY

In general, spine IG-HSRT is delivered by either multileaf collimator (MLC) linear accelerator (LINAC) treatment units or CyberKnife technology (Adler et al., 1997, 1999; Ryu et al., 2001). Key differences between treatment delivery systems have been previously described and are beyond the scope of this chapter (Sahgal et al., 2011). MLC-based LINAC treatment delivery entails the use of 7 to 11 coplanar beams with 5 to 15 MLC apertures or segments per beam. Intensity-modulated beam delivery can be delivered via "step-and-shoot" technique with a fixed gantry position or dynamic MLC approach. Additionally, CT image guidance and/or stereoscopic kilovoltage x-rays allow for treatment setup verification and interfraction shifts; however, there is no intrafractional imaging during the course of treatment. Alternatively, the CyberKnife system is nonisocentric, and treatment is delivered via a mobile robotic arm that is mounted with a compact LINAC. Highly conformal dose delivery is achieved by using a large number of beam angles (~100–200) with circular collimators of varying size to maximize target coverage and spare OARs. A distinct difference between these technologies is that CyberKnife uses near real-time imaging during treatment to make intrafractional adjustments (Sahgal et al., 2011). BrainLAB ExacTrac is another system that offers the ability to detect intrafractional tumor motion during stereotactic treatments using in-room x-ray imaging.

12.3.6 SPECIAL CONSIDERATIONS

Treatment planning and delivery in the postoperative spine setting pose several unique challenges. Target delineation and visualization of residual tumor are complicated by imaging artifact created by hardware (Pekmezci et al., 2006; Sahgal et al., 2011). Postsurgical blood products and debris are often difficult to differentiate from residual tumor, which introduces uncertainty into contouring. As previously mentioned, a CT myelogram may be helpful for spinal cord delineation in this setting. An additional challenge is the impact of spinal instrumentation and surgical hardware upon radiation dosimetry. The amount of artifact and image distortion increases as the amount of hardware increases for both CT and MRI (Sahgal et al., 2011). High Z (atomic number) materials such as titanium can result in significant dose perturbation. Dosimetric modeling studies on the effect of hardware on spine IG-HSRT reported that in relation to the prescribed dose, the dose in front of the hardware was approximately 6% higher due to electron back scatter, while it was approximately 7% lower beyond the hardware due to photon attenuation (Wang et al., 2013). Indeed, other studies have observed up to 13% attenuation with 6 MV photon beams where hardware material consisted of both titanium rods and pedicle screws (Liebross et al., 2002). These dosimetric inconsistencies are clinically relevant as they may result in hot spots within OARs or in underdosing of the target volume. It is important to communicate the composition of the surgical hardware to radiation dosimetrists so that an appropriate CT number to electron density conversion and density override is used for treatment planning. Furthermore, utilizing multiple-beam arrangements may reduce the impact of high Z material dose perturbation (Wang et al., 2013).

Table 12.8 Reasonable organ at risk (OAR) dosimetric constraints for spine IG-HSRT

OAR	Max critical volume above threshold (cc)[c]	MAX POINT DOSE (GY)[a]		
		1 FRACTION	3 FRACTIONS	5 FRACTIONS
Esophagus		15.4	25.2	35
Heart/pericardium		22	30	38
Great vessels		37	45	53
Trachea		20.2	30	40
Stomach		12.4	22.2	32
Bowel		12.4	22.2	32
Skin[b]		26	33	39.5
		Threshold dose (Gy)		
Combined kidneys	200	8.4	16	17.5
Combined lung	1000	7.4	12.4	13.5
Liver	700	9.1	19.2	21

Source: Adapted from Benedict, S.H. et al., Med. Phys., 39(1), 63, 2010.
[a] Max point dose.
[b] Unless intentionally included as CTV.
[c] For parallel tissues, the volume–dose constraints are based on critical minimum volume of tissue that should receive a dose equal to or less than the indicated threshold dose.

12.4 OUTCOMES

12.4.1 REPORTED CLINICAL OUTCOMES IN THE LITERATURE

The literature to date evaluating postoperative spine IG-HSRT is evolving but suggests promising clinical outcomes. There is currently no level 1 randomized data comparing conventional radiotherapy and IG-HSRT in the postoperative spine setting. The current data reporting clinical outcomes using postoperative spine IG-HSRT are summarized in Table 12.9.

Early reports by Gerszten et al. (2005) described outcomes of 26 patients with symptomatic pathological compression fracture treated with kyphoplasty-based closed fracture reduction followed by SRS with doses ranging from 16 to 20 Gy. With a median follow-up of 16 months (range, 11–24 months), 92% of patients reported long-term improvement in back pain, and there were no neurological symptoms attributable to radiosurgery. Given concerns regarding kyphoplasty alone, the same group from the University of Pittsburgh Medical Center subsequently evaluated minimally invasive surgery involving tumor debulking and kyphoplasty followed by adjuvant SRS (Gerszten and Monaco, 2009). In this study, 11 patients with pain secondary to pathological compression fracture with moderate (20% to 50%) spinal canal compromise underwent transpedicular coblation corpectomy combined with closed fracture reduction and fixation and were then treated with conformal radiosurgery with doses ranging from 16 to 22.5 Gy delivered over one session. Similarly, all patients experienced some degree of improvement in back pain after treatment by 10-point visual analog scale. There were no documented acute radiation toxicities nor new neurological deficits during the median follow-up period of 11 months (range, 7–44 months).

Using comparatively modest radiosurgical doses, Rock et al. (2006) retrospectively reviewed a series of 18 patients treated with a variety of surgical approaches followed by SRS. The operative procedure varied based upon the extent of osseous involvement and need for stabilization after decompression/resection. The radiosurgery dose ranged between 6 and 16 Gy. Among the patients with neurological deficits preceding therapy, 62% of these patients demonstrated improvement in neurological status, and 30% had stable neurological deficits during the median follow-up period of 7 months. One patient demonstrated

Table 12.9 Clinical studies evaluating postoperative spine IG-HSRT

STUDY (AUTHOR AND YEAR)		HSRT CHARACTERISTICS					CLINICAL OUTCOMES							
	NUMBER	SURGICAL TECHNIQUE	MEDIAN TOTAL DOSE (GY/NO. FRACTIONS)	TOTAL DOSE (GY/NO. FRACTIONS)	NBED (GY₂/₂)ᵃ	CTV DESCRIPTION	MEDIAN FOLLOW-UP (MONTHS)	LOCAL CONTROL (LAST FOLLOW-UP)	LOCAL CONTROL (1-YEAR)	PATTERNS OF LOCAL FAILURE	MEDIAN SURVIVAL (MONTHS)	PAIN CONTROL	TOXICITY
Gerszten et al., 2005	n = 26	Kyphoplasty	18 Gy/1 Fx	16–20 Gy/1 Fx	72–110 Gy	Entire VB and any adjacent tumor extension	16 (11–124)	92%	NR	NR	NR	92% long-term improvement in pain by VAS	NoRT-induced SCI
Rock et al., 2006	n = 18	Variable, open surgery	12 Gy/1 Fx	6–16 Gy/1 Fx	12–72 Gy	Involved VB and anterior one-third of pedicles, including paravertebral and epidural soft tissue component	7 (4–36)	94%	NR	NR	NR	33% with complete pain relief	6% possible RT-induced toxicity 92% with stable or improved deficits
Gerszten, 2009	n = 11	Percutaneous trans pedicular corpectomy with closed fracture reduction	19 Gy/1 Fx	16–22.5 Gy/1 Fx	72–138 Gy	Entire VB and any adjacent tumor extension	11 (7–44)	100%	NR	NR	NR	100% long-term improvement in back pain	No RT-induced SCI
Moulding et al., 2010	n = 21	Posterolateral decompression and instrumentation	24 Gy/1 Fx	18–24 Gy/1 Fx	90–156 Gy	Preoperative GTV, including entire VB, accounting for microscopic tumor	10.3	81%	90.5%	NR	10.2	NR	Acute neuritis (n = 1), grade 4 esophagitis (n = 1)
Massicotte et al., 2012	n = 10	Minimal access spine surgery (MASS)	24 Gy/2 Fx	20 Gy/1 Fx (20%)	110 Gy	Visible tumor on imaging, areas at risk of microscopic disease, and ipsilateral trajectory of tube	13 (3–18)	70%	NR	67% epidural, 33% "in-field" within VB	NR	80% with improvement of pain by VAS	Acute pain flare (n = 2), no reports of RM
				18–24 Gy/2–3 Fx (70%)	36–84 Gy								
				35 Gy/5 Fx (10%)	80 Gy								
Laufer et al., 2013	n = 186	Separation surgery with circumferential epidural resection	24 Gy/1 Fx	24 Gy/1 Fx (22%)	156 Gy	Preoperative GTV, including entire VB, accounting for microscopic tumor	7.6 (1–66.4)	NR	83.6%	NR	54% alive at last follow-up	NR	NR
			27Gy/3 Fx	24–30 Gy/3 Fx (20%)	60–90 Gy								
			30 Gy/5–6 Fx	18–36 Gy/5–6 Fx (58%)	24–72 Gy								

(*Continued*)

Table 12.9 (Continued) Clinical studies evaluating postoperative spine IG-HSRT

STUDY (AUTHOR AND YEAR)	NUMBER	HSRT CHARACTERISTICS					CLINICAL OUTCOMES						
		SURGICAL TECHNIQUE	MEDIAN TOTAL DOSE (GY/NO. FRACTIONS)	TOTAL DOSE (GY/NO. FRACTIONS)	NBED (GY$_{2/2}$)[a]	CTV DESCRIPTION	MEDIAN FOLLOW-UP (MONTHS)	LOCAL CONTROL (LAST FOLLOW-UP)	LOCAL CONTROL (1-YEAR)	PATTERNS OF LOCAL FAILURE	MEDIAN SURVIVAL (MONTHS)	PAIN CONTROL	TOXICITY
Al-Omair, Masucci et al., 2013	n = 80	Stabilization alone OR decompression ± stabilization	24 Gy/2 Fx	18–26 Gy/1–2 Fx (44%)	50–156 Gy	Circumferential "donut" (90%)	8.3	74%	84%	71% epidural, 5% posterior elements, 24% "mixed" in-field bony + epidural	64%	NR	Acute pain flare (n = 7); no RM
				18–40 Gy/3–5 Fx (56%)	36–100 Gy								
Puvanesarajah et al., 2015	n = 32	Decompression with stabilization +/- instrumentation	21 Gy/3 fx	12–30 Gy/1–5 fx	42–60 Gy	Preoperative GTV, accounting for microscopic tumor	NR	92%	100%	NR	43.7%	55.8%	No cases of RM
Bate et al., 2015	n = 21	Posterolateral decompression with stabilization	16 Gy/1 fx	16–22 Gy/1 fx	72–132 Gy	Postoperative residual based on CT myelogram, contiguous elements of involved VB	13.7	NR	90.5% (100% for single-fraction; 71.4% for multifraction)	66% in-field, 33% out-of-field (paraspinal musculature)	NR	100%	NR
				20–30 Gy/2–5 fx	60–74.25 Gy								
Harel et al., 2016	n = 22	Decompression and stabilization	Mean 14.6 Gy/ 1 fx	12–16 Gy/1 fx	42–72 Gy	NR	12.6	88.3%	NR	50% epidural	NR	88.3%	NR
Ito et al., 2018	n = 28	Primarily posterior decompression and stabilization	24 Gy/2 fx	24 Gy/ 2 fx	84 Gy	Postoperative residual, preoperative extent of bony epidural disease, spinal instrumentation, surgical incision, immediately adjacent bony anatomical compartments	13	NR	70%	NR	NR; 63% overall survival at 1 year	NR	RM (n = 1), VCF (n = 3)
Redmond et al., 2020	n = 35	Gross total resection (17%), subtotal resection (71%), or biopsy (11%)	30 Gy/5 fx	30 Gy/ 5 fx	60 Gy	Preoperative GTV and postoperative residual disease, accounting for microscopic tumor	10.5	90%	90%	100% epidural	14.3	90%	No grade 3+ toxicity

[a] nBED using α/β ratio = 2, delivered in 2 Gy per fraction.

Abbreviations: IG-HSRT, image-guided hypofractionated stereotactic radiotherapy; Gy, Gray; GTV, gross tumor volume; CTV, clinical target volume; nBED, normalized biologically effective dose; Fx, fractions; VB, vertebral body; NR, not reported; VAS, 10-point visual analog scale of pain; RT, radiation therapy; SCI, spinal cord injury; RM, radiation myelopathy; VCF, vertebral compression fracture.

neurological deterioration after radiosurgery that was attributed to rapid tumor progression. Although pain control was not comprehensively reported, 33% of patients experienced complete resolution of pre-treatment pain.

The MSKCC experience with decompressive surgery and adjuvant radiosurgery was initially described by Moulding et al. In this study, 21 patients were treated with posterolateral decompression and fixation followed by high-dose SRS of 18–24 Gy. They reported an excellent 1-year local control rate of 90.5% and a median survival of 10.2 months after IG-HSRT (Moulding et al., 2010). In general, toxicities were considered acceptable with a single self-limited episode of acute neuritis immediately following radiosurgery, three cases of grade 2 esophagitis, and a single case of grade 4 esophagitis requiring surgical intervention. Subsequently, Laufer and colleagues reported the largest series of postoperative spine IG-HSRT cases to date with an overall median follow-up of 7.6 months. In this retrospective analysis, 186 patients underwent separation surgery consisting of posterolateral laminectomy and circumferential epidural tumor resection followed by adjuvant hypofractionated or single-fraction SRS (Laufer et al., 2013). IG-HSRT was delivered as either single-fraction treatment of 24 Gy, high-dose hypofractionated treatment of 24–30 Gy delivered over 3 fractions, or low-dose hypofractionated treatment of 18–36 Gy delivered over 5–6 fractions. The cumulative incidence of local progression was 16.4% at 1 year. On multivariate analysis, a statistically significant increase in local control using high-dose hypofractionated treatment compared with low-dose hypofractionated regimens was noted; however, no significant difference in local control was observed between single-fraction and high-dose hypofractionated treatments.

MASS was explored as a neoadjuvant therapy to spine stereotactic radiosurgery by Massicotte et al. at the University of Toronto. Ten consecutive patients with mechanical pain due to single-level metastatic spine involvement with variable degrees of epidural disease (Bilsky grade 1a–3) were treated with MASS followed by SRS that was delivered over a range of doses and fractionation schedules (18–35 Gy over 1–5 fractions; Massicotte et al., 2012). With a median follow-up of 13 months, the local control rate was favorable at 70%, and 80% of patients reported some degree of improvement in pain by 10-point visual analog scale. With regards to toxicity, there were no reports of RM, and two patients experienced transient acute pain flares within a week of treatment completion. Of note, an improvement in disability and quality of life following decompression and radiosurgery was also observed.

Al-Omair, Masucci et al. (2013) reported on a series of 80 patients that underwent postoperative spine IG-HSRT with a median follow-up of 8.3 months. Surgical technique was heterogeneous ranging from stabilization alone to epidural disease resection/decompression with or without stabilization. Procedure selection was largely driven by preoperative epidural disease (55% of patients with high-grade Bilsky 2–3 epidural disease) extent and instability. Similarly, high-dose single-fraction and hypofractionated radiosurgery (18–26 Gy over 1–2 fractions) as well low-dose hypofractionated radiosurgery (18–40 Gy over 3–5 fractions) treatments were utilized; the median total dose was 24 Gy over 2 fractions. Local control rate at 1 year was 84%, and overall survival rate at 1 year was 64%. Isolated progression within the epidural space was the most common pattern of local failure. The toxicity analysis identified seven patients who experienced transient acute pain flare events, similar to previous reports (Moulding et al., 2010; Massicotte et al., 2012; Chiang et al., 2013), and there were no reported cases of RM.

Puvanesarajah et al. (2015) retrospectively analyzed pain palliation in 32 patients with spinal malignancy who had undergone surgical decompression followed by IG-HSRT using fairly moderate radiation doses (median nBED of 47.25 Gy). Nevertheless, there were no local failures within 1 year, and significantly fewer patients reported local spine pain at 3, 6, and 12 months post-radiation. Interestingly, only a history of prior radiation to the treated vertebral levels was associated with persistent post-treatment pain on multivariable analysis.

The effects of IG-HSRT fractionation were explored by Bate et al. (2015) in a series of 57 consecutive patients with 69 treated lesions, of which 21 lesions had undergone prior separation surgery. IG-HSRT doses were 16–22 Gy in a single fraction or 20–30 Gy in 2–5 fractions. Among patients who underwent surgery followed by IG-HSRT, local control at 1 year was excellent at 90.5%. There were no failures within 1 year for 14 patients who underwent surgery followed by single-fraction radiotherapy, as compared to failures in 2 out of 7 (28.6%) patients who underwent surgery followed by multifraction radiotherapy; while intriguing, these differences were not statistically significant.

Harel et al. (2016) reported on the oncologic and surgical outcomes of 22 patients who underwent spinal decompression, fusion, or both followed by single-fraction radiotherapy to a dose of 12–16 Gy. With a median follow-up of 12.6 months, local control was 88.3%, and there were no cases of wound infection or dehiscence arising from SRS. Two symptomatic local failures occurred at 5 and 18 months post-radiation, respectively, both at doses of 14 Gy and the latter of which involving the epidural compartment. No instances of RM were noted.

Further insight into postoperative reirradiation using IG-HSRT was provided by Ito et al. (2018). Twenty-eight patients who had undergone prior conventional radiotherapy with a median dose of 30 Gy in 10 fractions received IG-HSRT with a dose of 24 Gy in 2 fractions to local recurrences following posterior decompression and fixation surgery. With a median follow-up of 13 months, 1 year local control was 70% in all patients but 100% in the subset of patients with a favorable Rades score (Rades et al., 2006). Toxicities included three vertebral compression fractures and one case of RM resulting in complete paraplegia; notably, the patient who experienced RM had received 70.4 Gy equivalents in 16 fractions using carbon ion radiotherapy on the same vertebral levels 7 years prior to decompression and IG-HSRT.

Most recently, Redmond et al. (2020) reported the results of a phase 2 study of postoperative IG-HSRT to 35 lesions in 33 patients, with a median follow-up of 10.5 months. No patients had previously received radiotherapy, and targets were limited to ≤3 contiguous vertebral levels. Surgical technique varied from biopsy only to gross total resection, but all patients received a standardized radiation dose of 30 Gy in 5 fractions. Local control and overall survival at 1 year were 90% and 54%, respectively. Of the three local failures, all were epidural in nature, consistent with prior data. No patients experienced grade 3 or higher toxicities attributable to radiotherapy.

12.4.2 PREDICTORS OF LOCAL CONTROL

Local control rates with postoperative spine IG-HSRT based upon imaging and pain control are consistently superior to those reported following conventional radiotherapy. Acknowledging differences in definitions of local control, median follow-up, technique, dose, and fractionation, local control rates of 70% to 100% (84% to 90.5% at 1 year) have been achieved with postoperative spine IG-HSRT. By way of comparison, Klekamp and Samii reported local control rates of 40.1% at 6 months and 30.7% at 1 year with postoperative low-dose conventional radiotherapy (Klekamp and Samii, 1998).

There may be several factors that impact local control in postoperative spine IG-HSRT. Dose and fractionation have been evaluated in several studies. In the Moulding et al. series, a cumulative 1 year local control rate of 90.5% was observed. However, a cumulative local failure rate of 60% among patients undergoing low-dose IG-HSRT prompted further investigation (Moulding et al., 2010). Stratifying local failure status by IG-HSRT dose, there was a significant difference in risk of local failure among patients treated with 24 Gy (6.3%) and those treated with 18 or 21 Gy (20%). As mentioned earlier, Laufer et al. reported a significant improvement in local control among patients treated with high-dose hypofractionated regimens (24–30 Gy in 3 fractions) compared to low-dose hypofractionated regimens (18–36 Gy in 5–6 fractions). However, no differences were observed between high-dose single-fraction and high-dose hypofractionated regimens (Laufer et al., 2013). Similarly, Al-Omair et al. identified high-dose regimens of 18–26 Gy in 1–2 fractions as a significant predictor of local control when compared to low-dose regimens (18–40 Gy in 3–5 fractions). Again, there were no differences in local control among patients treated with high-dose IG-HSRT when comparing single-fraction and hypofractionated regimens. Conversely, the single-arm phase 2 trial by Redmond et al. reported an excellent 1-year local control rate of 90% using a low-dose IG-HSRT regimen of 30 Gy in 5 fractions, suggesting that appropriate patient selection and target delineation may be of comparable importance in achieving optimal outcomes (Redmond et al., 2020).

The extent of postoperative residual epidural disease also plays a role, as one series identified ESCC grade 0–1 disease (e.g., as shown in Figure 13.1) as a significant predictor of local control. Further subgroup analysis of patients with preoperative ESSC grade 2–3 disease demonstrated that successful decompression to postoperative ESCC grade 0–1 disease resulted in a significantly improved local control rate compared to patients with postoperative ESCC grade 2 disease (Al-Omair, Masucci et al., 2013).

12.4.3 PATTERNS OF FAILURE

Emerging data suggest that patterns of failure following postoperative spine SBRT are similar to those following SBRT for intact vertebral bodies. Specifically, local failure within the epidural space appears to be the most common (Massicotte et al., 2012; Al-Omair, Masucci et al., 2013; Redmond et al., 2020), likely because of the decreased dose at the spinal cord–tumor interface required to achieve OAR constraints. In one series, 21 patients experienced local failure of which 71% developed isolated progression within the epidural space, 5% developed failure in the posterior elements alone when outside of the CTV, and 24% developed a mixed pattern of infield bony and/or epidural space progression. Approximately 90% of patients in this study were treated with circumferential "donut" fields, which may explain this predominant pattern of failure (Al-Omair, Masucci et al., 2013).

Chan et al. (2016) further characterized patterns of epidural failure in 24 patients who presented preoperatively with epidural disease and underwent surgery followed by postoperative IG-HSRT. In patients with preoperative epidural disease confined to the anterior compartment (sectors 1, 2, and 6 per Cox et al., 2012), 100% of recurrences involved the anterior compartment while only 56% of recurrences involved the posterior compartment (sectors 3, 4, and 5 per Cox) and none involved sector 4, abutting the spinous process. Comparatively, in patients with preoperative anterior and posterior disease, 73% of failures involved the anterior compartment and 93% of failures involved the posterior compartment. These data suggest that circumferential epidural coverage (the "donut" field) may be unnecessary in selected patients with preoperative epidural disease limited to the anterior compartment.

While a more complete discussion is outside the scope of this chapter, intraoperative seed or plaque brachytherapy has also been investigated as a potential means to treat local failure within the epidural space, particularly in patients for whom systemic therapy or external beam radiotherapy is not possible. We direct the reader to a comprehensive recent review on the topic by Zuckerman et al. (2018) for additional information.

12.4.4 SINGLE-FRACTION VERSUS MULTIFRACTION HSRT

Much remains to be investigated regarding the use of postoperative spine IG-HSRT as a therapeutic modality, and there are no prospective randomized data to compare outcomes with single-fraction versus multifraction technique. Work by Al-Omair, Masucci et al. (2013) and Laufer, Iorgulescu et al. (2013) suggests that higher-dose regimens are associated with increased local control; however, there does not appear to be a difference when comparing single-fraction and multifraction high-dose IG-HSRT regimens. This raises the possibility that cumulative BED is more predictive of local control than fractionation schema. Conversely, Redmond et al. (2020) demonstrated excellent outcomes for post-operative IG-HSRT using a relatively low-dose regimen of 30 Gy in 5 fractions, suggesting that local control is influenced by factors beyond radiation dose and fractionation alone. It should be noted that hypofractionation should be considered in situation where single-fraction IG-HSRT is not able to achieve OAR or target coverage parameters.

12.5 TOXICITIES

Toxicities associated with postoperative spine IG-HSRT mostly mirror those following IG-HSRT for intact vertebral bodies and relate to injury of the spinal cord resulting in RM, injury to nearby OARs (e.g., esophageal stricture), and VCF.

Although exceedingly uncommon, RM is the most serious toxicity associated with spine IG-HSRT as it can result in irreversible paralysis, permanent neurological deficits, or death. Adherence to appropriate spinal cord dose constraints results in a very low number of RM events. The current evidence suggests that high doses to small volumes of the spinal cord are predictive of RM, and therefore point max (P_{max}) is an important dosimetric parameter. Sahgal and colleagues developed a predictive probability model of RM based upon P_{max} to the thecal sac, which is referenced in Table 12.7 (Sahgal, Weinberg et al., 2012). Furthermore, dose constraints for thecal sac P_{max} for reirradiation with IG-HSRT have been previously described in Section 13.3.3 and suggests that reirradiation is reasonably safe if cumulative normalized BED (nBED) does not exceed 70 $Gy_{2/2}$ (e.g., as shown in Table 12.6; Sahgal, Ma et al., 2012). There have been few cases of RM documented in the spine IG-HSRT literature (Ryu et al., 2007; Gibbs et al., 2009). Among the postoperative IG-HSRT series, cases of true RM have been exceedingly rare, though Ito and colleagues documented a case of dense paraplegia following IG-HSRT reirradiation in a patient who received a cumulative prescription nBED of nearly 200 $Gy_{2/2}$ to T10-T12 (Ito et al., 2018).

The presence of spinal instrumentation likely reduces the risk for VCF, but it remains a possibility. Extrapolating from the intact spine IG-HSRT literature and accounting for differences in dose and fractionation, the risk of VCF is approximately 11% to 39% (Sahgal, Whyne et al., 2013). Factors such as lower location within the spine, lytic vertebral body disease, kyphotic/scoliotic deformity, high dose per fraction, age greater than 55 years, and baseline VCF confer an increased risk of VCF (Rose et al., 2009; Boehling et al., 2012; Cunha et al., 2012; Sahgal, Whyne et al., 2013; Sahgal, Atenafu et al., 2013). Stabilization alone does not offset radiation-induced collagen damage and osteoradionecrosis, and therefore postoperative spine IG-HSRT patients remain at risk for VCF (Sahgal, Atenafu et al., 2013). The Al-Omair series reported a total of nine VCF events, five of which were *de novo* and four progressing from preexisting fractures. This corresponded to a crude VCF rate of 11%, which is consistent with the existing literature and supports that surgical stabilization alone does decrease risk of VCF related to osteoradionecrosis (Al-Omair, Masucci et al., 2013; Al-Omair, Smith et al., 2013).

Acute pain flare is a transient common adverse reaction that occurs within the first 10 days after spine IG-HSRT (Chiang et al., 2013). It is usually responsive to steroid administration, and dexamethasone prophylaxis is a consideration to reduce to incidence and/or severity of acute pain flare. Other spine IG-HSRT-related toxicities include damage to nearby OARs such as the esophagus, bowel, and kidneys, and care should be taken to minimize dose to these critical structure in order to reduce risk of acute and long-term side effects.

There are also risks that are unique to the postoperative setting, specifically wound-healing complications and hardware failure. Earlier work suggests that the rates of hardware failure may be approximately 1% to 2%, which is similar to the rate following surgery alone (Al-Omair, Masucci et al., 2013; Amankulor et al., 2014). In a recent prospective trial of postoperative IG-HSRT, there were no instances of wound dehiscence or hardware failure in 35 treated lesions (Redmond et al., 2020). Taken together, these data suggest that postoperative spine IG-HSRT does not increase the risk of hardware failure.

12.6 CONTROVERSIES

Important questions regarding optimal dose and fractionation, target volumes, and dose constraints remain unanswered in the postoperative spine IG-HSRT setting. Early data suggest that higher nBED regimens may result in higher local control rates; however, it remains unclear if tumoricidal doses delivered in single-fraction or multifraction impact local control. Furthermore, dose escalation can increase the incidence of serious late toxicities, such as RM and VCF, which were significantly less common in the pre-spine IG-HSRT era. Discerning the balance between dose escalation and minimizing toxicity is critical for optimizing patient outcomes. Insufficient distance between the spinal cord–tumor interface is not an uncommon scenario that requires a compromise in either target coverage or prescription dose to meet normal tissue constraints. The superiority of compromising coverage with a higher prescription dose over improved coverage with a lower prescription dose is unknown and merits further investigation. Additionally, there are no specific consensus guidelines for target delineation in the postoperative setting. The need to include the entire preoperative extent of disease or only residual and high-risk areas is another area of uncertainty. In the end, it is unknown whether more conservative dose constraints for OARs should be adopted in consideration of surgical manipulation and disruption of spinal vasculature as this may be associated with an increased incidence of RM and other toxicities, although initial studies do not suggest an increased risk (Sahgal et al., 2011).

12.7 FUTURE DIRECTIONS

Prospective and randomized trials will be critical in gaining a better understanding of the nuances of postoperative spine IG-HSRT. While the existing literature is promising from a local control, pain control, and toxicity perspective, it remains inadequate at present. As IG-HSRT and surgical technique continually evolve, it will be important to further understand which patients will maximally benefit from open decompressive surgeries versus those which can be managed with minimally invasive procedures followed by IG-HSRT. Minimally invasive approaches have the advantage of reducing surgical morbidity and improving patient quality of life while maximizing local control. Future investigations will further bolster the concept of surgical intervention as a neoadjuvant modality to postoperative spine IG-HSRT.

CHECKLIST: KEY POINTS FOR CLINICAL PRACTICE

KEY POINTS FOR POSTOPERATIVE SPINE IG-HSRT

- Postoperative spine IG-HSRT is an increasingly relevant treatment modality for the management of spinal metastases.
- Current literature suggests excellent local control rates (70% to 100%) can be achieved with postoperative spine IG-HSRT. While no level 1 randomized evidence exists, results are consistently superior to conventional radiotherapy.
- Surgical hardware introduces complexities in target delineation and treatment planning. Strict adherence to normal tissue constraints can reduce the risk of RM and other toxicities.

✓	ACTIVITY	SOME CONSIDERATIONS
	Patient selection	*Is the patient appropriate for surgery followed by HSRT?* • Medically fit for surgery • Patients with spinal instability (SINS ≥ 13) • Neurological symptoms due to MESCC (Bilksy grade 2–3) • Recurrent disease following prior radiotherapy • Oligometastatic disease and good KPS
	SRS vs HSRT	*Is the lesion amenable to single-fraction SRS?* • In general, HSRT should be considered in situations where SRS is not able to achieve OAR constraints while maintaining target coverage • Some prefer HSRT and not single fraction and is an area of controversy
	Simulation	*Immobilization* • Supine, positioning, and immobilization device are dependent on location of spinal lesion • For C1–T3, consider long thermoplastic mask *or* custom head mold and open face mask • For T4–L3, consider wing board setup with arms positioned above the head • For L4—sacrum, consider Hevezi pad or Vac Lok
		Imaging • Both CT scan and MRI both with/without contrast should be performed in treatment position • MRI should be thin-slice (≤2 mm slices), T1 ± gadolinium sequence for tumor delineation, and T2/STIR sequence for spinal cord and tumor delineation • Use preoperative MRI to assist CTV delineation • CT myelogram for spinal cord delineation, if there is significant metallic artifact from instrumentation

(Continued)

✓	ACTIVITY	SOME CONSIDERATIONS

	Treatment planning	*Contours*
		• GTV: gross residual disease as delineated on postoperative imaging
		• CTV includes GTV plus
		• Entire anatomic compartment of preoperative disease plus immediately adjacent spinal segment
		• Do not need to include surgical hardware or operative scar unless clinically indicated
		• PTV is a geometric expansion of ~1–2 mm from the CTV
		• Contour all OARs at least one vertebral body superior/inferior to PTV and to a further extent for nonisocentric planning due to potential for remote hot spots
		• Consider subtracting PRVs (planning organ at risk volume) from PTV
		Treatment planning
		• Isocentric or nonisocentric
		• Must account for location of instrumentation in selection of beam angles and use a density override in dose calculations
		• Dose (variable depending on clinical scenario)
		  • 18–20 Gy in 1 fraction
		  • 24 Gy in 2 fractions
		  • 27–30 Gy in 3 fractions
		  • 30–40 Gy in 5 fractions
		• Coverage
		  • 85% to 95% target coverage by 90% to 100% of prescription dose is generally acceptable
		  • Compromise PTV coverage to meet OAR constraints (or consider hypofractionation)

	Treatment delivery	*Technology/imaging*
		• Generally, the two types of treatment delivery systems are
		  1. MLC-based LINAC
		    • 7–11 coplanar beams
		    • 5–15 MLC apertures or segments per beam
		    • "Step-and-shoot" technique with fixed gantry position or dynamic MLC
		  2. CyberKnife
		    • Nonisocentric
		    • Mobile robotic arm mounted with compact LINAC
		    • Large number of beam angles (~100–200)
		    • Circular collimators of varying size to maximize target coverage and spare OARs
		    • Cone-beam CT and/or stereoscopic kV x-rays for treatment setup verification and interfraction shifts
		    • CyberKnife permits near real-time imaging during treatment to make intrafractional adjustments
		Concurrent systemic therapy
		• Caution should be utilized to limit concurrent administration of targeted agents and systemic therapy unless clinically essential

REFERENCES

Adler JR, Jr, Chang SD, Murphy MJ, Doty J, Geis P, Hancock SL (1997) The cyberknife: A frameless robotic system for radiosurgery. *Stereotactic and Functional Neurosurgery* 69:124–128.

Adler JR, Jr, Murphy MJ, Chang SD, Hancock SL (1999) Image-guided robotic radiosurgery. *Neurosurgery* 44:1299–1307.

Al-Omair A, Masucci L, Masson-Cote L, Campbell M, Atenafu EG, Parent A, Letourneau D et al. (2013) Surgical resection of epidural disease improves local control following postoperative spine stereotactic body radiotherapy. *Neuro-Oncology* 15(10):1413–1419.

Al-Omair A, Smith R, Kiehl TR, Lao L, Yu E, Massicotte EM, Keith J, Fehlings MG, Sahgal A (2013) Radiation-induced vertebral compression fracture following spine stereotactic radiosurgery: Clinicopathological correlation. *Journal of Neurosurgery Spine* 18(5):430–435.

Amankulor NM, Xu R, Iorgulescu JB, Chapman T, Reiner AS, Riedel E, Lis E, Yamada Y, Bilsky M, Laufer I (2014) The incidence and pattern of hardware failure after separation surgery in patients with spinal metastatic tumors. *Journal of Neurosurgery Spine* 14(9):1850–1859.

Bate BG, Khan NR, Kimball BY, Gabrick K, Weaver J (2015) Stereotactic radiosurgery for spinal metastases with or without separation surgery. *Journal of Neurosurgery Spine* 22(4):409–415.

Benedict SH, Yenice KM, Followill D, Galvin JM, Hinson W, Kavanagh B, Keall P et al. (2010) Stereotactic body radiation therapy: The report of AAPM Task Group 101. *Medical Physics* 39(1):563.

Bilsky MH, Boland PJ, Panageas KS, Woodruff JM, Brennan MF, Healey JH (2001) Intralesional resection of primary and metastatic sarcoma involving the spine: Outcome analysis of 59 patients. *Neurosurgery* 49(6):1277–1286.

Bilsky MH, Laufer I, Fourney DR, Groff M, Schmidt MH, Varga PP, Vrionis FD, Yamada Y, Gerszten PC, Kuklo TR (2010) Reliability analysis of the epidural spinal cord compression scale. *Journal of Neurosurgery Spine* 13(3):324–328.

Bishop AJ, Tao R, Guadagnolo BA, Allen PK, Rebueno NC, Wang XA, Amini B, Tatsui CE, Rhines LD, Li J, Chang EL, Brown PD, Ghia AJ (2017) Spine stereotactic radiosurgery for metastatic sarcoma: Patterns of failure and radiation treatment volume considerations. *Journal of Neurosurgery Spine* 27(3):303–311.

Boehling NS, Grosshans DR, Allen PK, McAleer MF, Burton AW, Azeem S, Rhines LD, Chang EL (2012) Vertebral compression fracture risk after stereotactic body radiotherapy for spinal metastases. *Journal of Neurosurgery Spine* 16(4):379–386.

Chan MW, Thibault I, Atenafu EG, Yu E, John Cho BC, Letourneau D, Lee Y, Yee A, Fehlings MG, Sahgal A (2016) Patterns of epidural progression following postoperative spine stereotactic body radiotherapy: Implications for clinical target volume delineation. *Journal of Neurosurgery Spine* 24(4):652–659.

Chi JH, Gokaslan Z, McCormick P, Tibbs PA, Kryscio RJ, Patchell RA (2009) Selecting treatment for patients with malignant epidural spinal cord compression—Does age matter?: Results from a randomized clinical trial. *Spine* (1976) 34(5):431–435.

Chiang A, Zeng L, Zhang L, Lochray F, Korol R, Loblaw A, Chow E, Sahgal A (2013) Pain flare is a common adverse events in steroid-naïve patients after spine stereotactic body radiation therapy: A prospective clinical trial. *International Journal of Radiation Oncology, Biology, Physics* 86(4):638–642.

Cox BW, Spratt DE, Lovelock M, Bilsky MH, Lis E, Ryu S, Sheehan J (2012) International spine radiosurgery consortium consensus guidelines for target volume definition in spinal stereotactic radiosurgery. *International Journal of Radiation Oncology, Biology, Physics* 83(5):e597–605.

Cunha MV, Al-Omair A, Atenafu EG, Masucci GL, Letourneau D, Korol R, Yu E et al. (2012) Vertebral compression fracture (VCF) after spine stereotactic body radiation therapy (SBRT): Analysis of predictive factors. *International Journal of Radiation Oncology, Biology, Physics* 84(3):e343–349.

Findlay GF (1984) Adverse effects of the management of malignant spinal cord compression. *J Neurol Neurosurg Psychiatry* 47(8):761–768.

Fisher CG, DiPaola CP, Ryken TC, Bilsky MH, Shaffrey CI, Berven SH, Harrop JS et al. (2010) A novel classification system for spinal instability in neoplastic disease: An evidence based-approach and expert consensus from the Spine Oncology Study Group. *Spine* (1976) 35(22):E1221–1229.

Fisher CG, Schouten R, Versteeg AL, Boriani S, Varga PP, Rhines LD, Kawahara N et al. (2014) Reliability of the Spinal Instability Neoplastic Score (SINS) among radiation oncologists: An assessment of instability secondary to spinal metastases. *Radiat Oncol* 9:69.

Fisher CG, Versteeg AL, Schouten R, Boriani S, Varga PP, Rhines LD, Heran MK et al. (2014) Reliability of the spinal instability neoplastic scale among radiologists: An assessment of instability secondary to spinal metastases. *Am J Roentgenol* 203(4):869–874.

Fornasier VL, Horne JG (1975) Metastases to the vertebral column. *Cancer* 36(2):590–594.

Fourney DR, Frangou EM, Ryken TC, Dipaola CP, Shaffrey CI, Berven SH, Bilsky MH et al. (2011) Spinal instability neoplastic score: An analysis of reliability and validity from the spine oncology study group. *J Clin Oncol* 29(22):3072–3077.

Gerszten PC, Germanwala A, Burton SA, Welch WC, Ozhasoglu C, Vogel WJ (2005) Combination kyphoplasty and spinal radiosurgery: A new treatment paradigm for pathological fractures. *J Neurosurg Spine* 3:296–301.

Gerszten PC, Monaco EA, III (2009) Complete percutaneous treatment of vertebral body tumors causing spinal canal compromise using transpedicular cavitation, cement augmentation, and radiosurgical technique. *Neurosurg Focus* 27(6):E9.

Ghia AJ, Guha-Thakurta N, Hess K, Yang JN, Settle SH, Sharpe HJ, Li J, McAleer M, Chang EL, Tatsui CE, Brown PD, Rhines LD (2018) Phase 1 study of spinal cord constraint relaxation with single session spine stereotactic radiosurgery in the primary management of patients with inoperable, previously unirradiated metastatic epidural spinal cord compression. *International Journal of Radiation Oncology, Biology, Physics* 102(5):1481–1488.

Gibbs IC, Patil C, Gerszten PC, Adler JR, Jr., Burton SA (2009) Delayed radiation-induced myelopathy after spinal radiosurgery. *Neurosurgery* 64(2 Suppl):A67–72.

Gilbert RW, Kim JH, Posner JB (1978) Epidural spinal cord compression from metastatic tumor: Diagnosis and treatment. *Ann Neurol* 3(1):40–51.

Greenberg HS, Kim JH, Posner JB (1980) Epidural spinal cord compression from metastatic tumor: Results from a new treatment protocol. *Ann Neurol* 8:361–366.

Harel R, Emch T, Chao S, Elson P, Krishnaney A, Djemil T, Suh J, Angelov L (2016) Quantitative evaluation of local control and wound healing following surgery and stereotactic spine radiosurgery for spine tumors. *World Neurosurg* 87:48–54.

Harrington KD (1993) Metastatic tumors of the spine: Diagnosis and treatment. *J Am Acad Orthop Surg* 1:76–86.

Ito K, Nihei K, Shimizuguchi T, Ogawa H, Furuya T, Sugita S, Hozumi T; Keisuke Sasai, MD, PhD, Karasawa K (2018) Postoperative reirradiation using stereotactic body radiotherapy for metastatic epidural spinal cord compression. *J Neurosurg Spine* 29(3):332–338

Kirshblum SC, Burns SP, Biering-Sorensen F, Donovan W, Graves DE, Jha A, Johansen M et al. (2011) International standards for neurological classification of spinal cord injury (revised 2011). *J Spinal Cord Med* 34(6):535–546.

Kirshblum SC, Waring W, III (2014) Updates for the international standards for neurological classification of spinal cord injury. *Phys Med Rehabil Clin North Am* 25(3):505–517.

Klekamp J, Samii M (1998) Surgical results for spinal metastases. *Acta Neurochir* 140:957–967.

Klimo P, Jr., Schmidt MH (2004) Surgical management of spinal metastases. *Oncologist* 9(2):188–196.

Laufer I, Iorgulescu JB, Chapman T, Lis E, Shi W, Zhang Z, Cox BW, Yamada Y, Bilsky MH (2013) Local disease control for spinal metastases following "separation surgery" and adjuvant hypofractionated or high-dose single-fraction stereotactic radiosurgery: Outcome analysis in 186 patients. *J Neurosurg Spine* 18(3):207–214.

Laufer I, Rubin DG, Lis E, Cox BW, Stubblefield MD, Yamada Y, Bilsky MH (2013) The NOMS framework: Approach to treatment of spinal metastatic tumors. *Oncologist* 18(6):744–751.

Liebross RH, Starkschall G, Wong PF, Horton J, Gokaslan ZL, Komaki R (2002) The effect of titanium stabilization rods on spinal cord radiation dose. *Med Dosim* 27(1):21–24.

Maranzano E, Latini P (1995) Effectiveness of radiation therapy without surgery in metastatic spinal cord compression: Final results from a prospective trial. *Int J Radiat Biol Phys* 32:959–967.

Massicotte E, Foote M, Reddy R, Sahgal A (2012) Minimal access spine surgery (MASS) for decompression and stabilization performed as an out-patient procedure for metastatic spinal tumors followed by spine stereotactic body radiotherapy (SBRT): First report of technique and preliminary outcomes. *Technol Cancer Res Treat* 11(1):15–25.

Masucci GL, Yu E, Ma L, Chang EL, Letourneau D, Lo S, Leung E et al. (2011) Stereotactic body radiotherapy is an effective treatment in re-irradiating spinal metastases: Current status and practical considerations for safe practice. *Expert Rev Anticancer Ther* 11(12):1923–1933.

Moulding HD, Elder JB, Lis E, Lovelock DM, Zhang Z, Yamada Y, Bilsky MH (2010) Local disease control after decompressive surgery and adjuvant high-dose single-fraction radiosurgery for spine metastases. *J Neurosurg Spine* 13(1):87–93.

Moussazadeh N, Laufer I, Yamada Y, Bilsky MH (2014) Separation surgery for spinal metastases: Effect of spinal radiosurgery on surgical treatment goals. *Cancer Control* 21(2):168–174.

Patchell RA, Tibbs PA, Regine WF, Payne R, Saris S, Kryscio RJ, Mohiuddin M, Young B (2005) Direct decompressive surgical resection in the treatment of spinal cord compression caused by metastatic cancer: A randomized trial. *Lancet* 366(9486):643–648.

Pekmezci M, Dirican B, Yapici B, Yazici M, Alanay A, Gürdalli S (2006) Spinal implants and radiation therapy: The effect of various configurations of titanium implant systems in a single-level vertebral metastasis model. *J Bone Joint Surg Am* 88(5):1093–1100.

Puvanesarajah V, Lo SL, Aygun N, Liauw JA, Jusué-Torres I, Lina IA, Hadelsberg U, Elder BD, Bydon A, Bettegowda C, Sciubba DM, Wolinsky JP, Rigamonti D, Kleinberg LR, Gokaslan ZL, Witham TF, Redmond KJ, Lim M (2015) Prognostic factors associated with pain palliation after spine stereotactic body radiation therapy. *J Neurosurg Spine* 23(5):620–629.

Redmond KJ, Lo SS, Fisher C, Sahgal A (2016) Postoperative stereotactic body radiation therapy (SBRT) for spine metastases: A critical review to guide practice. *International Journal of Radiation Oncology, Biology, Physics* 95(5):1414–1428.

Redmond KJ, Lo SS, Soltys SG, Yamada Y, Barani IJ, Brown PD, Chang EL, Gerszten PC, Chao ST, Amdur RJ, De Salles AA, Guckenberger M, Teh BS, Sheehan J, Kersh CR, Fehlings MG, Sohn MJ, Chang UK, Ryu S, Gibbs IC, Sahgal A (2017) Consensus guidelines for postoperative stereotactic body radiation therapy for spinal metastases: Results of an international survey. *J Neurosurg Spine* 26(3):299–306.

Redmond KJ, Robertson S, Lo SS, Soltys SG, Ryu S. McNutt T, Chao ST, Yamada Y, Ghia A, Chang EL, Sheehan J, Sahgal A (2017) Consensus contouring guidelines for postoperative stereotactic body radiation therapy for metastatic solid tumor malignancies to the spine. *International Journal of Radiation Oncology, Biology, Physics* 97(1):64–74.

Redmond KJ, Sciubba D, Khan M, Gui C, Lo SL, Gokaslan ZL, Leaf B, Kleinberg L, Grimm J, Ye X, Lim M (2020) A Phase 2 study of post-operative stereotactic body radiation therapy (SBRT) for solid tumor spine metastases. *International Journal of Radiation Oncology, Biology, Physics* 106(2):261–268.

Rock JP, Ryu S, Shukairy MS, Yin FF, Sharif A, Schreiber F, Abdulhak M, Kim JH, Rosenblum ML (2006) Postoperative radiosurgery for malignant spinal tumors. *Neurosurgery* 58(5):891–898.

Rose PS, Laufer I, Boland PJ, Hanover A, Bilsky MH, Yamada J, Lis E (2009) Risk of fracture after single fraction image-guided intensity-modulated radiation therapy to spinal metastases. *J Clin Oncol* 27(30):5075–5079.

Ryu SI, Chang SD, Kim DH, Murphy MJ, Le QT, Martin DP, Adler JR, Jr. (2001) Image-guided hypo-fractionated stereotactic radiosurgery to spinal lesions. *Neurosurgery* 49(4):838–846.

Ryu SI, Jin JY, Jin R, Rock J, Ajlouni M, Movsas B, Rosenblum M, Kim JH (2007) Partial volume tolerance of the spinal cord and complications of single-dose radiosurgery. *Cancer* 109(3):628–636.

Sahgal A, Atenafu EG, Chao S, Al-Omair A, Boehling N, Balagamwala EH, Cunha M et al. (2013) Vertebral compression fracture after spine stereotactic body radiotherapy: A multi-institutional analysis with a focus on radiation dose and spinal instability neoplastic score. *J Clin Oncol* 31(27):3426–3431.

Sahgal A, Bilsky M, Chang EL, Ma L, Yamada Y, Rhines LD, Létourneau D et al. (2011) Stereotactic body radiotherapy for spinal metastases: Current status, with a focus on its application in the post-operative patient. *J Neurosurg Spine* 14:151–166.

Sahgal A, Ma L, Gibbs I, Gerszten PC, Ryu S, Soltys S, Weinberg V et al. (2010) Spinal cord tolerance for stereotactic body radiotherapy. *International Journal of Radiation Oncology, Biology, Physics* 77:548–553.

Sahgal A, Ma L, Weinberg V, Gibbs IC, Chao S, Chang UK, Werner-Wasik M et al. (2012) Reirradiation human spinal cord tolerance for stereotactic body radiotherapy. *International Journal of Radiation Oncology, Biology, Physics* 82(1):107–116.

Sahgal A, Weinberg V, Ma L, Chang E, Chao S, Muacevic A, Gorgulho A et al. (2012) Probabilities of radiation myelopathy specific to stereotactic body radiation therapy to guide safe practice. *International Journal of Radiation Oncology, Biology, Physics* 85(2):341–347.

Sahgal A, Whyne CM, Ma L, Larson DA, Fehlings MG (2013) Vertebral compression fracture after stereotactic body radiotherapy for spinal metastases. *Lancet Oncol* 14(8):e310–e320.

Sciubba DM, Gokaslan ZL (2006) Diagnosis and management of metastatic spine disease. *Surg Oncol* 15(3):141–151.

Sciubba DM, Petteys RJ, Dekutoski MB, Fisher CG, Fehlings MG, Ondra SL, Rhines LD, Gokaslan ZL (2010) Diagnosis and management of metastatic spine disease. A review. *J Neurosurg Spine* 13(1):94–108.

Sharan AD, Szulc A, Krystal J, Yassari R, Laufer I, Bilsky MH (2014) The integration of radiosurgery for the treatment of patients with metastatic spine diseases. *J Am Acad Orthop Surg* 22(7):447–454.

Tatsui CE, Stafford RJ, Li J, Sellin JN, Amini B, Rao G, Suki D, Ghia AJ, Brown P, Lee SH, Cowles CE, Weinberg JS, Rhines LD (2015) Utilization of laser interstitial thermotherapy guided by real-time thermal MRI as an alternative to separation surgery in the management of spinal metastasis. *J Neurosurg Spine*23(4):400–411.

Wang X, Yang JN, Li X, Tailor R, Vassilliev O, Brown P, Rhines L, Chang E (2013) Effect of spine hardware on small spinal stereotactic radiosurgery dosimetry. *Phys Med Biol* 58(19):6733–6747.

Wong DA, Fornasier VL, MacNab I (1990) Spinal metastases: The obvious, the occult, and the impostors. *Spine* 15(1):1–4.

Yong RF, Post EM, King GA (1980) Treatment of spinal epidural metastases. Randomised prospective comparison of laminectomy and radiotherapy. *J Neurosurg* 53(6):741–748.

Zuckerman SL, Lim J, Yamada Y, Bilsky MH, Laufer I (2018) Brachytherapy in spinal tumors: A systematic review. *World Neurosurg* 118:e235–e244.

13 Postoperative Cavity Image-Guided Stereotactic Radiotherapy Outcomes

Yaser Hasan, Paul D. Brown, and Mary Frances McAleer

Contents

13.1 INTRODUCTION

Up to 30% of cancer patients will be diagnosed with brain metastases during their illness. Surgery can serve multiple purposes including providing pathological diagnosis and relieving mass effect and has been associated with improved survival in some patients. Local tumor recurrence in the surgical bed is still common despite complete resection, occurring in over half of patients during the first year following resection (Mahajan et al., 2017). For decades, whole-brain radiotherapy (WBRT) has been considered the standard of care therapy following resection of brain metastases based on positive results of prospective study by Patchell et al. (1998), which was confirmed by a trial conducted by the European Organization for Research and Treatment of Cancer (Kocher et al., 2011). Addition of WBRT significantly improved local control (LC) in patients randomized to receive the adjuvant radiation compared with those followed by observation after surgical resection of one (Patchell et al., 1998) and up to two (Kocher et al., 2011) brain lesions in these trials. However, WBRT is associated with increased adverse effects on patient neurocognitive function and quality of life (QOL) as well as delay of potentially life-prolonging systemic therapies (see, e.g., Soffietti et al., 2013; Brown, 2017).

There is an emerging body of evidence in support of limited field, single-fraction, or short hypofractionated courses of stereotactic radiotherapy to the at-risk intracranial surgical cavity for improved local tumor control for select patients with brain metastases. Radiosurgery has the additional advantage of avoiding delays in systemic therapy, but with increased risk of local radionecrosis. Utilization of such therapy has become more widely accepted, given the advances in neuroradiographic imaging, neurosurgical, and radiotherapeutic techniques. Here, we present the current data and indications for postoperative image-guided single-fraction or hypofractionated stereotactic radiotherapy (IG-HSRT) for resected brain metastases.

13.2 INDICATIONS

The indications for post-operative cavity stereotactic radiosurgery (SRS) are derived from the recently published, multi-institutional, Alliance cooperative group phase III trial that compared cavity SRS to WBRT for resected brain metastases (North Central Cancer Treatment Group [NCCTG] N107C/CEC.3). This study specifies the following criteria for adult subjects to be eligible to receive focal SRS to the surgical cavity with a 2 mm margin:

- A maximum of four brain metastases, with resection of one of these lesions
- Confirmed non-CNS, metastatic solid tumor, excluding germ cell, small cell, and lymphoma
- less than 3.0 cm, which is the maximum diameter of any unresected metastasis on postcontrast MRI
- less than 5.0 cm, which is the maximum diameter of resection cavity on postoperative imaging
- All lesions more than 5 mm from the optic chiasm and outside the brainstem
- Good performance status (Karnofsky Performance Scale [KPS] ≥ 70)
- Non-pregnant, non-nursing females
- No prior cranial RT
- No concurrent cytotoxic systemic therapy during postoperative SRS
- No leptomeningeal metastatic disease

13.3 APPROACH

The recommended approach described here is guided by the Alliance N107C/CEC.3 trial of post-operative SRS compared to WBRT for resected metastatic brain disease, using the indications above for patient selection. Patients identified as appropriate candidates for post-operative SRS underwent volumetric MRI of brain within 7 days of SRS. The SRS treatment was frame based and was delivered using either the Gamma Knife (GK) Perfexion unit (Elekta, Stockholm, Sweden) or a linear accelerator (LINAC) with arcs or mini-multileaf collimation. Real-time treatment planning was employed. The single-fraction dose to be delivered was defined based on the surgical cavity volume (Table 13.1). The maximum point dose to the optic apparatus was limited to less than 9 Gy, and up to 1.0 cubic centimeters (cc) of brainstem could receive 12 Gy. The surgical cavity was treated with a 2-mm margin.

After treatment, patients were monitored every 6–9 weeks for 1 year, then every 3–4 months during year 2, then every 6 months thereafter with MRI of brain to assess LC. The treatment approach to administer adjuvant SRS to the surgical bed in the trial specified the following:

- Technique: GK or ≥4 MV x-rays using LINAC
- Mandatory patient immobilization/localization system
- Target: MRI-defined surgical cavity +2 mm margin, excluding anatomic barriers
- <5.0 cm maximum diameter of cavity
- Dose: Prescribed to highest isodose line covering cavity + margin
- Dose based on cavity volume (without margin) (Table 13.1)

Table 13.1 Prescription dose for postoperative SRS as defined by N107C/CEC.3 trial

CAVITY VOLUME (CC)	DOSE (GY)
<4.2	20
≥4.2 to <8.0	18
≥8.0 to <14.4	17
≥14.4 to <20	15
≥20.0 to <30	14
≥30 to <5.0 cm max diam	12

Abbreviations: cc, cubic cm; max diam, maximum diameter.

13.4 OUTCOME DATA

13.4.1 PROSPECTIVE TRIALS

Until recently, there was only one prospective, single-institution, phase II trial from Memorial Sloan Kettering Cancer Center (MSKCC) that has reported results of LC using single-fraction SRS following resection of intracranial metastases (Brennan et al., 2014). Now two landmark phase III randomized controlled trials have been published. The first one is the multi-institutional Alliance trial led by Brown et al., described earlier. The other is a single-institution MD Anderson Cancer Center randomized trial of cavity SRS vs observation for completely resected brain metastases (Mahajan et al., 2017). These prospective studies are summarized in Table 13.2.

In the MSKCC trial, 49 evaluable patients with 50 surgical cavities were enrolled. Patients had median age of 59 years (range, 23–81), median KPS of 90 (range, 70–100), and the majority (57%) had histologically confirmed diagnosis of non-small cell lung cancer (NSCLC), followed by 20% with breast cancer and 8% with melanoma. Sixty-five percent of patients had primary tumor site controlled, and 45% had extracranial metastatic disease. Ninety-eight percent of patients had single metastases, with median tumor diameter of 2.9 cm (range, 1.0–5.2), and GTR was reported in 92% of cases. Of the 49 evaluable patients, 39 received adjuvant SRS to a total of 40 surgical cavities, with median cavity diameter of 2.8 cm (range, 1.7–5.4) that included the surgical track. A 2-mm margin was added to each cavity, as identified on postcontrast MRI and CT head, and single-fraction SRS was delivered based on the RTOG 90–05 dose guidelines (Shaw et al., 1996, 2000) and LINAC delivery of 8–12 noncoplanar, static beams with micro-multileaf collimation a median of 31 days (range, 7–56) from surgery.

One-year local failure (LF) was identified in 15% of patients treated with postoperative SRS (Brennan et al., 2014). This percentage increased to 53% in those patients with cavities measuring at least 3 cm and having dural invasion. DBF at 1 year was seen in 44% at a median of 4.4 months (range, 1.1–17.9), and median OS was 14.7 months (range, 1–94.1). WBRT was used as salvage in approximately two-thirds of patients who had intracranial disease progression. The results of this study revealed comparable cavity LC as was observed in the WBRT arms of the prospective randomized controlled studies of adjuvant WBRT following surgery or SRS, and the rate of DBF was similar to the observation arms of these studies (Patchell et al., 1998; Kocher et al., 2011). Of note, however, OS in the MSKCC study was almost 4 months longer than in the older trials, and WBRT was delayed or omitted in the majority of these patients.

Following the MSKCC trial, a smaller, randomized trial of single or multi-fraction cavity stereotactic radiotherapy (SRT) versus WBRT was published (Kepka et al., 2016). A total of 59 patients with total or subtotal resection were randomized to receive WBRT ($n = 30$) or SRT ($n = 29$), with either 15 Gy or 25 Gy in 1 or 5 fractions, respectively. After median follow-up of 29 months with 15 patients still alive, the primary endpoint of neurologic/cognitive failure at 6 months was 8% worse in the SRT group. Neurologic/cognitive failure at 2 years was 75% in the SRT arm and 62% in the WBRT arm ($p = 0.31$). The study authors concluded that, while not sufficiently powered, non-inferiority of SRT was not demonstrated and could only be considered hypothesis-generating.

The landmark phase III Alliance trial (Brown et al., 2017) is the only adequately powered multi-institutional randomized study directly comparing SRS with WBRT, assessing both quality of life (QOL) and cognitive function. Over a 4-year period, 194 patients were randomized between receiving postoperative WBRT ($n = 96$; 49 patients received 30 Gy in 10 fractions, 43 received 37.5 Gy in 15 fractions, 4 did not receive treatment; fractionation regimen used was predetermined by each institution) and SRS ($n = 98$; 5 did not receive treatment) to the surgical cavity, with doses ranging from 12 to 20 Gy, based on cavity volume (Table 13.1). Median follow-up was 11.1 months for all patients and 22.6 months (range, 13.8–34.6) for surviving patients. Patients were allowed to have up to three unresected metastases that were to be treated with SRS in either study arm.

Most patients enrolled had a single metastasis (77%). The overall survival, a co-primary endpoint, was similar in both arms (11.6 months WBRT versus 12.2 months SRS, HR 1.07, $p = 0.7$). The other co-primary endpoint of this study was cognitive-deterioration-free survival. As expected, this period was shorter after WBRT versus SRS (median 3.0 versus 3.7 months, HR 0.47, $p < 0.0001$). Cognitive deterioration at 6 months was less frequent (52% versus 85%, $p = 0.00031$) in patients receiving SRS. Interestingly, 6-month and 12-month surgical bed control were worse with SRS than WBRT (80.4% versus 87.1%, and 60.5% versus 80.6%, $p = 0.00068$, respectively). However, there was no difference in the development of leptomeningeal

Table 13.2 Summary of prospective studies using postoperative single-fraction stereotactic radiosurgery

STUDY	NO.	AGE (YEARS)[b]	KPS	% HISTOLOGY	% ONE LESION	% GTR	CAVITY (CC)	MARGIN (MM)	DOSE (GY)	% LC	% DBF	OS (MONTHS)	% NECROSIS	% LMD	% WBRT SALVAGE
NCCTG-N107C/CEC.3 (Brown et al., 2017)	98	61 (54–66)	100% ≥70	59 L 30 other 11 radioresistant	77	92	NR	2	12–20	62	35.7	12.2	4	7.2	20
MD Anderson (Mahajan et al., 2017)	63	59 (20–80)	100% ≥70	22 M 21 L 14 B	60	100	8.9 (0.9–28.6)	1	12–16	72	58	17	0	28	38
MSKCC[a] (Brennan et al., 2014)	39	59 (23–81)	90 (70–100)	57 L 20 B 8 M	97	92	2.8[c] (1.7–5.4) +track	2 + track	18 (15–22)	85	44[d]	14.7 (1–94.1)	17.5	NR	65

Note: Data are presented as median (range) unless otherwise indicated.

[a] Phase 2 study.
[b] Median (range).
[c] Size in cm.
[d] At 1 year.

Abbreviations: KPS, Karnofsky performance status; Histology L, non-small cell lung cancer; B, breast cancer; M, melanoma; GTR, gross total resection; cc, cubic cm; preop, preoperative; LC, local control; DBF, distant brain failure; OS, overall survival; postop, postoperative; LMD, leptomeningeal disease; WBRT, whole-brain radiotherapy; NR, not reported.

disease (7.2% with SRS versus 5.4% with WBRT, $p = 0.62$) or frequency of requiring local salvage therapy (32% with SRS versus 21% with WBRT, $p = 0.12$). Overall, intracranial control was greater in patients treated with WBRT than with SRS (72% versus 36.6% at 12 months, $p < 0.0001$). QOL, functional independence and physical wellbeing measured by patients were significantly better in those treated with SRS. Median duration of stable or improved functional independence was longer after SRS than WBRT (not reached with SRS versus 14 months for WBRT, HR 0.56, $p = 0.034$). There were 54 long-term survivors who had evaluation of cognitive function at 12 months or more from time of randomization. In these patients, long-term cognitive deterioration was less frequent after SRS than WBRT (37% versus 89% at 3 months, $p = 0.00016$; 46% versus 88% at 6 months, $p = 0.0025$; 48% versus 81% at 9 months, $p = 0.02$; and 60% versus 91% at 12 months, $p = 0.0188$). In these patients, intracranial tumor control was better with WBRT than SRS (92.6% versus 70.4% at 6 months and 81.5% versus 40.7% at 12 months; overall HR 3.12, $p = 0.0033$). The combined evidence from the trial suggests that SRS to the surgical cavity results in no significant difference in survival but improved preservation of QOL and cognitive outcomes compared with WBRT.

The other practice-changing, phase III randomized controlled study investigating post-operative SRS to the surgical cavity versus observation for completely resected brain metastases was published contemporaneously by MD Anderson Cancer Center (Mahajan et al., 2017). The median patient age was 59 years, and median dose of post-operative radiation was 16 Gy (range, 12–18) to the 50% isodose line, depending on cavity volume. Prescription doses and SRS target volumes were 16 Gy (for ≤10 cc), 14 Gy (for 10.1–15 cc), and 12 Gy (for >15 cc). To be eligible for this study, patients had to have a KPS more than 70 and a complete resection of one to three brain metastases (maximum diameter of cavity up to 4 cm). Patients were randomly assigned to either SRS to the resection cavity within 30 days of surgery or observation. Further stratification was performed according to the primary tumor histology, tumor size, and number of metastases. The primary endpoint was time to local recurrence in the resection cavity.

A total of 132 patients were randomly assigned between the two groups, with 128 patients eligible for analysis. Freedom from local recurrence at 12 months was better in the SRS group than the observation arm (72% versus 43%, HR 0.46, $p = 0.015$). Median time to local recurrence was 7.6 months in the observation arm and was not reached in the SRS cohort. Median OS was similar between the two arms (18 months with observation and 17 months with SRS, $p = 0.24$). Freedom from distant brain recurrence at 12 months was similar between the two groups (33% with observation and 42% with SRS, $p = 0.35$). Lesions with diameter ≤2.5 cm were associated with better LC at 12 months (91%) when compared to larger lesions (40% for lesions 2.6–3.5 cm; 46% if >3.5 cm). In the multivariate analysis for time to local recurrence, factors found to be significant predictors of local recurrence were SRS and metastasis size. Primary cancer histology, systemic disease status, graded prognostic assessment (GPA), and number of brain metastases were not associated with time to local recurrence. The results of this study reinforce that complete surgical resection alone is insufficient to provide durable local control and suggest that adjuvant SRS could be an alternative to WBRT after surgical resection of up to three brain metastases.

An essential limitation of the Alliance trial raised by its authors involves the relatively worse surgical bed control after SRS than reported in previous series. Surgical bed control rate was only 62% in this trial (Brown et al., 2017), compared to 72% in the single-institution MD Anderson trial (Mahajan et al., 2017), despite the use of lower SRS doses and tighter margins around the cavity in the latter trial. The differences in surgical bed control might have resulted from multiple factors, including inclusion of patients with subtotal resection, high rate of piecemeal resection, lack of central review for determining local control, and variations between the patient populations between the two trials.

These two trials establish post-operative SRS to the surgical cavity as the standard of care, offering improved local control compared to observation and better cognitive function preservation compared to WBRT. Questions remain as to the best radiation regimen (dose/fraction) and optimal timing of the SRS treatment.

13.4.2 RETROSPECTIVE STUDIES

Since 2008, there have been numerous retrospective analyses of postoperative focal brain radiotherapy published, including 13 reports of outcomes using single-fraction SRS (Table 13.3), 18 with multifraction IG-HSRT regimens (Table 13.4), and another 8 that combined both single- and multiple-fraction treatments (Table 13.5). Four of these reports also compared the outcomes of patients receiving focal

Table 13.3 Summary of retrospective studies using postoperative single-fraction stereotactic radiosurgery

STUDY	NO.	AGE (YEARS)	KPS	% HISTOLOGY	% ONE LESION	% GTR	CAVITY(CC)	MARGIN (MM)	DOSE (GY)	% LC	% DBF	OS (MONTHS)	% NECROSIS	% LMD	% WBRT SALVAGE
Dartmouth (Hartford et al., 2013)	47	64 (24–85)	80 (50–100)	49 L 11 B 15 M	70	76	3.0[b] (1.3–4.6) preop	2	10 (8–20)	85.5[c]	56.2[c]	52.5%[c]	NR	NR	45
Tufts (Hwang et al., 2010)	25	59.5[a] (48–71)	NR	NR	NR	95	NR	NR	NR	100	28	15 (6.0–35.8)	NR	NR	NR
Sherbrooke (Iorio-Morin et al., 2014)	110	58 (37–84)	90 (50–100)	50 L 13 B 10 M	30	81	12 (0.6–43)	1	18 (10–20)	73[c]	54	11 (1.4–84)	0.9	11	28
Osaka (Iwai et al., 2008)	21	61 (41–80)	88[b] (70–100)	24 L 10 B	76	86	10.7 (3.4–26.9)	NR	17 (13–23)	76	48	20 postop	NR	24	NR
UVA (Jagannathan et al., 2009)	47	61 (37–88)	90 (60–100)	40 L 15 B 21 M	13	100	10.5 (1.75–35.45)	2–3	19 (6–22)	94	6	11 (7–36)	0	NR	28
Wake Forest (Jensen et al., 2011)	106	56.1 (22.6–88.0)	NR	47 L 14 B 10 M	57.5	96.4	8.0 (0.32–33.4)	0	17 (11–23)	80.3[c]	NR	10.9	3	7.5	37
Barrow (Kalani et al., 2010)	68	60[a] (28–89)	90 (40–100)	44 L 15 B 13 M	100	NR	10.35 (0.9–45.4)	1–3	15 (14–30)	79.5	39.7	13.2	NR	NR	NR
Allegheny (Karlovits et al., 2009)	52	61 (31–85)	NR	46 L 17 B	34	92.3	3.85 (0.08–22)	2	15 (8–18)	93	44	15	NR	NR	31
Wash U (Limbrick et al., 2009)	15	56.8 (41–85)	93% ≥70	40 L 27 B 20 R	80	80	0.18–16.0	NR	(16–24)	73.3 at 11 m	60	20 (5–68)	NR	NR	40
U Pitt (Luther et al., 2013)	120	58[a]	NR	40 L 21 B 16M	NR	100	7.3 PTV	NR	16	85.8	40	NR	NR	NR	16

(Continued)

Table 13.3 (*Continued*) Summary of retrospective studies using postoperative single-fraction stereotactic radiosurgery

STUDY	NO.	AGE (YEARS)	KPS	% HISTOLOGY	% ONE LESION	% GTR	CAVITY(CC)	MARGIN (MM)	DOSE (GY)	% LC	% DBF	OS (MONTHS)	% NECROSIS	% LMD	% WBRT SALVAGE
U Pitt and Sherbrooke (Mathieu et al., 2008)	40	59.5	80 (60–100)	40 L 10 B 20 M	67.5	80	9.1 (0.6–39.9)	1	16 (11–20)	73	54	13 (2–56)	0	NR	16
U Penn (Ojerholm et al., 2014b)	91	60 (22–82)	96% >70	43 L 13 B 14 M	57	82	9.2 (0.6–34.7)	0	16 (12–21)	82	64	22.3	7	14	33
Heny Ford (Robbins et al., 2012)	85	58 (38–83)	80 (60–100)	59 L 11 B 13 M	62	68	13.96	2–3	16 (12–20)	81.2	55	12.1	8	8	35

Note: Data are presented as median (range) unless otherwise indicated.
[a] Mean (range).
[b] Size in cm.
[c] At 1 year.
Abbreviations: KPS, Karnofsky performance status; Histology L, non-small cell lung cancer; B, breast cancer; M, melanoma; R, renal cell carcinoma; GTR, gross total resection; cc, cubic cm; preop, preoperative; LC, local control; DBF, distant brain failure; OS, overall survival; postop, postoperative; LMD, leptomeningeal disease; WBRT, whole-brain radiotherapy; NR, not reported.

Table 13.4 Summary of studies using postoperative multiple-fraction stereotactic radiotherapy

STUDY	NO.	AGE (YEARS)	KPS	% HISTOLOGY	% ONE LESION	% GTR	CAVITY (CC)	MARGIN (MM)	DOSE (GY/FX)	% LC	% DBF	OS (MONTHS)	% NECROSIS	% LMD	% WBRT SALVAGE
MD Anderson CC (Traylor et al., 2019)	67	62 (18–79)	100% ≥70	32.8 L 16.4 C	80.6	76.8	6.4 (0.2–61.4)	1	24–30/3–5	85.1	34.3	63.3% 12-month 51.5% 18-month	13.4	16.4	NR
Sunnybrook (Soliman et al., 2019)	137	62 (22–92)	100% ≥70	44 L 21 B 8 R	73	89	30.1[a]	2 + track	25–35/5	84	71	17	30	26	33
MSKCC (Lockney et al., 2017)	143	61.3 (22–87.5)	94% ≥70	35.7 L 16.8 B 16.8 M	73.8	84	NR	2–5	30/5	83	45	13.9	4.2	NR	NR
UCLA (Lima et al., 2017)	41	56.6 (22–78)	100% ≥80	26.8 L 19.5 B 22 M	50	100	26.4 (14.1–38.4)[a]	2	25/5 or 30/10	77.1	22.4	28.27	0	14.6	NR
Strasbourg (Keller et al., 2017)	181	60.6 (26–82.5)	75% >70	45 L 11 B 10 C	73	94	14.15 (0.8–65.8)[a]	2	33/3	86.5	48.6	17	18.5	14.6	30
Saint-Herblain (Dore et al., 2017)	95	60 (36.4–82.5)	73% >70	39 L 16.5 R 11.3 B	86.6	94.8	11.45 (1.0–67.2)	2	23.1/3 (to 70% IDL)	84	44	25	7.2	17	33
Vanderbilt (Cleary et al., 2017)	85	58.9 (30.9–85.3)	NR	40 L 21 M 10.6 R	73	NR	9.8 (1.1–43.1)	0–2	30/5 (18–35/2–5)	87	52	13	2.4	11.7	NR
Milan (Pessina et al., 2016)	69	51 (33–77)	100% ≥0	34.8 B 30.4 L 21.7 M	74	100	55.2 (17.2–282.9)[a]	3	30/3	100	35	24	2.9	NR	28
Freiburg (Bilgera et al., 2016)	60	62 (32–84)	87% >70	31.7 L 18 B 15 M	85	88	NR	3	30/6 or 35/7	89	40	15	NR	3	50
Munich (Specht et al., 2016)	46	56 (19–84)	100% ≥70	26 L 24 B 21 C	87	63	14.16 (1.44–38.68)	2	35/7	82	43	25	10	NR	43

(Continued)

Table 13.4 (*Continued*) **Summary of studies using postoperative multiple-fraction stereotactic radiotherapy**

STUDY	NO.	AGE (YEARS)	KPS	% HISTOLOGY	% ONE LESION	% GTR	CAVITY (CC)	MARGIN (MM)	DOSE (GY)/FX	% LC	% DBF	OS (MONTHS)	% NECROSIS	% LMD	% WBRT SALVAGE
Moffitt (Abuodeh et al., 2016)	77	63 (18–81)	97% ≥70	31.2 L 29.9 M 14.3 R	63	95	13.8 (1.93–128.43)[a]	1–2	25/5	88.8	43	73.1%[b]	3	2.7	13
Sunnybrook (Al-Omair et al., 2013)	20	70 (41–90)	100% ≥70	50 L 15 B	NR	85	23.6[c] (3.1–42.1)	2	25–37.5/5	79[b]	NR	23.6	NR	NR	NR
NYU (Connolly et al., 2013)	33	56.6 (27–82)	90 (70–90)	39 L 27 B 24 M	100	NR	3.3[c] (1.7–5.7) preop	10	40.05/15	85	39.3[b]	65.6%[b]	0	NR	14.3
Emory (Eaton et al., 2013)	22[d]	58 (23–81)	88% ≥60	24 L 29 B 21 M	NR	50	24.5[a] (0.8–122.0)	2 (0–10)	21/3[c] (67%) 24/4[e] (14%) 30/5[e] (12%)	61[b]	71	Not reached	9.5	NR	40
Umeå (Lindvall et al., 2009)	47	64.9[f]	87% >70	45 L 21 B 2 M	100	100[g]	6 (0.6–26) 74% <10 cc	NR	35–40/5	84	19	5	2.1	NR	NR
Sapienza (Minniti et al., 2013)	101	57	80 (60–100)	22.8 L 18.8 B 27.8 M	100	100	29.5[a] (18.5–52.7)	3	27/3	92	53	17	9	NR	24
Hannover (Steinmann et al., 2012)	33	58 (33–73)	100% ≥70	42 L 27 B 9 M	100	75	22.6[a] (4.9–93.6)	4	40/4 (67%) 35/7 (21%) 30/6 (12%)	73	47	20	NR	NR	39
Beth Isreal Deaconess (Wang et al., 2012)	37	73% <65	97% ≥70	27 L 24 B 32 M	24	NR	28.8[f] (11.1–81.0)	2–3	24/3	80[b]	20	5.5	2.9	NR	14.3

Note: Data are presented as median (range) unless otherwise indicated.
[a] Planning target volume.
[b] At 1 year.
[c] Size in cm.
[d] Patients with resected brain metastases (total 44 patients in study).
[e] Additional 1–1.5 Gy simultaneous integrated boost for subtotally resected lesions.
[f] Mean (range).
[g] Based on neurosurgeon's report.
Abbreviations: KPS, Karnofsky performance status; Histology L, non-small cell lung cancer; B, breast cancer; M, melanoma; C, colorectal; R, renal cell carcinoma; GTR, gross total resection; cc, cubic cm; preop, preoperative; fx, fraction; LC, local control; DBF, distant brain failure; OS, overall survival; LMD, leptomeningeal disease; WBRT, whole-brain radiotherapy; NR, not reported.

Table 13.5 Summary of studies using both postoperative single- and multiple-fraction stereotactic radiotherapy

STUDY	NO.	AGE (YEARS)	KPS	% HISTOLOGY	%ONE LESION	% GTR	CAVITY (CC)	MARGIN (MM)	DOSE (GY/FX)	% LC	% DBF	OS (MONTHS)	% NECROSIS	% LMD	% WBRT SALVAGE
Birmingham (Foreman et al., 2018)	91	>18	NR	42.8 L 18.7 M 12 B	70	77	3.6 cm (1.4–7.3 cm)	0	16 (10–20)/1 25/5–30/6	78	NR	83.5% 12-month	6	35	NR
Stanford (Choi et al., 2012b)	112	61 (18–86)	92% ≥70	43 L 16 B 16M	63	90	8.5 (0.08–66.8)	0 (48%) 2 (52%)	12–30/1–5 86% 1 fx, 76% 3 fx	89.2	46[a]	17 (2–114)	3.5	NR	28
City of Hope (Do et al., 2009)	30	61.5 (40–93)	96.7% ≥70	47 L 20 B 20 M	NR	NR	NR	1[b] 2–3[c]	15–18/1 22–27.5/4–6	82	63	51%[a]	6.6	NR	47
Dana Farber (Kelly et al., 2012)	17	61.8 (38–81)	80 (70–100)	35 L 35 M	NR	94.4	3.49 (0.53–10.8)	0	18 (15–18)/1 (82%) 25/5 (12%) 30/10 (6%)	89	35	Not reached	NR	NR	41
U Pitt (Ling et al., 2015)	99	64 (39–81)	80 (60–100)	40 L 18 B 17 M	61	81	12.9[d] (0.6–51.1)	0–1[c]	15–21/1 (26%) 20–24/2 18–27/3 (56%) 24/4 20–28/5	72[a]	36[a]	12.7	9[g]	6	50
Emory (Patel et al., 2014)	96	56 (20–83)	96% ≥70	47 L	71	74	7.19 (0.90–35.70)	1	21 Gy ≤ 2 cm 18 Gy 2.1–3cm 15 Gy 3.1–4cm 3–5 fx >4 cm	83[a]	50	12.7	27	31[f]	14
Emory (Prabhu et al., 2012)	62	55 (20–75)	85% ≥70	41 L 11 B 23 M	71	81	8.5 (0.7–57)	>1 (95%)	18 (15–24)/1 (86%) 3–4 fx (14%)	78[a]	49[a]	13.4 (9.3–17.5)	8	NR	26[a]
U Pitt (Rwigema et al., 2011)	77	63 (39–83)	80 (60–100)	43 L 14 B 12M	85.7	NR	7.6 (1.1–59)	1	18 (12–27)/1–3	76.1[a]	53.3[a]	14.5 (1.6–51.4)	2.6	NR	26

Note: Data are presented as median (range) unless otherwise indicated.

[a] At 1 year.
[b] Frame-based, preoperative lesion.
[c] Mask-based, preoperative lesion ≥3 cm.
[d] Planning target volume.
[e] Physician preference.
[f] Compared with 13% incidence LMD in patients receiving WBRT.
[g] "Radiation injury".

Abbreviations: KPS, Karnofsky performance status; Histology L, non-small cell lung cancer; B, breast cancer; M, melanoma; GTR, gross total resection; cc, cubic cm; fx, fraction; LC, local control; DBF, distant brain failure; OS, overall survival; LMD, leptomeningeal disease; WBRT, whole-brain radiotherapy; NR, not reported.

radiotherapy to those receiving WBRT following resection of intracranial metastases (Lindvall et al., 2009; Hwang et al., 2010; Al-Omair et al., 2013; Patel et al., 2014). The patient populations of these investigations were similar to those included in the prospective trials described earlier. Specifically, the median age of patients was 55–65 years, virtually all had KPS of at least 70, and the most common tumor histologies were NSCLC, breast cancer, and melanoma. The majority of subjects in these studies had solitary brain metastases, and the extent of resection was considered gross total in approximately 70% of cases or more, with exception of one of the multifractionation studies where only half of patients had GTR (Eaton et al., 2013). While the reported local control of the surgical cavity in the prospective trials ranged from 62% to 85%, this percentage in the retrospective series ranged from approximately 61% to 100%.

The majority of patients in the retrospective series received their adjuvant treatment approximately 4 weeks from time of surgery. The surgical cavity size ranged from 3.0 to 13 cc in the studies that included singe-fraction SRS (Tables 13.3 and 13.5) and ranged larger at 6.0 to 55 cc in the multi-fraction series (Table 13.4). The median volume treated in the MD Anderson trial was 8.9 cc, approximately equal to the median volume of 11.5 cc in the MSKCC series (i.e., maximum target diameter converted to spherical volume). For the series that had only fractionated treatments, the target often included the setup margin and was reported as planning target volume. Consequently, the target size in these reports often exceeded 20 cc (Table 13.4). The margin added to the surgical cavity, identified by postcontrast MRI in the majority of the retrospective analyses, ranged from 0 to 3 mm for the investigations that included single-fraction SRS, up to 10 mm in the multifraction series. The margin applied around the cavity in the prospective single fraction cavity SRS trials was 2 mm in NCCTG-N107C/CEC.3 (Brown et al., 2017) and 1 mm in the MD Anderson trial (Mahajan et al., 2017).

There was considerable variability in technique and dose regimens utilized in the various retrospective series. The majority (10/13) of single-fraction-only studies were conducted using GK radiosurgery, whereas the analyses that included any patients treated with multiple fractions all utilized LINAC-based approaches, including CyberKnife and helical TomoTherapy. In the series that included only one-fraction adjuvant treatments (Table 13.3), the median marginal dose was 15–18 Gy, but the dose range was wider compared to the prospective trials (8–30 in the retrospective series vs 15–22 Gy (MSKCC), 12–18 Gy (MD Anderson), and 12–20 Gy (Alliance N107/CEC.3)). The total dose delivered in the hypofractionated regimens was 21–40 Gy administered using various regimens of 3–15 fractions (Table 13.4). For the studies that combined single- and multiple-fraction courses, the median marginal dose was 12–30 Gy delivered in 1–5 fractions (Table 13.5). While the debate continues as to the applicability of the commonly accepted models to determine biologic equivalent dose of hypofractionated and single-fraction radiotherapy courses to standard 1.8–2 Gy per fraction treatments delivered over several weeks, Eaton et al. (2013) estimated that the most common hypofractionated regimens used would have comparable tumor control as 17.4–22.8 Gy single-fraction SRS.

Despite the lack of uniformity in approach to focal adjuvant treatment of resected brain metastases, the outcomes are remarkably similar across the spectrum of treatment employed. The mean LC rate for single-fraction cavity SRS studies was 81% (range, 73%–100%), 84% (range, 61%–100%) for multifraction IG-HSRT, and 81% (range, 72%–89%) for the cohorts that included both single and multiple fractions (Tables 13.3 through 13.5). Restricting the analysis to only that subset of studies that reported LC at 1 year, the mean result was 75% for the SRS group (n = 5; Jensen et al., 2011; Hartford et al., 2013; Iorio-Morin et al., 2014; Brown et al., 2017; Mahajan et al., 2017), 80% for the IG-HSRT group (n = 7; Wang et al., 2012; Al-Omair et al., 2013; Eaton et al., 2013; Specht et al., 2016; Bilgera et al., 2016; Soliman et al., 2018; Traylor et al., 2019), and 77% for the combination group (n = 4; Rwigema et al., 2011; Prabhu et al., 2012; Patel et al., 2014; Ling et al., 2015). When only addressing the larger multifraction studies (>60 patients), 1-year LC ranged from 83 to 100%, which is comparable to the LC in the single fraction MSKCC trial (85%, n = 39), but superior to the more recent larger randomized trials (72% Mahajan et al., 2017 and 62% Brown et al., 2017). The mean DBF rate for the retrospective single-fraction, multifraction, and combination cohorts was 47%, 41.5%, and 48%, respectively. These findings are comparable to the DBF reported at 1 year for the subjects in the MSKCC trial (44%) and fall between the ranges of the NCCTG-N107C/CEC.3 trial and MD Anderson trial (35.7% and 58%, respectively). The average median OS time was 14.9 months based on 11 of 13 retrospective studies of the one-fraction SRS patients

(Table 13.3). This value was 17.8 months based on 14 of 18 studies in the multifraction IG-HSRT, excluding three studies where the OS at 1 year was around 67% and one study where the median OS was not yet reached (Table 13.4). Similarly, the average median OS was 14.4 months for subjects in four of seven of the combined single-fraction and multifraction reports and one report excluded with 93% 1-year OS and median OS not yet reached (Kelly et al., 2012; Table 13.5). These results are comparable to the median OS of 14.7 months reported in the MSKCC trial (Brennan et al., 2014), fall between the range reported by NCCTG-N107C/CEC.3 (12.2 months) and MD Anderson's trial (17 months), and are superior to the less than 12 months median OS of all subjects in the prospective studies of adjuvant WBRT by Patchell et al. (1998) and Kocher et al. (2011). In those retrospective studies where salvage WBRT was reported, fewer than half of patients received this treatment (Tables 13.3 through 13.5), which is comparable to NCCTG-N107C/CEC.3 (20%) and MD Anderson's trial (38%) and compares favorably to the 65% of subjects in the MSKCC trial receiving this treatment.

As always, when comparing the results of any retrospective analysis to those of prospective clinical trials, the issue of selection bias in the retrospective cohorts must be taken into consideration. Nevertheless, it is encouraging that the outcomes of the many retrospective studies that employed various techniques of postoperative partial brain radiotherapy following resection of intracranial metastases are not inferior to the prospective trials investigating either the adjuvant focal brain radiation or WBRT approach. Given concerns of adverse effects of WBRT on neurocognitive function and functional independence, the observation that, where reported in these retrospective analyses, over 50% of subjects treated with SRS or IG-HSRT to the surgical cavity of resected brain metastases were spared this treatment.

13.5 TOXICITIES

Postoperative radiotherapy delivered to the resection cavity of a metastatic brain lesion is well tolerated with very little toxicity. The most commonly cited adverse event following this therapy is brain radionecrosis. In the prospective phase II trial from MSKCC of single-fraction SRS following metastatectomy, radiation necrosis was reported in 17.5% of subjects (Brennan et al., 2014). The rates of radionecrosis were much lower in the more modern phase III trials: NCCTG-N107C/CEC.3 reported 4% and 0% in the MD Anderson trial. The incidence of necrosis following single-fraction SRS in retrospective studies ranged from 0% to 8% (Table 13.3) and from approximately 3% to 30% in the studies that included multifraction IG-HSRT (Tables 13.4 and 13.5). While median volume of normal brain receiving 24 Gy was identified as the most significant factor associated with necrosis in one study of multifraction IG-HSRT (Minniti et al., 2013), most of the studies identifying necrosis following adjuvant focal tumor bed radiotherapy failed to identify specific risk factors for this treatment-associated injury (see, e.g., Choi et al., 2012a; Eaton et al., 2013). Dosimetric factors that have been reported to affect radionecrosis include worse conformality index and larger volume irradiated. For multifraction regimens, a volume receiving 24 Gy (V24) more than 16.8 cc was associated with 16% risk of radionecrosis compared to only 2% if less than 16.8 cc. For single fraction SRS, V12 more than 20.0 cc has been shown to increase risk of radionecrosis (14% versus 4%; Minniti et al., 2013). One of the large retrospective series of multifraction radiosurgery reported 30% risk of radionecrosis, likely due to the predilection for larger cavity size in this series (Soliman et al., 2019). Notably, only 6% of those with necrosis in this series were symptomatic (Soliman et al., 2018). Another changing paradigm that may influence rates of radionecrosis is preoperative radiosurgery, where lower rates of radionecrosis were estimated at 0–1.5% (Patel et al., 2016). When reported, necrosis was most often managed by steroid administration and/or surgical resection (Jensen et al., 2011; Choi et al., 2012a; Robbins et al., 2012; Wang et al., 2012; Eaton et al., 2013; Gans et al., 2013; Patel et al., 2014; Ling et al., 2015). Bevacizumab to treat radionecrosis is also increasingly common to substitute or delay surgery, hyperbaric oxygen or prolonged steroid use.

13.6 CONTROVERSIES

With only two prospective, randomized controlled trials addressing postoperative IG-HSRT in the management of brain metastases, there remain many unanswered questions and topics for debate. A few of the outstanding controversial issues related to this are highlighted below.

13.6.1 PATIENT-RELATED ISSUES

As with any clinical therapeutic decision, the risk versus benefit of treatment must be carefully measured and weighed against anticipated patient survival, particularly in this setting of metastatic cancer to the brain. Thus, proper patient selection is critical. Extrapolating from the RPA (Gaspar et al., 1997, 2000) and the more recent GPA index (Sperduto et al., 2008) that have been developed to identify those patients with brain metastases likely to have the best survival, younger patients (<50–65 years) with excellent performance status (≥70), limited number of brain lesions (1–3), and controlled primary site of disease with no or limited extracranial metastases would be the ones anticipated to benefit most from aggressive therapy including surgical resection and postoperative SRS or IG-HSRT. In the two prospective randomized trials of cavity SRS, only patients having KPS ≥70 were eligible for enrollment. The MSKCC prospective study and most of the retrospective analyses highlighted earlier included subjects with more favorable RPA/GPA scores. Therefore, the precise optimal patient profile for adjuvant cavity radiotherapy is still unknown.

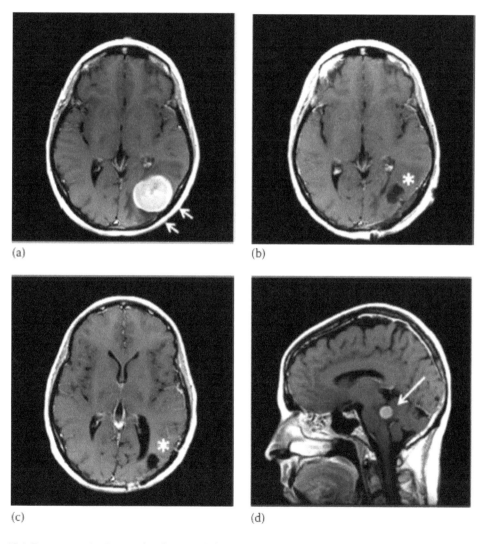

Figure 13.1 Representative T1-weighted postgadolinium MRI images of 38-year-old female with breast cancer. (a) Axial image demonstrating 4.3 cm left occipital metastasis abutting the dura (short arrows). (b) Status of post gross total resection followed by 16 Gy single-fraction SRS to cavity + 1 mm margin on protocol (*cavity). (c) Five months after stereotactic radiosurgery, there is no evidence of local recurrence (*cavity), but (d) sagittal image shows leptomeningeal disease with metastasis in fourth ventricle (long arrow).

Related to the prognostic factor of number of brain metastases, another undefined patient-specific issue is the maximal number of lesions (resected and not) that can be treated with postoperative SRS/IG-HSRT. The Alliance N107C/CEC.3 trial allowed up to four brain lesions (77% single, 23%, two to four lesions) while the MD Anderson trial allowed up to three brain metastases (60% single, 29% two lesions, and 11% three lesions). Most patients in the other analyses above had only one resected brain metastasis, although several studies included subjects with more than one resection cavity and/or up to a total of 10 intracranial lesions treated using the focal adjuvant radiation treatment (Jensen et al., 2011; Brennan et al., 2014).

13.6.2 SURGICAL ISSUES

The majority of lesions treated with postoperative SRS/IG-HSRT in the reported studies were considered to be GTR. How the extent of resection was established in the various retrospective analyses was often not specified, and in one study, this determination was based on self-reporting by the neurosurgeon (Lindvall et al., 2009). Uniform criteria to measure completeness of resection are as yet not defined but likely necessary in the determination of appropriateness of adjuvant radiotherapy to the surgical cavity in these cases. In the MD Anderson trial mandating GTR, the completeness of resection was verified on postoperative imaging review by a dedicated neuroradiologist.

While the dose to the completely resected tumor bed is still indeterminate (see Section 13.6.4), the question of how to clinically manage STR brain metastases is another matter for debate. In two of the retrospective studies, a higher radiation dose was delivered to STR lesions than to the GTR ones (Robbins et al., 2012; Eaton et al., 2013). What the dose should be to residual tumor in the resection cavity and benefit of dose escalation in that circumstance are presently unknown.

Another ill-defined area likely to impact outcomes of patients treated with this approach is the skill of the neurosurgeon, as mentioned in the review of postoperative tumor bed radiosurgery by Roberge and Souhami (2010). In surgical series, lesions resected with piecemeal surgery had worse LC than tumors resected en bloc. In the MD Anderson trial, complete resection alone resulted in 45% 12-month freedom from local recurrence in the observation arm, compared to 54% in the Patchell study (Patchell et al., 1998). Given anticipated wide variability on this front and no easily applied algorithm to normalize surgical outcomes, establishing a uniform system to report extent of resection is again underscored.

13.6.3 TUMOR-SPECIFIC ISSUES

While it is known that a wide variety of tumor histologies are able to metastasize to the brain, not all are amenable to resection and/or focal treatment options given predilection for microscopic dissemination (e.g., small cell lung cancer; Ojerholm et al., 2014a). The most common diagnoses of patients in the prospective and retrospective studies of cavity SRS/IG-HSRT summarized earlier include NSCLC, breast cancer, and melanoma. Whether the same radiation dose delivered adjuvantly to the tumor bed is equally efficacious regardless of histology is not known. The relationship between histology and local recurrence has been mixed in the literature. In the MD Anderson trial, histology was not found to be a predictor for local recurrence on multivariate analysis. However, other retrospective series identified certain histologies to be associated with increased local recurrence after surgery and cavity radiation, such as colorectal, melanoma (Soliman et al., 2019), and breast histology (Ojerholm et al., 2014b).

Also unknown is the maximal lesion size that should be considered amenable to postoperative radiotherapy. When reported in the studies earlier, the median tumor size ranged from 2.7 to 3.9 cm, with largest tumor diameter measuring 6.9 cm in the analysis by Ojerholm et al. (2014b). The maximum tumor diameter treated in the dose escalation trial RTOG 90–05 of SRS for intact, previously irradiated brain tumors (either primary or metastatic) was 4 cm (Shaw et al., 1996, 2000). Since the dose-limiting factor is the amount of "normal" brain receiving radiation, it can be reasoned that the dose–volume constraints would still apply in the postoperative setting, but without ample prospective data, this statement must be considered purely conjecture. In surgical series, without adjuvant radiation, tumors larger than 3 cm were associated with higher rates of local failure (Patel et al., 2010). In the MD Anderson trial, lesions with diameter ≤2.5 cm were associated with better LC at 12 months (91%) compared to larger lesions (40% if diameter was between 2.6 and 3.5 cm; 46% if diameter was more than 3.5 cm). In the multivariate analysis for time to local recurrence, the only factors found to be significant predictors of local recurrence were cavity SRS and metastasis size. One unanswered question is whether the increased local recurrence associated with larger irradiated cavities is merely a function of reduced adjuvant SRS doses delivered to the cavity to mitigate risk of radiation necrosis. In one retrospective series of multifraction SRS, preoperative tumor diameter, of which 57% were more than 3 cm, was not predictive of LC (n = 137, Soliman et al., 2019).

Another as-yet undefined matter related, in part, to the target size is the optimal timing of cavity SRS/IG-HRST following resection. Several investigators have reported on the dynamics of the surgical cavity volume over time. Jarvis et al. (Hartford et al., 2013) analyzed the volume change of 43 resection cavities in 41 patients from first postoperative day T1-weighted postcontrast MRI until similar imaging prior to planned SRS over an average of 24 days (range, 2–104 days). The majority of tumor beds remained stable, but one-quarter collapsed and another third increased by more than 2 cc. In a separate analysis of 63 subjects with 68 resection cavities, Atalar et al. (2013a) observed no significant difference in immediate postoperative target volume within 3 days of surgery and that measured at the time of SRS treatment planning up to 33 days following craniotomy. In this series, the largest tumors (>4.2 cc and diameter of 2 cm) were associated with the greatest reduction in tumor bed volume (median change—35%, $p < 0.001$), whereas those smaller than 4.2 cc had larger cavity volumes (median change +46%, $p = 0.001$). Another series of cavity dynamics evaluated 61 cavities and found a significant volume reduction compared to tumor volume (4 cc; $p = 0.03$). While larger tumors showed a volume shrinkage of 11.6%, smaller tumors showed an increase in size by up to 34.4%. Factors predictive of cavity volume reduction were tumor size greater than 3 cm, dural involvement, and time more than 21 days from surgery (Alghamdi et al., 2018). Figure 13.2 demonstrates a case of cavity "shrinkage" 1 month following resection of a brain metastasis. In a separate analysis of 37 patients with 39 resection cavities, Ahmed et al. (2014) identified the presence of at least 15 mm vasogenic edema surrounding the tumor bed immediately following surgery to be predictive of at least a 10% reduction of cavity size when imaged within 30 days after resection. Besides cavity size dynamics, other issues surrounding the timing of adjuvant SRS/IG-HRST following surgery include the competing risk of adequate wound healing and that of tumor recurrence. Historically, postoperative radiotherapy is typically delayed for a period of 10–14 days to allow wound repair, with treatment to be initiated ideally within 4–6 weeks of surgery (see, e.g., Patchell et al., 1998; Kocher et al., 2011). The issue of tumor recurrence and timing of SRS following surgery was commented upon in the work by Jarvis et al. (2012), since 2 of the 13 subjects found to have an increase in tumor bed size prior to SRS had clear evidence of disease regrowth and another two potential recurrences by 19 days (range, 4–76 days) after initial postoperative MRI, making up 23% recurrence rate with a median interval of 30 days post-operatively. Cavity

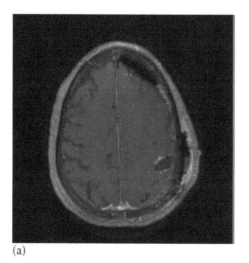
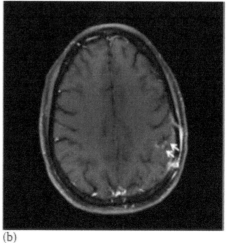

(a)   (b)

Figure 13.2 Representative T1-weighted postgadolinium MRI images of 48-year-old female with breast cancer. Axial image at (a) 24 hours and (b) 1 month following gross total resection of a left parietal metastasis. Note collapse of the cavity (short arrows) at the later postoperative time point.

SRS was delivered within 30 days of resection in the MD Anderson trial and between 2 and 8 weeks in the MSKCC trial.

Often reported in analyses of postoperative SRS for resected brain metastases is the development of leptomeningeal disease (LMD). The rate of LMD was 7.2% in NCCTG- N107C/CEC.3, 28% in the MD Anderson trial, and none observed in the MSKCC trial. In a recent study looking at risk of LMD in patients treated with SRS to surgical cavity versus intact brain metastases, the risk of LMD after cavity SRS was reported to be 16.9% (n = 112; Johnson et al., 2016). In another series comparing the outcomes of patients treated with pre- versus post-operative SRS, the rates of LMD were found to be higher in patients treated after surgery (16.6% versus 3.2% at 2 years, p = 0.01; Patel et al., 2016). A more recent analysis of predictors of LMD following IG-HSRT for intact and resected brain metastases found that radiosensitive tumors were associated with higher risk of LMD among treated cavities, and the rate for all cavities was 24.1% (Nguyen et al. 2020). Interestingly, the use of targeted agents or immunotherapy in this study was associated with lower risk of LMD (Nguyen et al., 2020). LMD was noted in five of the retrospective single-fraction SRS studies, occurring at a mean incidence of 13% (range, 7.5%–24%; Table 13.3) and in 6% to 31% of subjects in two of the seven combination single-/multiple-fraction reports (Table 13.5). LMD was noted in the more recent multifraction series (Table 13.4) at a mean incidence of 12.6% (range, 3%–22%). In the one study with the highest occurrence of LMD following postoperative cavity radiation, this incidence was more than halved in subjects treated at the same institution with adjuvant WBRT (Patel et al., 2014). In the NCCTG-N107C/CEC.3 trial, no difference was observed in the rate of LMD between cavity SRS and WBRT. This is an important finding of the trial, since prior retrospective series hypothesized that WBRT may have lower rates of LMD due to treating widespread microscopic disease. Although not universal among all of these investigations, the use of adjuvant SRS alone (Patel et al., 2014; Hsieh et al., 2015), breast cancer histology (Jensen et al., 2011; Atalar et al., 2013b; Ojerholm et al., 2014b; Soliman et al., 2019), infratentorial tumor location (Iwai et al., 2008; Jensen et al., 2011; Ojerholm et al., 2014b), and subtotal resection (STR; Soliman et al., 2019) have been identified as potential risk factors for LMD. The variability in rates of LMD between series is likely related to multiple factors, including difficulty defining and identifying LMD on imaging, inconsistent imaging follow-up, and lack of radiology review. There is also emerging evidence to suggest that the pattern of tumor progression following surgical resection and focal treatment to the cavity may be distinct from conventional LMD. In a recent retrospective cohort study of patients with brain metastases treated with either surgical resection followed by focal radiation or focal radiation alone, there was no difference in development of LMD between the two groups; however, pachymeningeal seeding was only observed in patients undergoing neurosurgical resection (Cagney et al., 2019). The incidence of pachymeningeal seeding was found to be 8.4%, with a higher incidence (13.7%) observed in patients with controlled extracranial disease. Of note, the rate of neurological death was 72% in patients with pachymeningeal seeding, and less than half of the patients with this pattern of recurrence treated with salvage radiation survived 1 year (Cagney et al., 2019). Figure 13.1 demonstrates a case of LMD 5 months following postoperative SRS in a patient with breast cancer.

13.6.4 RADIOTHERAPY ISSUES

What is the optimal dose, fractionation regimen, planning margin, and sequencing of SRS? All these questions remain to be answered, and some are subjects of ongoing clinical trials.

The dose regimens utilized in the prospective and many of the retrospective studies of cavity SRS/ IG-HSRT were based on RTOG 90–05 dose escalation trial of SRS for intact, previously irradiated brain tumors (either primary or metastatic) (Shaw et al., 1996, 2000). There are two distinctions that must be highlighted between SRS in the postoperative setting and the treatment used in RTOG 90–05: (1) altered blood supply leading to hypoxic area along the resection cavity and (2) radiation-naïve brain in patients receiving the cavity SRS/IG-HSRT treatment. Hypoxic tissue is well recognized as being resistant to standard fractionated doses of radiation, and a recent *in silico* analysis of radiosurgery for brain metastases further suggests the benefit of hypofractionation for large tumors as a means to overcome hypoxia (Toma-Dasu et al., 2014). Following resection of a brain metastasis, the blood vessels "feeding" the tumor are severed, thus rendering the potential residual microscopic malignant cells hypoxic. The optimal adjuvant radiation treatment approach in this setting, namely, single-fraction versus hypofractionated radiotherapy, is unknown, as underscored by the *in silico* study of unresected metastases.

The other factor that distinguishes radiation dose selection in the postoperative state from the dose-finding RTOG 90–05 study is that in the former case, the tumor has been removed and the target is a rind of "normal tissue," whereas in the latter case, the tumor was intact and previously irradiated and hence, arguably, radiation resistant. Because RTOG 90–05 defined the maximum safe, tolerated radiosurgical dose to be delivered to the target (based on maximum diameter) in the setting of prior brain radiation, use of these established dose–volume criteria could be considered conservative for unirradiated brain tissue. Use of the RTOG 90–05 dose regimen has been criticized, for this reason, since LC following cavity SRS/IG-HSRT has been found to be only approximately 80% in the studies reporting outcomes using this technique, and since more than 90% of failures were identified as occurring within the prescription isodose line in the analysis of patients treated with postoperative SRS at Emory University by Prabhu et al. (2012) (same patient population updated in Patel et al. 2014) in Table 13.5. Although RTOG 90–05 dose levels are still widely used in the setting of cavity SRS, the more recent phase III trials migrated to volume-based dose determination, NCCTG-N107C/CEC.3 using doses ranging from 12 to 20 Gy, and MD Anderson trial delivering 12–16 Gy, depending on cavity volume, instead of preoperative tumor or cavity diameter.

While these two randomized trials utilized single-fraction SRS, multifraction regimens are increasing in use. Interestingly, on average, multifraction regimens have been showing overall better local control at 1 year, averaging approximately 85%, compared to that observed in the two single fractions SRS trials (61% and 72%). One potential reason for the lower LC observed in the single fraction SRS trials could be the lower radiation BED used to treat larger cavities (>3 cm), where the dose used was reduced to 12 Gy to mitigate the risk of radiation necrosis. For larger cavity size lesions, multifraction regimens have been hypothesized to offer superior therapeutic ratio, providing higher BED while still maintaining low rate of radionecrosis. This particular question is being asked in the ongoing Alliance A071801 (ClinicalTrials. gov Identifier: NCT04114981) randomized trial, assessing post-surgical single fraction SRS compared with fractionated stereotactic radiotherapy in treating patients with resected brain metastases. The study limits the total number of unresected brain lesions to three or less, with unresected lesions to be less than 4 cm so they can be targeted with SRS, and mandates the preoperative tumor size be more than 2 cm and the lesions have GTR. The single-fraction SRS doses are similar to the volume-based dose levels used in NCCTG-N107C/CEC.3, while the multi-fraction doses allowed are 27 Gy/3 fractions for cavities less than 30 cc, or 30 Gy in 5 fractions for cavities ≥30 cc up to 5.0 cm maximum diameter. The primary outcome measure is surgical bed recurrence-free survival, and secondary outcomes include change in functional assessment, performance status, overall survival, incidence of adverse events, including radiation necrosis, and time until WBRT.

Another yet unanswered question relates to the optimal sequencing of SRS for brain metastases. One dreaded outcome among patients with resected brain metastases is the development of LMD, with rates ranging in post-operative trials from 3% to as high as 30%. It has been hypothesized that one potential cause could be tumor seeding at the time if surgical resection and, therefore, preoperative SRS may sterilize the surgical margin and reduce that risk. Preoperative SRS may also reduce radionecrosis as potentially smaller volumes are irradiated due to the lack of margin applied compared to when treating a cavity. The recent retrospective phase II study by Patel et al. (2016) looked at the outcomes of patients treated pre- and post-operatively with SRS. While the rate of LC and salvage WBRT was similar in both arms, the rate of LMD was lower in patients treated preoperatively (2-year LMD risk 3.2% versus 16.6%, $p = 0.01$). Similar trends were found for radiation necrosis and symptomatic radiation necrosis, with 1- and 2-year cumulative incidence of symptomatic radiation necrosis of 14.6% versus 1.5% and 16.4% versus 4.9%, respectively ($p = 0.01$). Similar outcomes were demonstrated in a separate series of 117 patients (125 lesions) treated with preoperative SRS, where 2-year rates of LMD, radionecrosis, and symptomatic radionecrosis were favorably low at 4.3%, 4.8%, and 2.6%, respectively (Prabhu et al., 2018). Potential concerns related to the preoperative SRS approach include the lack of pathological confirmation at the time of SRS and the possibility of subtotal resection after SRS. Phase III randomized trials are needed to compare the two approaches, and many are currently being conducted at multiple institutions.

Another area of great contention is the target definition for radiotherapy following resection of brain metastases. In the studies of adjuvant SRS/IG-HSRT tabulated above, the radiation target included a margin on the surgical cavity of 0–3 mm for single-fraction treatments and of up to 10 mm for the multifraction regimens. In the prospective phase III trials of cavity SRS, a 2-mm margin was applied in the Alliance N107C/CEC.3 trial and 1-mm margin in the MD Anderson trial. The Alliance N107C/CEC.3 trial did not cover surgical access tracks for deep lesions, and the patterns of failure relative to the CTV were not reported (Brown et al., 2018). For the MD Anderson trial, a separate analysis of patterns of local failure was performed. Of the 64 patients randomized to cavity SRS, 12 developed local failure, of which one-quarter were marginal and all had preoperative dural involvement. The addition of a 2-mm margin may obviate the benefit of postoperative cavity shrinkage by increasing the target volume by 2 to 200%, as was demonstrated by Atalar et al. (2013a). The need for, and extent of, margin will also vary depending upon the immobilization (frame based or frameless) and technique (GK or LINAC-based, use of intrafractional image guidance) employed to deliver the adjuvant radiotherapy. However, since the incidence of marginal failures is low (Prabhu et al., 2012; Eaton et al., 2013), the extent of marginal expansion of the cavity and consequent treatment of more "normal" brain must be carefully considered. Other debated issues related to target definition for cavity SRS include the need to cover the surgical track (Brennan et al., 2014; Kelly et al., 2012; Minniti et al., 2013; Patel et al., 2014) and the optimal timing and sequence of imaging used to delineate the target (discussed in Section 13.6.3). The recent publication of the consensus contouring guidelines for postoperative completely resected cavity SRS for brain metastases (Soliman et al., 2018) provides direction for treatment planning. Recommendations from the consensus of 10 experts include fusion of preoperative MRI to aid volume delineation, contouring the entire surgical tract regardless of tumor location, extension of the CTV 5–10 mm along the dura when preoperative dural contact is present and a margin less than 5 mm into adjacent sinus when preoperative venous sinus contact is present. Whether to include the surgical tract as part of the target for adjuvant focal brain radiation or not has been an area of debate. While the omission of the tract may offer the benefit of yielding a smaller SRS volume allowing dose escalation to the cavity to improve local control, some raise concerns related to tract recurrence. A recent retrospective review of 66 deep brain metastases (defined as located >1.0 cm from the pial surface) treated with postresection SRS with or without covering the tract showed no difference in 12-month progression, local failure, tract failure, cavity failure, or adverse radiation effect (Shi et al., 2020). The findings of this analysis support that the tract may be safely omitted from the postoperative SRS target for resected deep tumors.

13.6.5 OTHER ISSUES

In addition to the myriad unresolved questions regarding optimal patient, target, dose, or technique of radiotherapy to be delivered adjuvantly following resection of solid brain metastases, there remain other outstanding issues as to the utility of cavity SRS/IG-HSRT. One such issue is the timing of cavity radiotherapy and systemic therapy. Despite being closed early due to poor accrual, the phase III trial by the RTOG (RTOG 0320) investigating potential OS benefit of adding temozolomide or erlotinib to WBRT and SRS for subjects with NSCLC and one to three brain metastases revealed statistically higher incidence of grade 3–5 toxicity in the concurrent systemic therapy arms (Sperduto et al., 2013). While the use of single-fraction or short-course hypofractionated radiotherapy has often been advocated to reduce delays in systemic chemo-, biologic, targeted, or immunotherapy for patients with metastatic disease, there are limited prospective data on the optimal interval from this type of radiation delivery (either before or after planned systemic therapy) in the setting of unresected brain metastases and no data in the setting of postoperative SRS/IG-HSRT with respect to potential exacerbation of toxicity versus either therapy alone. Recently, there has been an expanding interest in the role of targeted and immunotherapy in treating brain metastases. The use of targeted and immunotherapy has increased in use in the metastatic setting across many tumor histologies, including melanoma, NSCLC, and breast cancer. Immunotherapy has also been demonstrated to have effects across the blood–brain barrier on untreated brain metastases. A recent trial from MD Anderson (Tawbi et al., 2018) showed that nivolumab combined with ipilimumab had clinically meaningful intracranial efficacy, concordant with extracranial activity, in patients with melanoma and untreated brain metastases. The rate of clinical benefit was 57% (47%–68%): 26% of patients showing complete response of untreated brain metastases, 30% partial response, and 2% stable disease. Similar trends were illustrated in another randomized trial (Long et al., 2018) showing that intracranial response rate was achieved in 46% patients, 17% with complete response, when nivolumab combined with ipilimumab was used to treat patients with metastatic melanoma and untreated brain metastases. In this context, targeted and immunotherapy may have similar, if not greater, effects on brain metastases in the post-operative setting, since the blood–brain barrier is already disrupted. Larger trials are needed to better define the role of targeted immunotherapy in the post-operative setting and its interaction with post-operative cavity SRS in terms of efficacy and toxicity.

13.7 FUTURE DIRECTIONS

While the two recent phase III trials defined cavity SRS as a new standard of care versus observation or WBRT, ongoing research is still needed to further refine its details. Ongoing trials will address utility of single-fraction vs multi-fraction radiation regimens, pre- versus post-operative SRS, and target volume delineation. Advancements in imaging of brain metastases, including PET and functional imaging, may aid in postoperative target delineation, treatment planning, and more accurate definition of recurrence versus radiation necrosis. The interaction between cavity SRS and targeted/immunotherapy is also yet to be defined.

CHECKLIST: KEY POINTS FOR CLINICAL PRACTICE

✓	ACTIVITY	SOME CONSIDERATIONS
	Patient selection	*Is the patient appropriate for postoperative HSRT?* • Consider enrollment on clinical trial
	SRS versus HSRT	*Is the surgical cavity amenable to single-fraction SRS?* • Current dose–volume constraints based on extrapolation from studies of unresected brain metastases • Consider enrollment on clinical trial
	Simulation	*Immobilization* • Rigid frame versus thermoplastic mask *Imaging* • CT and/or MRI should be performed in treatment position • MRI should ideally include volumetric (1 mm) images, T1 sequence with IV contrast • Accurate image co-registration should be verified by clinician and used for contouring targets
	Treatment planning	*Contours* • Surgical cavity = clinical target volume • No consensus on margin, dose, fractionation, timing • Consider enrollment on clinical trial
	Treatment delivery	*Technique* • Both Gamma Knife and linear accelerator–based approaches acceptable • Patient localization system versus near- to real-time image guidance is mandatory *Concurrent systemic therapy* • Limited prospective data • Consider enrollment on clinical trial

REFERENCES

Abuodeh Y, Ahmad KA, Naghavi AO, Venkat PS, Sarangkasiri S, Johnstone PA, Estame AB, Yu HM (2016) Postoperative stereotactic radiosurgery using 5-Gy × 5 sessions in the management of brain metastases. *World Neurosurgery* 90:58–65.

Ahmed S, Hamilton J, Colen R, Schellingerhout D, VuT, Rao G, McAleer MF, Mahajan A (2014) Change in postsurgical cavity size within the first 30 days correlates with extent of surrounding edema: Consequences for postoperative radiosurgery. *Journal of Computer Assisted Tomography* 38:457–460.

Alghamdi M, Hasan Y, Ruschin M, Atenafu EG, Myrehaug S, Tseng, CL, Spears J, Mainprize T, Sahgal A, Soliman H (2018) Stereotactic radiosurgery for resected brain metastasis: Cavity dynamics and factors affecting its evolution. *Journal of Radiosurgery & SBRT* 5(3):191–200.

Al-Omair A, Soliman H, Xu W, Karotki A, Mainprize T, Phan N, Das S et al. (2013) Hypofractionated stereotactic radiotherapy in five daily fractions for post-operative surgical cavities in brain metastases patients with and without prior whole brain radiation. *Technology in Cancer Research and Treatment* 12:493–499.

Atalar B, Choi CY, Harsh GR, 4th, Chang SD, Gibbs IC, Adler JR, Soltys SG (2013a) Cavity volume dynamics after resection of brain metastases and timing of postresection cavity stereotactic radiosurgery. *Neurosurgery* 72:180–185; discussion 185.

Atalar B, Modlin LA, Choi CY, Adler JR, Gibbs IC, Chang SD, Harsh GR, IV, et al. (2013b) Risk of leptomeningeal disease in patients treated with stereotactic radiosurgery targeting the postoperative resection cavity for brain metastases. *International Journal of Radiation Oncology, Biology, Physics* 87:713–718.

Barnholtz-Sloan JS, Sloan AE, Davis FG, Vigneau FD, Lai P, Sawaya RE (2004) Incidence proportions of brain metastases in patients diagnosed (1973 to 2001) in the metropolitan Detroit cancer surveillance system. *Journal of Clinical Oncology* 22:2865–2872.

Bilgera A, Milanovica D, Lorenza H, Oehlke O, Urbach H, Schmucker M, Weyerbroch A, Nieder C, Grosu A (2016) Stereotactic fractionated radiotherapy of the resection cavity in patients with one to three brain metastases. *Clinical Neurological Neurosurgery* 142:81–86.

Brennan C, Yang TJ, Hilden P, Zhang Z, Chan K, Yamada Y, Chan TA et al. (2014) A phase 2 trial of stereotactic radiosurgery boost after surgical resection for brain metastases. *International Journal of Radiation Oncology, Biology, Physics* 88:130–136.

Brown PD, Ballman KV, Cerhan JH, Anderson SK, Carrero XW, Whitton AC, Greenspoon J et al. (2017) Postoperative stereotactic radiosurgery compared with whole brain radiotherapy for resected metastatic brain disease (NCCTG N107C/CEC·3: A multicenter, randomised, controlled, phase 3 trial. *Lancet Oncology* 18(8):1049–1060.

Cagney DN, Lamba N, Sinha S, Catalano PJ, Bi WL, Alexander BM, Aizer AA (2019) Association of neurosurgical resection with development of pachymeningeal seeding in patients with brain metastases. *JAMA Oncology* 5(5):703–709.

Chang EL, Selek U, Hassenbusch SJ, III, Maor MH, Allen PK, Mahajan A, Sawaya R, Woo SY (2005) Outcome variation among "radioresistant" brain metastases treated with stereotactic radiosurgery. *Neurosurgery* 56:936–945.

Chang EL, Wefel JS, Hess KR, Allen PK, Lang FF, Kornguth DG, Arbuckle RB et al. (2009) Neurocognition in patients with brain metastases treated with radiosurgery or radiosurgery plus whole-brain irradiation: A randomised controlled trial. *Lancet Oncology* 10:1037–1044.

Choi CY, Chang SD, Gibbs IC, Adler JR, Harsh GR IV, Atalar B, Lieberson RE, Soltys SG (2012a) What is the optimal treatment of large brain metastases? An argument for a multidisciplinary approach. *International Journal of Radiation Oncology, Biology, Physics* 84:688–693.

Choi CY, Chang SD, Gibbs IC, Adler JR, Harsh GR IV, Lieberson RE, Soltys SG (2012b) Stereotactic radiosurgery of the postoperative resection cavity for brain metastases: Prospective evaluation of target margin on tumor control. *International Journal of Radiation Oncology, Biology, Physics* 84:336–342.

Cleary R K, Meshman J, Dewan M, Du L, Cmelak AJ, Luo G, Morales-Paliza M et al. (2017) Postoperative fractionated stereotactic radiosurgery to the tumor bed for surgically resected brain metastases. *Cureus* 9(5):e1279.

Connolly EP, Mathew M, Tam M, King JV, Kunnakkat SD, Parker EC, Golfinos JG, Gruber ML, Narayana A (2013) Involved field radiation therapy after surgical resection of solitary brain metastases—mature results. *Neuro Oncology* 15:589–594.

Do L, Pezner R, Radany E, Liu A, Staud C, Badie B (2009) Resection followed by stereotactic radiosurgery to resection cavity for intracranial metastases. *International Journal of Radiation Oncology, Biology, Physics* 73:486–491.

Doré M, Martin S, Delpon G, Clément K, Campion L, Thillays F (2017) Stereotactic radiotherapy following surgery for brain metastasis: Predictive factors for local control and radionecrosis. *Cancer Radiotherapy* 21:4–9.

Eaton BR, Gebhardt B, Prabhu RS, Shu HK, Curran WJ, Jr., Crocker I (2013) Hypofractionated radiosurgery for intact or resected brain metastases: Defining the optimal dose and fractionation. *Radiation & Oncology* 8:135.

Foreman PM, Jackson BE, Singh KP, Romeo AK, Guthrie BL, Fisher WS, O Riley K et al. (2018) Postoperative radiosurgery for the treatment of metastatic brain tumor: Evaluation of local failure and leptomeningeal disease. *Journal of Clinical Neuroscience* 49:48–55.

Gans JH, Raper DM, Shah AH, Bregy A, Heros D, Lally BE, Morcos JJ, Heros RC, Komotar RJ (2013) The role of radiosurgery to the tumor bed after resection of brain metastases. *Neurosurgery* 72:317–325; discussion 325–316.

Gaspar LE, Scott C, Murray K, Curran W (2000) Validation of the RTOG recursive partitioning analysis (RPA) classification for brain metastases. *International Journal of Radiation Oncology, Biology, Physics* 47:1001–1006.

Gaspar LE, Scott C, Rotman M, Asbell S, Phillips T, Wasserman T, McKenna WG, Byhardt R (1997) Recursive partitioning analysis (RPA) of prognostic factors in three Radiation Therapy Oncology Group (RTOG) brain metastases trials. *International Journal of Radiation Oncology, Biology, Physics* 37:745–751.

Hartford AC, Paravati AJ, Spire WJ, Li Z, Jarvis LA, Fadul CE, Rhodes CH (2013) Postoperative stereotactic radiosurgery without whole-brain radiation therapy for brain metastases: Potential role of preoperative tumor size. *International Journal of Radiation Oncology, Biology, Physics* 85:650–655.

Hsieh J, Elson P, Otvos B, Rose J, Loftus C, Rahmathulla G, Angelov L, Barnett GH, Weil RJ, Vogelbaum MA (2015) Tumor progression in patients receiving adjuvant whole-brain radiotherapy vs localized radiotherapy after surgical resection of brain metastases. *Neurosurgery* 76:411–420.

Hwang SW, Abozed MM, Hale A, Eisenberg RL, Dvorak T, Yao K, Pfannl R et al. (2010) Adjuvant Gamma Knife radiosurgery following surgical resection of brain metastases: A 9-year retrospective cohort study. *J Neuro-Oncology* 98:77–82.

Iorio-Morin C, Masson-Cote L, Ezahr Y, Blanchard J, Ebacher A, Mathieu D (2014) Early Gamma Knife stereotactic radiosurgery to the tumor bed of resected brain metastasis for improved local control. *Journal of Neurosurgery* 121(Suppl):69–74.

Iwai Y, Yamanaka K, Yasui T (2008) Boost radiosurgery for treatment of brain metastases after surgical resections. *Surgical Neurology* 69:181–186; discussion 186.

Jagannathan J, Yen CP, Ray DK, Schlesinger D, Oskouian RJ, Pouratian N, Shaffrey ME, Larner J, Sheehan JP (2009) Gamma Knife radiosurgery to the surgical cavity following resection of brain metastases. *Journal of Neurosurgery* 111:431–438.

Jarvis LA, Simmons NE, Bellerive M, Erkmen K, Eskey CJ, Gladstone DJ, Hug EB, Roberts DW, Hartford AC (2012) Tumor bed dynamics after surgical resection of brain metastases: Implications for postoperative radiosurgery. *International Journal of Radiation Oncology, Biology, Physics* 84:943–948.

Jensen CA, Chan MD, McCoy TP, Bourland JD, deGuzman AF, Ellis TL, Ekstrand KE et al. (2011) Cavity-directed radiosurgery as adjuvant therapy after resection of a brain metastasis. *Journal of Neurosurgery* 114:1585–1591.

Johnson MD, Avkshtol V, Baschnagel AM, Meyer K, Ye H, Grills IS, Chen PY et al. (2016) Surgical resection of brain metastases and the risk of leptomeningeal recurrence in patients treated with stereotactic radiosurgery. *International Journal of Radiation Oncology, Biology, Physics* 94(3):537–543.

Kalani MY, Filippidis AS, Kalani MA, Sanai N, Brachman D, McBride HL, Shetter AG, Smith KA (2010) Gamma Knife surgery combined with resection for treatment of a single brain metastasis: Preliminary results. *Journal of Neurosurgery* 113(Suppl):90–96.

Karlovits BJ, Quigley MR, Karlovits SM, Miller L, Johnson M, Gayou O, Fuhrer R (2009) Stereotactic radiosurgery boost to the resection bed for oligometastatic brain disease: Challenging the tradition of adjuvant whole-brain radiotherapy. *Neurosurgical Focus* 27:E7.

Keller A, Doré M, Cebula H, Thillays F, Proust F, Darie I, Martin SA et al. (2017) Hypofractionated stereotactic radiation therapy to the resection bed for intracranial metastases. *International Journal of Radiation Oncology, Biology, Physics* 99(5):1179–1189.

Kelly PJ, Lin YB, Yu AY, Alexander BM, Hacker F, Marcus KJ, Weiss SE (2012) Stereotactic irradiation of the postoperative resection cavity for brain metastasis: A frameless linear accelerator-based case series and review of the technique. *International Journal of Radiation Oncology, Biology, Physics* 82:95–101.

Kepka L, Tyc-Szczepaniak D, Bujko K, Olszyna-Serementa M, Michalski W, Sprawka A, Trabska-Kluch B, Komosinska K, Wasilewska-Tesluk E, Czeremszynska B (2016) Stereotactic radiotherapy of the tumor bed compared to whole brain radiotherapy after surgery of single brain metastasis: Results from a randomized trial. *Radiotherapy & Oncology* 121(2):217–224.

Kocher M, Soffietti R, Abacioglu U, Villà S, Fauchon F, Baumert BG, Fariselli L et al. (2011) Adjuvant whole-brain radiotherapy versus observation after radiosurgery or surgical resection of one to three cerebral metastases: Results of the EORTC 22952–26001 study. *J Clin Oncol Off J Am Soc Clin Oncol* 29:134–141.

Lima LC, Sharim J, Levin-Epstein R, Tenn S, Teles AR, Kaprealian T, Pouratian N (2017) Hypofractionated stereotactic radio Surgery and radiotherapy to large resection cavity of metastatic brain tumors. *World Neurosurgery* 97:571–579.

Limbrick DD, Jr., Lusis EA, Chicoine MR, Rich KM, Dacey RG, Dowling JL, Grubb RL et al. (2009) Combined surgical resection and stereotactic radiosurgery for treatment of cerebral metastases. *Surgical Neurology* 71:280–288; disucssion 288–289.

Lindvall P, Bergstrom P, Lofroth PO, Tommy Bergenheim A (2009) A comparison between surgical resection in combination with WBRT or hypofractionated stereotactic irradiation in the treatment of solitary brain metastases. *Acta Neurochirurgica* 151:1053–1059.

Ling DC, Vargo JA, Wegner RE, Flickinger JC, Burton SA, Engh J, Amankulor N, Quinn AE, Ozhasoglu C, Heron DE (2015) Postoperative stereotactic radiosurgery to the resection cavity for large brain metastases: Clinical outcomes, predictors of intracranial failure, and implications for optimal patient selection. *Neurosurgery* 76:150–157.

Lockney NA, Wang DG, Gutin PH, Brennan C, Tabar V, Ballangrud A, Pei X et al. (2017) Clinical outcomes of patients with limited brain metastases treated with hypofractionated (5 x 6 Gy) conformal radiotherapy. *Radiotherapy & Oncology* 123(2):203–208.

Long GV, Atkinson V, Lo S, Sandhu S, Guminski AD, Brown MP, Wilmott JS et al. (2018) Combination nivolumab and ipilimumab or nivolumab alone in melanoma brain metastasis: A multicenter randomised phase 2 study. *Lancet Oncology* 19(5):672–681.

Luther N, Kondziolka D, Kano H, Mousavi SH, Engh JA, Niranjan A, Flickinger JC, Lunsford LD (2013) Predicting tumor control after resection bed radiosurgery of brain metastases. *Neurosurgery* 73:1001–1006; discussion 1006.

Mahajan A, Wang X, Ahmed S, McAleer MF, Weinberg JS, Li J, Brown PD et al. (2017) Local recurrence pattern of patients enrolled on a randomized study of post-operative stereotactic radiosurgery vs observation for completely resected brain metastases. *Neuro-Oncology* 19(Suppl 6):226–226.

Mathieu D, Kondziolka D, Flickinger JC, Fortin D, Kenny B, Michaud K, Mongia S, Niranjan A, Lunsford LD (2008) Tumor bed radiosurgery after resection of cerebral metastases. *Neurosurgery* 62:817–824.

Minniti G, Esposito V, Clarke E, Scaringi C, Lanzetta G, Salvati M, Raco A, Bozzao A, Maurizi Enrici R (2013) Multidose stereotactic radiosurgery (9 Gy ×3) of the postoperative resection cavity for treatment of large brain metastases. *International Journal of Radiation Oncology, Biology, Physics* 86:623–629.

Nguyen TK, Sahgal A, Detsky J, Atenafu EG, Myrehaug S, Tseng CL, Husain Z et al. (2020) Predictors of leptomeningeal disease following hypofractionated stereotactic radiotherapy for intact and resected brain metastases. *Neuro-Oncology* 22(1):84–93.

Ojerholm E, Alonso-Basanta M, Simone CB II (2014a) Stereotactic radiosurgery alone for small cell lung cancer: A neurocognitive benefit? *Radiation Oncology* 9:218.

Ojerholm E, Lee JY, Thawani JP, Miller D, O'Rourke DM, Dorsey JF, Geiger GA et al. (2014b) Stereotactic radiosurgery to the resection bed for intracranial metastases and risk of leptomeningeal carcinomatosis. *Journal of Neurosurgery* 121(Suppl):75–83.

Patchell RA, Tibbs PA, Regine WF, Dempsey RJ, Mohiuddin M, Kryscio RJ, Markesbery WR, Foon KA, Young B (1998) Postoperative radiotherapy in the treatment of single metastases to the brain: A randomized trial. *JAMA* 280:1485–1489.

Patel AJ, Suki D, Hatiboglu MA, Abouassi H, Shi W, Wildrick DM, Lang FF, Sawaya R (2010) Factors influencing the risk of local recurrence after resection of a single brain metastasis. *Journal of Neurosurgery* 113(2):181–189.

Patel KR, Burri SH, Asher AL, Crocker IR, Fraser RW, Zhang C, Chen Z et al. (2016) Comparing preoperative with postoperative stereotactic radiosurgery for resectable brain metastases: A multi-institutional analysis. *Neurosurgery* 79(2):279–285.

Patel KR, Prabhu RS, Kandula S, Oliver DE, Kim S, Hadjipanayis C, Olson JJ et al. (2014) Intracranial control and radiographic changes with adjuvant radiation therapy for resected brain metastases: Whole brain radiotherapy versus stereotactic radiosurgery alone. *Journal of Neuro-Oncology* 120:657–663.

Pessina F, Navarria P, Cozzi L, Ascolese AM, Maggi G, Riva M, Masci G et al. (2016) Outcome evaluation of oligometastatic patients treated with surgical resection followed by hypofractionated stereotactic radiosurgery (HSRS) on the tumor bed, for single, large brain metastases. *PLoS One* 11(6):1–13.

Prabhu RS, Miller KR, Asher AL, Heinzerling JH, Moeller BJ, Lankford SP, McCammon RJ et al. (2018) Preoperative stereotactic radiosurgery before resection of brain metastases: Updated analysis of efficacy and toxicity of a novel treatment paradigm. *Journal of Neurosurgery* 2018 Dec 1:1–8.

Prabhu RS, Patel KR, Press RH, Soltys SG, Brown PD, Mehta MP, Asher AL, Burri SH (2019) Preoperative vs postoperative radiosurgery for resected brain metastases: A review. *Neurosurgery* 84:19–29.

Prabhu RS, Shu HK, Hadjipanayis C, Dhabaan A, Hall W, Raore B, Olson J, Curran W, Oyesiku N, Crocker I (2012) Current dosing paradigm for stereotactic radiosurgery alone after surgical resection of brain metastases needs to be optimized for improved local control. *International Journal of Radiation Oncology, Biology, Physics* 83:e61e66.

Robbins JR, Ryu S, Kalkanis S, Cogan C, Rock J, Movsas B, Kim JH, Rosenblum M (2012) Radiosurgery to the surgical cavity as adjuvant therapy for resected brain metastasis. *Neurosurgery* 71:937–943.

Roberge D, Souhami L (2010) Tumor bed radiosurgery following resection of brain metastases: A review. *Technology in Cancer Research and Treatment* 9:597–602.

Rwigema JC, Wegner RE, Mintz AH, Paravati AJ, Burton SA, Ozhasoglu C, Heron DE (2011) Stereotactic radiosurgery to the resection cavity of brain metastases: A retrospective analysis and literature review. *Stereotactic and Functional Neurosurgery* 89:329–337.

Schouten LJ, Rutten J, Huveneers HA, Twijnstra A (2002) Incidence of brain metastases in a cohort of patients with carcinoma of the breast, colon, kidney, and lung and melanoma. *Cancer* 94:2698–2705.

Shaw E, Scott C, Souhami L, Dinapoli R, Bahary JP, Kline R, Wharam M et al. (1996) Radiosurgery for the treatment of previously irradiated recurrent primary brain tumors and brain metastases: Initial report of radiation therapy oncology group protocol (90–05). *International Journal of Radiation Oncology, Biology, Physics* 34:647–654.

Shaw E, Scott C, Souhami L, Dinapoli R, Kline R, Loeffler J, Farnan N (2000) Single dose radiosurgical treatment of recurrent previously irradiated primary brain tumors and brain metastases: Final report of RTOG protocol 90–05. *International Journal of Radiation Oncology, Biology, Physics* 47:291–298.

Shi S, Sandhu N, Jin M, Wang E, Liu E, Jaoude JA, Schofield K et al. (2020) Stereotactic radiosurgery for resected brain metastases: Does the surgical corridor need to be targeted? *Practice in Radiation Oncology*. https://doi.org/10.1016/j.prro.2020.04.009.

Soffietti R, Kocher M, Abacioglu UM, Villa S, Fauchon F, Baumert BG, Fariselli L et al. (2013) A European organisation for research and treatment of cancer phase III trial of adjuvant whole-brain radiotherapyversus observation in patients with one to three brain metastases from solid tumors after surgical resection or radiosurgery: Quality-of-life results. *J Clin Oncol Off J Am Soc Clin Oncol* 31:65–72.

Soliman H, Myrehaug S, Tseng CL, Ruschin M, Hashmi A, Mainprize T, Spears J et al. (2019) Image-guided, linac-based, surgical cavity-hypofractionated stereotactic radiotherapy in 5 daily fractions for brain metastases. *Neurosurgery* 85:E860–869.

Soliman H, Ruschin M, Angelov L, Brown PD, Chiang VL, Kirkpatrick JP, Lo SS et al. (2018) Consensus contouring guidelines for postoperative completely resected cavity stereotactic radiosurgery for brain metastases. *International Journal of Radiation Oncology, Biology, Physics* 100:436–442.

Specht HM, Kessel KA, Oechsner M, Meyer B, Zimmer C, Combs SE (2016) HFSRT of the resection cavity in patients with brain metastases. *Strahlenther Onkol* 192:368–376.

Sperduto PW, Berkey B, Gaspar LE, Mehta M, Curran W (2008) A new prognostic index and comparison to three other indices for patients with brain metastases: An analysis of 1,960 patients in the RTOG database. *International Journal of Radiation Oncology, Biology, Physics* 70:510–514.

Sperduto PW, Wang M, Robins HI, Schell MC, Werner-Wasik M, Komaki R, Souhami L et al. (2013) A phase 3 trial of whole brain radiation therapy and stereotactic radiosurgery alone versus WBRT and SRS with temozolomide or erlotinib for non-small cell lung cancer and 1 to 3 brain metastases: Radiation Therapy Oncology Group 0320. *International Journal of Radiation Oncology, Biology, Physics* 85:1312–1318.

Steinmann D, Maertens B, Janssen S, Werner M, Frühauf J, Nakamura M, Christiansen H, Bremer M (2012) Hypofractionated stereotactic radiotherapy (hfSRT) after tumour resection of a single brain metastasis: Report of a single-centre individualized treatment approach. *Journal of Cancer Research in Clinical Oncology* 138:1523–1529.

Tawbi HA, Forsyth PA, Algazi A, Hamid O, Hodi FS, Moschos SJ, Khushalani NI et al. (2018) Combined nivolumab and ipilimumab in melanoma metastatic to the brain. *The New England Journal of Medicine* 379:722–730.

Toma-Dasu I, Sandstrom H, Barsoum P, Dasu A (2014) To fractionate or not to fractionate? That is the question for the radiosurgery of hypoxic tumors. *Journal of Neurosurgery* 121(Suppl):110–115.

Traylor JI, Habib A, Patel R, Muir M, Gadot R, Briere T, Yeboa DN, Li J, Rao G (2019) Fractionated stereotactic radiotherapy for local control of resected brain metastases. *Journal of Neuro-Oncology* 144:343–350.

Wang CC, Floyd SR, Chang CH, Warnke PC, Chio CC, Kasper EM, Mahadevan A, Wong ET, Chen CC (2012) Cyberknife hypofractionated stereotactic radiosurgery (HSRS) of resection cavity after excision of large cerebral metastasis: Efficacy and safety of an 800 cGy ×3 daily fractions regimen. *Journal of Neuro-Oncology* 106:601–610.

14 Brain Metastases Image-Guided Hypofractionated Radiation Therapy: Rationale, Approach, Outcomes

Michael J. Moravan, John M. Boyle, Jordan A. Torok, Peter E. Fecci, Carey Anders, Jeffrey M. Clarke, April K.S. Salama, Justus Adamson, Scott R. Floyd, and Joseph K. Salama

Contents

14.1 INTRODUCTION

Image-guided hypofractionated stereotactic radiation therapy (HSRT) has been increasingly used for the treatment of brain metastases. This technique employs noninvasive immobilization methods and modern imaging techniques at the time of treatment delivery for precise localization, allowing for the delivery of highly conformal radiation therapy. HSRT differs from commonly employed methods for delivering stereotactic radiosurgery (SRS) in the use of noninvasive immobilization and in the delivery of fractionated courses of radiation. In addition, HSRT aims to maximize the therapeutic ratio, permitting safe and effective treatment of large intracranial lesions or those in close proximity to critical normal tissues that would normally be unsuitable for single-fraction radiosurgery.

In this chapter, we briefly review the epidemiology of brain metastases to define the scope of the problem for which HSRT is well suited. Subsequently, we present the data and rationale for the use of SRS in this setting, followed by a discussion of the rationale for HSRT. Next, we present our institutional techniques for HSRT with a focus on clinical scenarios where HSRT may be beneficial. In the end, we will review the available data supporting the use of HSRT with a focus on treated metastasis control and normal tissue toxicity. The chapter closes with a brief discussion of controversies and future directions.

14.1.1 EPIDEMIOLOGY AND BACKGROUND

Brain metastases are a significant cause of morbidity and mortality among cancer patients. It is estimated that 20% to 40% of cancer patients will develop brain metastases in the course of their disease, translating to an incidence of between 100,000 and 300,000 cases annually in the United States (Johnson and Young, 1996; Mehta et al., 2005). Lung, breast, melanoma, renal cell, and colorectal cancers represent the majority of BM (Nayak et al., 2012). Evidence suggests that the incidence of brain metastases may be on the rise. This may be due to increased detection of clinically occult disease through staging magnetic resonance scans (MRI) (Sundermeyer et al., 2005). Alternatively, the increased incidence of brain metastases may be due to the improved efficacy of systemic therapies, which unmask the true incidence of brain metastases in a population of cancer patients with controlled extracranial disease. For example, a number of retrospective breast cancer series demonstrate a trend toward higher rates of brain metastases in women with HER2 overexpression and treatment with trastuzumab (Crivellari et al., 2001; Slimane et al., 2004; Lin and Winer, 2007). Additionally, the extracranial efficacy of erlotinib in patients with epidermal growth factor receptor (EGFR)-mutant non-small cell lung cancer has led to a similar hypothesis (Patel et al., 2014).

A number of prognostic indices have been developed to predict survival in patients with newly diagnosed brain metastases. The Radiation Therapy Oncology Group's (RTOG's) recursive partitioning analysis (RPA) was first published in 1997 and identified higher Karnofsky Performance Status (KPS), younger age, control of the primary tumor, and absence of extracranial metastases as prognostic factors for overall survival (OS) (Gaspar et al., 1997). This work was further developed, taking into account the number of brain metastases, histology of the primary tumor, and, for non-small cell lung cancer (NSCLC) and melanoma, the presence of molecular markers (Patrikidou et al., 2020; Sperduto et al., 2008, 2010, 2012, 2017a, 2017b, 2018, 2019). The resulting graded prognostic assessment (GPA) as well as the subsequent diagnosis-specific GPAs and molGPAs provide useful tools for predicting the prognosis of patients with brain metastases. These prognostic indices can also aid in the selection of patients for whom HSRT is an appropriate treatment strategy.

For oncologists treating patients with brain metastases, there is an increasing arsenal of therapeutic options including best supportive care, surgery, radiation therapy, and systemic therapies (Achrol et al., 2019). In particular, recent evidence demonstrating intracranial activity of systemic agents used for patients with breast cancer, melanoma, and non-small cell lung cancer has challenged the idea that upfront brain metastasis-directed therapy is appropriate for all patients with asymptomatic brain metastases who are candidates of these novel targeted agents. Selecting the best option for the patient often involves weighing the potential toxicities and benefits of the treatments, while keeping in mind the devastating implications of uncontrolled intracranial disease (Moravan et al., 2020). Unfortunately, both aggressive metastasis-directed treatment and uncontrolled disease can cause a decline in neurocognition, performance status, and even increase mortality (DeAngelis et al., 1989; Regine et al., 2001). Therefore, the optimal choice of therapy is best made on an individual patient level in a multidisciplinary setting. Thankfully, a growing body of level 1 evidence and consensus guidelines are aiding oncologists and patients in selecting the optimal treatment strategy.

14.1.2 RANDOMIZED TRIALS AND RATIONALE FOR STEREOTACTIC RADIOSURGERY

Multiple randomized trials have been conducted to establish the appropriate roles of surgery and radiation therapy in management of brain metastases. Three early randomized controlled trials investigated the benefit of adding surgical metastasectomy to whole-brain radiation therapy (WBRT) for patients with single-brain metastases (Patchell et al., 1990; Noordijk et al., 1994; Mintz et al., 1996). The first two of these trials found improvement in OS in the surgical arm (Patchell et al., 1990; Noordijk et al., 1994). The third trial by Mintz et al. (1996), while failing to establish a benefit for surgery, clarified the importance of systemic disease status. Indeed, it was found that a lack of extracranial disease control resulted in increased mortality (risk ratio 2:3; Mintz et al., 1996). The benefit of surgery for patients with controlled extracranial disease was also seen in the study by Noordijk et al. (1994). Median survival for patients in either arm with uncontrolled extracranial disease was 5 months. Conversely, survival was improved by 5 months for patients with controlled extracranial disease (median 12 versus 7 months). These studies established the benefit of aggressive metastasis-directed therapy with surgery in appropriately selected patients.

Based in part on the results of the surgical trials, similar attempts have also been made to improve treated metastasis control by adding radiosurgery to WBRT. RTOG 9805 selected patients with one to three brain metastases (stratified a priori) and a KPS ≥ 70 and randomized to WBRT or WBRT followed by aggressive metastasis-directed therapy using radiosurgery (Andrews et al., 2004). The addition of radiosurgery to WBRT failed to meet the primary endpoint of improved OS for the entire cohort; however on subset analysis, there was a survival benefit favoring SRS for patients with a single metastasis (4.9 vs 6.5 months). In addition, KPS was significantly improved for *all* patients receiving SRS, and steroid requirements were significantly reduced when compared with patients receiving WBRT alone. In summary, both surgical studies and subset analysis of the RTOG trials suggest that aggressive metastasis-directed therapy of a single-brain lesion improves survival. Furthermore, treatment of up to three metastases with radiosurgery following WBRT was safe and improved performance status.

The utility of WBRT following craniotomy for metastases was established by Patchell et al. (1998) in a trial investigating the role of WBRT following surgical resection of a single-brain metastases. In patients undergoing surgery alone, there were substantially higher rates of both treated metastasis (46% versus 10%) and other brain recurrences (37% versus 14%). Despite a dramatic reduction in the risk of in-brain recurrence with the addition of WBRT (70% versus 18%), this did not translate into an OS benefit. However, the percentage of deaths due to neurologic causes was reduced with the addition of WBRT (44% versus 14%), presumably due to a reduction in intracranial recurrence. In the modern MRI era, the European Organisation for Research and Treatment of Cancer (EORTC) conducted a similar study, addressing the benefit of immediate WBRT following surgical resection or radiosurgery. This trial randomized WBRT following intracranial metastasis-directed therapy for one to three metastases, allowing both complete surgical resection and SRS at the discretion of the treating surgeon/radiation oncologist (Kocher et al., 2011). The primary endpoint of the study, time to decline in WHO performance status more than 2, was no different with or without WBRT (median 9.5 versus 10 months). The secondary endpoint, median OS, also was not affected by the addition of WBRT (10.7 versus 10.9 months). Similar to the study by Patchell et al., there was a significant reduction in intracranial progression with WBRT (78% versus 48%, *p* < 0.02), observed as a decrease in the rate of both treated metastasis and distant in-brain recurrence. Intriguingly, rates of 2-year-treated metastasis progression were markedly lower with SRS as compared to surgery, either with WBRT (19% versus 37%) or without WBRT (31% versus 69%).

These studies demonstrate that the addition of WBRT immediately following ablative therapy with surgery or radiation results in decreased intracranial progression and neurocognitive death rates while failing to alter OS. The benefits of WBRT must be weighed against the toxicity of treatment. While WBRT is associated with mild acute symptoms, long-term effects on cognitive and cerebellar function may be worse than with intracranial metastasis-directed therapy alone (Regine et al., 2001; Tallet et al., 2012). As a result, SRS alone has been increasingly utilized to deliver targeted high-dose radiation to intact and resected brain metastases, minimizing dose to surrounding normal brain tissue, and potentially resulting in less adverse neurocognitive effects than WBRT.

In the setting of surgically resected disease, a single-center randomized phase 3 study compared post-operative single-fraction SRS (SF-SRS) to observation following brain metastasis resection (Mahajan et al., 2017). The primary endpoint was the time to local recurrence, which was significantly increased with the delivery of post-op SRS (12-month freedom from local recurrence of 72% with post-op SRS vs 43% with observation, HR 0.46, *p* = 0.015). No serious adverse events were noted in the SRS group, and no difference in OS was noted between groups, although the study was not powered for OS. Most local recurrences received salvage treatment with SRS, WBRT, surgery, or some combination of these modalities. Importantly, there was a high rate (~60%) of new brain metastases in both treatment groups over 1 year. Another prospective randomized study compared postoperative SRS versus WBRT with co-primary endpoints of cognitive-deterioration-free survival and OS (Brown et al., 2017). Compared to WBRT, treatment with SRS was associated with superior cognitive-deterioration-free survival (median 3 versus 3.7 months, HR 0.47, *p* < 0.0001) and less frequent cognitive deterioration at 6 months (85% versus 52%). Median OS was similar in both groups (~12 months), yet WBRT resulted in significantly higher rates of intracranial tumor control (72% versus 40%) and surgical bed control at 12 months (80% versus 60%) compared to SRS. Despite the reported inferior-treated metastasis control rate, the authors concluded that SRS should be considered a standard of care in this setting due to the reduction in neurotoxicity and similar OS. More contemporary studies have therefore sought to clarify the role of WBRT in patients with intact brain metastases treated with SRS.

A Japanese study enrolled patients with good performance status (Eastern Cooperative Oncology Group [ECOG] ≥ 2) and up to four brain metastases and randomized to WBRT followed by SRS or SRS alone (Aoyama et al., 2006). Consistent with other trials, the primary endpoint of OS was not significantly different between treatment arms despite a reduction of in-brain recurrence with WBRT. The authors of the trial hypothesized that the efficacy of salvage therapy, which was more often used in patients treated initially with SRS alone, could explain the findings. A similar study randomized patients with up to three metastases to SRS followed by WBRT or SRS alone (Chang et al., 2009). The primary endpoint of the study was neurocognitive function, as measured by a standard battery of neurocognitive function tests. The study was stopped early by the institutions data and safety monitoring board when interim analysis suggested a high probability (96%) that the addition of WBRT resulted in a decline in learning and memory function (52%) at 4 months following treatment compared to those receiving SRS alone (24%). Furthermore, a recent pooled analysis was conducted of three trials comparing SRS with and without WBRT. It was found that for patients younger than 50, SRS alone led to improved survival compared to that with the addition of WBRT (Sahgal et al., 2015).

To better address this question, a multi-institutional phase III study randomized patients with one to three brain metastases to receive WBRT and SRS versus SRS alone, with the primary endpoint of cognitive deterioration at 3 months (Brown et al., 2016). At 3 months, patients receiving SRS alone had significantly less cognitive decline (63.5% versus 91.7%, respectively) and improved quality of life compared to patients receiving WBRT and SRS. When long-term survivors were analyzed separately, the addition of WBRT resulted in a significantly higher incidence of cognitive deterioration at both 3 and 12 months. Although patients receiving WBRT and SRS had improved rates of intracranial tumor control at 12 months (85% versus 51%, $p < 0.001$), no significant difference in overall survival was observed compared to those receiving SRS alone. Based on these results and the similar findings in patients undergoing surgical resection, SRS alone is a reasonable treatment option for patients with one to four brain metastases.

The aforementioned trials investigated whether or not upfront WBRT could be safely omitted in patients with limited brain metastases (one to four) and of limited size (<4 cm) treated with single-fraction SRS. However, the safety of omitting upfront WBRT for patients with more extensive intracranial metastatic disease remains an open question. A prospective, observational, non-inferiority study (JLGK0901) enrolled newly diagnosed, good performance status patients with 1 to 10 brain metastases (largest tumor <10 mL and <3 cm in maximal diameter with an aggregate tumor volume ≤15 mL) to receive SRS to all intracranial disease (Yamamoto et al., 2014). Patients were divided into three groups (those with a single-brain metastasis, two to four metastases, and 5 to 10 metastases) with the primary endpoint of overall survival. Surprisingly, the median OS for patients with 5 to 10 brain metastases was non-inferior to those with 2 to 4 brain metastases, with no difference in the rates of treatment-related adverse events seen between the groups. The results of this study support consideration of SRS alone for patients with good performance status and up to 10 untreated brain metastases, who meet JLGK0901 inclusion criteria. Notably, only 17% of patients on this study had 5 to 10 brain metastases, and the median number of brain metastases was 6. This finding is consistent with multiple retrospective studies showing that the total volume of brain metastases was a better predictor of OS than the total number of brain metastases (Baschnagel et al., 2013; Bhatnagar et al., 2006; Likhacheva et al., 2013; Shultz et al., 2015). Importantly, the superiority of SRS versus WBRT (and vice versa) has not been established in the setting of more than four brain metastases. Currently, there is a phase III cooperative group trial (NCT03550391) comparing hippocampal-avoidant WBRT to single-fraction SRS for patients with 5 to 15 brain metastases, with the primary endpoints of both OS and neurocognitive PFS, which should shed light on this issue. However, these data suggest that the omission of WBRT does not compromise patient survival and may actually reduce morbidity and improve quality of life.

In summary, there are a number of radiation treatment strategies for patients with intact or resected brain metastases. Given the results of the above trials, SRS alone has become widely adopted as the preferred management for patients with a limited number of brain metastases, as reflected by recently published guidelines (Graber et al., 2019).

14.1.3 RATIONALE FOR HYPOFRACTIONATED STEREOTACTIC RADIOSURGERY

HSRT has been proposed as an alternative to single-fraction SRS for both intact and resected brain metastases. This technique uses a noninvasive stereotactic head frame system to deliver ablative doses of radiation to intracranial targets in 2 to 10 fractions with precision and conformity. There are several hypothetical advantages to this approach including (1) reduced toxicity of treatment, particularly for large targets or those located in eloquent regions of the brain, (2) improved control through delivery of total radiation doses with greater biological effectiveness, and (3) use in radioresistant histologies, such as melanoma and renal cell carcinoma.

The initial trials of SRS employed doses of 15–24 Gy delivered in a single fraction to metastases of less than 4 cm in size. The doses selected in these studies were based on the results of the RTOG 90–05 trial (Shaw et al., 2000). In this trial, previously irradiated patients with recurrent brain tumors (primary or metastatic) measuring up to 4 cm were enrolled and underwent single-fraction treatment with either Gamma Knife or linear accelerator-based SRS. Based on clinically determined toxicity, the maximally tolerated doses were 24, 18, and 15 Gy, for tumors measuring ≤20, 21–30, and 31–40 mm, respectively. The target volume included the contrast-enhancing volume on MRI or CT. The dose was prescribed to the 50%–90% isodose line with the entire target volume covered by the prescription isodose line without a margin. As a result of this trial, enrollment into the randomized trials of SRS was limited to tumors less than 4 cm. Subsequently, there is an appropriate hesitation in treating larger tumors with single-fraction SRS, given that these doses were associated with clinically significant rates of grade 3 or higher chronic toxicity: 10% for metastases less than 2 cm treated with 24 Gy, 20% for metastases 21–30 mm treated with 18 Gy, and 14% for metastases 31–40 mm treated with 15 Gy. This initial trial demonstrated the maximum acceptable toxicity for a given size and dose; however, efficacy of treatment was not an evaluated endpoint.

There is a large body of work investigating the dose–volumetric tolerance of normal brain tissue to single-fraction SRS. A number of metrics predict radiation injury, defined as neuroradiological changes, pathologic findings, and subsequent clinical side effects. Parameters such as volume of tissue receiving 8 Gy (Flickinger et al., 1997), 10 Gy (Voges et al., 1996; Flickinger et al., 1997; Levegrun et al., 2004; Blonigen et al., 2010; Minniti et al., 2011), or 12 Gy (Flickinger et al., 1997, 1998, 2000; Korytko et al., 2006; Blonigen et al., 2010; Lawrence et al., 2010; Minniti et al., 2011) have been evaluated. A representative series of 206 patients with 310 brain metastases treated with single-fraction SRS found that the patients treated with a V10 Gy more than 12.6 cm^3 and V12 Gy more than 10.9 cm^3 developed radiographic changes on MRI consistent with radionecrosis at a rate of 47% (Minniti et al., 2011). The correlation between V10 Gy and V12 Gy was more significant for symptomatic rather than asymptomatic radionecrosis. These studies demonstrate a dose–volume relationship for single-fraction SRS toxicity observed with long-term follow-up.

Beyond dose–volume constraints, the location of the treated lesions has also been shown to be predictive of radiation injury following single-fraction SRS (Flickinger et al., 1992, 1998, 2000). In a series of 422 patients treated with single-fraction SRS for AVM, of whom 85 experienced radiation injury, multivariate logistic regression analysis was used to identify risk factors predicting injury (Flickinger et al., 2000). In addition to V12 Gy, intracranial location was also found to be predictive of radiation injury, with the lowest risks for lesions in the cerebral and cerebellar hemispheres, followed by midbrain structures, and the highest risk in the brainstem. For example, a V12 Gy of 20 cc predicted an approximate 40% risk of radiation injury for a lesion in the basal ganglia compared to a less than 5% risk if the lesion were in the frontal lobe. These studies further refined the understanding of risks in patients undergoing SRS.

There is a limited amount of data reporting the risk of brainstem toxicity with single-fraction SRS, and interpretation of these studies is complicated by the differing dose prescription and volumetric information reported. An early study by Foote et al. (2001) analyzed outcomes among 149 patient treated with SRS for vestibular schwannomas using doses from 10 to 22.5 Gy. The reported outcomes were stratified by years of treatment as before 1994, when planning was largely CT based, and after 1994 when MRI was primarily used. For patients treated in the more contemporary cohort, the 2-year rates of facial and trigeminal neuropathies were 5% and 2%, respectively, as opposed to 29% and 7% in the older patient cohort. A multivariate analysis of risk factors found that a maximum dose (D_{max}) of 17.5 Gy and proximity of the tumor to the brainstem predicted for subsequent development of cranial nerve neuropathy. Available data suggest greater risks for higher V12 Gy, prescription doses more than 15 Gy, and volumes more than 4 cc (Spiegelmann et al., 2001; Kano et al., 2012). One early study of 43 patients with metastases in the brain treated with single-fraction SRS (Koyfman et al., 2010), using a median dose of 15 Gy, demonstrated no grade 3 or 4 toxicities. More recently, the efficacy and toxicity for 547 patients with 596 brainstem metastases treated with single-fraction Gamma Knife SRS using marginal doses ranging from 8 to 25 Gy, median 16 Gy, was reported (Trifiletti et al., 2016). Grade 3 to 4 SRS-related toxicities were observed in 7.4% of patients and consistent with earlier data, increasing odds of SRS-related toxicity were associated with increasing tumor volume, increasing margin dose, and a history of WBRT. Surprisingly, the V12 Gy was not associated with delayed toxicity in this population, although it was postulated by the authors that this was due to the relatively small volume of the brainstem metastases in their study (median brain metastasis volume = 0.8 mL). Based on the available data, it seems that doses of 12.5–13 Gy are associated with a less than 5% risk of clinically observed toxicity.

The optic nerves and optic chiasm are other critical structures that are often dose limiting for patients receiving single-fraction SRS. Most studies reporting on radiation-induced optic neuropathy report the maximum point dose. Early published data supported maximum doses of less than 8 Gy as safe, with increasing risk from 8 to 12 Gy (Leber et al., 1998; Stafford et al., 2003; Pollock et al., 2008). More recently, however, an analysis of published reports for radiation-naïve patients receiving SRS and HSRT near the optic nerves and/or chiasm with dosimetric and visual endpoints found that the risk of radiation-induced optic nerve/chiasm neuropathy (RION) with maximum point doses of less than 12 Gy in a single-fraction was less than 1%, although the authors recommended a limit of 10 Gy (Milano et al., 2018). Importantly, this study reported a crude 10-fold increased risk of developing RION in patients who had previously received radiation therapy. Similarly, low rates of RION following doses of less than 10 Gy were also reported by Hiniker et al. (2016). For maximum point doses more than 12 Gy, the risks of RION are generally felt to be unacceptably high (Mayo et al., 2010; Milano et al., 2018).

As described earlier, single-fraction SRS has dose limitations based on the size of the metastasis and its location, which may not make SRS the most appropriate for all intact or resected brain metastases. Importantly, toxicity data from RTOG 9005 found that tumors 3–4 cm in size were only able to be treated safely with 15 Gy in a single fraction. Theoretically, reducing dose as the tumor diameter increases less than ideal because as the number of cancer cells inside the tumor increases, potentially more dose is required to achieve equivocal cell killing compared to smaller tumors. It has been shown that large volume metastases, where dose is most limited by toxicity, SRS doses of less than 15–18 Gy are predictive of worse tumor control (Shiau et al., 1997; Mori et al., 1998; Vogelbaum et al., 2006; Yang et al., 2011). One would also expect similar reductions in tumor control for lesions in which the dose had to be reduced because of their proximity to critical structures of the brain.

Another setting where HSRT may be more appropriate than SRS is in the treatment of patients with resected brain metastases who require adjuvant radiation to their surgical cavities. Typically, this results in large and irregular tumor volumes, posing the same technical and clinical challenges for the use of single-fraction SRS in patients with large intact brain metastases in order to limit toxicity. The aforementioned prospective trials utilizing single-fraction SRS in this setting provide insight into this challenging clinical situation. In the study by Mahajan et al. (2016), following surgical resection for brain metastases, patients were randomized to either observation or single-fraction SRS (maximum resection cavity diameter ≤4 cm). The SRS dose was determined by the target volume: 16 Gy for ≤10 cc, 14 Gy for 10.1–15 cc, and 12 Gy for more than 15 ccs. The use of single-fraction SRS significantly reduced local recurrence compared to observation, yet tumor size (even after resection) was an independent predictor for local recurrence, suggesting decreasing the dose of single-fraction SRS resulted in sub-optimal surgical bed control. A similar phenomenon was seen in NCCTG N107C/CEC.3 (Brown et al., 2017) that compared the efficacy and toxicity of post-operative WBRT versus single-fraction SRS (maximum allowed resection cavity diameter $\leq$5 cm). SRS dose was based on cavity size in a manner similar to Mahajan et al. (2016). Single-fraction SRS resulted in 1-year surgical bed control of 60%, which was inferior to that of WBRT at 81%. As discussed earlier, patients receiving SRS demonstrated improved cognitive outcomes and equivalent overall survival as those receiving WBRT; however the lower rate of surgical bed control seen following single-fraction SRS suggests that further study is required to determine the optimal dose and fractionation of targeted radiation therapy to maximize surgical bed control while minimizing toxicity. HSRT may represent one method for accomplishing this goal.

A radiobiologic argument for HSRT, presented in work by Hall and Brenner (1993), highlights the benefits of a fractionated approach to SRS. It is recognized that many tumors contain a population of hypoxic and, therefore, radioresistant cells. Fractionated SRS takes advantage of the reoxygenation that occurs between treatments. Each dose of radiation will primarily kill the aerobic cell population. It has been documented that following each dose, the tumor will reestablish its original proportion of oxygenated cells, maximizing the effect of the subsequent dose. Another potential advantage of HSRT lies in the fact that mitotic cells are known to be the most sensitive to radiation and that fractionating treatment may allow for surviving tumor cells to enter mitosis, where they may succumb to a subsequent fraction. Additionally, fractionation takes advantage of the differing dose–response curves of early responding tissues (including tumors) and late responding tissues (such as brain). Radiobiological principles dictate that total dose, rather than fractional dose, most affects the amount of cell killing for early responding tissues. This is, to a degree, in opposition to late responding normal tissues that are more sensitive to changes in fractionation. Thus, more fractionated schedules will preferentially spare normal tissues and reduce late effects without compromising tumor control probability.

Employing the linear–quadratic model of cell kill and extrapolating from experience with low-dose rate brachytherapy for recurrent gliomas, Brenner et al. (1991) described alternative dose fractionations and their equivalent single-fraction dose. For example, alternative fractionation schemes equivalent to 18 Gy in a single fraction would be 27.4 Gy in 3 fractions or 32.6 Gy in 5 fractions. These doses have been employed in the setting of brain metastases, resulting in high rates of treated metastasis control and low rates of late toxicity (Manning et al., 2000). Concerns regarding the applicability of the linear–quadratic model to SRS and HSRT are discussed in Section 15.5.

14.2 RADIATION TECHNIQUE

The following section will detail the radiation technique in place at Duke University since 2008, having been employed in more than 600 patients. While other centers may utilize slightly different hardware and software solutions, the basic technique will be similar for most centers using a linear accelerator to deliver HSRT.

14.2.1 SIMULATION

Unless contraindicated, all patients we plan to treat with HSRT at Duke are immobilized in a thermoplastic face mask that is either frameless or attaches to a rigid U-frame system (BrainLAB, Munich, Germany) as shown in Figure 15.1. Subsequently, all patients undergo CT imaging, and for those patients without contraindication to MRI, axial fine-cut (1 mm) contrast-enhanced spoiled gradient (3D SPGR) MRI is obtained. The MR and CT images are fused using iPlan treatment planning software (BrainLAB), and the target lesion is contoured on the axial 3D SPGR images. For patients who are unable to undergo an MRI, a fine-cut (1 mm) CT with contrast enhancement is obtained.

14.2.2 CONTOURS

For patients with intact brain metastases, we define the gross tumor volume (GTV) as the contrast-enhancing lesion on the 3D SPGR MRI. A margin of 1 mm is applied to the GTV to generate a planning target volume (PTV), based on our prior analyses demonstrating equivalent treated metastasis control between 1 and 3 mm PTV expansions with a trend toward reduced rates of radionecrosis favoring the smaller expansion (Kirkpatrick et al., 2015). In the situation where only a fine-cut CT is available, we may utilize a

Figure 14.1 Example of a frameless thermoplastic immobilization system.

slightly larger margin expansion (typically 2 mm), as it is more difficult to visualize the boundary between tumor and normal brain.

For patients with resected brain metastases, a GTV is only identified if there is a residual tumor in the resection cavity post-operatively on 3D SPGR MRI. The clinical target volume (CTV) is defined as GTV (if applicable) plus the surgical bed and tract on contrast-enhanced MRI. A PTV margin of 2 mm is applied to the CTV based on work from Choi et al. (2012) that demonstrated improved local control without an apparent increase in toxicity. Patterns of recurrence after HSRT suggest that larger margins may be necessary especially when dural contact is present (Gui et al., 2018). Recently, expert consensus SRS/HSRT contouring guidelines for completely resected brain metastases have been published (Soliman et al., 2018). In both the intact and resected settings, normal tissues contoured include the whole brain, cochlea, optic nerves, optic chiasm, lenses, and brainstem delineated on fine-cut CT and MR images.

14.2.3 TREATMENT PLANNING

At our institution, we employ a number of different treatment techniques, all utilizing linear accelerator SRS systems equipped with a high-definition multileaf collimator with 2.5 mm wide leaves in the proximal ±4 cm from the isocenter and 5 mm wide leaves at more than 4 cm, as well as kV-cone-beam CT-based image guidance and a robotic couch capable of applying a 6D correction (translation + rotation) to match the planning CT (Novalis Tx & Truebeam STX, Varian Medical Systems, Palo Alto, CA). Dynamic conformal arc therapy (DCAT) and volumetric modulated arc therapy (VMAT) both take advantage of rotational radiation therapy to deliver highly conformal treatment. DCAT is a technique where the aperture of the MLC leaves in the beam's eye view dynamically conforms to the PTV with an added margin during gantry rotation. We typically utilize a 1–2 mm margin between the projection of the PTV and the MLC aperture in order to achieve a balance between dose conformity and falloff, and acceptable dose heterogeneity. A full description of the dosimetric trade-offs associated with the choice of PTV to MLC margin is given by Zhao et al. (2014). VMAT consists of rotational arcs where the MLC leaf positions, dose, and dose rate of the linear accelerator are varied as a function of gantry angle, the combination of which is determined via an inverse optimization to achieve a desired dose distribution. DCAT is typically ideal for smaller or more uniformly shaped targets, while VMAT may be preferred for larger or irregularly shaped targets. Intensity-modulated radiation therapy (IMRT) employs nonrotational "static" beams where the MLC leaves are again modified via optimization to allow "dose painting." Similar to VMAT, IMRT is typically best suited for larger or nonuniform target volumes, especially when the concave surface of a target volume abuts a sensitive normal structure. In this instance, it may be desirable to employ static beams to tightly constrain the angles from which dose is delivered, allowing preferential sparing via a nonuniform dose distribution. At our institution, DCAT planning is performed using iPlan (BrainLAB), while VMAT and IMRT are planned using Eclipse (Varian).

In our experience, either a DCAT or VMAT technique may be used when treating a single intracranial target. However, the length of time required for treatment increases with the number of intracranial targets, such that using a separate treatment plan for each target becomes cumbersome. A newer linear-accelerator-based technique that has been applied clinically is the use of a single isocenter VMAT technique to treat multiple intracranial targets. Recent developments have included reports of treatment techniques (Clark et al., 2010, 2012; Morrison et al., 2016; Audet et al., 2011; Ballangrud et al., 2018; Ohira et al., 2018), immobilization and quality assurance (Thomas et al., 2013; Roper et al., 2015; Faught et al. 2016; Stanhope et al., 2016; Adamson et al., 2019; Pant et al., 2020), and clinical outcomes (Lau et al., 2015; Alongi et al., 2019; Gregucci et al., 2019). Typically these plans include four to six noncoplanar VMAT arcs with the isocenter placed either near the center of the brain or the combined PTV and with particular attention to collimator angle and normal tissue constraints in the inverse optimization to minimize dose bridging in the healthy brain tissue located between targets (Clark et al., 2010; Kang et al., 2010; Morrison et al., 2016; Stanhope et al., 2016; Wu et al., 2016). In adopting a single isocenter VMAT technique, careful attention should be paid regarding the risk of geometric miss due to rotational uncertainties; in this regard, studies have shown that interfractional rotational uncertainties need be corrected prior to treatment via a patient support assembly capable of 6D motion (Roper et al., 2015; Stanhope et al., 2016). In addition, tightening of QA tolerances for patient support assembly and collimator angles beyond those outlined by recommending bodies may also be warranted (Faught et al., 2016).

Our standard dose and fractionation for HSRT is 25–27.5 Gy delivered in 5 fractions on consecutive days, allowing breaks for weekends. Dose selection is guided by the size, location, and histology of the primary tumor as well as whether the treatment is to an intact metastasis versus a resection cavity. Dose is prescribed to the 100% isodose line, and dose homogeneity is typically maintained with D_{max} less than 115%, with certain exceptions to maximize tumor control and minimize normal tissue dose. When using DCAT, hotspots of 120% may be acceptable for similar reasons as mentioned previously. Our goal is for a minimum of 99% coverage of the PTV by the prescription dose. Conformity index is an additional metric of dose conformity that is calculated by dividing the volume of tissue receiving the prescription dose by the volume of the PTV. Our goal is to achieve a conformity index of less than 2.0, though this may not be feasible with DCAT when treating very small or highly irregularly shaped lesions.

More recently, our department has shifted to using 27.5 Gy delivered in 5 fractions more regularly, as the BED2 is similar for this dose to the widely accepted constraints of 54 Gy in 30 fractions for the brainstem and optic apparatus. Other dose constraints for 5 fraction therapy have been published as well to help guide physicians (Benedict et al., 2010; Marcrom et al., 2017). The goal is to minimize dose to the noninvolved normal tissues such as brainstem, optic nerves, and optic chiasm. In cases of prior treatment, whether surgical or radiation therapy, constraints may be needed as judged clinically applicable. We do not typically make dose reductions in the setting of prior WBRT, though we find it useful to calculate the biologically equivalent dose (BED) for the HSRT course and the previous treatment, particularly as regards the total BED versus the maximum tolerated BED observed in conventionally fractionated radiotherapy.

14.2.4 TREATMENT DELIVERY

The time interval between simulation and treatment delivery should optimally be less than 5–7 days. Of note, for patients who have recently received systemic therapy, time is typically allowed between the patient's last dose of systemic therapy and HSRT to provide a washout period (3–5 half lives) in order to limit toxicity. Immediately before each treatment, patients are first positioned on the table using the in-room lasers. All patients then undergo image guidance with kV orthogonal imaging followed by kV cone-beam CT while on the treatment table. Appropriate adjustment of the isocenter is made, including translational and rotational corrections (6D) when necessary, after which cone-beam CT imaging is repeated. This technique ensures a translation position deviation of less than 1 mm in any direction and less than 1.0 degrees of rotation (Ma et al., 2009). Imaging is independently checked by both the physicist and physician who must agree on the accuracy of image guidance. The final position is verified visually in the room. For plans with high monitor units (MU), we often utilize either a 6X flattening filter free photon energy (Truebeam STX) or an SRS-specific 6X photon energy with a smaller flattening filter (Novalis Tx), allowing dose rates of up to 1400 and 1000 MU per minute, respectively. Following the delivery of the final treatment, patients are often discharged on a short course of steroids (typically dexamethasone) with dose and duration guided by patient-reported symptoms.

14.3 INDICATIONS

Single-fraction SRS remains a reasonable option for tumors that are small and in noneloquent areas of the brain. However, for larger tumors or tumors that are located within or in close proximity to critical areas of the brain, HSRT should be considered. Such critical areas include the optic chiasm, optic nerves, thalamus, basal ganglia, corpus callosum, or brainstem. In general, HSRT should be considered for lesions that are more than 2 cm in any axis, more than 3–4 cc in total volume, or for lesions located within or immediately adjacent to critical areas. Included below are clinical scenarios where HSRT was deemed appropriate and employed.

14.3.1 CASE 1 (FIGURE 14.2)

A 70-year-old woman presented with a 2-month history of progressive dizziness and disequilibrium. Her primary care physician ordered a CT of the head to rule out a stroke that revealed a hypodense region in the brainstem. A follow-up MRI revealed an enhancing lesion in the pons. A subsequent staging CT scan

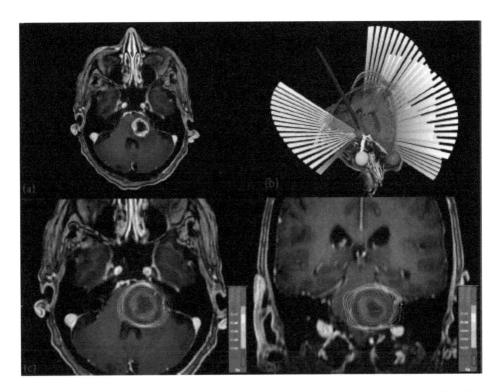

Figure 14.2 Case example of brainstem metastasis. Representative axial image from planning MRI with gross tumor volume (red contour) and planning target volume (PTV) (magenta contour) (a). Diagram demonstrating the orientation and degree of the five dynamic conformal arcs. Notice the avoidance of the optic nerves and optic chiasm (b). Axial image with isodose lines demonstrating steep dose gradient. Prescription isodose line (orange contour) and PTV (magenta volume) (c). Coronal image with isodose lines again demonstrating steep dose gradient and coverage of the PTV (d).

of the chest, abdomen, and pelvis showed nodularity in the right lung apex, but no clear lesion amenable to biopsy. Thus, a biopsy was performed of the lesion in the brainstem returning metastatic adenocarcinoma consistent with primary non-small cell lung cancer. Given a KPS of 90 and a low burden of disease, SRS was recommended. Given the location in the brainstem, HSRT was selected.

For this case of HSRT, a five dynamic conformal arc HSRT plan was selected. A dose of 2500 cGy was delivered in 5 fractions over 5 days. The dose was prescribed to the 100% isodose line yielding a maximum dose of 27.9 Gy and 99.2% coverage of the PTV with a conformity index of 1.4. A short course of dexamethasone was prescribed at discharge.

14.3.2 CASE 2 (FIGURE 14.3)

A 65-year-old male with a history of T1bN0 squamous cell carcinoma of the left upper lobe underwent lobectomy 2 years ago. On routine follow-up, he complained of ataxia and multiple episodes of falling down. A brain MRI was ordered by his oncologist revealing a solitary brain metastasis in the medulla. The patient was restaged with positron emission tomography and CT scans (PET/CT) showing no evidence of extracranial disease. Given his reasonable performance status and absence of extracranial disease, SRS was recommended. Due to the location in the brainstem, HSRT was selected.

For this case, a five dynamic conformal arc HSRT plan was selected. A dose of 2500 cGy in 5 fractions was delivered over 7 days. The dose was prescribed to the 100% isodose line yielding a maximum dose of 29.3 Gy with 99.5% coverage of the PTV, and a conformity index of 2.0. SRS was tolerated without difficulty, and the patient was discharged on a short course of dexamethasone.

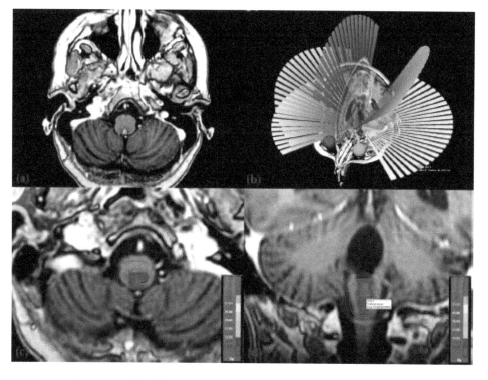

Figure 14.3 A second case example of brainstem metastasis. Representative axial image from planning MRI with gross tumor volume (red contour) and planning target volume (PTV) (magenta contour) (a). Diagram demonstrating the orientation and degree of rotation in the five dynamic conformal arcs. Notice the avoidance of the optic nerves and optic chiasm (b). Axial image with isodose lines demonstrating steep dose gradient. Prescription isodose line (orange contour) and PTV (magenta volume) (c). Coronal image with isodose lines again demonstrating steep dose gradient and a good coverage of the PTV (d).

14.3.3 CASE 3 (FIGURE 14.4)

A 68-year-old male presented to his primary care physician with complaints of memory loss and word-finding difficulties. He reported trouble recalling his children's and pets' names. An MRI of the brain was ordered revealing a 5.5 ×4 ×3.5 cm enhancing lesion involving the left splenium of the corpus callosum. He underwent staging PET/CT showing a mass in the right lower lobe and mediastinal adenopathy. He underwent bronchoscopy and biopsy of the lung mass revealing adenocarcinoma. He was staged clinically as T2N2M1 lung cancer. At the patient's request, the plan was to proceed with primary chemotherapy following treatment of the brain metastases. He was evaluated by neurosurgery and deemed inappropriate for surgical management. After discussing WBRT, SRS, and HSRT, the latter was selected. This approach was chosen to maximize the chance of treated metastasis control while minimizing the risk of treatment toxicity. This approach also prevented excessive delay prior to initiation of systemic therapy.

For this case, a three-arc VMAT HSRT plan was selected. A dose of 2500 cGy was delivered in 5 fractions on consecutive days. The prescription was made to the 100% isodose line yielding a maximum dose of 27.7 Gy and 99% coverage of the PTV and a conformity index of 99%. Therapy was tolerated well, and he was discharged to proceed with systemic therapy.

14.3.4 CASE 4 (FIGURE 14.5)

A 42-year-old woman with a history of HER2+ pathologic T2N2a breast cancer was diagnosed 7 years ago. She was treated with mastectomy followed by adjuvant chemotherapy as well as adjuvant chest wall and nodal radiation therapy. Five years ago, she developed a chest wall recurrence for which she was

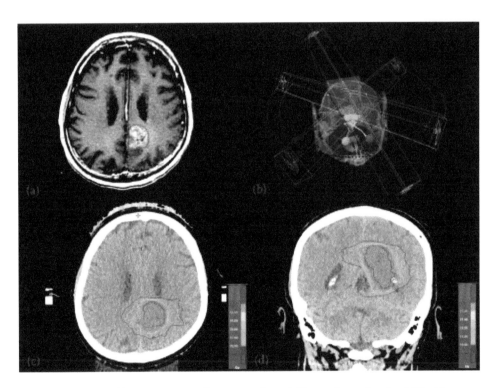

Figure 14.4 Case example of large metastasis. Representative axial image from planning MRI with gross tumor volume (GTV) (red contour) and planning target volume (PTV) (magenta contour) (a). Diagram demonstrating the orientation and degree of rotation of arcs in the five arc volumetric modulated arc therapy plan. These angles were selected to avoid the optic structures anteriorly and brainstem inferiorly (b). Axial image with isodose lines demonstrating steep dose gradient. Prescription isodose line (orange contour), GTV (red volume), and PTV (magenta contour) (c). Coronal image with isodose lines again demonstrating a steep dose gradient with a good coverage of the GTV and PTV (d).

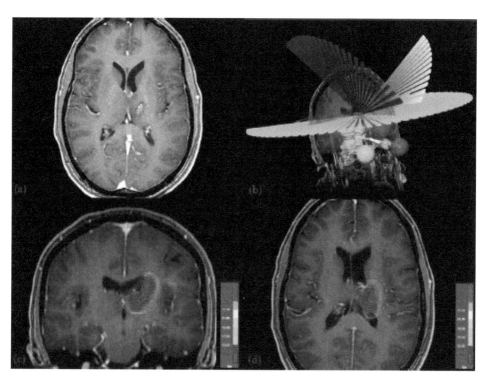

Figure 14.5 Case example of thalamic metastasis. Representative axial image from the planning MRI with gross tumor volume (red contour) and planning target volume (PTV) (magenta contour) (a). Diagram demonstrating the orientation and degree of the three arcs in the volumetric modulated arc therapy plan (b). Axial image with isodose lines demonstrating steep dose gradient. Prescription isodose line (orange contour) and PTV (magenta volume) (c). Coronal image with isodose lines again demonstrating steep dose gradient and a good coverage of the PTV (d).

treated with reirradiation followed by trastuzumab chemotherapy. Four years ago, she developed new onset disequilibrium. An MRI of the brain was ordered, concerning for leptomeningeal disease. A spine MRI showed no spinal leptomeningeal disease. She underwent a course of WBRT with a dose of 30 Gy in 10 fractions. She presents now with new onset ataxia and occasional falls. A repeat brain MRI showed a solitary enhancing lesion in the left thalamus. She has no current evidence of extracranial disease. Treatment options including repeat WBRT, SRS, and HSRT were discussed with the patient. In an effort to maximize control while minimizing toxicity, HSRT was selected.

For this case, a three-arc VMAT HSRT plan was selected to deliver a dose of 2500 cGy over a 7-day period. A VMAT plan was selected given the abnormal shape of the target volume. In this instance, VMAT allows for high conformality despite the nonuniform volume. The prescription was made to the 100% isodose line yielding a maximum dose of 27.4 Gy with 99% coverage of the PTV and a conformity index of 1.2. Treatment was tolerated well, and she was discharged to continue systemic therapy.

14.3.5 CASE 5 (FIGURE 14.6)

A 53-year-old male presented 1 year ago to the emergency room of an outside hospital with intense headaches associated with nausea and vomiting. A CT scan showed a hyperdense lesion in the left frontal lobe. An MRI showed an enhancing mass in the left frontal lobe measuring 4.5 × 4.0 × 3.5 cm. Staging studies showed a lesion in the left kidney. He underwent craniotomy and resection of a tumor with pathology returning metastatic renal cell carcinoma. He was then lost to follow-up and received no adjuvant therapy but, recently, established care at our institution. A repeat brain MRI showed an enhancing lesion posterior to the resection cavity involving the left caudate head and left lentiform nucleus. A separate enhancing

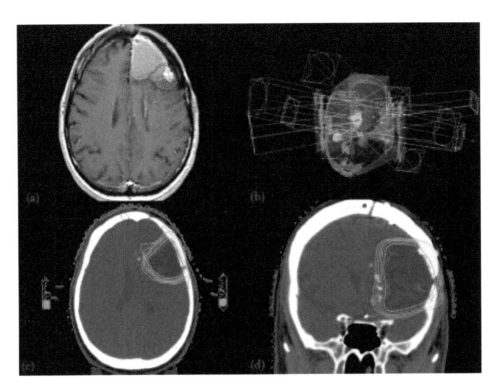

Figure 14.6 Case example of large target volume adjacent to critical normal tissues. Representative axial image from planning MRI with gross tumor volume (GTV) (red contour) and planning target volume (PTV) (magenta contour) (a). Diagram demonstrating the orientation and degree of rotation of arcs in the five-field IMRT plan. These angles were selected to avoid the optic structures inferiorly (b). Axial image with isodose lines demonstrating steep dose gradient. Prescription isodose line (orange contour), GTV (red volume), and PTV (magenta contour) (c). Coronal image with isodose lines again demonstrating a steep dose gradient with a good coverage of the GTV and PTV. Inferior to the PTV is the left optic nerve (yellow contour) (d).

lesion was seen at the lateral periphery of the resection cavity. He was evaluated by neurosurgery but deemed not to be a surgical candidate. Treatment options were discussed including WBRT and HSRT to the new enhancing lesions, as well as to the entire resection bed. Given the size of the projected treatment volume, as well as the left optic nerve, and the desire to maximize treated metastasis control, HSRT was selected.

For this case, a five-field IMRT HSRT plan was selected to deliver 2500 cGy in five consecutive daily treatments. The prescription was made to the 100% isodose line yielding a maximum dose of 27.5 Gy and 99% coverage of the PTV with a conformity index of 1.3. He received a single dose of dexamethasone and was discharged to the care of urologic oncology.

14.4 OUTCOMES DATA

There is a no level 1 evidence for HSRT with regard to toxicity or treated metastasis control or resection cavity control. Additionally, to date, no prospective randomized studies have been published comparing outcomes of HSRT to SRS for patients with intact or resected brain metastases. However, there is a growing body of retrospective and prospective evidence to support the feasibility, safety, and efficacy of using HSRT in the treatment of patients with brain metastases (Aoyama et al., 2003; Choi et al., 2012; Eaton et al., 2013; Ernst-Stecken et al., 2006; Fahrig et al., 2007; Faruqi et al., 2020; Feuvret et al., 2014; Fokas et al., 2012; Follwell et al. 2012; Hasegawa et al., 2017; Inoue et al., 2014a, 2014b; Jeong et al., 2015; Jiang et al., 2012; Kim et al., 2011, 2016; Kwon et al., 2009; Lim et al., 2018; Ling et al., 2015; Manning et al., 2000; Minniti et al., 2013, 2014a, 2014b, 2016, 2017; Murai et al., 2014; Navarria et al., 2016;

Pessina et al., 2016; Rajakesari et al., 2014; Serizawa et al., 2018; Soliman et al., 2019; Vogel et al., 2015; Wang et al., 2012; Wegner et al., 2015; Yomo et al., 2012, 2014; Zhong et al., 2017). Some of these studies will be discussed in more detail below, but they generally show high rates of treated metastasis or surgical bed control and favorable rates of toxicity.

14.4.1 TREATED METASTASIS CONTROL

Universally high rates of treated metastasis control have been reported with the use of HSRT for intact brain metastases. Rates are typically between 70% and 90% at 1 year (Table 14.1). Caution must be made when interpreting these rates and comparing them to outcomes in patients treated with SRS, where patients are more likely to have smaller metastases in less sensitive areas of the brain. When employing HSRT for intact metastases, a number of factors have been identified as predictive of treated metastasis control including tumor volume, tumor dimensions, use of concurrent chemotherapy, histology, and RPA class (Manning et al., 2000; Kwon et al., 2009; Fokas et al., 2012;Minniti et al., 2014). Interestingly, various analyses of treated metastasis control by dose and fractionation have yielded mixed results with some studies showing a dose–response (Aoyama et al., 2003), whereas others have found no such relationship (Fahrig et al., 2007; Kwon et al., 2009).

An early series by Manning et al. (2000) included 57 metastases in 32 patients treated with WBRT (30 Gy in 10 fractions) followed by HSRT. Patients were treated with 3 fractions of 6–12 Gy and prescribed to the 80%–90% isodose line. With a median follow-up of 37 months in surviving patients, only two experienced treated metastasis progression. Of note, this series included relatively small metastases (median volume 2.16 cm^3 corresponding to a diameter of 1–2 cm) and excluded patients with brainstem involvement or metastases within 5 mm of the chiasm.

Consistent with a broader trend, Aoyama et al. explored the efficacy of HSRT without WBRT in an attempt to minimize toxicity. Patients with up to four metastases were included (Aoyama et al., 2003). The analysis included 159 metastases in 87 patients with a median tumor volume of 3.3 cc (approximately equivalent to 2 cm spherical tumor diameter). A dose of 35 Gy in 4 fractions was prescribed to the treatment isocenter. The acceptable minimum dose was 28 Gy, though 32 Gy was the median minimum dose. For lesions in the brainstem, dose was reduced by 10%–20%, whereas for targets less than 1 cc, the dose was increased by 10%–20%. Treated metastasis control was 85%, 81%, and 69% at 6 months, 1 year, and 2 years, respectively. A univariate analysis identified a tumor volume of more than 3 cc, a minimum dose of less than 32 Gy, and dose at the isocenter of less than 35 Gy as predictive of progression. Only tumor volume remained significant on multivariable analysis. Thirty patients developed metachronous brain metastasis, of whom 22 had sufficient performance status to receive additional therapy.

As early experiences of HSRT have demonstrated acceptable rates of treated metastasis control, interest has grown in the use of HSRT to treat lesions where SRS was seen as ineffective or overly risky. As discussed previously, such situations include large tumor or those involving sensitive structures of the brain. To this end, a German phase II study enrolled 51 patients with 72 brain metastases, which involved the brainstem, mesencephalon, basal ganglia, or capsula interna or with a sum tumor volume of more than 3 cc (Ernst-Stecken et al., 2006). Patients with controlled extracranial disease were treated with HSRT alone (35 Gy in 5 fractions to the 90% isodose line), and those without extracranial disease control were treated with WBRT followed by HSRT (30 Gy in 5 fractions to the 90% isodose line). This approach resulted in treated metastasis control of 89% and 76% at 6 months and 1 year, respectively, despite a median tumor volume of 6 cc. A total of 14 patients (27%) experienced neurologic symptoms largely (11/14) due to metachronous brain metastases. This seminal study demonstrates that even large tumors, or those in sensitive areas, can be effectively controlled utilizing HSRT, notably without employing WBRT.

Further support for these findings comes from a German study where HSRT was routinely employed in patients with tumors in the brainstem, mesencephalon, basal ganglia, or capsula interna, or with a total tumor volume of more than 3 cc (Fahrig et al., 2007). A total of 243 brain metastases in 150 patients were treated at one of three treatment centers, which employed various dose fractionations schedules including 30–35 Gy in 5 fractions (*n* = 51), 40 Gy in 4 fractions (*n* = 36), or 35 Gy in 7 fractions (*n* = 63). Of note, 72 patients had prior WBRT though outcomes were not stratified to identify its effect. With a median follow-up of 28 months, the crude rate of treated metastasis control was 93% with no differences observed

Table 14.1 Treated metastasis control in select series of hypofractionated stereotactic radiation therapy for patients with intact brain metastases

STUDY	NO. OF PATIENTS/ METASTASES	PRIOR WBRT (DOSE/ FRACTIONS)	HSRT (DOSE/FRACTIONS)	MEDIAN TUMOR VOLUME (CC)	LOCAL CONTROL (%)	TIME POINT	COMMENT
Manning et al. (2000)	32/57	30 Gy/10	18–36 Gy/3 marginal dose	2.16	91	Crude Median F/U: 37 months	—
Aoyama et al. (2003)	87/159	None	35 Gy/4 Isocentric dose	3.3	81	1 year	Tumor volume >3 cm^3 predictive of local failure
Ernst-Strecken et al. (2006)	51/72	None 40 Gy/20	35 Gy/5 30 Gy/5 Marginal dose	6	76	1 year	Phase II study, including large tumors or those in critical brain regions not amenable to SRS
Fahrig et al. (2007)	150/243	72 of 150 patients treated with 40 Gy/20	30–35 Gy/5 (n = 51) 40 Gy/10 (n = 36) 35Gy/7 (n = 63) Marginal dose	Median PTV = 6.1	93	Crude	No difference in tumor response with fractionation
Kwon et al. (2009)	27/52	45 of 52 metastases treated with 37.5 Gy/15	20–36 Gy/4–6 25 Gy/5 most common	0.52	68.2	1 year	Smaller tumor dimension, tumor volume, & concurrent chemotherapy predictive of local control
Lindvall et al. (2009)	47/47	None	35–40 Gy/5 Marginal dose	6	84	Crude	—
Kim et al. (2011)	40/49	16 of 40 patients 30 Gy/10	30–42 Gy/6 Marginal dose Median dose = 36 Gy	Median PTV = 5.0	69	1 year	—
Fokas et al. (2012)	122/not reported	None	35 Gy/7 40 Gy/10	2.04 5.93	75 71	1 year	—

(Continued)

Table 14.1 (*Continued*) Treated metastasis control in select series of hypofractionated stereotactic radiation therapy for patients with intact brain metastases

STUDY	NO. OF PATIENTS/ METASTASES	PRIOR WBRT (DOSE/ FRACTIONS)	HSRT (DOSE/FRACTIONS)	MEDIAN TUMOR VOLUME (CC)	LOCAL CONTROL (%)	TIME POINT	COMMENT
Minniti et al. (2014)	135/170	None	27 Gy/3 if ≤2 cm 36 Gy/3 if >2 cm	10.1	72	1 year	Melanoma histology predictive of failure (HR 6.1)
Murai et al. (2014)	54/61	None	Group 1: 18–22 Gy/3 or 21–25 Gy/5 Group 2: 22–27 Gy/3 or 25–31 Gy/5 Group 3: 27–30 Gy/3 or 31–35 Gy/5	Median Diameter ≥2.5 cm	Group 1: 66% Group 2: 65% Group 3: 68%	1 year	—
Rajakesari et al. (2014)	70/not reported	40/70 patients	25 Gy/5 in 61/70 of patients	2.4 cm³	56%	1 year	—
Navarria et al. (2016)	102/102 Group 1: 51 Group 2: 51	None	Group 1: 27 Gy/3 if 2.1–3 cm Group 2: 32 Gy/4 if 3.1–5 cm	16.3 cm³	96% Group 1: 100% Group 2: 91%	1 year	—
Minniti et al. (2017)	60/70 Melanoma brain metastasis patients	None	9 Gy/3	GTV: 11.2 (3.1–37.1) PTV: 15.5 (5.6–44.6)	72%	1 year	Tumor size ≥3 cm predicted for increased local failure
Minniti et al. (2019)	127	None	9 Gy/3	GTV: 10.3 (3.1–37.1) TV: 15.6 (5.6–44.6)	92%	1 year	All patients had brain metastases from NSCLC
Faruqi et al. (2020)	187/250 targets Intact metastasis: 132/250	42/132 Intact patients	20–35 Gy/5 median dose: 30 Gy	Intact: 7.7	Intact: 78.2%	1 year	—

with the differing dose regimens. More recently, a Canadian study was published describing the outcomes from 187 consecutively treated patients with both intact (53%) and resected (47%) metastases (Faruqi et al., 2020). The median total prescribed dose was 30 Gy (20–35 Gy), and in all cases it was delivered daily, over 5 fractions. At 1 year, the local control rate was 78.2% for patients with intact metastases treated with HSRT.

In one of the few prospective studies investigating HF-SRS, Murai et al. performed a dose escalation study utilizing 3- and 5-fraction regimens. Patients with tumors ≥2.5 cm were included, those with tumors in the 2.5 to 4 cm range were treated with 3 fractions while those with tumors ≥4 cm were treated with 5 fractions. A total of 54 patients with 61 large brain metastases were included with the dose safely being escalated to the highest dose levels of 27–30 Gy in three fractions and 31–35 Gy in five fractions. One-year local control was 69%, and no grade 3 toxicities were reported. At least three institutions have compared outcomes between patients treated with single-fraction SRS and HSRT. One such series from Korea included 130 metastases in 98 patients treated with SRS ($n = 58$) or HSRT ($n = 40$) (Kim et al., 2011). Prior WBRT had been employed in 12 out of 58 and 16 out of 40 patients treated with SRS and HSRT, respectively. The groups differed in size of the PTV, with median volumes of 2.21 cc for SRS as compared to 5.0 cc for HSRT ($p = 0.02$ for comparison). The SRS dose was 18–22 Gy in all patients. Patients treated with HSRT received 30–42 Gy (median 36 Gy) in 6 fractions daily. Despite the HSRT group having significantly larger tumor volumes, treated metastasis control between the two groups was no different at 1 year (71% versus 69%), and there was no statistical difference in toxicity. Further support for their findings comes from a German series (Fokas et al., 2012). As was the institutional policy, patients with lesions more than 3 cm or those with involvement of the brainstem, mesencephalon, basal ganglia, or capsula interna were treated with HSRT. In their analysis, which included 260 patients with one to three brain metastases treated with SRS ($n = 138$) or HSRT at 35 Gy in 7 fractions ($n = 61$) or 40 Gy in 10 fractions ($n = 61$), they found no difference in treated metastasis control (71%–75%). This was despite the respective differences in median tumor volumes of patients receiving SRS (0.87 cc), HSRT at 35 Gy in 7 fractions (2.04 cc), and HSRT at 40 Gy in 10 fractions (5.93 cc). In the end, an Italian experience (Minniti et al., 2014) using HSRT (9–12 Gy × 3 fractions) in patients with one to three brain metastases demonstrated a 1-year-treated metastasis control rate of 88%. This group (Minniti et al., 2016) published their institutional experience of treating patients with brain metastases more than 2 cm in diameter and compared outcomes for those receiving SRS to those receiving HSRT. In 289 patients, the cumulative treated metastasis control rates for SRS and HSRT were 77% and 91%, respectively. Importantly, this difference in disease control persisted after propensity score adjustment.

14.4.2 SURGICAL BED CONTROL

Due to the challenges and limitations of utilizing single-fraction SRS for large-target volumes and the toxicity associated with conventional WBRT, the use of HSRT for patients with resected brain metastases has increased. Published studies have demonstrated high rates of resection cavity control, with reported rates of surgical bed control typically between 70% and 90% at 1 year (Table 14.2) and acceptable toxicity. When employing HSRT for the treatment of resection cavities, margin size, PTV, graded prognostic assessment, and meningeal contact have been shown to be associated with tumor bed control (Choi et al., 2012; Keller et al., 2017).

The first study published reporting on the use of HSRT in patients with resected brain metasteses was from Do et al. (2009). It consisted of seven patients treated with 22–27.5 Gy in 4–6 fractions, but it did include patients treated with single-fraction SRS. For entire cohort of resected patients, local recurrence occurred in 12% of patients; however, specifc data for the HSRT cohort was not provided. The study by Steinmann et al. (2012) was the first study consisting solely of patients with resected brain metastases receiving HSRT and utilized multiple dose fractionation schemes. The study included 33 patients with a singular resected brain metastases who were treated in 5–10 fractions to doses of 30–40 Gy. Local control at 1 year was 71%, and no grade 2 of higher toxicity was reported.

Further support of the use of HSRT for resected brain metastases came from a 101-patient study by Minniti et al. (2013). All patients had a single brain metastasis treated with surgical resection (cavity >3 cm) followed by HSRT (9 Gy × 3), with none of the patients having received prior WBRT. Local control was reported to be 93% at 1 year and 84% at 2 years. Importantly, the 1-year local control of HSRT in this setting did not appear to be influenced by tumor histology. A later analysis by Minniti et al. (2017) focusing on patients with resected or intact brain metastases from melanoma treated with HSRT in a similar fashion also showed a high rate of local control with surgical resection followed by HSRT at 12 months (88%). Keller et al. (2017) demonstrated a high local control rate at 1 year (88.2%) in 181 patients with 189 brain metastases resection cavities following HSRT. In this study, the HSRT dose was 33 Gy, prescribed to the isocenter with the 70% isodose line covering the PTV, and was delieered over three fractions, every other day. A more recent study by Minniti et al. (2019) that included 95 non-small cell lung cancer patients with resected or intact brain metastases treated with HSRT (9Gy × 3) also demonstrated a high rate of local control at 1 year (83%). More recently, Faruqi et al. (2020) published the outcomes of 187 consecutive patients treated with HSRT for both intact (53%) and resected (47%) metastases. The median total prescribed dose was 30 Gy (20–35 Gy), and in all cases it was delivered in 5 daily fractions. At 1 year, the local control rate of patients with resected brain metastases treated with HSRT was 84%. In the end, a large study by Shi et al. (2020) involving 442 patients with 501 resected brain metastases treated with single-fraction SRS or HSRT. There were a total of 425 resection cavities treated with HSRT in 2, 3, or 5 fractions, with the most common dose and fractionation being 24 Gy in 3 fractions. For all patients treated with HSRT, the 12-month local control rate was 93%.

Although no prospective randomized trials comparing the efficacy and toxicity of single-fraction SRS and HSRT exist, recently a systematic review and meta-analysis of post-operative SRS and HSRT studies was conducted by Akanda et al. (2020). The authors analyzed 50 studies including 3458 patients and found that the local control at 12 months was significantly better in patients treated with HSRT than those treated with single-fraction SRS (87.3% versus 80.0%, $p = 0.021$). Surprisingly, the analysis did not find a significant difference in local control between patients treated with or without an additional margin (84.3% versus 83.1%, respectively, $p = 0.714$), although the authors admit that they were not analyzing the difference in margin size across studies, just the addition of a specialized margin. Interestingly, local control at 12 months was not found to be different between patients treated within 30 days of resection or greater than 30 days following resection (85.1% versus 82.7%) although heterogeneity across studies makes the significance of this finding difficult to assess but suggests that further investigation may be warranted to determine the optimum interval.

In sum, these data provide ample support for the efficacy of HSRT in the treatment of patients with intact or resected brain metastases, especially in the setting of larger intracranial targets (>2 cm) or those located near sensitive organs at risk. The reported rates of treated metastasis control and surgical bed control are high and consistent with those reported for patients with brain metastases treated with single-fraction SRS. Furthermore, these data demonstrate that HSRT maintains favorable control rates despite being used near critical structures and for larger target volumes. A discussion on the safety of HSRT is presented below.

14.4.3 TOXICITY

One hypothetical benefit of HSRT is the ability to treat larger intracranial targets or those in critical areas of the brain without excess acute and late toxicity. As the vast majority of data is retrospective in nature, caution should again be taken when interpreting the data. That being said, the literature does indeed support the hypothesis. Patients treated with SRS are at risk of acute toxicity such as nausea, vomiting, seizures, and headaches, as well as late toxicity including alopecia, chronic headaches, visual or hearing impairment, chronic seizures, and radionecrosis. Acute toxicity is typically mild and rates are consistently less than 5%, whereas rates of grade 3–4 late toxicity are consistently less than 10%. Table 14.3 is a summary of selected studies of acute and late toxicities.

In the study by Manning et al. (2000), there was no acute toxicity. Of the 32 patients, 4 (12.5%) experienced seizures up to 12 months following treatment. Aoyama et al. documented very low rates of both acute (4.6%) and late (2.7%) toxicity (Aoyama et al., 2003). It is worth noting that one patient who received a dose of 40 Gy experienced brain edema 2 weeks following therapy resulting in hemiparesis, not responsive to steroids. Both late toxicities were due to biopsy-proven radionecrosis. The phase II study by Ernst-Stecken et al. (2006), while including somewhat higher-risk patients, demonstrated a significantly

Table 14.2 Resection cavity control in select series with hypofractionated stereotactic radiation therapy for patients with resected brain metastases

STUDY	NO. OF PATIENTS/ TARGETS	PRIOR WBRT (DOSE/ FRACTIONS)	HSRT (DOSE/FRACTIONS)	MEDIAN TARGET VOLUME (CC)	LOCAL CONTROL (%)	TIME POINT	COMMENT
Choi et al. (2012)	112/120 Total 63 lesions received HSRT	None	Median: 8 Gy/3	10.1 / 14.8 for those treated with 2-mm PTV margin	96.6% for patients treated with PTV margin (89% received HSRT)	year	12-month toxicity rates without and with margin were 8% and 3%, respectively
Minniti et al. (2013)	101/101	None	9 Gy/3	GTV: 17.5 (12.6–35.7) PTV: 29.5 (18.5–52.7)	93%	1 year	Radionecrosis by radiology or pathology was seen in 9% of patients
Ling et al. (2015)	99/100	4 patients Dose not reported	10–28 Gy in 1–5 fractions Median dose: 22 Gy 56.6% of patients treated in 3 fractions	PTV: 12.9	71.8% (74.3% of patients treated with HSRT)	1 year	9% of patients developed radiographic or histological radiation injury
Vogel et al. (2015)	30/33	None	79% treated in 5 fractions Median dose = 30 Gy	5-fraction PTV: 4.2 (3.2–6.7)	68.5%	1 year	10% radionecrosis rate
Pessina et al. (2016)	69/69	NR	10 Gy/3	CTV: 29.0 (4.1–203.1) PTV: 55.2 (17.2–282.9)	100%	1 year	Symptomatic radionecrosis rate was 0%
Keller et al. (2017)	181	None	11 Gy/3 to isocenter PTV covered by 70% IDL (23.10 Gy)	CTV: 7.6 (0.20–48.81) PTV: 14.15 (0.8–65.8)	88.2%	1 year	
Minniti et al. (2017)	60/67	None	9 Gy/3	GTV: 12.1 (3.3–49.9) PTV: 20.6 (6.1–66.8)	88%	1 year	Melanoma brain metastasis patients

(Continued)

Table 14.2 (*Continued*) Resection cavity control in select series with hypofractionated stereotactic radiation therapy for patients with resected brain metastases

STUDY	NO. OF PATIENTS/ TARGETS	PRIOR WBRT (DOSE/ FRACTIONS)	HSRT (DOSE/FRACTIONS)	MEDIAN TARGET VOLUME (CC)	LOCAL CONTROL (%)	TIME POINT	COMMENT
Minniti et al. (2019)	95	None	9 Gy/3	GTV: 10.8 (3.5–46.3) PTV: 22.4 (6.3–67.4)	83%	1 year	All patients had brain metastases from NSCLC
Soliman et al. (2019)	122/137	14%	All 5-fraction HSRT 62% received 30 Gy Dose range: 25–35 Gy	Mean PTV: 30.1	84%	1 year	Median OS 17 months OS at 1 year: 62%
Faruqi et al. (2020)	187/250 targets Cavities: 118/250	15/118 cavity patients Dose not reported	20–35 Gy/5 Median dose: 30 Gy	Cavity: 24.9	Cavity: 84%	1 year	—
Shi et al. (2020)	Total: 442/501 2-, 3-, or 5-fraction HSRT: 425 cavities	Whole cohort: 32 (7%) Dose not reported	24 Gy (24–27 Gy) 3-fraction HSRT most common: 389 cavities	PTV: 15.67 (10.68–23.84)	93%	1 year	

Table 14.3 **Toxicity in select series of hypofractionated stereotactic radiation therapy for intact and/or resected brain metastases**

STUDY	NUMBER OF PATIENTS	ACUTE TOXICITY[A] (%)	LATE TOXICITY[A] (%)	COMMENT
Manning et al. (2000)	32	0	12	Seizures from 3 weeks to 12 months post-HSRT
Aoyama et al. (2003)	87	4.6	2.7	Both late toxicities due to radionecrosis (RN)
Ernst-Stecken et al. (2006)	72	0	34 symptomatic RN	Radiographic RN correlated to V4 Gy per fraction (5 fractions)
Fahrig et al. (2007)	150	0	10	Various dose fractionations
				No toxicities with 40 Gy/10 fractions
Kwon et al. (2009)	27	0	14.8 Grade 2	1 Grade 3 toxicity (headache)
			3.7 Grade 3	
Lindvall et al. (2009)	47	0	2.1	—
Kim et al. (2011)	40	0	0	—
Fokas et al. (2012)	122	0	1–2% Grade 3	One case of RN
Minniti et al. (2013)	101/101	Grade 3–4 RTOG neurologic toxicity in five patients	Radionecrosis occurred in nine patients	Acutarial RN risk at 1 year was 7%
Minniti et al. (2014)	135	Not reported	4% Grade 3–4	9% rate of radiographic RN, predicted by increasing V18 Gy and V21 Gy
Murai et al. (2014)	54	Grade ≤ 2 alopecia in all patients without cavities	No Grade 3 or greater toxicity noted	Median overall survival was 6 months
Rajakesari et al. (2014)	70	16% total 9% Grade 1 4% Grade 2 3% Grade 3	4.3% rate ofsymptomatic RN	-
Keller et al. (2017)	181	No Grade 3+ toxicities	No Grade 3 or 4 late toxicities	18.5% of cavities developed RN Median time = 15 months Infratentorial location was associated with radionecrosis development

(Continued)

STUDY	NUMBER OF PATIENTS	ACUTE TOXICITY[a] (%)	LATE TOXICITY[a] (%)	COMMENT
Minniti et al. (2017)	120 patients 137 targets	Grade 2/3 in 14 patients Symptomatic RN patients: 7/60 surgery + HSRT 4/60 HSRT alone	RN cumulative incidence: 13% after surgery + HSRT 8% after HSRT alone	Intact tumor size ≥3 cm and the volume of normal brain receiving ≥18 Gy predicted for increased risk of RN
Faruqi et al. (2020)	118	38% of adverse radiation effects (ARE) occurred ≤6 months	ARE in 21.2% of lesions (51% symptomatic) Symptomatic ARE rates: 6-month: 4% 1 year: 9% 2 years: 12%	For intact metastases, total brain volume minus GTV receiving 30 Gy (BMC30) ≥10.5 cm³ predicted symptomatic ARE (OR = 7.2)
Shi et al. (2020)	Patients: 442 Cavities: 501 425 cavities received HSRT	78 cavities with ARE Grade 1: 36 (46%) Grade 2: 16 (21%) Grade 3: 5 (6%) Grade 4: 21 (27%)	1-year ARE rate: 8.5% 1-year symptomatic ARE rate: 5.5%	Higher single fraction equivalent dose using an alpha/beta ratio of 2 (SFED₂) associated with ARE development

[a] Grade included if reported.

higher rate of radionecrosis (34%). Of 72 lesions trethoated, 48 (61%) experienced increasing edema (enlargement of T2-weighted signal) of which 25 (34%) required an increase in steroid medication. A test of correlation identified increasing V4 Gy per fraction to be predictive of radiographic necrosis, with a threshold of 23 cc (14% if <23 cc and 70% if >23 cc). The analysis of different dose fractionation schemes by Fahrig et al. (2007) found no acute or late toxicity in patients treated with 40 Gy in 10 fractions, suggesting that increased fractionation may reduce complications. The Dana Farber/Harvard Cancer Center commonly utilized HSRT (~90% received 5 Gy × 5) for tumors of size more than 3 cm, cases wherein single fraction SRS would have carried a high V_{12Gy}, and/or for those in close proximity to critical structures. They reported their experience of 70 patients treated with HSRT and observed symptomatic radiation-induced treatment changes occurring in only 4% of patients (Rajakesari et al., 2016).

In their comparison of outcomes for patients with intact metastases treated with SRS and HSRT, Kim et al. (2011) showed a reduction in toxicity with the latter without compromising treated metastasis control. Despite significantly larger target volumes and higher rates of WBRT in the HSRT cohort (5.0 versus 2.2 cc), there was no grade ≥2 toxicity. In comparison, there was a 2% rate of grade 3 and a 7% rate of grade 4 toxicity in the SRS cohort. Similarly, Fokas et al. (2012) found higher rates of grade 1–3 toxicity in patients treated with SRS (14%) as compared to HSRT at 35 Gy in 7 fractions (6%) or HSRT at 40 Gy in 10 fractions (2%). As noted earlier, this is despite the significantly larger target volumes in the latter groups. In the institutional experience of patients with brain metastases more than 2 cm being treated with SRS versus HSRT by Minniti et al. (2016), patients receiving HSRT had radionecrosis rates of 9% compared to 18% in those receiving SRS, which remained significantly different even after propensity score adjustment.

Toxicity for patients with resected brain metastases following HSRT also appears to be low. In the study by Minniti et al. (2013), neurologic complications were recorded in nine patients, but only five patients experienced grade 3–4 toxicity. The actuarial risk of radionecrosis in this study at 1 and 2 years was 7% and 16%, respectively, and the volume of normal brain receiving 24 Gy ($V24_{Gy}$) was the most significant factor associated with the development of radionecrosis. Patients with a $V_{24Gy} \geq 16.8$ cm^3 had a crude risk of radionecrosis of 16%, while patients with a V_{24Gy} less than 16.8 cm^3 had a crude risk of 2%. There were no other factors associated with the development of radionecrosis.

Similar incidences of toxicity for HSRT in the treatment of resected brain metastases were seen in the study from Keller et al. (2017). No grade 3+ acute or late side effects were seen. Additionally, only 18.5% of cavities developed radionecrosis following HSRT, with a median time of 15 months. On multivariate analysis, infratentorial location of the brain metastasis was the only factor predictive of the development of radionecrosis.

The study by Faruqi et al. (2020) analyzed predictive factors for the development of any adverse radiation effect (ARE) and symptomatic ARE following HSRT for both intact and resected brain metastases. For the entire cohort, 53 targets developed ARE with a little over half being symptomatic. The 1-year rate and 2-year rates of ARE and symptomatic ARE were 17% and 21%, and 9% and 12%, respectively.

For patients treated with HSRT for intact brain metastases, the rate of ARE at 1 year was 12%. ARE developed in 34 (26%) of intact metastases treated with HSRT, of which 18 (53%) were symptomatic ARE. For patients with intact brain metastases treated with HSRT, a single metastasis was associated with a higher risk of developing ARE than patients treated for multiple metastases in the same treatment. Additionally, the volume of brain minus the GTV receiving 30 Gy (BMC30) was found to be a significant risk factor in the development of symptomatic ARE. More specifically, it was determined that in their cohort of patients, a BMC30 of ≥ 10.5 cm^3 versus less than 10.5 cm^3 increased the likelihood of developing symptomatic ARE with OR of 7.2. The investigators also found that the percentage of targets developing symptomatic ARE was 61% for patients with BMC30 values ≥ 10.5 cm^3 compared to only 13% in patients with a BMC30 less than 10.5 cm^3.

For patients receiving HSRT following resection of their brain metastases, the rate of ARE development at 1 year was 22%. ARE was seen in 19 (16%) of resection cavities treated with HSRT (16%), of which 9 (47%) were symptomatic. In patients in which resection cavities were treated with HSRT, only the receipt of targeted or immunotherapies within 1 month of radiation was a significant risk factor for developing any ARE. Both prior WBRT and prior SRS were found to be associated with symptomatic ARE.

Shi et al. (2020) also found low rates of ARE and symptomatic ARE in their large cohort of patients with resected brain metastases treated with SRS/HSRT. ARE developed in 78 cavities with 36 (46%) being grade 1, 16 (21%) being grade 2, 5 (6%) being grade 3, and 21 (27%) being grade 4. In this study, the only factor that was associated with an increased risk of ARE on multivariate analysis was a higher single fraction equivalent dose using an alpha/beta ratio of 2 (SFED$_2$).

In the prospective HSRT trial by Murai et al. (2014), a dose escalation study was performed investigating 3- and 5-fraction HSRT regimens in patients with brain metastsases ≥ 2.5 cm. Patients with brain metastases in the 2.5 to 4 cm range were treated with 3 fractions, while those with brain metastsases ≥ 4 cm were treated with 5 fractions. In 54 patients with 61 brain metastases (≥ 2.5 cm), the dose was safely escalated to 27–30 Gy in 3 fractions and 31–35 Gy in 5 fractions and no grade 3 toxicities were reported.

Another potential advantage of HSRT is potentially enhancing the therapeutic ratio for lesions close to OAR. Recently, Milano et al. (2018) published recommended dose tolerances for single-fraction SRS and HSRT based on an analysis of published reports of radiation-naïve patients receiving SRS and HSRT near the optic nerves and/or chiasm with dosimetric and visual endpoints. Thirty-four studies describing outcomes for 1578 patients receiving treatment for pituitary adenoma, cavernous sinus meningioma, craniopharyngioma, and malignant skull base tumors were included in the analysis. This data was used to investigate normal tissue complication probability (NTCP) for RION in patients with no prior history of radiation therapy from 1997 and beyond using an α/β ratio for the optic apparatus of 1.6. For patients receiving HSRT, it was calculated that there was a less than 1% risk of RION for maximum doses to the optic apparatus of 20 Gy in 3 fractions and 25 Gy in 5 fractions. As was the case for SRS, similar low rates of RION (<1%) were reported following doses of 20 Gy in 3 fractions and 25 Gy in 5 fractions were reported in a study byHiniker et al. (2016), which was not included in the analysis by Milano et al. (2018).

The available retrospective data suggests that HSRT can be delivered safely without excess acute and late toxicity and may possibly be less toxic than single-fraction SRS. Importantly, this is despite the treatment of larger tumors and/or those in close approximation to sensitive normal tissues.

14.5 CONTROVERSIES AND FUTURE DIRECTIONS

While there is a growing body of literature analyzing the efficacy and safety of HSRT, there is much room for future investigations (see Table 14.4). Ideally, future studies would prospectively enroll patients with limited brain metastases and randomize them to receive HSRT versus single-fraction SRS or WBRT, with endpoints of disease control, neurocognitive function, and neurotoxicity. Recently, the Alliance cooperative group activated a randomized phase III trial (NCT04114981) comparing single-fraction SRS versus HSRT in patients with zero to three unresected brain metastases (<4 cm in diameter) and one completely resected brain metastasis (<5 cm cavity diameter) with a primary endpoint of surgical bed recurrence-free survival. Secondary endpoints will investigate differences in performance status, quality of life, overall survival, and neurotoxicity. The results of this should inform practitioners as to whether HSRT provides a benefit over single-fraction SRS with respect to surgical bed control and/or neurotoxicity.

Although the intracranial activity of systemic agents for melanoma, NSCLC, and breast cancer patients with brain metastases is exciting, the optimal sequencing of systemic therapy and SRS/HSRT has not been established. Sequencing is likely dependent on a combination of patient and disease characteristics such as the presence of neurologic symptoms, tumor histology, tumor genetics, the total volume of intracranial metastases, prior systemic therapies, and the patient's candidacy for targeted therapies. Though upfront systemic therapy can address both intracranial and extracranial disease, sparing them from SRS/HRST neurotoxicity until intracranial disease progression, others argue that the risk of neurotoxicity from SRS/HSRT is low and that earlier SRS/HSRT in the course of treatment may result in better outcomes. Future research is necessary to determine how best to integrate these therapies and identifying which patient and disease factors may influence the optimal sequencing of therapies. Until these studies have been performed, patients with brain metastases who are candidates for intracranially active systemic therapies should be discussed in a multidisciplinary setting to determine the optimal integration of therapies.

There is a paucity of data on the optimal dose fractionation schemes for different clinical scenarios, focusing both on disease control and normal tissue toxicity. First principles dictate that higher doses should improve disease control. However, a much more nuanced understanding of the dose–response curve for different dose–fractionation schedules is needed for selecting an optimal schedule. As a correlate to this, similar models are needed to predict normal tissue damage. As much has been done for single-fraction SRS, work should focus on establishing the dose–volume–location relationships for lesions treated with HSRT. In particular, prospective toxicity data are needed for treatment of lesions in the brainstem or within close proximity to the optic nerves and optic chiasm.

Furthermore, it is imperative to further investigate the mechanisms of tissue damage when large doses of radiation are used. There is a great deal of controversy regarding the utility of the linear–quadratic model in the setting of SRS or HSRT (Kirkpatrick et al., 2008, 2009; Hanin and Zaider, 2010). Traditional understanding of radiation biology principles does not explain the excellent clinical outcomes when using these techniques. Standard principles of radiation biology dictate that cell killing occurs largely through DNA damage via double-strand breaks. Alternative mechanisms of cell killing have been proposed such as vascular damage (Kirkpatrick et al., 2008; Park et al., 2012; Song et al., 2013) or radiation-induced immune response (Finkelstein et al., 2011). In the absence of clinical data, preclinical models may be needed for mechanistic studies that will inform future clinical investigations.

HSRT has also been utilized following laser interstitial thermal therapy (LITT) in the setting of potentially recurrent brain metastases previously treated with SRS. LITT utilizes focal laser energy delivered through a small fiberoptic catheter to cause interstitial hyperthermia and coagulate surrounding tissue, and it is increasingly being used to treat patients with recurrent brain metastases following SRS. Retrospective studies have shown treated metastases control rates of 60%–100% (Hong et al., 2019; Ahluwalia et al., 2018; Ali et al., 2016). Notably, one multi-institutional study found that all instances of tumor recurrence following LITT occurred in patients where less than 80% of the lesion was ablated (Ali et al., 2016). A subset of patients in this study received post-LITT HSRT (5 Gy × 5 daily fractions) to the treatment area. Interestingly, none of the patients receiving HSRT demonstrated treated metastasis progression, despite some patients being treated with an ablation efficiency of less than 80%, suggesting that adjuvant HSRT may enhance the efficacy of LITT. At our institution, HSRT (5 Gy × 5 daily fractions) is recommended for all patients with biopsy-proven recurrent brain metastases following SRS who undergo LITT therapy.

Table 14.4 Select clinical trials of hypofractionated stereotactic radiation therapy (HSRT) in patients with intact brain metastases

STUDY	PHASE	CLINICAL TRIALS.GOV IDENTIFIER	BRAIN METASTASES: NUMBER AND CHARACTERISTICS	PRIMARY OUTCOME
Phase I/II Study of Fractionated Stereotactic Radiosurgery to Treat Large Brain Metastases	/2	NCT00928226	1–4 brain metastases or resection cavities with a volume of 4.2–33.5 cm³	Maximum tolerated dose (MTD) of 3 fraction HSRT for brain metastases or cavities
Fractionated Stereotactic Radiotherapy (FSRT) in Treatment of Brain Metastases	1	NCT02187822	1–3 brain metastases Max tumor diameter ≤ 5 cm or max tumor volume ≤ 120 cc	MTD of TPI-287 given concurrently with 5 fraction HSRT
Hypofractionated Stereotactic Radiosurgery in Treating Patients With Large Brain Metastasis	1	NCT01705548	1 intact or resected brain metastases, 3–6 cm in diameter	MTD of 5-fraction SRS
Perfexion Brain Metastasis (HF-SRT)	1	NCT00805103	1–5 recurrent brain metastases after WBRT with ≥1 metastasis >2 cm in max diameter	MTD of HSRT for recurrent brain metastases
SIMT Stereotactic Radiosurgery Outcomes Study	NA	NCT02886572	4–10 brain metastases None > 4 cm in diameter	Proportion of patients living longer than predicted by GPA
Fractionated Stereotactic Radiosurgery with Concurrent Bevacizumab for Brain Metastases: A Phase I Dose-escalation Trial	1	NCT02672995	1–3 brain metastases Group 1: 1.5–2.5 cm in diameter Group 2: 2.5–3.5 cm in diameter	MTD of 3-fraction SRS + bevacizumab
Frameless Fractionated Stereotactic Radiation Therapy (FSRT) for Brain Mets	2	NCT02798029	1–4 brain metastases Max diameter ≤5 cm	Local control, intracranial progression-free survival, overall survival, and cost associated with 3–5 fraction HSRT
Fractionated Stereotactic Radiosurgery for Large Brain Metastases	1	NCT02054689	1–3 brain metastases 3–5 cm in diameter Other lesions treated with SRS	MTD and patient toxicities for 3-fraction HSRT
Hypofractionated Stereotactic Radiation Therapy of Brain Metastases: Evaluation of WBRT	NA	NCT02913534	1–3 brain metastases	Overall survival of patients with 1–3 brain metastases treated with HSRT

(Continued)

Table 14.4 *(Continued)* **Select clinical trials of hypofractionated stereotactic radiation therapy (HSRT) in patients with intact brain metastases**

STUDY	PHASE	CLINICAL TRIALS.GOV IDENTIFIER	BRAIN METASTASES: NUMBER AND CHARACTERISTICS	PRIMARY OUTCOME
Single-Fraction Stereotactic Radiosurgery Compared With Fractionated Stereotactic Radiosurgery in Treating Patients With Resected Metastatic Brain Disease	3	NCT04114981	1 completely resected brain metastasis with cavity <5 cm in diameter 0–3 unresected brain metastases, <4 cm in diameter	Surgical bed recurrence-free survival
A Study Evaluating the Efficacy of the Combination of Hypofractionated Stereotactic Radiation Therapy With the Anti-PDL1 Immune Checkpoint Inhibitor Durvalumab in NSCLC Patients With 1 to 4 Brain Metastases (SILK BM)	2	NCT03955198	1–4 brain metastases <3.5 cm in diameter, all amenable to hFSRT, with ≥1 metastasis ≥1 cm (RANO-BM criteria evaluation)	Time to intracranial progression according to Response Assessment in Neuro-Oncology Brain Metastases (RANO-BM) criteria
HFSRT With Concurrent TMZ for Large BMs	3	NCT03778541	1–3 brain metastases Tumor volume ≥6 cm^3 Maximum diameter ≥3 cm	Intracranial progression-free survival rate
HA-WBRT versus SRS in Patients With Multiple Brain Metastases (HipSter)	3	NCT04277403	4–15 brain metastases Total volume <25 cm^3	Intracranial progression-free survival

HSRT is typically delivered 2–3 weeks following LITT. Future studies should focus on patient, tumor, and LITT characteristics that can be used to better identify those who would benefit from post-LITT HSRT.

In the end, there is interesting evidence that the treatment of tumors with radiation may release tumor-derived antigens, which can stimulate the immune system, potentially improving local and distant disease control (Demaria et al., 2015). Although the high doses utilized for single-fraction SRS are effective at damaging vasculature to enhance tumor killing, this phenomenon may limit the ability of tumor-derived antigens and/or immune cells to interact, thereby inhibiting the overall immunomodulatory effect of radiation (Park et al., 2012). It has been postulated that HSRT may be able to generate tumor antigens without impairing their transport, which would result in a more robust immune response (Demaria et al., 2012, 2015; Park et al., 2012). This effect may be further amplified if combined with one or more immunomodulatory cancer therapies (such as ipilumimab, pembrolizumab, or nivolumab), although the interaction between radiation and these immune checkpoint inhibitors is not well understood and is the subject of current clinical studies.

CHECKLIST: KEY POINTS FOR CLINICAL PRACTICE

✓	ACTIVITY	SOME CONSIDERATIONS
	Patient selection	*Is the patient appropriate for SRS/HSRT?* • Is the tumor causing symptoms, mass effect, and/or the patient has no tissue diagnosis of cancer? If so, consider resection • One to four metastases present? (Unless SIMT is available) • KPS ≥ 70? • Recent systemic therapy? (Typically allow for 3–5 half-lives of washout prior to HSRT delivery) • Is the patient asymptomatic and a candidate for a CNS penetrating targeted therapeutic option available?
	SRS versus HSRT	*Might the patient benefit from HSRT versus single-fraction SRS?* • Is the lesion >3 cm in size? • Is the lesion in the brainstem or in close proximity to the brainstem, optic nerves, or optic chiasm? • Has the patient had previous radiation? • If yes, consider HSRT
	Simulation	*Immobilization* • Rigid frameless system with <1 mm of translational or <1.0 degree of rotation • Consider patient comfort and replicability of setup imaging • Both CT scan and MRI should be performed in treatment position • CT simulation ideally should be fine cut (1 mm slices) • MRI ideally should be fine cut (1 mm slices) in axial orientation, T1 sequence with IV contrast • Accurate MRI fusion to treatment planning CT should be verified by clinician and used for contouring targets
	Treatment planning	*Contours* • GTV should include contrast-enhancing lesions • PTV created by using institutionally appropriate expansion (1–2 mm) • Organs at risk should be contoured on CT *Treatment planning* • For uniform lesions, consider DCAT • For nonuniform lesions or those in close proximity to sensitive normal tissues, consider VMAT or IMRT

(Continued)

✓	ACTIVITY	SOME CONSIDERATIONS
		• Evidence to support multiple-dosefractionation schemes • Consider 21–27 Gy in 3 fractions or 25–30 Gy in 5 fractions; marginal dose prescription • >99% PTV coverage • Conformity index <2.0 • Normal tissues but should be kept as low as reasonably achievable (ALARA), although multiple dose constraints published for the various fractionation schemes • Dose constraint exceptions in the setting or prior radiation or surgical manipulation
	Treatment delivery	*Equipment* Micromultileaf collimator is recommended 6-degree couch is recommended *Imaging* At a minimum, orthogonal kV imaging should be performed to verify isocenter prior to treatment delivery If available, cone-beam CT should be performed to assess translational and rotational accuracy of setup Reimage if shifts are made *HSRT discharge* For large lesions or those near brainstem, consider a short course of low-dose dexamethasone For patients already on steroids, dose adjustments should be symptom-based Repeat MRI at 2–3 months post-HSRT

REFERENCES

Achrol AS et al. (2019) Brain metastases. *Nat Rev Dis Primers* 5(1):1–5.

Adamson J et al. (2019) Delivered dose distribution visualized directly with onboard kV-CBCT: Proof of principle. *Int J Radiat Oncol Biol Phys* 103(5):1271–1279.

Ahluwalia M et al. (2018) Laser ablation after stereotactic radiosurgery: A multicenter prospective study in patients with metastatic brain tumors and radiation necrosis. *J Neurosurg* 130:804–811.

Akanda ZZ, Hong W, Nahavandi S, Haghighi N, Phillips C, Kok DL (2020) Post-operative stereotactic radiosurgery following excision of brain metastases: A systematic review and meta-analysis. *Radiother Oncol* 142:27–35.

Ali MA et al. (2016) Stereotactic laser ablation as treatment for brain metastases that recur after stereotactic radiosurgery: A multiinstitutional experience. *Neurosurg Focus* 41: E11.

Alongi F et al. (2019) First experience and clinical results using a new non-coplanar monoisocenter technique (HyperArc™) for linac-based VMAT radiosurgery in brain metastases. *J Cancer Res Clin Oncol* 145:193–200.

Andrews DW et al. (2004) Whole brain radiation therapy with or without stereotactic radiosurgery boost for patients with one to three brain metastases: Phase III results of the RTOG 9508 randomised trial. *Lancet* 363(9422):1665–1672.

Angelov L et al. (2018) Impact of 2-staged stereotactic radiosurgery for treatment of brain metastases >/= 2 cm. *J Neurosurg* 129(2):366–382.

Aoyama H et al. (2003) Hypofractionated stereotactic radiotherapy alone without whole-brain irradiation for patients with solitary and oligo brain metastasis using noninvasive fixation of the skull. *Int J Radiat Oncol Biol Phys* 56(3):793–800.

Aoyama H et al. (2006) Stereotactic radiosurgery plus whole-brain radiation therapy vs stereotactic radiosurgery alone for treatment of brain metastases: A randomized controlled trial. *JAMA* 295(21):2483–2491.

Audet C et al. (2011) Evaluation of volumetric modulated arc therapy for cranial radiosurgery using multiple noncoplanar arcs. *Med Phys* 38(11):5863–5872.

Ballangrud Å et al. (2018) Institutional experience with SRS VMAT planning for multiple cranial metastases. *J Appl Clin Med Phys* 19(2):176–183.

Baschnagel AM et al. (2013) Tumor volume as a predictor of survival and local control in patients with brain metastases treated with Gamma Knife surgery. *J Neurosurg.* 119(5):1139–1144.

Benedict SH et al. (2010) Stereotactic body radiation therapy: The report of AAPM task group 101. *Med Phys* 37(8):4078–4101.

Bhatnagar AK, Flickinger JC, Kondziolka D, Lunsford LD. (2006) Stereotactic radiosurgery for four or more intracranial metastases. *Int J Radiat Oncol Biol Phys* 64(3):898–903.

Blonigen BJ, Steinmetz RD, Levin L, Lamba MA, Warnick RE, Breneman JC (2009) Irradiated volume as a predictor of brain radionecrosis after linear accelerator stereotactic radiosurgery. *Int J Radiat Oncol Biol Phys* 77(4):996–1001.

Brenner DJ, Martel MK, Hall EJ (1991) Fractionated regimens for stereotactic radiotherapy of recurrent tumors in the brain. *Int J Radiat Oncol Biol Phys* 21(3):819–824.

Brown PD et al. (2016) Effect of radiosurgery alone vs radiosurgery with whole brain radiation therapy on cognitive function in patients with 1 to 3 brain metastases: A randomized clinical trial. *JAMA* 316(4):401–409.

Brown PD et al. (2017) Postoperative stereotactic radiosurgery compared with whole brain radiotherapy for resected metastatic brain disease (NCCTG N107C/CEC.3): A multicentre, randomised, controlled, phase 3 trial. *Lancet Oncol* 18(8):1049–1060.

Chang EL et al. (2009) Neurocognition in patients with brain metastases treated with radiosurgery or radiosurgery plus whole-brain irradiation: A randomised controlled trial. *Lancet Oncol* 10(11):1037–1044.

Choi CY et al. (2012) Stereotactic radiosurgery of the postoperative resection cavity for brain metastases: Prospective evaluation of target margin on tumor control. *Int J Radiat Oncol Biol Phys* 84(2):336–342.

Clark GM, Popple RA, Young PE, Fiveash JB (2010) Feasibility of single-isocenter volumetric modulated arc radiosurgery for treatment of multiple brain metastases *Int J Radiat Oncol Biol Phys* 76:296–302.

Clark GM et al. (2012) Plan quality and treatment planning technique for single isocenter cranial radiosurgery with volumetric modulated arc therapy. *Pract Radiat Oncol* 2:306–313.

Crivellari Det al. (2001) High incidence of centralnervous system involvement in patients with metastatic or locally advanced breast cancer treated with epirubicinand docetaxel. *Ann Oncol* 12(3):353–356.

DeAngelis LM, Delattre JY, Posner JB (1989) Radiation-induced dementia in patients cured of brain metastases. *Neurology*39(6):789–796.

Demaria S, Formenti SC (2012) Radiation as an immunological adjuvant: Current evidence on dose and fractionation. *Front Oncol* 2:153.

Demaria S, Golden EB, Formenti SC (2015) Role of local radiation therapy in cancer immunotherapy. *JAMA Oncol* 1(9):1325–1332.

Do L et al. (2009) Resection followed by stereotactic radiosurgery to resection cavity for intracranial metastases. *Int J Radiat Oncol Biol Phys* 73(2):486–491.

Dohm AE et al. (2018a) Staged stereotactic radiosurgery for large brain metastases: Local control and clinical outcomes of a one-two punch technique. *Neurosurgery* 83(1):114–121.

Dohm AE et al. (2018b) Surgical resection and postoperative radiosurgery versus staged radiosurgery for large brain metastases. *J Neurooncol* 140(3):749–756.

Eaton BR, Gebhardt B, Prabhu R, Shu HK, Curran WJ, Jr, Crocker I. (2013) Hypofractionated radiosurgery for intact or resected brain metastases: Defining the optimal dose and fractionation. *Radiat Oncol* 8:135.

Ernst-Stecken A, Ganslandt O, Lambrecht U, Sauer R, Grabenbauer G (2006) Phase II trial of hypofractionated stereotactic radiotherapy for brain metastases: Results and toxicity. *Radiother Oncol* 81(1):18–24.

Fahrig A et al. (2007) Hypofractionated stereotactic radiotherapy for brain metastases—Results from three different dose concepts. *Strahlenther Onkol* 183(11):625–630.

Faruqi S et al. (2020) Adverse radiation effect after hypofractionated stereotactic radiosurgery in 5 daily fractions for surgical cavities. *Int J Radiat Oncol Biol Phys* 106(4):772–779.

Faught AM, Trager M, Yin F-F, Kirkpatrick J, Adamson J (2016) Re-examining TG-142 recommendations in light of modern techniques for linear accelerator based radiosurgery. *Med Phys* 43(10):5437–5441.

Feuvret L et al. (2014) Stereotactic radiotherapy for large solitary brain metastases. *Cancer Radiother* 18(2):97–106.

Finkelstein SE et al. (2011) The confluence of stereotactic ablative radiotherapy and tumor immunology. *Clin Dev Immunol* 2011:439752.

Flickinger JC, Kondziolka D, Maitz AH, Lunsford LD (1998) Analysis of neurological sequelae from radiosurgery of arteriovenous malformations: How location affects outcome. *Int J Radiat Oncol Biol Phys* 40(2):273–278.

Flickinger JC, Kondziolka D, Pollock BE, Maitz AH, Lunsford LD (1997) Complications from arteriovenous malformation radiosurgery: Multivariate analysis and risk modeling. *Int J Radiat Oncol Biol Phys* 38(3):485–490.

Flickinger JC, Lunsford LD, Kondziolka D, Maitz AH, Epstein AH, Simons SR, Wu A (1992) Radiosurgery and brain tolerance: An analysis of neurodiagnostic imaging changes after Gamma Knife radiosurgery for arteriovenous malformations. *Int J Radiat Oncol Biol Phys* 23(1):19–26.

Flickinger JC et al. (2000) Development of a model to predict permanent symptomatic postradiosurgery injury for arteriovenous malformation patients: Arteriovenous malformation radiosurgery study group. *Int J Radiat Oncol Biol Phys* 46(5):1143–1148.

Fokas E, Henzel M, Surber G, Kleinert G, Hamm K, Engenhart-Cabillic R (2012) Stereotactic radiosurgery and fractionated stereotactic radiotherapy: Comparison of efficacy and toxicity in 260 patients with brain metastases. *J Neurooncol* 109(1):91–98.

Follwell MJ et al. (2012) Volume specific response criteria for brain metastases following salvage stereotactic radiosurgery and associated predictors of response. *Acta Oncol* 51(5):629–635.

Foote KD, Friedman WA, Buatti JM, Meeks SL, Bova FJ, Kubilis PS (2001) Analysis of risk factors associated with radiosurgery for vestibular schwannoma. *J Neurosurg* 95(3):440–449.

Gaspar L et al. (1997) Recursive partitioning analysis (RPA) of prognostic factors in three radiation therapy oncology group (RTOG) brain metastases trials. *Int J Radiat Oncol Biol Phys* 37(4):745–751.

Graber JJ, Cobbs CS, Olson JJ (2019) Congress of neurological surgeons systematic review and evidence-based guidelines on the use of stereotactic radiosurgery in the treatment of adults with metastatic brain tumors. *Neurosurgery* 84(3):E168–170.

Gregucci F et al. (2019) Linac-based radiosurgery or fractionated stereotactic radiotherapy with flattening filter-free volumetric modulated arc therapy in elderly patients. *Strahlentherapie und Onkologie* 195(3):218–225.

Gui C et al. (2018) Local recurrence patterns after postoperative stereotactic radiation surgery to resected brain metastases: A quantitative analysis to guide target delineation. *Pract Radiat Oncol* 8:388–396.

Hall EJ, Brenner DJ (1993) The radiobiology of radiosurgery: Rationale for different treatment regimes for AVMs and malignancies. *Int J Radiat Oncol Biol Phys* 25(2):381–385.

Han JH, Kim DG, Chung HT, Paek SH, Park CK, Jung HW (2012a) Radiosurgery for large brain metastases. *Int J Radiat Biol Oncol Phys* 83(1):113–120.

Han JH, Kim DG, Kim CY, Chung HT, Jung HW (2012b) Stereotactic radiosurgery for large brain metastases. *Prog Neurol Surg* 25:248–260.

Hanin LG, Zaider M (2010) Cell-survival probability at large doses: An alternative to the linear-quadratic model. *Phys Med Biol* 55(16):4687–4702.

Hasegawa T et al. (2017) Multisession Gamma Knife surgery for large brain metastases. *J Neurooncol* 131(3):517–524.

Higuchi Y et al. (2009) Three-staged stereotactic radiotherapy without whole brain irradiation for large metastatic brain tumors. *Int J Radiat Oncol Biol Phys* 74(5):1543–158.

Hiniker SM et al. (2016) Dose-response modeling of the visual pathway tolerance to single-fraction and hypofractionated stereotactic radiosurgery. *Semin Radiat Oncol* 26:97–104.

Hong CS, Deng D, Vera A, Chiang VL. (2019) Laser-interstitial thermal therapy compared to craniotomy for treatment of radiation necrosis or recurrent tumor in brain metastases failing radiosurgery. *J Neurooncol* 142:309–317.

Inoue HK et al. (2014a) Five-fraction CyberKnife radiotherapy for large brain metastases in critical areas: Impact on the surrounding brain volumes circumscribed with a single dose equivalent of 14 Gy (V14) to avoid radiation necrosis. *J Radiat Res* 55(2):334–342.

Inoue HK et al. (2014b) Optimal hypofractionated conformal radiotherapy for large brain metastases in patients with high risk factors: A single-institutional prospective study. *Radiat Oncol* 9:231.

Jeong WJ, Park JH, Lee EJ, Kim JH, Kim CJ, Cho YH (2015) Efficacy and safety of fractionated stereotactic radiosurgery for large brain metastases. *J Korean Neurosurg Soc* 58(3):217–224.

Jiang X et al. (2012) Hypofractionated stereotactic radiotherapy for brain metastases larger than three centimeters. *Radiat Oncol* 7(36).

Johnson JD, Young B (1996) Demographics of brain metastasis. *Neurosurg Clin North Am* 7(3):337–344.

Kang J, Ford EC, Smith K, Wong J, McNutt TR (2010). A method for optimizing LINAC treatment geometry for volumetric modulated arc therapy of multiple brain metastases. *Med Phys* 37(8):4146–4154.

Kano H et al. (2012) Stereotactic radiosurgery for arteriovenous malformations, part 5: Management of brainstem arteriovenous malformations. *J Neurosurg* 116(1):44–53.

Keller A et al. (2017) Hypofractionated stereotactic radiation therapy to the resection bed for intracranial metastases *Int J Radiat Oncol Biol Phys* 99(5):1179–1189.

Kim JW et al. (2016) Fractionated stereotactic Gamma Knife radiosurgery for large brain metastases: A retrospective, single center Study. *PLoS One* 11(9):e0163304.

Kim YJ et al. (2011) Single-dose versus fractionated stereotactic radiotherapy for brain metastases. *Int J Radiat Oncol Biol Phys* 81(2):483–489.

Kirkpatrick JP, Brenner DJ, Orton CG (2009) Point/counterpoint: The linear-quadratic model is inappropriate to model high dose per fraction effects in radiosurgery. *Med Phys* 36(8):3381–3384.

Kirkpatrick JP, Meyer JJ, Marks LB (2008) The linear-quadratic model is inappropriate to model high dose per fraction effects in radiosurgery. *Semin Radiat Oncol* 18(4):240–243.

Kirkpatrick JP et al. (2015) Defining the optimal planning target volume in image-guided stereotactic radiosurgery of brain metastases: Results of a randomized trial. *Int J Radiat Oncol Biol Phys* 91(1):100–108.

Kocher M et al. (2011) Adjuvant whole-brain radiotherapy versus observation after radiosurgery or surgical resection of one to three cerebral metastases: Results of the EORTC 22952–26001 study. *J Clin Oncol* 29(2):134–141.

Korytko T et al. (2006) 12 Gy Gamma Knife radiosurgical volume is a predictor for radiation necrosis in non-avm intracranial tumors. *Int J Radiat Oncol Biol Phys* 64(2):419–424.

Koyfman SA et al. (2010) Stereotactic radiosurgery for single brainstem metastases: The cleveland clinic experience. *Int J Radiat Oncol Biol Phys* 78(2):409–414.

Kwon AK, Dibiase SJ, Wang B, Hughes SL, Milcarek B, Zhu Y (2009) Hypofractionated stereotactic radiotherapy for the treatment of brain metastases. *Cancer* 115(4):890–898.

Lau SKM et al. (2015) Single-isocenter frameless volumetric modulated arc radiosurgery for multiple intracranial metastases. *Neurosurgery* 77(2):233–240.

Lawrence YR et al. (2010) Radiation dose-volume effects in the brain. *Int J Radiat Oncol Biol Phys* 76(3 Suppl):S20–27.

Leber KA, Bergloff J, Pendl G (1998) Dose-response tolerance of the visual pathways and cranial nerves of the cavernous sinus to stereotactic radiosurgery. *J Neurosurg* 88(1):43–50.

Levegrun S, Hof H, Essig M, Schlegel W, Debus J (2004) Radiation-induced changes of brain tissue after radiosurgery in patients with arteriovenous malformations: Correlation with dose distribution parameters. *Int J Radiat Oncol Biol Phys* 59(3):796–808.

Likhacheva A et al. (2013) Predictors of survival in contemporary practice after initial radiosurgery for brain metastases. *Int J Radiat Oncol Biol Phys* 85(3):656–661.

Lim TK, Kim WK, Yoo CJ, Kim EY, Kim MJ, Yee GT (2018) Fractionated stereotactic radiosurgery for brain metastases using the novalis Tx(R) system. *J Korean Neurosurg Soc* 61(4):525–529.

Lin NU, Winer EP (2007) Brain metastases: The HER2 paradigm. *Clin Cancer Res* 13(6):1648–1655.

Lindvall P, Bergstrom P, Lofroth PO, Tommy Bergenheim A (2009) A comparison between surgical resection in combination with WBRT or hypofractionated stereotactic irradiation in the treatment of solitary brain metastases. *Acta Neurochir* 151(9):1053–1059.

Ling DC et al. (2015) Postoperative stereotactic radiosurgery to the resection cavity for large brain metastases: Clinical outcomes, predictors of intracranial failure, and implications for optimal patient selection. *Neurosurgery* 76(2):150–156; discussion 6–7; quiz 7.

Ma J, Chang Z, Wang Z, Wu QJ, Kirkpatrick JP, Yin FF (2009) ExacTrac X-ray 6 degree-of-freedom image-guidance for intracranial non-invasive stereotactic radiotherapy: Comparison with kilo-voltage cone-beam CT. *Radiother Oncol* 93(3):602–608.

Mahajan A et al. (2017) Post-operative stereotactic radiosurgery versus observation for completely resected brain metastases: A single-centre, randomised, controlled, phase 3 trial *Lancet Oncol* 18(8):1040–1048.

Manning MA et al. (2000) Hypofractionated stereotactic radiotherapy as an alternative to radiosurgery for the treatment of patients with brain metastases. *Int J Radiat Oncol Biol Phys* 47(3):603–608.

Marcrom SR et al. (2017) Fractionated stereotactic radiation therapy for intact brain metastases. *Adv Radiat Oncol* 2:564–571.

Mayo C, Martel MK, Marks LB, Flickinger J, Nam J, Kirkpatrick J (2010) Radiation dose-volume effects of optic nerves and chiasm. *Int J Radiat Oncol Biol Phys* 76(3 Suppl):S28–35.

Mehta MP et al. (2005) The American society for therapeutic radiology and oncology (ASTRO) evidence-based review of the role of radiosurgery for brain metastases. *Int J Radiat Oncol Biol Phys* 63(1):37–46.

Milano MT et al. (2018) Single- and multi-fraction stereotactic radiosurgery dose tolerances of the optic pathways. *Int J Radiat Oncol Biol Phys* https://doi.org/10.1016/j.ijrobp.2018.01.053

Minniti G et al. (2011) Stereotactic radiosurgery for brain metastases: Analysis of outcome and risk of brain radionecrosis. *Radiat Oncol* 6:48.

Minniti G et al. (2013) Multidose stereotactic radiosurgery (9 Gy x 3) of the postoperative resection cavity for treatment of large brain metastases. *Int J Radiat Oncol Biol Phys* 86(4):623–629.

Minniti G et al. (2014a) Fractionated stereotactic radiosurgery for patients with brain metastases. *J Neurooncol* 117(2):295–301.

Minniti G et al. (2014b) Fractionated stereotactic radiosurgery for patients with skull base metastases from systemic cancer involving the anterior visual pathway. *Radiat Oncol* 9:110.

Minniti G et al. (2016) Single-fraction versus multifraction (3 x 9 Gy) stereotactic radiosurgery for large (>2 cm) brain metastases: A comparative analysis of local control and risk of radiation-induced brain necrosis. *Int J Radiat Biol Phys* 95(4):1142–1148.

Minniti G et al. (2017) Outcomes of postoperative stereotactic radiosurgery to the resection cavity versus stereotactic radiosurgery alone for melanoma brain metastases. *J Neurooncol* 132(3):455–462.

Minniti G et al. (2019)

Mintz AH et al. (1996) A randomized trial to assess the efficacy of surgery in addition to radiotherapy in patients with a single cerebral metastasis. *Cancer* 78(7):1470–1476.

Moravan MJ et al. (2020) Current multidisciplinary management of brain metastases. *Cancer* 126(7):1390–1406.

Mori Y, Kondziolka D, Flickinger JC, Kirkwood JM, Agarwala S, Lunsford LD (1998) Stereotactic radiosurgery for cerebral metastatic melanoma: Factors affecting local disease control and survival. *Int J Radiat Oncol Biol Phys* 42(3):581–589.

Morrison J et al. (2016) Is a single isocenter sufficient for volumetric modulated arc therapy radiosurgery when multiple intracranial metastases are spatially dispersed? *Med Dosim* 41(4):285–289.

Muacevic A, Wowra B, Siefert A, Tonn JC, Steiger HJ, Kreth FW. (2008) Microsurgery plus whole brain irradiation versus Gamma Knife surgery alone for treatment of single metastases to the brain: A randomized controlled multicentre phase III trial. *J Neurooncol* 87(3):299–307.

Murai T et al. (2014) Fractionated stereotactic radiotherapy using CyberKnife for the treatment of large brain metastases: A dose escalation study. *Clin Oncol (R Coll Radiol)* 26(3):151–158.

Navarria P et al. (2016) Hypofractionated stereotactic radiotherapy alone using volumetric modulated arc therapy for patients with single, large brain metastases unsuitable for surgical resection. *Radiat Oncol* 1:76.

Nayak L, Lee EQ, Wen PY (2012) Epidemiology of brain metastases. *Curr Oncol Rep* 14(1):48–54.

Noordijk EM et al. (1994) The choice of treatment of single brain metastasis should be based on extracranial tumor activity and age. *Int J Radiat Oncol Biol Phys* 29(4):711–717.

Ohira S et al. (2018) HyperArc VMAT planning for single and multiple brain metastases stereotactic radiosurgery: A new treatment planning approach. *Radiat Oncol* 13:13–21.

Pant K, Umeh C, Oldham M, Floyd S, Giles W, Adamson J (2020) Comprehensive radiation and imaging isocenter verification using NIPAM kV-CBCT dosimetry. *Med Phys* 47(3):927–936.

Park HJ, Griffin RJ, Hui S, Levitt SH, Song CW (2012) Radiation-induced vascular damage in tumors: Implications of vascular damage in ablative hypofractionated radiotherapy (SBRT and SRS). *Radiat Res* 177(3):311–327.

Patchell RA et al. (1990) A randomized trial of surgery in the treatment of single metastases to the brain. *N Engl J Med* 322(8):494–500.

Patchell RA et al. (1998) Postoperative radiotherapy in the treatment of single metastases to the brain: A randomized trial. *JAMA* 280(17):1485–1489.

Patel S et al. (2014) Risk of brain metastasis in EGFR-mutant NSCLC treated with erlotinib: A role for prophylactic cranial irradiation? *Int J Radiat Oncol Biol Phys* 90(1s):S643–644.

Patrikidou A et al. (2020) Development of a disease-specific graded prognostic assessment index for the management of sarcoma patients with brain metastases (Sarcoma-GPA). *BMC Cancer* 20(1):117.

Pessina F et al. (2016) Outcome evaluation of oligometastatic patients treated with surgical resection followed by hypofractionated stereotactic radiosurgery (HSRS) on the tumor bed, for single, large brain metastases. *PLoS One* 11(6):e0157869.

Prabhu RS et al. (2017) Single-fraction stereotactic radiosurgery (SRS) alone versus surgical resection and SRS for large brain metastases: A multi-institutional analysis. *Int J Radiat Oncol Biol Phys* 99(2):459–467.

Pollock BE, Cochran J, Natt N, Brown PD, Erickson D, Link MJ, Garces YI, Foote RL, Stafford SL, Schomberg PJ (2008) Gamma Knife radiosurgery for patients with nonfunctioning pituitary adenomas: Results from a 15-year experience. *Int J Radiat Oncol Biol Phys* 70(5):1325–1329.

Rajakesari S et al. (2014) Local control after fractionated stereotactic radiation therapy for brain metastases. *J Neurooncol* 120(2):339–346.

Regine WF, Scott C, Murray K, Curran W (2001) Neurocognitive outcome in brain metastases patients treated with accelerated-fractionation vs. accelerated-hyperfractionated radiotherapy: An analysis from radiation therapy oncology group study 91–04. *Int J Radiat Oncol Biol Phys* 51(3):711–717.

Roper J, Chanyavanich V, Betzel G, Switchenko J, Dhabaan A (2015) Single-isocenter multiple-target stereotactic radiosurgery: Risk of compromised coverage. *Int J Radiat Oncol Biol Phys* 93(3):540–546.

Sahgal A et al. (2015) Phase 3 trials of stereotactic radiosurgery with or without whole-brain radiation therapy for 1 to 4 brain metastases: Individual patient data meta-analysis. *Int J Radiat Oncol Biol Phys* 91(4):710–717.

Serizawa T et al. (2018) Comparison of treatment results between 3- and 2-stage Gamma Knife radiosurgery for large brain metastases: A retrospective multi-institutional study. *J Neurosurg* 1–11.

Shaw E et al.(2000) Single dose radiosurgical treatment of recurrent previously irradiated primary brain tumors and brain metastases: Final report of RTOG protocol 90–05. *Int J Radiat Oncol Biol Phys* 47(2):291–298.

Shi S et al. (2020) Stereotactic radiosurgery for resected brain metastsases: Single-institutional experience of over 500 cavities. *Int J Radiat Oncol Biol Phys* 106(4):764–771.

Shiau CY et al. (1997) Radiosurgery for brain metastases: Relationship of dose and pattern of enhancement to local control. *Int J Radiat Oncol Biol Phys* 37(2):375–383.

Shultz DB et al. (2015) Repeat courses of stereotactic radiosurgery (SRS), deferring whole-brain irradiation, for new brain metastases after initial SRS. *Int J Radiat Oncol Biol Phys* 92(5):993–999.

Slimane K et al. (2004) Risk factors for brain relapse in patients with metastatic breast cancer. *Ann Oncol* 15(11):1640–1644.

Soliman H et al. (2017) Consensus contouring guidelines for postoperative completely resected cavity stereotactic radiosurgery for brain metastases. *Int J Radiat Oncol Biol Phys* 100(2):436–442.

Song CW, Cho LC, Yuan J, Dusenbery KE, Griffin RJ, Levitt SH (2013) Radiobiology of stereotactic body radiation therapy/stereotactic radiosurgery and the linear-quadratic model. *Int J Radiat Oncol Biol Phys* 87(1):18–19.

Sperduto PW, Berkey B, Gaspar LE, Mehta M, Curran W (2008) A new prognostic index and comparison to three other indices for patients with brain metastases: An analysis of 1,960 patients in the RTOG database. *Int J Radiat Oncol Biol Phys* 70(2):510–514.

Sperduto PWet al. (2010) Diagnosis-specific prognostic factors, indexes, and treatment outcomes for patients with newly diagnosed brain metastases: A multi-institutional analysis of 4,259 patients. *Int J Radiat Oncol Biol Phys* 77(3):655–661.

Sperduto PW et al. (2012) Summary report on the graded prognostic assessment: An accurate and facile diagnosis-specific tool to estimate survival for patients with brain metastases. *J Clin Oncol* 30(4):419–425.

Sperduto PW et al. (2017a) Estimating survival in melanoma patients with brain metastases: An update of the graded prognostic assessment for melanoma using molecular markers (melanoma-molGPA). *Int J Radiat Oncol Biol Phys* 99(4):812–816.

Sperduto PW et al. (2017b) Estimating survival in patients with lung cancer and brain metastases: An update of the graded prognostic assessment for lung cancer using molecular markers (lung-molGPA). *JAMA Oncol* 7(3):827–831.

Sperduto PW et al. (2018) Estimating survival for renal cell carcinoma patients with brain metastases: An update of the renal graded prognostic assessment tool. *Neuro Oncol* 20(12):1652–1660.

Sperduto PW et al. (2019) Estimating survival in patients with gastrointestinal cancers and brain metastases: An update of the graded prognostic assessment for gastrointestinal cancers (GI-GPA). *Clin Transl Radiat Oncol* 27(18):39–45.

Spiegelmann R, Lidar Z, Gofman J, Alezra D, Hadani M, Pfeffer R (2001) Linear accelerator radiosurgery for vestibular schwannoma. *J Neurosurg* 94(1):7–13.

Stafford SL et al. (2003) A study on the radiation tolerance of the optic nerves and chiasm after stereotactic radiosurgery. *Int J Radiat Oncol Biol Phys* 55(5):1177–1181.

Stanhope C et al. (2016) Physics considerations for single-isocenter, volumetric modulated arc radiosurgery for treatment of multiple intracranial targets. *Pract Radiat Oncol* 6(3):207–213.

Steinmann D et al. (2012) Hypofractionated stereotactic radiotherapy (hfSRT) after tumour resection of a single brain metastasis: Report of a single-centre individualized treatment approach. *J Cancer Res Clin Oncol* 138:1523–1529.

Sundermeyer ML, Meropol NJ, Rogatko A, Wang H, Cohen SJ (2005) Changing patterns of bone and brain metastases in patients with colorectal cancer. *Clin Colorectal Cancer* 5(2):108–113.

Tallet AV et al. (2012) Neurocognitive function impairment after whole brain radiotherapy for brain metastases: Actual assessment. *Radiat Oncol* 7:77.

Thomas A et al. (2013) A comprehensive investigation of the accuracy and reproducibility of a multitarget single isocenter VMAT radiosurgery technique. *Med Phys* 40(12):121725.

Trifiletti DM et al. (2016) Stereotactic radiosurgery for brainstem metastases: An international cooperative study to define response and toxicity. *Int J Radiat Oncol Biol Phys* 96(2):280–288.

Vogel J et al. (2015) Intracranial control after Cyberknife radiosurgery to the resection bed for large brain metastases. *Radiat Oncol* 10:221.

Vogelbaum MA, Angelov L, Lee SY, Li L, Barnett GH, Suh JH (2006) Local control of brain metastases by stereotactic radiosurgery in relation to dose to the tumor margin. *J Neurosurg* 104(6):907–912.

Voges J et al. (1996) Risk analysis of linear accelerator radiosurgery. *Int J Radiat Oncol Biol Phys* 36(5):1055–1063.

Wang CC et al. (2012) Cyberknife hypofractionated stereotactic radiosurgery (HSRS) of resection cavity after excision of large cerebral metastasis: Efficacy and safety of an 800 cGy x 3 daily fractions regimen. *J Neurooncol* 106(3):601–610.

Wegner RE et al. (2015) Fractionated stereotactic radiosurgery for large brain metastases. *Am J Clin Oncol* 38(2):135–159.

Wu Q et al. (2016). Optimization of treatment geometry to reduce normal brain dose in radiosurgery of multiple brain metastases with single-isocenter volumetric modulated arc therapy. *Sci Rep* 6:34511.

Yamamoto M et al. (2014) Stereotactic radiosurgery for patients with multiple brain metastases (JLGK0901): A multi-institutional prospective observational study. *Lancet Oncol* 15(4):387–395.

Yang HC, Kano H, Lunsford LD, Niranjan A, Flickinger JC, Kondziolka D (2011) What factors predict the response of larger brain metastases to radiosurgery? *Neurosurgery* 68(3):682–690; discussion 690.

Yomo S, Hayashi M. (2014) A minimally invasive treatment option for large metastatic brain tumors: Long-term results of two-session Gamma Knife stereotactic radiosurgery. *Radiat Oncol* 9:132.

Yomo S, Hayashi M, Nicholson C (2012) A prospective pilot study of two-session Gamma Knife surgery for large metastatic brain tumors. *J Neurooncol* 109(1):159–165.

Zhao B et al. (2014) Prescription to 50–75% isodose line may be optimum for linear accelerator based radiosurgery of cranial lesions. *J Radiosurg SBRT* 3(2):139.

Zhong J et al. (2017) Postoperative stereotactic radiosurgery for resected brain metastases: A comparison of outcomes for large resection cavities. *Pract Radiat Oncol* 7(6):e419–425.

Zimmerman AL et al. (2016) Treatment of large brain metastases with stereotactic radiosurgery. *Technol Cancer Res Treat* 15(1):186–195.

15

Image-Guided Hypofractionated Stereotactic Whole-Brain Radiotherapy and Simultaneous Integrated Boost for Brain Metastases

Alan Nichol

Contents

15.1 INTRODUCTION

The Radiation Therapy Oncology Group (RTOG) 9508 study showed that whole-brain radiotherapy (WBRT) followed by a stereotactic radiosurgery (SRS) boost, compared to WBRT alone, resulted in better survival for patients with solitary brain metastases and for patients with good prognosis (Andrews et al., 2004; Sperduto et al., 2014). The development of helical tomotherapy and volumetric modulated arc therapy (VMAT) led to an interest in using these technologies to deliver the WBRT and boost the metastases simultaneously (Bauman et al., 2007; Gutierrez et al., 2007; Hsu et al., 2010). This technique for treating the whole brain with a concurrent boost to the brain metastases is often referred to as WBRT and simultaneous integrated boost (SIB). This chapter will review the issue of using WBRT for selected patients, summarize publications about the various WBRT and SIB prescriptions, and describe the delivery of WBRT and SIB in 5 fractions.

WBRT has been shown to reduce the risk of local and distant recurrences of brain metastases when combined with SRS compared to SRS alone (Aoyama et al., 2006; Chang et al., 2009; Kocher et al., 2011; Sahgal et al., 2015; Churilla et al., 2017). According to the Cochrane Database of Systematic Review of WBRT+SRS versus SRS alone, WBRT decreased the relative risk of any intracranial disease progression at 1 year by 53% (HR 0.47, 95% CI 0.34–0.66, $p < 0.0001$), but there was no significant difference in overall

291

survival (HR 1.11, 95% CI 0.83–1.48, p = 0.47) (Soon et al., 2014). The randomized clinical trials also showed that focal treatment was associated with less alopecia, less fatigue, less cognitive impairment, and better quality of life as compared to treatment including WBRT, although the quality of the evidence was judged to be low (Patchell et al., 1998; Aoyama et al., 2006, 2007; Roos et al., 2006; Chang et al., 2009; Kocher et al., 2011; McDuff et al., 2013; Soffietti et al., 2013).

The best evidence for the neurocognitive effects of WBRT comes from studies of prophylactic cranial irradiation (PCI) because they isolate the effects of WBRT from those of palliative chemotherapy and progression or recurrence of brain metastases (McDuff et al., 2013). Gondi et al. reported on a combined analysis of the RTOG 0212 and 0214 studies. Of the patients who completed the baseline cognitive testing, 410 received PCI and 173 were observed (Gondi et al., 2013). This analysis showed that PCI was significantly associated with declines in self-reported cognitive functioning (p < 0.0001) and Hopkins Verbal Learning Test–Recall (p = 0.002) at 6 and 12 months. The RTOG 0212 study compared two PCI prescriptions (Wolfson et al., 2011). Significant declines on one or more cognitive tests had an incidence of 62% in the 25 Gy arm and 85% in the 36 Gy arm. Age was shown to predispose patients to a higher risk of new neurocognitive impairment with 36 Gy, compared to 25 Gy PCI in the RTOG 0212 trial (Wolfson et al., 2011). In a multivariable model, age over 60 predicted for a decline in the Hopkins verbal learning test-delayed recall at 12 months in Gondi's RTOG 0212 and 0214 analysis (Gondi et al., 2013).

WBRT diminishes neurocognitive function, but so too does poor control of brain metastases (Regine et al., 2001; Meyers et al., 2004; Aoyama et al., 2007; Li et al., 2007). Regine et al. reported a significant drop in Mini-Mental Status Examination (MMSE) scores in patients whose brain metastases were not controlled by WBRT (Regine et al., 2001). Progression of treated metastases was shown to diminish neurocognitive function in an exploratory analysis of a randomized controlled trial of WBRT with or without motexafin gadolinium. Meyers et al. (2004) demonstrated that patients with partial responses were more likely to have improvement of their cognition on a battery of eight tests than patients with progressive disease. Li et al. (2007) assessed volumetric response of brain metastases treated with WBRT and found that better volumetric response correlated with better neurocognitive function. The median time to cognitive impairment was longer in good responders than in poor responders on all eight cognitive tests used in the study (p = 0.008). In their randomized controlled trial, Aoyama et al. observed that three-point declines in MMSE were delayed from 7.6 months in the SRS alone arm to 16.5 months in the WBRT+SRS arm (p = 0.05), which they attributed to better local and distant control of brain metastases, most evident in the first 2 years after treatment (Aoyama et al., 2007).

The American Society for Radiation Oncology issued a Choosing Wisely® statement: "Don't routinely add whole-brain radiotherapy to stereotactic radiosurgery for limited brain metastases" because of the absence of a survival benefit with WBRT and its known side effects (American Society of Radiation Oncology, 2014). The Choosing Wisely® campaign is intended to encourage more detailed conversations between physicians and patients about treatment options. The conversation about WBRT and SIB should include a frank discussion about the patient's prognosis and the likelihood of side effects from WBRT, compared to the likelihood of complications that could arise from intracranial relapse. There are a number of characteristics of patients that predispose them to a high risk of new brain metastases and need for salvage treatment, both of which can be reduced with adjuvant WBRT. Given its known toxicities, the use of WBRT must be selective (Abe and Aoyama, 2012; Mehta, 2015; Sahgal, 2015). The following section presents a discussion about the patients for whom the risk/benefit ratio may favor WBRT.

15.2 INDICATIONS FOR WBRT

Selected patients at high risk of new and early new metastases may be suited to adjuvant WBRT. Rodrigues et al. (2014) and Ayala-Peacock et al. (2014) published nomograms that identify patients at high risk of intracranial relapse. These tools can be employed to make individualized treatment recommendations to patients about adjuvant WBRT regarding their personal risk of new metastases and the need for salvage therapies.

Rodrigues et al. (2014) studied the incidence of new brain metastases at 1 year in a cohort of patients treated with SRS alone for one to three metastases. Their univariate analysis showed that younger patients with smaller gross tumor volumes (GTV), two to three brain metastases, and better performance status were more likely to have new metastases at 1 year. The odds ratio for new metastases was 0.78 (95% CI, 0.63–0.95) for each increase in age of 10 years, indicating that older patients had a lower risk of new metastases. Worse performance status also conferred a lower risk of new metastases: World Health Organization Performance Status (WHO-PS) = 0 versus 1, odds ratio = 0.50 (95% CI, 0.29–0.87), and 2 versus 3, odds ratio = 0.24 (95% CI, 0.11–0.56). On multivariable analysis, patients with larger GTVs also had a lower risk of new metastases: odds ratio = 0.66 (95% CI, 0.44–0.99). Patients with two to three lesions had a higher risk of new metastases than patients with only one lesion: odds ratio = 2.27 (95% CI, 1.37–3.75). Thus, the risk of new metastases at 1 year generally increased with predictors of longer survival, presumably because the incidence of new brain metastases competes with the risk of death from extracranial disease. Their recursive partitioning analysis identified a high-risk subgroup of patients with WHO-PS = 0 and two to three metastases, with a cumulative incidence of new metastases at 1 year of almost 70%, and an intermediate risk group with a cumulative incidence of new metastases of almost 50%.

Ayala-Peacock et al. (2014) performed a similar retrospective study on 464 patients treated with SRS alone for up to 13 brain metastases. The risk of new brain metastases was significantly increased with a higher baseline number of brain metastases, progressive extracranial disease, detection of new metastases on the SRS planning MRI, and high-risk histologies: melanoma and Her2-negative breast cancer. Although salvage WBRT was only required in 30% of the patients before death, there were some patients who had a high risk of needing early WBRT. For example, salvage WBRT was required at a median of 3.3 months for patients with melanoma and at a median of 3.0 months for patients with poorly differentiated lung cancer.

Patients' long-term risk of dementia from whole-brain radiotherapy rises with advancing age and better prognosis from their extracranial disease. Considering the relative risks of dementia and new metastases suggests that patients with a median life expectancy of about 6 months and a high risk of new metastases are good candidates for WBRT with SRS or WBRT and SIB. WBRT and SIB should not be used for patients who are eligible for targeted systemic therapies because of their prolonged survival (Nichol et al., 2016.

15.3 STUDIES OF WBRT AND SIB

There are several radiotherapy systems that can deliver WBRT and SIB. Helical tomotherapy has been used in both planning studies and clinical trials (Bauman et al., 2007; Gutierrez et al., 2007; Sterzing et al., 2009; Edwards et al., 2010; Rodrigues et al., 2011, 2019). VMAT can also deliver WBRT and SIB (Hsu et al., 2010; Lagerwaard et al., 2009; Weber et al., 2011; Awad et al., 2013; Oehlke et al., 2015; Nichol et al., 2016). Reports of clinical experience with WBRT and SIB are summarized in Table 15.1.

Sterzing et al. (2009) retreated two patients with WBRT and SIB using helical tomotherapy for relapsed brain metastases following prior 40 Gy in 20 fractions of WBRT. One patient had 8 new metastases and the other had 11 new metastases. The prescriptions were 15 Gy in 10 fractions to the whole brain and 30 Gy in 10 fractions to a 2-mm planning target volume (PTV) on the brain metastases. The authors reported local control in both patients at 6 and 12 months of follow-up.

Edwards et al. (2010) treated 11 patients with WBRT and SIB using helical tomotherapy for one to four brain metastases measuring up to 8 cm in diameter. The prescriptions were 30 Gy in 10 fractions to the whole brain and 40 Gy in 10 fractions to the metastases. The whole-brain PTV was an expansion of the brain by 3 mm. The brain metastases were treated with a 0 mm GTV–PTV margin. All patients received corticosteroids during the treatment. All the lesions exhibited 1-month local responses, and the patients experienced no complications.

Lagerwaard et al. (2009) treated three patients with WBRT and SIB using VMAT for one to three new brain metastases. The prescriptions employed were 20 Gy in 5 fractions to the whole-brain PTV and 40 Gy in 5 fractions to the metastasis PTVs. Both the brain and the GTVs were expanded by a 2-mm margin to create the PTVs. The patients were positioned using cone-beam CT (CBCT). All head rotations over 0.8° were corrected clinically by repositioning the patients and repeating the CBCT. All translational corrections were applied. A follow-up publication reported on 50 patients treated using a robotic couch capable

Table 15.1 Summary of whole-brain radiotherapy and simultaneous integrated boost results

STUDY	NUMBER OF SUBJECTS	NUMBER OF METASTASES MEDIAN (RANGE)	DIAMETER OR / VOLUME OF METASTASES	GTV–PTV MARGIN (MM)	STUDY ENDPOINT	RESULT	MEDIAN SURVIVAL (MONTHS)	WBRT DOSE (GY)/ FRACTIONS	SIB DOSE (GY)/ FRACTIONS
Lagerwaard (Lagerwaard et al., 2009)	3	3 (1–3)	1.5–25.8 cm^3	2	NR	NR	NR	20/5	40/5
Sterzing (Sterzing et al., 2009)	2	8–11	NR	2	Reirradiation	NR	NR	15/10	30/10
Awad (Awad et al., 2013)[a]	30	2 (1–8)	Mean 6.9 cm^3	2	3.5 month LC	81%	9.4	28.6–37.5/15	50–70.8/15
Edwards (Edwards et al., 2010)	11	(1–4)	2.5–8 cm	3	3 month LC	100%	>4	30/10	40/10
RodriguesPhase I (Rodrigues et al., 2011)	48	<3 cm	<3 cm	0	60/10	Maximum tolerated dose	5.3	30/10	35–60/10
Rodrigues and Lagerwaard (Rodrigues et al., 2012)[b]	120	2 (1–6)	<3 cm	Various	1 year	~67%[c]	5.9	30/10 20/5	35–60/10 40/5
Weber (Weber et al., 2011)	29	(1–4)	<40 cm^3	NR	6-month PFS	78%	>6	30/10	40/10
Oehlke (Oehlke et al., 2015)[a]	20	5 (2–13)	0.78 ±1.17 cm^3	1	1-year local control	~64%[c]	16.5	30/12	51/12
Kim (Kim et al., 2015)[a]	11	4 (2–15)	<10.1 cm^3	0	1-year brain control	67%	14.5	25–28/10–14	40–48/10–14
Nichol Phase II (Nichol et al., 2016)	60	3 (1–10)	<3 cm	2	3-month freedom from progression	90%	10.1	20/5	38/5

(Continued)

Table 15.1 (*Continued*) **Summary of whole-brain radiotherapy and simultaneous integrated boost results**

STUDY	NUMBER OF SUBJECTS	NUMBER OF METASTASES MEDIAN (RANGE)	DIAMETER OR / VOLUME OF METASTASES	GTV— PTV MARGIN (MM)	STUDY ENDPOINT	RESULT	MEDIAN SURVIVAL (MONTHS)	WBRT DOSE (GY)/ FRACTIONS	SIB DOSE (GY)/ FRACTIONS
Ferro Phase I (2017)	30	1 (2–4)	<3 cm	3	50/10	Maximum tolerated dose	12 months	30/10	35–50/10
Yang et al. (2017)	24	2 (1–3)	2.9 (0.6–5.9) cm	3	Overall response rate	66.7%	8.0	30/10	40/10
Rodrigues Phase II (2019)	60	2 (1–3)	1.3 (0.2–3.0) cm	0	6-month actuarial local control	89%	5.4	30/10	60/10
Zhong et al. (2020)	13	6 (2–10)	2.4 (0.6–49.9) cm³	2	1-year local control	92%	8.6	25/10 or 37.5/15	45/10 or 52.5/15

[a] Hippocampal avoidance.
[b] Combined cohorts—including all 48 patients from Rodrigues' phase I study.
[c] Estimated from a figure.
Abbreviations: NR, not reported; LC, local control; PFS, progression-free survival; GTV, gross tumor volume; PTV, planning target volume.

of correcting patient positioning in six degrees of freedom. The crude rate of radionecrosis reported in this subgroup was 6% (3/50) (Rodrigues et al., 2012).

Rodrigues et al. (2011) reported a phase I trial of 30 Gy/10 WBRT and increasing SIB doses from 35 to 60 Gy in 5 Gy increments to the brain metastasis GTVs with no margin. Dose-limiting toxicity was not observed at the final dose level of 60 Gy. A 60-patient phase II study of the final dose level (60 Gy/10 SIB) from the phase I study has been conducted (Rodrigues et al., 2019). The prescription dose was delivered to the GTVs with no margin using helical tomotherapy. With a median survival of 5.4 months, only two patients developed asymptomatic imaging changes consistent with radiation necrosis.

Weber et al. (2011) treated 29 patients with one to four brain metastases with 30 Gy in 10 fractions to the whole brain and 40 Gy in 10 fractions to the metastases using VMAT. The 6-month overall survival was 55%.

Awad et al. (2013) reported on 30 patients treated with VMAT using a variety of planning techniques and fractionation schedules: 17 received hippocampal avoidance (HA)-WBRT and SIB and 5 received WBRT and SIB. Twenty-six of the patients had melanoma. The hippocampus was contoured according to the RTOG 0933 study guidelines (Gondi et al., 2010, 2014) and was expanded to create an HA volume. The median dose to the whole brain was 30 Gy in 15 fractions and to the brain metastases was 50 Gy in 15 fractions. Overall survival for the 22 patients treated with WBRT and SIB was not separately reported, but it was 9.4 months for the whole cohort.

Oehlke et al. (2015) reported on the first 20 patients treated with HA-WBRT and SIB using VMAT in a single-institution study. The whole-brain prescription was 30 Gy in 12 fractions, and the boost prescription was 51 Gy in 12 fractions. The hippocampus was contoured according to the RTOG 0933 study guidelines (Gondi et al., 2010, 2014). The whole-brain PTV was an expansion of the brain parenchyma by 3 mm; however, the PTV for brain metastases was an expansion of the GTVs by 1 mm. The treatment was well tolerated. Grade 1 and 2 treatment-related toxicities were commonly observed, but only one patient developed radionecrosis requiring surgery (grade 4 toxicity). The median intracranial progression-free survival was 9.2 months, and the median overall survival was 16.6 months.

Kim et al. (2015) reported on 11 patients with metastatic lung cancer treated with HA-WBRT and SIB using helical tomotherapy for 70 brain metastases (Kim et al., 2015). They delivered 25–28 Gy to the whole brain and 40–48 Gy to metastases with no margin in 10–14 fractions. The mean dose to the hippocampus was 13.7 Gy. The intracranial control rate at 1 year was 67%.

At BC Cancer, a phase 2 study with 60 patients was conducted using VMAT to deliver 20 Gy/5 WBRT and 47.5 Gy/5 SIB to brain metastases without a margin and 38 Gy/5 SIB to PTVs with a 2 mm GTV to PTV margin (Nichol et al., 2016). The median survival was 10.1 months. Freedom from in-brain progression was 90% at 3 months. The crude incidence of symptomatic radionecrosis was 12% in the 60 patients. The incidence of radionecrosis with this boost prescription was unacceptable for deep-brain metastases (Flickinger et al., 1998; Wegner et al., 2011). The SIB prescription of 47.5 Gy/5 to metastases with no margin that was used in our study is not recommended for metastases in the brainstem, thalamus, and basal ganglia.

Ferro et al. treated 30 patients with 30 Gy/10 WBRT in a phase I study, which determined that 50 Gy/10 SIB was the maximum tolerated dose for brain metastases treated with a 3-mm margin (Ferro et al., 2017). Yang et al. treated 24 RPA class 3 patients with 40 Gy/10 fractions SIB to brain metastases with a 3-mm margin, along with 30 Gy/10 WBRT and found that this SIB dose was well tolerated (Yang et al., 2017). Rodrigues et al. used helical tomotherapy to deliver 60 Gy/10 SIB with no margin along with 30 Gy/10 WBRT to 87 patients. They determined that this treatment was well tolerated and offered excellent 6-month local control. (Rodrigues et al., 2019). Zhong et al. treated 11 out of 13 patients with 25 Gy/10 WBRT and 45 Gy/10 fractions SIB as well as two patients with 37.5 Gy/15 WBRT and 52.5 Gy/15 SIB (Zhong et al., 2020). They reported a remarkable 1-year local control of 92%.

15.4 CHOOSING A PRESCRIPTION FOR WBRT AND SIB

A 5-fraction course of WBRT and SIB was selected at our center because it was more convenient for patients than longer courses, and it could be delivered during the second week of a 3-week cycle of

Table 15.2 **Equivalent doses in 2 Gy fractions (EQD2) for whole-brain radiotherapy prescriptions**

WHOLE-BRAIN RADIOTHERAPY DOSE (GY)	FRACTIONS	EQD2 (GY$_3$)	EQD2 (GY$_{10}$)
20	5	28	23
25	10	28	26
30	10	36	33
37.5	15	41	39
40	20	40	40

Table 15.3 **Equivalent doses in 2 Gy fractions (EQD2) for simultaneous integrated boosts, compared to the radiosurgery dosing schema of the Radiation Therapy Oncology Group 9508 study**

INDICATIONS	BOOST (GY)	FRACTIONS	EQD2 (GY$_3$)	EQD2 (GY$_{10}$)
Brainstem metastases, brain metastases 3–4 cm, and 2–3 cm metastases in deep, eloquent brain	15	1	54	31
	30	5	54	40
	40	10	56	47
Brain metastases 2–3 cm and <2 cm metastases in deep, eloquent brain	18	1	76	42
	35	5	70	50
	45	10	68	54
Brain metastases <2 cm in non-eloquent brain	24	1	130	68
	40	5	88	60
	50	10	80	63

Note: Five-fraction courses should be given with six degree-of-freedom setup correction. Ten-fraction courses can be given with three-degree-of-freedom setup correction.

chemotherapy without delaying the systemic therapy schedule. The WBRT prescription of 20 Gy in 5 fractions has a long history of use for treating brain metastases with local control results that are slightly lower than prescriptions like 30 Gy in 10 fractions, 37.5 Gy in 15 fractions, and 40 Gy in 20 fractions (Table 15.2) (Tsao et al., 2012). However, when focal high-dose treatment is delivered to the GTVs, the WBRT is adjuvant therapy, intended only to control subclinical microscopic disease. In this setting, we regard doses like those used for PCI as being sufficient to manage the risk of subclinical disease in the brain. Based on calculations of biologically equivalent doses, the WBRT prescription of 20 Gy in 5 fractions would be expected to have similar effectiveness and late side effects to the commonly used PCI prescription of 25 Gy in 10 fractions.

The WBRT+SRS arm of the RTOG 9508 study employed a WBRT prescription of 37.5 Gy/15 and SRS prescribing based on the diameter of metastases: 24 Gy for less than 2 cm, 18 Gy for 2–3 cm, and 15 Gy for 3–4 cm (Andrews et al., 2004). Table 15.3 presents the biologically equivalent doses from RTOG 9508 for comparison with possible 5-fraction and 10-fraction boost prescriptions. There are a number of studies of 5- to 10-fraction schedules of stereotactic radiotherapy that can inform the choice of an SIB prescription (Lindvall et al., 2005; Aoki et al., 2006; Ernst-Stecken et al., 2006; Fahrig et al., 2007; Narayana et al., 2007; Tomita et al., 2008; Kwon et al., 2009; Scorsetti et al., 2009; Chen et al., 2011; Fokas et al., 2012; Jiang et al., 2012; Martens et al., 2012; Ogura et al., 2012; De Potter et al., 2013; Eaton et al., 2013; Ahmed et al., 2014; Murai et al., 2014; Rajakesari et al., 2014). In addition, there are investigations of the biological equivalence of various boost schedules for hypofractionated stereotactic radiotherapy (Yuan et al., 2008; Wiggenraad et al., 2011).

All of the phase 3 trials of WBRT+SRS or SRS alone for patients with brain metastases excluded patients with brainstem metastases. There are a number of retrospective reports that describe treatment

Table 15.4 **Dose constraints and typical volumetric modulated arc therapy planning priorities for 20 Gy/5 whole-brain radiotherapy and 30–40 Gy/5 simultaneous integrated boost delivered with translational and rotational setup correction to 1 mm planning target volumes**

	DOSE CONSTRAINTS		
STRUCTURE	LIMIT	VOLUME (%)	DOSE (GY)
Anterior chambers (no margin)	Upper	<1	10
Cochleas and middle ears (no margin)	Upper	<1	10
Retinas (no margin)	Upper	<1	25
Optics_PRV (1 mm margin)	Upper	<1	25
Spinal cord_PRV (1 mm margin)	Upper	<1	25
Whole-brain PTV	Upper	<5	25
(1 mm margin)	Lower	98	19
Brain metastasis PTVs (1-mm margin)	Upper	0	< 150% of PTV prescription
	Lower	99	PTV prescription

of brainstem metastases with SRS (Kased et al., 2008; Samblas et al., 2009; Koyfman et al., 2010; Valery et al., 2011; Leeman et al., 2012; Kilburn et al., 2014; Hsu et al., 2015). For patients with multiple brain metastases, the likelihood of one of them being deep and within eloquent brain increases with the number of metastases at baseline, so it is important to have a safe and effective SIB prescription for brainstem metastases (Leeman et al., 2012). Kilburn et al. (2014) reported that the risk of toxicity rises when treating metastases with volumes over 1 cm^3 using 18 Gy to the GTV with no margin. Our standard WBRT and SIB prescribing is found in Table 15.4.

15.5 SIMULATION

Patients require highly reproducible immobilization for WBRT and SIB. The treatment can be delivered in about 12 minutes, and the chosen mask system should ensure intrafraction stability for this treatment duration. At our center, we used a thick thermoplastic mask with a patient-specific mouthpiece attached inside the shell. If only translational image-guided setup correction is available, the use of a mouthpiece offers the advantage of reducing rotation (Dhabaan et al., 2012; Dincoglan et al., 2012; Theelen et al., 2012).

The planning CT scan must be performed with high resolution in both the axial plane and the craniocaudal direction. To optimize contouring of small structures like the chiasm and metastases, the field of view needs to be as small as possible, but the field of view needs to be large enough to include the entire immobilization device. A 35-cm field of view covers most head immobilization devices and offers 0.7 mm resolution in the axial plane. The CT reconstruction resolution in the craniocaudal direction must be ≤1 mm for targeting metastases with high-dose radiotherapy.

The MRI sequence used for radiotherapy planning should be a gadolinium-enhanced T1, high-resolution 3D acquisition with axial and craniocaudal resolution of ≤1 mm to permit accurate segmentation of the metastases in three dimensions. T1 with gadolinium sequences sometimes exhibit high-signal artifacts that can resemble metastases, often near blood vessels. Thus, in addition to the high-resolution 3D sequence, it is important to acquire another T1 with gadolinium sequence in a different plane to avoid misinterpretation of the radiotherapy planning sequence and the inadvertent targeting of MRI artifacts. In addition, coregistrations of CT and MRI are imperfect: MRIs can have geometric distortions and artifact due to movement of the patients during acquisition. At our center, we use contrast enhancement for the planning CT scan because it is the fiducial image set on which the dose calculations are performed and the setup correction is executed. It is reassuring to have another contrast-enhanced imaging modality to verify the presence of metastases and to confirm the accuracy of the MRI-CT fusion in the regions of the high-dose boosts. CT imaging with contrast also helps to prevent the targeting of MRI artifacts.

15.6 STRUCTURE AND TARGET DELINEATION

Organs at risk must be contoured to ensure that the dose they receive is controlled in the planning optimization. Organs at risk include anterior chambers (lenses), retinas, brainstem, and spinal cord. In addition, the optic nerves and chiasm should be contoured as a single structure to avoid gaps that can be left between separate chiasm and optic nerve structures. Figure 16.1 illustrates the contours of a patient who was treated in our study. The whole-brain contour can be automatically generated by the planning software. If present, skull metastases can be added to the whole-brain clinical target volume to ensure that they receive the whole-brain prescription dose. This whole-brain and skull metastasis clinical target volume can be expanded by 2 mm to create a whole-brain PTV. The GTVs (or PTVs) of the metastases should be Boolean subtracted from the whole-brain PTV, so it can be used for dose–volume histogram calculations on the "normal" (or nontarget) brain.

Studies comparing different GTV to PTV margin widths for SRS showed that wider margins increase the risk of radionecrosis. In a study by Nataf et al. (2008), the incidence of radionecrosis was 20% for a 2-mm margin in contrast to 7% for a 0-mm margin. This study suggested that the PTV margin employed for SIB likely has a clinically significant effect on the risk of radionecrosis. The simplest prescription is to the GTVs, so multiple GTVs do not need margins added to create multiple PTVs, but it is important to select a dose and a margin that are appropriate for the accuracy of the treatment planning process and the setup correction that is employed. With single-isocenter, multiple-target SIB, a 1° rotation around the plan isocenter could cause a 2-mm displacement of metastases near the skull (Lagerwaard et al., 2009; Peng et al., 2011). Thus, for treatment delivered with only translation setup correction, prescribing to a PTV

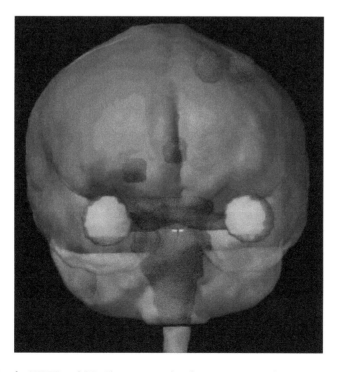

Figure 15.1 Contouring for WBRT and SIB. The anterior chambers (green) are the anterior one-third of the globes and the retinas (orange) are the posterior two-thirds of the globes. The Optics_PRV is the optic chiasm and optic nerves expanded by 1 mm (brown). The brainstem is represented by magenta. The Spinal_PRV is the spinal cord expanded by 1 mm (cyan). Blue PTVs surround the GTVs (not visible) with a 1-mm margin. The whole-brain PTV (yellow) is the brain expanded by 1 mm.

defined by a 2-mm margin from the GTV is recommended (Table 15.4). When using six-degree of freedom setup correction with correction of translational and rotational uncertainties to less than 1 mm and less than 1 degree, a 1-mm GTV–PTV margin can be used to ensure full dose to the metastases. To minimize the risk of radionecrosis, there is also the option of trusting that the whole-brain prescription will deal with the potential for minor geographic miss when using a 0 mm GTV–PTV margin (Rodrigues et al., 2019).

15.7 RADIOTHERAPY PLANNING

In our study, WBRT and SIB was delivered on Varian linear accelerators with multi-leaf collimators that had 5-mm central leaves using two 360° coplanar arcs at a dose rate of 600 monitor units/min, with the collimator set at 45° to 30° for the clockwise arc and 315° to 330° for the counterclockwise arc. However, we now use a high-definition MLC and a dose-rate of 1,200 monitor units/min. For 5-fraction WBRT and SIB, dose constraints for the organs at risk and target doses are found in Table 15.4. This table illustrates two possible SIB prescriptions: either to GTVs with no margin or to PTVs with a margin on GTVs.

Noncoplanar beam arrangements prolong treatment times and diminish setup accuracy unless a patient-tracking system is employed at non-zero couch angles. They are very useful for reducing the dose to normal brain when planning VMAT SRS to metastases alone but rarely offer an advantage for WBRT and SIB, where the whole brain is a target volume. However, for patients with multiple brain metastases in the same axial plane, noncoplanar arcs may diminish the likelihood of bridges of high-dose forming between the boost volumes. There are other strategies for eliminating high-dose bridges between adjacent metastases, including the use of multiple shells around the target volumes and the segmentation of avoidance structures between the target volumes (Thomas et al., 2014). The simplest approach to optimizing the gradient index around the boost target volumes is to combine them into single structure using Boolean OR, create a multi-shell structure around all of them with a 3 to 4 mm margin on the target volumes, and constrain the mean dose of the boost and whole-brain prescriptions using the multi-shell structure. Individual shells around certain targets may need to be added in challenging cases where lesions are in close proximity.

15.8 TREATMENT DELIVERY

The BC Cancer phase 2 study of WBRT and SIB used daily online translational image-guided setup correction with orthogonal kV imaging. However, we now use CBCT for daily setup correction of translations and rotations. Automated 6-degree-of-freedom matching can detect smaller rotations than the radiation therapists can perceive with orthogonal kV imaging. Even when using a mask with a mouthpiece, rotations of over 1°, particularly in the sagittal plane, are sometimes detected by CBCT. A 6-degree-of-freedom robotic couch to correct for both translations and rotations is recommended for delivery of 5-fraction WBRT and SIB, whereas 10-fraction WBRT and SIB can be delivered with daily translational setup correction.

15.9 TOXICITY

An oral antinauseant is recommended 1 hour before each fraction. For patients who are taking dexamethasone before treatment for symptom management, individualized tapering instructions are provided for after the treatment. The acute side effects of WBRT and SIB are the same as would be expected from WBRT alone.

Alopecia occurred after WBRT and SIB with a similar time course as the alopecia seen after conventional WBRT, so VMAT delivery did not reduce acute alopecia. This observation was similar to the observations of De Puysseleyr et al. who reported that alopecia, assessed at 1 month, was not diminished in the 10 patients treated in their study (De Puysseleyr et al., 2014). However, the hair grew back better at the vertex and occiput of the scalp than after WBRT with open beams. Thus, the use of IGRT setup for a 2-mm PTV margin around the brain can prevent the formation of the traditional permanent regions of alopecia caused by tangential dose in open beams.

In our study, acute and subacute serous otitis media occurred in several patients. The middle ears were not contoured as organs at risk. Because they are located between the inferior temporal lobes and the posterior fossa, the VMAT dose optimization delivered the full WBRT prescription dose to the middle ears. The likelihood of serous otitis media might have been reduced by contouring the middle ear and Eustachian tubes as an organ at risk and limiting the dose using VMAT (Wang et al., 2009). The cochleas could have been spared at the same time by including them in "L Ear" and "R Ear" structures. For the middle ear and cochlea, recommended threshold doses are not known, so a dose that is as low as reasonably achievable below the WBRT dose would be a suitable goal (Bhandare et al., 2010; Theunissen et al., 2014).

15.10 CONTROVERSIES

There is ongoing debate about the relative cost of treatment with SRS alone, as compared to WBRT+SRS. Patients treated with SRS alone require many more SRS treatments for salvage because of their elevated risk of new brain metastases. With WBRT+SRS, up-front WBRT in 10 to 15 fractions is delivered to every patient, whereas with SRS alone, WBRT for salvage was required in only about 30% of patients (Ayala-Peacock et al., 2014). A study by Hall et al. (2014) showed that SRS alone was more cost effective than WBRT+SRS when long courses of WBRT were employed. It remains to be seen whether 5-fraction or 10-fraction WBRT and SIB might be more cost effective than either SRS alone or WBRT+SRS, because it reduces the duration of up-front treatment compared to WBRT+SRS and the need for salvage therapies compared to SRS alone.

15.11 FUTURE DIRECTIONS

A clinical trial comparing WBRT plus salvage SRS, if needed, WBRT and SIB, and SRS/stereotactic radiotherapy alone plus salvage WBRT, if needed, for patients with 2 to 10 metastases would help clarify the risks and benefits of these three approaches for patients with multiple brain metastases and could be designed to shed light on the controversy about the risks and benefits of WBRT. The 5-fraction and 10-fraction WBRT and SIB regimens could also be compared in a trial to determine which one offers the best balance of disease control and risk of radionecrosis.

HA-WBRT can be delivered with helical tomotherapy or VMAT (Gutierrez et al., 2007; Hsu et al., 2010; Awad et al., 2013; Oehlke et al., 2015). Following promising results of the RTOG 0933 study (Gondi et al., 2014), which showed less neurocognitive decline with HA-WBRT compared to historical controls, a randomized study (NRG CC001, clinicaltrials.gov identifier: NCT02360215) determined that hippocampal avoidance during WBRT reduces the risk of subacute and long-term memory impairment compared to conventional WBRT (Brown et al., 2020). HA-WBRT and SIB haves been shown to be technically and clinically feasible (Awad et al., 2013; Kim et al., 2015; Oehlke et al., 2015).

CHECKLIST: KEY POINTS FOR CLINICAL PRACTICE

✓	ACTIVITY	SOME CONSIDERATIONS
	Patient selection	• Up to 10 brain metastases • Largest metastasis <3 cm diameter for 5-fraction WBRT and SIB • Any diameter for 40 Gy/10-fraction WBRT and SIB • Patient with good enough prognosis to justify a boost to brain metastases • Patient with high enough risk of new brain metastases to justify the use of WBRT
	Simulation	• Immobilization using a head shell designed to prevent rotation • CT and MRI, both with 1-mm resolution in three dimensions • CT with contrast in the immobilization device • MRI with contrast for fusion with CT • Verify the accuracy of the CT–MRI fusion

(Continued)

✓	ACTIVITY	SOME CONSIDERATIONS
	Treatment planning	• Metastases segmented on MRI and verified on CT • Whole brain or inner skull table segmented on CT with automated tools, if available • Organs at risk segmented on MRI and verified on CT • Inverse-planned volumetric modulated arc therapy using two coplanar arcs • Employ shells around the metastases to control dose falloff • With 6-degree-of-freedom setup correction, use 1-mm GTV–PTV margins and prescribe 35 Gy in 5 fractions to brain PTVs. Use 1-mm PRV margins. Prescribe 30 Gy in 5 fractions to brainstem PTVs • With 3-degree-of-freedom setup correction, use 2 mm GTV–PTV margins and prescribe 45 Gy/10 fractions to brain PTVs. Use 2-mm PRV margins. Prescribe 40 Gy in 10 fractions to brainstem PTVs
	Treatment delivery	• Corticosteroids and antinauseants daily 1 hour before treatment • Clinical setup using lasers in head shell • For translational setup correction, cone-beam CT is used to assess rotation in head shell; if head rotation is more than 1°, reposition head clinically and repeat cone-beam CT to verify correction of head rotation, then apply translational setup correction • For translational and rotational setup correction, obtain cone-beam CT and apply the translational and rotational setup correction

REFERENCES

Abe E, Aoyama H (2012) The role of whole brain radiation therapy for the management of brain metastases in the era of stereotactic radiosurgery. *Current Oncology Reports* 14:79–84.

Ahmed KA, Sarangkasiri S, Chinnaiyan P, Sahebjam S, Yu HH, Etame AB, Rao NG (April 21, 2014) Outcomes following hypofractionated stereotactic radiotherapy in the management of brain metastases. *American Journal of Clinical Oncology* 11 [Epub ahead of print].

American Society of Radiation Oncology (September 14, 2014) *ASTRO releases second list of five radiation oncology treatments to question, as part of National 'Choosing Wisely®' Campaign, 2015.* ABIM Foundation, Philadelphia.

Andrews DW, Scott CB, Sperduto PW, Flanders AE, Gaspar LE, Schell MC, Werner-Wasik M et al. (2004) Whole brain radiation therapy with or without stereotactic radiosurgery boost for patients with one to three brain metastases: Phase III results of the RTOG 9508 randomised trial. *Lancet* 363:1665–1672.

Aoki M, Abe Y, Hatayama Y, Kondo H, Basaki K (2006) Clinical outcome of hypofractionated conventional conformation radiotherapy for patients with single and no more than three metastatic brain tumors, with noninvasive fixation of the skull without whole brain irradiation. *International Journal of Radiation Oncology Biology Physics* 64:414–418.

Aoyama H, Shirato H, Tago M, Nakagawa K, Toyoda T, Hatano K, Kenjyo M et al. (2006) Stereotactic radiosurgery plus whole-brain radiation therapy vs stereotactic radiosurgery alone for treatment of brain metastases: A randomized controlled trial. *JAMA* 295:2483–2491.

Aoyama H, Tago M, Kato N, Toyoda T, Kenjyo M, Hirota S, Shioura H et al. (2007) Neurocognitive function of patients with brain metastasis who received either whole brain radiotherapy plus stereotactic radiosurgery or radiosurgery alone. *International Journal of Radiation Oncology Biology Physics* 68:1388–1395.

Awad R, Fogarty G, Hong A, Kelly P, Ng D, Santos D, Haydu L (2013) Hippocampal avoidance with volumetric modulated arc therapy in melanoma brain metastases—The first Australian experience. *Radiation Oncology* 8:62.

Ayala-Peacock DN, Peiffer AM, Lucas JT, Isom S, Kuremsky JG, Urbanic JJ, Bourland JD (2014) A nomogram for predicting distant brain failure in patients treated with Gamma Knife stereotactic radiosurgery without whole brain radiotherapy. *Neuro-Oncology* 16:1283–1288.

Bauman G, Yartsev S, Fisher B, Kron T, Laperriere N, Heydarian M, VanDyk J (2007) Simultaneous infield boost with helical tomotherapy for patients with 1 to 3 brain metastases. *American Journal of Clinical Oncology* 30:38–44.

Bhandare N, Jackson A, Eisbruch A, Pan CC, Flickinger JC, Antonelli P, Mendenhall WM (2010) Radiation therapy and hearing loss. *International Journal of Radiation Oncology Biology Physics* 76:S50–57.

Brown PD, Gondi V, Pugh S, Tome WA, Wefel JS, Armstrong TS, Bovi JA (April 1, 2020) Hippocampal avoidance during whole-brain radiotherapy plus memantine for patients with brain metastases: Phase III trial NRG oncology CC001. *Journal of Clinical Oncology*38(10):1019–1029.

Chang EL, Wefel JS, Hess KR, Allen PK, Lang FF, Kornguth DG, Arbuckle RB et al. (2009) Neurocognition in patients with brain metastases treated with radiosurgery or radiosurgery plus whole-brain irradiation: A randomised controlled trial. *Lancet Oncology* 10:1037–1044.

Chen XJ, Xiao JP, Li XP, Jiang XS, Zhang Y, Xu YJ, Dai JR, Li YX (2011) Risk factors of distant brain failure for patients with newly diagnosed brain metastases treated with stereotactic radiotherapy alone. *Radiation Oncology* 6:175.

Churilla TM, Ballman KV, Brown PD, Twohy EL, Jaeckle K, Farace E et al. (December 1, 2017) Radiosurgery with or without whole-brain radiation therapy for limited brain metastases: A secondary analysis of the North Central cancer treatment group N0574 (alliance) randomized controlled trial. *International Journal of Radiation Oncology Biology Physics*99(5):1173–1178.

De Potter B, De Meerleer G, De Neve W, Boterberg T, Speleers B, Ost P (2013) Hypofractionated frameless stereotactic intensity-modulated radiotherapy with whole brain radiotherapy for the treatment of 1–3 brain metastases. *Neurological Sciences*34:647–653.

De Puysseleyr A, Van De Velde J, Speleers B, Vercauteren T, Goedgebeur A, Van Hoof T, Boterberg T, De Neve W, De Wagter C, Ost P (2014) Hair-sparing whole brain radiotherapy with volumetric arc therapy in patients treated for brain metastases: Dosimetric and clinical results of a phase II trial. *Radiation Oncology* 9:170.

Dhabaan A, Schreibmann E, Siddiqi A, Elder E, Fox T, Ogunleye T, Esiashvili N, Curran W, Crocker I, Shu HK (2012) Six degrees of freedom CBCT-based positioning for intracranial targets treated with frameless stereotactic radiosurgery. *Journal of Applied Clinical Medical Physics* 13:3916.

Dincoglan F, Beyzadeoglu M, Sager O, Oysul K, Sirin S, Surenkok S, Gamsiz H, Uysal B, Demiral S, Dirican B (2012) Image-guided positioning in intracranial non-invasive stereotactic radiosurgery for the treatment of brain metastasis. *Tumori*98:630–635.

Eaton BR, Gebhardt B, Prabhu R, Shu HK, Curran WJ, Jr, Crocker I (2013) Hypofractionated radiosurgery for intact or resected brain metastases: Defining the optimal dose and fractionation. *Radiation Oncology* 8:135.

Edwards AA, Keggin E, Plowman PN (2010) The developing role for intensity-modulated radiation therapy (IMRT) in the non-surgical treatment of brain metastases. *British Journal of Radiology* 83:133–136.

Ernst-Stecken A, Ganslandt O, Lambrecht U, Sauer R, Grabenbauer G (2006) Phase II trial of hypofractionated stereotactic radiotherapy for brain metastases: Results and toxicity. *Radiotherapy and Oncology* 81:18–24.

Fahrig A, Ganslandt O, Lambrecht U, Grabenbauer G, Kleinert G, Sauer R, Hamm K (2007) Hypofractionated stereotactic radiotherapy for brain metastases—results from three different dose concepts. *Strahlentherapie und Onkologie* 183:625–630.

Ferro M, Chiesa S, Macchia G, Cilla S, Bertini F, Frezza G et al. (2017) Intensity modulated radiation therapy with simultaneous integrated boost in patients with brain oligometastases: A phase 1 study (ISIDE- BM-1). *International Journal of Radiation Oncology Biology Physics*97:82–90.

Flickinger JC, Kondziolka D, Maitz AH, Lunsford LD (1998) Analysis of neurological sequelae from radiosurgery of arteriovenous malformations: How location affects outcome. *International Journal of Radiation Oncology Biology Physics* 40:273–278.

Fokas E, Henzel M, Surber G, Kleinert G, Hamm K, Engenhart-Cabillic R (2012) Stereotactic radiosurgery and fractionated stereotactic radiotherapy: Comparison of efficacy and toxicity in 260 patients with brain metastases. *Journal of Neuro-Oncology* 109:91–98.

Gondi V, Paulus R, Bruner DW, Meyers CA, Gore EM, Wolfson A, Werner-Wasik M, Sun AY, Choy H, Movsas B (2013) Decline in tested and self-reported cognitive functioning after prophylactic cranial irradiation for lung cancer: Pooled secondary analysis of radiation therapy oncology group randomized trials 0212 and 0214. *International Journal of Radiation Oncology Biology Physics* 86:656–664.

Gondi V, Pugh SL, Tome WA, Caine C, Corn B, Kanner A, Rowley H et al. (2014) Preservation of memory with conformal avoidance of the hippocampal neural stem-cell compartment during whole-brain radiotherapy for brain metastases (RTOG 0933): A phase II multi-institutional trial. *Journal of Clinical Oncology* 32:3810–3816.

Gondi V, Tolakanahalli R, Mehta MP, Tewatia D, Rowley H, Kuo JS, Khuntia D, Tome WA (2010) Hippocampal-sparing whole-brain radiotherapy: A "how-to" technique using helical tomotherapy and linear accelerator-based intensity-modulated radiotherapy. *International Journal of Radiation Oncology Biology Physics* 78:1244–1252.

Gutierrez AN, Westerly DC, Tome WA, Jaradat HA, Mackie TR, Bentzen SM, Khuntia D, Mehta MP (2007) Whole brain radiotherapy with hippocampal avoidance and simultaneously integrated brain metastases boost: A planning study. *International Journal of Radiation Oncology Biology Physics* 69:589–597.

Hall MD, McGee JL, McGee MC, Hall KA, Neils DM, Klopfenstein JD, Elwood PW (2014) Cost-effectiveness of stereotactic radiosurgery with and without whole-brain radiotherapy for the treatment of newly diagnosed brain metastases. *Journal of Neurosurgery* 121(Suppl):84–90.

Hsu F, Carolan H, Nichol A, Cao F, Nuraney N, Lee R, Gete E, Wong F, Schmuland M, Heran M, Otto K (2010) Whole brain radiotherapy with hippocampal avoidance and simultaneous integrated boost for 1–3brain metastases: A feasibility study using volumetric modulated arc therapy. *International Journal of Radiation Oncology Biology Physics* 76(5):1480–1485.

Hsu F, Nichol A, Ma R, Kouhestani P, Toyota B, McKenzie M (2015) Stereotactic radiosurgery for metastases in eloquent central brain locations. *Canadian Journal of Neurological Sciences* 42(5):333–337.

Jiang XS, Xiao JP, Zhang Y, Xu YJ, Li XP, Chen XJ, Huang XD, Yi JL, Gao L, Li YX (2012) Hypofractionated stereotactic radiotherapy for brain metastases larger than three centimeters. *Radiation Oncology* 7:36.

Kased N, Huang K, Nakamura JL, Sahgal A, Larson DA, McDermott MW, Sneed PK (2008) Gamma Knife radiosurgery for brainstem metastases: The UCSF experience. *Journal of Neuro-Oncology* 86:195–205.

Kilburn JM, Ellis TL, Lovato JF, Urbanic JJ, Daniel Bourland J, Munley MT, Deguzman AF et al. (2014) Local control and toxicity outcomes in brainstem metastases treated with single fraction radiosurgery: Is there a volume threshold for toxicity? *Journal of Neuro-Oncology* 117:167–174.

Kim KH, Cho BC, Lee CG, Kim HR, Suh YG, Kim JW, Choi C, Baek JG, Cho J (January 18, 2015) Hippocampus-sparing whole-brain radiotherapy and simultaneous integrated boost for multiple brain metastases from lung adenocarcinoma: Early response and dosimetric evaluation. *Technology in Cancer Research & Treatment* pii, doi:10.1177/1533034614566993 [Epub ahead of print].

Kocher M, Soffietti R, Abacioglu U, Villa S, Fauchon F, Baumert BG, Fariselli L et al. (2011) Adjuvant whole-brainradiotherapy versus observation after radiosurgery or surgical resection of one to three cerebral metastases: Results of the EORTC 22952–26001 study. *Journal of Clinical Oncology* 29:134–141.

Koyfman SA, Tendulkar RD, Chao ST, Vogelbaum MA, Barnett GH, Angelov L, Weil RJ, Neyman G, Reddy CA, SuhJH (2010) Stereotactic radiosurgery for single brainstem metastases: The Cleveland clinic experience. *International Journal of Radiation Oncology Biology Physics* 78:409–414.

Kwon AK, Dibiase SJ, Wang B, Hughes SL, Milcarek B, Zhu Y (2009) Hypofractionated stereotactic radiotherapy for the treatment of brain metastases. *Cancer* 115:890–898.

Lagerwaard FJ, van der Hoorn EA, Verbakel WF, Haasbeek CJ, Slotman BJ, Senan S (2009) Whole-brain radiotherapy with simultaneous integrated boost to multiple brain metastases using volumetric modulated arc therapy. *International Journal of Radiation Oncology Biology Physics* 75:253–259.

Leeman JE, Clump DA, Wegner RE, Heron DE, Burton SA, Mintz AH (2012) Prescription dose and fractionation predict improved survival after stereotactic radiotherapy for brainstem metastases. *Radiation Oncology* 7:107.

Li J, Bentzen SM, Renschler M, Mehta MP (2007) Regression after whole-brain radiation therapy for brain metastases correlates with survival and improved neurocognitive function. *Journal of Clinical Oncology* 25:1260–1266.

Lindvall P, Bergstrom P, Lofroth PO, Henriksson R, Bergenheim AT (2005) Hypofractionated conformal stereotactic radiotherapy alone or in combination with whole-brain radiotherapy in patients with cerebral metastases. *International Journal of Radiation Oncology Biology Physics* 61:1460–1466.

Martens B, Janssen S, Werner M, Fruhauf J, Christiansen H, Bremer M, Steinmann D (2012) Hypofractionatedstereotactic radiotherapy of limited brain metastases: A single-centre individualized treatment approach. *BMCCancer*12, doi:10.1186/1471-2407-12-497497-2407-12-497.

McDuff SG, Taich ZJ, Lawson JD, Sanghvi P, Wong ET, Barker FG II, Hochberg FH et al. (2013) Neurocognitive assessment following whole brain radiation therapy and radiosurgery for patients with cerebral metastases. *Journal of Neurology, Neurosurgery, and Psychiatry* 84:1384–1391.

Mehta MP (2015) The controversy surrounding the use of whole-brain radiotherapy in brain metastases patients. *Neuro-Oncology* 17:919–923.

Meyers CA, Smith JA, Bezjak A, Mehta MP, Liebmann J, Illidge T, Kunkler I et al. (2004) Neurocognitive function and progression in patients with brain metastases treated with whole-brain radiation and motexafin gadolinium: Results of a randomized phase III trial. *Journal of Clinical Oncology* 22:157–165.

Murai T, Ogino H, Manabe Y, Iwabuchi M, Okumura T, Matsushita Y, Tsuji Y, Suzuki H, Shibamoto Y (2014) Fractionated stereotactic radiotherapy using CyberKnife for the treatment of large brain metastases: A dose escalation study. *Clinical Oncology (The Royal College of Radiologists)* 26:151–158.

Narayana A, Chang J, Yenice K, Chan K, Lymberis S, Brennan C, Gutin PH (2007) Hypofractionated stereotactic radiotherapy using intensity-modulated radiotherapy in patients with one or two brain metastases. *Stereotactic and Functional Neurosurgery* 85:82–87.

Nataf F, Schlienger M, Liu Z, Foulquier JN, Gres B, Orthuon A, Vannetzel JM et al. (2008) Radiosurgery with or without A 2-mm margin for 93 single brain metastases. *International Journal of Radiation Oncology Biology Physics* 70:766–772.

Nichol AM, Ma R, Hsu F, Gondara L, Carolan H, Olson R et al. (February 1, 2016) Volumetric radiosurgery for 1 to 10 brain metastases: A multicenter, single-arm, phase 2 study. *International Journal of Radiation Oncology Biology Physics* 94(2):312–321.

Oehlke O, Wucherpfennig D, Fels F, Frings L, Egger K, Weyerbrock A, Prokic V, Nieder C, Grosu AL (2015) Whole brain irradiation with hippocampal sparing and dose escalation on multiple brain metastases: Local tumour control and survival. *Strahlentherapie und Onkologie* 191(6):461–469.

Ogura K, Mizowaki T, Ogura M, Sakanaka K, Arakawa Y, Miyamoto S, Hiraoka M (2012) Outcomes of hypofractionated stereotactic radiotherapy for metastatic brain tumors with high risk factors. *Journal of Neuro-Oncology* 109:425–432.

Patchell RA, Tibbs PA, Regine WF, Dempsey RJ, Mohiuddin M, Kryscio RJ, Markesbery WR, Foon KA, Young B (1998) Postoperative radiotherapy in the treatment of single metastases to the brain: A randomized trial. *JAMA* 280:1485–1489.

Peng JL, Liu C, Chen Y, Amdur RJ, Vanek K, Li JG (2011) Dosimetric consequences of rotational setup errors with direct simulation in a treatment planning system for fractionated stereotactic radiotherapy. *Journal of Applied Clinical Medical Physics* 12:3422.

Rajakesari S, Arvold ND, Jimenez RB, Christianson LW, Horvath MC, Claus EB, Golby AJ et al. (2014) Local control after fractionated stereotactic radiation therapy for brain metastases. *Journal of Neuro-Oncology* 120:339–346.

Regine WF, Scott C, Murray K, Curran W (2001) Neurocognitive outcome in brain metastases patients treated with accelerated-fractionation vs. accelerated-hyperfractionated radiotherapy: An analysis from radiation therapy oncology group study 91–04. *International Journal of Radiation Oncology Biology Physics* 51:711–717.

Rodrigues G, Eppinga W, Lagerwaard F, de Haan P, Haasbeek C, Perera F, Slotman B, Yaremko B, Yartsev S, Bauman G (2012) A pooled analysis of arc-based image-guided simultaneous integrated boost radiation therapy for oligometastatic brain metastases. *Radiotherapy and Oncology* 102:180–186.

Rodrigues G, Warnemr A, Zindler J, Slotman B, Lagerwaard F (2014) A clinical nomogram and recursive partitioning analysis to determine the risk of regional failure after radiosurgery alone for brain metastases. *Radiotherapy and Oncology* 111:52–58.

Rodrigues G, Yartsev S, Roberge D, MacRae R, Tay KY, Pond GR et al. (December 16, 2019) A phase II multi-institutional clinical trial assessing fractionated simultaneous in-field boost radiotherapy for brain oligometastases. *Cureus* 11(12):e6394.

Rodrigues G, Yartsev S, Yaremko B, Perera F, Dar AR, Hammond A, Lock M et al. (2011) Phase I trial of simultaneous in-field boost with helical tomotherapy for patients with one to three brain metastases. *International Journal of Radiation Oncology Biology Physics* 80:1128–1133.

Roos DE, Wirth A, Burmeister BH, Spry NA, Drummond KJ, Beresford JA, McClure BE (2006) Whole brain irradiation following surgery or radiosurgery for solitary brain metastases: Mature results of a prematurely closed randomized trans-tasman radiation oncology group trial (TROG 98.05). *Radiotherapy and Oncology* 80:318–322.

Sahgal A (2015) Point/counterpoint: Stereotactic radiosurgery without whole-brain radiation for patients with a limited number of brain metastases: The current standard of care? *Neuro-Oncology* 17:916–918.

Sahgal A, Aoyama H, Kocher M, Neupane B, Collette S, Tago M, Shaw P, Beyene J, Chang EL (2015) Phase 3 trialsof stereotactic radiosurgery with or without whole-brain radiation therapy for 1 to 4 brain metastases: Individualpatient data meta-analysis. *International Journal of Radiation Oncology Biology Physics*91:710–717.

Samblas JM, Sallabanda K, Bustos JC, Gutierrez-Diaz JA, Peraza C, Beltran C, Samper PM (2009) Radiosurgery andwhole brain therapy in the treatment of brainstem metastases. *Clinical and Translational Oncology* 11:677–680.

Scorsetti M, Facoetti A, Navarria P, Bignardi M, De Santis M, Ninone SA, Lattuada P, Urso G, Vigorito S, Mancosu P, Del Vecchio M (2009) Hypofractionated stereotactic radiotherapy and radiosurgery for the treatment of patientswith radioresistant brain metastases. *Anticancer Research* 29:4259–4263.

Soffietti R, Kocher M, Abacioglu UM, Villa S, Fauchon F, Baumert BG, Fariselli L et al. (2013) A European organisation for research and treatment of cancer phase III trial of adjuvant whole-brain radiotherapy versus observation inpatients with one to three brain metastases from solid tumors after surgical resection or radiosurgery: Quality-of-life results. *Journal of Clinical Oncology* 31:65–72.

Soon YY, Tham IW, Lim KH, Koh WY, Lu JJ (2014) Surgery or radiosurgery plus whole brain radiotherapy versus surgery or radiosurgery alone for brain metastases. *Cochrane Database of Systematic Reviews* 3:CD009454.

Sperduto PW, Shanley R, Luo X, Andrews D, Werner-Wasik M, Valicenti R, Bahary JP, Souhami L, Won M, Mehta M (2014) Secondary analysis of RTOG 9508, a phase 3 randomized trial of whole-brain radiation therapy versus WBRT plus stereotactic radiosurgery in patients with 1–3 brain metastases; poststratified by the graded prognostic assessment (GPA). *International Journal of Radiation Oncology Biology Physics* 90:526–531.

Sterzing F, Welzel T, Sroka-Perez G, Schubert K, Debus J, Herfarth KK (2009) Reirradiation of multiple brain metastases with helical tomotherapy: A multifocal simultaneous integrated boost for eight or more lesions. *Strahlentherapie und Onkologie* 185:89–93.

Theelen A, Martens J, Bosmans G, Houben R, Jager JJ, Rutten I, Lambin P, Minken AW, Baumert BG (2012) Relocatable fixation systems in intracranial stereotactic radiotherapy: Accuracy of serial CT scans and patient acceptance in a randomized design. *Strahlentherapie und Onkologie* 188:84–90.

Theunissen EA, Zuur CL, Yurda ML, van der Baan S, Kornman AF, de Boer JP, Balm AJ, Rasch CR, Dreschler WA (2014) Cochlea sparing effects of intensity modulated radiation therapy in head and neck cancers patients: A long-term follow-up study. *Journal of Otolaryngology – Head & Neck Surgery* 43:30.

Thomas EM, Popple RA, Wu X, Clark GM, Markert JM, Guthrie BL, Yuan Y, Dobelbower MC, Spencer SA, Fiveash JB (2014) Comparison of plan quality and delivery time between volumetric arc therapy (RapidArc) and Gamma Knife radiosurgery for multiple cranial metastases. *Neurosurgery* 75:409–417; discussion 417–418.

Tomita N, Kodaira T, Tachibana H, Nakamura T, Nakahara R, Inokuchi H, Shibamoto Y (2008) Helical tomo-therapy for brain metastases: Dosimetric evaluation of treatment plans and early clinical results. *Technology in Cancer Research & Treatment* 7:417–424.

Tsao MN, Lloyd N, Wong RK, Chow E, Rakovitch E, Laperriere N, Xu W, Sahgal A (2012) Whole brain radio-therapy for the treatment of newly diagnosed multiple brain metastases. *Cochrane Database of Systematic Reviews* 4:CD003869.

Valery CA, Boskos C, Boisserie G, Lamproglou I, Cornu P, Mazeron JJ, Simon JM (2011) Minimized doses for linear accelerator radiosurgery of brainstem metastasis. *International Journal of Radiation Oncology Biology Physics* 80(2):362–368.

Wang SZ, Li J, Miyamoto CT, Chen F, Zhou LF, Zhang HY, Yang G et al. (2009) A study of middle ear function in the treatment of nasopharyngeal carcinoma with IMRT technique. *Radiotherapy & Oncology* 93:530–533.

Weber DC, Caparrotti F, Laouiti M, Malek K (2011) Simultaneous in-field boost for patients with 1 to 4 brain metas-tasis/es treated with volumetric modulated arc therapy: A prospective study on quality-of-life. *Radiotherapy & Oncology* 6:79.

Wegner RE, Oysul K, Pollock BE, Sirin S, Kondziolka D, Niranjan A, Lunsford LD, Flickinger JC (2011) A modified radiosurgery-based arteriovenous malformation grading scale and its correlation with outcomes. *International Journal of Radiation Oncology Biology Physics* 79:1147–1150.

Wiggenraad R, Verbeek-de Kanter A, Kal HB, Taphoorn M, Vissers T, Struikmans H (2011) Dose-effect relation in stereotactic radiotherapy for brain metastases. A systematic review. *Radiotherapy & Oncology* 98:292–297.

Wolfson AH, Bae K, Komaki R, Meyers C, Movsas B, Le Pechoux C, Werner-Wasik M, Videtic GM, Garces YI, Choy H (2011) Primary analysis of a phase II randomized trial radiation therapy oncology group (RTOG) 0212: Impact of different total doses and schedules of prophylactic cranial irradiation on chronic neurotoxic-ity and quality of life for patients with limited-disease small-cell lung cancer. *International Journal of Radiation Oncology Biology Physics* 81:77–84.

Yang J, Zhan W, Zhang H, Song T, Jia Y, Xu H et al. (October2017) Intensity-modulated radiation therapy for patients with 1 to 3 brain metastases in recursive partitioning analysis class 3. *Medicine (Baltimore)* 96(40):e7715.

Yuan J, Wang JZ, Lo S, Grecula JC, Ammirati M, Montebello JF, Zhang H, Gupta N, Yuh WT, Mayr NA (2008) Hypofractionation regimens for stereotactic radiotherapy for large brain tumors. *International Journal of Radiation Oncology Biology Physics* 72:390–397.

Zhong J, Waldman AD, Kandula S, Eaton BR, Prabhu RS, Huff SB Shu HK(March 2020) Outcomes of whole-brain radiation with simultaneous in-field boost (SIB) for the treatment of brain metastases. *Journal of Neuro-Oncology* 147(1):117–123.

16 Image-Guided Hypofractionated Stereotactic Radiation Therapy for High-Grade Glioma

Luke Pike and Kathryn Beal

Contents

16.1 INTRODUCTION

Glioblastoma is the most common high-grade glioma and the most common primary brain tumor in adults. It is characterized by an extremely aggressive clinical course and poor prognosis. Even with optimal treatment of maximal resection followed by concurrent chemoradiation and adjuvant chemotherapy, median progression-free survival and overall survival for patients from contemporary studies are on the order of 11 months and 17 months, respectively (Gilbert et al. 2014; Chinot et al. 2014).

The role of adjuvant radiation therapy (RT) following maximal surgical resection was established in the 1970s in the BTSG studies (Walker et al. 1978, 1980). Radiation therapy in these studies was prescribed with a dose of 50 to 60 Gy to midplane, with opposed laterals to the whole brain. Since that time, although significant advances have been made in the technical application of RT in the treatment of glioblastoma, these approaches have not yet yielded significant improvements in progression-free or overall survival for these patients. Specifically, studies of dose escalation with conventional fractionation, hyperfractionation, radiosurgery, and brachytherapy boosts have not improved disease control.

However, technological advances in the delivery of RT enable the safe delivery of larger doses of radiation via image-guided hypofractionated stereotactic radiotherapy (IG-HSRT). It is postulated that such approaches might improve tumor control in this aggressive disease. IG-HSRT may improve the therapeutic ratio with increased effectiveness, decreased side effects, and abbreviated treatment courses for patients. This chapter will review (1) current indications and evidence supporting IG-HSRT for both newly diagnosed and recurrent glioblastoma, (2) technical approaches to the planning and delivery of IG-HSRT, (3) potential toxicities and their management, and (4) controversies and future directions for the use of IG-HSRT for malignant glioma.

16.2 INDICATIONS

16.2.1 NEWLY DIAGNOSED GLIOBLASTOMA

The current standard of care for the treatment of newly diagnosed glioblastoma was defined by the EORTC/NCIC trial (Stupp et al. 2009, 2005). In this study, 573 patients were randomized between postoperative RT alone and RT with concurrent and adjuvant temozolomide. Combined modality treatment significantly improved median overall survival from 12.1 to 14.6 months. More recently, the addition of tumor-treating fields (TTFs) or lomustine to adjuvant temozolomide has yielded a modest additional survival benefit in the first-line setting (Herrlinger et al. 2019; Stupp et al. 2017), while the addition of bevacizumab in the first line setting has not demonstrated an overall survival benefit (Gilbert et al. 2014; Chinot et al. 2014).

Efforts to improve outcomes through various radiotherapy techniques and fractionation schemes have largely been unsuccessful. The addition of a 10–20 Gy radiosurgical boost following conventional radiation resulted in significantly increased levels of radionecrosis, with nearly 50% of patients continuing to require steroids at 1 year following radiosurgery, without evidence of improved disease-specific outcomes (Loeffler et al. 1992). Likewise, analyses of hyperfractionation schema used in RTOG protocols have failed to demonstrate a benefit over conventional fractionation (Scott et al. 1998). In the end, no improvement in overall survival was seen with the use of I-125 interstitial brachytherapy boosts at the time of surgery (Selker et al. 2002) nor with dose escalation utilizing conventional fractionation (Tsien et al. 2009; Watkins et al. 2009; Graf et al. 2005). Recent preliminary data from the prospective phase III study BN001 unfortunately did not show a potential benefit to dose escalation to 75 Gy with intensity-modulated radiotherapy as compared to 60 Gy, and in fact the MGMT-methylated cohort in this study showed statistically inferior overall survival with dose escalation (Gondi et al. 2020).

These data have called into question the efficacy of conventionally fractionated radiation regimens. By contrast, hypofractionation has several theoretical advantages. First, delivering a higher dose per fraction has potential radiobiological advantages: cell line models have demonstrated increased direct tumor cell kill and inhibiting cellular repopulation in glioblastoma (Kaaijk et al. 1997). Estimates of the alpha/beta ratio of glioblastoma range from 2 to 5, which argues for a higher dose per fraction than conventionally fractionated treatments. From a practical standpoint, a hypofractionated approach is more convenient for patients by reducing treatment courses from six or more weeks to potentially to a week, thereby reducing the physical and financial burden of travel into a radiation oncology department. This is of particular consideration in a patient population whose prognosis is guarded and whose life expectancy can be measured in a small number of months in many instances.

With these considerations in mind, early studies of hypofractionated RT were conducted with frail and elderly patients. The first published Canadian study randomized patients over 60 years of age to either 60 Gy in 30 fractions or to 40 Gy in 15 fractions, without chemotherapy, after maximal surgical resection. Although not powered for non-inferiority, median overall survival appeared similar between the treatment groups (5.1 versus 5.1 months) (Roa et al. 2004). In follow-on studies, the use of concurrent temozolomide with this fractionation schedule demonstrated a clear survival benefit over radiation alone (Perry et al. 2017). A further study in the same vein compared 40 Gy in 15 fractions to 25 Gy in 5 fractions in patients who were frail and/or elderly and showed similar overall survival with either approach (7.6 versus 7.9 months, $p = 0.988$) (Roa et al. 2020). In the end, in the NORDIC trial, patients were randomized to treatment with either temozolomide alone or radiation with 34 Gy in 10 fractions or 60 Gy in 30 fractions. Intriguingly, similar outcomes were seen between temozolomide alone and 34 Gy in 10 fractions, while those who received 60 Gy in 30 fractions appeared to fare worse (Malmström et al. 2012). It is important to note, however, that although these studies suggest a potential role of the use of hypofractionated regimens in older patients with poor performance status, most of whom were not powered for non-inferiority, they do not necessarily suggest a clear benefit of hypofractionated treatment in younger, healthier patients.

As such, recent prospective efforts have taken advantage of intensity modulation and simultaneous integrated boost capabilities to study hypofractionation in less selected populations. Floyd et al. (2004) reported on 20 patients with good performance status treated with a dose of 30 Gy in 10 fractions to the area of edema with simultaneous integrated boost to a total dose of 50 Gy in 10 fractions to the enhancing primary disease. Disease control in this cohort was poor, however, with median time to disease progression of 6 months following completion of radiation, with half of patients incurring radionecrosis requiring surgical resection (Floyd et al. 2004).

In provocative results from the University of Colorado, 24 patients with good performance status and a median age of 60.5 years underwent hypofractionated IGRT with dose painting to a dose of 60 Gy in 10 fractions to the area of enhancement and 30 Gy in 10 fractions to the area of T2 abnormality with concurrent temozolomide (Reddy et al. 2012). Median survival was 16.6 months, with no reported grade 3 or greater non-hematologic toxicity. It is notable, however, that six patients underwent salvage resection for suspected recurrence, of which four specimens contained predominantly (>80%) necrosis. Similarly promising results were reported in a single-arm study by Iuchi and colleagues, in which 46 patients with newly diagnosied GBM received a total of 68 Gy in 8 fractions, with a three-layered dose painting technique with concurrent and adjuvant temozolomide (Iuchi et al. 2014). Median overall survival was 20 months following treatment, with radionecrosis observed in nearly half of patients. Notably, disease failure was predominately outside the high-dose field, and radionecrosis was associated with prolonged survival, but ultimately resulted in significant deterioration in performance status in survivors. More recently, investigators here at Memorial Sloan Kettering Cancer Center conducted a phase II trial investigating concurrent bevacizumab and temozolomide with dose-painted image-guided hypofractionated stereotactic radiotherapy (Omuro et al. 2014). Forty patients were treated with 36 Gy in 6 fractions to the area of enhancement and with 24 Gy in 6 fractions to the area of T2 hyperintensity, with dose-painting, and followed for a median of 42 months among surviving patients. Median overall survival was 19 months, and the overall survival was 93% at 1 year. Notably, both the clinical and radiographic incidence of radionecrosis was lower in this study compared to the other studies investigating hypofractionation in the up-front setting, and no pseudoprogression was identified, suggesting a potential role for the use of bevacizumab in attenuating radionecrosis in this setting. More recently, a multi-institutional phase I/II dose-finding study was conducted, with increasing doses from 25 to 40 Gy using 5-fraction radiosurgical technique with concurrent temozolomide, intriguingly, utilizing a 5-mm CTV expansion. The maximum tolerated dose was determined to be 40 Gy with dose-limiting toxicity occurring in two patients, one of whom experienced significant post-treatment cerebral edema. Median survival in this study was at least comparable to that obtained via conventionally fractionated approaches—14.8 months (Azoulay et al. 2020). Studies investigating the use of hypofractionation with 10 or less fractions in newly diagnosed glioma are summarized in Table 16.1. Overall, these studies demonstrate IG-HSRT to be a safe, effective, and convenient therapy approach for newly diagnosed glioma that requires further study in the randomized trial setting.

16.2.2 RECURRENT GLIOBLASTOMA

Although the addition of temozolomide to conventional radiotherapy has resulted in an improvement in outcomes, more than 90% of patients will succumb to their disease in the first 5 years after diagnosis (Stupp et al. 2009, 2005). Importantly, more than 75% of patients who received conventional chemoradiation for GBM will recur within 2 cm of the original area of enhancement, squarely in the middle of the high-dose region (Sherriff et al. 2013). As such, many patients will return for consideration of salvage therapy. Options in this setting include surgical resection, chemotherapy, and reirradiation. Due to proximity to eloquent brain structures, further resection is often not possible. Although brachytherapy and single-fraction stereotactic radiosurgery (SRS) have been employed in the recurrent setting, they have been associated with significant toxicity, and thus IG-HSRT is often the preferred approach.

The use of IG-HSRT in the recurrent setting affords the ability to deliver higher fractional doses of radiation to tumor that has demonstrated itself to be radioresistant. Mitigating treatment-related toxicity while palliating tumor-related symptoms is usually the primary goal of care in this setting, and thus IMRT/VMAT technique with CT- and MRI-based image guidance is often used to avoid excessive cumulative doses to sensitive brain structures.

There is a growing body of retrospective and prospective data demonstrating safety and efficacy of reirradiation with hypofractionated RT (using 10 fractions or less) for glioblastoma, summarized in Table 16.2. In an early phase I/II study, 22 patients with recurrent glioma who had previously received RT were treated

Table 16.1 Studies of hypofractionated stereotactic radiation therapy in newly diagnosed high-grade glioma in patients with good performance status

AUTHOR	PATIENTS	STUDY TYPE	MEDIAN DOSE	NUMBER OF FRACTIONS	BED (A/B = 3)	MEDIAN TUMOR VOLUME (RANGE)	CONCURRENT SYSTEMIC AGENT	TOXICITY REQUIRING SURGICAL INTERVENTION	MEDIAN OVERALL SURVIVAL (MONTHS)
Floyd et al. (2004)	20	Prospective phase I/II	50/30[a]	10	133.3/60	NR[b]	None	15%	7
Reddy et al. (2012)	24	Prospective phase II	60/30[a]	10	180/60	97.87 (53.9– 145.09)/258.04 (126–452.49) [c]	TMZ	17%	16.6
Iuchi et al. (2014)	46	Prospective phase II	68/40/32[a]	8	260.7/ 106.7/74.7	80.9 (26.5–267.2)/ 160.7 (78.5–374.3)/ 0.6 (0–65.9)[d]	TMZ	NR[b]	20
Omuro et al. (2014)	40	Prospective phase II	36/24[a]	6	108/56	NR[b]	TMZ + bevacizumab	5%	19
Azoulay et al. (2020)	30	Prospective phase I/II	25–40	5	67–147	60 (14.7–137.3)	TMZ	NR	13.8

[a] Dose painting.
[b] Not reported.
[c] PTV_1/PTV_2.
[d] $PTV_1/PTV_2/PTV_3$.

on a dose escalation study, with a starting dose of 30 Gy in 6 fractions, increasing to 50 Gy in 10 fractions (Laing et al. 1993). Median survival from SRT was 9.8 months, though those patients who received 40 Gy or more required steroids for the control of radionecrosis/edema. Other studies have shown median overall survival rates after reirradiation ranging from 7.0 to 18.0 months. A recent phase I dose-escalation study treated patients with bevacizumab and escalating three-fraction regimens, from 27 Gy to 33 Gy, with the latter being declared the maximum-tolerated dose (Clarke et al. 2017). Median overall survival in this cohort was 13 months. These rates compare favorably to series utilizing stereotactic radiotherapy with standard fraction sizes: Cho et al. reported a median survival of 12 months after the delivery of 37.5 Gy in 2.5 Gy fractions (Cho et al. 1999). In a cohort of 565 patients with recurrent high-grade glioma treated with a median of 36 Gy in 2 Gy fractions, Combs and colleagues reported a median survival of 7.5 months (Combs et al. 2018). Outcomes of IG-HSRT for recurrent glioma also compare favorably to the use of both cytotoxic and molecularly targeted chemotherapy. Series investigating treatment approaches with chemotherapy alone have reported median overall survival times ranging from 4.4 to 9.8 months (Bambury and Morris 2014). The rates of radiation necrosis requiring surgical intervention after IG-HSRT appear to be acceptably low (ranging from 0% to 15%) with the exception of the Voynov and Kohshi studies, which reported rates of 20% and 28%, respectively (Voynov et al. 2002; Kohshi et al. 2007).

Recent efforts have focused on the addition of novel agents such as immune checkpoint inhibitors (ICIs) to the treatment paradigm for recurrent glioblastoma, though to date, these have not been successful. For example, a recent phase III trial, CHECKMATE-143, of 369 patients with recurrent glioblastoma showed no survival benefit gleaned from single agent nivolumab as compared to bevacizumab (Reardon et al. 2020). A phase II study comparing neoadjuvant to adjuvant pembrolizumab has shown interesting results, with patients in the neoadjuvant arm faring better than those in the adjuvant arm, though phase III data are forthcoming (Cloughesy et al. 2019). There are numerous ongoing studies in the first line or recurrent setting, evaluating the activity of single agent ICIs or in combination with salvage radiotherapy (typically 35 Gy in 10 fractions). However, preclinical data have suggested a specific pro-inflammatory effect of hypofractionated RT (Vanpouille-Box et al. 2017; Twyman-Saint Victor et al. 2015), and thus it will be crucial to assess the combination of IG-HSRT and ICIs or other novel therapeutics in prospective trials. One such study taking place at MSKCC is 18–400 in which patients with recurrent MGMT methylated glioblastoma will receive hypofractionated reirradiation concurrently with bevacizumab and nivolumab.

16.3 SIMULATION, TARGET DELINEATION, AND TREATMENT DELIVERY

At our institution, prior to simulation, patients undergo an MRI of the brain with contrast with thin cuts, ideally 1 mm in size. Patients are simulated in the supine position utilizing a three-point thermoplastic mask. More recently, immobilization is achieved with a custom head mold and open-face mask (CDR Systems, Calgary, Alberta, Canada). A CT scan utilizing 1-mm slice thickness through the entire brain and brainstem and with IV contrast is obtained, and images are fused to the MRI. Every effort is made to obtain an MRI within 1 week of SIM and 2 weeks of treatment initiation.

For patients with newly diagnosed glioblastoma, we employ a dose-painting technique. Enhancing tumor with a small margin is dose escalated, while also ensuring lower dose delivery to areas of T2 hyperintensity consistent with tumor. Variable approaches and doses are used, and careful consideration is made with regards to specific anatomy and patters of infiltration. In all cases, careful review with neuroradiology in a multidisciplinary setting is conducted prior to defining targets. The eyes, optic nerves, optic chiasm, brainstem, and spinal cord are contoured as organs at risk, with consideration given to the prior radiation course in terms of cumulative dose limits to each of these structures.

Inverse planning is typically utilized, and at this institution we make use of VMAT technique for delivery. However, hypofractionated RT may be delivered safely with conventional IMRT or 3D conformal approaches or with the use of Gamma Knife or CyberKnife platforms. It is crucial that image guidance, with daily cone-beam CT and real time motion tracking be utilized.

At our institution, treatment is delivered daily or every other day with a Varian TrueBeam Linear Accelerator with onboard imaging capability. Patients are first set up conventionally using room lasers and

Table 16.2 Studies of hypofractionated stereotactic radiation therapy in recurrent high-grade glioma

AUTHOR	PATIENTS	STUDY TYPE	MEDIAN DOSE	MEDIAN NUMBER OF FRACTIONS	BED (A/B = 3)	MEDIAN TUMOR VOLUME (CM³)	CONCURRENT SYSTEMIC AGENT	TOXICITY REQUIRING SURGICAL INTERVENTION	MEDIAN OS AFTER SALVAGE RT
Laing et al. (1993)	22	Prospective phase I/II	40 (20–50)	8 (4–10)	106.7	25 (1–93)	None	NR	9.8 months
Glass et al. (1997)	20	Prospective phase I/II	42 (37.5–42)	7 (7–10)	126	14.3 (1.76–122)	Cisplatin	15%	12.7 months
Shepherd et al. (1997)	29	Retrospective	35 (20–50)	7 (4–10)	93.3	24 (3–93)	None	6%	10.7 months
Hudes et al. (1999)	19	Prospective phase I	30 (21–35)	10 (7–10)	60	12.66 (0.89–47.5)	None	0%	10.5 months
Lederman et al. (2000)	88	Prospective phase I/II	24 (18–36)	4	72	32.7 (1.5–150.3)	Paclitaxel	8%	7 months
Voynov et al. (2002)	10	Retrospective	30 (25–40)	4 (2–5)	105	34.69 (4.29–75.23)	Various	20%	10.1 months
Vordermark et al. (2005)	19	Retrospective	30 (20–30)	5 (2–6)	90	15 (4–70)[a]	None	0%	9.3 months
Grosu et al. (2005)	44	Prospective phase I/II	30	6	80	19	Temozolomide	7%	8 months
Kohshi et al. (2007)	25 (11 GBM, 14 AA)	Retrospective	22 (18–27)	8	42.2	8.7(1.7–159.3)	None	28%	11 months (GBM)
Ernst-Stecken et al. (2007)	15	Prospective phase I/II	35	7	93.3	5.75 (0.77–21.94)	None	NR	12 months
Patel et al. (2009)	10	Retrospective	36	6	108	51.1 (16.1–123.3)	None	10%	7.4 months
Gutin et al. (2009)	25	Prospective phase I/II	30	5	90	34 (2–62)[a]	Bevacizumab	4%	12.5 months
Henke et al. (2009)	31	Retrospective	20 (20–25)	5 (4–5)	46.7	55 (0.9–277)[a]	Various	6%	10.2 months

(Continued)

Table 16.2 (Continued) Studies of hypofractionated stereotactic radiation therapy in recurrent high-grade glioma

AUTHOR	PATIENTS	STUDY TYPE	MEDIAN DOSE	MEDIAN NUMBER OF FRACTIONS	BED (A/B = 3)	MEDIAN TUMOR VOLUME (CM³)	CONCURRENT SYSTEMIC AGENT	TOXICITY REQUIRING SURGICAL INTERVENTION	MEDIAN OS AFTER SALVAGE RT
Fokas et al. (2009)	53	Retrospective	32.5 (20–60)	10 (5–30)	67.7	35.01 (3–204)[a]	TMZ, ACNU/VM-26, or PCV	0%	9 months
Fogh et al. (2010)	147	Retrospective	35	10	75.8	22 (0.6–104)	Various	0%	11 months (GBM)
Kim et al. (2011)	8	Retrospective	25	5	66.7	69.5[a]	None	13%	7.6 months
Minniti et al. (2011)	54 (42 GBM, 12 AA)	Prospective phase I/II	25	5	66.7	13.1 (1–35.3)	Bevacizumab or fotemustine	5%	13.0 months, AA 9 months, GBM
McKenzie et al. (2013)	35	Retrospective	30 (8–30)	5 (1–5)	90	8.54 (0.4–46.56)[a]	Various	0%	8.6 months
Ciammella et al. (2013)	15	Retrospective	25	5	66.7	NR[b]	None	0%	9.5 months
Wuthrick et al. (2014)	11	Prospective phase I/II	35 (30–42)	10 (10–15)	75.8	6.75 (0.05–72.01)	Sunitinib	0%	11 months
Miwa et al. (2014)	21	Prospective phase I/II	25–35	5	66.7–116.7	27.4 (3.4–102.9)[a]	None	4.8%	11 months
Navarria et al. (2015)	25	Retrospective	25 (20–50)	5 (5–10)	66.7	35 (2.46–116.7)	Various	0%	18 months
Clarke et al. (2017)	15	Prospective phase I	27, 30, and 33	3	108–154	<40	Bevacizumab	0%	13 months

[a] Planning target volume.
[b] Not reported.

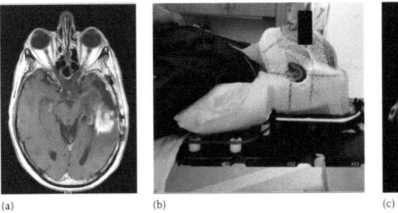
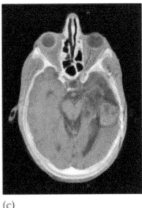

(a)   (b)   (c)

Figure 16.1 Fifty-eight-year-old male with WHO Grade IV glioblastoma who underwent resection and concurrent chemoradiation therapy. He experienced a recurrence 2 years later and was treated with partial resection of the tumor. He was recommended to undergo postoperative hypofractionated radiation therapy. An MRI of the brain (a) was used for treatment planning purposes. The patient was simulated in a CDR head frame (b). Images from the planning MRI were fused with CT images from the simulation in order to aid in target delineation (c). GTV—blue; PTV—red.

manual shifts. If the CDR head frame is used, an optical surface imaging program (AlignRT, VisionRT, London, UK) is used in conjunction with a treatment couch with six degrees of freedom to correct translational and rotational shifts.

The timing of concurrent chemotherapy is variable and should be instituted according to institutional practice and experience.

16.4 TOXICITY AND MANAGEMENT

Expected acute side effects from treatment to the brain include localized skin erythema, alopecia, fatigue, and mild headache. Skin effects can be managed with nonscented, noncolored moisturizers. Mild headache can be managed with over-the-counter medication. In the reirradiation setting, permanent hair loss may occur.

Unlikely acute side effects include headache requiring narcotic medication, nausea, vomiting, exacerbation of existing neurological symptoms, and the development of new focal neurological symptoms. The pathophysiology surrounding these effects is almost always due to an acute radiation-induced inflammatory reaction. In such cases, dexamethasone bolus followed by a taper should be considered. When used concurrently, the effects of bevacizumab may offset the signs and symptoms of acute inflammation and radionecrosis and decrease the need for corticosteroids.

Possible but unlikely long-term side effects include changes in memory or cognitive function, white matter changes, and radiation necrosis. The risk of symptomatic radiation necrosis is estimated to be less than 10%. In certain cases of symptomatic radiation necrosis that is refractory to steroid medication, surgical excision should be considered. The risk of radionecrosis is typically proportional to the irradiated treatment volume, and thus clinical judgment should be practiced in selecting appropriate patients and target volumes for IG-HSRT.

16.5 CONTROVERSIES AND FUTURE DIRECTIONS

Despite these important data demonstrating IG-HSRT for newly diagnosed glioblastoma to be well tolerated for small-volume disease, with evidence of treatment response and prolonged survival in some cases, the prognosis remains poor. While there have been promising results in a phase I trial combining hypofractionated

RT with bevacizumab, recently published phase III trials of utilizing standard chemoradiation with bevacizumab failed to show an overall survival benefit (Chinot et al. 2020; Gilbert et al. 2014) and have curbed enthusiasm for the use of this expensive agent in the up-front setting. The use of hypofractionation in the up-front setting may have the added the disadvantage of lessening the time that patients are treated concurrently with temozolomide, from 6 weeks with a standard regimen to only 1–2 weeks with a hypofractionated approach. Conversely, however, the efficacy of novel therapeutics such as immune checkpoint inhibitors might be enhanced when combined with hypofractionated radiation therapy, or, conversely, lymphopenia related to prolonged courses of radiotherapy may blunt immune-mediated tumor killing (Pike et al. 2019).

In the recurrent setting, the use of IG-HSRT for limited volume disease has a firm rationale in using high doses per fraction to treat radioresistant disease, image guidance to ensure accurate treatment delivery, and intensity modulation to help spare previously treated structures at risk. It appears to be well tolerated with low rates of radiation necrosis requiring surgical intervention, especially for small-volume disease. Recent advances have been made in imaging modalities that may help discriminate between radionecrosis and true tumor progression, such as FDG or choline-PET, perfusion mapping, and contrast subtraction mapping.

However, controversies still exist in the management of recurrent patients. Further study is needed to clarify which patients should undergo surgical resection, definitive reirradiation, chemotherapy, or observation. Similarly, the role of targeted inhibitors, cancer vaccines, oncolytic viruses, and immune checkpoint inhibitors in the recurrent setting has yet to be defined, either in combination with radiotherapy or as single modality treatments.

CHECKLIST: KEY POINTS FOR CLINICAL PRACTICE

✓	ACTIVITY	SOME CONSIDERATIONS
	Patient selection	*Is the patient appropriate for HSRT?* • Consider for patients who have a unifocal tumor or unifocal recurrence • KPS > 50 • No recent cytotoxic chemotherapy • Up-front setting: GTV < 60 cc • Recurrent setting: PTV < 40 cc
	SRS vs HSRT	*Is the lesion amenable to single-fraction SRS?* • In general, patients with HGG are treated with either HSRT or conventionally fractionated RT. The role of single-fraction SRS is limited in this disease
	Simulation	*Immobilization* • Rigid fixation, such as with a thermoplastic mask *Imaging* • CT with contrast with fused thin-cut MRI with contrast and T2 FLAIR sequences
	Treatment planning	*Contours* Newly diagnosed glioblastoma • Dose-painting technique with dose escalation to the area of enhancement • CTV and PTV expansions should be based on anatomic and technical parameters *Treatment planning* • Inverse treatment planning with VMAT is an option, but other IG-HSRT approaches may be used including IMRT, CyberKnife, or Gamma Knife • Dose constraints should consider cumulative dose from prior radiation courses in the recurrent setting

(Continued)

✓	ACTIVITY	SOME CONSIDERATIONS
	Treatment delivery	*Imaging* • Patients are first set up conventionally using room lasers and manual shifts • Real-time motion monitoring with devices such as ExactTrack or AlignRT should be utilized • Treatment couch with six degrees of freedom to correct translational and rotational shifts • Cone-beam CT to match to bony anatomy • Intrafraction motion monitoring *Concurrent systemic therapy* • Should be instituted per protocol based on center practice and experience

REFERENCES

Azoulay, Melissa, Steven D. Chang, Iris C. Gibbs, Steven L. Hancock, Erqi L. Pollom, Griffith R. Harsh, John R. Adler, et al. 2020. "A Phase I/II Trial of 5-Fraction Stereotactic Radiosurgery With 5-Mm Margins with Concurrent Temozolomide in Newly Diagnosed Glioblastoma: Primary Outcomes." *Neuro-Oncology* 22 (8): 1182–1189. https://doi.org/10.1093/neuonc/noaa019.

Bambury, Richard M., and Patrick G. Morris. 2014. "The Search for Novel Therapeutic Strategies in the Treatment of Recurrent Glioblastoma Multiforme." *Expert Review of Anticancer Therapy* 14 (8): 955–964. https://doi.org/10.1586/14737140.2014.916214.

Chinot, Olivier L., Wolfgang Wick, Warren Mason, Roger Henriksson, Frank Saran, Ryo Nishikawa, Antoine F. Carpentier, et al. 2014. "Bevacizumab Plus Radiotherapy—Temozolomide for Newly Diagnosed Glioblastoma." *New England Journal of Medicine* 370 (8): 709–722. https://doi.org/10.1056/NEJMoa1308345.

Chinot, Olivier L., Wolfgang Wick, Warren Mason, Roger Henriksson, Frank Saran, Ryo Nishikawa, Magalie Hilton, Lauren Abrey, and Timothy Cloughesy. 2020. "Bevacizumab Plus Radiotherapy—Temozolomide for Newly Diagnosed Glioblastoma." *New England Journal of Medicine* 709–722. https://doi.org/10.1056/NEJMoa1308345.

Cho, Kwan H., Walter A. Hall, Bruce J. Gerbi, Patrick D. Higgins, Warren A. McGuire, and H. Brent Clark. 1999. "Single Dose Versus Fractionated Stereotactic Radiotherapy for Recurrent High-Grade Gliomas." *International Journal of Radiation Oncology Biology Physics* 45 (5): 1133–1141. https://doi.org/10.1016/S0360-3016(99)00336-3.

Clarke, Jennifer, Elizabeth Neil, Robert Terziev, Philip Gutin, Igor Barani, Thomas Kaley, Andrew B. Lassman, et al. 2017. "Multicenter, Phase 1, Dose Escalation Study of Hypofractionated Stereotactic Radiation Therapy With Bevacizumab for Recurrent Glioblastoma and Anaplastic Astrocytoma." *International Journal of Radiation Oncology Biology Physics* 99 (4): 797–804. https://doi.org/10.1016/j.ijrobp.2017.06.2466.

Cloughesy, Timothy F., Aaron Y. Mochizuki, Joey R. Orpilla, Willy Hugo, Alexander H. Lee, Tom B. Davidson, Anthony C. Wang, et al. 2019. "Neoadjuvant Anti-PD-1 Immunotherapy Promotes a Survival Benefit With Intratumoral and Systemic Immune Responses in Recurrent Glioblastoma." *Nature Medicine* 25 (3): 477–486. https://doi.org/10.1038/s41591-018-0337-7.

Combs, Stephanie E., Maximilian Niyazi, Sebastian Adeberg, Nina Bougatf, David Kaul, Daniel F. Fleischmann, Arne Gruen, et al. 2018. "Re-Irradiation of Recurrent Gliomas: Pooled Analysis and Validation of an Established Prognostic Score—Report of the Radiation Oncology Group (ROG) of the German Cancer Consortium (DKTK)." *Cancer Medicine* 7 (5): 1742–1749. https://doi.org/10.1002/cam4.1425.

Floyd, Nathan S., Shiao Y. Woo, Bin S. Teh, Charlotte Prado, Wei Yuan Mai, Todd Trask, Philip L. Gildenberg, et al. 2004. "Hypofractionated Intensity-Modulated Radiotherapy for Primary Glioblastoma Multiforme." *International Journal of Radiation Oncology Biology Physics* 58 (3): 721–726. https://doi.org/10.1016/S0360-3016(03)01623-7.

Gilbert, Mark R., James J. Dignam, Terri S. Armstrong, Jeffrey S. Wefel, Deborah T. Blumenthal, Michael A. Vogelbaum, Howard Colman, et al. 2014. "A Randomized Trial of Bevacizumab for Newly Diagnosed Glioblastoma." *New England Journal of Medicine* 370 (8): 699–708. https://doi.org/10.1056/NEJMoa1308573.

Gondi, V., S. Pugh, C. Tsien, T. Chenevert, M. Gilbert, A. Omuro, J. Mcdonough, et al. 2020. "Radiotherapy (RT) Dose-Intensification (DI) Using Intensity-Modulated RT (IMRT) Versus Standard-Dose (SD) RT with Temozolomide (TMZ) in Newly Diagnosed Glioblastoma (GBM): Preliminary Results of NRG Oncology BN001." *Radiation Oncology Biology* 108 (3): S22–23. https://doi.org/10.1016/j.ijrobp.2020.07.2109.

Graf, Reinhold, Bert Hildebrandt, Wolfgang Tilly, Geetha Sreenivasa, Renate Ullrich, Klaus Maier-Hauff, Roland Felix, and Peter Wust. 2005. "Dose-Escalated Conformal Radiotherapy of Glioblastomas—Results of a Retrospective Comparison Applying Radiation Doses of 60 and 70 Gy." *Onkologie* 28 (6–7): 325–330. https://doi.org/10.1159/000085574.

Herrlinger, Ulrich, Theophilos Tzaridis, Frederic Mack, Joachim Peter Steinbach, Uwe Schlegel, Michael Sabel, Peter Hau, et al. 2019. "Lomustine-Temozolomide Combination Therapy Versus Standard Temozolomide Therapy in Patients with Newly Diagnosed Glioblastoma with Methylated MGMT Promoter (CeTeG/NOA-09): A Randomised, Open-Label, Phase 3 Trial." *Lancet* 393 (10172): 678–688. https://doi.org/10.1016/S0140-6736(18)31791-4.

Iuchi, Toshihiko, Kazuo Hatano, Takashi Kodama, Tsukasa Sakaida, Sana Yokoi, Koichiro Kawasaki, Yuzo Hasegawa, and Ryusuke Hara. 2014. "Phase 2 Trial of Hypofractionated High-Dose Intensity Modulated Radiation Therapy with Concurrent and Adjuvant Temozolomide for Newly Diagnosed Glioblastoma." *International Journal of Radiation Oncology Biology Physics* 88 (4): 793–800. https://doi.org/10.1016/j.ijrobp.2013.12.011.

Kaaijk, P., D. Troost, P. Sminia, M. C. C. M. Hulshof, A. H. W. Van Der Kracht, S. Leenstra, and D. A. Bosch. 1997. "Hypofractionated Radiation Induces a Decrease in Cell Proliferation But No Histological Damage to Organotypic Multicellular Spheroids of Human Glioblastomas." *European Journal of Cancer Part A* 33 (4): 645–651. https://doi.org/10.1016/S0959-8049(96)00503-5.

Kohshi, Kiyotaka, Haruaki Yamamoto, Ai Nakahara, Takahiko Katoh, and Masashi Takagi. 2007. "Fractionated Stereotactic Radiotherapy Using Gamma Unit after Hyperbaric Oxygenation on Recurrent High-Grade Gliomas." *Journal of Neuro-Oncology* 82 (3): 297–303. https://doi.org/10.1007/s11060-006-9283-1.

Laing, Robert W., Alan P. Warrington, John Graham, Juliet Britton, Frances Hines, and Michael Brada. 1993. "Efficacy and Toxicity of Fractionated Stereotactic Radiotherapy in the Treatment of Recurrent Gliomas (Phase I/II Study)." *Radiotherapy and Oncology* 27 (1): 22–29. https://doi.org/10.1016/0167-8140(93)90040-F.

Loeffler, Jay S., Eben Alexander, W. Michael Shea, Patrick Y. Wen, Howard A. Fine, Hanne M. Kooy, and Peter Mc L. Black. 1992. "Radiosurgery as Part of the Initial Management of Patients With Malignant Gliomas." *Journal of Clinical Oncology* 10 (9): 1379–1385. https://doi.org/10.1200/JCO.1992.10.9.1379.

Malmström, Annika, Bjørn Henning Grønberg, Christine Marosi, Roger Stupp, Didier Frappaz, Henrik Schultz, Ufuk Abacioglu, and Björn Tavelin. 2012. "Temozolomide Versus Standard 6-Week Radiotherapy Versus Hypofractionated Radiotherapy in Patients Older Than 60 Years With Glioblastoma : The Nordic Randomised, Phase 3 Trial." *Lancet Oncology* 13 (9): 916–926. https://doi.org/10.1016/S1470-2045(12)70265-6.

Omuro, Antonio, Kathryn Beal, Philip Gutin, Sasan Karimi, Denise D. Correa, Thomas J. Kaley, Lisa M. DeAngelis, et al. 2014. "Phase II Study of Bevacizumab, Temozolomide, and Hypofractionated Stereotactic Radiotherapy for Newly Diagnosed Glioblastoma." *Clinical Cancer Research* 20 (19): 5023–5031. https://doi.org/10.1158/1078-0432.CCR-14-0822.

Perry, James R., Normand Laperriere, Christopher J. O'Callaghan, Alba A. Brandes, Johan Menten, Claire Phillips, Michael Fay, et al. 2017. "Short-Course Radiation Plus Temozolomide in Elderly Patients With Glioblastoma." *New England Journal of Medicine* 376 (11): 1027–1037. https://doi.org/10.1056/NEJMoa1611977.

Pike, Luke R. G., Andrew Bang, Brandon A. Mahal, Allison Taylor, Monica Krishnan, Alexander Spektor, Daniel N. Cagney, et al. 2019. "The Impact of Radiation Therapy on Lymphocyte Count and Survival in Metastatic Cancer Patients Receiving PD-1 Immune Checkpoint Inhibitors." *International Journal of Radiation Oncology Biology Physics* 103 (1): 142–151. https://doi.org/10.1016/j.ijrobp.2018.09.010.

Reardon, David A., Alba A. Brandes, Antonio Omuro, Paul Mulholland, Michael Lim, Antje Wick, Joachim Baehring, et al. 2020. "Effect of Nivolumab vs Bevacizumab in Patients with Recurrent Glioblastoma: The CheckMate 143 Phase 3 Randomized Clinical Trial." *JAMA Oncology* 6 (7): 1003–1010. https://doi.org/10.1001/jamaoncol.2020.1024.

Reddy, Krishna, Denise Damek, Laurie E. Gaspar, Douglas Ney, Allen Waziri, Kevin Lillehei, Kelly Stuhr, Brian D. Kavanagh, and Changhu Chen. 2012. "Phase II Trial of Hypofractionated IMRT With Temozolomide for Patients with Newly Diagnosed Glioblastoma Multiforme." *International Journal of Radiation Oncology Biology Physics* 84 (3): 655–660. https://doi.org/10.1016/j.ijrobp.2012.01.035.

Roa, Wilson, P. M. A. Brasher, G. Bauman, M. Anthes, E. Bruera, A. Chan, B. Fisher, et al. 2004. "Abbreviated Course of Radiation Therapy in Older Patients With Glioblastoma Multiforme: A Prospective Randomized Clinical Trial." *Journal of Clinical Oncology* 22 (9): 1583–1588. https://doi.org/10.1200/JCO.2004.06.082.

Roa, Wilson, Lucyna Kepka, Narendra Kumar, Valery Sinaika, Juliana Matiello, and Darejan Lomidze. 2020. "International Atomic Energy Agency Randomized Phase III Study of Radiation Therapy in Elderly and/or Frail Patients With Newly Diagnosed Glioblastoma Multiforme." *Journal of Clinical Oncology* 33 (35). https://doi.org/10.1200/JCO.2015.62.6606.

Scott, Charles B., Charles Scarantino, Raul Urtasun, Benjamin Movsas, Christopher U. Jones, Joseph R. Simpson, A. Jennifer Fischbach, and Walter J. Curran. 1998. "Validation and Predictive Power of Radiation Therapy Oncology Group (RTOG) Recursive Partitioning Analysis Classes for Malignant Glioma Patients: A Report Using RTOG 90–06." *International Journal of Radiation Oncology Biology Physics* 40 (1): 51–55. https://doi.org/10.1016/S0360-3016(97)00485-9.

Selker, Robert G., William R. Shapiro, Peter Burger, Margaret S. Blackwood, Melvin Deutsch, Vincent C. Arena, John C. Van Gilder, et al. 2002. "The Brain Tumor Cooperative Group NIH Trial 87–01: A Randomized Comparison of Surgery, External Radiotherapy, and Carmustine Versus Surgery, Interstitial Radiotherapy Boost, External Radiation Therapy, and Carmustine." *Neurosurgery* 51 (2): 343–357. https://doi.org/10.1227/00006123-200208000-00009.

Sherriff, J., J. Tamangani, L. Senthil, G. Cruickshank, D. Spooner, B. Jones, C. Brookes, and P. Sanghera. 2013. "Patterns of Relapse in Glioblastoma Multiforme Following Concomitant Chemoradiotherapy with Temozolomide." *British Journal of Radiology* 86 (1022). https://doi.org/10.1259/bjr.20120414.

Stupp, Roger, Monika E. Hegi, Warren P. Mason, Martin J. van den Bent, Martin J. B. Taphoorn, Robert C. Janzer, Samuel K. Ludwin, et al. 2009. "Effects of Radiotherapy With Concomitant and Adjuvant Temozolomide Versus Radiotherapy Alone on Survival in Glioblastoma in a Randomised Phase III Study: 5-Year Analysis of the EORTC-NCIC Trial." *The Lancet Oncology* 10 (5): 459–466. https://doi.org/10.1016/S1470-2045(09)70025-7.

Stupp, Roger, Warren P. Mason, Martin J. Van Den Bent, Michael Weller, Barbara Fisher, Martin J.B. Taphoorn, Karl Belanger, et al. 2005. "Radiotherapy Plus Concomitant and Adjuvant Temozolomide for Glioblastoma." *New England Journal of Medicine* 352 (10): 987–996. https://doi.org/10.1056/NEJMoa043330.

Stupp, Roger, Sophie Taillibert, Andrew Kanner, William Read, David M. Steinberg, Benoit Lhermitte, Steven Toms, et al. 2017. "Effect of Tumor-Treating Fields Plus Maintenance Temozolomide vs Maintenance Temozolomide Alone on Survival in Patients With Glioblastoma a Randomized Clinical Trial." *JAMA—Journal of the American Medical Association* 318 (23): 2306–2316. https://doi.org/10.1001/jama.2017.18718.

Tsien, Christina, Jennifer Moughan, Jeff M. Michalski, Mark R. Gilbert, James Purdy, Joseph Simpson, John J. Kresel, Walter J. Curran, Aidnag Diaz, and Minesh P. Mehta. 2009. "Phase I Three-Dimensional Conformal Radiation Dose Escalation Study in Newly Diagnosed Glioblastoma: Radiation Therapy Oncology Group Trial 98–03." *International Journal of Radiation Oncology Biology Physics* 73 (3): 699–708. https://doi.org/10.1016/j.ijrobp.2008.05.034.

Twyman-Saint Victor, Christina, Andrew J. Rech, Amit Maity, Ramesh Rengan, Kristen E. Pauken, Erietta Stelekati, Joseph L. Benci, et al. 2015. "Radiation and Dual Checkpoint Blockade Activate Non-Redundant Immune Mechanisms in Cancer." *Nature* 520 (7547): 373–377. https://doi.org/10.1038/nature14292.

Vanpouille-Box, Claire, Amandine Alard, Molykutty J. Aryankalayil, Yasmeen Sarfraz, Julie M. Diamond, Robert J. Schneider, Giorgio Inghirami, C. Norman Coleman, Silvia C. Formenti, and Sandra Demaria. 2017. "DNA Exonuclease Trex1 Regulates Radiotherapy-Induced Tumour Immunogenicity." *Nature Communications* 8 (June): 15618. https://doi.org/10.1038/ncomms15618.

Voynov, George, Seth Kaufman, Theodore Hong, Arthur Pinkerton, Richard Simon, and Robert Dowsett. 2002. "Treatment of Recurrent Malignant Gliomas With Stereotactic Intensity Modulated Radiation Therapy." *American Journal of Clinical Oncology: Cancer Clinical Trials* 25 (6): 606–611. https://doi.org/10.1097/00000421-200212000-00017.

Walker, Michael D., E. Alexander, W. E. Hunt, C. S. MacCarty, M. S. Mahaley, J. Mealey, H. A. Norrell, et al. 1978. "Evaluation of BCNU and/or Radiotherapy in the Treatment of Anaplastic Gliomas: A Cooperative Clinical Trial." *Journal of Neurosurgery* 49 (3): 333–343. https://doi.org/10.3171/jns.1978.49.3.0333.

Walker, Michael D., Sylvan B. Green, David P. Byar, Eben Alexander, Ulrich Batzdorf, William H. Brooks, William E. Hunt, et al. 1980. "Randomized Comparisons of Radiotherapy and Nitrosoureas for the Treatment of Malignant Glioma After Surgery." *New England Journal of Medicine* 303 (23): 1323–1329. https://doi.org/10.1056/NEJM198012043032303.

Watkins, John M., David T. Marshall, Sunil Patel, Pierre Giglio, Amy E. Herrin, Elizabeth Garrett-Mayer, and Joseph M. Jenrette. 2009. "High-Dose Radiotherapy to 78 Gy With or Without Temozolomide for High Grade Gliomas." *Journal of Neuro-Oncology* 93 (3): 343–348. https://doi.org/10.1007/s11060-008-9779-y.

17

Hypofractionated Stereotactic Radiotherapy for Meningiomas, Vestibular Schwannomas, and Pituitary Adenomas

Adomas Bunevicius, Eric J. Lehrer, Einsley-Marie Janowski, and Jason P. Sheehan

Contents

17.1 INTRODUCTION

Skillful microsurgical resection that heavily relies on detailed anatomical knowledge, technical expertise, manual dexterity, and experience continues to hold a central role in the management of skull base tumors. Ongoing perfection of microsurgical techniques and operative approaches together with technological advances of preoperative brain imaging, intraoperative brain mapping, endoscopic and microscopic techniques, refinements of skull base reconstruction, and postoperative care have substantially contributed toward improved safety and efficacy of microsurgical resection of complex skull base tumors. Nevertheless, the learning curve of skull base neurosurgery remains steep, and postoperative morbidity continues to challenge open complete surgical removal of certain skull base tumors. Substantial experience is required for optimized efficacy and safety of skull base surgery, thus largely limiting the availability of open skull base neurosurgical treatments to experienced clinical centers with adequate patient volume [1, 2]. Even in experienced hands, the incidence of temporary and permanent cranial nerve deficits after skull base tumor resection has been reported to be as high as 44% and 56%, respectively [3–10]. Postoperative mortality rates after skull base surgery has been reported as high as 9%, with lower rates in more recent series [4, 6, 11–20].

Benign tumors comprise a vast majority of all primary CNS tumors [21, 22]. Meningiomas are the most common primary intracranial neoplasm, comprising 38% of all CNS tumors, followed by pituitary tumors (17%) [21, 22]. Vestibular schwannomas are the most common non-malignant nerve sheath tumors, comprising 9% of all primary intracranial tumors [21, 23]. Benign brain tumors are often slow-growing neoplasms that can remain asymptomatic for a prolonged period of time, often being diagnosed incidentally [24–26]. Given the high incidence rate and generally favorable long-term prognosis of benign brain tumors, preservation of neurological function and quality of life is imperative. Therefore, potential benefits of attempted complete microsurgical resection of benign skull base tumors should be carefully weighed and balanced against the risks of treatment. Clinicians must choose wisely between gross total surgical resection, subtotal resection followed by radiation therapy, definitive radiation therapy, or imaging surveillance. Advantages of microsurgical resection include immediate decompression of the brain, neural and/or vascular structures, alleviation of mass effect, and confirmation of histological diagnosis. However, complete microsurgical resection of some meningiomas, pituitary adenomas, and vestibular schwannomas is often not technically possible due to their infiltrative tumor growth and intimate contact with manipulation-sensitive vital neural structures and blood vessels. Incomplete surgical resection of meningiomas without adjuvant radiation therapy can be associated with accelerated tumor progression after surgery [27]. A 4-mm mean annual growth rate of partially resected petroclival meningiomas has been reported [28]. The most common complications of surgical resection of vestibular schwannomas include hearing loss, facial weakness, and CSF leak [29–32], and the recurrence rate of incompletely resected vestibular schwannomas approaches 10% to 15% [33–35]. Local control of benign CNS tumors can be significantly improved with adjuvant radiation therapy. For example, a 5-year progression-free survival of 98% was reported for incompletely resected meningiomas treated with adjuvant conventionally fractionated external beam radiotherapy [36, 37]. Adjuvant radiation therapy was also consistently shown to improve local control and overall survival of patients harboring atypical and malignant meningiomas [38–40]. Therefore, adjuvant radiation therapy is often considered for WHO grade II and III meningiomas as well as for incompletely resected and/or progressing WHO grade I meningiomas [2, 41–48]. Similarly, radiation therapy for incompletely resected or infiltrative pituitary adenomas results in local tumor control of more than 90%–95% as well as endocrine remission of the majority of functioning pituitary adenomas [49–52].

Cytoreductive surgery followed by radiosurgery for residual tumors allows for a less aggressive surgical resection with the goal of functional preservation, and this paradigm is becoming increasingly used for the contemporary management of benign skull base tumors [53, 54]. This treatment approach entails the combination of planned subtotal surgical debulking of a tumor followed by radiosurgery to the tumor remnant(s) that are intentionally left adjacent to the most critical brain structures. For many patients, combining surgery with radiosurgery offers the best and most durable tumor control, while offering neurological preservation or improvement.

Radiation therapy can be used for adjuvant treatment and as a primary treatment approach for benign cranial base tumors [55–58]. Stereotactic radiosurgery (SRS) allows spatially precise treatment of intracranial lesions using high doses of radiation with submillimeter accuracy. SRS was introduced more than four decades ago, and its safety and efficacy for skull base tumors are well documented [59–62]. Increasing experience, technological advances in SRS treatment planning and treatment delivery, and improved imaging quality for imaging-guided radiation have contributed towards the improved safety and efficacy of SRS [63]. The treatment envelope of SRS has also expanded, making lesion size and location less technically restrictive than they once were. For example, the first-generation, frame-based techniques initially allowed for only limited access to low-lying tumors, such as skull base and upper neck. Newer frames allow better access to these low-lying tumors. However, as the treatment volume increases, so does the area of normal brain and other healthy tissue that is irradiated, thereby increasing the risk of radiation-induced complications. Because of such limitations, radiosurgery has been traditionally used primarily to treat "smaller" lesions, defined as less than 3 cm in largest diameter (or smaller than 10 to 15 cm^3 in volume). Fractionated radiosurgery has allowed for larger treatment volumes while minimizing the risk of both radionecrosis and normal structure toxicity [64–66].

17.2 RADIOBIOLOGY OVERVIEW

Delivery of therapeutic radiation is a sophisticated modality that requires the knowledge of both radiobiology and physics. Radiation therapy damages cellular DNA by both direct and indirect mechanisms, thereby destroying the cell's ability to divide and grow. Radiation exposure can cause damage to both normal tissues and tumors. Radiation oncology has evolved over the course of decades to optimize the therapeutic ratio in radiation treatments, thus improving tumor cellular killing while minimizing critical structure toxicity. Radiation therapy can produce both lethal damage, which is irreversible and irreparable, or sublethal damage, which, under normal conditions, can be repaired in hours after exposure. Cellular sensitivity to radiation has many factors, including which phase the cell is in the cell cycle, with the S phase being relatively radiation resistant and the G2M phase being radiation sensitive. The 4R's of radiobiology include repair, reassortment, repopulation, and reoxygenation. Repair refers to the cellular repair of sublethal damage within hours of radiation exposure. Reassortment refers to the progression of cells through the cell cycle during the interval between radiation doses. If the interval between doses exceeds the length of the cell cycle, repopulation, or an increase of the surviving fraction resulting from cell division, can occur, a scenario only typically seen in rapidly dividing cells. In the end, reoxygenation refers to the process by which hypoxic cells become better oxygenated during periods after radiation, which subsequently improves the cellular lethality to radiation. A final factor to be considered is radiosensitivity, which is often thought of as an intrinsic factor to the tissue being treated. These factors can influence the therapeutic ratio in independent ways, and modern fractionation schemes have been developed based on manipulating these effects to maximize tumor cell kill while avoiding acute and late toxicity in normal tissue. The linear–quadratic equation has been developed as a model to predict cell killing by radiation, with the important parameters required in this calculation being the alpha/beta ratio, or the dose where cell killing due to the linear and quadratic components are equal. The alpha/beta ratio is typically high for tumors and acutely responding tissue and low for late-responding tissue. Ultimately, late tissues are typically more sensitive to changes in radiation fraction sizes [67].

17.3 CONVENTIONAL RADIATION THERAPY VERSUS RADIOSURGERY

External beam radiation may be delivered using either linear accelerators, which produce therapeutic X-rays, or with Gamma Knife units, which utilize Cobalt-60 radioactive decay to produce gamma rays. Technological innovations in radiation delivery have allowed optimization of the therapeutic ratio, with increasingly sophisticated improvements in immobilization, treatment delivery systems and treatment planning. These innovations have allowed more precise targeting of tumors and reduced exposure to normal tissues. As more precise tumor targeting has been achieved, radiation dosing has also evolved, including the development of radiosurgery.

Universally, whenever radiation therapy is being considered as a therapeutic option, great care must be taken in minimizing the dose delivered to radiosensitive normal intracranial structures, such as the visual pathways and brainstem. This issue becomes of even greater concern when the target involves any of these structures. Thus, with the goal of protecting radiosensitive intracranial structures and preserving or enhancing neurological function, radiosurgery has had a number of innovations. The optimization of patient immobilization with frame-based devices and the improvement in target/critical structure delineation with image-guided radiosurgery have both improved the accuracy of treatment delivery, thereby allowing for tighter radiation margins and less critical structure dose exposure. In addition, the use of nonisocentric beam delivery and development of inverse planning approaches have allowed for a higher degree of dose homogeneity. As a result, hot spots (defined as regions exceeding the prescription dose and typically kept to less than 110% of the dose prescription) can be kept away from nearby radiosensitive structures at risk [68].

However, despite these important advances, single-fraction radiosurgery tends to be less applicable for some cases than conventionally fractionated radiation therapy (CFRT), particularly with respect to the development of late radiotoxicities, which are more sensitive to higher doses of radiation. Hypofractionated radiosurgery (HFRS), which is typically defined as treating targets in 5 fractions or less at a dose of 5 Gy/fraction, has emerged as a viable treatment alternative in a variety of clinical settings.

The major advantages of frame-based radiosurgical techniques over CFRT are improved treatment accuracy and an enhanced ability to spare nearby radiosensitive structures, due to optimized patient immobilization, tighter PTV margins, and image-guided treatment delivery [69, 70]. Gradient index is typically improved with Gamma Knife radiation, which would improve the low-dose spray to surrounding structures, at least when treating multiple targets. While conventional radiation delivery techniques have undergone a great deal of innovation over the past two decades, which have resulted in markedly improved dose conformality, such as intensity modulated radiation therapy (IMRT) and volumetric modulated arc therapy (VMAT), radiosurgery still provides a superior dose conformality and selectivity [69–72]. In contrast, a major advantage to fractionated approaches during the delivery of ionizing radiation with respect to the development of late toxicity is interfraction tissue repair, or sublethal damage repair, which allows delivery of high total doses of radiation while maintaining low levels of critical structure late toxicity. The relationship between radiation dose and tumor cell survival may be represented by the linear–quadratic model, at least below 10 Gy per fraction [73]. However, while the linear–quadratic model is frequently used to compare radiation treatment regimens, use of this model in radiosurgical settings remains controversial. Indeed, some argue that higher doses of radiation per day are associated with other radiobiologic mechanisms besides classic DNA damage, including vascular damage and antigen expression, which might both achieve an enhanced dose response beyond what is estimated using the linear–quadratic model and increase toxicity to nearby critical structures [74]. Furthermore, the increased number of days of treatment and larger treatment fields associated with CFRT may increase the risk of radiotherapy-induced lymphopenia [75–77].

17.4 SINGLE SESSION VERSUS HYPOFRACTIONATED RADIOSURGERY

One of the major radiobiologic differences between CFRT and SRS is that the former is generally most effective in killing rapidly dividing cells, while the high radiation doses delivered using the latter technique may achieve lethality by more than just DNA damage (i.e., vascular damage and antigen presentation) [78]. With 5-year tumor control rates exceeding 90%, SRS has been shown to be safe and effective at treating many benign parasellar lesions [68, 79, 80]. However, single-session radiosurgery does have limitations, such as when treating targets that are close or within eloquent intracranial structures. The dose gradient achieved with all forms of single-fraction radiosurgery can at times be inadequate for the treatment of lesions abutting optical and brainstem structures. For instance, multiple studies have shown that exposure of the optic chiasm or nerves to more than 8–12 Gy in a single fraction is associated with a higher risk of visual impairment [81, 82], but a maximal dose of 20–25 Gy in 5 fractions is generally well tolerated to the anterior optic pathways [68, 83, 84]. In addition, the ability to effectively delineate the optic apparatus that is effaced or displaced by a tumor is often difficult, despite the use of imaging modalities, such as volumetric MRI. Thus, single-session radiosurgery may be relatively contraindicated in cases when the distance between the target and anterior visual pathways is less than 3mm, as the delivery of a curative dose to the target while maintaining an acceptable low risk of damage to the optic apparatus is often not possible in this setting. Hypofractionated radiosurgery might also be the prudent choice for reirradiation cases. While recent studies have suggested that the risk of radiation-induced optic nerve/chiasm neuropathy after single-fraction SRS is less than 1% in patients treated with 10–12 Gy who have no had previous radiation therapy [84], the ability of the optic structures to tolerate radiation therapy can be further compromised by pre-existing visual deficits and prior course of radiation therapy in this area [85].

Hypofractionated radiosurgery is aimed at integrating the short treatment time and high degree of target conformality of SRS with the benefits of normal tissue repair between fractions, as well as the hypothetical benefit of enhanced tumor cell killing due to re-oxygenation and redistribution to more radiosensitive phases of the cell cycle between fractions that is seen with CFRT [86]. Thus, hypofractionated radiosurgery is able to mitigate the limitations of single-session SRS in the setting of larger volume targets and suboptimal dose falloff [68, 87]. The utilization of hypofractionated radiosurgery has increased in tandem with the number of available linear accelerator-based radiosurgical modalities [68, 88]. Reflecting these developments, the definition of "radiosurgery" has been revised to encompass treatment delivery in 1 to 5 fractions [89].

17.5 RADIOBIOLOGY OF HYPOFRACTIONATED SRS

The reason for using larger doses per fraction of radiation stems from fundamental radiological principles. While there is a paucity of data demonstrating the benefit of larger fraction size in treating benign, as opposed to malignant brain tumors, there is a strong theoretical foundation for this conclusion [90]. High rates of tumor control have been observed in studies comparing SRS to CFRT; however, the higher dose per fraction in SRS is associated with a higher biologically effective dose (BED), which has been correlated with greater tumor shrinkage on follow-up imaging [91]. This has been observed in doses of more than 5 Gy per fraction, which appear to activate cellular mechanisms and induce vascular changes that differ by doses used in CFRT [28].

When using hypofractionated approaches, higher total doses of radiation are required to produce the same physiologic and biologic effects than what would be necessary to achieve the same effect with single-fraction radiation (e.g., SRS). This is largely due to both the cellular sublethal damage repair that occurs between hypofractionated SRS doses of radiation as well as the likely synergistic aspects of high-dose single-fraction SRS in cell killing, which also cause vascular damage and antigen expression. For example, a patient treated with Gamma Knife radiosurgery with a margin dose of 19.7 Gy prescribed to the 50% isodose line given over three to five treatment sessions for a parasellar or perioptic lesion has a mean BED (α/β = 2.5) of 60.9 Gy; however, when the same target is treated with single-fraction SRS to a margin dose of 12 Gy, the mean BED (α/β = 2.5) is 69.6 Gy.

The linear–quadratic model, which is used to calculate BED, is a widely used principle in radiation therapy to compare different fraction schemes. Despite the popularity of the linear–quadratic model, which likely stems from its ease of use, many have raised concerns that this model becomes less valid, particularly when regimens are delivered at high doses per fraction [28]. Furthermore, this is particularly true when the α/β ratio is dose range dependent and when high doses per fraction are administered, as the linear–quadratic curve bends continuously on the log-linear plot. This behavior is not observed in the experimental setting, as many clonogenic cell survival studies at high doses of radiation therapy have shown a dose–response relationship that more closely resembles a straight line [92, 93]. While alternative models have been proposed [73, 93], they require further validation in both the in vitro and clinical setting.

The universal survival curve, which was proposed in 2008, is a hybrid model that combines principles of both the linear–quadratic and multi-target models, which can be used to compare the dose fractionation schemes of both CFRT and SRS [94]. This model is said to provide an empirically and clinically well-justified rationale for the use of hypofractionated radiation, while preserving the valid strengths of the linear–quadratic model for CFRT. Additional models, such as the modified linear–quadratic model, which incorporate a new parameter into the traditional linear–quadratic model that is derived from in vitro cell survival data of several human tumor cell lines and in vivo animal iso-effect curves, have also been proposed [73]. It is thought that the modified linear–quadratic model describes cell survival better after radiosurgery and hypofractionated surgery than the traditional linear–quadratic model [73]. This continues to remain a point of controversy.

17.6 TREATMENT PLANNING

Hypofractionated stereotactic radiotherapy is often considered for treatment of large tumors (>10 cm^3), tumors residing in close spatial proximity or abutting critical neural structures, such as the optic nerve or brainstem, and for patients who failed prior single-session SRS. Meningiomas and vestibular schwannomas are extra-axial brain tumors that usually vividly enhance with sharp margins of demarcation on contrast-enhanced T1-weighted MRI sequences, thus presenting themselves as ideal targets for highly conformal image-guided radiation therapy techniques.

SRS is minimally invasive (frame-based) or noninvasive (thermoplastic mask) therapy that can be safely performed in nearly all patients, including patients who are not fit for open surgical resection or do not wish to pursue open surgery. HFSRT is usually performed using a frameless approach with a thermoplastic mask, thus allowing safe treatment of nearly all patients, including those in need of continuous anticoagulation therapy. On the other hand, HFSRT can be more challenging for patients who are uncooperative and for children because of the need for conscious sedation or general anesthesia.

Radiation dose selection for treatment of benign lesions is influenced by numerous factors that include tumor type, volume, and proximity to sensitive and critical neural structures, extent of irradiated optic nerve and/or brainstem, as well as a previous history of radiation therapy and surgical treatment. Although biologically effective dose (BED) formulas can be used to guide optimal dose section (for temporally fractionated radiosurgery), the initial choice of number of sessions and dose is to a large extent empirically and experience based. Historically, protocols were derived from an earlier experience with hypofractionated frame-based radiosurgery in patients with no other treatment options [68]. With hypofractionation, the radiation tolerance to the anterior optic apparatus can be extended to 15.3–17.4 Gy for 3 fractions and 23–25 Gy for 5 fractions [93]. We believe that the maximal number of sessions (most conservative approach) should be reserved for patients with the longest involvement of the optic apparatus or where the optic nerve or chiasm are most displaced and, as such, cannot be clearly contoured.

There are few adequately powered studies to guide us in choosing multi-fraction radiosurgical parameters for the treatment of different benign lesions. Hypofractionated, multistage (i.e., temporal and spatial fractionation) treatments, and/or a combination of both have been performed for years using frame-based Gamma Knife radiosurgery to treat selected large cranial base meningiomas and brain arteriovenous malformations [95–97]. There also remains between-center variations in fractionation schemes, number of fractions used per lesion, and radiation dose selection (per fraction and total treatment dose). Furthermore, radiation planning software and radiation delivery devices (GK versus CK versus LINAC) also vary across studies. Center experience and era of treatment should also be considered when interpreting published series, given technological evolution of imaging and SRS techniques. Most centers tend to be more conservative in their initial dose selection prior to established safety with their earliest-treated patients, and these centers sequentially transition to gradual dose escalation in order to maximize the likelihood of long-term tumor control with preserved safety profile. However, whether or not a further dose escalation and resulting increase in BED are warranted remains uncertain.

17.7 TEMPORAL VERSUS SPATIAL FRACTIONATION

Increasing experience and better understanding of the radiobiology of fractionated SRS have substantially expanded the treatment envelope of SRS for intracranial malignancies [68, 98, 99]. Hypofractionated treatment regimens can improve the therapeutic ratio and in selected cases better avoid normal tissue toxicity compared to single-fraction SRS regimens [93]. As such, patients with multiple lesions may be treated with repeated single-fraction SRS to control the local progression of each lesion without potentially deleterious effects of less conformal radiation therapy techniques, such as whole-brain radiation therapy [100, 101]. However, it is important to distinguish a multisession hypofractionated treatment of the same index lesion from multiple single-fraction treatments of different lesions in patients with multiple lesions or distant recurrences. Multiple serial sessions of single-fraction SRS are often used for patients with brain metastases, where new lesions (not seen at the initial treatment scans) emerge during the disease course on follow-up brain imaging and are treated with SRS.

The multi-staging approach (volume-staged fractionation) most often refers to the segmentation of a large volume lesions to smaller volumes that are treated with single-session SRS at different points in time. This approach has been employed in SRS for large AVMs for decades [95, 102, 103] and is also sometime used for large brain metastases [104] and meningiomas [105]. Volume-staged radiosurgery has shown tumor control rates of 90% with low morbidity for intracranial meningiomas exceeding 3 cm in diameter [106].

17.8 MENINGIOMAS

17.8.1 OPTIC NERVE SHEATH MENINGIOMAS

Optic nerve sheath meningiomas (ONSMs) are extremely rare tumors, representing 2% of all orbital tumors, 1%–2% of intracranial meningiomas, and one-third of optic nerve lesions [81, 107, 108]. ONSMs arise from the meninges surrounding the optic nerve and usually grow circumferentially along the nerve. The most frequent symptom of ONSMs are visual decline (reduction in visual acuity and visual fields or both), resulting from either direct compression of the optic nerve (and resultant optic atrophy) or from vascular rearrangement (the so-called optociliary shunts, which are usually a late and rare sign, resulting from the direct compressive optic neuropathy). This triad of visual disturbances, optic nerve atrophy, and optociliary shunts are pathognomonic of ONSM [109]. Given their intimate contact with the optic nerve, management of ONSM requires special considerations. Treatment options for ONSMs includes observation, surgery, CFRT, or SRS. Conservative treatment with observation, whose reasoning stems from the benign nature of the tumor, invariably leads to visual deterioration and complete blindness [108, 110], and thus has fallen out of favor. Surgical resection was historically considered as the treatment of choice; however, it is associated with a high risk of vision loss in the affected eye due to a shared blood supply by the tumor and the optic nerve [111, 112, 108]. Furthermore, a non-negligible fact is that surgery may lead to very poor aesthetic results. Decompression of the optic sheath is sometimes considered to allow decompression of the optic nerve and confirm tissue diagnosis, but this can be associated with tumor seeding [108, 113]. A surgical approach best suits those instances where histological confirmation is required, an immediate reduction in tumor burden in the frame of deteriorating vision is needed to preserve vision, or when patients have no useful vision.

Given radiation tolerance of the optic nerve, single-session SRS has seldom if ever been proposed as an ONSM treatment unless the patient's vision is substantially compromised in the affected eyes [82, 83, 108, 114, 115]. However, HFSRT can be used to treat ONSM [116–118] (Table 17.1). Marchetti et al. reported their experience with 21 patients treated with radiosurgery using a frameless CyberKnife system, with 5 fractions of 5 Gy each prescribed to the 75%–85% isodose line [117]. During a mean follow-up of 30 months, the majority of patients responded well, with only one patient developing transient steroid-responsive optic neuropathy. Similar results were reported by Romanelli with colleagues in five patients with ONSM treated with HFSRT delivering 5 Gy in 4 fractions, with follow-up ranging from 36 to 74 months [118]. Local tumor control was achieved in all patients, and four patients experienced restoration of normal vision 6 to 12 months after the treatment. A fractionation schedule of 5 Gy delivered over 4 fractions, according to the linear–quadratic model, provides an equivalent dose at least comparable with conventional fractionated regimens, which deliver 50.4 to 56 Gy in 1.8 to 2 Gy fractions [117]. CFRT is traditionally considered for management of ONSMs because it achieves a durable local tumor control and visual preservation in the majority of patients with minimal risk of late toxicity [86, 119, 120]. HFSRT appears to offer comparable tumor control rates and safety profile as CFRT [86, 120].

17.8.2 PERIOPTIC MENINGIOMAS

Meningiomas residing within 2 mm of the optic apparatus were coined as perioptic meningiomas [68, 121]. Surgical resection of perioptic lesions can be technically challenging, and manipulation of the optic nerves can cause visual deterioration. CFRT is most commonly used for the management of perioptic lesions; however, CFRT can cause irradiation of larger regions of surrounding brain, optic apparatus, pituitary gland, and hypothalamus, which can subsequently increase the risk for radiation necrosis, secondary malignancies, and cerebrovascular events [122–125]. Single-session SRS allows for highly conformal delivery of high doses of radiation and can be considered for the management of perioptic lesions; however, 2–3 mm distance from the optic apparatus is traditionally accepted for optimized risk–benefit ratio of SRS (Figure 17.1).

Adler with colleagues reported the Stanford University Medical Center experience with hypofractionated CyberKnife-based SRS for treatment of 49 consecutive perioptic lesions (27 meningiomas) situated within 2 mm of a "short segment" of the optic apparatus, with some tumors displacing or obscuring the optic apparatus [68]. The primary sites of meningiomas were cavernous sinus alone (n = 9), cavernous sinus with posterior orbital involvement (n = 6), tuberculum sella (n = 6), medial sphenoid wing (n = 3), orbital apex (n = 2), and petroclival region (n = 1). Patients were treated using the CyberKnife system, delivering radiation in two to five sessions with average cumulative marginal dose of 20.3 Gy (range, 15–30 Gy). In the total sample of brain tumor patients, local tumor control was achieved in 94% of patients during a median imaging follow-up of 46 months (range, 13–100 months). Two meningioma patients experienced disease recurrence close to or within the treatment field. During visual follow-up of 46 months (range, 6–96 months), visual fields remained stable or improved in 94% of patients. Aside from transient and steroid responsive diplopia, two patients with cavernous sinus meningiomas experienced visual deterioration in the ipsilateral eye that was attributed to tumor progression. Similarly, Conti with colleagues reported 100% tumor control rate without permanent complications during a median follow-up of 58 months in 25 patients with perioptic meningiomas treated using HFSRT (median dose: 25 Gy; range: 18–40 Gy) [84]. Demiral with colleagues used a LINAC-based HFSRT for treatment of 22 anterior clinoid meningiomas located within 2 mm of the optic nerve, delivering 25 Gy in 5 fractions [126]. PFS at 3 years was 89%, and two (11%) patients experienced new visual field deficits. Kim J. W. et al. reported their experience with HFSRT for 13 perioptic meningiomas all within 1 mm of the optic apparatus [98]. Radiation was delivered in 3 to 4 fractions with a median cumulative marginal dose of 20 Gy. Mean follow-up time was 29 months, during which tumor control was achieved in 100% (n = 13). Visual function was stable or improved in all patients. No other complications were observed. In their report focusing on HFSRT for the treatment of orbital tumors, Kim M. S. et al. [127] reported on their experience with five patients harboring orbital meningiomas near the optic nerve. Treatment was delivered in 3 or 4, and median marginal dose was 20 Gy (15–20 Gy). During a follow-up period ranging from 7 to 43 months, tumor control was confirmed in four of five patients. Of note, one patient who suffered tumor recurrence had a WHO-II rhabdoid meningioma, which required surgical resection. No other complications were reported.

17.8.3 SKULL BASE MENINGIOMAS

Skull base meningiomas are a heterogeneous group with respect to their anatomic location, behavior, and treatment strategy. However, in some published series, skull base meningiomas of different anatomic locations are often analyzed together.

In the German and Italian pooled cohort study, 205 patients with skull base meningiomas were treated using HFSRT, and 136 patients were treated using CFRT [128]. Local tumor control rates were similar between the two patient groups of 99% at 1 year and 80% at 10 years. Eight percent of patients treated in the HFSRT group experienced mild or moderate cranial neuropathy, with one case of visual disturbance. Park with colleagues presented their experience with HFSRT (Gamma Knife) for large (≥10 cm³) skull base meningiomas in 23 patients with mean tumor volume of 21.2 ± 15.63 cm³ (range, 10.09–71.42 cm³) [129]. Treatment was delivered in 3–4 fractions with a median dose per fraction and margin dose of 6 Gy and 18 Gy, respectively [130]. During a mean follow-up of 38 months, tumor control was achieved in all patients, and mean tumor volume decreased from 21.63 ± 15.46 cm³ to 19.75 ± 15.70 cm³. Two patients developed new or worsening peritumoral edema that was asymptomatic and successfully managed with steroids, and five patients experienced transient new cranial neuropathy. Navarria with colleagues treated

26 patients for skull base meningiomas delivering 30 Gy over 5 fractions using the volumetric modulated arc therapy (RapidArc) with Varian linear accelerators (Varian Medical Systems, Palo Alto, the United States) [130]. During a median follow-up of 24.5 months (range, 5–57 months) there were no cases of tumor progression nor severe adverse radiation toxicity [130]. Han with colleagues presented their experience with 22 skull base meningioma patients treated using LINAC-based hFSRT delivering 25 Gy over 5 treatment fractions, and found that, during a median follow-up of 32 months (range, 7–97 months), progression-free survival was 94% with no serious morbidity [131].

17.8.4 UNSPECIFIED OR OTHER INTRACRANIAL LOCATIONS FOR MENINGIOMAS

A handful of groups published their experience with HFSRT for meningiomas not discriminating by anatomic locations. Meniai-Merzouki with colleagues combined experience with HFSRT from three institutions in France [132]. Their cohort included 126 patients who harbored convexity (n = 42), skull base (n = 27), and posterior fossa (n = 27) meningiomas. Patients were treated using a median prescription dose of 25 Gy (range, 12–40 Gy) delivered over a median of 5 fractions (range, 3–10 fractions). Actuarial local tumor control rates at 1 year and at 2 years were 78% and 70%, respectively. Histories of CFRT were associated with inferior local tumor control rates after HFSRT. Higher total dose, lower fraction dose, higher number of fractions, and lower WHO grade were significant favorable prognostic factors of progression-free survival in the primary HFSRT group, while none of the examined prognostic factors was associated with progression-free survival in the reirradiated HFSRT cohort. Greater hypofractionation was also associated with inferior local tumor control in the primary HFSRT subgroup with 2-year progression-free survival of 62% in patients treated using 21–23 Gy in 3 fractions versus 92% in patients treated with 25–40 Gy in 5–10 fractions. There were no cases of severe early radiation toxicities, and 34% of patients experienced delayed grade 1–2 toxicities. Seven patients (two in the primary HFSRT group and five in reirradiated group) developed radiation necrosis at a median of 4.4 months after the SRS that was successfully managed with steroids; anti-vascular endothelial growth factor treatment was required in one patient. Yazici with colleagues treated 48 patients for supratentorial (64%) and infratentorial (33%) meningiomas with HFSRT (CyberKnife) delivering 18–25 Gy over 3 fractions or 20–33 Gy over 5 fractions [133]. Five patients experienced disease progression during a median follow-up of 58 months. Manabe with colleagues presented their experience with 32 meningioma patients treated with 25 Gy over 2–10 fractions [134]. Five-year progression-free survival was 86% in the total cohort of patients who included patients treated using HFSRT (n = 32) and SRS (n = 9). Five patients in the HFSRT cohort experienced greater than grade 2 toxicities that included optic neuropathy, cerebral necrosis requiring necroctomy, stroke due to cerebral artery occlusion, hydrocephalus, and cerebral edema causing pyramidal tract syndrome. All patients who experienced grade 2 toxicities had tumors larger than 13.5 mL, and authors suggested that CFRT should be considered for large meningiomas. Morimoto with colleagues used HFSRT (CyberKnife) to treat 31 patients for intracranial meningiomas delivering 27.8 Gy (range, 21–36 Gy) in 3–5 fractions [135]. Five-year progression-free survival was 83%. Six (31%) patients developed marked peritumoral edema that was associated with a larger treatment volume.

17.9 VESTIBULAR SCHWANNOMAS

A vestibular schwannoma (more commonly known as an acoustic neuroma) is a benign Schwann-cell derived tumor of the vestibular division of cranial nerve VIII. These tumors are diagnosed in approximately 1 among 100,000 people each year and account for 5–9% of all brain tumors [62]. The incidence has been on the rise in recent years, likely due to greater availability of imaging (e.g., MRI), which has resulted in earlier detection [136]. Patients commonly present with hearing loss, tinnitus, vertigo, and gait disturbance. Pain, which is due to involvement of the trigeminal ganglion and its branches, and hemifacial spasms are much less common clinical manifestations. If left untreated, tumor progression can lead to mass effect on the brainstem, cranial neuropathies, and hydrocephalus. Approximately 90% of vestibular schwannomas are unilateral; however, they can occur bilaterally and are associated with genetic conditions such as

Neurofibromatosis Type 2. Available treatment options include observation, microsurgical resection, CFRT, and SRS.

17.9.1 TUMOR CONTROL

Single-fraction SRS is associated with excellent rates of tumor control in the treatment of vestibular schwannomas [137–149] (Table 17.2). Tumor control rates with single-fraction SRS have been reported upwards of 90% [1]. A meta-analysis of recent individual institutional experiences comparing tumor control in over 2,000 patients treated with single-fraction SRS or observation found that tumor control was 96.9% in the former and 65% in the latter [150]. Despite excellent tumor control rates seen in earlier studies, hearing preservation was often compromised and was seen in only 51%–60% of patients treated with single-fraction SRS [151–154]. Due to the high rates of tumor control seen with single-fraction SRS, efforts have been focused on minimizing the risk of cranial nerve deficits post-treatment. As a result, the single-fraction SRS margin dose has decreased from 16–20 Gy to 11–13 Gy in most settings [155–158].

In the hypofractionation setting, dose de-escalation has also been observed. While earlier studies typically utilized doses ranging from 21 Gy in 3 fractions to 18 Gy in 3 fractions, the rate of tumor control with the latter compared with the former was shown to be equivalent [159]. An early report by Williams et al. that was published in 2002, where 90% of patients were treated with 25 Gy delivered over 5 fractions via linear accelerator (LINAC), found a 1-year tumor control rate of 100% [160]. Meijer et al. published a series of 80 patients treated to 20–25 Gy in 5 fractions via LINAC, which demonstrated a tumor control rate of 94% [161]. However, these two studies utilized radiosurgical techniques that are less conformal and less accurate than those used in current practice and should therefore be viewed with caution.

Chang et al. reported their experience using staged CyberKnife-based SRS treatment of unilateral vestibular schwannomas in 61 patients over 3 sessions to a total dose of 18–21 Gy [151]. Of these 61 patients, 13% (*n* = 8) had previous surgery, and residual tissue was targeted. The mean pretreatment maximal tumor dimension was 18.5 mm (range, 5–32 mm). After radiosurgery, 29 (48%) of 61 tumors decreased in size, and 31 tumors (50%) were stable, resulting in a tumor control rate of 98% [151].

Hansasuta et al. reported on their experience of treating 383 patients with CyberKnife fractionated radiosurgery [162]. Patients were treated to a total dose of 18 Gy delivered over 3 fractions with a median follow-up of 43 months. Of these patients, 10 experienced progressive growth, necessitating additional treatment (microsurgery [*n* = 9] or repeat SRS [*n* = 1]). The 3- and 5-year Kaplan–Meier resection/repeat SRS-free tumor control rates were 99% and 96%, respectively. Of note, NF2-associated tumors had worse tumor control compared with sporadic tumors, which is in accordance with other published reports. The 3- and 5-year Kaplan–Meier tumor control rates for sporadic vs NF-2-associated tumors were 99% and 96% compared with 93% and 84%, respectively (*p* = 0.03).

In 2014, Anderson et al. published a single-institutional experience, where 37 patients were treated with a dose of 20 Gy delivered over 5 fractions via LINAC [163]. The 5-year tumor control rate was 90.5% ± 5.2%. A meta-analysis was published in 2017 that evaluated outcomes in 228 patients with a median total dose of 25 Gy treated with via LINAC found a 96% tumor control rate post-hypofractionated radiosurgery [164].

17.9.2 HEARING PRESERVATION

Preservation of hearing is a major concern when utilizing radiosurgery to treat vestibular schwannomas. Early series noted hearing preservation rates of 51%–60% [151–154]. However, more recent advances in SRS, such as using additional isocenters, have permitted enhanced target conformality. Studies utilizing these radiosurgical techniques have demonstrated hearing preservation rates of 71%–73% [151, 165–167]. A literature review of 45 published articles published in 2010 that represented 4,234 patients treated with Gamma Knife radiosurgery demonstrated an overall hearing preservation rate of 51%; the authors found dose to be a statistically significant factor associated with hearing preservation (hearing preservation rate of 60.5% at less than 13 Gy versus 50.4% at >13 Gy: *p* = .001) [168]. However, the effect of fractionation on hearing preservation is less clear. Although the 70%–100% hearing preservation rates reported with CFRT or hypofracted SRS are promising [169–173], no prospective randomized trial has evaluated the effects of fractionation (with either CFRT or 2- to 5-session SRS) on hearing preservation.

Williams et al., in their series published in 2002, observed that at 1 year, Gardner–Robertson Grade 1 or 2 hearing was maintained in 70% of patients with tumors of diameters less than 3cm, who were treated with 25 Gy delivered over 5 fractions [160]. Similarly, the series published by Meijer et al. in 2000 noted a 66% hearing preservation rate at 5 years [161]. However, as mentioned previously, these studies should not serve as benchmarks, as they used less than contemporary radiosurgical techniques.

Chang et al. observed 90% hearing preservation rate in the 48 patients who completed audiograms both before and after radiosurgery [151]. The follow-up period in this study was 48 months (range, 36–62 months). Hansasuta et al. observed a crude serviceable hearing preservation rate of 76% with a median follow-up period of 3 years for hearing [162]. Additionally, smaller tumor volumes were associated with a higher hearing preservation rate after SRS ($p = 0.001$, with tumor volume as a continuous variable). Tumors of volume less than 3 cm^3 had a serviceable hearing preservation rate of 80% compared with 59% for tumors of volume more than 3 cm^3 ($p = 0.009$) [162].

In 2011, Collen et al. published a single-institution experience, where 41 patients were treated with fractionated radiosurgery for vestibular schwannomas [174]. Three different fractionation schemes were utilized: (1) 2 Gy × 25 fractions ($n = 10$); (2) 4 Gy × 10 fractions ($n = 11$); (3) 3 Gy × 10 fractions ($n = 20$). With a median follow-up of 52 months, no statistically significant difference in hearing preservation probability between the single-fraction SRS and hypofractionated SRS was observed [174]. Anderson et al. observed a 5-year rate of serviceable hearing preservation of 63.2 ± 11.4% in the hypofractionated radiosurgery group [163].

17.9.3 ADVERSE EVENTS

Facial nerve palsy and trigeminal nerve dysfunction are also potential sequelae when radiosurgery is utilized in the management of vestibular schwannoma. Anderson et al. reported a 0% incidence of facial nerve injury; however, it should be noted that the median follow-up for these patients was 12 months [163]. Meijer et al. reported a 5-year facial nerve preservation rate of 96% [163]. Hansasuta et al. analyzed facial nerve function at time of presentation and after hypofractionated SRS using the House–Brackmann scale [162, 175]. Non-auditory complications were associated with larger tumor volume, where the largest quartile tumors (3.4 cm^3) had a complication rate of 9.6% compared with 3.5% for the remainder ($p = 0.03$). Similarly, the rate of non-auditory complications for Koos stage IV tumors was 9.3% compared with 3.8% for Koos stage I, II, and III tumors ($p = 0.05$) [162].

Collen et al. reported significant facial nerve toxicity (an increase of 2 or more points on the House–Brackmann scale) in 0 patients [174]. Similarly, Anderson et al. reported worsening of trigeminal neuralgia in a single patient (2.7%) who was treated with hypofractionated SRS [163]. A meta-analysis published in 2017 found rates of facial nerve and trigeminal nerve preservation to be 100% and 98%, respectively [164].

17.10 PITUITARY ADENOMAS

Pituitary adenomas are relatively common among the general population and represent approximately 10% to 20% of all central nervous system tumors [176]. Approximately 12,000 new cases are diagnosed each year in the United States [21]. Pituitary adenomas are classically divided by their size and endocrine status. Microadenomas are defined as measuring less than 1 cm in size, while macroadenomas are defined as measuring 1 cm in size or greater. Functional or endocrine status refers to hormone hypersecretion from functioning lesions and lack of abnormal hormonal production from nonfunctioning or null cell lesions. Approximately 70% to 75% of newly diagnosed pituitary adenomas are functional in nature [177].

Surgical resection achieves tumor control in only approximately 50% to 80% of cases [178]. Therefore, these lesions frequently require additional treatment, such as SRS. Radiosurgery is an excellent treatment modality in the postoperative, progressive, and recurrent setting. However, a major concern is the close proximity of the pituitary gland to the anterior optic pathways, particularly the optic chiasm (Figure 17.2).

Current experience and published literature on hypofractionated radiosurgical approaches to pituitary adenomas is limited as well as inconsistent in terms of methodology and definition. However, there are several single-institution published series that exist, which have studied hypofractionated approaches in this setting [179–181] (Table 17.3). Ultimately, the paucity of available data in this setting should be noted and is an important consideration when considering hypofractionated radiosurgery in this setting.

In 2011, Iwata et al. published a prospective study that was conducted from 2000 to 2009, where 100 patients with nonfunctioning pituitary adenomas were treated with hypofractionated radiosurgery in doses of 17–21 Gy in 3 fractions or 22–25 Gy in 5 fractions, which was administered via CyberKnife [179]. Of note, 94% of these patients were being treated for recurrent disease or were undergoing adjuvant therapy. With a median follow-up of 33 months, 5-year local tumor control was 98%. One patient developed a grade 2 visual disorder, where the optic nerve and chiasm received a D_{max} of 20.8 Gy and 20.7 Gy in 3 fractions, respectively. Three patients went on to develop hypopituitarism after treatment, with each receiving a D_{max} of 20.4 Gy, 20.5 Gy, and 20.9 Gy to the pituitary gland in 3 fractions, respectively.

In 2014, Liao et al. published a retrospective study of 34 patients with pituitary adenomas (13 functional/21 non-functional) within 3 mm of the optic apparatus, who were treated with LINAC-based hypofractionated radiosurgery [180]. Mean follow-up was 36.8 ± 15.7 months. Of these patients, 16 had immediate residual tumor after a resection, and 18 had a disease recurrence. Patients were treated to a dose of 21 Gy delivered over 3 fractions. There was no worsening or development of new visual field deficits after undergoing hypofractionated radiosurgery.

In 2017, a single-institution retrospective study, where 11 patients with pituitary adenomas (10 nonfunctioning/1 functioning) underwent hypofractionated Gamma Knife radiosurgery with a dose of 20 Gy in 4 fractions, each of the 11 pituitary adenomas demonstrated an interval response to hypofractionated radiosurgery [181]. A meta-analysis of 2,671 patients published in 2020 evaluating the effect of radiosurgery on nonfunctioning pituitary adenomas reported a 5-year control pooled control rate in hypofractionated studies of 97% [51].

17.11 CONCLUSIONS AND FUTURE DIRECTIONS

HFSRT is increasingly being used for adjuvant and upfront treatment of benign intracranial tumors residing in close proximity to critical or eloquent neural structures, such as the optic pathway and brainstem, or for larger tumors that cannot be safely treated in a single fraction with SRS. HFSRT reduces normal tissue toxicity while maintaining a high local tumor control rate. HFSRT takes advantage of the beneficial effect that fractionation has on early and late-responding tissues seen in conventional radiotherapy. HFSRT should be considered for tumors that are outside of the single-session SRS treatment envelope due to their size and proximity to vital neural structures. An accumulating body of evidence indicates that HFSRT is an effective and safe treatment option for meningiomas, vestibular schwannomas, and pituitary adenomas and should be considered in the armamentarium of treatment options.

The integration of onboard imaging such as cone-beam CT and even MRI to SRS devices will only further expand the role and ease of implementation of hypofractionated SRS for intracranial pathologies. Integration of advancing imaging modalities, such as 7T ultra-high field MRI and cranial nerve tractography, into SRS treatment planning could further optimize treatment accuracy and precision and improve safety of the SRS. Further studies are needed to clarify the most appropriate candidates and optimal dose fractionation schedules of HFSRT. Robust clinical trials comparing the safety and efficacy of HFSRT with single-session SRS and CFRT for larger tumors and tumors residing near the critical structures could potentially provide more definitive evidence about optimal treatment approaches. Well-powered retrospective institutional series and multi-institutional registries with adequate clinical and imaging follow-up are also valuable to better understand the long-term efficacy and safety of HFSRT for benign brain tumors.

Figure legends:

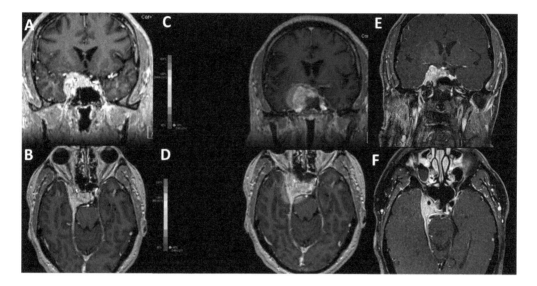

Figure 17.1 Fifty-six-year-old male presenting with worsening right lid ptosis and was diagnosed with homogenously enhancing right-sided sphenoid wing lesion elevating the right optic nerve and involving the cavernous sinus and sella. The tumor was most consistent with a meningioma (A and B). The tumor was treated using fractionated SRS 4.5 Gy × 4 fractions to the 50% isodose (C and D). Follow-up brain MRI at 6 years after SRS documented decrease in tumor volume (E and F).

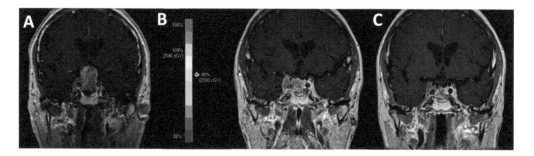

Figure 17.2 Sixty-five-year-old female who presented with bitemporal hemianopsia and was diagnosed with pituitary macroadenoma infiltrating the right cavernous sinus, with suprasellar extension and compression of the optic chiasm (A). The patient underwent subtotal transsphenoidal resection of the adenoma that was positive for LH, FSH, and alpha subunit. Follow-up MRI revealed stable residual adenoma centered in the right sella and cavernous sinus in close proximity to right optic nerve and chiasm (B). The patient was treated using fractionated stereotactic radiosurgery delivering 25 Gy in 5 fractions with maximal dose to right optic nerve of 3.9 Gy. Follow-up brain MRI 18 months after radiosurgery documented decreased size of the residual pituitary adenoma (C).

Table 17.1 Clinical series of hypofractionated SRS for meningiomas

AUTHORS	NUMBER OF PATIENTS	WHO GRADE II OR III	TUMOR VOLUME (ML)	SRS DEVICE	NUMBER OF FRACTIONS	MARGIN DOSE: MEDIAN (RANGE)	FOLLOW-UP (MONTHS)	TUMOR CONTROL (%)	PERMANENT COMPLICATIONS (%)
Optic Nerve Sheath Meningiomas									
Marchetti et al., 2011 [117]	21	0%	Median: 2.8 Range: 0.3–23	CK	5	Per fraction: 5 Gy Total dose: 25 Gy	Mean: 30 Range: 11–68	100%	0%
Perioptic meningiomas									
Adler et al., 2008 [68]	27	n/r	Mean: 7.7 [A] Range: 1.2–42 [A]	CK	2–5 [A]	Total dose: 20.3 Gy (15–30 Gy) [A]	Mean: 46 [A] Range: 13–100	94% [A]	7%
Demiral et al., 2016 [126]	22	0%	Median: 36 Range: 4–103	LINAC	5	Per fraction: 5 Gy Total dose: 25 Gy	Median: 53 Range: 36–63	1-year: 100% 3-years: 89%	11%
Conti et al., 2015 [37]	25	n/r	Median: 7.4 Range 0.3–44.4	CK	2–5	Total dose: 25 Gy (18–40 Gy)	Median: 57.5 Range: 48–82	100%	0%
Kim JW et al., 2008 [98]	13	n/r	Mean: 3.9 [B] Range: 0.3–11.4 [B]	GK	3–4	Total dose: 20 Gy (15–20 Gy) [B]	Means: 29 Range: 14–44	100%	0%
Kim MS et al., 2008	5 (orbital meningiomas)	N=1	Mean: 3.7 Range: 0.7–13	GK	3–4	Total dose: 20 Gy (15–20 Gy)	Median: 24 Range: 7–43	80%	0%
Skull base meningiomas									
Conti et al., 2019	205	n/r	25.1 ± 31.2 cm³	CK	5	Per fraction: 5 Gy	Median: 33 Range: 2–135 months	1-year: 99% 3-year: 97% 10-year: 80%	0.5
Park et al., 2018 [129]	23	n/r	Mean: 21.2 ± 15.63 cm³ Range: 10.09–71.42 cm³	GK	3–4	Total dose: 18 Gy (15–20 Gy) Per fraction: 6 Gy (5–6 Gy)	Mean: 38 months Range: 17–78 months	100%	0%
Navarria et al., 2015 [130]	26	n/r	Mean: 13.00 ± 19.1 cm³ Range:1.8–93.4 cm³	LINAC	5	Total dose: 30 Gy	Median: 24.5 months Range: 5–57 months	100%	0%

(Continued)

Table 17.1 (Continued) Clinical series of hypofractionated SRS for meningiomas

AUTHORS	NUMBER OF PATIENTS	WHO GRADE II OR III	TUMOR VOLUME (ML)	SRS DEVICE	NUMBER OF FRACTIONS	MARGIN DOSE: MEDIAN (RANGE)	FOLLOW-UP (MONTHS)	TUMOR CONTROL (%)	PERMANENT COMPLICATIONS (%)
Han et al., 2014 [131]	22	0%	Median: 4.8 cm³ Range: 0.88–20.38 cm³	LINAC	5	Total dose: 25 Gy	Median: 32 months Range: 7–97 months	94%	0%
Various locations									
Meniai-Merzouki et al., 2018 [132]	126	14%	Median: 4.84 cm³ Range: 0.31–44.70 cm³	CK	Median: 5 Range: 3–10	Total dose: 25 Gy (12–40 Gy)	Median: 20.3 Range: 1–77	2 years: 70%	0%
Yazici et al., 2018 [133]	2018	48	Median: 7.5 cm³ Range: 0.4–23.5 cm³	CK	3–5	Total dose: 25 Gy (18–33 Gy)	Median: 58 months Range: 6–144 months	At 5 years: 89.5%	Not reported
Manabe et al., 2017 [134]	32	n/r	Median: 11.3 cm³ Range: 1.4–56.9 cm³	CK	2–10	Total dose: 25 Gy (14–38 Gy)	Median: 49 months Range: 7–138 months	At 5 years: 86% [D]	16%
Morimoto et al., 2011 [135]	2011	31	Median: 6.3 cm³ Range: 1.4–27.1 cm³	CK	3–5	Median: 27.8 Gy Range: 21–36 Gy	Median: 48 months	5 year: 83%	15% (symptomatic peritumoral edema)

Abbreviations: CK, CyberKnife; GK, Gamma Knife.
[A] For total sample: meningiomas (n = 27), pituitary adenomas (n = 19), craniopharyngiomas (n = 2) and mixed germ cell tumor (n = 1).
[B] For total cohort of patients (n = 22)
[C] For the total cohort of 802 patients
[D] For total cohort of SRS (n = 9) and hFSRT (n = 32)

Table 17.2 Clinical series of hypofractionated SRS for vestibular schwannomas

AUTHOR, YEAR	NUMBER OF PATIENTS	RADIOSURGERY PLATFORM	NUMBER OF FRACTIONS	TOTAL MEDIAN MARGIN DOSE (GY)	MEDIAN FOLLOW-UP (MONTHS)	TUMOR CONTROL (%)	PERMANENT COMPLICATIONS (%)
Williams et al., 2002 [160]	150	LINAC	5	25 Gy (90% of patients)	12	100%	70% hearing preservation; 0% facial weakness
Meijer et al., 2000 [161]	80	LINAC	5	20–25 Gy	60	94%	4% facial nerve; 34% hearing loss
Chang et al., 2005 [151]	61	CK	3	18–21	48	98%	10% hearing loss
Hansasuta et al., 2011 [162]	383	CK	3	18	43	99%; 96% at 3 and 5 years	3.8%–9.3% non-auditory, 24% loss of hearing
Collen at al., 2011 [174]	41	LINAC	2–4	30–50	52	NR	59% hearing preservation at 4 years; 3% facial nerve damage; 3% trigeminal nerve dysfunction
Anderson et al., 2014 [163]	37	LINAC	5	20	43.1	90.5 ± 5.2% at 5 years	0% worsened facial sensation; 2.7% trigeminal neuralgia; 0% facial weakness; 2.7% worsened vestibular function; 2.7% worsened tinnitus

Abbreviations: CK: CyberKnife; Gy: gray; LINAC: linear accelerator; NR: not reported

Table 17.3 Hypofractionated radiosurgery for pituitary adenomas

AUTHOR, YEAR	NUMBER OF PATIENTS	RADIOSURGERY PLATFORM	NUMBER OF FRACTIONS	TOTAL MEDIAN MARGIN DOSE (GY)	MEDIAN FOLLOW-UP (MONTHS)	TUMOR CONTROL (%)	PERMANENT COMPLICATIONS (%)
Iwata et al., 2011 [179]	100	CK	3 or 5	17–21 or 22–25	33	98% at 5 years	Grade 2 visual disorder in 1 patient and hypopituitarism in 3 patients
Liao et al, 2014 [180]	34	LINAC	3	21	36.8 ± 15.7 (mean)	NR	No development of visual field deficits after radiosurgery
McTyre et al., 2017 [37]	11	GK	4	20	NR	All tumors noted to have interval response	NR

Abbreviations: CK, CyberKnife; GK, Gamma Knife; Gy, gray; LINAC, linear accelerator; NR, not reported.

CHECKLIST: KEY POINTS FOR CLINICAL PRACTICE

✓	ACTIVITY	SOME CONSIDERATIONS
	Patient selection	**Is the patient appropriate for HSRT?** • Consider for patients who have a unifocal tumor or unifocal recurrence • KPS ≥ 70 • No recent cytotoxic chemotherapy • Primary or recurrent setting: typically PTV <40 cc
	SRS vs HSRT	**Is the lesion amenable to single-fraction SRS?** • In general, patients with benign intracranial tumors are treated with single session SRS. Patients harboring larger tumors (>10 cc), those near radiation-sensitive critical structures, or those who have failed prior SRS are typical candidates for HSRT
	Simulation	**Immobilization** • Supine position with three-point thermoplastic mask, a custom head mold and open face mask, or an open vacuum-assisted mouthpiece **Imaging** • Both CT scan and MRI should be performed in the treatment position • MRI ideally should be a thin slice study (1 mm slices), including T1 sequence with IV contrast as well as a T2 or FLAIR sequence • Further image sequences as per clinical preference and optimized for tumor histology (e.g., fat saturation for post-resection pituitary adenomas) • Accurate image fusion should be verified by a physicist and the treating clinician and used for contouring targets

	Treatment planning	**Contours**
		Newly diagnosed benign intracranial tumor
		• Dose painting technique
		• GTV_1 should include post-operative residual contrast tumor
		• No margin is added to create a clinical target volume (CTV_1)
		• Additional 1–2 mm are added to create the planning target volume (PTV_1). This is not always done in Gamma Knife-based systems
		• Organs at risk should be contoured on CT and verified via MRI
		Recurrent benign intracranial tumor
		• GTV includes all enhancing disease on the T1 post contrast sequence (or enhancing post-operative cavity if disease is previously resected)
		• PTV is directly generated by expanding the GTV by 1–2 mm
		Treatment planning
		• Forward or Inverse treatment planning depending upon the SBRT platform
		• Planning techniques vary by SBRT platform but can include noncoplanar IMRT or multiple VMAT arcs (linac), multiple isocenters (Gamma Knife), or non-isocentric beams (CyberKnife)
		• Dose
		  • Varies depending upon the underlying tumor histology. Please refer to the chapter's text for further guidance
		• Coverage
		  • D95% $\geq$ Rx
		• Dose constraints
		  • Optic structures: max point dose (MPD) 23–25 Gy in up to five fractions.
		  • Brainstem: D_{05} of 30 Gy in up to 5 fractions

(*Continued*)

✓	ACTIVITY	SOME CONSIDERATIONS
	Treatment delivery	**Setup Verification Imaging (Varies depending upon the SBRT platform)** • Patients are first set up conventionally using room lasers and manual shifts • OSI program if the CDR head frame is used • Cone-beam CT to match to bony anatomy • Treatment couch with six degrees of freedom to correct translational and rotational shifts • Position confirmed using 2D kilo-voltage imaging **Manual Setup Verification (for SBRT platforms with no onboard imaging)** • Patient position measured via manual measurements or by matching field templates to light field **Treatment delivery** • Couch shifts are automated for some systems (Gamma Knife) and manual on others (some linacs). Shifts on some platforms may require imaging for position verification

REFERENCES

1. Kondziolka D, Mousavi SH, Kano H, Flickinger JC, Lunsford LD (2012) The newly diagnosed vestibular schwannoma: Radiosurgery, resection, or observation? *Neurosurgerical Focus* 33(3):E8
2. Kondziolka D, Patel AD, Kano H, Flickinger JC, Lunsford LD (2016) Long-term outcomes after Gamma Knife radiosurgery for meningiomas. *American Journal of Clinical Oncology* 39(5):453–457
3. Arnautović KI, Al-Mefty O, Husain M (2000) Ventral foramen magnum meninigiomas. *Journal of Neurosurgery* 92(1 Suppl):71–80
4. Barker FG, Klibanski A, Swearingen B (2003) Transsphenoidal surgery for pituitary tumors in the United States, 1996–2000: Mortality, morbidity, and the effects of hospital and surgeon volume. *Journal of Clinical Endocrinology & Metabolism* 88(10):4709–4719
5. Bricolo AP, Turazzi S, Talacchi A, Cristofori L (1992) Microsurgical removal of petroclival meningiomas: A report of 33 patients. *Neurosurgery* 31(5):813–828; discussion 828
6. Burton BN, Hu JQ, Jafari A, Urman RD, Dunn IF, Bi WL, DeConde AS, Gabriel RA (2018) An updated assessment of morbidity and mortality following skull base surgical approaches. *Clinical Neurological Neurosurgery* 171:109–115
7. De Jesús O, Sekhar LN, Parikh HK, Wright DC, Wagner DP (1996) Long-term follow-up of patients with meningiomas involving the cavernous sinus: Recurrence, progression, and quality of life. *Neurosurgery* 39(5):915–919; discussion 919–920
8. Fliss DM, Gil Z (2016) Complications after skull base surgery. In: Fliss DM, Gil Z (eds) *Atlas of Surgical Approaches to Paranasal Sinuses and the Skull Base*. Springer, Berlin, Heidelberg, pp. 295–303
9. Kassam AB, Prevedello DM, Carrau RL, et al. (2011) Endoscopic endonasal skull base surgery: Analysis of complications in the authors' initial 800 patients: A review. *Journal of Neurosurgery* 114(6):1544–1568
10. Singh R, Siddiqui SH, Choi Y, Azmy MC, Patel NM, Grube JG, Hsueh WD, Baredes S, Eloy JA (2019) Morbidity and mortality associated with ventral skull base surgery: Analysis of the national surgical quality improvement program. *International Forum of Allergy & Rhinology* 9(12):1485–1491
11. Betka J, Zvěřina E, Balogová Z, Profant O, Skřivan J, Kraus J, Lisý J, Syka J, Chovanec M (2014) Complications of microsurgery of vestibular schwannoma. *BioMed Research International* 2014:e315952
12. Couldwell WT, Fukushima T, Giannotta SL, Weiss MH (1996) Petroclival meningiomas: Surgical experience in 109 cases. *Journal of Neurosurgery* 84(1):20–28
13. DeMonte F, Smith HK, al-Mefty O (1994) Outcome of aggressive removal of cavernous sinus meningiomas. *Journal of Neurosurgery* 81(2):245–251
14. George B, Lot G, Boissonnet H (1997) Meningioma of the foramen magnum: A series of 40 cases. *Surgical Neurology* 47(4):371–379
15. Halvorsen H, Ramm-Pettersen J, Josefsen R, Rønning P, Reinlie S, Meling T, Berg-Johnsen J, Bollerslev J, Helseth E (2014) Surgical complications after transsphenoidal microscopic and endoscopic surgery for pituitary adenoma: A consecutive series of 506 procedures. *Acta Neurochirurgica (Wien)* 156(3):441–449
16. McClelland S, Guo H, Okuyemi KS (2011) Morbidity and mortality following acoustic neuroma excision in the United States: Analysis of racial disparities during a decade in the radiosurgery era. *Neuro Oncology* 13(11):1252–1259

17. Natarajan SK, Sekhar LN, Schessel D, Morita A (2007) Petroclival meningiomas: Multimodality treatment and outcomes at long-term follow-up. *Neurosurgery* 60(6):965–979; discussion 979–981

18. Samii M, Carvalho GA, Tatagiba M, Matthies C (1997) Surgical management of meningiomas originating in Meckel's cave. *Neurosurgery* 41(4):767–774; discussion 774–775

19. Samii M, Klekamp J, Carvalho G (1996) Surgical results for meningioma of the craniocervical junction. *Neurosurgery* 39(6):1086–1094; discussion 1094–1095

20. Samii M, Tatagiba M (1992) Experience with 36 surgical cases of petroclival meningiomas. *Acta Neurochirurgica (Wien)* 118(1–2):27–32

21. Ostrom QT, Cioffi G, Gittleman H, Patil N, Waite K, Kruchko C, Barnholtz-Sloan JS (2019) CBTRUS Statistical Report: Primary Brain and Other Central Nervous System Tumors Diagnosed in the United States in 2012–2016. *Neuro Oncology* 21(Suppl_5):v1—v100

22. Wiemels J, Wrensch M, Claus EB (2010) Epidemiology and etiology of meningioma. *Journal of Neuro-Oncology* 99(3):307–314

23. Babu R, Sharma R, Bagley JH, Hatef J, Friedman AH, Adamson C (2013) Vestibular schwannomas in the modern era: Epidemiology, treatment trends, and disparities in management. *Journal of Neurosurgery* 119(1):121–130

24. Chamoun R, Krisht KM, Couldwell WT (2011) Incidental meningiomas. *Neurosurgical Focus* 31(6):E19

25. Ezzat S, Asa SL, Couldwell WT, Barr CE, Dodge WE, Vance ML, McCutcheon IE (2004) The prevalence of pituitary adenomas. *Cancer* 101(3):613–619

26. Schmidt RF, Boghani Z, Choudhry OJ, Eloy JA, Jyung RW, Liu JK (2012) Incidental vestibular schwannomas: A review of prevalence, growth rate, and management challenges. *Neurosurgical Focus* 33(3):E4

27. Condra KS, Buatti JM, Mendenhall WM, Friedman WA, Marcus RB, Rhoton AL (1997) Benign meningiomas: Primary treatment selection affects survival. *International Journal of Radiation Oncology, Biology, Physics* 39(2):427–436

28. Kirkpatrick JP, Meyer JJ, Marks LB (2008) The linear-quadratic model is inappropriate to model high dose per fraction effects in radiosurgery. *Seminars in Radiation Oncology* 18(4):240–243

29. Ansari SF, Terry C, Cohen-Gadol AA (2012) Surgery for vestibular schwannomas: A systematic review of complications by approach. *Neurosurgical Focus* 33(3):E14

30. Huang X, Xu J, Xu M, Chen M, Ji K, Ren J, Zhong P (2017) Functional outcome and complications after the microsurgical removal of giant vestibular schwannomas via the retrosigmoid approach: A retrospective review of 16-year experience in a single hospital. *BMC Neurology.* doi: 10.1186/s12883-017-0805-6

31. Samii M, Matthies C (1997) Management of 1000 vestibular schwannomas (acoustic neuromas): The facial nerve-preservation and restitution of function. *Neurosurgery* 40(4):684–694; discussion 694–695

32. Sughrue ME, Yang I, Aranda D, Rutkowski MJ, Fang S, Cheung SW, Parsa AT (2011) Beyond audiofacial morbidity after vestibular schwannoma surgery. *Journal of Neurosurgery* 114(2):367–374

33. Cerullo L, Grutsch J, Osterdock R (1998) Recurrence of vestibular (acoustic) schwannomas in surgical patients where preservation of facial and cochlear nerve is the priority. *British Journal of Neurosurgery* 12(6):547–552

34. Ohta S, Yokoyama T, Nishizawa S, Uemura K (1998) Regrowth of the residual tumour after acoustic neurinoma surgery. *British Journal of Neurosurgery* 12(5):419–422

35. Roche P-H, Ribeiro T, Khalil M, Soumare O, Thomassin J-M, Pellet W (2008) Recurrence of vestibular schwannomas after surgery. *Progress in Neurological Surgery* 21:89–92

36. Goldsmith BJ, Wara WM, Wilson CB, Larson DA (1994) Postoperative irradiation for subtotally resected meningiomas. A retrospective analysis of 140 patients treated from 1967 to 1990. *Journal of Neurosurgery* 80(2):195–201

37. Maire JP, Caudry M, Guérin J, Célérier D, San Galli F, Causse N, Trouette R, Dautheribes M (1995) Fractionated radiation therapy in the treatment of intracranial meningiomas: Local control, functional efficacy, and tolerance in 91 patients. *International Journal of Radiation Oncology, Biology, Physics* 33(2):315–321

38. Rogers CL, Won M, Vogelbaum MA, et al. (2020) High-risk Meningioma: Initial Outcomes From NRG Oncology/RTOG 0539. *International Journal of Radiation Oncology, Biology, Physics* 106(4):790–799

39. Sughrue ME, Sanai N, Shangari G, Parsa AT, Berger MS, McDermott MW (2010) Outcome and survival following primary and repeat surgery for World Health Organization Grade III meningiomas. *Journal of Neurosurgery* 113(2):202–209

40. Wang C, Kaprealian TB, Suh JH, Kubicky CD, Ciporen JN, Chen Y, Jaboin JJ (2017) Overall survival benefit associated with adjuvant radiotherapy in WHO grade II meningioma. *Neuro-oncology* 19(9):1263–1270

41. Cohen-Inbar O, Tata A, Moosa S, Lee C-C, Sheehan JP (2018) Stereotactic radiosurgery in the treatment of parasellar meningiomas: Long-term volumetric evaluation. *Journal of Neurosurgery* 128(2):362–372

42. Forbes AR, Goldberg ID (1984) Radiation therapy in the treatment of meningioma: The joint center for radiation therapy experience 1970 to 1982. *Journal of Clinical Oncology* 2(10):1139–1143

43. Glaholm J, Bloom HJ, Crow JH (1990) The role of radiotherapy in the management of intracranial meningiomas: The Royal Marsden Hospital experience with 186 patients. *International Journal of Radiation Oncology, Biology, Physics* 18(4):755–761

44. Goldbrunner R, Minniti G, Preusser M, et al. (2016) EANO guidelines for the diagnosis and treatment of meningiomas. *Lancet Oncology* 17(9):e383–391

45. Hasegawa T, Kida Y, Yoshimoto M, Koike J, Iizuka H, Ishii D (2007) Long-term outcomes of Gamma Knife surgery for cavernous sinus meningioma. *Journal of Neurosurgery* 107(4):745–751

46. NCCN Clinical practice Guidelines in Oncology Central Nervous System Cancers. Version 2.2019 — September 16, 2019. www.nccn.org/professionals/physician_gls/pdf/cns.pdf. Accessed 30 September 2019

47. Rogers L, Barani I, Chamberlain M, Kaley TJ, McDermott M, Raizer J, Schiff D, Weber DC, Wen PY, Vogelbaum MA (2015) Meningiomas: Knowledge base, treatment outcomes, and uncertainties. A RANO review. *Journal of Neurosurgery* 122(1):4–23

48. Taylor BW, Marcus RB, Friedman WA, Ballinger WE, Million RR (1988) The meningioma controversy: Postoperative radiation therapy. *International Journal of Radiation Oncology, Biology, Physics* 15(2):299–304

49. Bunevicius A, Kano H, Lee C-C, et al. (2020) Early versus late Gamma Knife radiosurgery for Cushing's disease after prior resection: Results of an international, multicenter study. *Journal of Neurosurgery* 1–9

50. Castinetti F, Nagai M, Morange I, et al. (2009) Long-term results of stereotactic radiosurgery in secretory pituitary adenomas. *Journal of Clinical Endocrinology & Metabolism* 94(9):3400–3407

51. Kotecha R, Sahgal A, Rubens M, et al. (2020) Stereotactic radiosurgery for non-functioning pituitary adenomas: Meta-analysis and International Stereotactic Radiosurgery Society practice opinion. *Neuro Oncology* 22(3):318–332

52. Sheehan JP, Pouratian N, Steiner L, Laws ER, Vance ML (2011) Gamma Knife surgery for pituitary adenomas: Factors related to radiological and endocrine outcomes. *Journal of Neurosurgery* 114(2):303–309

53. van de Langenberg R, Hanssens PEJ, van Overbeeke JJ, Verheul JB, Nelemans PJ, de Bondt B-J, Stokroos RJ (2011) Management of large vestibular schwannoma. Part I. Planned subtotal resection followed by Gamma Knife surgery: Radiological and clinical aspects. *Journal of Neurosurgery* 115(5):875–884

54. Starnoni D, Daniel RT, Tuleasca C, George M, Levivier M, Messerer M (2018) Systematic review and meta-analysis of the technique of subtotal resection and stereotactic radiosurgery for large vestibular schwannomas: A "nerve-centered" approach. *Neurosurgical Focus* 44(3):E4

55. Combs SE, Kessel K, Habermehl D, Haberer T, Jäkel O, Debus J (2013) Proton and carbon ion radiotherapy for primary brain tumors and tumors of the skull base. *Acta Oncology* 52(7):1504–1509

56. Malouff TD, Mahajan A, Krishnan S, Beltran C, Seneviratne DS, Trifiletti DM (2020) Carbon ion therapy: A modern review of an emerging technology. *Frontiers in Oncology* 10:82

57. Murphy ES, Suh JH (2011) Radiotherapy for vestibular schwannomas: A critical review. *International Journal of Radiation Oncology, Biology, Physics* 79(4):985–997

58. Wattson DA, Tanguturi SK, Spiegel DY, et al. (2014) Outcomes of proton therapy for patients with functional pituitary adenomas. *International Journal of Radiation Oncology, Biology, Physics* 90(3):532–539

59. Bunevicius A, Laws ER, Vance ML, Iuliano S, Sheehan J (2019) Surgical and radiosurgical treatment strategies for Cushing's disease. *Journal of Neuro-Oncology* 145(3):403–413

60. Kondziolka D, Flickinger JC, Lunsford LD (2008) The principles of skull base radiosurgery. *Neurosurgical Focus* 24(5):E11

61. Trifiletti DM, Dutta SW, Lee C-C, Sheehan JP (2019) Pituitary tumor radiosurgery. *Progress in Neurological Surgery* 34:149–158

62. Tsao MN, Sahgal A, Xu W, et al. (2017) Stereotactic radiosurgery for vestibular schwannoma: International stereotactic radiosurgery society (ISRS) practice guideline. *Journal of Radiosurgery & SBRT* 5(1):5–24

63. Bunevicius A, Sheehan D, Lee Vance M, Schlesinger D, Sheehan JP (2020) Outcomes of Cushing's disease following Gamma Knife radiosurgery: Effect of a center's growing experience and era of treatment. *Journal of Neurosurgery* 1–8

64. Akanda ZZ, Hong W, Nahavandi S, Haghighi N, Phillips C, Kok DL (2020) Post-operative stereotactic radiosurgery following excision of brain metastases: A systematic review and meta-analysis. *Radiotherapy and Oncology* 142:27–35

65. Masucci GL (2018) Hypofractionated Radiation therapy for large brain metastases. *Frontiers in Oncology*. doi: 10.3389/fonc.2018.00379

66. Navarria P, Pessina F, Clerici E, et al. (2019) Surgery Followed by Hypofractionated Radiosurgery on the Tumor Bed in Oligometastatic Patients With Large Brain Metastases. Results of a Phase 2 Study. *International Journal of Radiation Oncology, Biology, Physics* 105(5):1095–1105

67. Hall EJ, Giaccia AJ (2018) *Radiobiology for the Radiologist*. Lippincott Williams & Wilkins

68. Adler JR, Gibbs IC, Puataweepong P, Chang SD (2006) Visual field preservation after multisession CyberKnife radiosurgery for perioptic lesions. *Neurosurgery* 59(2):244–254; discussion 244–254

69. Han EY, Wang H, Luo D, Li J, Wang X (2019) Dosimetric comparison of fractionated radiosurgery plans using frameless Gamma Knife ICON and CyberKnife systems with linear accelerator-based radiosurgery plans for multiple large brain metastases. *Journal of Neurosurgery* 1–7

70. Schmitt D, Blanck O, Gauer T, et al. (2020) Technological quality requirements for stereotactic radiotherapy. *Strahlenther Onkol* 196(5):421–443

71. Descovich M, Sneed PK, Barbaro NM, McDermott MW, Chuang CF, Barani IJ, Nakamura JL, Lijun M (2010) A dosimetric comparison between Gamma Knife and CyberKnife treatment plans for trigeminal neuralgia. *Journal of Neurosurgery* 113(Suppl):199–206

72. Dutta D, Balaji Subramanian S, Murli V, Sudahar H, Gopalakrishna Kurup PG, Potharaju M (2012) Dosimetric comparison of Linac-based (BrainLAB®) and robotic radiosurgery (CyberKnife®) stereotactic system plans for acoustic schwannoma. *Journal of Neuro-Oncology* 106(3):637–642

73. Guerrero M, Li XA (2004) Extending the linear-quadratic model for large fraction doses pertinent to stereotactic radiotherapy. *Physics in Medicine and Biology* 49(20):4825–4835

74. Kirkpatrick JP, Soltys SG, Lo SS, Beal K, Shrieve DC, Brown PD (2017) The radiosurgery fractionation quandary: Single fraction or hypofractionation? *Neuro-oncology* 19(Suppl_2):ii38–ii49

75. Hughes MA, Parisi M, Grossman S, Kleinberg L (2005) Primary brain tumors treated with steroids and radiotherapy: Low CD4 counts and risk of infection. *International Journal of Radiation Oncology, Biology, Physics* 62(5):1423–1426

76. Pike LRG, Bang A, Mahal BA, et al. (2019) The impact of radiation therapy on lymphocyte count and survival in metastatic cancer patients receiving PD-1 immune checkpoint inhibitors. *International Journal of Radiation Oncology, Biology, Physics* 103(1):142–151

77. Rudra S, Hui C, Rao YJ, et al. (2018) Effect of Radiation Treatment Volume Reduction on Lymphopenia in Patients Receiving Chemoradiotherapy for Glioblastoma. *International Journal of Radiation Oncology, Biology, Physics* 101(1):217–225

78. Niranjan A, Gobbel GT, Kondziolka D, Flickinger JC, Lunsford LD (2004) Experimental radiobiological investigations into radiosurgery: Present understanding and future directions. *Neurosurgery* 55(3):495–504; discussion 504–505

79. Kondziolka D, Levy EI, Niranjan A, Flickinger JC, Lunsford LD (1999) Long-term outcomes after meningioma radiosurgery: Physician and patient perspectives. *Journal of Neurosurgery* 91(1):44–50

80. Stafford SL, Perry A, Suman VJ, Meyer FB, Scheithauer BW, Lohse CM, Shaw EG (1998) Primarily resected meningiomas: Outcome and prognostic factors in 581 Mayo Clinic patients, 1978 through 1988. *Mayo Clinical Proceedings* 73(10):936–942

81. Leber KA, Berglöff J, Pendl G (1998) Dose-response tolerance of the visual pathways and cranial nerves of the cavernous sinus to stereotactic radiosurgery. *Journal of Neurosurgery* 88(1):43–50

82. Tishler RB, Loeffler JS, Lunsford LD, Duma C, Alexander E, Kooy HM, Flickinger JC (1993) Tolerance of cranial nerves of the cavernous sinus to radiosurgery. *International Journal of Radiation Oncology, Biology, Physics* 27(2):215–221

83. Conti A, Pontoriero A, Midili F, Iatì G, Siragusa C, Tomasello C, La Torre D, Cardali SM, Pergolizzi S, De Renzis C (2015) CyberKnife multisession stereotactic radiosurgery and hypofractionated stereotactic radiotherapy for perioptic meningiomas: Intermediate-term results and radiobiological considerations. *Springerplus* 4:37

84. Milano MT, Grimm J, Soltys SG, et al. (2018) Single- and Multi-fraction stereotactic radiosurgery dose tolerances of the optic pathways. *International Journal of Radiation Oncology, Biology, Physics.* doi: 10.1016/j.ijrobp.2018.01.053

85. Nguyen JH, Chen C-J, Lee C-C, Yen C-P, Xu Z, Schlesinger D, Sheehan JP (2014) Multisession Gamma Knife radiosurgery: A preliminary experience with a noninvasive, relocatable frame. *World Neurosurgery* 82(6):1256–1263

86. Milker-Zabel S, Huber P, Schlegel W, Debus J, Zabel-du Bois A (2009) Fractionated stereotactic radiation therapy in the management of primary optic nerve sheath meningiomas. *Journal of Neuro-Oncology* 94(3):419–424

87. Killory BD, Kresl JJ, Wait SD, Ponce FA, Porter R, White WL (2009) Hypofractionated CyberKnife radiosurgery for perichiasmatic pituitary adenomas: Early results. *Neurosurgery* 64:A19–25

88. Barnett GH, Linskey ME, Adler JR, Cozzens JW, Friedman WA, Heilbrun MP, Lunsford LD, Schulder M, Sloan AE (2007) Stereotactic radiosurgery—an organized neurosurgery-sanctioned definition. *Journal of Neurosurgery* 106(1):1–5

89. Brenner DJ, Hall EJ (1994) Stereotactic radiotherapy of intracranial tumors—an ideal candidate for accelerated treatment. *International Journal of Radiation Oncology, Biology, Physics* 28(4):1039–41; discussion 1047

90. Metellus P, Regis J, Muracciole X, Fuentes S, Dufour H, Nanni I, Chinot O, Martin PM, Grisoli F (2005) Evaluation of fractionated radiotherapy and Gamma Knife radiosurgery in cavernous sinus meningiomas: Treatment strategy. *Neurosurgery* 57(5):873–886; discussion 873–886

91. Garcia-Barros M, Paris F, Cordon-Cardo C, Lyden D, Rafii S, Haimovitz-Friedman A, Fuks Z, Kolesnick R (2003) Tumor response to radiotherapy regulated by endothelial cell apoptosis. *Science* 300(5622):1155–1159

92. Choi CY, Soltys SG, Gibbs IC, Harsh GR, Sakamoto GT, Patel DA, Lieberson RE, Chang SD, Adler JR (2011) Stereotactic radiosurgery of cranial nonvestibular schwannomas: Results of single- and multisession radiosurgery. *Neurosurgery* 68(5):1200–8; discussion 1208

93. Park C, Papiez L, Zhang S, Story M, Timmerman RD (2008) Universal survival curve and single fraction equivalent dose: Useful tools in understanding potency of ablative radiotherapy. *International Journal of Radiation Oncology, Biology, Physics* 70(3):847–852

94. Benedict SH, Yenice KM, Followill D, et al. (2010) Stereotactic body radiation therapy: The report of AAPM task group 101. *Medical Physics* 37(8):4078–4101

95. Firlik AD, Levy EI, Kondziolka D, Yonas H (1998) Staged volume radiosurgery followed by microsurgical resection: A novel treatment for giant cerebral arteriovenous malformations: Technical case report. *Neurosurgery* 43(5):1223–1228

96. Sirin S, Kondziolka D, Niranjan A, Flickinger JC, Maitz AH, Lunsford LD (2008) Prospective staged volume radiosurgery for large arteriovenous malformations: Indications and outcomes in otherwise untreatable patients. *Neurosurgery* 62(Suppl 2):744–754

97. Tanaka T, Kobayashi T, Kida Y (1996) Growth control of cranial base meningiomas by stereotactic radiosurgery with a Gamma Knife unit. *Neurol Medico Chirurgica (Tokyo)* 36(1):7–10

98. Kim J-W, Im Y-S, Nam D-H, Park K, Kim J-H, Lee J-I (2008) Preliminary report of multisession Gamma Knife radiosurgery for benign perioptic lesions: Visual outcome in 22 patients. *Journal of Korean Neurosurgical Society* 44(2):67–71

99. Tuniz F, Soltys SG, Choi CY, Chang SD, Gibbs IC, Fischbein NJ, Adler JR (2009) Multisession CyberKnife stereotactic radiosurgery of large, benign cranial base tumors: Preliminary study. *Neurosurgery* 65(5):898–907; discussion 907

100. Brown PD, Jaeckle K, Ballman KV, et al. (2016) Effect of radiosurgery alone vs radiosurgery with whole brain radiation therapy on cognitive function in patients with 1 to 3 brain metastases: A randomized clinical trial. *JAMA* 316(4):401–409

101. Umansky F, Shoshan Y, Rosenthal G, Fraifeld S, Spektor S (2008) Radiation-induced meningioma. *Neurosurgical Focus* 24(5):E7

102. Ilyas A, Chen C-J, Ding D, Taylor DG, Moosa S, Lee C-C, Cohen-Inbar O, Sheehan JP (2018) Volume-staged versus dose-staged stereotactic radiosurgery outcomes for large brain arteriovenous malformations: A systematic review. *Journal of Neurosurgery* 128(1):154–164

103. Pollock BE, Link MJ, Stafford SL, Lanzino G, Garces YI, Foote RL (2017) Volume-staged stereotactic radiosurgery for intracranial arteriovenous malformations: Outcomes based on an 18-year experience. *Neurosurgery* 80(4):543–550

104. Dohm A, McTyre ER, Okoukoni C, et al. (2018) Staged stereotactic radiosurgery for large brain metastases: Local control and clinical outcomes of a one-two punch technique. *Neurosurgery* 83(1):114–121

105. Iwai Y, Yamanaka K, Shimohonji W, Ishibashi K (2019) Staged Gamma Knife radiosurgery for large skull base meningiomas. *Cureus* 11(10):e6001

106. Haselsberger K, Maier T, Dominikus K, Holl E, Kurschel S, Ofner-Kopeinig P, Unger F (2009) Staged Gamma Knife radiosurgery for large critically located benign meningiomas: Evaluation of a series comprising 20 patients. *Journal of Neurological and Neurosurgery Psychiatry* 80(10):1172–1175

107. Marchetti M, Conti A, Beltramo G, Pinzi V, Pontoriero A, Tramacere I, Senger C, Pergolizzi S, Fariselli L (2019) Multisession radiosurgery for perioptic meningiomas: Medium-to-long term results from a CyberKnife cooperative study. *Journal of Neuro-Oncology* 143(3):597–604

108. Parker RT, Ovens CA, Fraser CL, Samarawickrama C (2018) Optic nerve sheath meningiomas: Prevalence, impact, and management strategies. *Eye Brain* 10:85–99

109. Sibony PA, Krauss HR, Kennerdell JS, Maroon JC, Slamovits TL (1984) Optic nerve sheath meningiomas. Clinical manifestations. *Ophthalmology* 91(11):1313–1326

110. Shields JA, Shields CL, Scartozzi R (2004) Survey of 1264 patients with orbital tumors and simulating lesions: The 2002 Montgomery lecture, part 1. *Ophthalmology* 111(5):997–1008

111. Alper MG (1981) Management of primary optic nerve meningiomas. Current status--therapy in controversy. *Journal of Clinical Neuroophthalmology* 1(2):101–117

112. Dutton JJ (1992) Optic nerve sheath meningiomas. *Survey of Ophthalmology* 37(3):167–183

113. Wright JE, McNab AA, McDonald WI (1989) Primary optic nerve sheath meningioma. *British Journal of Ophthalmology* 73(12):960–966

114. Kwon Y, Bae JS, Kim JM, Lee DH, Kim SY, Ahn JS, Kim JH, Kim CJ, Kwun BD, Lee JK (2005) Visual changes after Gamma Knife surgery for optic nerve tumors. Report of three cases. *Journal of Neurosurgery* 102(Suppl):143–146

115. Stafford SL, Pollock BE, Leavitt JA, Foote RL, Brown PD, Link MJ, Gorman DA, Schomberg PJ (2003) A study on the radiation tolerance of the optic nerves and chiasm after stereotactic radiosurgery. *International Journal of Radiation Oncology, Biology, Physics* 55(5):1177–1181

116. Klink DF, Miller NR, Williams J (1998) Preservation of residual vision 2 years after stereotactic radiosurgery for a presumed optic nerve sheath meningioma. *Journal of Neuroophthalmology* 18(2):117–120

117. Marchetti M, Bianchi S, Milanesi I, Bergantin A, Bianchi L, Broggi G, Fariselli L (2011) Multisession radiosurgery for optic nerve sheath meningiomas—an effective option: Preliminary results of a single-center experience. *Neurosurgery* 69(5):1116–1122; discussion 1122–1123

118. Romanelli P, Bianchi L, Muacevic A, Beltramo G (2011) Staged image guided robotic radiosurgery for optic nerve sheath meningiomas. *Computer Aided Surgery* 16(6):257–266

119. Eddleman CS, Liu JK (2007) Optic nerve sheath meningioma: Current diagnosis and treatment. *Neurosurgical Focus* 23(5):E4

120. Saeed P, Blank L, Selva D, Wolbers JG, Nowak PJCM, Geskus RB, Weis E, Mourits MP, Rootman J (2010) Primary radiotherapy in progressive optic nerve sheath meningiomas: A long-term follow-up study. *British Journal of Ophthalmology* 94(5):564–568

121. Ellenberger C (1976) Perioptic meningiomas. Syndrome of long-standing visual loss, pale disk edema, and optociliary veins. *Archives of Neurology* 33(10):671–674

122. Agha A, Sherlock M, Brennan S, O'Connor SA, O'Sullivan E, Rogers B, Faul C, Rawluk D, Tormey W, Thompson CJ (2005) Hypothalamic-pituitary dysfunction after irradiation of nonpituitary brain tumors in adults. *Journal of Clinical Endocrinology & Metabolism* 90(12):6355–6360

123. Hoshi M, Hayashi T, Kagami H, Murase I, Nakatsukasa M (2003) Late bilateral temporal lobe necrosis after conventional radiotherapy. *Neurol Medico Chirurgica (Tokyo)* 43(4):213–216

124. Minniti G, Traish D, Ashley S, Gonsalves A, Brada M (2005) Risk of second brain tumor after conservative surgery and radiotherapy for pituitary adenoma: Update after an additional 10 years. *Journal of Clinical Endocrinology & Metabolism* 90(2):800–804

125. van Varsseveld NC, van Bunderen CC, Ubachs DHH, Franken A a. M, Koppeschaar HPF, van der Lely AJ, Drent ML (2015) Cerebrovascular events, secondary intracranial tumors, and mortality after radiotherapy for nonfunctioning pituitary adenomas: A subanalysis from the Dutch national registry of growth hormone treatment in adults. *Journal of Clinical Endocrinology & Metabolism* 100(3):1104–1112

126. Demiral S, Dincoglan F, Sager O, Gamsiz H, Uysal B, Gundem E, Elcim Y, Dirican B, Beyzadeoglu M (2016) Hypofractionated stereotactic radiotherapy (HFSRT) for who grade I anterior clinoid meningiomas (ACM). *Japanese Journal of Radiology* 34(11):730–737

127. Kim M-S, Park K, Kim JH, Kim Y-D, Lee J-I (2008) Gamma Knife radiosurgery for orbital tumors. *Clinical Neurology & Neurosurgery* 110(10):1003–1007

128. Alfredo C, Carolin S, Güliz A, et al. (2019) Normofractionated stereotactic radiotherapy versus CyberKnife-based hypofractionation in skull base meningioma: A German and Italian pooled cohort analysis. *Radiation Oncology* 14(1):201

129. Park HR, Lee JM, Park K-W, Kim JH, Jeong SS, Kim JW, Chung H-T, Kim DG, Paek SH (2018) Fractionated Gamma Knife radiosurgery as initial treatment for large skull base meningioma. *Experimental Neurobiology* 27(3):245–255

130. Navarria P, Pessina F, Cozzi L, et al. (2015) Hypofractionated stereotactic radiation therapy in skull base meningiomas. *Journal of Neuro-Oncology* 124(2):283–289

131. Han J, Girvigian MR, Chen JCT, Miller MJ, Lodin K, Rahimian J, Arellano A, Cahan BL, Kaptein JS (2014) A comparative study of stereotactic radiosurgery, hypofractionated, and fractionated stereotactic radiotherapy in the treatment of skull base meningioma. *American Journal of Clinical Oncology* 37(3):255–260

132. Meniai-Merzouki F, Bernier-Chastagner V, Geffrelot J, Tresch E, Lacornerie T, Coche-Dequeant B, Lartigau E, Pasquier D (2018) Hypofractionated stereotactic radiotherapy for patients with intracranial meningiomas: Impact of radiotherapy regimen on local control. *Scientific Reports* 8(1):1–8

133. Ragos V, Yazici O, Guzle Adas Y, et al. (2018) Intracranial meningioma: Experience with stereotactic radiotherapy. *Journal of BUON* 23(4):1169–1173

134. Manabe Y, Murai T, Ogino H, Tamura T, Iwabuchi M, Mori Y, Iwata H, Suzuki H, Shibamoto Y (2017) CyberKnife stereotactic radiosurgery and hypofractionated stereotactic radiotherapy as first-line treatments for imaging-diagnosed intracranial meningiomas. *Neurol Medico Chirurgica (Tokyo)* 57(12):627–633

135. Morimoto M, Yoshioka Y, Shiomi H, et al. (2011) Significance of tumor volume related to peritumoral edema in intracranial meningioma treated with extreme hypofractionated stereotactic radiation therapy in three to five fractions. *Japanese Journal of Clinical Oncology* 41(5):609–616

136. Propp JM, McCarthy BJ, Davis FG, Preston-Martin S (2006) Descriptive epidemiology of vestibular schwannomas. *Neuro Oncology* 8(1):1–11

137. Boari N, Bailo M, Gagliardi F, Franzin A, Gemma M, del Vecchio A, Bolognesi A, Picozzi P, Mortini P (2014) Gamma Knife radiosurgery for vestibular schwannoma: Clinical results at long-term follow-up in a series of 379 patients. *Journal of Neurosurgery* 121(Suppl):123–142

138. Bowden G, Cavaleri J, Monaco E, Niranjan A, Flickinger J, Lunsford LD (2017) Cystic vestibular schwannomas respond best to radiosurgery. *Neurosurgery* 81(3):490–497

139. Breivik CN, Nilsen RM, Myrseth E, Pedersen PH, Varughese JK, Chaudhry AA, Lund-Johansen M (2013) Conservative management or Gamma Knife radiosurgery for vestibular schwannoma: Tumor growth, symptoms, and quality of life. *Neurosurgery* 73(1):48–56; discussion 56–57

140. Chopra R, Kondziolka D, Niranjan A, Lunsford LD, Flickinger JC (2007) Long-term follow-up of acoustic schwannoma radiosurgery with marginal tumor doses of 12 to 13 Gy. *International Journal of Radiation Oncology, Biology, Physics* 68(3):845–851

141. Friedman WA, Bradshaw P, Myers A, Bova FJ (2006) Linear accelerator radiosurgery for vestibular schwannomas. *Journal of Neurosurgery* 105(5):657–661

142. Fukuoka S, Takanashi M, Hojyo A, Konishi M, Tanaka C, Nakamura H (2009) Gamma Knife radiosurgery for vestibular schwannomas. *Progress in Neurological Surgery* 22:45–62

143. Hasegawa T, Kida Y, Kobayashi T, Yoshimoto M, Mori Y, Yoshida J (2005) Long-term outcomes in patients with vestibular schwannomas treated using Gamma Knife surgery: 10-year follow up. *Journal of Neurosurgery* 102(1):10–16

144. Hsu PW, Chang CN, Lee ST, Huang YC, Chen HC, Wang CC, Hsu YH, Tseng CK, Chen YL, Wei KC (2010) Outcomes of 75 patients over 12 years treated for acoustic neuromas with linear accelerator-based radiosurgery. *Journal of Clinical Neuroscience* 17(5):556–560

145. Jacob JT, Carlson ML, Schiefer TK, Pollock BE, Driscoll CL, Link MJ (2014) Significance of cochlear dose in the radiosurgical treatment of vestibular schwannoma: Controversies and unanswered questions. *Neurosurgery* 74(5):466–474; discussion 474

146. Kalogeridi MA, Georgolopoulou P, Kouloulias V, Kouvaris J, Pissakas G (2009) Long-term results of LINAC-based stereotactic radiosurgery for acoustic neuroma: The Greek experience. *Journal of Cancer Research & Therapy* 5(1):8–13

147. Kim YH, Kim DG, Han JH, et al. (2013) Hearing outcomes after stereotactic radiosurgery for unilateral intracanalicular vestibular schwannomas: Implication of transient volume expansion. *International Journal of Radiation Oncology, Biology, Physics* 85(1):61–67

148. Regis J, Carron R, Park MC, Soumare O, Delsanti C, Thomassin JM, Roche PH (2010) Wait-and-see strategy compared with proactive Gamma Knife surgery in patients with intracanalicular vestibular schwannomas. *Journal of Neurosurgery* 113(Suppl):105–111

149. Tamura M, Carron R, Yomo S, Arkha Y, Muraciolle X, Porcheron D, Thomassin JM, Roche PH, Regis J (2009) Hearing preservation after Gamma Knife radiosurgery for vestibular schwannomas presenting with high-level hearing. *Neurosurgery* 64(2):289–296; discussion 296

150. Leon J, Lehrer EJ, Peterson J, et al. (2019) Observation or stereotactic radiosurgery for newly diagnosed vestibular schwannomas: A systematic review and meta-analysis. *Journal of Radiosurgery & SBRT* 6(2):91–100

151. Chang SD, Gibbs IC, Sakamoto GT, Lee E, Oyelese A, Adler JR (2005) Staged stereotactic irradiation for acoustic neuroma. *Neurosurgery* 56(6):1254–1261; discussion 1261–1263

152. Hirato M, Inoue H, Nakamura M, Ohye C, Hirato J, Shibazaki T, Andou Y (1995) Gamma Knife radiosurgery for acoustic schwannoma: Early effects and preservation of hearing. *Neurol Medico Chirurgica (Tokyo)* 35(10):737–741

153. Hirato M, Inoue H, Zama A, Ohye C, Shibazaki T, Andou Y (1996) Gamma Knife radiosurgery for acoustic schwannoma: Effects of low radiation dose and functional prognosis. *Stereotactic and Functional Neurosurgery* 66(Suppl 1):134–141

154. Kondziolka D, Flickinger JC, Perez B (1998) Judicious resection and/or radiosurgery for parasagittal meningiomas: Outcomes from a multicenter review. Gamma Knife meningioma study group. *Neurosurgery* 43(3):405–413; discussion 413–414

155. Flickinger JC, Kondziolka D, Niranjan A, Lunsford LD (2001) Results of acoustic neuroma radiosurgery: An analysis of 5 years' experience using current methods. *Journal of Neurosurgery* 94(1):1–6

156. Flickinger JC, Kondziolka D, Niranjan A, Maitz A, Voynov G, Lunsford LD (2004) Acoustic neuroma radiosurgery with marginal tumor doses of 12 to 13 Gy. *International Journal of Radiation Oncology, Biology, Physics* 60(1):225–230

157. Flickinger JC, Kondziolka D, Pollock BE, Lunsford LD (1996) Evolution in technique for vestibular schwannoma radiosurgery and effect on outcome. *International Journal of Radiation Oncology, Biology, Physics* 36(2):275–280

158. Lunsford LD, Niranjan A, Flickinger JC, Maitz A, Kondziolka D (2005) Radiosurgery of vestibular schwannomas: Summary of experience in 829 cases. *Journal of Neurosurgery* 102(Suppl):195–199

159. Poen JC, Golby AJ, Forster KM, Martin DP, Chinn DM, Hancock SL, Adler JR (1999) Fractionated stereotactic radiosurgery and preservation of hearing in patients with vestibular schwannoma: A preliminary report. *Neurosurgery* 45(6):1299–1305; discussion 1305–1307

160. Williams JA (2002) Fractionated stereotactic radiotherapy for acoustic neuromas. *Acta Neurochirurgica (Wien)* 144(12):1249–1254; discussion 1254

161. Meijer OW, Vandertop WP, Baayen JC, Slotman BJ (2003) Single-fraction vs. fractionated linac-based stereotactic radiosurgery for vestibular schwannoma: A single-institution study. *International Journal of Radiation Oncology, Biology, Physics* 56(5):1390–1396

162. Hansasuta A, Choi CY, Gibbs IC, et al. (2011) Multisession stereotactic radiosurgery for vestibular schwannomas: Single-institution experience with 383 cases. *Neurosurgery* 69(6):1200–1209

163. Anderson BM, Khuntia D, Bentzen SM, et al. (2014) Single institution experience treating 104 vestibular schwannomas with fractionated stereotactic radiation therapy or stereotactic radiosurgery. *Journal of Neuro-Oncology* 116(1):187–193

164. Nguyen T, Duong C, Sheppard JP, Lee SJ, Kishan AU, Lee P, Tenn S, Chin R, Kaprealian TB, Yang I (2018) Hypofractionated stereotactic radiotherapy of five fractions with linear accelerator for vestibular schwannomas: A systematic review and meta-analysis. *Clinical Neurology & Neurosurgery* 166:116–123

165. Flickinger JC, Kondziolka D, Lunsford LD (1999) Dose selection in stereotactic radiosurgery. *Neurosurgical Clinical of North America* 10(2):271–280

166. Niranjan A, Lunsford LD, Flickinger JC, Maitz A, Kondziolka D (1999) Dose reduction improves hearing preservation rates after intracanalicular acoustic tumor radiosurgery. *Neurosurgery* 45(4):753–762; discussion 762–765

167. Spiegelmann R, Gofman J, Alezra D, Pfeffer R (1999) Radiosurgery for acoustic neurinomas (vestibular schwannomas). *Isr Med Assoc J* 1(1):8–13

168. Yang I, Sughrue ME, Han SJ, Aranda D, Pitts LH, Cheung SW, Parsa AT (2010) A comprehensive analysis of hearing preservation after radiosurgery for vestibular schwannoma. *Journal of Neurosurgery* 112(4):851–859

169. Andrews DW, Suarez O, Goldman HW, Downes MB, Bednarz G, Corn BW, Werner-Wasik M, Rosenstock J, Curran WJ (2001) Stereotactic radiosurgery and fractionated stereotactic radiotherapy for the treatment of acoustic schwannomas: Comparative observations of 125 patients treated at one institution. *International Journal of Radiation Oncology, Biology, Physics* 50(5):1265–1278

170. Chan AW, Black P, Ojemann RG, Barker FG, Kooy HM, Lopes VV, McKenna MJ, Shrieve DC, Martuza RL, Loeffler JS (2005) Stereotactic radiotherapy for vestibular schwannomas: Favorable outcome with minimal toxicity. *Neurosurgery* 57(1):60–70; discussion 60–70

171. Combs SE, Volk S, Schulz-Ertner D, Huber PE, Thilmann C, Debus J (2005) Management of acoustic neuromas with fractionated stereotactic radiotherapy (FSRT): Long-term results in 106 patients treated in a single institution. *International Journal of Radiation Oncology, Biology, Physics* 63(1):75–81

172. Kopp C, Fauser C, Muller A, Astner ST, Jacob V, Lumenta C, Meyer B, Tonn JC, Molls M, Grosu AL (2011) Stereotactic fractionated radiotherapy and LINAC radiosurgery in the treatment of vestibular schwannoma-report about both stereotactic methods from a single institution. *International Journal of Radiation Oncology, Biology, Physics* 80(5):1485–1491

173. Sawamura Y, Shirato H, Sakamoto T, Aoyama H, Suzuki K, Onimaru R, Isu T, Fukuda S, Miyasaka K (2003) Management of vestibular schwannoma by fractionated stereotactic radiotherapy and associated cerebrospinal fluid malabsorption. *Journal of Neurosurgery* 99(4):685–692

174. Collen C, Ampe B, Gevaert T, Moens M, Linthout N, De Ridder M, Verellen D, D'Haens J, Storme G (2011) Single fraction versus fractionated linac-based stereotactic radiotherapy for vestibular schwannoma: A single-institution experience. *International Journal of Radiation Oncology, Biology, Physics* 81(4):e503–509

175. House JW, Brackmann DE (1985) Facial nerve grading system. *Otolaryngol Head Neck Surgery* 93(2):146–147

176. Ezzat S, Asa SL, Couldwell WT, Barr CE, Dodge WE, Vance ML, McCutcheon IE (2004) The prevalence of pituitary adenomas: A systematic review. *Cancer* 101(3):613–619

177. Mehta GU, Lonser RR (2017) Management of hormone-secreting pituitary adenomas. *Neuro-oncology* 19(6):762–773

178. Hoybye C, Rahn T (2009) Adjuvant Gamma Knife radiosurgery in non-functioning pituitary adenomas; low risk of long-term complications in selected patients. *Pituitary* 12(3):211–216

179. Iwata H, Sato K, Tatewaki K, Yokota N, Inoue M, Baba Y, Shibamoto Y (2011) Hypofractionated stereotactic radiotherapy with CyberKnife for nonfunctioning pituitary adenoma: High local control with low toxicity. *Neuro Oncology* 13(8):916–922

180. Liao H-I, Wang C-C, Wei K-C, Chang C-N, Hsu Y-H, Lee S-T, Huang Y-C, Chen H-C, Hsu P-W (2014) Fractionated stereotactic radiosurgery using the Novalis system for the management of pituitary adenomas close to the optic apparatus. *Journal of Clinical Neuroscience* 21(1):111–115

181. McTyre E, Helis CA, Farris M, et al. (2017) Emerging indications for fractionated Gamma Knife radiosurgery. *Neurosurgery* 80(2):210–216

18 Radiation Necrosis

Kenneth Y. Usuki, Kajsa Mayo, Susannah Ellsworth, Sara Hardy,
Steven J. Chmura, and Michael T. Milano

Contents

18.1 INTRODUCTION

Single-fraction stereotactic radiosurgery (SRS) and 2–5 fraction stereotactic radiotherapy (hSRT) are widely used and efficacious treatments for brain/spine metastases, meningiomas, arteriovenous malformations, vestibular schwannomas (also referred to as acoustic neuromas), trigeminal neuralgia, and recurrent gliomas. Modern treatment planning, in conjunction with stereotactic techniques discussed elsewhere in this book, achieves a sharp-dose gradient allowing an ablative dose of radiation to the target volumes while minimizing significant dose delivery to adjacent healthy critical structures. While stereotactic techniques can be used in conjunction with conventionally fractionationated radiotherapy (1.8–2 Gy), hypofractionation (larger fractional dose) is more convenient to the patient and likely affords a different and relatively greater radiobiologic effect on the target.

A general tenet of radiation oncology is that with increased fraction size (i.e., greater dose per fraction), there is a greater risk of late normal tissue complication. While penumbra is not a clinically significant issue for low-dose standard fractionated therapy, at the high doses of SRS/hSRT, the penumbra dose becomes clinically significant and can be damaging to healthy tissue. With SRS/hSRT, stereotactic techniques are used to minimize the volume of normal tissue exposed to deleterious doses, mitigating but not eliminating late toxicity risk. The historically used linear–quadratic model may be less predictive of normal tissue effects after SRS/hSRT, likely due to different biologic mechanisms with high dose per fraction schedules (Milano et al., 2011), although this is controversial (Shuryak et al., 2015). The linear–quadratic model is derived from in vitro cell survival assays of cancer cell lines and has been shown to be clinically predictive at low dose per fraction treatment. This model is not necessarily expected to predict *in vivo* toxicity with increased fraction sizes to normal tissues for which injury of different cell types and varied intracellular components may be of increased relevance (Glatstein, 2008). The doses needed to sterilize tumor cells in vitro are much higher than the doses to attain long-term control of metastasis of similar size using SRS.

Looking at cell survival curves of EMT6 cells, Miyakawa et al. (2013) concluded that the linear–quadratic model is not applicable to single-fraction and hypofractionated irradiation. In the cell line investigated, the LQ model was considered applicable to 7- to 20-fraction irradiation or doses per fraction of 2.57 Gy or smaller. Szeifert et al. (2006) found that lesions that were excised post-SRS had parenchymal changes, stromal alterations, and vasculopathies not commonly seen in normal fractionated radiation and that these changes were correlated with length of local control. Thus, the importance of clinical outcome data for assessing the risk of late complications including parenchymal brain radiation necrosis after SRS/hSRT is paramount.

18.2 SYMPTOMS AND ENDPOINTS

Brain radiation necrosis is the most significant late complication after cranial SRS/hSRT, resulting from tissue damage/breakdown involving vascular endothelial injury and glial injury (Schultheiss et al., 1995; Sheline et al., 1980; Chao et al., 2013), with blood–brain barrier (BBB) disruption likely a key component (Brown et al., 2005; Li et al., 2004). This normal tissue parenchymal breakdown is generally associated with surrounding edema. Edema may also occur in the absence of normal tissue necrosis. Brain edema and/or necrosis can be symptomatic or asymptomatic.

There have been many studies describing the frequency of symptomatic and asymptomatic brain parenchymal necrosis with variation partially attributable to differences in endpoints used and the length of follow-up. There has been a wide range of the reported frequency of necrosis from 2% to 24% (Andrews et al., 2004; Chin et al., 2001; Lutterbach et al., 2003; Minniti et al., 2011; Petrovich et al., 2002; Sneed et al., 2015). Symptoms of edema and/or necrosis include headache, nausea, seizure, ataxia, and localized neurologic deficits (from necrotic brain parenchymal changes, its associated edema, and/or normal brain compression). These focal symptoms depend upon the region(s) of brain affected and may be completely or partially reversible.

The RTOG/EORTC LENT SOMA scale published in 1995 graded central nervous system injury into categories of fully functional with minor neurologic findings (grade 1), neurologic findings requiring home care, nursing assistance, and/or medications (grade 2), neurologic symptoms requiring hospitalization (grade 3), and serious impairment that includes paralysis, coma, mediation-resistant seizures with hospitalization required (grade 4) (Cox et al., 1995). More detailed grading, grouped by symptomatic, objective, management, and analytic criteria, were also provided in the organ-specific RTOG/EORTC LENT SOMA reviews (Schultheiss et al., 1995). Limited peri-lesional necrosis (grade 2), focal necrosis with mass effect (grade 3), and pronounced mass effect requiring surgical intervention (grade 4) were described as MRI criteria. The CTC-AE versions 3, 4, and 5 (Trotti et al., 2003) scored CNS necrosis as asymptomatic (grade 1), moderate symptoms requiring corticosteroids (grade 2), severe symptoms requiring medical intervention (grade 3), life-threatening symptoms requiring urgent intervention (grade 4), and death related to adverse event (grade 5).

18.3 FACTORS AFFECTING NECROSIS RISK

Several studies have correlated risk of necrosis with treatment-related factors. In the RTOG 90–05 study of SRS dose escalation for retreated primary and metastatic malignant brain tumors, the maximal tolerated marginal dose was 24 Gy for ≤2.0 cm lesions (dose tolerance was not met as there was doubt to the clinical relevance of continued dose escalation), 18 Gy for 2.1–3.0 cm lesions, and 15 Gy for 3.1–4.0 cm lesions. In this study, tumor volume more than 8.2 ml and a ratio of maximum dose to prescription dose more than 2 were associated with unacceptable toxicity. In a multivariate analysis, maximum tumor diameter was one variable associated with a significantly increased risk of grade 3, 4, or 5 neurotoxicity. Tumors between the diameters 21 and 40 mm were 7.3 to 16 times more likely to develop grade 3–5 neurotoxicity compared to tumors of diameters less than 20 mm. The overall actuarial incidence of radionecrosis was 5%, 8%, 9%, and 11% at 6, 12, 18, and 24 months, respectively, following radiosurgery (Shaw et al., 2000).

The RTOG 95–08 study randomized 333 patients with one to three brain metastases to whole-brain radiation with or without SRS, using the recommended RTOG 90–05 phase I dose levels (Andrews et al., 2004). Acute grade 3 (severe neurologic symptoms requiring medication) to grade 4 (life-threatening neurologic symptoms) toxicities occurred in 3% of those receiving SRS versus none among those not receiving SRS; late grade 3–4 toxicities occurred in 6% versus 3%, respectively. These differences were not significant, and among patients with solitary brain metastases, the risk of toxicity did not significantly differ between the three dose/size levels. Interestingly, a University of Kentucky study of 160 patients with 468 metastases of diameter 62 cm were treated with SRS and followed for 1–82 (median 7) months. The study found that peripheral doses less than 20 versus 20 Gy resulted in inferior tumor control, while doses more than 20 Gy versus equal to 20 Gy did not result in improved tumor control, and there was a trending association with greater grade 3–4 neurologic toxicity risk (5.9% versus 1.9%, $p = 0.078$) (Shehata et al., 2004).

Sneed et al. found prior SRS to the same lesion to be an important factor with a 20% 1-year risk of symptomatic necrosis, 4% for prior whole-brain radiotherapy, 8% for concurrent WBRT, versus 3% for no prior treatment. When they excluded lesions treated previously with SRS, the 1-year probabilities of symptomatic radiation necrosis were less than 1% for 0.3–0.6 cm diameter target, 1% for 0.7–1.0 cm, 3% for 1.1–1.5 cm, 10% for 1.6–2.0 cm, and 14% for 2.1–5.1 cm for maximum diameter (Sneed et al., 2015)

The most studied treatment-related factor related to brain radiation necrosis is the volume of tissue irradiated at or greater than a specific dose. Many of these studies are summarized in Table 18.1; select studies are discussed later. Different studies have examined different definitions of tissue volumes (i.e., "tissue", normal tissue, normal brain tissue, treated tissue including target volume). It is also important to note that most dosimetric variables are highly collinear with each other and with target volume size, complicating efforts to model dose–volume effects on radionecrosis risk.

Flickinger and colleagues have published several studies looking at the relationship between volume and dose in regards to risk of radiation necrosis (Flickinger, 1989; Flickinger et al., 1990; Flickinger and Steiner, 1990; Flickinger, Lunsford, and Kondziolka, 1991; Flickinger, Lunsford, and Wu et al., 1991). In a study of neurodiagnostic imaging changes after Gamma Knife SRS for arteriovenous malformations (AVM), the only factor that correlated with imaging changes indicative of radiation necrosis was treatment volume (mean 3.75 cubic cm) (Flickinger et al., 1992). Similarly, in a more recent study of patients treated with proton radiosurgery for brain metastases, target volume was the only significant predictor of radionecrosis risk on multivariable analysis (Atkins K et al., 2018). The AVM Study Group specifically found, using multivariate analysis, the effects of AVM location and volume of tissue receiving 12 Gy or more were significant in predicting permanent sequelae (Flickinger et al., 2000). In addition, AVM location impacted the risk of necrosis, with the lowest to increasing risk as follows: frontal, temporal, intraventricular, parietal, cerebellar, corpus callosum, occipital, medulla, thalamus, basal ganglia, and pons/midbrain. The authors created a statistical model to predict risk of permanent symptomatic sequelae using both location and 12 Gy volume (which did not exclude target volume). Marginal 12 Gy volume (target volume excluded) did not significantly improve the risk-prediction model for permanent sequelae.

Sneed et al. (2015) found the volume parameters that correlated with symptomatic radiation necrosis included target, prescription isodose, V12, and V10. Excluding lesions treated with repeat SRS, the 1-year probability of symptomatic radiation necrosis leveled off at 13%–14% for brain metastases with maximum diameter more than 2.1 cm, target volume more than 1.2 cm³, prescription isodose volume more than 1.8 cm³, 12-Gy volume more than 3.3 cm³, and 10-Gy volume more than 4.3 cm³. Interestingly, capecitabine was the only systemic therapy within 1 month of SRS that appeared to increase symptomatic radiation necrosis risk.

A study by Blonigen et al. (2010) from the University of Cincinnati looked at the relationship between dose and volume in a group of patients where 63% had received previous whole-brain irradiation with a mean prescribed SRS dose of 18 Gy. Symptomatic radiation necrosis was observed in 10% and asymptomatic radiation necrosis in 4% of lesions treated. Multivariate regression analysis showed V8–V16 to be most predictive of symptomatic radiation necrosis. For V10 and V12, they showed that the threshold volumes for which radionecrosis significantly increased were between the 75th and 90th percentiles. These percentiles corresponded to a V10 between 6.4 and 14.5 cm³ and a V12 distribution between 4.8 and 10.9 cm³.

Table 18.1 Select studies analyzing brain dose–volume metrics for complications after SRS

STUDY	STUDY COHORT	FOLLOW-UP (MEDIAN)	ENDPOINT/OUTCOME	ONSET	DOSE-VOLUME PARAMETER	RISK ¶	p-VALUE
U. Cologne	135 tumors or AVMs	9–59 (28) months	Radiation-induced	4–35 months	Tissue V10 †		<0.0001
(Voges et al., 1996)			Tissue reactions		≤10 ml	0%	
			(edema ± ring enhancement)		>10 ml	24%	
Case Western	198 tumors	Not reported	Symptomatic necrosis	Not reported	Tissue V12		Significant
(Korytko et al., 2006)	(not AVMs)		(decline in neurologic function that is		0–5 ml	23%	
			associated with imaging changes)		5–10 ml	20%	
					10–15 ml	54%	
					>15 ml	57%	
			Asymptomatic necrosis	Not applicable	Tissue V12		Not significant
					<10 ml	19%	
					>10 ml	19%	
AVM radiosurgery	422 AVMs	9–140 (34) months	Symptomatic	Not reported	Tissue V12	‡	0.0001
Study Group		for those with complications	Complications/necrosis				
(Flickinger et al., 2000)		24–92 (45) months					
		for those without complications					
U. Maryland	243 tumors £	Until death or	Radiation necrosis:	2–14 (median 4)	Tissue V10	–	

(Continued)

Table 18.1 (Continued) **Select studies analyzing brain dose–volume metrics for complications after SRS**

STUDY	STUDY COHORT	FOLLOW-UP (MEDIAN)	ENDPOINT/OUTCOME	ONSET	DOSE–VOLUME PARAMETER	RISK ¶	p-VALUE
(Chin et al., 2001)		>15 months	*based upon MRI + pathology*	months	median of 28.4 versus 7.8 ml	-	0.007
			or necrotic lesion that resolved				
					Normal brain V10	-	
					Median of 19.8 versus 7.1 ml		0.005
U. Florida	269 AVMs	Not reported	Permanent radiation-induced	Not reported	Tissue V12	–	0.047–0.080
(Fredman et al., 2003)			Complications				
			Transient radiation-induced	Not reported	Tissue V12	–	0.052–0.145
			complications				
U. Cincinnati	173 brain metastases	3.5–51 (14) months	Asymptomatic or	*symptomatic:*	Tissue V10 and V12*	–	<0.0001
(Blonigen et al., 2010)		*>6 months unless*	Symptomatic necrosis*	2–41 months	<2.2 ml <1.6 ml	5%	
		developed necrosis		*asymptomatic:*	2.2–6.3 ml 1.6–4.7 ml	12%	
				3–19 months	6.4–14.5 ml 4.8–10.8 ml	35%	
					>14.5 ml >10.8 ml	69%	
U. Maryland	243 tumors £	Until death or	Radiation necrosis:	2–14 (median 4)	Tumor volume	-	
(Chin et al., 2001)		>15 months	*see above*	months	median of 4.4 versus 1.5 ml		0.04

(Continued)

Table 18.1 (Continued) Select studies analyzing brain dose-volume metrics for complications after SRS

STUDY	STUDY COHORT	FOLLOW-UP (MEDIAN)	ENDPOINT/OUTCOME	ONSET	DOSE-VOLUME PARAMETER	RISK[¶]	p-VALUE
U. Pittsburgh	208 brain	12–122 (18) months	Neurologic	10–25 months	Treatment volume		0.009
(Varlotto et al., 2003)	metastases		complications ¥		≤2 ml	2% @ 1 year	
					≤2 ml	4% @ 5-years	
					>2 ml	3 % @ 1 year	
					>2 ml	16% @ 5 years	
UCSF	73 AVMs	3–93 months	Neurologic complications	3–62 months	Treatment volume		0.04
(Miyawaki et al., 1999)			requiring steroids,		<1.0 ml	0%	
			anti-convulsants, or surgery		1.0–3.9 ml	15%	
					4.0–13.9 ml	14%	
					≥14 ml	27%	
			Necrosis requiring resection		<14 ml	0%	0.01
					≥14 ml	13%	
			MRI T2 abnormalities		<1.0 ml	13%	
					1.0–3.9 ml	31%	
					4.0–13.9 ml	50%	
					≥14 ml	69%	
UCSF	1181 AVMs or tumors	>2 months	Grade ≥3 neurologic	0.3–17.6 months	Prescription volume		0.009
(Miyawaki et al., 1999)			complications		0.05–0.66 ml	0% @ 1.5-years	

(Continued)

Table 18.1 (Continued) Select studies analyzing brain dose–volume metrics for complications after SRS

STUDY	STUDY COHORT	FOLLOW-UP (MEDIAN)	ENDPOINT/OUTCOME	ONSET	DOSE–VOLUME PARAMETER	RISK [1]	p-VALUE
					0.67–3.0 ml	3% @ 1.5 years	
					3.1–8.6 ml	7% @ 1.5 years	
					8.7–95.1 ml	9% @ 1.5 years	
U. Rome 310 brain metastases 2–42 (9.4) months neurologic complications	*symptomatic:*	Brain V10 and V12					
(Minniti et al., 2011)			RTOG grade 3–4 median 11 months < 4.5 ml <3.3 ml 2.6%				
		MRI	*asymptomatic:*	4.5–7.7 ml, 3.3–5.9 ml, 11%			
Median 10 months 7.8–12.6 ml, 6.0–10.9 ml 24%							
>12.6 ml >10.9 ml 47%							
>19.1 ml >15.4 ml 62%							
V10 and V12		-	p = 0.001				

(Continued)

Table 18.1 (Continued) Select studies analyzing brain dose–volume metrics for complications after SRS

STUDY	STUDY COHORT	FOLLOW-UP (MEDIAN)	ENDPOINT/OUTCOME	ONSET	DOSE-VOLUME PARAMETER	RISK ¶	p-VALUE
Gifu U. 131 brain metastases 7–45.9 (18.2) months clinical symptoms or *symptomatic*:	Non-WBRT cases						
(*Ohtakara et al., 2012*)			MRI 2.2–24.2 (median 3.7) V12 cutoff-values				
				Asymptomatic:	*Symptomatic*: 8.87 m	-	0.006
				2.5–8.4 (median 6.9) *all*: 8.62 ml	-	0.008	
					V22 cutoff-values		
					Symptomatic: 2.62 ml	0.001	
					All: 2.14 ml	-	<0.001
					WBRT cases		
					V12 cutoff-values		
					Symptomatic: 8.39 ml	-	0.009
					All: 8.39 ml	-	<0.001
					V15 cutoff-values		
					Symptomatic: 5.20 ml	-	0.006
					All: 2.14 ml	-	<0.001
					V18 cutoff-values		
					Symptomatic: 1.72 ml	-	0.088
					All: 1.72 ml	-	0.063
U. Toronto							

(*Continued*)

Table 18.1 (Continued) Select studies analyzing brain dose–volume metrics for complications after SRS

STUDY	STUDY COHORT	FOLLOW-UP (MEDIAN)	ENDPOINT/OUTCOME	ONSET	DOSE-VOLUME PARAMETER	RISK ¶	p-VALUE
(Faruqi et al., 2020) 250 brain metastases 12 months clinical symptoms and or symptomatic: intact symptomatic:		MRI median = 7 months Brain-GTV >= 10.5 ccm −0.02					
		Any ARE: 1 year: 13% BMC 30 < 10.5 ccm					
				Median = 7.9 months		61% bmc 30 > 10.5 ccm	

Source: Table modified from reference Milano et al., 2011.

¶ crude risk unless other specified.

"tissue" implies target + normal brain

† Minimal and maximal target dose and target volume were not significant. The V10 of the tissue minus target (i.e., normal brain V10) was not significant.

‡ Eighty-five patients developed symptomatic necrosis, and 38/85 were classified as having permanent symptomatic necrosis, with unchanged symptoms ≥2 years after SRS.

¥ Of 11 complications, mostly included necrosis (n = 4) and persistent or symptomatic edema (n = 4).

£ Matched-pair analyses.

* Symptomatic necrosis defined as requiring steroids, hyperbaric oxygen, vitamin E, or pentoxifylline therapy, or with new neurologic complaints. Tissue V8–18 was significant for symptomatic necrosis and tissue V8–14 significant for asymptomatic necrosis. Symptomatic necrosis was observed in 10% and asymptomatic necrosis in 4% of lesions treated.

Abbreviations: AVM = arteriovenous malformation; U.= University.

The midpoint of each interval was 10.45 cm^3 and 7.85 cm^3, respectively. The risk of radionecrosis for V10 volume less than 2.2 cm^3 was 4.7%, for 2.2–6.3 cm^3 was 11.9%, for 6.4–14.5 cm^3 was 34.6%, and for more than 14.5 cm^3 was 68.8%. The risk of radionecrosis was the same for the V12 volumes of less than 1.6 cm^3, 1.6–4.7 cm^3, 4.8–10.8 cm^3, and more than 10.8 cm^3, respectively. There were no cases of radionecrosis below the 25th percentile for either V10 (<0.68 cm^3) or V12 (<0.5 cm^3). The data demonstrates that the risk of radiation necrosis exists over a relatively wide range of dose/volume exposures, although risks are low at smaller volumes; risks are unlikely to be zero at therapeutic doses and likely impacted by other (non-dosimetric) factors.

Minniti et al. (2011) studied patients who received SRS as their primary and only brain metastases treatment. Brain radionecrosis occurred in 24% of treated lesions and was symptomatic in 10% and asymptomatic in 14%. V10 Gy and V12 Gy were the most predictive. For V10 Gy more than 12.6 cm^3 and V12 Gy more than 10.9 cm^3, the risk of radionecrosis was 47%. Lesions with V12 Gy more than 8.5 cm^3 carried a risk of radionecrosis more than 10%.

In another volume-based study, Ohtakara et al. (2012) investigated whether a superficial location, which could lead to spillage of dose into the extra-parenchymal tissue outside of the brain, might decrease the risk of brain radiation necrosis. They evaluated 131 lesions, 43.5% of which received prior whole-brain radiotherapy. A three-tiered location grade was defined with grade 1 involving less than or equal to 5 mm depth from the brain surface; grade 2 located at greater than 5 mm; and grade 3 located in the brainstem, cerebellar peduncle, diencephalon, or basal ganglion. Symptomatic radiation necrosis and asymptomatic radiation necrosis were observed in 8.4% and 6.9% of cases, respectively. Multivariate analysis indicated that the significant factors for both types of necrosis were location grade, V12 Gy, and V22 Gy. In all cases, V12 was the most significant dosimetric variable for radionecrosis. Looking at non-WBRT cases, V22 and location grade were the most significant. For WBRT cases, V15 and location grade had the strongest correlation. For the non-WBRT cases, the cutoff values of V22 Gy were 2.62 and 2.14 cm^3 for symptomatic necrosis and combined (symptomatic and asymptomatic), respectively. For the WBRT cases, the cutoff values of V15 Gy were 5.61 and 5.20 cm^3 for symptomatic necrosis and combined, respectively. In addition to the dose–volume data, location grade helped predict the risk of radionecrosis.

From the recent HyTEC analyses, pooled data from published studies were used to create normal tissue complication probability (NTCP) models to predict risks of radiation necrosis (Milano et al. In Press). Table 18.2 summarizes risks of symptomatic radiation necrosis from this report.

In addition to the different dosimetric variables for treating superficial and deep brain lesions, different anatomic regions of the central nervous system may possess varying susceptibilities to injury. If this is the case, the difference may be related to tissue vascularity, glial cell population, or repair capacity. Permanent, late clinical manifestations of radiation injury are likely impacted not only by location, but also by tumor-related factors (factors released by treated tumor on surrounding normal tissue, histology, genetic morphology), patient-related factors (comorbid conditions, prior surgery, prior radiation, sex), and redundancy in brain function and plasticity (repair and recruitment of function from other regions of the brain). As previously mentioned, the University of Pittsburgh analysis also suggests that occipital, parietal, cerebellar, corpus callosum, and intraventricular AVMs are at greater risk of symptomatic necrosis versus frontal or temporal locations (Flickinger et al., 2000). Interestingly, a recent study also reported decreased symptomatic radionecrosis risk in patients treated with a 5-fraction stereotactic regimen to resection cavities, compared with those who underwent the same regimen for intact tumors, despite a considerably smaller target volume on average in non-operated patients (Faruqi et al., 2020). Although caution is needed in interpreting these findings, given imbalances in the rate of prior whole-brain radiotherapy, SRS therapy and anti-angiogenic/immunotherapy agent use in the intact versus operated patients, the apparently greater tolerance of large-volume hypofractionated regimens in the surgical patients is an intriguing finding.

In a study from Case Western Reserve University, occipital and temporal lobe non-AVM tumors had a greater likelihood of symptomatic necrosis. In addition to increased fractional dose (Ganz et al., 1996; Kalapurakal et al., 1997) and increased tumor size (Kalapurakal et al., 1997; Kondziolka et al., 1998), several meningioma studies have implicated para-sagittal/midline location as a significant variable with greater risks of symptomatic perilesional edema after SRS for meningiomas (Chang et al., 2003; Ganz et al., 1996; Kalapurakal et al., 1997; Patil et al., 2008). In the AVM studies from the University of Pittsburgh group,

Table 18.2 Brain necrosis NTCP risks from HyTEC pooled analyses*

TARGET	OAR	ENDPOINT	#	MEASURE		RISK
Brain metastases	Total brain including GTV	Symptomatic necrosis‡	1	V_{12}	≤ 5 cc	≤ 10%
				V_{12}	≤ 10 cc	≤ 15%
				V_{12}	≤ 15 cc	≤ 20%
Brain metastases	Total brain including GTV	Symptomatic necrosis‡	3	$V_{19.6}$	≤ 10 cc	≤ 10%
				$V_{19.6}$	≤ 15 cc	≤ 15%
			5	$V_{24.4}$	≤ 10 cc	≤ 10%
				$V_{24.4}$	≤ 15 cc	≤ 15%
Brain metastases	Total brain including GTV	Symptomatic necrosis requiring surgery†	3	$V_{23.1}$	≤ 10 cc	≤ 1%
				$V_{23.1}$	≤ 20 cc	≤ 5%
				$V_{23.1}$	≤ 30 cc	≤ 10%
			5	$V_{28.8}$	≤ 10 cc	≤ 1%
				$V_{28.8}$	≤ 20 cc	≤ 5%
				$V_{28.8}$	≤ 30 cc	≤ 10%
AVMs	Total brain including target	Symptomatic necrosis¥	1	V_{12}	≤ 10 cc	≤ 10%

‡ Data from Figure 4 in HyTEC paper* which used binned NTCP data from selected studies.
† Data from Figure 6B in HyTEC paper* which used individual patient NTCP data from selected studies.
¥ Data from Figure 2 in HyTEC paper* which used binned NTCP data from selected studies.
* Milano MT et al. Single and multi-fraction stereotactic radiosurgery dose tolerances of the brain. *Int J Radiat Oncol Biol Phys.* 2020; In Press.
Abbreviations: NTCP = normal tissue complication probability; GTV = gross target volume; AVM = arteriovenous malformation; OAR = organ at risk.

post-SRS imaging changes and symptomatic radiation necrosis (Flickinger et al., 1997, 1998, 2000) after SRS for AVM were increased among patients with brainstem targets. These findings raise the question of whether or not the brainstem is more susceptible to radiation injury, or whether radiation injury to this structure results in more symptomatic injury versus other areas of the central nervous system (Loeffler et al., 2008).

Several papers have reported toxicity outcomes after SRS for brainstem metastases, with a broad range of peripheral doses (9–30 Gy), with median peripheral doses of 15–20 Gy (Fuentes et al., 2006; Huang et al., 1999; Hussain et al., 2007; Kased et al., 2008; Koyfman et al., 2010; Lorenzoni et al., 2009; Shuto et al., 2003; Yen et al., 2006). Symptomatic brainstem radionecrosis is relatively uncommon after brainstem SRS with good reported local tumor control (76%–96%) (Hatiboglu et al., 2011; Sengöz et al., 2013), likely attributable to the smaller size of brainstem lesions at presentation with cone usage decreasing geometric penumbra, allowing lower peripheral brainstem doses while still maintaining clinically significant doses within the target. In addition, the prescribed dose used by practitioners tends to be lower due to the critical nature of the structure. Unfortunately, these patients suffer from poor survival (median 4–11 months) and may not have the opportunity to manifest late radiation toxicity. When the possibility of radiation toxicity occurs, it is often difficult to discern from symptomatic progression. Reported toxicities include hemiparesis, ataxia, cranial nerve deficits, headaches, nausea/vomiting, and seizures (Hatiboglu et al. 2011; Huang et al. 1999; Hussain et al., 2007; Kased et al., 2008).

A study from the University of Pittsburgh using imaging findings and neurologic deficits after SRS for 38 patients with benign tumors followed 6–84 (median 41) months (Sharma et al., 2008). Not all patients with adverse imaging findings developed deficits arising within the brainstem long tracts or adjacent cranial nerve. Some patients developed neurologic deficits in the absence of adverse imaging findings. Interestingly, there was no correlation between marginal dose and adverse imaging findings or neurologic deficits.

Interestingly, marginal doses more than 18 Gy were associated with less neurologic deficits than marginal doses of 15–17 Gy (16.6% versus 19.1%, not significant), which the authors attribute to differences in distribution of target types between different dose groups (i.e., the 15–17 Gy group represented mostly cavernomas and the >18 Gy group represented mostly arteriovenous malformations).

MD Anderson reported that 20% of their patients developed complications likely related to SRS as early as 1 month after radiosurgery and that multivariate analysis showed pre-SRS tumor volume and male sex were associated with a significantly shorter overall survival interval (Hatiboglu et al., 2011). Sengöz et al. (2013) found peri-tumoral changes were detected radiologically in 4% of the metastatic lesion sites treated with Gamma Knife radiosurgery but none of the patients exhibited symptoms. Female gender, KPS 70, mesencephalon tumor location, and response to treatment were associated with longer survival (Hatiboglu et al., 2011; Sengöz et al., 2013). Based upon the published studies, a brainstem maximum dose of 10–12 Gy is expected to result in a minimal (<1–2%) risk of brainstem toxicity.

The QUANTEC review for single-fraction doses of 12.5, 14.2, 16.0, and 17.5 Gy: partial volume irradiation to one-third of the brainstem results in normal tissue complication probabilities (NTCP) of 1%, 13%, 61%, and 94%, respectively; and brainstem maximal doses results in NTCPs of 0.2%, 3.2%, 26%, and 68%, respectively (Mayo et al., 2010). For benign tumors compressing the brainstem, for which patients are already symptomatic, or are at risk of becoming symptomatic, marginal doses of 13 Gy appear to be well tolerated (Nakaya et al., 2010). The generally accepted dose constraint to the brainstem is considered to be on the order of 10–12 Gy, and efforts to minimize brainstem dose are recommended. In general, the brainstem maximum should be maintained below 12 Gy if feasible, though when therapeutic dose to the target is compromised with such a constraint, particularly when the risks of treatment failure outweighs the risks of treatment toxicity, higher maximal doses to the brainstem should be considered. Poor local control has been shown to decrease survival and higher doses improve local control. If tumor control is compromised with lower SRS doses, progression of the metastatic lesion or primary tumor in the brainstem could be fatal. Even when exceeding tolerance to small volumes, the risk of brainstem toxicity is low and is less likely to cause a poor outcome than a failure to achieve brainstem tumor control. In the treatment of trigeminal neuralgia, we know that negligible volumes of brainstem may even safely receive doses more than 20–50 Gy. When selecting the aggressiveness of the prescription and brainstem dose, factors such as histology, KPS, primary tumor/other metastatic disease control, and the presence of extracranial metastatic disease should be taken into consideration.

18.3.1 SYSTEMIC TARGETED THERAPY WITH SRS

Targeted therapy can impact radionecrosis risk when combined with SRS. Investigators at Emory University found that among 87 patients with intracranial metastatic melanoma who were treated with a combination of BRAF inhibitors (vemurafenib and dabrafenib) and SRS had a higher incidence of radiographic and symptomatic radionecrosis than those treated with SRS alone (Patel et al., 2016). Smaller studies have found very few incidents of radionecrosis among patients treated with combined SRS and BRAF inhibition, with no difference in overall toxicity (Narayana et al., 2014; Ahmed et al., 2015). However, short median survival in these studies (and 13.7 and 7.2 months, respectively) may have precluded detection of radionecrosis that would have been evident on a longer timescale.

The safety of SRS in combination with T-DM1 (trastuzumab emtansine) for HER-2 positive metastatic breast cancer has not been extensively studied with respect to radionecrosis. A small retrospective study of 12 HER-2 positive patients at the Institut Curie in France found radionecrosis rates among patients treated with T-DM1 and SRS of 50% (concurrently) and 26.8% (sequentially). Comparison of these rates to average radionecrosis incidence for SRS alone supports the idea that T-DM1 may augment the risk of radionecrosis (Geraud et al., 2016).

When combined with upfront WBRT, a combined SRS and targeted therapy regimen has been correlated with higher rates of radiographic radiation necrosis. A group in Cleveland retrospectively analyzed radionecrosis rates among 445 brain metastasis patients who were treated with combined SRS and at least 1 systemic therapy (including cytotoxic, hormone, cytokine, and targeted systemic therapies). They found that the group treated with upfront SRS and WBRT along with systemic targeted therapy had significantly more radionecrosis events over 12 months (8.8 versus 5.3%, $p < 0.01$). In particular, VEGFR tyrosine kinase inhibitors (14.3 versus 6.6%, $p = 0.04$) and EGFR tyrosine kinase inhibitors (15.6 versus 6.0%, $p = 0.04$) were implicated (Kim et al., 2017).

18.3.2 SYSTEMIC IMMUNOTHERAPY WITH SRS

Multiple studies indicate that combining immunotherapeutic checkpoint inhibitors with SRS may increase radionecrosis risk. University of Toronto study showed targeted or immunotherapies within 1 month before or after cavity treatment with hSRT to be a significant risk factor for symptomatic radionecrosis (Faruqi et al., 2020).

A retrospective study at the University of Southern California compared treatment for metastatic melanoma with SRS and ipilimumab (an anti CTLA-4 monoclonal antibody) to SRS alone. Among 91 patients, 51 were treated with the combination therapy, and 40 were treated with SRS alone. There were only five radiation necrosis events in this study, but four of the five were in the group that received ipilimumab. However, survival among the Ipilimumab group was longer (15.1 months compared to 7.8 months in patients who did not receive ipilimumab), which may partially explain the difference in observed radionecrosis events (Diao et al., 2018).

A retrospective study at the Dana Farber Cancer Institute of 480 patients with metastatic melanoma, non-small cell lung cancer, and renal cell carcinoma compared incidence of radiation necrosis with treatment using immunotherapeutic checkpoint inhibitors (ipilimumab, pembrolizumab, or nivolumab) in combination with SRS to SRS alone. Ipilimumab and SRS were correlated with increased incidence of radionecrosis, and the association was especially strong in melanoma patients. PD-1 inhibition with SRS was also associated with increased radionecrosis events, but this finding was not statistically significant (Martin et al., 2018). In a study from Yale University of 180 patients with metastatic melanoma treated with Gamma Knife Radiosurgery, those who also received immunotherapy (Anti-CD 137, anti-CLTA-4, Anti-PD-1, Interferon, or Interleukin 2) showed an increased risk of radionecrosis compared to those treated with cytotoxic chemotherapy or targeted therapy (Colaco et al., 2016). Another study from Italy evaluated treatment of 80 patients with metastatic melanoma treated with combination ipilimumab and SRS or nivolumab and SRS. The 12-month incidence of radiologic RN was 25% and 17% in the ipilimumab and nivolumab groups, respectively (Minniti et al., 2019).

At least one study has reported little or no significant increase in radionecrosis risk with anti-CTLA-4 and anti-PD-1 therapy. A retrospective study of 182 metastatic melanoma patients by a group in Vienna found that the occurrence of radiation necrosis after Gamma Knife Radiosurgery did not show any statistically significant difference between groups treated with anti-CTLA-4, anti-PD-1, neither, or both (Gatterbauer et al., 2020).

Many immunotherapy and targeted therapy regimens have been correlated with increased overall survival and time to recurrence when combined with SRS. In addition, the morbidity and mortality of uncontrolled brain metastases is high. As discussed previously, the local control of SRS and hSRT is extremely high. However, the effectiveness and safety of SRS or hSRT are inversely related to the size of the lesions being treated. Giving a therapy with a response rate usually well below 50% of a therapy, risking progression, while delaying a therapy with local control around 90% needs to be considered. In addition, there may be some additional local and/or abscopal immune stimulation from giving these therapies concurrently or in succession. Concerns regarding the possible increased risk of radionecrosis with combined therapy must be weighed against these considerations.

18.4 CORTICOSTEROID USE IN THE MANAGEMENT OF BRAIN NECROSIS

Radiation necrosis following SRS/hSRT tends to be a temporary problem with most of the symptoms typically caused by brain edema and not from parenchymal damage (although this may occur as well). SRS/hSRT induced necrosis can usually be managed conservatively with observation in asymptomatic or minimally symptomatic patients. The first-line treatment of symptomatic radiation necrosis is corticosteroids due to the drug's rapid ability to reduce cerebral edema. Corticosteroids are often well tolerated, particularly when using moderate doses over a short period of time. In patients who are suffering from symptomatic edema, corticosteroids usually briskly alleviate or improve symptoms. Once symptoms are controlled, corticosteroids should be gradually tapered with the speed of the taper dependent on the length of steroid dependency. If steroids were given for less than 2 weeks, a taper is likely not necessary. One should try to avoid the use of corticosteroids for an extended time period, as their deleterious effects tend to be a function of length of use and the dose. Patients should be prescribed the lowest possible dose to alleviate their symptoms. Steroids may be difficult to tolerate in patients with a psychiatric history who may be more prone to steroid-induced psychosis and diabetics who may have a difficult time controlling blood sugar levels. Careful attention needs to be paid to these two groups. Diabetics may benefit from recommendations by their primary care physician or endocrinologist concerning any changes in diabetic therapies and monitoring that may help the patient tolerate the temporary steroid treatment. The possible side effects of corticosteroids are many and include anxiety, depression, psychosis, sleeplessness, aggressive behavior, infection, diabetes, GI irritation, increased appetite, osteoporosis, and facial swelling.

At least one study has indicated that lower dose corticosteroids may be an acceptable treatment for radionecrosis. One hundred sixty-nine nasopharyngeal carcinoma patients with radiographic and symptomatic radionecrosis were treated with either high-dose methylprednisone (500 mg for 3 days, 80 mg for 4 days, 40 mg for 4 days, then oral prednisone 30 mg/day, gradually tapering by 5 mg/week to a maintenance dose of 10 mg daily for 3 months) or low-dose methylprednisone (1 mg/kg/day for 5 consecutive days, then 40 mg for 5 days, then oral prednisone 30 mg per day, gradually tapering by 5 mg/week to a maintenance dose of 10 mg daily for 3 months). At 12 months, there was no significant difference in response rate based on MRI or clinical symptoms (Xhou et al., 2019).

18.5 WARFARIN, PENTOXYFYLLINE, AND VITAMIN E

Vitamin E and pentoxifylline given in combination are reported to benefit other body sites afflicted with radiation damage. The University of Pittsburgh group performed a pilot study looking at oral pentoxifylline and vitamin E therapy given to 11 patients with suspected adverse radiation effects after SRS (Williamson et al., 2008). These patients were followed and their MR FLAIR volume changes were plotted over time. The change in edema volume varied from 59.6 ml in one patient (worse edema) to 324.2 ml (improvement). The average change in edema from pre- to post-treatment was 72.3 ml. One patient had more edema despite treatment; this patient was found to have tumor recurrence. Two patients discontinued pentoxifylline because of persistent nausea and abdominal discomfort (Williamson et al., 2008). Since radiation necrosis is usually a transient occurrence after SRS/hSRT, more extensive data are needed before any conclusion on the efficacy of vitamin E and pentoxifylline can be drawn.

Glantz et al. (1994) evaluated anticoagulation with heparin and warfarin on radiation necrosis in 11 patients with late radiation-induced nervous system injuries (8 with cerebral radionecrosis, 1 with a myelopathy, and 2 with plexopathies, all unresponsive to dexamethasone and prednisone). In five of the eight patients with cerebral radionecrosis, some recovery of function occurred. Anticoagulation was continued for 3 to 6 months. In one patient with cerebral radionecrosis, symptoms recurred after discontinuation of anticoagulation and disappeared again after reinstitution of treatment. The group hypothesized that anticoagulation may improve small vessel endothelial injury leading to clinical improvement.

18.6 BEVACIZUMAB IN THE TREATMENT AND PREVENTION OF CENTRAL NERVOUS SYSTEM RADIATION NECROSIS

Vascular endothelial growth factor (VEGF) is released in the setting of radiation injury to the blood–brain barrier and is a key mediator of radiation-induced white matter toxicity and central nervous system radiation necrosis (Kim et al., 2004). Bevacizumab, a humanized monoclonal VEGF antibody, therefore represents a promising targeted agent for brain radionecrosis. It has been studied as a treatment for known brain radionecrosis and as an adjunct to radiation dose escalation or dose intensification with the goal of preventing the development of brain radionecrosis.

Despite considerable clinical interest in bevacizumab as a treatment for brain radionecrosis, most reports on its use in this setting have been case reports or retrospective series. Tye et al. (2014) conducted a pooled analysis of the 16 studies published through 2012 (including 71 cases) of brain radionecrosis treatment with bevacizumab. The overall radiographic response rate was 97%, and the clinical improvement rate (measured in terms of performance status) was 79%. The median decreases in T1 contrast-enhancing area and FLAIR signal abnormality were 63% and 59%, respectively, and dexamethasone dose decreased by a median of 6 mg.

Additionally, a small randomized trial was published in 2012 by Levin et al. This single-institution study randomized 14 patients with known brain radionecrosis (diagnosed via MRI and/or biopsy) to bevacizumab versus placebo and permitted cross-over in patients who had neurologic or radiographic progression. Ultimately, all the patients in the placebo arm received bevacizumab, rendering this essentially a trial of early versus delayed bevacizumab for brain radionecrosis. The study reported a radiographic response rate of 100% in both arms. Symptomatically, no differences were observed between the two groups, and four of the five patients who were on dexamethasone at the time of study enrollment were able to decrease their steroid doses. Observed adverse events included sagittal sinus thrombosis, pulmonary embolism, and pneumonia (in one patient each).

One issue with bevacizumab treatment is the phenomenon of recurrence of radionecrosis following cessation of treatment. This was illustrated in a retrospective assessment of bevacizumab as a monotherapy for 50 nasopharyngeal carcinoma patients with temporal radionecrosis in Guangdong, China. Those who received bevacizumab showed a robust initial response, with an average decrease of 72.6% shown on T2-weighted MRI. However, among the 38 patients (76%) who were responsive to bevacizumab, 15 exhibited recurrence following termination of treatment. The strongest predictors of recurrence were the duration between radiation therapy and bevacizumab treatment and the duration between radiation therapy and RN diagnosis (Li et al., 2018).

Another study raised similar concerns regarding recurrence of radionecrosis following cessation of bevacizumab. Thirteen of 14 patients included in the retrospective analysis initially responded to bevacizumab; however, 10 of the 13 responsive patients (76.9%) exhibited a recurrence on T1-weighted MRI over a median follow-up time of 10 months (Zhuang et al., 2016).

Studies of bevacizumab for the prevention of radionecrosis in brain tumor patients undergoing dose-escalated radiation therapy or reirradiation have produced mixed results. Gutin et al. (2009) reported a prospective trial of bevacizumab in conjunction with hypofractionated stereotactic reirradiation for recurrent malignant glioma. Patients without evidence of disease progression after one cycle of bevacizumab (10 mg/kg every 14 days of a 28-day cycle) went on to receive 30 Gy in 5 fractions. Bevacizumab was subsequently continued until disease progression or unacceptable toxicity occurred. The study excluded patients with tumors larger than 3.5 cm in diameter. No instances of radionecrosis occurred in this cohort, even though all but two of the patients received reirradiation to areas that had previously received 60 Gy. However, three patients (12%) stopped therapy due to toxicities that included brain hemorrhage, intestinal perforation, and wound dehiscence. A fourth patient developed gastrointestinal bleeding 3 weeks after discontinuing study treatment due to tumor progression.

Most recently, Ney et al. (2015) reported a prospective clinical trial of an aggressively hypofractionated combination regimen in previously untreated glioblastoma. Dose-painting was used to treat the surgical cavity or enhancing tumor (plus a 1-cm margin) to 60 Gy in 10 fractions; areas of T2 abnormality (also plus a 1-cm margin) received 30 Gy in 10 fractions. Patients received radiation from Monday through Friday for 2 weeks without a scheduled break. Concurrent temozolomide was administered at standard doses (75 mg/m^2) during radiation, and patients received bevacizumab (10 mg/kg) on days 1 and 15. Adjuvant bevacizumab and temozolomide (200 mg/m^2) were re-started 4 to 6 weeks after radiation. This study was stopped early when interim analysis revealed that 50% of patients had developed radionecrosis (pathologically proven by surgical samples or autopsy in six individuals). Other potentially treatment-related toxicities included stroke ($n = 1$), grade 3 wound dehiscence ($n = 2$), and pulmonary embolism ($n = 2$). The unexpectedly high rate of radionecrosis in this study was thought to be due to a combination of large target volumes and a high total RT dose: the median 30-Gy volume was 342.6 cm^3 and the median 60-Gy volume was 131.1 cm^3.

For comparison, a previous prospective trial of concurrent hypofractionated RT/temozolomide/bevacizumab in newly diagnosed glioblastoma prescribed a total dose of 36 Gy in 6 fractions to enhancing tumor and 24 Gy in 6 fractions to the T2 abnormality (Omuro et al., 2014). Additionally, that study excluded patients with a tumor volume more than 60 cm^3 and reported a much lower rate of radionecrosis, with two cases of biopsy-proven radionecrosis (5%) observed among the 40 participants at a median follow-up of 42 months.

Overall, bevacizumab appears to be a promising treatment for brain radionecrosis. Although the available data are relatively sparse, studies published to date and clinical experience report impressively high radiographic and clinical response rates in patients with radiographic or pathologic evidence of radiation necrosis (Figure 18.1). The efficacy of bevacizumab as prophylaxis against radionecrosis for patients treated with intensified radiation regimens remains in question and requires further study. Considerable caution must be exercised in selecting patients for treatment with bevacizumab, which is associated with significant toxicity, particularly hemorrhage, thrombosis, and wound-healing difficulties. The authors therefore recommend that patients with known brain radionecrosis either demonstrate resistance to steroid therapy or have intolerable steroid toxicity before being considered for treatment with bevacizumab. Contraindications to bevacizumab use include the following: coagulation disorders (bleeding diathesis or hypercoagulability); use of antiplatelet agents or anticoagulants; pregnancy/nursing; a history of significant cardiovascular disease (stroke, cerebral hemorrhage, heart failure, uncontrolled hypertension, recent myocardial infarction, peripheral vascular disease, or aortic aneurysm/dissection); hemoptysis; and a recent history of GI bleeding, perforation, abscess, or fistula (Hershman et al., 2013). In patients who have undergone surgery, bevacizumab treatment should be delayed for at least 28 days postoperatively to allow adequate wound healing (Genentech/Roche).

It has been hypothesized that hypoxia-induced inflammation may also contribute to late-onset RN. As such, there is interest in using HIF-1α and CXCR4 inhibition to mitigate the development of RN following SRS. Early trials in mice have shown significantly less RN in mice treated with either topotecan or AMD3100 following SRS (Yang et al., 2018, 2015).

18.7 HYPERBARIC OXYGEN

Hyperbaric oxygen (HBO) is a treatment where patients enter a sealed chamber with 100% oxygen that is up to three times greater than the atmospheric pressure. This allows dissolving of oxygen into the plasma component, bypassing the need for hemoglobin saturation to deliver oxygen to tissue. This aids in delivering oxygen through the plasma to necrotic tissue or fibrotic tissue where the damaged vasculature may not allow effective hemoglobin passage, theoretically reducing fibrosis by stimulating angiogenesis and allowing stem cells to be recruited to the irradiated tissue. The treatment schedule is usually 5 days a week, and a patient may be prescribed between 20 and 40 treatments with each session taking 120 minutes.

Data for HBO therapy are paltry, and the reported effectiveness in CNS radionecrosis is mostly limited to case studies (Kohshi et al., 2003; Leber et al., 1998; Pérez-Espejo et al., 2009; Valadao et al., 2014). There is legitimate, though unproven, concern that HBO may stimulate malignant cells. In addition, there are side effects associated with hyperbaric oxygen although severe, life-threatening side effects are rare. Side effects may include seizure in 1%–2%, pulmonary symptoms in 15%–20%, and reversible myopia in 20% of patients treated (Leach et al., 1998).

HBO may have use as prophylaxis against radiation necrosis. Ohguri et al. (2007) looked at 78 patients presenting with 101 brain metastases treated with SRS. Of these, 32 patients with 47 brain metastases were treated with prophylactic HBO, which included all 21 patients who underwent subsequent or prior radiotherapy and 11 patients with common predictors of longer survival, such as inactive extracranial tumors and younger age. The other 46 patients with 54 brain metastases did not undergo HBO. The radiation-induced brain injuries were divided into two categories, white matter injury and radiation necrosis on the basis of imaging findings. Radiation-induced brain injury occurred in five lesions (white matter injury in two, necrosis in three) (11%) in the HBO group. In the non-HBO group, 11 lesions occurred (white matter injury in nine, necrosis in two) (20%). While the data is interesting, more research is needed into the subject of HBO for prophylaxis of brain radiation injury. HBO is an option for corticosteroid refractory radiation necrosis after SRS, particularly if bevacizumab or surgery is not option for treatment.

18.8 SURGICAL REMOVAL OF BRAIN NECROSIS

When corticosteroids and antiangiogenics fail to alleviate the pressure created by the radiation necrosis-induced edema with neurological deficits, surgical debulking of a radiation necrosis lesion can relieve increased intracranial pressure immediately and quickly improve or halt progression of a neurologic disability. The surgical removal of the necrosis may provide significant reversal of the edema-created neurological deficits, and many patients can be weaned from corticosteroids.

If bevacizumab has been given, most surgeons will delay surgery by at least a month to avoid surgical morbidity and wound-healing issues. Surgical resection also has the benefit of giving information as to whether radiographic changes and neurologic symptoms are from tumor progression or radiation necrosis. Surgery is associated with risk of complications including neurologic deficit, infection, anesthesia risk, and the risk of an inpatient hospital stay.

McPherson et al. (2004) looked at 11 patients who had malignant brain tumors and underwent surgery for radiation necrosis. The diagnosis of radiation necrosis was based primarily on MRI and clinical suspicion. Frameless stereotaxis was used in all patients, and intra-operative MRI was used in nine. Optimal resection was confirmed by intra-operative MRI and achieved in all patients by the use of frameless stereotaxis with no additional resection performed in any patient. All the nine patients taking steroids before treatment of necrosis had a substantial reduction in steroid dosage (pre- to postoperative dose 24 to 8 mg/day) after surgical treatment. Postoperatively, KPS improved in four patients, remained stable in four, and worsened in three. Complications from surgery included wound infection, asymptomatic carotid dissection, and pulmonary embolism. The authors concluded that morbidity including both surgical complications and neurologic deterioration was 54% and that given the success of medical therapies, they recommended that surgical treatment of radiation necrosis should be reserved for symptomatic patients in whom medical therapy has failed. Surgery should be considered in moderately to severely symptomatic patients in whom medical therapy fails and/or pathological examination could affect the treatment course.

18.9 FRACTIONATION

Fractionation with hSRT, delivering biologically effective doses similar to single-fraction SRS, theoretically reduces risks of normal tissue injury. hSRT may allow for equivalent or possibly superior (Shuryak et al., 2015) local control while conferring less toxicity treating large targets, multiple targets close by, or when the target is near a sensitive critical structure. Groups have treated brain metastases with hSRT and reported promising tumor control and toxicity, even for lesions larger than 4 cm or 3–5 ml (which are generally less amenable to single-fraction SRS due to greater risk of necrosis) and/or lesions involving or in close proximity to eloquent brain or optic nerve/chiasm (Aoyama et al., 2003; Ernst-Stecken et al., 2006; Fahrig et al., 2006; Feuvret et al., 2010; Kim et al., 2011; Kwon et al., 2009; Manning et al., 2000; Marchetti et al., 2011; Narayana et al., 2007). Retrospective studies have demonstrated comparable tumor control and neurologic toxicity of hSRT versus SRS, even though the hSRT group comprised patients with larger tumors (Feuvret et al., 2010; Kim et al., 2011) and/or adverse locations (Kim et al., 2011). Studies of hypo-fractionated hSRT have also been undertaken in patients with brain metastases who are potentially eligible for single-fraction SRS (i.e., smaller lesions not involving or abutting eloquent structures) (Aoyama et al., 2003; Kwon et al., 2009; Manning et al., 2000).

Minniti et al. in 2016 published a large retrospective series looking at the local control and radiation-induced brain necrosis in patients with brain metastases >2 cm in size who received single-fraction SRS or multifraction stereotactic radiosurgery (hSRT, fSRS). For patients who received SRS, doses were 18 Gy for metastases of 2–3 cm in size and 15–16 Gy for metastases greater than or equal to 3 cm, similar to RTOG 90–05. The hSRT was given as 3 × 9 Gy. There was a bias toward hSRT being used in tumors greater than or equal to 3 cm and closer to critical areas. The 1-year cumulative local control rates were 77% in the single-fraction SRS group and 91% in the multifraction hSRT (fSRS) group (p =.01). Of patients undergoing SRS, 20% versus 8% treated with hSRT experienced brain radionecrosis (p = .004); the 1-year cumulative incidence rate of radionecrosis was 18% and 9% (p = .01), respectively. These significant differences between SRS and hSRT in terms of local control and risk of radionecrosis maintained after propensity score adjustment. In the SRS group, univariate analysis showed that tumor size, GTV, and volume of normal brain that received doses of 12–16 Gy were predictive of brain necrosis. The V12-Gy was the most significant variable associated with the development of RN; at a median radiologic follow-up of 10 months, the incidence of RN was 13% for V12-Gy 13.2 cm^3 and 28% for V12-Gy more than 13.2 cm^3 (p =.02). In the hSRT group, GTV and volume of normal brain receiving doses of 15–24 Gy were predictive of RN. The brain volume receiving 18 Gy (V18-Gy) was the most significant prognostic factor for radiation necrosis; the incidence was 5% for V18-Gy less or equal to 30.2 cm^3 and 14% for V18-Gy more than 30.2 cm^3 (p =.04). (Minniti et al., 2016).

A study from Sunnybrook Odette Cancer Centre of 5-fraction hSRT for 132 intact brain metastases showed that the V30 of brain minus target volume of was associated with a significantly greater risk of symptomatic adverse radiation effects, with a more than seven fold relative-risk on multivariable analysis with a V30 more than 10.5 cc (Faruqi et al., 2020).

The aforementioned HyTEC analyses of NTCP separately analyzed risk for single- versus multi-fraction radiosurgery (Milano et al. In Press). Table 18.2 summarizes these risk estimates.

The data that exists for hypofractionated hSRT for meningiomas shows that it is safe and effective treatment even for large tumors (Gorman et al., 2008; Henzel et al., 2006; Shrieve et al., 2004; Trippa et al., 2009; Onodera, 2011; Park, 2018). Due to the biologic mechanism of obliterating the nidus and AVM's low alpha:beta ratio, all favoring high single doses of radiation, the radiobiologic utility of hSRT for AVMs is controversial (Kocher et al., 2004; Hall et al., 1993; Qi et al., 2007; Vernimmen et al., 2010; Wigg 1999). Several studies have demonstrated some efficacy of hSRT for larger (>2.5–4 cm and/or >10–14ml) AVMs where single-fraction ablative doses would be prohibitive (Chang et al., 2004; Lindvall et al., 2003; Silander et al., 2004; Veznedaroglu et al., 2004, 2008). Due to the previously mentioned radiobiologic advantages in using large single fractions, some studies have taken a volume staged approach to large, even combining volume staging with surgery with good and possibly superior results to dose staging (Abla et al., 2015; Moosa et al., 2014).

Thus, size and location are valid reasons to consider fractionated hSRT. However, radiobiologically, hSRT may result in less effective tumor control and/or greater risk of normal tissue toxicity as compared to regular conventional fractionation (i.e., 1.8–2Gy), particularly with targets near dose-limiting critical structures or low alpha:beta ratios (Shrieve et al., 2004; Vernimmen et al., 2010). In a study from the University of Erlangen, analyzing 51 patients (with 72 brain metastases) not considered candidates for single-fraction SRS (due to volume >3 ml or proximity/involvement of eloquent brain), a volume of normal brain ≥ 23 ml receiving 4 Gy/fraction (over 5 fractions) was associated with a significantly increased risk of brain necrosis (70% versus 14% for those with 4 Gy volume <23 ml, p = 0.001) (Ernst-Stecken et al., 2006). In another study from this same group of 150 patients with 228 brain metastases, a PTV of more than 17 ml (corresponding to approximately 3+ cm in diameter) was associated with increased neurologic toxicity; furthermore, as the number of fractions increased (5 × 6–7 Gy versus 7 × 5 Gy versus 10 × 4 Gy), the risk of toxicity decreased (22%, 7%, and 0%, respectively), though the less protracted regimens were associated with greater tumor response (Fahrig et al., 2007).

18.10 CONCLUSION

Some uncertainty remains with the lack of randomized data, about the risks for radiation necrosis and how best to utilize the varied brain dose–volume metrics. Complicated underlying genetic factors certainly play a role in determining radiation sensitivity and susceptibility to radiation-induced brain injury (Wang et al., 2019). There is current interest in using genomic data to inform the degree of radionecrosis risk. However, insufficient data at this time precludes application in clinical practice. What is the relevance of V10–12 to an individual lesion site versus a composite V10–12 for the entire SRS treatment? The number of lesions was not significant in one study (Varlotto et al., 2003). Based on our understanding of radiation-induced fibrosis and necrosis, it is reasonable to expect that a given volume more than 10–12 Gy is more likely to result in necrosis if that volume is not spread over multiple lesions but rather confined to one confluent location in the brain. Is the V10–12 Gy of the treatment volume versus normal tissue versus normal brain more clinically relevant?

There is a difference in how studies defined V12, and some studies did not specify a definition for V12. For studies of necrosis after treatment of AVMs, V12 tends to represent treated tissue receiving ≥12 Gy. Compared to benign and malignant tumors, the AVM nidus is more difficult to precisely define, as it is characterized by nidus and feeder vessels that interdigitate with normal brain. The size of the AVM may also be relevant in that with larger AVMs, it is possible that there is a greater impact of radiation in affecting regional blood flow and thus increasing necrosis risk. Flickinger et al. report that subtracting the target from V12 adds no additional benefit in correlating necrosis risk with V12. This finding may be partially attributable to the high doses used to treat AVMs, resulting in a significant component of the V12 volume being outside of the target, thus minimizing the predictive benefit of removing the target from V12. In some studies, the V12 is defined as 'normal brain' receiving ≥ to 12 Gy. This approach is logical for brain metastasis since normal brain is displaced by the target. In the treatment of brain metastases (and other tumors), the relative importance of necrosis within the target versus within the surrounding tissue is unclear, as both scenarios would likely result in surrounding vasogenic edema. Bulkier targets treated in excess of 12 Gy are likely to be at risk of tumor necrosis and surrounding transient edema, even if the V12 was conformed within the target volume. This edema caused by tumor necrosis is likely more temporary as the normal brain does not experience significant necrosis. The reported decreased symptomatic radionecrosis risk in patients treated with a 5-fraction stereotactic regimen to resection cavities, compared with those who underwent the same regimen for intact tumors, despite a considerably smaller target volume on average in non-operated patients (Faruqi et al., 2020) suggests an important intratumoral role in symptomatic radionecrosis.

Interestingly, the aforementioned Japanese study (Ohtakara et al. 2012) evaluated the depth of the target lesion and concluded that adding depth grade to dose-volumetric indices is predictive of necrosis risk, as superficial metastasis treated utilizing normal tissue definition of V12 (normal brain + dura, skull, skin) would deposit some of the radiation necrosis correlative dose–volume into non-brain tissue. In addition to dose being absorbed in non-brain tissue for superficial lesions, the depth of lesion also alters the relative size distribution of different radiation dose–volumes through the normal brain. As depth increases, more monitor units and geometric access (number/length of arcs and arc placement) become necessary to safely cover the target with a particular prescribed dose. Which raises another question: what is the relevance of the dose–volume metrics of lower doses on necrosis risk? Perhaps brain V4–8 correlates with late neurologic deficits or perhaps the volume receiving even lower doses (perhaps V1–2) correlates with second malignancy risk. In addition, how does the depth of the treated lesion relate to anatomical differences in tissue vascularity and glial cell population repair capacity?

SRS-induced malignancy is reportedly very rare (Balasubramaniam et al., 2007; Loeffler et al., 2003; McIver and Pollock, 2004; Niranjan et al., 2009; Rowe et al., 2007) as it is estimated to occur in fewer than 1 in 1,000–10,000 patients (McIver and Pollock, 2004; Muracciole and Regis, 2008; Niranjan et al., 2009; Pollock, 2003), though it is unclear if this risk is actually higher, and will be better appreciated as more cases are ascertained (Loeffler et al., 2003), or if the risk is no different than malignant tumors in the general population (Rowe et al., 2007).

We concur with QUANTEC that "toxicity increases rapidly once the volume of the brain exposed to >12 Gy is >5–10 cm^3" (Lawrence et al., 2010) and agree with the HyTEC risk estimates (Table 18.2). More stringent constraints should be considered in certain clinical contexts including eloquent brain regions and previous irradiation. Some non-dosimetric factors are also predictive for the development of radionecrosis including tumor location, previous irradiation, and male sex. The technique of brain SRS/hSRT is a very heterogeneous treatment with different dosimetric planning techniques, physical delivery modalities, definitions of V12, definitions of target volume, definitions of toxicity, length of follow-up, depth of lesions, histology of lesions, and variability of individual brain volumes. More investigation is needed to understand the risks and indicators that predict for radiation necrosis in SRS and hSRT patients.

REFERENCES

Abla AA, et al. (2015) A treatment paradigm for high-grade brain arteriovenous malformations: volume-staged radiosurgical downgrading followed by microsurgical resection. *Journal of Neurosurgery.* 122:419–432.

Ahmed KA, Freilich JM, Sloot S, et al. (2015) LINAC-based stereotactic radiosurgery to the brain with concurrent vemurafenib for melanoma metastases. *Journal of Neuro-Oncology.* 122(1):121–126.

Andrews DW, et al. (2004) Whole brain radiation therapy with or without stereotactic radiosurgery boost for patients with one to three brain metastases: phase III results of the RTOG 9508 randomized trial. *Lancet.* 363:1665–1672.

Aoyama H, et al. (2003) Hypofractionated stereotactic radiotherapy alone without whole-brain irradiation for patients with solitary and oligo brain metastasis using noninvasive fixation of the skull. *International Journal of Radiation Oncology, Biology, Physics.* 56:793–800.

Atkins K, et al. (2018) Proton stereotactic radiosurgery for brain metastases: a single-institution analysis of 370 patients. *International Journal of Radiation Oncology, Biology, Physics.* 101(4):820–829.

Balasubramaniam A, Shannon P, Hodaie M, Laperriere N, Michaels H, Guha A. (2007) Glioblastoma multiforme after stereotactic radiotherapy for acoustic neuroma: case report and review of the literature. *Neuro-Oncology.* 9:447–453.

Blonigen BJ, Steinmetz RD, Levin L, Lamba MA, Warnick RE, Breneman JC. (2010) Irradiated volume as a predictor of brain radionecrosis after linear accelerator stereotactic radiosurgery. *International Journal of Radiation Oncology, Biology, Physics.* 77:996–1001.

Brown WR, et al. (2005) Vascular damage after fractionated whole-brain irradiation in rats. *Radiat Res.* 164:662–668.

Chang JH, Chang JW, Choi JY, Park YG, Chung SS. (2003) Complications after Gamma Knife radiosurgery for benign meningiomas. *J Neurol Neurosurg Psychiatry.* 74:226–230.

Chang TC, et al. (2004) Stereotactic irradiation for intracranial arteriovenous malformation using stereotactic radiosurgery or hypofractionated stereotactic radiotherapy. *International Journal of Radiation Oncology, Biology, Physics.* 60:861–870.

Chao ST, et al. (2013) Challenges with the diagnosis and treatment of cerebral radiation necrosis. *International Journal of Radiation Oncology, Biology, Physics.* 87:449–457.

Chin LS, Ma L, DiBiase S. (2001) Radiation necrosis following Gamma Knife surgery: a case-controlled comparison of treatment parameters and long-term clinical follow up. *Journal of Neurosurgery.* 94:899–904.

Colaco RJ, Martin P, Kluger HM, Yu JB, Chiang VL. (2016) Does immunotherapy increase the rate of radiation necrosis after radiosurgical treatment of brain metastases? *Journal of Neurosurgery.* 125(1):17–23.

Cox JD, Stetz J, Pajak TF. (1995) Toxicity criteria of the radiation therapy oncology group (RTOG) and the European organization for research and treatment of cancer (EORTC). *International Journal of Radiation Oncology, Biology, Physics.* 31:1341–1346.

Diao K, Bian SX, Routman DM, et al. (2018) Stereotactic radiosurgery and ipilimumab for patients with melanoma brain metastases: clinical outcomes and toxicity. *J Neurooncol.* 139(2):421–429.

Ernst-Stecken A, Ganslandt O, Lambrecht U, Sauer R, Grabenbauer G. (2006) Phase II trial of hypofractionated stereotactic radiotherapy for brain metastases: results and toxicity. *Radiotherapy and Oncology.* 81:18–24.

Fahrig A, et al. (2007) Hypofractionated stereotactic radiotherapy for brain metastases—results from three different dose concepts. *Strahlenther Onkol.* 183:625–630.

Faruqi S, Ruschin M, Soliman H, Myrehaug S, Zeng KL, Husain Z, Atenafu E, Tseng CL, Das S, Perry J, Maralani P, Heyn C, Mainprize T, Sahgal A. (2020) Adverse radiation effect after hypofractionated stereotactic radiosurgery in 5 daily fractions for surgical cavities and intact brain metastases. *International Journal of Radiation Oncology, Biology, Physics.* 106(4):772–779.

Feuvret L, et al. (2010) Trifractionated stereotactic radiotherapy for large single brain metastases. *International Journal of Radiation Oncology, Biology, Physics.* 78:S284.

Flickinger JC.(1989) An integrated logistic formula for prediction of complications from radiosurgery. *International Journal of Radiation Oncology, Biology, Physics.* 17:879–885.

Flickinger JC, Schell MC, Larson DA. (1990) Estimation of complications for linear accelerator radiosurgery with the integrated logistic formula. *International Journal of Radiation Oncology, Biology, Physics.* 19:143–148.

Flickinger JC, Steiner L. (1990) Radiosurgery and the double logistic product formula. *Radiotherapy and Oncology.* 17:229–237.

Flickinger JC, Lunsford LD, Kondziolka D. (1991) Dose–volume considerations in radiosurgery. *Stereotact Funct Neurosurg.* 57:99–105.

Flickinger JC, Lunsford LD, Wu A, Kalend A. (1991) Predicted dose–volume isoeffect curves for stereotactic radiosurgery with the 60Co gamma unit. *Acta Oncol.* 30:363–367.

Flickinger JC, et al. (1992) Radiosurgery and brain tolerance: an analysis of neurodiagnostic imaging changes after Gamma Knife radiosurgery for arteriovenous malformations. *International Journal of Radiation Oncology, Biology, Physics.* 23:19–26.

Flickinger JC, Kondziolka D, Pollock BE, Maitz AH, Lunsford LD. (1997) Complications from arteriovenous malformation radiosurgery: multivariate analysis and risk modeling. *International Journal of Radiation Oncology, Biology, Physics.* 38:485–490.

Flickinger JC, Kondziolka D, Maitz AH, Lunsford LD. (1998) Analysis of neurological sequelae from radiosurgery of arteriovenous malformations: how location affects outcome. *International Journal of Radiation Oncology, Biology, Physics.* 40:273–278.

Flickinger JC, et al. (2000) Development of a model to predict permanent symptomatic postradiosurgery injury for arteriovenous malformation patients. Arteriovenous malformation radiosurgery study group. *International Journal of Radiation Oncology, Biology, Physics.* 46:1143–1148.

Fuentes S, Delsanti C, Metellus P, Peragut JC, Grisoli F, Regis J. (2006) Brainstem metastases: management using Gamma Knife radiosurgery. *Neurosurgery.* 58:37–42.

Ganz JC, Schrottner O, Pendl G. (1996) Radiation-induced edema after Gamma Knife treatment for meningiomas. *Stereotact Funct Neurosurg.* 66 (Suppl):129–133.

Gatterbauer B, Hirschmann D, Eberherr N, et al. (2020) Toxicity and efficacy of Gamma Knife radiosurgery for brain metastases in melanoma patients treated with immunotherapy or targeted therapy—A retrospective cohort study. *Cancer Medicine.* 9(11):4026–4036.

Genentech/Roche. Avastin (bevacizumab) prescribing information. Retrieved 24 May 2015 from www.avastin.com.

Geraud, A., Xu, H.P., Beuzeboc, P. et al. (2017) Preliminary experience of the concurrent use of radiosurgery and T-DM1 for brain metastases in HER2-positive metastatic breast cancer. *J Neurooncol.* 131:69–72.

Glantz MJ, Burger PC, Friedman AH, Radtke RA, Massey EW, Schold SC Jr. (1994) Treatment of radiation-induced nervous system injury with heparin and warfarin. *Neurology.* 44:2020.

Glatstein E. (2008) Hypofractionation, long-term effects, and the alpha/beta ratio. *International Journal of Radiation Oncology, Biology, Physics.* 72:11–12.

Gorman L, Ruben J, Myers R, Dally M. (2008) Role of hypofractionated stereotactic radiotherapy in treatment of skull base meningiomas. *J Clin Neurosci.* 15:856–862.

Gutin PH, et al. (2009) Safety and efficacy of bevacizumab with hypofractionated stereotactic irradiation for recurrent malignant gliomas. *International Journal of Radiation Oncology, Biology, Physics.* 75:156–163.

Hall EJ, Brenner DJ. (1993) The radiobiology of radiosurgery: rationale for different treatment regimes for AVMs and malignancies. *International Journal of Radiation Oncology, Biology, Physics.* 25:381–385.

Hatiboglu MA, Chang EL, Suki D, Sawaya R, Wildrick DM, Weinberg JS. (2011) Outcomes and prognostic factors for patients with brainstem metastases undergoing stereotactic radiosurgery. *Neurosurgery.* 69:796–806.

Henzel M, et al. (2006) Stereotactic radiotherapy of meningiomas: symptomatology, acute and late toxicity. *Strahlenther Onkol.* 182:382–388.

Hershman D, et al. (2013) Contraindicated use of bevacizumab and toxicity in elderly patients with colorectal cancer. *Journal of Clinical Oncology.* 31:3592–3599.

Huang CF, Kondziolka D, Flickinger JC, Lunsford LD. (1999) Stereotactic radiosurgery for brainstem metastases. *Journal of Neurosurgery.* 91:563–568.

Hussain A, Brown PD, Stafford SL, Pollock BE. (2007) Stereotactic radiosurgery for brainstem metastases: survival, tumor control, and patient outcomes. *International Journal of Radiation Oncology, Biology, Physics.* 67:521–524.

Kalapurakal JA, et al. (1997) Intracranial meningiomas: factors that influence the development of cerebral edema after stereotactic radiosurgery and radiation therapy. *Radiology.* 204:461–465.

Kased N, et al. (2008) Gamma Knife radiosurgery for brainstem metastases: the UCSF experience. *J Neurooncol.* 86:195–205.

Kim JH, et al. (2004) Upregulation of VEGF and FGF2 in normal rat brain after experimental intraoperative radiation therapy. *J Korean Med Sci.* 19:879–886.

Kim JM, Miller JA, Kotecha R, et al. (2017) The risk of radiation necrosis following stereotactic radiosurgery with concurrent systemic therapies. *J Neurooncol.* 133(2):357–368.

Kim YJ, et al. (2011) Single-dose versus fractionated stereotactic radiotherapy for brain metastases. *International Journal of Radiation Oncology, Biology, Physics.* 8:483–489.

Kocher M, et al. (2004) Alpha/beta ratio for arteriovenous malformations estimated from obliteration rates after fractionated and single-dose irradiation. *Radiotherapy and Oncology.* 71:109–114.

Kohshi K, et al. (2003) Successful treatment of radiation-induced brain necrosis by hyperbaric oxygen therapy. *J Neurol Sci.* 209:115–117.

Kondziolka D, Flickinger JC, Perez B. (1998) Judicious resection and/or radiosurgery for parasagittal meningiomas: outcomes from a multicenter review. Gamma Knife meningioma study group. *Neurosurgery.* 43:405–413.

Koyfman SA, et al. (2010) Stereotactic radiosurgery for single brainstem metastases: the Cleveland clinic experience. *International Journal of Radiation Oncology, Biology, Physics.* 78:409–414.

Kwon AK, Dibiase SJ, Wang B, Hughes SL, Milcarek B, Zhu Y. (2009) Hypofractionated stereotactic radiotherapy for the treatment of brain metastases. *Cancer.*115:890–898.

Lawrence YR, et al. (2010) Radiation dose—volume effects in the brain. *International Journal of Radiation Oncology, Biology, Physics.* 76:S20–S27.

Leach RM, Rees PJ, Wilmshurst P. (1998) Hyperbaric oxygen therapy. *BMJ.* 317:1140–1143.

Leber KA, et al. (1998) Treatment of cerebral radionecrosis by hyperbaric oxygen therapy. *Stereotact Funct Neurosurg.* 70(Suppl 1):229–236.

Levin VA, et al. (2011) Randomized double-blind placebo-controlled trial of bevacizumab therapy for radiation necrosis of the central nervous system. *International Journal of Radiation Oncology, Biology, Physics.* 79: 1487–1495.

Li YQ, et al. (2004) Early radiation-induced endothelial cell loss and blood-spinal cord barrier breakdown in the rat spinal cord. *Radiat Res.* 161:143–152.

Lindvall P, et al. (2003) Hypofractionated conformal stereotactic radiotherapy for arteriovenous malformations. *Neurosurgery.* 53:1036–1042.

Loeffler JS.(2008) Radiation tolerance limits of the brainstem. *Neurosurgery.* 63:733.

Loeffler JS, Niemierko A, Chapman PH. (2003) Second tumors after radiosurgery: tip of the iceberg or a bump in the road? *Neurosurgery.* 52:1436–1440.

Lorenzoni JG, et al. (2009) Brainstem metastases treated with radiosurgery: prognostic factors of survival and life expectancy estimation. *Surg Neurol.* 71:188–195.

Lutterbach J, et al. (2003) Radiosurgery followed by planned observation in patients with one to three brain metastases. *Neurosurgery.* 52:1066–1074.

Marchetti M, et al. (2011) Hypofractionated stereotactic radiotherapy for oligometastases in the brain: a single-institution experience. *Neurol Sci.* 32:393–399.

Manning MA, et al. (2000) Hypofractionated stereotactic radiotherapy as an alternative to radiosurgery for the treatment of patients with brain metastases. *International Journal of Radiation Oncology, Biology, Physics.* 47:603–608.

Martin AM, Cagney DN, Catalano PJ, et al. (2018) Immunotherapy and symptomatic radiation necrosis in patients with brain metastases treated with stereotactic radiation. *JAMA Oncology.* 4(8):1123.

Mayo C, Yorke E, Merchant TE. (2010) Radiation associated brainstem injury. *International Journal of Radiation Oncology, Biology, Physics.* 76:S36–S41.

McIver JI, Pollock BE. (2004) Radiation-induced tumor after stereotactic radiosurgery and whole brain radiotherapy: case report and literature review. *J Neurooncol.* 66:301–305.

McPherson CM, Warnick RE. (2004) Results of contemporary surgical management of radiation necrosis using frameless stereotaxis and intraoperative magnetic resonance imaging. *J Neurooncol.* 68:41–47.

Milano MT, Usuki KY, Walter KA, Clark D, Schell MC. (2011) Stereotactic radiosurgery and hypofractionated stereotactic radiotherapy: normal tissue dose constraints of the central nervous system. *Cancer Treat Rev.* 37:567–578.

Milano MT, Grimm J, Niemierko A, Soltys SG, Moiseenko V, Redmond KJ, Yorke E, Sahgal A, Xue J, Mahadevan A, Muacevic A, Marks LB, Kleinberg LR. (2020) Single- and multifraction stereotactic radiosurgery dose/volume tolerances of the brain. *International Journal of Radiation Oncology, Biology, Physics.* S0360–3016(20)34101–8. Epub ahead of print.

Minniti G, et al. (2011) Stereotactic radiosurgery for brain metastases: analysis of outcome and risk of brain radionecrosis. *Radiat Oncol.* 6:48.

Minniti G, Scaringi C, Paolini S, et al. (2016) Single-fraction versus multifraction (3 x 9 Gy) Stereotactic radiosurgery for large (>2 cm) brain metastases: a comparative analysis of local control and risk of radiation-induced brain necrosis. *Int J Radiation Oncol Biol Phs.* 95:1142–1148.

Minniti G, Anzellini D, Reverberi C, et al. (2019) Stereotactic radiosurgery combined with nivolumab or Ipilimumab for patients with melanoma brain metastases: evaluation of brain control and toxicity. *Journal for Immunotherapy of Cancer.* 7:102.

Miyakawa A, Shibamoto Y, Otsuka S, Iwata H. (2014) Applicability of the linear—quadratic model to single and fractionated radiotherapy schedules: an experimental study. *J Radiat Res.* 55:451–454.

Moosa S, et al. (2014) Volume-staged versus dose-staged radiosurgery outcomes for large intracranial arteriovenous malformations. *Neurosurgery Focus.* 37:E18.

Muracciole X, Regis J. (2008) Radiosurgery and carcinogenesis risk. *Prog Neurol Surg.* 21:207–213.

Narayana A, et al. (2007) Hypofractionated stereotactic radiotherapy using intensity-modulated radiotherapy in patients with one or two brain metastases. *Stereotact Funct Neurosurg.* 85:82–87.

Narayana A, Mathew M, Tam M, Kannan R, Madden KM, Golfinos JG, et al. (2013) Vemurafenib and radiation therapy in melanoma brain metastases. *J Neurooncol.* 113:411–416.

Nakaya K, et al. (2010) Gamma Knife radiosurgery for benign tumors with symptoms from brainstem compression. *International Journal of Radiation Oncology, Biology, Physics.* 77:988–995.

Ney DE, et al. (2015) Phase II trial of hypofractionated intensity-modulated radiotherapy combined with temozolomide and bevacizumab for patients with newly diagnosed glioblastoma. *J Neurooncol.* 122:135–143.

Niranjan A, Kondziolka D, Lunsford LD. (2009) Neoplastic transformation after radiosurgery or radiotherapy: risk and realities. *Otolaryngol Clin North Am.* 42:717–729.

Ohguri T, et al. (2007) Effect of prophylactic hyperbaric oxygen treatment for radiation-induced brain injury after stereotactic radiosurgery of brain metastases. *International Journal of Radiation Oncology, Biology, Physics.* 67:248–255.

Ohtakara K, et al. (2012) Significance of target location relative to the depth from the brain surface and high-dose irradiated volume in the development of brain radionecrosis after micromultileaf collimator-based stereotactic radiosurgery for brain metastases. *J Neurooncol.* 108:201–209.

Omuro A, et al. (2014) Phase II study of bevacizumab, temozolomide, and hypofractionated stereotactic radiotherapy for newly diagnosed glioblastoma. *Clinical Cancer Research.* 20:5023–5031.

Onodera S, Aoyama H, Katoh N, et al. (2011) Long—term outcomes of fractionated stereotactic radiotherapy for intracranial skull base benign meningiomas in single institution. *Japanese Journal of Clinical Oncology.* 4:462–468.

Park HR, Lee JM, Park KW, Kim JH, et al. (2018) Fractionated Gamma Knife radiosurgery as initial treatment for large skull base meningioma. *Exp Neurobiol.* 3:245–255.

Patel, K.R., Chowdhary, M., Switchenko, J.M., Kudchadkar, R., Lawson, D.H., Cassidy, R.J., Prabhu, R.S., Khan, M.K. (2016) BRAF inhibitor and stereotactic radiosurgery is associated with an increased risk of radiation necrosis. *Melanoma Res.* 26(4):387–394.

Patil CG, et al. (2008) Predictors of peritumoral edema after stereotactic radiosurgery of supratentorial meningiomas. *Neurosurgery.* 63:435–440.

Pérez-Espejo MA, et al. (2009) Usefulness of hyperbaric oxygen in the treatment of radionecrosis and symptomatic brain edema after LINAC radiosurgery. *Neurosurgery (Astur).* 20:449–453. See comment in PubMed Commons below

Petrovich Z, et al. (2002) Survival and pattern of failure in brain metastases treated with stereotactic Gamma Knife radiosurgery. *Journal of Neurosurgery.* 97:499–506.

Pollock B.(2003) Second tumors after radiosurgery: tip of the iceberg or a bump in the road? *Neurosurgery.* 52:1436–1440.

Qi XS, Schultz CJ, Li XA. (2007) Possible fractionated regimens for image-guided intensity-modulated radiation therapy of large arteriovenous malformations. *Physics in Medicine and Biology.* 52:5667–5682.

Rowe J, Grainger A, Walton L, Silcocks P, Radatz M, Kemeny A. (2007) Risk of malignancy after Gamma Knife stereotactic radiosurgery. *Neurosurgery.* 60:60–65.

Schultheiss TE, Kun LE, Ang KK, Stephens LC. (1995) Radiation response of the central nervous system. *International Journal of Radiation Oncology, Biology, Physics.* 31:1093–1112.

Sengöz M, Kabalay IA, Tezcanlı E, Peker S, Pamir N. (2013) Treatment of brainstem metastases with Gamma-Knife radiosurgery. *J Neurooncol.* 113:33–38.

Sharma MS, et al. (2008) Radiation tolerance limits of the brainstem. *Neurosurgery.* 63:728–732.

Shaw E, et al. (2000) Single dose radiosurgical treatment of recurrent previously irradiated primary brain tumors and brain metastases: final report of RTOG protocol 90–05. *International Journal of Radiation Oncology, Biology, Physics.* 47:291–298.

Shehata MK, et al. (2004) Stereotactic radiosurgery of 468 brain metastases < or =2 cm: implications for SRS dose and whole brain radiation therapy. *International Journal of Radiation Oncology, Biology, Physics.* 59:87–93.

Sheline GE, Wara WM, Smith V. (1980) Therapeutic irradiation and brain injury. *International Journal of Radiation Oncology, Biology, Physics.* 6:1215–1228.

Shrieve DC, Hazard L, Boucher K, Jensen RL. (2004) Dose fractionation in stereotactic radiotherapy for parasel-lar meningiomas: radiobiological considerations of efficacy and optic nerve tolerance. *Journal of Neurosurgery.* 101(Suppl 3):390–395.

Shuryak I, et al. (2015) High-dose and fractionation effects in stereotactic radiation therapy: analysis of tumor control data from 2965 patients. *Radiotherapy Oncol.* 8:339–348.

Shuto T, Fujino H, Asada H, Inomori S, Nagano H. (2003) Gamma Knife radiosurgery for metastatic tumours in the brainstem. *Acta Neurochir (Wien).* 145:755–760.

Silander H, et al. (2004) Fractionated, stereotactic proton beam treatment of cerebral arteriovenous malformations. *Acta Neurol Scand.* 109:85–90.

Sneed PK, et al. (2015) Adverse radiation effect after stereotactic radiosurgery for brain metastases: incidence, time course, and risk factors. *Journal of Neurosurgery.* 15:1–14.

Szeifert GT, Atteberry DS, Kondziolka D, Levivier M, Lunsford LD. (2006) Cerebral metastases pathology after radiosurgery: a multicenter study. *Cancer.* 106:2672–2681.

Trippa F, Maranzano E, Costantini S, Giorni C. (2009) Hypofractionated stereotactic radiotherapy for intracranial meningiomas: preliminary results of a feasible trial. *Journal of Neurosurgery Science.* 53:7–11.

Trotti A, et al. (2003) CTCAE v3.0: development of a comprehensive grading system for the adverse effects of cancer treatment. *Semin Radiat Oncol.* 13:176–181 . . . NIH Publication No. 09–5410: https://ctep.cancer.gov/proto-coldevelopment/electronic_applications/docs/CTCAE_v5_Quick_Reference_5x7.pdf

Tye K, et al. (2014) An analysis of radiation necrosis of the central nervous system treated with bevacizumab. *J Neurooncol.* 117:321–327.

Valadão J, Pearl J, Verma S, Helms A, Whelan H. (2003) Hyperbaric oxygen treatment for post-radiation central nervous system injury: a retrospective case series. *J Neurol Sci.* 209:115–117.

Varlotto JM, et al. (2003) Analysis of tumor control and toxicity in patients who have survived at least one year after radiosurgery for brain metastases. *International Journal of Radiation Oncology, Biology, Physics.* 57:452–464.

Vernimmen FJ, Slabbert JP. (2010) Assessment of the alpha/beta ratios for arteriovenous malformations, meningiomas, acoustic neuromas, and the optic chiasma. *Int J Radiat Biol.* 86:486–498.

Veznedaroglu E, et al. (2004) Fractionated stereotactic radiotherapy for the treatment of large arteriovenous malformations with or without previous partial embolization. *Neurosurgery.* 55:519–530.

Veznedaroglu E, et al. (2008) Fractionated stereotactic radiotherapy for the treatment of large arteriovenous malformations with or without previous partial embolization. *Neurosurgery.* 62(Suppl. 2):763–775.

Wang TM, Shen GP, Chen MY, et al. (2019) Genome-wide association study of susceptibility loci for radiation-induced brain injury. *J Natl Cancer Inst.* 111(6):620–628.

Wigg DR.(1999) Is there a role for fractionated radiotherapy in the treatment of arteriovenous malformations? *Acta Oncol.* 38:979–986.

Williamson R, Kondziolka D, Kanaan H, Lunsford LD, Flickinger JC. (2008) Adverse radiation effects after radiosurgery may benefit from oral vitamin E and pentoxifylline therapy: a pilot study. *Stereotact Funct Neurosurg.* 86:359–366.

Yen CP, Sheehan J, Patterson G, Steiner L. (2006) Gamma Knife surgery for metastatic brainstem tumors. *Journal of Neurosurgery.* 105:213–219.

Zhuo X, Huang X, Yan M, et al. (2019) Comparison between high-dose and low-dose intravenous methylprednisolone therapy in patients with brain necrosis after radiotherapy for nasopharyngeal carcinoma. *Radiotherapy and Oncology.* 137:16–23.

19 Vertebral Compression Fracture Post-Spine SBRT

Michael Hardisty, Joel Finkelstein, Jay Detsky, Isabelle Thibault, and Arjun Sahgal

Contents

19.1 INTRODUCTION

A decade of literature has established that vertebral compression fracture (VCF) is a relatively common adverse side effect following spine stereotactic body radiation therapy (SBRT). A recent review (Faruqi *et al.*, 2018) tabulated a pooled average incidence rate of VCF of 13.9% following SBRT treatments. However, the rate has been as high as 40% in some series (Rose *et al.*, 2009; Sahgal, Atenafu *et al.*, 2013), highlighting the importance in understanding patient, tumor, and treatment factors that impact the risk of VCF. VCF may lead to pain, neurological compromise, and the need for further therapy, including invasive procedures such as biopsy or cement augmentation. Patient selection as well as consideration of prophylactic stabilization are critical for ensuring the safe use of SBRT in the spine. The aim of this chapter is to focus on the reported literature to guide the definition, assessment, and management of SBRT-induced VCF. A summary of the specific risk factors (lytic tumor burden, baseline VCF, spinal deformity, dose per fraction) that have been shown to be predictive of VCF and the need for stabilization is also provided.

19.2 DEFINITION

VCF is defined as a collapse of the vertebral body (VB). Spine SBRT-induced VCF can occur as *de novo* (no existing baseline VB fracture) or as progression of an existing VCF within the treated vertebral segment (fracture progression). Those series reporting on spine SBRT-induced VCF have typically excluded segments that concomitantly experience local tumor progression, arguing that tumor progression itself might destabilize the spinal segment and predispose to a pathologic VCF rather than an iatrogenic VCF. Similarly, reporting of SBRT-induced VCF is typically restricted to patients without a history of prior invasive surgery (which limits the risks of any new fracture or progression).

19.3 RADIOLOGIC ASSESSMENT OF VCF AND SURVEILLANCE IMAGING

19.3.1 VERTEBRAL BODY HEIGHT MEASUREMENTS

The initial step in the radiologic assessment of VCF is to determine the treated VB height, based on the endplates, according to the baseline treatment planning CT and MRI. Ideally, that baseline VB height is compared to prior imaging to determine whether there is any baseline VCF or progression of an existing one. If prior imaging is not available, then the height is compared to the average VB height of the vertebrae located immediately above and below the treated segment.

Following SBRT, the baseline VB height is compared to subsequent follow-up imaging studies, preferably using the same imaging modality. Both sagittal and coronal views should be used if available; however, the key is to be consistent in the method. A difference of ≤5% in VB height is considered insignificant and could represent a measurement error, and, therefore, a radiographic score of VCF is based on more than 5% height loss.

19.3.2 IMAGING MODALITY

The preferred imaging modality post-SBRT for follow-up assessment is MRI (T1 nonenhanced sequences). As compared to CT, MRI has superior diagnostic accuracy (Buhmann (Kirchhoff) *et al.*, 2009) for spinal metastases and allows visualization of any epidural and/or paraspinal soft tissue extension. Given that VCF tends to occur shortly after spine SBRT, on average 2.5 months after radiation (Sahgal, Whyne *et al.*, 2013), the first follow-up MRI should be performed within the first 2 to 3 months following treatment. Most spine SBRT experts perform routine follow-up imaging at 2 to 3-month intervals for at least the initial years and at 3 to 6 month intervals thereafter.

If VCF is observed, then further imaging is often required to characterize the VCF. These include plain film x-rays and most importantly a CT spine. CT is useful in the assessment of the integrity of the bony anatomy with respect to the detection of cortical and trabecular bone destruction. CT further characterizes the tumor as osteolytic, osteosclerotic, or mixed type lesions (which cannot be reliably determined with MRI alone); this may have implications for treatment as lytic-based VCF may be treated with cement augmentation procedures as opposed to purely sclerotic-based VCF. CT helps determine the spinal alignment and degree of instability which has implications on surgical decision-making. For example, translation beyond a physiologically expected level is a high-risk situation often requiring surgical stabilization, and kyphotic deformity may also influence the type of intervention. Figure 19.1 presents a case of SBRT-induced VCF resulting in kyphotic deformity and mechanical pain.

19.3.3 CLASSIFICATION OF FRACTURE

The morphology and classification that best describes these fractures is a Type A fracture according to the AO spine classification of thoracolumbar injuries (Vaccaro *et al.*, 2013). These are compression injuries that involve the anterior portion of the vertebral column. The stability of the fracture is dependent on and further characterized based on factors such as: (1) involving just the anterior half of the vertebral body and a single endplate, (2) involving both endplates without the involvement of the posterior wall creating a "pincer" fracture, (3) single endplate and posterior wall involvement of the vertebral body that is compressed

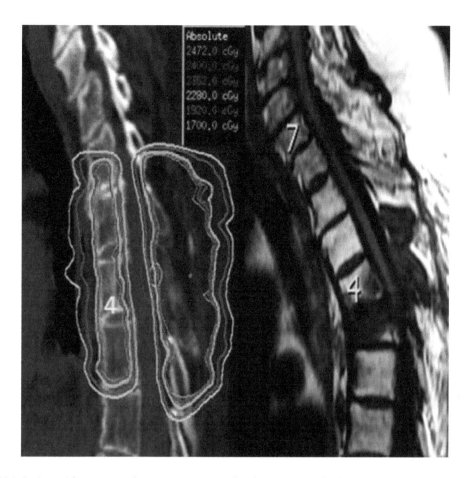

Figure 19.1 Patient with metastatic breast cancer treated with stereotactic body radiation therapy (SBRT) (24 Gy in 2 fractions) to T3–T4–T5. The sagittal dose distribution is shown on the left panel. Patient responded well to pain and 4 months later developed a sudden mechanical type of pain. Investigations showed fracture at T5 and resulting kyphotic deformity, and the sagittal T1-weighted magnetic resonance image is shown on the right panel, with numbered 7th cervical and 4th thoracic vertebral segments. The patient's pain settled without intervention, and the patient remained pain-free at the last follow-up (12 months post SBRT).

and retropulsed into the spinal canal (incomplete burst fracture), and (4) both end plates along with the posterior vertebral wall with retropulsion into the spinal canal (complete burst fracture). The instability and risk of neurological injury progress with each successive subtype. A similar classification for cervical spine fractures can be used as it relates to the understanding of VCF and potential instability patterns (Vaccaro *et al.*, 2016). A comprehensive assessment of spinal instability is required in metastatic patients and will be further discussed in this chapter.

19.4 PATHOPHYSIOLOGY OF SBRT-INDUCED VCF

SBRT-induced VCF may occur as a sub-acute or late radiation effect on the vertebral bone. The high-dose treatment causes collateral damage deriving from the ablation of adjacent (within the treated spinal segment) non-pathologically involved bone tissue, which results in damage to the bone matrix and local resorption of bone (Barth *et al.*, 2010; Al-Omair *et al.*, 2013; Steverink *et al.*, 2018). Early failure, occurring within several months of treatment, is likely related to a combination of necrosis, inflammation, and damage to the underlying bone matrix, with later stage VCF occurring following local bone remodeling, resulting in a degradation of the bone's ability to resist mechanical loading.

SBRT induces radiation necrosis, fibrosis, and disruption to bone remodeling. This phenomenon was first observed after spine SBRT by Al-Omair et al. (Al-Omair *et al.*, 2013), who identified radiation osteonecrosis and fibrosis in biopsy specimens of two cases with VCF suspected to be tumor progression. A larger (n = 23) histopathology study by Roerster et al. of tumor-involved vertebral specimens obtained from salvage surgery following SBRT showed that fractures were associated with disturbances to bone remodeling and bone marrow fibrosis (Foerster *et al.*, 2019). Prevalence of osteonecrosis (85%) and soft tissue (77%) necrosis was high in both fractured and unfractured vertebrae. Osteoradionecrosis is a well-known late, but rare, toxicity after conventional radiotherapy and has been observed in the mandible (Marx and Johnson, 1987), hand (Walsh, 1897), and femoral head (Tai *et al.*, 2000). It is characterized by osteolysis, altered collagen fibrils, and loss of minerals, occurring most commonly in a hypoxic and avascular environment.

Inflammation is thought to also play a role in the destabilization of treated vertebrae, leading to mechanical failure. Inflammation is suspected in acute cases of VCF because of the common occurrence of pain flair (incidence reported as high as 68%), combined with the observation that treatment with the anti-inflammatory drug, dexamethasone, alleviates the pain.

The effect of radiation is only one component contributing to the pathophysiology of VCF (Sahgal, Whyne *et al.*, 2013). The structural integrity and quality of the vertebral bone rely both on its material properties (tissue mineralization and collagen) and architecture. Strength, stiffness, and toughness are afforded by the hydroxyapatite content, arrangement of minerals within the collagen, and the collagen fiber network itself (Wang *et al.*, 2001; Burr, 2002; Whyne, 2014). Bone toughness has been shown to be decreased by ablation at radiation doses on the order of those seen in SBRT in *ex vivo* mechanical testing of crack propagation (Barth *et al.*, 2010). Bone quality is known to be affected by metastatic involvement of the bony spine, beyond the resorption of bone mass by lytic lesions. Basic science research has shown that the organic phase and mineral phase are both affected by the presence of tumor leading to changes in the material properties. These changes occur at multiple length scales, affecting trabecular architecture (Harwdisty *et al.*, 2012), mineralization (He *et al.*, 2017), and collagen crosslinking (Burke *et al.*, 2017). Metastatic involvement has been shown to affect tissue level (Nazarian *et al.*, 2008) and apparent mechanical properties (Nazarian *et al.*, 2008).

19.5 INCIDENCE OF VCF

A summary of the published papers reporting on VCF after spine SBRT is presented in Table 19.1. The first major study identifying VCF as a serious toxicity following spine SBRT was performed by the Memorial Sloan Kettering Cancer Center and reported a 39% risk of VCF following the dose of 18 to 24 Gy delivered in a single-fraction SBRT (Rose *et al.*, 2009). Many investigations have reported similarly high incidence of VCF, ranging from 4% to 46% following SBRT regimens of 16 to 45 Gy in 1–5 fractions, corresponding to a single equivalent mean dose of 21 Gy (Sung and Chang, 2014). The time to VCF has a wide range from 1.2 to 15.4 months) (Gong *et al.*, 2019).

To summarize the risk of VCF, a multi-institutional study was reported by Sahgal et al., pooling data from the MD Anderson Cancer Center (MDACC), Cleveland Clinic, and the University of Toronto. Based on 410 spinal segments treated, the crude rate of VCF was 14%, and the actuarial 1-year cumulative incidence was 12.4% (Sahgal, Atenafu *et al.*, 2013). The median and mean time to VCF was 2.46 and 6.33 months, respectively, with two-thirds of VCFs developing within the first 4 months post-SBRT.

19.6 PREDICTIVE FACTORS OF VCF

19.6.1 EFFECT OF DOSE FRACTIONATION

Dose per fraction has been identified as a predictor of VCF by multiple investigators (Thibault *et al.*, 2014; Jawad *et al.*, 2016; Lee *et al.*, 2016; Chen *et al.*, 2020). High-dose per fraction SBRT (exceeding 19 Gy) has been identified as a major predictor of VCF. Of 410 treated spinal segments reported in a multi-institutional study (Sahgal, Atenafu *et al.*, 2013), those receiving ≥24 Gy/fraction (HR 5.25; 95% CI 2.29–12.01) or 20–23 Gy/fraction (HR 4.91; 95% CI 1.96–12.28) had higher VCF risk compared to those treated with

SBRT of ≤19 Gy/fraction. SBRT dose per fraction of ≥20 Gy was also a significant predictor of VCF in the study by Cunha et al. (Cunha *et al.*, 2012). A further reduction in VCF incidence to 8.5% was noted for a dose of 24 Gy in 2 fractions (12 Gy per fraction) (Faruqi *et al.*, 2018). A recent review of radiosurgery for the treatment of renal cell carcinoma found similar results with the rate of VCF ranging from 16% to 27.5% in the literature and identified single-fraction therapy as having a greater risk of fracture (Smith *et al.*, 2018).

A need for SBRT regimens ≥20 Gy per fraction has been suggested for radioresistant tumor types. For example, a series of spinal metastases from GI primary cancers, known to be radioresistant, showed improved local control with high single-fraction equivalent dose greater than 20 Gy compared to lower dose per fraction less than 20 Gy (Sandhu *et al.*, 2020). The investigators reported a VCF incidence of 13% which is consistent with other investigations. In contrast, a recent systematic review found that the incidence of VCF was only modestly increased (10.7% versus 10.1%) in patients with a single fraction versus multiple fractions (Gong *et al.*, 2019). Consistent with the concept of dose fractionation being related to VCF risk, normal tissue complication probability modelling has shown that dose levels D90% and D80% parameters of dose–volume histograms have also been found to be significantly related to VCF occurrence (Sandhu *et al.*, 2020).

Further research is needed to determine optimal clinical practice and how the risk compares to conventional palliative radiotherapy. Currently, a trial is underway that examines a dose of 24 Gy in 2 fraction scheme against conventional radiotherapy (Canadian SC24, Phase 3 RCT, NCT02512965) which will provide an accurate incidence of the VCF risk in the setting of a phase 3 trial in both arms. However, no RCT has examined the importance of dose fractionation specific to spine SBRT.

19.6.2 BONE TUMOR TYPE

Osteolytic tumor type has also been shown to be a significant predictor of VCF. This factor has been identified by many published studies evaluating risk factors for VCF (Table 19.2). An early multi-institutional series reported by Sahgal et al. with 48 VCFs (18.8%) after SBRT among 256 spinal metastases noted an increased risk with lytic tumor type (HR 3.53; 95% CI 1.58–7.93) with more recent reports confirming this finding (HR = 2.5–3) (Boyce-Fappiano *et al.*, 2017). Consistent with this finding, volumetric quantification of lytic tumor burden was shown to greatly increase fracture risk (HR 38) (Thibault *et al.*, 2017). However, lytic tumor burden remains challenging to quantify volumetrically. Klein et al. have recently presented methods to automate the assessment of tumor type using a combination of deep learning methods that allow for automatic segmentation of the vertebral body and histogram-based identification of tumor type (Klein *et al.*, 2020).

U net Convolutional Neural Network architectures (left) have been shown to be fast and highly accurate for image segmentation problems. Developing fracture risk prediction in the metastatic spine has

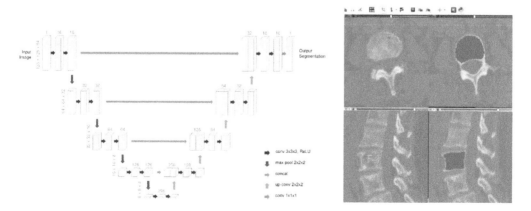

Figure 19.2 Vertebral compression fracture risk prediction: A deep learning approach.

initially focused on vertebral body segmentation (right). The approach uses deep networks for segmentation and deep feature extraction in combination with automation of calculation of existing stability features (vertebral collapse, tumor burden, mal-alignment, etc.). Future work will combine CT and MR images using a latent representation to enable multimodal assessment of mechanical stability and integration with a clinical scoring system.

Lytic tumor causes local bone resorption and degradation for the material properties of adjacent bone tissue (Burke *et al.*, 2017). Lytic tumor has reduced bone mineralization, compromised inherent bone structure, and increased propensity for pressurization (Whyne, 2014). Thus, lytic tumor as a predictor of VCF makes biophysical sense as it is predisposed to fracture even prior to SBRT.

19.6.3 PREEXISTING FRACTURE

The presence of a VCF at baseline has also been shown to predict for a higher risk of radiation-induced VCF by at least six investigations including multiple series (Boehling *et al.*, 2012) and multi-institutional reports (Boehling *et al.*, 2012; Sahgal, Atenafu *et al.*, 2013; Boyce-Fappiano *et al.*, 2017). This finding is also consistent with the trauma literature that has shown a well-established risk for future fracture being prior fracture (Klotzbuecher *et al.*, 2000). The elevated risk of post SBRT VCF due to an existing fracture is substantial but with a wide variation reported in the literature (HR=1.69–9.25) (Sahgal, Atenafu *et al.*, 2013; Boyce-Fappiano *et al.*, 2017).

The mechanical stability of the pre-existing fractures can be assessed by imaging. Hardisty et al. examined CT-based quantitative measures of stability of fractured vertebrae treated with SBRT (Hardisty *et al.*, 2020). Specifically fractured vertebral body morphology (volume loss, height reduction, anterior column, and posterior column collapse) was examined. Patients with fractured vertebral bodies that lost volume (>3%) between imaging prior to and after SBRT were more likely to go on to need stabilization procedures.

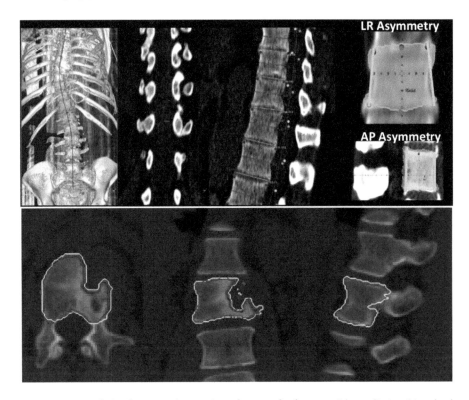

Figures 19.3 Automatic analysis of metastatic vertebrae features for fracture risk prediction: Vertebral stability can be informed by automated methods that identify vertebrae, segment vertebral bodies, place landmarks on endplates, and quantify vertebral body collapse and changes in left–right (LR) height asymmetry and anterior–posterior (AP) height asymmetry. Automated segmentation of fractured vertebrae showed good agreement (DSC = 88% ± 5%) with manually defined contours (Hardisty *et al.*, 2020).

Table 19.1 Comparison of studies reporting on VCF after spine SBRT

AUTHOR, SITE (YEAR)	NO. OF PTS.	NO. OF SEGMENTS	LYTIC SEGMENTS	BASELINE FRACTURED SEGMENTS[3]	SBRT TOTAL DOSE/FX	INCIDENCE OF VCF	MEDIAN TIME TO VCF (MONTHS)	SALVAGE INTERVENTION (%); TYPE
Sandhu et al. (2020)	74	114	29%	NA	18–24/1–3	13%	NA	3/15 (20%) 2 CAP
Rose et al. (2009)	62	71	65%	28%	18–24/1	39%	25	3/27 (11%); 2 S, 1 CAP
Boehling et al. (2012)	93	123	58%	28%	18–30/1–5	20%	3	10/25 (40%); 10 CAP
Cunha et al. (2012)	90	167	48%	17%	8–35/1–5	11%	2 (mean 3.3)	9/19 (47%); 3 S, 6 CAP
Balagamwala et al. (2012)	57	88	NA	30%	8–16/1	14%	NA	NA
Sahgal, Whyne et al. (2013)	252	410	62%	20%	8–35/1–5	14%	2.5 (mean 6.3)	24/57 (42%); 7 S, 17 CAP
Thibault et al. (2014)	37	61[b]	95%	21%	18–30/1–5	16%	1.6	4/10 (40%); 1 S, 3 CAP
Sung and Chang (2014)	72	72	NA	NA	18–45/1–5	36%	(mean: 1.5)	15/26 (58%); 5 S, 10 CAP
Guckenberger et al. (2014)	301	387	72%	20%	8–60/1–20	7.8%	NA	NA
Ito et al. (2019)	20	20	NA	NA	24/2	10%	15	NA
Miller et al. (2016)	100	232	NA	29%	10–24/1–3	21%	NA	5/30 (17%) 5S
Sellin et al. (2015)	37	40	93%	NA	24–30/1–5	35%	NA	6/14 (43%); 4 S, 2 CAP
(Sohn et al., 2014)	13	NA	NA	NA	Average 38/1–5	15%	NA	None
Boyce-Fappiano et al. (2017)	448	1070	27%	42%	10–60/1–5	11.9%	2.7	37/127 (29%), 16 S, 21 CAP
Germano et al. (2016)	79	143	39%	42%	10–18/1	21%	5	8/17 (30%), NA
Jawad et al. (2016)	NA	594	71%	24%	8–40/1–5	5.7%	3	NA
Lee et al. (2016)	79	100	NA	22%	16–27/1–3	32%	3.3	15/32 (47%) 5 S, 10 CAP
Thibault et al. (2016)	55	100	56%	24%	12–35/1–5	17%	1.68	NA

Note: The results of the multi-institutional study by Sahgal et al. are reported in bold text.

[a] Indicates percentage of segments with a preexisting vertebral compression fracture at baseline, before spine stereotactic body radiation therapy.

[b] Of a cohort of 71 renal cell cancer spinal segments, 61 were analyzed for vertebral compression fracture risk as 10 were postoperative stereotactic body radiation therapy cases.

Abbreviations: VCF, vertebral compression fracture; pts, patients; fx, fractions; S, surgery; CAP, cement augmentation procedure; NA, not available.

The question that arises from this data is whether patients with preexisting VCF should be routinely treated with prophylactic stabilization surgery. With minimally invasive techniques consisting of cement augmentation or percutaneous instrumentation, some types of stabilization can be performed as an outpatient procedure prior to or shortly after SBRT with significantly fewer adverse events as compared to those expected with traditional open surgery.

Gerszten et al. were one of the first to successfully combine kyphoplasty (Medtronic, Minnesota, MA) followed by SBRT for patients with painful pathologic VCF deemed eligible for SBRT (Gerszten et al., 2009). However, the risk of routine prophylaxis is that a significant proportion of patients might be over-treated, and there remains the risk of significant adverse events despite the minimally invasive nature of these interventions. This highlights the importance of scoring systems, such as the spinal instability neoplastic score (SINS) (Fisher et al., 2010), which may help identify those patients a priori who are mechanically unstable and are at risk of VCF. Prophylactic stabilization can be considered for patients with mechanical instability or mechanical pain. At this time, treatment with SBRT and intervention only upon VCF development or progression is the standard of care, until robust patient selection methods are determined and validated specific to radiation-induced VCF risk. An additional benefit to this approach is to treat an undisturbed target for SBRT planning, and the potential for tumor extravasation following cement injection has been reported (Cruz et al., 2014). Following SBRT, stabilization with cement or surgical instrumentation can be considered for painful VCF post-SBRT. Evidence-based algorithms for stabilization of VCF are needed both pre- and post-SBRT. These aspects will be discussed further in the chapter.

Table 19.2 **Summary of the literature on risk factors for vertebral compression fracture after spine stereotactic body radiation therapy**

AUTHOR (YEAR)	SIGNIFICANT PREDICTORS ON MVA
Rose et al. (2009)	Lytic tumor (HR 3.8); 41%–60% vertebral body involved (HR 3.9)
Boehling et al. (2012)	Age > 55 years (HR 5.7); baseline VCF[a] (HR 4.12); lytic tumor (HR 2.8)
Cunha et al. (2012)	Lytic tumors (HR 12.2); malalignment (HR 11.1); ≥20 Gy/fraction (HR 6.8); lung histology (HR 4.3); liver histology (HR 34)
Sahgal et al. (2013)	Baseline VCF[a] (HR 8.9 if <50% VCF, HR 6.9 if ≥50% VCF); lytic tumor (HR 3.5); ≥20 Gy/fraction (HR 4.9 if 20–23 Gy, HR 5.3 if ≥24 Gy); spinal malalignment (HR 3.0)
Sung and Chang (2014)	Vertebral body osteolysis rate of ≥60%
Thibault et al. (2014)	Baseline VCF (HR 9.25), single-fraction SBRT (5)
Thibault et al. (2016)	≥11.6% lytic volume (OR 52), baseline VCF (OR 37, SBRT dose/fraction ≥20 Gy (OR 12)
Jawad et al. (2016)	Solitary metastasis (OR 3.5), prescription dose to target volume ≥38.4 Gy EQD_2 (OR 2.3), preexisting VCF (OR 2.82)
Lee et al. (2016)	SINS ≥7 (HR 5.6), age (≥ 65)
Boyce-Fappiano et al. (2017)	<3 vs 3+ levels treated (HR 4–4.3), lytic lesion (HR 2.5–3), prior VCF (HR 1.69)
Chen (2020)	BMI (HR 0.90 per unit increase); total SINS (HR 2.44 unstable vs stable); PTV D80% (HR 1.11 per Gy increase)
Ehresman et al. (2020)	SINS; vertebral bone quality (VBQ, an MRI-derived normalized vertebral signal intensity, combined model AUC = 89%)

[a] Indicates preexisting vertebral compression fracture at baseline, prior to spine stereotactic body radiation therapy.
Abbreviations: VCF, vertebral compression fracture; MVA, multivariate analysis; HR, hazard ratio; Multi, multi-institutions, BMI, Body Mass Index; SINS Spinal Instability Neoplastic Score; PTV, Planning Target Volume.

19.6.4 SPINAL MALALIGNMENT

The presence of baseline spinal misalignment (kyphotic/scoliotic deformity or subluxation/translation) prior to SBRT was identified as an independent predictor of VCF increasing risk (HR = 3.0–11.1) by the large U of T series (Cunha *et al.*, 2012) and confirmed by a multi-institutional report (Sahgal, Atenafu *et al.*, 2013). Evaluation of spinal alignment and stability is one key aspect as to why multidisciplinary management involving the radiation oncologist and spinal surgeon is essential for joint decision-making and personalized patient care.

19.6.5 TUMOR HISTOLOGY

Cunha et al. identified lung and liver cancer histologies at increased risk of radiation-induced VCF (Cunha *et al.*, 2012); however, these findings have not been reproduced in other studies (Table 19.2). Consistent with tumor histology affecting VCF risk is the systematic review of SBRT for renal cell metastases to the spine showing a modestly higher incidence (14–27.5%) (Balagamwala *et al.*, 2012; Smith *et al.*, 2018) compared to those found for other cancer histologies (14%) (Faruqi *et al.*, 2018) and shown in Table 19.1. Limited confirmatory evidence exists from investigations that have noted the importance of histology in univariate analysis but without inclusion in multivariate analysis (Jawad *et al.*, 2016; Boyce-Fappiano *et al.*, 2017). Therefore, currently, there are insufficient data to identify tumor histological type as affecting risk for post-SBRT VCF.

19.6.6 OTHER FACTORS

Patient-specific factors such as patient age (HR 2.15–5.67), BMI (HR 0.90 per unit increase), and female sex (in univariate analysis HR 1.54) (Boyce-Fappiano *et al.*, 2017) have been shown to increase risk of VCF and could be considered as factors for patient management (Table 19.2). The presence of osteoporosis, use of bisphosphonate, and single versus multiple tumors treated within a single clinical target volume have been investigated but found not to impact on post-SBRT VCF risk (Rose *et al.*, 2009; Boehling *et al.*, 2012; Cunha *et al.*, 2012; Sahgal, Atenafu *et al.*, 2013; Sung and Chang, 2014).

Imaging-derived features have been identified by three different investigations as relevant to the prediction and management of VCF following SBRT (Thibault *et al.*, 2017; Ehresman *et al.*, 2020; Hardisty *et al.*, 2020). Thibault et al. identified the lytic tumor volume size as highly predictive (HR = 37, with lesions greater than 11.6% of vertebral body size) of vertebral compression fracture (Thibault *et al.*, 2017). Ehresman et al. evaluated trabecular bone quality in MRI scans and combined the metric with SINS for 89% accuracy for the prediction of VCF (Ehresman *et al.*, 2020). However, the patients studied by Ehresman had a high rate of VCF, 53%, and approximately half the patients in the study were excluded because of inadequate imaging. Hardisty et al. in a small pilot study observed that loss of vertebral body volume before and after SBRT was related to those patients with VCF, who went on to be stabilized (Hardisty *et al.*, 2020). These studies illustrate the potential of imaging-derived factors for predicting VCF and the need for stabilization; however, more research is needed to demonstrate these findings in larger patient populations. Further translation of these tools to clinical software is needed for these approaches to gain widespread adoption. This is an active area of research with new tools from artificial intelligence being applied to assess spinal metastases.

19.7 IDENTIFYING SPINAL INSTABILITY

Spine instability has been defined by the Spine Oncology Study Group (SOSG) as the "loss of spinal integrity as a result of a neoplastic process that is associated with movement-related pain, symptomatic or progressive deformity and/or neural compromise under physiological loads" (Fisher *et al.*, 2010). Of note, pain with movement or axial loading of the spine and/or pain relieved with recumbence is referring to mechanical pain. As mechanical pain is a key symptom to assess; some clinicians prefer to use the term mechanical instability rather than spinal instability (Laufer *et al.*, 2013).

The SOSG developed the SINS, based on expert consensus, to aid clinicians in determining which spinal tumor is "stable", "potentially unstable", or "unstable". This classification system has been validated as a reliable tool, with respect to intraobserver and interobserver reliability by surgeons, radiation oncologists, and radiologists (Fourney *et al.*, 2011; Fisher, Schouten *et al.*, 2014; Fisher, Versteeg *et al.*, 2014).

Table 19.3 **Spinal instability neoplastic score (SINS) classification**

6 SINS FACTOR	DESCRIPTION	POINTS
Location of tumor	SJunction: occiput–C2, C7–T2, T11–L1, L5–S1	3
	Mobile spine: C3–C6, L2–L4	2
	Semirigid T-spine: T3–T10	1
	Rigid sacrum: S2–S5	0
Pain	Mechanical pain[a]	3
	Occasional and nonmechanical	1
	None	0
Bone tumor type	Osteolytic	2
	Mixed	1
	Osteoblastic	0
Spinal malalignment	Subluxation or translation	4
	Kyphosis or scoliosis *de novo* deformity	2
	None	0
Vertebral body height collapse	≥50%	3
	<50%	2
	No compression fracture but >50% body involved	1
	None of the above	0
Tumoral involvement of posterior elements	Bilateral	3
	Unilateral	1
	None	0

Classification: Total SINS ranges from 0 to 18 points. According to SINS, stable segments have a score of 0–6, indeterminate/potentially unstable a score of 7–12, and unstable segments a score of 13–18.

[a] Movement-related or axial loading pain and/or pain improvement with recumbency.
Abbreviation: SINS, spinal instability neoplastic score.

The classification system is based on the assessment of six factors: location of the metastases, type of pain, bone lesion type, radiologic spinal alignment, VB collapse, and posterolateral involvement of spinal elements. Table 19.3 details the SINS classification.

Although the SINS criteria were originally developed to identify potentially unstable patients with spinal metastases, they were recently investigated for their predictive capacity with respect to developing SBRT-induced VCF. SINS as well as the individual SINS criteria (baseline VCF, osteolytic tumor, and malalignment) have been shown to be predictive of SBRT-induced VCF (Rose *et al.*, 2009; Boehling *et al.*, 2012; Cunha *et al.*, 2012; Sahgal, Atenafu*et al.*, 2013; Jawad *et al.*, 2016; Lee *et al.*, 2016; Boyce-Fappiano *et al.*, 2017; Chen *et al.*, 2020; Ehresman *et al.*, 2020). SINS is an important tool with three independent studies highlighting its utility to identify patients at increased risk for SBRT-induced VCF combined with other factors discussed in this chapter.

19.8 GUIDELINES

With respect to VCF risk, patient and tumor factors are essential. There is greater attention to determining which patients should undergo a stabilization procedure prior to SBRT and which patients should be closely monitored following SBRT. It is our recommendation that patients presenting at baseline with a spinal subluxation or translation should be referred for consultation with a spine surgeon prior to radiation. Patients with a baseline VCF, who also have mechanical pain and/or spinal malalignment, should also be referred to a spine surgeon prior to SBRT. What is more controversial, as to prophylactic stabilization, are those patients pain-free with a baseline VCF and those with significant osteolytic tumor with no fracture. In these patients, close monitoring within the first 6 months following SBRT should be maintained as the risk of VCF is greatest during this period, and the risk may be as high as 35%. Similarly, monitoring in the short term is needed for single-fraction SBRT of ≥20 Gy as the risk of VCF exceeds 20%.

19.9 SURGICAL MANAGEMENT OF SBRT-INDUCED VCF

Approximately 40% (range, 0%–58%) of SBRT-induced VCFs will be treated with a surgical salvage stabilization procedure due to the development of mechanical instability (Table 19.1). Most commonly, vertebral cement augmentation procedures (e.g., kyphoplasty or vertebroplasty) have been performed; however, approximately 25% will be treated with a more invasive instrumented surgical procedure (Table 19.1) (Ross et al., 2007; Boehling et al., 2012; Cunha et al., 2012; Sahgal, Atenafuet al., 2013; Sung and Chang, 2014; Thibault et al., 2014).

Traditionally, spine stabilization surgery was an open invasive surgery requiring a long incision, muscle dissection of paraspinal tissues, and instrumentation of levels above and below the injured segment. Epidural decompression or vertebrectomy could be performed within the same intervention due to the wide exposure. Such open surgery is associated with longer rehabilitation time, greater morbidity, and increased delay before initiating adjuvant therapies such as radiation and chemotherapy, compared to noninvasive surgical intervention.

Frailty is an age-related syndrome characterized by declined physiological reserve across multiple organ systems causing an inability to respond to provoked stress. The concept of frailty has increasingly been embraced by spine surgeons as a factor that can have a clinically significant role in predicting postoperative outcomes (Moskven et al., 2018). Similarly, sarcopenia, related to frailty, has been shown to predict surgical outcomes in spine surgeries for metastatic disease (Bourassa-Moreau et al., 2020). Considering all factors including frailty and the unique risks involved in surgical treatment of metastatic spine disease, the decision-making process for surgery has evolved.

The NOMS decision framework was introduced in 2013, integrating neurologic, oncologic, mechanical, and systemic considerations into the treatment decision-making for spine metastases (Laufer et al., 2013). Mechanical instability, referring to severe mechanical pain and instability according to SOSG, was stipulated as an independent surgical indication, regardless of tumor epidural extension or radiosensitivity (Laufer et al., 2013). In patients with mechanical instability, a surgical stabilization intervention was recommended by the authors, based on radiation, being unable to restore a spinal malalignment and steroids often failing to palliate mechanical pain. Most recently, the LMNOP system was proposed by the University of Saskatchewan, integrating SINS into a multifactorial decision-based and individualized approach for the general management of spinal metastases (Ivanishvili and Fourney, 2014). In potentially unstable metastases (SINS total score of 7–12), the authors' general guidance is to first consider a vertebral cement augmentation procedure if no cord compression is present, while pedicle screw fixation or more invasive procedures are usually reserved for unstable metastases (SINS of 13–18).

Percutaneous spinal interventions, such as vertebral cement augmentation (balloon kyphoplasty or vertebroplasty) and other minimally invasive spinal surgery (MISS) procedures, are now increasingly used and allow spinal procedures to be applied to a wider spectrum of spinal oncology patients. MISS refers to the application of instrumentation percutaneously and/or allows for epidural decompression via a tubular retraction system (Massicotte et al., 2012). In appropriate situations, MISS with separation surgery is an emerging practice prior to spine SBRT.

Regarding cement augmentation procedures, vertebroplasty consists of a high-pressure cement injection (usually of polymethylmethacrylate) into the fractured vertebra under fluoroscopic guidance. In contrast, kyphoplasty utilizes balloon inflation within the fractured VB followed by a lower-pressure cement injection. Both procedures are associated with a risk of cement leakage, which can be of major consequence if cement leaks into the spinal canal causing cord compression or it leaks into the vasculature causing cement emboli. It has been reported that kyphoplasty is associated with fewer symptomatic cement leakage complications, although the risk is low at under 5% (Lee et al., 2009; Berenson et al., 2011). A multicenter randomized control trial evaluating kyphoplasty as compared to nonsurgical interventions has been reported in patients with cancer and a pathologic VCF. In those randomized to kyphoplasty, a significant relief of pain at the primary (1-month post-kyphoplasty) endpoint (mean score change from 17.6 to 9.1) was observed in addition to improved functional outcomes (Berenson *et al.*, 2011). Although these data are specific to patients presenting with a pathologic VCF, it is reasonable to expect similar outcomes after kyphoplasty for radiation-induced VCF.

Tumor extravasation following cement augmentation was recently reported as a possible iatrogenic complication (Cruz *et al.*, 2014). Observed venous tumor extravasation and anterior subligamentous spread observed in two cases were postulated to be a direct result of the increased intravertebral pressure during balloon inflation and cement injection (Cruz *et al.*, 2014). Importantly, the consequence of this complication impaired the feasibility of subsequent spine SBRT due to the difficulty in adequately delineating the target volume. Because of the risk of cement leakage and tumor extravasation, although rare, some spine SBRT experts prefer to treat with spine SBRT first followed by a cement augmentation procedure. The intent is to then perform the cement augmentation procedure at 6 to 8 weeks post-SBRT as a planned, prophylactic intervention.

19.10 NONSURGICAL MANAGEMENT OF SBRT-INDUCED VCF

For symptomatic patients not eligible for a stabilization surgical procedure and/or declining a surgical intervention, pharmacologic pain management strategies consist of narcotic analgesics, nonsteroidal anti-inflammatory drugs, corticosteroids, membrane stabilizers, bisphosphonates, and tricyclic antidepressants. Treatment consideration may also include spinal orthosis (e.g., brace or cervical collar application).

19.11 FUTURE DIRECTIONS

VCF following spine SBRT has been identified as a significant adverse effect. In fact, the risk can approach nearly 40% of treated patients depending on patient, tumor, and radiation dose fractionation factors. Predictors of VCF following spine SBRT have been identified such that radiation oncologists can start to rely on these predictors to help in daily clinical practice. In particular, by identifying those with a high SINS, baseline fracture, significant osteolytic tumor burden, and spinal malalignment, appropriate referral to spine surgeons for the consideration of a stabilization procedure prior to SBRT can be made. Further research to determine decision-making trees, recursive portioning analyses to stratify patients into risk groups, better tools for assessing quantitative imaging features, and a better understanding of the pathophysiology of SBRT-induced VCF will further our ability to choose the most appropriate patients for spine SBRT.

CHECKLIST: KEY POINTS FOR CLINICAL PRACTICE

- High-dose per fraction SBRT (≥20 Gy), osteolytic tumor, a preexisting VCF at baseline, and spinal malalignment are significant predictors of VCF following spine SBRT.
- VCF tends to occur shortly after spine SBRT at a median time of 2.5 months. The first follow-up spine MRI should be performed within 2 to 3 months of spine SBRT.
- The pathophysiology of VCF is complex and related to a combination of necrosis, inflammation, and damage to the bone matrix with later-stage VCF related to local bone remodeling.
- The SINS aims to identify patients who are unstable and has been shown to be a useful tool to help clinicians identify those at increased risk of VCF after spine SBRT.
- Patients with spinal malalignment, symptomatic baseline VCF, and/or mechanical pain should be evaluated by a spine surgeon prior to spine SBRT to consider a stabilization procedure.
- Surgical management of symptomatic SBRT-induced VCF includes cement augmentation procedures (kyphoplasty, vertebroplasty), minimally invasive surgeries, or invasive stabilization approaches with fixation.
- When a surgical procedure is required following spine SBRT, it is recommended to perform a biopsy to diagnose tumor progression versus necrosis as the cause of the VCF.

REFERENCES

Al-Omair, A. *et al.* (2013) 'Radiation-induced vertebral compression fracture following spine stereotactic radiosurgery: Clinicopathological correlation', *Journal of Neurosurgery: Spine*, 18(5), pp. 430–435. doi: 10.3171/2013.2.SPINE12739.

Balagamwala, E. H. *et al.* (2012) 'Single-fraction stereotactic body radiotherapy for spinal metastases from renal cell carcinoma: Clinical article', *Journal of Neurosurgery: Spine*, 17(6), pp. 556–564. doi: 10.3171/2012.8.SPINE12303.

Barth, H. D. *et al.* (2010) 'On the effect of X-ray irradiation on the deformation and fracture behavior of human cortical bone.', *Bone*, 46(6), pp. 1475–1485. doi: 10.1016/j.bone.2010.02.025.

Berenson, J. *et al.* (2011) 'Balloon kyphoplasty versus non-surgical fracture management for treatment of painful vertebral body compression fractures in patients with cancer: A multicentre, randomised controlled trial', *The Lancet Oncology*, 12(3), pp. 225–235. doi: 10.1016/S1470-2045(11)70008-0.

Boehling, N. S. *et al.* (2012) 'Vertebral compression fracture risk after stereotactic body radiotherapy for spinal metastases: Clinical article', *Journal of Neurosurgery: Spine*, 16(4), pp. 379–386. doi: 10.3171/2011.11.SPINE116.

Bourassa-Moreau, É. *et al.* (2020) 'Sarcopenia, but not frailty, predicts early mortality and adverse events after emergent surgery for metastatic disease of the spine', *Spine Journal*, 20(1), pp. 22–31. doi: 10.1016/j.spinee.2019.08.012.

Boyce-Fappiano, D. *et al.* (2017) 'Analysis of the factors contributing to vertebral compression fractures after spine stereotactic radiosurgery', *International Journal of Radiation Oncology Biology Physics*. Elsevier Inc., 97(2), pp. 236–245. doi: 10.1016/j.ijrobp.2016.09.007.

Buhmann (Kirchhoff), S. *et al.* (2009) 'Detection of osseous metastases of the spine: Comparison of high resolution multi-detector-CT with MRI', *European Journal of Radiology*, 69(3), pp. 567–573. doi: 10.1016/j.ejrad.2007.11.039.

Burke, M. *et al.* (2017) 'The impact of metastasis on the mineral phase of vertebral bone tissue', *Journal of the Mechanical Behavior of Biomedical Materials*, 69, pp. 75–84. doi: 10.1016/j.jmbbm.2016.12.017.

Burr, D. B. (2002) 'The contribution of the organic matrix to bone's material properties', *Bone*, 31(1), pp. 8–11. doi: 10.1016/S8756-3282(02)00815-3.

Chen, X. *et al.* (2020) 'Normal tissue complication probability of vertebral compression fracture after stereotactic body radiotherapy for de novo spine metastasis', *Radiotherapy and Oncology*. Elsevier Ireland Ltd, 150, pp. 142–149. doi: 10.1016/j.radonc.2020.06.009.

Cruz, J. P. *et al.* (2014) 'Tumor extravasation following a cement augmentation procedure for vertebral compression fracture in metastatic spinal disease', *Journal of Neurosurgery: Spine*, 21(3), pp. 372–377. doi: 10.3171/2014.4.SPINE13695.

Cunha, M. V. R. *et al.* (2012) 'Vertebral compression fracture (VCF) after spine stereotactic body radiation therapy (SBRT): Analysis of predictive factors', *International Journal of Radiation Oncology Biology Physics*. Elsevier Inc., 84(3), pp. e343–e349. doi: 10.1016/j.ijrobp.2012.04.034.

Ehresman, J. *et al.* (2020) 'A novel MRI-based score assessing trabecular bone quality to predict vertebral compression fractures in patients with spinal metastasis', *Journal of Neurosurgery: Spine*. American Association of Neurological Surgeons, 32(4), pp. 499–506. doi: 10.3171/2019.9.SPINE19954.

Faruqi, S. *et al.* (2018) 'Vertebral compression fracture after spine stereotactic body radiation therapy: A review of the pathophysiology and risk factors', *Clinical Neurosurgery*. Oxford University Press, 83(3), pp. 314–322. doi: 10.1093/neuros/nyx493.

Fisher, C. G. *et al.* (2010) 'A novel classification system for spinal instability in neoplastic disease', *Spine*, 35(22), pp. E1221–E1229. doi: 10.1097/BRS.0b013e3181e16ae2.

Fisher, C. G., Schouten, R.*et al.* (2014) 'Reliability of the spinal instability neoplastic score (SINS) among radiation oncologists: An assessment of instability secondary to spinal metastases', *Radiation Oncology*. BioMed Central Ltd., 9(1). doi: 10.1186/1748-717X-9-69.

Fisher, C. G., Versteeg, A. L.*et al.* (2014) 'Reliability of the spinal instability neoplastic scale among radiologists: An assessment of instability secondary to spinal metastases', *American Journal of Roentgenology*. American Roentgen Ray Society, 203(4), pp. 869–874. doi: 10.2214/AJR.13.12269.

Foerster, R. *et al.* (2019) 'Histopathological findings after reirradiation compared to first irradiation of spinal bone metastases with stereotactic body radiotherapy: A cohort study', *Clinical Neurosurgery*. Oxford University Press, 84(2), pp. 435–441. doi: 10.1093/neuros/nyy059.

Fourney, D. R. *et al.* (2011) 'Spinal instability neoplastic score: An analysis of reliability and validity from the spine oncology study group', *Journal of Clinical Oncology*, 29(22), pp. 3072–3077. doi: 10.1200/JCO.2010.34.3897.

Gerszten, P. C. *et al.* (2009) 'Combination kyphoplasty and spinal radiosurgery: A new treatment paradigm for pathological fractures', *Journal of Neurosurgery: Spine*. Journal of Neurosurgery Publishing Group (JNSPG), 3(4), pp. 296–301. doi: 10.3171/spi.2005.3.4.0296.

Gong, Y. *et al.* (2019) 'Efficacy and safety of different fractions in stereotactic body radiotherapy for spinal metastases: A systematic review', *Cancer Medicine*, 8(14), pp. 6176–6184. doi: 10.1002/cam4.2546.

Hardisty, M. *et al.* (2020) 'CT based quantitative measures of the stability of fractured metastatically involved vertebrae treated with spine stereotactic body radiotherapy', *Clinical and Experimental Metastasis*. doi: 10.1007/s10585-020-10049-9.

Harwdisty, M. R. *et al.* (2012) 'Quantification of the effect of osteolytic metastases on bone strain within whole vertebrae using image registration', *Journal of Orthopaedic Research*, 30(7), pp. 1032–1039. doi: 10.1002/jor.22045.

He, F. *et al.* (2017) 'Multiscale characterization of the mineral phase at skeletal sites of breast cancer metastasis', *Proceedings of the National Academy of Sciences of the United States of America*. National Academy of Sciences, 114(40), pp. 10542–10547. doi: 10.1073/pnas.1708161114.

Ivanishvili, Z. and Fourney, D. R. (2014) 'Incorporating the spine instability neoplastic score into a treatment strategy for spinal metastasis: LMNOP', *Global Spine Journal*. SAGE Publications, 4(2), pp. 129–135. doi: 10.1055/s-0034-1375560.

Jawad, M. S. *et al.* (2016) 'Vertebral compression fractures after stereotactic body radiation therapy: A large, multi-institutional, multinational evaluation', *Journal of Neurosurgery: Spine*. American Association of Neurological Surgeons, 24(6), pp. 928–936. doi: 10.3171/2015.10.SPINE141261.

Klein, G. *et al.* (2020) 'Metastatic vertebrae segmentation for use in a clinical pipeline', in Cai, Y. et al. (eds) *Computational Methods and Clinical Applications for Spine Imaging. CSI 2019. Lecture Notes in Computer Science*, pp. 15–28. doi: 10.1007/978-3-030-39752-4_2.

Klotzbuecher, C. M. *et al.* (2000) 'Patients with prior fractures have an increased risk of future fractures: A summary of the literature and statistical synthesis', *Journal of Bone and Mineral Research*. American Society for Bone and Mineral Research, 15(4), pp. 721–739. doi: 10.1359/jbmr.2000.15.4.721.

Laufer, I. *et al.* (2013) 'The NOMS framework: Approach to the treatment of spinal metastatic tumors', *The Oncologist*, 18(6), pp. 744–751. doi: 10.1634/theoncologist.2012-0293.

Lee, M. J. *et al.* (2009) 'Percutaneous treatment of vertebral compression fractures: A meta-analysis of complications', *Spine*, 34(11), pp. 1228–1232. doi: 10.1097/BRS.0b013e3181a3c742.

Lee, S. H. *et al.* (2016) 'Can the spinal instability neoplastic score prior to spinal radiosurgery predict compression fractures following stereotactic spinal radiosurgery for metastatic spinal tumor? A post hoc analysis of prospective phase II single-institution trials', *Journal of Neuro-Oncology*. Springer New York LLC, 126(3), pp. 509–517. doi: 10.1007/s11060-015-1990-z.

Marx, R. E. and Johnson, R. P. (1987) 'Studies in the radiobiology of osteoradionecrosis and their clinical significance', *Oral Surgery, Oral Medicine, Oral Pathology*, 64(4), pp. 379–390. doi: 10.1016/0030-4220(87)90136-8.

Massicotte, E. *et al.* (2012) 'Minimal access spine surgery (MASS) for decompression and stabilization performed as an out-patient procedure for metastatic spinal tumours followed by spine stereotactic body radiotherapy (SBRT): First report of technique and preliminary outcomes', *Technology in Cancer Research and Treatment*. Adenine Press, pp. 15–25. doi: 10.7785/tcrt.2012.500230.

Moskven, E. *et al.* (2018) 'The impact of frailty and sarcopenia on postoperative outcomes in adult spine surgery. A systematic review of the literature', *Spine Journal*. Elsevier Inc., pp. 2354–2369. doi: 10.1016/j.spinee.2018.07.008.

Nazarian, A. *et al.* (2008) 'Bone volume fraction explains the variation in strength and stiffness of cancellous bone affected by metastatic cancer and osteoporosis', *Calcified Tissue International*, 83(6), pp. 368–379. doi: 10.1007/s00223-008-9174-x.

Rose, P. S. *et al.* (2009) 'Risk of fracture after single fraction image-guided intensity-modulated radiation therapy to spinal metastases', *Journal of Clinical Oncology*, 27(30), pp. 5075–5079. doi: 10.1200/JCO.2008.19.3508.

Ross, C. F. *et al.* (2007) 'Modulation of mandibular loading and bite force in mammals during mastication', *The Journal of Experimental Biology*, 210(Pt 6), pp. 1046–1063. doi: 10.1242/jeb.02733.

Sahgal, A., Atenafu, E. G. *et al.* (2013) 'Vertebral compression fracture after spine stereotactic body radiotherapy: A multi-institutional analysis with a focus on radiation dose and the spinal instability Neoplastic score', *Journal of Clinical Oncology*, 31(27), pp. 3426–3431. doi: 10.1200/JCO.2013.50.1411.

Sahgal, A., Whyne, C. M., Ma, L., Larson, D. A. and Fehlings, M. G. (2013) 'Vertebral compression fracture after stereotactic body radiotherapy for spinal metastases', *The Lancet Oncology*. Elsevier Ltd, 14(8), pp. e310–e320. doi: 10.1016/S1470-2045(13)70101-3.

Sandhu, N. *et al.* (2020) 'Local control and toxicity outcomes of stereotactic radiosurgery for spinal metastases of gastrointestinal origin', *Journal of Neurosurgery Spine*, pp. 1–8. doi: 10.3171/2020.1.SPINE191260.

Smith, B. W. *et al.* (2018) 'Radiosurgery for Treatment of renal cell metastases to spine: A systematic review of the literature', *World Neurosurgery*. Elsevier Inc., 109, pp. e502–e509. doi: 10.1016/j.wneu.2017.10.011.

Sohn, S. *et al.* (2014) 'Stereotactic radiosurgery compared with external radiation therapy as a primary treatment in spine metastasis from renal cell carcinoma: A multicenter, matched-pair study', *Journal of Neuro-Oncology*, 119(1), pp. 121–128. doi: 10.1007/s11060-014-1455-9.

Steverink, J. G. *et al.* (2018) 'Early tissue effects of stereotactic body radiation therapy for spinal metasta-ses', *International Journal of Radiation Oncology Biology Physics*, 100(5), pp. 1254–1258. doi: 10.1016/j.ijrobp.2018.01.005.

Sung, S.-H. and Chang, U.-K. (2014) 'Evaluation of risk factors for vertebral compression fracture after stereotactic radiosurgery in spinal tumor patients', *Korean Journal of Spine*, 11(3), p. 103. doi: 10.14245/kjs.2014.11.3.103.

Tai, P. *et al.* (2000) 'Pelvic fractures following irradiation of endometrial and vaginal cancers—A case series and review of literature', *Radiotherapy and Oncology*, 56(1), pp. 23–28. doi: 10.1016/S0167-8140(00)00178-X.

Thibault, I. *et al.* (2014) 'Spine stereotactic body radiotherapy for renal cell cancer spinal metastases: Analysis of outcomes and risk of vertebral compression fracture', *Journal of Neurosurgery: Spine*, pp. 711–718. doi: 10.3171/2014.7.SPINE13895.

Thibault, I. *et al.* (2017) 'Volume of lytic vertebral body metastatic disease quantified using computed tomography—based image segmentation predicts fracture risk after spine stereotactic body radiation therapy', *International Journal of Radiation Oncology Biology Physics*. Elsevier Inc., 97(1), pp. 75–81. doi: 10.1016/j.ijrobp.2016.09.029.

Vaccaro, A. R. *et al.* (2013) 'AOSpine thoracolumbar spine injury classification system: Fracture description, neurological status, and key modifiers', *Spine*. Spine (Phila Pa 1976), 38(23), pp. 2028–2037. doi: 10.1097/BRS.0b013e3182a8a381.

Vaccaro, A. R. *et al.* (2016) 'AOSpine subaxial cervical spine injury classification system', *European Spine Journal*. Springer Verlag, 25(7), pp. 2173–2184. doi: 10.1007/s00586-015-3831-3.

Walsh, D. (1897) 'Deep tissue traumatism from roentgen rat exposure', *British Medical Journal*, 2(1909), pp. 272–273. doi: 10.1136/bmj.2.1909.272.

Wang, X. *et al.* (2001) 'The role of collagen in determining bone mechanical properties', *Journal of Orthopaedic Research: Official Publication of the Orthopaedic Research Society*, 19(6), pp. 1021–1026. doi: 10.1016/S0736-0266(01)00047-X.

Whyne, C. M. (2014) 'Biomechanics of metastatic disease in the vertebral column', *Neurological Research*. Maney Publishing, 36(6), pp. 493–501. doi: 10.1179/1743132814Y.0000000362.

20 Spinal Cord Dose Limits for Stereotactic Body Radiotherapy

Joe H. Chang, Ahmed Hashmi, Shun Wong, David Larson, Lijun Ma, and Arjun Sahgal

Contents

20.1 INTRODUCTION

Spine stereotactic body radiotherapy (SBRT), also known as spine stereotactic radiosurgery (SRS), is an emerging treatment option for patients with spinal bone metastases. There is accumulating evidence for its effectiveness in the setting of spinal oligometastases (Chang et al., 2017a; Zeng et al., 2019), radioresistant histologies (Zeng et al., 2019), prior spinal radiation therapy (Mantel et al., 2013), and in the postoperative setting (Redmond et al., 2020). The main benefit of spine SBRT is the ability to deliver high doses to the tumor volume while sparing the adjacent organs at risk (OARs).

The most important OAR to spare using a spine SBRT technique is the spinal cord, which is typically located in close proximity to the vertebral target volume. If the dose tolerance of the spinal cord is exceeded, the patient is put at risk of developing radiation myelopathy (RM) which is one of the most feared complications associated with spine SBRT. RM may lead to patients becoming permanently paralyzed, and it can even be fatal especially if it occurs in the cervical spinal cord (Schultheiss et al., 1986).

There has been a lot of variation in dose constraints applied to the spinal cord, especially among early adopters of spine SBRT. These variations are due to uncertainties such as whether or not calculating biologically effective doses (BEDs) at the high doses per fraction with SBRT is accurate, whether or not the volume of spinal cord irradiated makes any difference to tolerances, and how the spinal cord OAR volume is delineated.

While there are still many uncertainties in understanding safe spinal cord constraints, we now have over a decade of worldwide experience on spinal SBRT to guide our practice. This chapter summarizes the data on spinal cord tolerances to SBRT in both those who have never had prior radiotherapy (*de novo* SBRT) and those who received prior radiotherapy (reirradiation SBRT).

20.2 RADIATION MYELOPATHY

RM is a late effect of radiotherapy delivered to the spinal cord, with clinical effects ranging from minor sensory and/or motor deficits to complete paraplegia/quadriplegia and loss of autonomic functioning. It is a diagnosis of exclusion, based on neurologic signs and symptoms consistent with damage to the irradiated segment of the spinal, without evidence of recurrent or progressive tumor affecting the spinal cord.

Demyelination and necrosis of the spinal cord, typically confined to white matter, are the main histologic features of radiation-induced myelopathy, although they are not pathognomonic of radiation injury (Wong et al., 2015). Apart from white matter changes, varying degrees of vascular damage and glial reaction can often be seen. Injury of the microvasculature including disruption of the blood–spinal cord barrier (BSCB) has been implicated in the pathogenesis of RM, although vascular changes may be absent or inconspicuous histologically (Wong et al., 2015).

The underlying biologic mechanism of RM remains unclear. The prevalent model suggests that mitotic death of endothelial cells results in BSCB disruption. This leads to vasogenic edema, hypoxia, and an inflammatory cascade resulting in demyelination and necrosis. The pathobiology of RM was reviewed recently, and further discussions can be found in the review by Wong et al., 2015.

MRI is currently the most commonly used imaging tool in the diagnostic assessment of RM. As this is a rare complication, the literature consists of mainly case reports generally without histopathologic correlation (Sahgal et al., 2012, 2013). Characteristic MRI changes in the cord include areas of low signals on T1-weighted images, high signals on T2, and focal contrast enhancement (Wong et al., 2015). In the rat spinal cord, high signal intensity on T2 has been shown to correlate histopathologically with edema and confluent necrosis, and postcontrast enhancement has been shown to correlate with BSCB disruption (Wong et al., 2015). Advanced quantitative MRI techniques such as apparent diffusion coefficients, magnetization transfer, and diffusion tensor imaging may provide additional information regarding structural changes after radiation, particularly in white matter after radiation.

20.3 DEFINING THE SPINAL CORD ORGAN AT RISK

Segmenting the spinal cord OAR accurately is critical for safe spinal SBRT practice. On the one hand, if a larger than necessary volume is defined as the spinal cord OAR, one risks underdosing the epidural space, and it has been shown that progression within the epidural space is the most common pattern of failure following spine SBRT (Chang et al., 2016). On the other hand, if an inappropriately small or inaccurate volume is defined as the spinal cord OAR, one risks delivering a higher-than-intended dose to the true spinal cord and therefore increasing the risk of RM.

A technique that allows accurate visualization of the spinal cord is essential for spinal cord segmentation. A common approach is to fuse T1 and T2 MRI sequences to the radiotherapy planning CT scan and then to segment the spinal cord on the fused images. A drawback of this method is that it introduces uncertainties related to the accuracy of image fusion. Another approach is to use CT myelography to visualize the spinal cord. This method involves applying the myelogram contrast agent immediately before performing the radiotherapy planning CT scan, with the patient immobilized in the treatment position. This method eliminates the uncertainty associated with image fusion; however, it is an invasive procedure and can be associated with complications (Thariat et al., 2009). As a result, there are tradeoffs to either MRI- or CT-based approaches for visualizing the spinal cord. Importantly, CT alone (without myelogram contrast) is insufficient for visualizing the spinal cord.

Many factors can contribute to uncertainty about the spinal cord position, including image fusion accuracy as described above, inter- and intrafraction patient motion (Hyde et al., 2012; Ma et al., 2009), and motion of the spinal cord (Tseng et al., 2015). In order to account for this uncertainty, many clinicians use a safety margin around the imaging-defined spinal cord. This margin can be applied by segmenting the spinal cord using one of techniques described earlier, followed by applying a uniform planning OAR volume (PRV) expansion margin. Alternatively, a surrogate structure for the spinal cord that is larger than the spinal cord itself can be defined, such as the thecal sac or spinal canal. Using the spinal canal as the surrogate structure for the spinal cord, OAR is generally not advised, as this is larger than is usually necessary to account for uncertainty, and may lead to the epidural space and adjacent bone being underdosed (Sahgal et al., 2019). The thecal sac is a commonly used surrogate structure for the spinal cord OAR and represents a reasonable safety margin beyond the true spinal cord that can account for uncertainty in spinal cord position (Sahgal et al., 2019).

The most consistent and modern method of defining a spinal cord OAR may be contouring the spinal cord using one of the techniques described above and applying a PRV expansion margin. Some clinicians assume the uncertainty in spinal cord position to be negligible, and therefore, do not apply a PRV expansion margin at all. Others have used margins of 1 to 2 mm, citing the uncertainties discussed above and the fact that because the steepest dose gradients are almost always adjacent to the spinal cord, even small motions can be dosimetrically significant (Guckenberger et al., 2007; Wang et al., 2008). Ideally, institutions should determine the errors associated with their own image fusion, inter- and intrafraction motion, and contouring accuracy to determine center-specific PRV margins (Chang et al., 2017b; Guckenberger et al., 2011).

20.4 REVIEW OF OUTCOMES DATA

A review of the literature regarding spinal cord tolerance was performed for both *de novo* and reirradiation SBRT in the Hypofractionation Treatment Effects in the Clinic (HyTEC) review (Sahgal et al., 2019). Because of the heterogeneity of the data reported, several assumptions had to be made to be able to synthesize the data.

First, a wide variety of fractionation schedules were used in the various spine SBRT studies. In order to be able to make meaningful comparisons between the different fractionation schedules, doses were converted into BEDs. The linear–quadratic (LQ) model has traditionally been used for this purpose; however, its accuracy in modeling biological effects at the high doses per fraction used in SBRT (>10 Gy per fraction) has been questioned (Brenner, 2008; Kirkpatrick et al., 2008). Several new models have been proposed that aim to improve on the LQ model at these higher doses; however, none have been validated with clinical data with sufficient confidence to shift practice away from the LQ model (Huang et al., 2013; Park et al., 2008; Wang et al., 2010). At present, the LQ model is still based on the fewest number of assumptions, is easily calculated in the clinic, and the most commonly used model in current SBRT literature. For these reasons, the LQ model was used for comparing the doses between studies in the HyTEC review (Sahgal et al., 2019). The most important parameter for performing LQ calculations is the α/β of the spinal cord which ranges from about 1 to 4 in the literature (Kirkpatrick et al., 2010). The study with the largest number of cases of RM to date (Sahgal et al., 2013) used an α/β of 2; therefore, an α/β of 2 was used in the HyTEC review (Sahgal et al., 2019) wherever possible to maintain consistency with this analysis. The LQ model was used to convert reported doses into a common metric, the equivalent dose in 2 Gy fractions using an α/β of 2 ($EQD2_2$). These $EQD2_2$ values were used to compare doses among SBRT schedules.

Second, a variety of dose–volume histogram (DVH) parameters have been reported for dose to the spinal cord. The D_{max}, defined as the maximum absorbed dose as specified by a single calculation point (International Commission on Radiation Units and Measurements, 2017), was the most commonly reported parameter in the reviewed studies. Some studies have used a "near-max" dose (e.g., $D_{0.03cc}$), which may be associated with less uncertainty and may, therefore, be a more reliable metric for reporting the spinal cord dose (Benedict et al., 2010; Ma et al., 2019). Because most studies simply reported the D_{max} without a "near-max" dose, and because there is no way of reliably deducing one metric from the other (Ma et al., 2019), data was summarized only using the D_{max}.

Third, data was pooled even though slightly different structures were used for the spinal cord OAR among the different studies. Some studies used the spinal cord with no PRV expansion, some used the spinal cord with a 1.5 mm PRV expansion, and others used the thecal sac. Because there is no way of reliably determining the dose for one structure from the other, the data was pooled, with the caveat that this increases the uncertainty in the final recommendations.

The following two sections describes the main studies reviewed in the HyTEC review (Sahgal et al., 2019) and the main recommendations.

20.5 SPINAL CORD TOLERANCE TO *DE NOVO* SPINE SBRT

The data for *de novo* spine SBRT has been thoroughly reviewed in the HyTEC review (Sahgal et al., 2019). The main studies used to make the recommendations are summarized below.

Sahgal and colleagues performed a multi-institutional study where the dose–volume histogram (DVH) data for nine cases of RM from spinal SBRT were compared to a cohort of 66 spine SBRT patients with no RM (Sahgal et al., 2013). Notably, one case of RM in this series had been treated with SBRT as a boost 6 weeks after 30 Gy in 10 fractions delivered conventionally (SBRT thecal sac D_{max} was 15 Gy). This case was included in the series, as within a 6-week time period, one would not expect sufficient recovery to classify it as a reirradiation case. Six cases of RM followed single-fraction SBRT, one case following two fractions, and one case following three fractions. The median follow-up intervals among patients with and without RM were 23 months (range, 8–40 months) and 15 months (range, 4–64 months), respectively. The median time to development of RM was 12 months (range, 3–15 months). The mean and median $EQD2_2$ D_{max} in the RM cases were 70.6 Gy and 73.7 Gy, respectively, whereas the mean and median $EQD2_2$ D_{max} in the control cohort were 38.8 Gy and 35.7 Gy, respectively. Based on this analysis, recommendations were made for thecal sac D_{max} for a 5% or lower risk of RM. A thecal sac $EQD2_2$ D_{max} of 44.6 Gy was recommended, which according to the LQ model translates to 12.4, 17, 20.3, 23, and 25.3 in 1, 2, 3, 4, and 5 fractions, respectively.

Katsoulakis and colleagues reported a series of 228 patients treated to 259 sites with single-fraction SBRT to doses ranging from 18 to 24 Gy that included two RM cases (Katsoulakis et al., 2017). The D_{max} to the spinal cord (as defined using CT myelography) with no PRV expansion margin was reported for these cases. The spinal cord D_{max} ranged from 9.61 to 15.21 Gy, with the two RM cases receiving 13.4 Gy and 13.6 Gy (both of which were lower than 13.85 Gy, the median in the series). The conclusion was that a spinal cord D_{max} constraint of 14 Gy in one fraction carries a less than 1% rate of RM.

Gibbs and colleagues reported a series of 74 patients treated to 102 sites (Gibbs et al., 2007). One case developed RM following *de novo* SBRT, and two cases developed RM following reirradiation SBRT. The *de novo* SBRT data from Gibbs and colleagues (Gibbs et al., 2007) was pooled with the data from Katsoulakis and colleagues (Katsoulakis et al., 2017) to create a logistic model for RM as a function of spinal cord D_{max}, designated the Katsoulakis–Gibbs (KG) model (Sahgal et al., 2019). The modeled rate of myelitis at 14 Gy was consistent with the overall rate of myelitis of 1% for treatments limited to a D_{max} less than 14 Gy as per the Katsoulakis study (Katsoulakis et al., 2017). Using the LQ model, the dose of 14 Gy in one fraction corresponds to an $EQD2_2$ of 56 Gy, 19.3 Gy in 2 fractions, 23.1 Gy in 3 fractions, 26.2 Gy in 4 fractions, and 28.8 Gy in 5 fractions.

Considering the large uncertainties in synthesizing the data as discussed in Section 20.4, we have defined a range of doses that represent safe practice. The Sahgal study (Sahgal et al., 2013) likely represents conservative dose limits that form the lower limit for safe practice for 1–5 fraction SBRT. The KG model, when converted from single fraction to 2–5 fraction SBRT using the LQ model, has higher doses at each number of fractions and likely represents the upper limit for safe practice for 1–5 fraction SBRT (Sahgal et al., 2019). Based on these two models, the range of doses that represent safe practice are 12.4–14 Gy for 1-fraction SBRT, 17–19.3 Gy for 2-fraction SBRT, 20.3–23.1 Gy for 3-fraction SBRT, 23–26.2 Gy for 4-fraction SBRT, and 25.3–28.8 Gy for 5 fraction SBRT. The KG model and Sahgal model estimates for risk of RM at these doses are 1% and 5%, respectively (Sahgal et al., 2013, 2019). Considering the limitations of both models, 1% to 5% constitutes a reasonable estimation of

Table 20.1 Spinal cord and thecal sac D$_{max}$ values recommended in previous publications compared with model-derived limits

Number of fractions	EXISTING EXPERT-BASED RECOMMENDATIONS FOR D$_{max}$		MODEL-BASED LIMITS FOR D$_{max}$ DERIVED FROM CLINICAL DATA		
	AAPM TG101 (Benedict et al., 2010)	Kim et al., 2017	Sahgal Model*	Katsoulakis–Gibbs Model*	Approximate
			LQ, α/β = 2 Gy	LQ, α/β = 2 Gy	Risk
	(Gy)	(Gy)	(Gy)	(Gy)	of RM
1	14	14	12.4	14	1–5%
2		18.3	*17*	*19.3*	1–5%
3	21.9	22.5	*20.3*	*23.1*	1–5%
4		25.6	*23*	*26.2*	1–5%
5	30	28	*25.3*	*28.8*	1–5%

*, *The spinal cord itself (from CT myelogram or MRI) was used as the dose reporting structure in Katsoulakis–Gibbs model (Sahgal et al., 2019), and the thecal sac was used as a surrogate structure for the spinal cord in the Sahgal model (Sahgal et al., 2013). Numbers in italics denote LQ-based extrapolations from the single-fraction limit.*
Note: Due to the uncertainties involved, the decimal place may not be meaningful, and an approximately equivalent set of median-rounded limits from the recommendations/models would be 14, 18, 22, 26, and 28 Gy, for 1–5 fractions, respectively.
Source: Sahgal, A. et al., *Int. J. Radiat. Oncol. Biol. Phys.* 2019.
Abbreviations: D$_{max}$, maximum dose; RM, radiation myelopathy; AAPM TG101, American Association of Physicists in Medicine Task Group 101; LQ, linear–quadratic.

RM risk at these doses. These values can be compared with other published recommended limits in Table 20.1.

20.6 SPINAL CORD TOLERANCE TO REIRRADIATION SBRT

The data for reirradiation spine SBRT has been thoroughly reviewed in the HyTEC review (Sahgal et al., 2019). The main study used to make the recommendations is summarized later.

Sahgal and colleagues reported a DVH analysis of five cases of RM following reirradiation spinal SBRT and compared the DVH data with 16 retreatment spinal SBRT controls (Sahgal et al., 2012). The thecal sac EQD2$_2$ D$_{max}$ ranged from 18.3 to 52.5 Gy for the first course and from 44.1 to 104.9 Gy for the SBRT reirradiation. The median EQD2$_2$ D$_{max}$ for the SBRT component and cumulative EQD2$_2$ were 12.5 Gy (range, 1.9–58.7 Gy) and 52.4 Gy (range, 39.1–111.2 Gy), respectively, in the no-RM control cohort. In the no-RM cohort, there was a minimum duration of 5 months between initial radiation and reirradiation. With further analysis comparing the SBRT EQD2$_2$ D$_{max}$ to the cumulative EQD2$_2$ D$_{max}$ and time to re-treatment, the following recommendations were been made for reirradiation SBRT:

1. The cumulative thecal sac EQD2$_2$ D$_{max}$ should not exceed 70 Gy.
2. The reirradiation SBRT thecal sac EQD2$_2$ D$_{max}$ should not exceed 25 Gy.
3. The reirradiation SBRT thecal sac EQD2$_2$ D$_{max}$ to cumulative EQD2$_2$ D$_{max}$ ratio should not exceed 0.5.
4. The minimum time interval to reirradiation should be at least 5 months.

Based on these recommendations, spinal cord dose constraints are provided for reirradiation SBRT following commonly used radiotherapy schedules in Table 20.2. These likely represent conservative dose limits that have a very low risk of RM. Further data are required to quantify the risk and to define an upper limit of safe practice that still has an acceptably low risk of RM.

Table 20.2 Maximal spinal cord doses for reirradiation, associated with a low risk of RM

PRIOR RT		RECOMMENDED SPINAL CORD* D_{max} IN 1–5 FRACTIONS (GY)				
DOSE (GY)/ FRACTIONS	EQD2$_2$ (GY)	1 FRACTION	2 FRACTIONS	3 FRACTIONS	4 FRACTIONS	5 FRACTIONS
20/5	30	9	12.2	14.5	16.2	18
30/10	37.5	9	12.2	14.5	16.2	18
40/20	40	N/A	12.2	14.5	16.2	18
45/25	43	N/A	12.2	14.5	16.2	18
50/25	50	N/A	11	12.5	14	15.5

*, The thecal sac was used as a surrogate structure for the spinal cord in this study.
Source: Sahgal, A. et al., Int. J. Radiat. Oncol. Biol. Phys. 2019.
Abbreviations: RT, radiotherapy; D_{max}, maximum dose; EQD2$_2$, equivalent dose in 2 Gy fractions (α/β = 2 Gy).

20.7 VOLUME EFFECTS

The spinal cord has classically been described as an organ with serial functional architecture, and, as a result, damage to small volumes within the structure can have a major impact on neurologic function.

Animal studies have yielded conflicting results regarding the functional architecture of the spinal cord. Proton-based rat spinal cord irradiation experiments suggested that the spinal cord might also have a component of parallel architecture when irradiating with inhomogeneous dose distributions, such as those inherent to spine SBRT (Bijl et al., 2002, 2003). Photon-based SBRT studies on pigs, however, did not support these findings, with the data suggesting that the tolerance is similar to that expected with homogeneous irradiation (Medin et al., 2011).

Human data is mostly consistent with the spinal cord having serial functional architecture. This issue was explored by Sahgal and colleagues (Sahgal et al., 2013). Doses to volumes of thecal sac ranging from 0 (D_{max}) to 2 cm^3 were analyzed. Significant differences between RM cases and controls were observed up to 0.8 cm^3; however, the significance was greatest for D_{max}, suggesting a serial architecture to the spinal cord. Grimm and colleagues (Grimm et al., 2016) also noted that RM was best associated with doses to small volumes (D_{1cc} and D_{max}). Katsoulakis and colleagues (Katsoulakis et al., 2017) reported an RM rate of 1 in 13 (7.7%) in patients with cord V7 Gy greater than 5.8 cm^3, suggesting that there may be a parallel architecture component to spinal cord tolerance; however, this was not statistically significant. More data is required to resolve the issue of volume effects.

20.8 CONCLUSION

There are many limitations to the currently available data on RM after SBRT, due to RM being an extremely rare outcome and because of the heterogeneity in the reporting of the data. Despite these limitations, the data on RM after SBRT continues to expand, and our understanding of safe practice continues to grow. The data summarized in this chapter represents a detailed analysis of spinal cord tolerance for spine SBRT and can guide radiation oncologists with respect to safe practice.

CHECKLIST: KEY POINTS FOR CLINICAL PRACTICE

- Dose should be constrained to a spinal cord OAR structure that can be the spinal cord itself, the spinal cord PRV, or the thecal sac.
- For *de novo* SBRT delivered in 1 to 5 fractions, D_{max} values of 12.4 to 14 Gy in 1 fraction, 17 to 19.3 Gy in 2 fractions, 20.3 to 23.1 Gy in 3 fractions, 23 to 26.2 Gy in 4 fractions, and 25.3 to 28.8 Gy in 5 fractions appear to be associated with an estimated risk of RM ranging from 1% to 5%.

- For reirradiation SBRT delivered in 1 to 5 fractions, (1) the cumulative thecal sac EQD2$_2$ D$_{max}$ should not exceed 70 Gy; (2) the reirradiation SBRT thecal sac EQD2$_2$ D$_{max}$ should not exceed 25 Gy; (3) the reirradiation SBRT thecal sac EQD2$_2$ D$_{max}$ to cumulative EQD2$_2$ D$_{max}$ ratio should not exceed 0.5; and (4) the minimum time interval to reirradiation should be at least 5 months.

REFERENCES

Benedict, S. H., Yenice, K. M., Followill, D., Galvin, J. M., Hinson, W., Kavanagh, B., Keall, P., et al. (2010). Stereotactic body radiation therapy: The report of AAPM task group 101. *Medical Physics, 37*(8), 4078–4101.

Bijl, H. P., van Luijk, P., Coppes, R. P., Schippers, J. M., Konings, A. W., & van der Kogel, A. J. (2002). Dose-volume effects in the rat cervical spinal cord after proton irradiation. *International Journal of Radiation Oncology, Biology, Physics, 52*(1), 205–211.

Bijl, H. P., van Luijk, P., Coppes, R. P., Schippers, J. M., Konings, A. W., & van der Kogel, A. J. (2003). Unexpected changes of rat cervical spinal cord tolerance caused by inhomogeneous dose distributions. *International Journal of Radiation Oncology, Biology, Physics, 57*(1), 274–281.

Brenner, D. J. (2008). The linear-quadratic model is an appropriate methodology for determining isoeffective doses at large doses per fraction. *Seminars in Radiation Oncology, 18*(4), 234–239.

Chang, J. H., Gandhidasan, S., Finnigan, R., Whalley, D., Nair, R., Herschtal, A., Eade, T., et al. (2017a). Stereotactic ablative body radiotherapy for the treatment of spinal oligometastases. *Clinical Oncology, 29*(7), e119–e125.

Chang, J. H., Sangha, A., Hyde, D., Soliman, H., Myrehaug, S., Ruschin, M., Lee, Y., et al. (2017b). Positional accuracy of treating multiple versus single vertebral metastases with stereotactic body radiotherapy. *Technology in Cancer Research and Treatment, 16*(2), 231–237.

Chang, J. H., Shin, J. H., Yamada, Y. J., Mesfin, A., Fehlings, M. G., Rhines, L. D., & Sahgal, A. (2016). Stereotactic body radiotherapy for spinal metastases: What are the risks and how do we minimize them? *Spine (Phila Pa 1976), 41*(Suppl 20), S238–S245.

Gibbs, I. C., Kamnerdsupaphon, P., Ryu, M. R., Dodd, R., Kiernan, M., Chang, S. D., & Adler, J. R., Jr. (2007). Image-guided robotic radiosurgery for spinal metastases. *Radiotherapy and Oncology, 82*(2), 185–190.

Grimm, J., Sahgal, A., Soltys, S. G., Luxton, G., Patel, A., Herbert, S., Xue, J., et al. (2016). Estimated risk level of unified stereotactic body radiation therapy dose tolerance limits for spinal cord. *Seminars in Radiation Oncology, 26*(2), 165–171.

Guckenberger, M., Meyer, J., Wilbert, J., Baier, K., Bratengeier, K., Vordermark, D., & Flentje, M. (2007). Precision required for dose-escalated treatment of spinal metastases and implications for image-guided radiation therapy (IGRT). *Radiotherapy and Oncology, 84*(1), 56–63.

Guckenberger, M., Sweeney, R. A., Flickinger, J. C., Gerszten, P. C., Kersh, R., Sheehan, J., & Sahgal, A. (2011). Clinical practice of image-guided spine radiosurgery--results from an international research consortium. *Radiation Oncology, 6*, 172.

Huang, Z., Mayr, N. A., Yuh, W. T., Wang, J. Z., & Lo, S. S. (2013). Reirradiation with stereotactic body radiotherapy: Analysis of human spinal cord tolerance using the generalized linear-quadratic model. *Future Oncology, 9*(6), 879–887.

Hyde, D., Lochray, F., Korol, R., Davidson, M., Wong, C. S., Ma, L., & Sahgal, A. (2012). Spine stereotactic body radiotherapy utilizing cone-beam CT image-guidance with a robotic couch: Intrafraction motion analysis accounting for all six degrees of freedom. *International Journal of Radiation Oncology, Biology, Physics, 82*(3), e555–e562.

International Commission on Radiation Units and Measurements. (2017). Prescribing, recording, and reporting of stereotactic treatments with small photon beams. ICRU Report 91. *Journal of the International Commission on Radiation Units and Measurements, 14*(2), 1–160.

Katsoulakis, E., Jackson, A., Cox, B., Lovelock, M., & Yamada, Y. (2017). A detailed dosimetric analysis of spinal cord tolerance in high-dose spine radiosurgery. *International Journal of Radiation Oncology, Biology, Physics, 99*(3), 598–607.

Kim, D. W. N., Medin, P. M., & Timmerman, R. D. (2017). Emphasis on repair, not just avoidance of injury, facilitates prudent stereotactic ablative radiotherapy. *Seminars in Radiation Oncology, 27*(4), 378–392.

Kirkpatrick, J. P., Meyer, J. J., & Marks, L. B. (2008). The linear-quadratic model is inappropriate to model high dose per fraction effects in radiosurgery. *Seminars in Radiation Oncology, 18*(4), 240–243.

Kirkpatrick, J. P., van der Kogel, A. J., & Schultheiss, T. E. (2010). Radiation dose-volume effects in the spinal cord. *International Journal of Radiation Oncology, Biology, Physics, 76*(Suppl 3), S42–S49.

Ma, L., Sahgal, A., Hossain, S., Chuang, C., Descovich, M., Huang, K., Gottschalk, A., et al. (2009). Nonrandom intrafraction target motions and general strategy for correction of spine stereotactic body radiotherapy. *International Journal of Radiation Oncology, Biology, Physics, 75*(4), 1261–1265.

Ma, T. M., Emami, B., Grimm, J., Xue, J., Asbell, S. O., Kubicek, G. J., Lanciano, R., et al. (2019). Volume effects in radiosurgical spinal cord dose tolerance: How small is too small? *Journal of Radiation Oncology*, 1–9.

Mantel, F., Flentje, M., & Guckenberger, M. (2013). Stereotactic body radiation therapy in the re-irradiation situation--a review. *Radiation Oncology*, 8, 7.

Medin, P. M., Foster, R. D., van der Kogel, A. J., Sayre, J. W., McBride, W. H., & Solberg, T. D. (2011). Spinal cord tolerance to single-fraction partial-volume irradiation: A swine model. *International Journal of Radiation Oncology, Biology, Physics*, 79(1), 226–232.

Park, C., Papiez, L., Zhang, S., Story, M., & Timmerman, R. D. (2008). Universal survival curve and single fraction equivalent dose: Useful tools in understanding potency of ablative radiotherapy. *International Journal of Radiation Oncology, Biology, Physics*, 70(3), 847–852.

Redmond, K. J., Sciubba, D., Khan, M., Gui, C., Lo, S. L., Gokaslan, Z. L., Leaf, B., et al. (2020). A phase 2 study of post-operative stereotactic body radiation therapy (SBRT) for solid tumor spine metastases. *International Journal of Radiation Oncology, Biology, Physics*, 106(2), 261–268.

Sahgal, A., Chang, J. H., Ma, L., Marks, L. B., Milano, M. T., Medin, P., Niemierko, A., et al. (2019). Spinal cord dose tolerance to stereotactic body radiotherapy. *International Journal of Radiation Oncology, Biology, Physics*. Advance online publication. doi: 10.1016/j.ijrobp.2019.09.038

Sahgal, A., Ma, L., Weinberg, V., Gibbs, I. C., Chao, S., Chang, U. K., Werner-Wasik, M., et al. (2012). Reirradiation human spinal cord tolerance for stereotactic body radiotherapy. *International Journal of Radiation Oncology, Biology, Physics*, 82(1), 107–116.

Sahgal, A., Weinberg, V., Ma, L., Chang, E., Chao, S., Muacevic, A., Gorgulho, A., et al. (2013). Probabilities of radiation myelopathy specific to stereotactic body radiation therapy to guide safe practice. *International Journal of Radiation Oncology, Biology, Physics*, 85(2), 341–347.

Schultheiss, T. E., Stephens, L. C., & Peters, L. J. (1986). Survival in radiation myelopathy. *International Journal of Radiation Oncology, Biology, Physics*, 12(10), 1765–1769.

Thariat, J., Castelli, J., Chanalet, S., Marcie, S., Mammar, H., & Bondiau, P. Y. (2009). CyberKnife stereotactic radiotherapy for spinal tumors: Value of computed tomographic myelography in spinal cord delineation. *Neurosurgery*, 64(Suppl 2), A60–A66.

Tseng, C. L., Sussman, M. S., Atenafu, E. G., Letourneau, D., Ma, L., Soliman, H., Thibault, I., et al. (2015). Magnetic resonance imaging assessment of spinal cord and cauda equina motion in supine patients with spinal metastases planned for spine stereotactic body radiation therapy. *International Journal of Radiation Oncology, Biology, Physics*, 91(5), 995–1002.

Wang, H., Shiu, A., Wang, C., O'Daniel, J., Mahajan, A., Woo, S., Liengsawangwong, P., et al. (2008). Dosimetric effect of translational and rotational errors for patients undergoing image-guided stereotactic body radiotherapy for spinal metastases. *International Journal of Radiation Oncology, Biology, Physics*, 71(4), 1261–1271.

Wang, J. Z., Huang, Z., Lo, S. S., Yuh, W. T., & Mayr, N. A. (2010). A generalized linear-quadratic model for radiosurgery, stereotactic body radiation therapy, and high-dose rate brachytherapy. *Science Translational Medicine*, 2(39), 39ra48.

Wong, C. S., Fehlings, M. G., & Sahgal, A. (2015). Pathobiology of radiation myelopathy and strategies to mitigate injury. *Spinal Cord*, 53(8), 574–580.

Zeng, K. L., Tseng, C. L., Soliman, H., Weiss, Y., Sahgal, A., & Myrehaug, S. (2019). Stereotactic body radiotherapy (SBRT) for oligometastatic spine metastases: An overview. *Frontiers in Oncology*, 9, 337.

21 Summary of IG-HSRT: Serious Late Toxicities and Strategies to Mitigate Risk

Simon S. Lo, Balamurugan Vellayappan, Kristin J. Redmond,
Nina A. Mayr, William T. Yuh, Matthew Foote, Eric L. Chang,
Samuel T. Chao, John H. Suh, Bin S. Teh, and Arjun Sahgal

Content

21.1 INTRODUCTION

Serious late complications are one of the most feared aspects of image-guided hypofractionated radiotherapy (IG-HSRT). Albeit their relatively rare occurrence with modern technology and the experience gained in these procedures throughout these years, they have been observed and reported. Fortunately, many of these serious complications can be prevented by applying suitable strategies. This chapter will summarize the serious late complications associated with intracranial and spinal IG-HSRT and the strategies to mitigate the risks. In this chapter, the term "spinal HSRT" is used interchangeably with "spinal SBRT". Details pertaining to each disease condition have been discussed in previous chapters.

21.2 SERIOUS LATE TOXICITIES FROM INTRACRANIAL IG-HSRT AND STRATEGIES TO MITIGATE RISK

This section will cover the most important complications including optic neuropathy and other cranial nerve injuries, vascular injury, brainstem injury, and radiation necrosis.

21.2.1 OPTIC NEUROPATHY

Optic neuropathy is one of the most feared and devastating complications from SRS or HSRT. Overall, the incidence of optic neuropathy after SRS is low (Stafford et al., 2003; Leavitt et al., 2013; Pollock et al., 2014). The most commonly used constraint for the optic apparatus is 8 Gy (Tishler et al., 1993), but it is very likely to be overly conservative. Subsequent data from Mayo Clinic showed that the risk of developing a clinically significant radiation-induced optic neuropathy was 1.1% for patients receiving a single maximum point dose of 12 Gy or less to the optic apparatus (Stafford et al., 2003). Overall, radiation-induced optic neuropathy developed in less than 2% of patients, despite that 73% received dose more than 8 Gy to a short segment of the optic apparatus. In their follow-up study, they found that the risk of radiation-induced optic neuropathy was 0, 0, 0, and 10% when the maximum radiation doses received by the anterior visual pathway were ≤8 Gy, 8.1–10.0 Gy, 10.1–12.0 Gy, and more than 12 Gy, respectively (Leavitt et al., 2013). The overall risk of radiation-induced optic neuropathy in patients receiving dose more than 8 Gy to the anterior visual pathway was 1.0% (Leavitt et al., 2013). In another series from Mayo Clinic where 133 patients (266 sides) with pituitary adenomas were treated with SRS, no optic neuropathy was observed at a median follow-up of 32 months when the maximum point dose to the optic apparatus did not exceed 12 Gy (Pollock et al., 2014).

Data on the optic nerve tolerance for other hypofractionated regimens other than a single fraction are limited. In a study from Japan, a dose of 24 Gy or less in 6 fractions appeared to be safe for the optic apparatus (Kanesaka et al., 2011). In another study from Japan, grade 2 visual disorder was observed in one patient who received 20.8 Gy and 20.7 Gy in 3 fraction to the optic nerve and chiasm, respectively (Iwata et al., 2011). Iwata et al. reported a 1% risk of optic neuropathy using either 17–21 Gy in 3 fractions or 22–25 Gy in 5 fractions for HSRT for pituitary adenomas (Iwata et al., 2011). In a study from Italy where 25 patients with perioptic meningioma were treated with CyberKnife-based HSRT using regimens including 18 Gy in 2 fractions, 18–21 Gy in 3 fractions, 20–22 Gy in 4 fractions, and 23–25 Gy in 5 fractions, none of the patients developed visual deterioration with a median follow-up of 57.5 months (Conti et al., 2015). The optic pathway constraints used were 10 Gy in 2 fractions, 15 Gy in 3 fractions, 20 Gy in 4 fractions, and 25 Gy in 5 fractions. Using the same constraints for 2–5 fractions, a further group of 39 patients were treated prospectively with HSRT regimens including 18 Gy in 2 fractions, 18–21 Gy in 3 fractions, 20–22 Gy in 4 fractions, 25 Gy in 5 fractions, 27.5 Gy in 6 fractions, 30 Gy in 9 fractions, 34 Gy in 10 fractions, and 40 Gy in 15 fractions. With a median follow-up of 15 months, no visual toxicities were observed (Conti et al., 2015). In a study from Barrow Neurological Institute, 20 patients with pituitary adenomas were treated with CyberKnife-based HSRT with the optic pathway constraint set at 25 Gy in 5 fractions. With a median follow-up of 26.6 months, none of the patients developed visual deficits (Killory et al., 2009). The authors concluded that the tolerance of optic pathway was 25 Gy in 5 fractions (Killory et al., 2009). In a study from University of Virginia, 15 patients with meningioma, pituitary adenoma, or pilocytic astrocytoma were treated with Gamma Knife-based HSRT using a relocatable system (Nguyen et al., 2014). The doses to the optic pathway were tracked. The maximum doses delivered to the optic pathway were 3.6–14.4 Gy in 3 fractions, 2.8–22.8 Gy in 4 fractions, and 5–24.5 Gy in 5 fractions. With a median follow-up of 13.8 months (range, 4–44.3 months), no visual toxicities were observed (Nguyen et al., 2014).

Since the publication of the last edition of this book, additional studies reported the visual outcomes of patients treated with HSRT for tumors close to the optic apparatus. One study from Thailand reported no optic pathway toxicity after HSRT to a dose of 25 Gy (range, 20–28 Gy) in 5 fractions (range, 3–5) for perioptic pituitary adenoma (Puataweepong et al., 2016). The optic nerve/chiasm maximum doses ranged from 2.4 to 32.0 Gy. The authors reported that with strict adherence to tolerance doses, which were not specified, no visual toxicity was reported. In another study from Milan, Italy, 143 patients were treated with HSRT to a dose of 15–21 Gy in 3 fractions, 16–20 Gy in 4 fractions, and 20–25 Gy in 5 fractions for meningiomas (Marchetti et al., 2016). Visual deterioration occurred in 10 of 143 patients after optic nerve/chiasm maximum doses ranging from 2.5 to 34.0 Gy (range, 4.0–32.0 Gy among those with visual deterioration). The authors did not find any significant correlation between worsening vision and total or fractional maximum optic nerve or chiasm doses, though these maximum doses were not reported for the seven patients with radiation-induced optic neuropathy compared to the three with worsening vision from progressive disease.

Most recently, the American Association of Physicists in Medicine (AAPM) Working Group on Stereotactic Body Radiotherapy published the SRS and HSRT dose tolerance of optic pathway (Milano et al., 2018). Thirty-four studies were identified with data of 1578 patients included. In patients with no prior radiotherapy receiving SRS or HSRT, maximum point doses of ≤10 Gy in 1 fraction, 20 Gy in 3 fractions, and 25 Gy in 5 fractions resulted in less than 1% risk of radiation-induced optic neuropathy.

21.2.2 III, IV, V, AND VI CRANIAL NERVE PALSIES

The situations in which the III, IV, and VI cranial nerves are likely to be exposed to significant doses of radiation are when cavernous sinus meningiomas and pituitary adenomas are treated with SRS or HSRT. Data in the literature on SRS for skull base meningiomas seem to suggest that the III, IV, and VI nerves are able to tolerate a high single dose of radiation (Witt, 2003). Witt from Indiana University reviewed the data on 1255 patients who underwent SRS for pituitary adenoma with the marginal dose delivered ranging from 14 to 34 Gy in one fraction. Given the proximity of the sella to the cavernous sinus and the high doses used to treat pituitary adenomas, especially secretory tumors, the cranial nerves in the cavernous sinus can potentially be exposed to very high radiation doses. However, the overall incidence of permanent III, IV, or VI neuropathy was 0.4% (Witt, 2003). In a study from Israel, when 102 patients with cavernous sinus meningiomas were treated with LINAC-based SRS to a dose of 12–17.5 Gy in 1 fraction, the incidence of VI nerve palsy was less than 2% (Spiegelmann et al., 2010). Colleagues from University of Virginia reported a very low incidence of III and IV nerve palsies and zero incidence of IV nerve palsy in patients with prolactinomas with cavernous sinus invasion after Gamma Knife SRS to a median marginal dose of 25 Gy in 1 fraction (Cohen-Inbar et al., 2015). Park et al. reported III, IV, and VII nerve toxicity rates of 2.8%, 0%, and 1.7%, respectively, after Gamma Knife-based SRS for cavernous sinus meningiomas to a dose of 13 Gy (range, 10–20 Gy) in 1 fraction with a median imaging follow-up of 101 months (Park et al., 2018). For the V nerve, the incidence of neuropathy was 0.2 % based on the analysis of patients with pituitary adenoma treated with SRS, which can potentially expose the V nerve to a high single dose of radiation (Witt, 2003). In the study from Israel mentioned earlier, the incidence of V nerve injury was less than 2% (Spiegelmann et al., 2010). Other clinical scenarios where the V nerve can be exposed to a substantial level of radiation are when SRS is used for the treatment of vestibular schwannoma, cavernous sinus meningiomas, and trigeminal neuralgia. In the early series of SRS for vestibular schwannoma from University of Pittsburgh Medical Center (UPMC) where a higher dose of 14–20 Gy was used, the incidence of V nerve deficit was 16% (Kondziolka et al., 1998). Other series using a dose of 12–13 Gy, including a later series from UPMC, reported much lower V nerve deficit rates of 2%–8% (Petit et al., 2001; Flickinger et al., 2004; Murphy and Suh, 2011). The study on Gamma Knife-based SRS for cavernous meningiomas by Park et al. mentioned above reported a V nerve toxicity rate of 7.8% (Park et al., 2018). A study from University of Cologne showed that the incidence of V nerve deficit was 3.1% after SRS for vestibular schwannoma with a dose of 12.3 Gy (Ruess et al., 2020). Söderlund Diaz and Hallqvist reported a V nerve toxicity rate of 8% (grade 3 or higher rate of V neuropathy of 0%) after SRS for vestibular schwannoma with a dose of 12 Gy (Soderlund Diaz and Hallqvist, 2020). Based on the data from retrospective series, the risks of III, IV, V, and VI cranial nerve palsies are generally very low with SRS.

Much less data is available for HSRT. Wang and colleagues treated 14 patients with cavernous sinus hemangiomas with HSRT to a dose of 21 Gy in 3 fractions in a phase II trial. Among the six patients with cranial nerve deficits, they either showed complete recovery or improvement of function (Wang et al., 2012). Other patients did not experience cranial nerve palsies. In another report of HSRT for large cavernous sinus hemangiomas by Wang et al., the rates of III, IV, V, and VI nerve toxicities were zero with HSRT regimens of 19.5–21 Gy in 3 fractions and 18–22 Gy in 4 fractions (Wang et al., 2018). In a study from Johns Hopkins University where vestibular schwannomas were treated with HSRT to 25 Gy in 5 fractions, the incidence of V nerve deficit was 7% (Song and Williams, 1999). In another study from Japan, HSRT for vestibular schwannoma delivering 18–21 Gy in 3 fractions or 25 Gy in 5 fractions resulted in a zero incidence of V nerve deficit at a median follow-up of 80 months (Morimoto et al., 2013). In the Georgetown University series of HSRT for vestibular schwannoma delivering 25 Gy in 5 fractions, the rate of V nerve injury was 5.5% (Karam et al., 2013). A recent study from Göteborg showed that the rates of V nerve toxicities after HSRT for vestibular schwannoma were 13% and 15% with a dose of 18–21 Gy in 3 fractions and 25 Gy in 5 fractions, respectively (Soderlund Diaz and Hallqvist, 2020). The corresponding rates of grade 3 or higher V neuropathy were 0% and 3.3%.

21.2.3 VII AND VIII CRANIAL NERVE PALSIES

Most of the data on toxicities of VII and VIII cranial nerve palsies came from SRS series for vestibular schwannoma. In an early series reported by Kondziolka et al. where a dose of 14–20 Gy in one fraction was used for Gamma Knife SRS, the hearing retention rate was only 47%, and the risk of VII nerve injury was 15% (Kondziolka et al., 1998). In more recent studies where a dose of 12–13 Gy in 1 fraction was used, the incidence of retention of serviceable hearing ranged from 35% to 88% with Gamma Knife or LINAC-based SRS while local control was not jeopardized (Petit et al., 2001; Flickinger et al., 2004; Murphy and Suh, 2011; Ruess et al., 2020; Soderlund Diaz and Hallqvist, 2020). The risk of VII nerve deficit was also observed to be less than 5%--% (Petit et al., 2001; Flickinger et al., 2004; Murphy and Suh, 2011; Ruess et al., 2020; Soderlund Diaz and Hallqvist, 2020). Protons have been used to deliver SRS, but the hearing preservation rates were lower at approximately 30% if a low-dose regimen of 12 CGE was used (Murphy and Suh, 2011).

Less data is available with VII and VIII nerve deficits from HSRT. Song from Johns Hopkins University published his experience with the use of HSRT for vestibular schwannoma using a regimen of 25 Gy in 5 fractions (Song and Williams, 1999). With relatively short follow-up times of 6–44 months, the hearing preservation rate was 75% and the rate of VII nerve deficits was 0%. In a study from Japan where 25 patients with 26 vestibular schwannomas were treated with CyberKnife-based HSRT, using regimens of 18–21 Gy in 3 fractions or 25 Gy in 5 fractions, the overall VII and VIII nerve preservation rates were 92% and 50%, respectively (Morimoto et al., 2013). In a study from Taiwan, treatment with 18 Gy in 3 fractions was associated with a serviceable hearing retention rate of 81.5% at a mean follow-up of 61.1 months (Tsai et al., 2013). Using the same regimen of 18 Gy in 3 fractions, colleagues from UPMC reported a serviceable hearing retention rate of 53.5% (Vivas et al., 2014). Colleagues from Georgetown University used mainly the regimen of 25 Gy in 5 fractions and observed a hearing preservation rate of 73% at 5 years and a zero incidence of VII nerve palsy (Karam et al., 2013). A Dutch study comparing SRS and HSRT for vestibular schwannoma showed that with regimens of 20–25 Gy in 5 fractions, the rates of preservation of VII and VIII nerve function were 97% and 61%, respectively (Meijer et al., 2003). A recent study from Göteborg showed that the rates of VII nerve toxicities after HSRT for vestibular schwannoma were 18% for both with 18–21 Gy in 3 fractions and 25 Gy in 5 fractions, respectively (Soderlund Diaz and Hallqvist, 2020). The rates of grade 3 or higher VII neuropathy were 5.1% and 0% for 3 and 5 fractions, respectively. The corresponding rates of hearing preservation was 36% and 35%, respectively.

21.2.4 IX, X, XI, AND XII CRANIAL NERVE PALSIES

The toxicity data on IX, X, XI, and XII cranial nerves are mainly extracted from studies on SRS for glomus tumors, which are frequently located close to the internal jugular vein. There have been two meta-analyses examining the outcomes of SRS for glomus tumors (Guss et al., 2011; Ivan et al., 2011). In the meta-analysis from University of California San Francisco (UCSF), among 339 patients treated with SRS alone, the incidence of IX, X, XI, and XII nerve deficits was 9.7%, 9.7%, 12%, and 8.7%, respectively (Ivan et al., 2011). However, no dosimetric data were available. The meta-analysis from Johns Hopkins University did not report any toxicity data (Guss et al., 2011). The prescribed dose ranged from 12 Gy to 20.4 Gy (median 15.1 Gy) in one fraction.

Sharma et al. from Cleveland Clinic Foundation reported relatively low rates of cranial nerve toxicities after Gamma Knife-based SRS for glomus tumors with a dose of 15 Gy in one fraction (Sharma et al., 2018).

Data is even more limited on IX, X, XI, and XII nerve toxicities caused by HSRT. In a study from University of Texas Southwestern Medical Center, 31 patients with skull base glomus tumors were treated with HSRT to a dose of 25 Gy in 5 fractions, and no brainstem toxicities were observed with a median follow-up of 24 months (Chun et al., 2014).

21.2.5 VASCULAR INJURY

The circumstances in which major vessels such as internal carotid arteries are usually exposed to high doses of radiation occur when cavernous sinus meningiomas and pituitary adenomas are treated with SRS or HSRT. Witt from Indiana University performed a systematic review on SRS for pituitary adenomas and found that among the 1,255 patients evaluated in the studies, there were only 3 cases of internal carotid artery occlusion or stenosis (Witt, 2003). In one case, the dose to the internal carotid artery was estimated to be less than 20 Gy in 1 fraction. It was recommended that less than 50% of the circumference should receive the prescribed dose, and the internal carotid artery dose should be kept below 30 Gy in 1 fraction (Witt, 2003).

Colleagues from Mayo Clinic examined the risk of internal carotid artery stenosis or occlusion after SRS for cavernous sinus meningioma or growth hormone-secreting pituitary adenoma in 283 patients. Pre-SRS ICA involvement was graded based on Hirsch scale for cavernous sinus meningioma or Knosp scale for pituitary adenoma:

Hirsch category 1: Tumor abuts but does not encircle internal carotid artery
Hirsch category 2: Tumor encircles but does not narrow internal carotid artery
Hirsch category 3: Tumor encircles and narrows internal carotid artery
Knosp Grade 0: Adenoma does not cross medial internal carotid artery margin
Knosp Grade 1: Adenoma crosses medial internal carotid artery margin, embracing less than 50% of internal carotid artery
Knosp Grade 2: Adenoma crosses medial internal carotid artery margin, embracing less than 50% of internal carotid artery
Knosp Grade 3: Adenoma extends beyond lateral internal carotid artery margin
Knosp Grade 4: Adenoma fully envelops internal carotid artery

All tumors were classified using Hirsch scale for the ease of comparison. The median time to stenosis or occlusion was 4.8 years (range, 1.8–7.6 years). With a median follow-up of 6.6 years, none of the Hirsch category 1 cavernous sinus meningioma or acromegaly patients developed internal carotid artery stenosis or occlusion. For Hirsch category 2 and 3 cavernous sinus meningioma patients, the 5- and 10-year risks of new internal carotid artery stenosis or occlusion was 7.5% and 12.4%, respectively, but the corresponding risks of ischemic stroke were both only 1.2%. All new stenosis or occlusion occurred completely or predominantly within the cavernous sinus. Regarding prescribed marginal dose, the median values were 15 Gy (range, 14–16 Gy) and 25 Gy (range, 20–25 Gy) in one fraction for cavernous sinus meningiomas and growth hormone-secreting pituitary adenomas. Multivariate analysis showed that pre-SRS, internal carotid grade, age, and pathological diagnosis (with higher risk with cavernous sinus meningioma) predicted stenosis or occlusion. Interestingly, there were no significant association between any dosimetric parameter and stenosis or occlusion (Graffeo et al., 2019).

Pertaining to HSRT, among the studies on pituitary adenoma or cavernous sinus tumors, no vascular injury was observed when regimens of 19.5–21 Gy in 3 fractions or 25 Gy in 5 fractions were used (Killory et al., 2009; Wang et al., 2012; Nguyen et al., 2014; Huang et al., 2019). In a report of HSRT for giant cavernous sinus hemangiomas using regimens of 19.5 Gy in 3 fractions and in 18–22 Gy in 4 fractions, no vascular injury was observed (Wang et al., 2018).

21.2.6 BRAINSTEM INJURY

While the brainstem tolerance to conventional radiotherapy is quite well established, there is much less data on the brainstem tolerance to SRS and HSRT. Mayo and colleagues published a comprehensive review on radiation-induced brainstem injury as part of the Quantitative Analyses of Normal Tissue Effects in the Clinic (QUANTEC) project in 2010 (Mayo et al., 2010a). Based on literature review, the authors have concluded that a maximum brainstem dose of 12.5 Gy in one fraction is associated with a less than 5% risk of brainstem injury (Mayo et al., 2010a). There have been studies on SRS delivering doses up to 15–20 Gy to the brainstem with very low incidence of complication (Mayo et al., 2010a). This phenomenon may be explained in patients with brainstem metastases treated with SRS, where the survival is expected to be too short to develop complications.

Data on brainstem injury from HSRT largely derive from studies of HSRT for skull base tumors, mainly vestibular schwannoma. An early seminal study on HSRT for benign and malignant tumors at various locations from McGill University, Canada, utilizing a regimen of 42 Gy in 6 fractions observed late and serious brainstem complications in 4 of 77 patients treated (Clark et al., 1998). Based on the non-conformal isodose distribution shown in the publication, it was expected that a relatively large volume of the brainstem would be included in the prescribed isodose line (Clark et al., 1998). While the delivery of HSRT using dynamic rotation technique was regarded as state-of-the art at that time, with the quantum leap of technology, a highly conformal isodose distribution around the tumor volume is expected. As a result, the data presented in this Canadian study may not apply to modern HSRT.

In the phase II study of HSRT for cavernous sinus hemangiomas by Wang and colleagues, no brainstem injury was observed with a brainstem dose of 19.8 Gy (range, 12.4–22.8 Gy) in 3 fractions (Wang et al., 2012). In a Japanese study, no brainstem injury was observed when the maximum brainstem dose was limited to 35 Gy in 5 fractions or 27 Gy in 3 fractions (Morimoto et al., 2013). Karam et al. from Georgetown University reported zero incidence of brainstem injury when 25 Gy in 5 fractions or 21 Gy in 3 fractions was used as the HSRT regimen for vestibular schwannoma (Karam et al., 2013). Dosimetric details were not available. In the study from Johns Hopkins where a dose of 25 Gy in 5 fraction was used, no brainstem injury was observed (Song and Williams, 1999). Wang et al. reported zero incidence of brainstem injury after HSRT for large cavernous sinus hemangioma with maximum brainstem doses of 19.08 Gy in 3 fractions and 20.1 Gy in 4 fractions (Wang et al., 2018).

21.2.7 RADIATION NECROSIS

Both intracranial SRS and HSRT deliver ablative doses of radiation to the target volume as well as the brain parenchyma immediately adjacent to it and hence predisposing the treated volume to a risk of radiation necrosis. Colleagues from UPMC first developed a model for predicting symptomatic radiation necrosis from SRS for arteriovenous malformation (AVM). They have determined that the AVM location and the volume of tissue receiving ≥12 Gy (12 Gy-volume) were the best predictors and have used those parameters to construct a significant post-radiosurgery injury expression (SPIE) score, which was directly proportional to the risk of symptomatic necrosis (Flickinger et al., 2000). Locations of AVM in ascending order of risk and SPIE score (0–10) were frontal lobe, temporal lobe, intraventricular area, parietal lobe, cerebellum, corpus callosum, occipital lobe, medulla, thalamus, basal ganglia, and pons/midbrain (Flickinger et al., 2000). The 12 Gy-volume (V12) has since become the standard parameter to predict symptomatic necrosis for SRS for AVM. However, it was unclear whether this would apply to intracranial tumors. Korytko et al. from Case Western Reserve University did a retrospective review on their 129 patients with 198 non-AVM tumors treated with Gamma Knife-based SRS and found that, as in AVM, V12 correlated with risk of symptomatic necrosis (Figure 21.1). The risk was 23%, 20%, 54%, and 57% when V12 was 0–5 cc, 5–10 cc, 10–15 cc, and more than 15 cc, respectively. V12 was not predictive of the risk of asymptomatic necrosis (Korytko et al., 2006) (Figure 21.1).

The data on radiation necrosis from intracranial HSRT is even more limited. Studies with dosimetric correlation are lacking. In a study from Sweden where 56 patients with AVMs were treated with HSRT to 30–32.5 Gy in 5 fractions or 35 Gy in 5 fractions, all patients who developed symptomatic radiation necrosis received 35 Gy in 5 fractions. The target volumes ranged from 1.5 to 29 cc (Lindvall et al., 2010). Inoue et al. from Japan have attempted to estimate the risk of symptomatic necrosis in 78 patients with 85 large brain metastases treated with 5-fraction HSRT. The surrounding brain volumes encompassed by the 28.8 Gy isodose line (single dose equivalent to 14 Gy or V14) were measured to evaluate the risk of radiation necrosis (Inoue et al., 2014). The risk of brain necrosis increased in patients' long-term survival when V14 ≥ 7.0 cc, and none of the patients with a V14 of less than 7.0 cc experienced

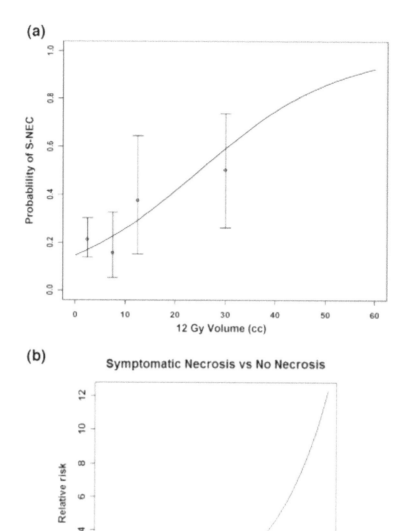

Figure 21.1 (a) Percentage of symptomatic radiation necrosis (S-NEC) versus 12-Gy Volume. Logistic regression of the probability of S-NEC versus 12-Gy Volume. The model was fitted with 12-Gy Volume as the sole covariate. The error bars are Clopper–Pearson 95% confidence interval for the proportion of patients with S-NEC. (b) Relative risk of S-NEC versus 12-Gy Volume. (Reprinted from *International Journal of Radiation Oncology, Biology and Physics*, 64 (2), Timothy Korytko, Tomas Radivoyevitch, Valdir Colussi, Barry W. Wessels, Kunjan Pillai, Robert J. Maciunas, Douglas B. Einstein, 12 Gy Gamma Knife radiosurgical volume is a predictor for radiation necrosis in non-AVM intracranial tumors, Pages 419–424, Copyright 2006, with permission from Elsevier.)

brain necrosis requiring surgical intervention (Inoue et al., 2014). Recent data from Sunnybrook Health Science Centre, Toronto, Canada showed that for HSRT in brain metastases, there was a higher risk with SRS for intact lesions compared to resection cavity, and the investigators suggested that the total brain minus gross tumor volume receiving more than 30 Gy in 5 fractions should be limited to 10.5 cc (Faruqi et al., 2020). (Figure 21.2).

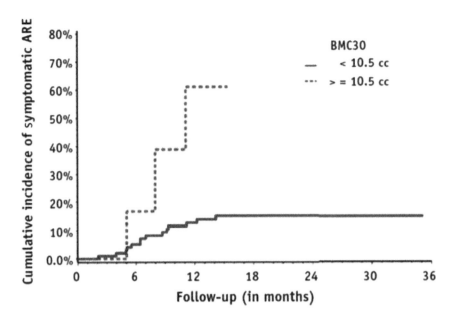

Figure 21.2 Cumulative incidence of symptomatic adverse radiation effect in intact metastases cohort based on brain minus gross tumor volume receiving 30 Gy volume (BMC30) <10.5 cc versus ≤10.5 cc (5 fractions). (Reprinted from *International Journal of Radiation Oncology, Biology*, 106(4), Salman Faruqi, Mark Ruschin, Hany Soliman, Sten Myrehaug, K Liang Zeng, Zain Husain, Eshetu Atenafu, Chia-Lin Tseng, Sunit Das, James Perry, Pejman Maralani, Chris Heyn, Todd Mainprize, Arjun Sahgal, Adverse Radiation Effect After Hypofractionated Stereotactic Radiosurgery in 5 Daily Fractions for Surgical Cavities and Intact Brain Metastases, Pages 772–779, Copyright 2020, with permission from Elsevier.)

In a most recent HyTEC analysis, 51 reports in the literature on radiation necrosis for SRS and HSRT were reviewed (Milano et al., 2020). In the setting of brain metastases, it was deemed to be difficult to distinguish between SRS- or HSRT- associated radionecrosis and tumor progression. For single-fraction SRS to brain metastases, normal brain parenchymal tissue volumes including target volumes receiving 12 Gy (V12) of 5 cc, 10 cc, or more than 15 cc were associated with risks of symptomatic radionecrosis of approximately 10%, 15%, and 20%, respectively. In the setting of AVM, SRS was associated with lower rates of symptomatic radionecrosis for the same V12. For HSRT using 3 fractions for brain metastases, normal brain parenchyma V18 less than 30 cc and V23 less than 7 cc were associated with less than 10% risk of radionecrosis (Milano et al., 2020).

21.2.8 STRATEGIES TO MITIGATE RISKS OF COMPLICATIONS FROM INTRACRANIAL IG-HSRT

21.2.8.1 Cranial Nerve Injury

For patients undergoing intracranial SRS or HSRT for targets close to the optic apparatus, it is crucial to have the optic nerves, optic chiasm, and optic tracks contoured to track the doses and to ascertain that the tolerance is not exceeded. It is prudent to limit the volume of the optic apparatus receiving the tolerance dose. If a Gamma Knife unit is used for SRS, several maneuvers can be used to decrease the dose to the optic apparatus. Witt has described that by orienting the stereotactic frame parallel to the long axis of the optic nerves and chiasm, the anteroposterior axis of the peripheral isodose curves will then be parallel to the optic apparatus in the sagittal plane, and the extremely steep falloff of radiation in the craniocaudal direction can be exploited (Witt, 2003). The use of multiple 4 mm shots can also steepen the dose gradient, potentially decreasing the dose delivered to the optic apparatus. The automatic beam-shaping feature of the Perfexion or ICON model can also facilitate sparing of the optic apparatus. If a linear accelerator-based or robotic system is used, the use of inverse treatment planning can steer the dose away from the optic structures, potentially decreasing the risk of complications.

Based on the data in the literature, there is data to suggest that a maximum point dose of 12 Gy in one fraction is safe for the optic apparatus (Stafford et al., 2003; Leavitt et al., 2013; Pollock et al., 2014). The Quantitative Analyses of Normal Tissue Effects in the Clinic (QUANTEC) has estimated that the tolerance of the optic apparatus is 8–12 Gy in one fraction but has not made recommendations for 2–5 fractions (Mayo et al., 2010b). There is a fair amount of data from retrospective studies from multiple radiosurgery centers suggesting a maximum point dose of 25 Gy in 5 fractions should be safe for the optic apparatus (Killory et al., 2009; Iwata et al., 2011; Nguyen et al., 2014; Conti et al., 2015). Much less data is available for 2–4 fractions. Although it is tempting for one to extrapolate toxicity data from conventional fractionation, the derived tolerance is not rigorously tested in a clinical setting. Most recently, the AAPM published their analysis of the risk of optic neuropathy after SRS or HSRT, and they have determined that maximum point doses of ≤10 Gy in 1 fraction, 20 Gy in 3 fractions, and 25 Gy in 5 fractions resulted in less than 1% risk of radiation-induced optic neuropathy (Milano et al., 2018). Taking all the information together, for SRS, it is prudent to limit the maximum point dose to the optic apparatus to ≤10 Gy, although in situations where the tumor is close to the optic apparatus and adequate high-dose coverage is necessary, for example, in a case of secretory pituitary adenoma, in which case a maximum point dose of 12 Gy is acceptable. Alternatively, the use of HSRT, limiting the dose to the apparatus to 20 Gy in 3 fractions or 25 Gy in 5 fractions based on AAPM recommendations, can be considered (Milano et al., 2018).

Studies of SRS and HSRT for cavernous sinus tumors and pituitary adenomas seem to suggest that the III, IV, V, and VI nerves are quite resistant to ablative doses of radiation (Witt, 2003; Spiegelmann et al., 2010; Park et al., 2018; Wang et al., 2018). However, dosimetric analysis is lacking in those studies, and it is unclear exactly how much radiation is actually being delivered to those cranial nerves, which can be difficult to visualize even on MRI. When the cavernous sinus is invaded by a meningioma or a macroadenoma, the III, IV, V, and VI nerves are likely to be included in the tumor volume to be treated and are likely receiving at least the prescribed dose. Based on the doses used in various studies, it appears that for SRS, the usual doses of 14–34 Gy in 1 fraction used for pituitary adenoma and 12–18 Gy in 1 fraction used for cavernous sinus meningioma are expected to result in a low risk of III, IV, and VI nerve damage (Witt, 2003; Spiegelmann et al., 2010; Park et al., 2018). One note of caution is that although a single dose of up to 34 Gy has been used for SRS for pituitary adenomas, this regimen is only used in secretory microadenomas, and the doses delivered to the III, IV, V, and VI nerves are likely to be much lower. For HSRT, the use of regimens of 21 Gy in 3 fractions and 25 Gy in 5 fractions appears to be safe for the III, IV, V, and VI nerves, although the exact tolerances of those nerves to HSRT are unknown (Killory et al., 2009; Wang et al., 2012; Karam et al., 2013; Morimoto et al., 2013; Wang et al., 2018). Given the lack of information regarding the exact tolerance of the III, IV, V, and VI nerves to SRS and HSRT, it is imperative that every effort should be made to minimize high-dose spillage by creating a highly conformal isodose distribution around the periphery of the tumor. Significant hotspots are likely to be present inside the prescribed isodose line, usually 50% for Gamma Knife-based SRS/ HSRT and ranging from 70% to 85% for LINAC-based SRS/ HSRT. By limiting the high-dose spillage, the III, IV, V, and VI nerves, which if involved by the tumor are expected to be at the periphery of the tumor, can be spared of an excessive dose of radiation from the hotspots.

Similar principles apply to the VII, VIII, IX, X, XI, and XII nerves. The use of highly conformal isodose plans can potentially spare cranial nerves (Figure 21.3). This can be accomplished in different ways depending on the device being used. For Gamma Knife, the use of smaller shots can create a highly conformal plan with a steep dose fall-off beyond the tumor. For other LINAC-based treatment planning systems, the use of intensity-modulated radiotherapy (IMRT) or volumetric modulated arc therapy (VMAT) planning can achieve the same goals. The VII and VIII nerves are typically encountered in the setting of SRS or HSRT for vestibular schwannoma. For SRS, data in the literature suggest that a peripheral dose of 12–13 Gy in 1 fraction will result in less than 5% risk of VII nerve palsy, and serviceable hearing rates of 44%–88% (Murphy and Suh, 2011). For vestibular schwannoma HSRT, regimens of 18 Gy in 3 fractions and 25 Gy in 5 fractions appear to be associated with good hearing preservation and low risk of VII nerve injury (Song and Williams, 1999; Meijer et al., 2003; Karam et al., 2013; Morimoto et al., 2013; Vivas et al., 2014). For the IX, X, XI, and XII nerves, the use of a dose of 15 Gy in 1 fraction for SRS or 25 Gy

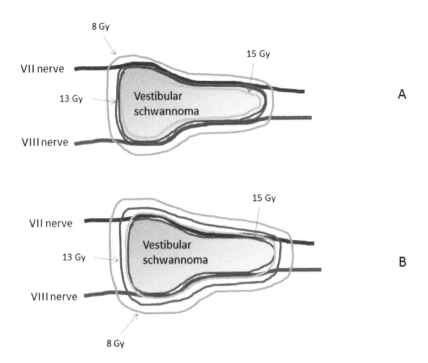

Figure 21.3 A dose of 13 Gy (red isodose line) in 1 fraction is prescribed to the vestibular tumor. The upper diagram (A) shows a highly conformal plan with the 15 Gy line located inside the tumor, and the VII and VIII nerves that are displaced by the tumor to the periphery are receiving a dose close to the prescribed dose of 13 Gy. By contrast, the lower diagram (B) shows a non-conformal plan with significant high-dose spillage, resulting in the encompassment of the VII and VIII nerves within the 15 Gy line, potentially increasing the risk of cranial nerve injury.

in 5 fractions for HSRT for glomus tumors appears to be associated with low-to-acceptable risk of injury (Guss et al., 2011; Ivan et al., 2011; Chun et al., 2014).

21.2.8.2 Vascular Injury, Brainstem Injury, and Radiation Necrosis

Detailed dosimetric data on vascular injury caused by SRS or HSRT are very scarce. Based on the experience in the treatment of pituitary adenomas with SRS, the risk of internal carotid artery stenosis is very low at the dose range used (Witt, 2003). However, it is prudent to avoid having more than 50% of the circumference receiving the prescribed dose and to limit the maximum dose to 30 Gy in 1 fraction as suggested in the limited literature (Witt, 2003). Recent data from Mayo Clinic showed that the dosimetric parameters were not predictive of stenosis/occlusion of internal carotid artery, but patients who had SRS for cavernous sinus meningiomas were at a higher risk of developing internal carotid artery stenosis/occlusion compared to those who had SRS for growth hormone-secreting pituitary adenomas (15 Gy in 1 fraction versus 25 Gy in 1 fraction) (Graffeo et al., 2019).

From HSRT, data from the literature showed that when pituitary adenomas and cavernous sinus lesions were treated with regimens of 19.5–21 Gy in 3 fractions, 18–22 Gy in 4 fractions, and 25 Gy in 5 fractions, no vascular injury was observed (Killory et al., 2009; Wang et al., 2012; Nguyen et al., 2014; Wang et al., 2018; Huang et al., 2019). Although data with detailed dosimetric correlation are limited, it is reasonable to consider using these values as reference when treating pituitary adenomas and cavernous sinus meningiomas. It is also prudent to avoid irradiating more than 50% of the circumference as in the case of SRS.

According to the literature and QUANTEC, the maximum point dose to the brainstem should be limited to 12.5 Gy in 1 fraction for SRS (Mayo et al., 2010a). For HSRT for tumors close to the brainstem, regimens of 21 Gy in 3 fractions and 25 Gy in 5 fractions appear to be safe for the brainstem (Song and Williams, 1999; Meijer et al., 2003; Karam et al., 2013; Morimoto et al., 2013; Vivas et al., 2014). More recently, Wang et al. found that maximum brainstem point doses of 19.1 Gy in 3 fractions and 20.1 Gy in 4 fractions resulted in zero brainstem toxicity (Wang et al., 2018). When planning SRS or HSRT for tumors close to the brainstem, it is crucial to contour the brainstem to ascertain that the maximum brainstem dose does not exceed the tolerance. For Gamma Knife-based SRS or HSRT, the isodose line can be shaped around the brainstem by manipulating the small shots as well as blocking one or more of the eight sectors for each shot. For LINAC-based or CyberKnife-based SRS or HSRT, inverse planning can achieve the same purpose.

The risk of symptomatic radiation necrosis is a function of target location, volume treated, and radiation dose delivered. Data from UPMC has provided guidelines for the estimation of risk of radiation necrosis from SRS for AVM, based on the SPIE and V12 (Flickinger et al., 2000). Based on these parameters, one can adjust the prescribed dose according to the estimated risk of symptomatic necrosis. This will also help one determine whether a volume-staged approach should be used. This has become one of the most important strategies to decrease the risk of symptomatic necrosis after SRS for AVM. The applicability of V12 as a parameter for risk estimation for non-AVM has been validated by the group from Case Western Reserve University. It appears that V12 is a reasonable parameter to be used for SRS for brain metastases (Korytko et al., 2006). Recent HyTEC study suggested that V12 values of 5 cc, 10 cc, or more than 15 cc were associated with risks of symptomatic radionecrosis of approximately 10%, 15%, and 20%, respectively (Milano et al., 2020). Based on these data, one can decide the risk level one is willing to accept based on the clinical scenario.

For HSRT, the HyTEC study also modeled the risk of radiation based on a 3-fracton regimen. As mentioned above, normal brain parenchyma V18 less than 30 cc and V23 less than 7 cc were associated with less than 10% risk of radionecrosis (Milano et al., 2020). When a 5-fraction regimen is used, limited data seem to suggest that the volume encompassed by 28.8 Gy is a reasonable parameter to use to estimate the risk of symptomatic necrosis, but this is subject to further validation (Inoue et al., 2014). In a recent LINAC-based HSRT series from Sunnybrook Health Science Centre, Toronto, Ontario, Canada, a brain minus gross tumor volume of ≥10.5 cc, based on a 5-fraction regimen for intact brain metastases, resulted in a much higher risk of symptomatic radiation necrosis (Faruqi et al., 2020). Given the above information, if this constraint cannot be met during treatment planning, a reduction of prescribed dose should be considered. What is unknown is whether the constraints for HSRT are applicable to both Gamma Knife-based and LINAC-based or CyberKnife-based treatments as the integral dose inside the prescribed isodose volume is higher for Gamma Knife-based HSRT. As Gamma Knife-based HSRT is a recent development, more data with longer-term follow-up are needed to determine whether the risk of radiation necrosis is higher with the same volume parameters.

21.3 SERIOUS LATE TOXICITIES FROM SPINAL IG-HSRT AND STRATEGIES TO MITIGATE RISK

This section will cover radiation myelopathy, vertebral compression fracture, radiation plexopathy, neuropathy, pain flare, and esophageal injury.

21.3.1 RADIATION MYELOPATHY

Rare occurrence of radiation myelopathy (RM) has been observed after spinal SBRT in both radiation-naïve and reirradiated patients. Sahgal et al. from Sunnybrook Health Science Centre at the University of Toronto have reported on nine cases of RM after spinal SBRT

in radiation-naïve patients and provided an unprecedented detailed dose–volume histogram (DVH) analysis (Sahgal, Weinberg et al., 2013). The DVH data for the 9 RM patients were compared to those of 66 controls. In the study, the thecal sac was contoured as a surrogate for spinal cord, taking into account potential sources of error such as physiologic spinal cord motion, intrafraction patient motion, variation in spinal cord delineation, potential errors in magnetic resonance imaging (MRI) and computed tomography (CT) image fusion, the treatment planning calculation algorithm, the image-guidance system, treatment couch motions, gantry rotation precision, and micromultileaf collimator (mMLC) leaf position calibration. Based on the data analysis to limit the risk of RM to below 5%, the thecal sac doses should be limited to 12.4 Gy in 1 fraction, 17 Gy in 2 fractions, 20.3 Gy in 3 fractions, 23 Gy in 4 fractions, and 25.3 Gy in 5 fractions recommended (Sahgal, Weinberg et al., 2013). Recent AAPM HyTEC guideline for spinal cord tolerance arrived at the same conclusions (Sahgal et al., 2019).

For reirradiation with SBRT, Sahgal et al. also performed analysis on five previously irradiated patients who developed RM after reirradiation with SBRT for spinal tumors and the dosimetric parameters were compared with 16 re-treatment spinal SBRT controls treated at the University of California, San Francisco (Sahgal et al., 2012). Given the different fractionation regimens used, normalized 2-Gy equivalent biologically effective dose (nBED) was used to facilitate comparison. The thecal sac cumulative nBED was then calculated by adding the nBED of the first course of radiotherapy to the point maximum nBED from the re-treatment SBRT course, using an α/β ratio of 2 for the spinal cord. They concluded that the cumulative nBED to the thecal sac point maximum should not exceed 70 Gy2/2 based on the conditions that the SBRT thecal sac re-treatment point maximum nBED did not exceed 25 Gy2/2, the thecal sac SBRT point maximum nBED/cumulative point maximum nBED ratio did not exceed 0.5, and the minimum time interval to re-treatment was at least 5 months (Sahgal et al., 2012). The AAPM HyTEC guideline for spinal cord tolerance also deemed the above constraints to be valid (Sahgal et al., 2019). Huang et al. reanalyzed these data using generalized linear–quadratic (LQ) model and concluded that the thecal sac point maximum nBED should not exceed 70 Gy2/2 (Huang et al., 2013). Readers are advised to refer to the radiation myelopathy chapter of this book for further details.

21.3.2 VERTEBRAL COMPRESSION FRACTURE

Vertebrae involved by metastases, especially lytic lesions, are predisposed to pathologic fractures due to the replacement of healthy bone with tumor. Radiotherapy, frequently used as a treatment for bone metastasis, can increase the risk of vertebral compression fracture (VCF), but the risk is deemed to be low in general with conventional palliative radiotherapy. However, SBRT delivers ablative doses of radiation to the clinical target volume (CTV), which typically encompasses the entire vertebral body. Up until the recent years, there were no detailed data in the literature pertaining to VCF caused by spinal SBRT.

Colleagues from Memorial Sloan-Kettering Cancer Center (MSKCC) first reported their observation of VCF after SBRT for spinal metastases to 18 to 24 Gy in one fraction, with a majority of patients receiving 24 Gy in one fraction (Rose et al., 2009). A progressive VCF rate of 39% was observed at a median time of 25 months. The location of spinal metastasis (above T10 versus T10 or below), the type of the spinal metastasis (lytic versus sclerotic and mixed), and the percentage of vertebral body involvement were identified as predictors of VCF (Rose et al., 2009). M.D. Anderson Cancer Center (MDACC) and University of Toronto also reported their series on VCF after spinal SBRT. However, they have reported a much lower rate of VCF occurring at 2–3.3 months after SBRT (Boehling et al., 2012; Cunha et al., 2012). In the MDACC study, the incidence of new or progressive VCF after SBRT in 93 patients with 123 spinal metastases was 20%. Approximately two-thirds of the patients in the MDACC series received either 27 Gy in 3 fractions or 20 to 30 Gy in 5 fractions. They have determined that factors predicting VCF were age more than 55 years, a pre-existing fracture, and baseline pain (Boehling et al., 2012). Similarly, the University of Toronto study of 90 patients with 167 spinal metastases treated also included patients treated with 2 to 5 fractions in addition to those treated with 1 fraction. The presence of kyphosis/ scoliosis, lytic appearance, primary lung and hepatocellular carcinoma, and a dose per fraction ≥ 20 Gy were identified as risk factors for VCF (Cunha et al., 2012). The crude rate of VCF was 11%, and the 1-year fracture-free probability was 87.3%.

Sahgal et al. pooled data of 252 patients with 410 spinal segments treated with SBRT treated at University of Toronto, M.D. Anderson Cancer Center (MDACC), and Cleveland Clinic Foundation (CCF) to identify the risk factors of VCF. The Spinal Instability Neoplastic Scoring (SINS) system was also applied to determine predictive value (Sahgal, Atenafu et al., 2013). There were 57 fractures (57 of 410, 14%), with 47% (27 of 57) new fractures, and 53% (30 of 57) fracture progression, and the median time to VCF was 2.46 months with 65% of VCFs occurring within the first 4 months. The 1- and 2-year cumulative incidences of fracture were 12.4% and 13.5%, respectively (Sahgal, Atenafu et al., 2013). On multivariable analysis, dose per fraction (greatest risk for ≥ 24 Gy vs 20 to 23 Gy vs ≤ 19 Gy) and three of the six original SINS criteria, namely, baseline VCF, lytic tumor, and spinal deformity, were identified as significant predictors of VCF (Sahgal, Atenafu et al., 2013).

The Elekta Spine Radiosurgery Research Consortium analyzed the factors predicting an increased risk for VCF following SBRT for spinal tumors. The prescribed dose ranged from 8 to 40 Gy in 1–5 fractions (median 20 Gy in 1 fraction). With a median follow-up of 10.1 months, 5.7% of the patients developed VCF. On multivariate analysis, the factors that predicted VCF included solitary metastasis, pre-SBRT compression fracture, and prescription dose to high-dose target volume (EQD2 of 38.4 Gy or higher). For new fractures, the factors predicting VCF on multivariate analysis were solitary metastasis and not using MRI for target delineation (Jawad et al., 2016).

Chen et al. from Johns Hopkins Cancer Center analyzed clinical and radiation planning characteristics of 193 patients with 302 vertebral segments predicting post-SBRT VCF using a novel normal tissue complication probability (NTCP) model (Chen et al., 2020). The 1 and 2-year cumulative rates were 4.6% and 6.7%, respectively. Based on NTCP modeling, there was a steep response of risk of VCF to the dose to 80% and 50% volume of the planning target volume (PTV D80% and D50%), but not to maximum dose or dose to 1 cc or 10% of the PTV. The D80% of 25 Gy and D50% of 28 Gy, based on 3 fractions, corresponded to a 10% risk of VCF. Other significant risk factors based on multivariate analysis included lower body mass index, total spinal instability neoplastic score, and PTV D80%. It appeared that the larger volume of spine PTV receiving lower doses were more closely associated with post-SBRT VCF compared to the high-dose regions.

21.3.3 RADIATION PLEXOPATHY/NEUROPATHY

The spinal nerves and nerve plexuses can potentially be injured by ablative doses of radiation delivered to the vertebrae via SBRT. Fortunately, radiation radiculopathy or plexopathy is uncommonly encountered after SBRT for spinal tumors. In a phase I/II trial of single-dose SBRT for radiation-naïve spinal metastases conducted at MDACC, 10 of 61 patients treated developed mild (grade 1 or 2) numbness and tingling, and one developed grade 3 radiculopathy at L5 after a dose of 16–24 Gy (Garg et al., 2012). In a study from Beth Israel Deaconess Hospital, 4 cases of persistent or new radiculopathy were observed among their 60 patients treated with SBRT for recurrent epidural spinal metastases. However, it is uncertain whether the complications were caused by tumor progression, radiation injury of the spinal nerves, or a combination of both (Mahadevan et al., 2011) since all those patients had evidence of radiological progression of disease. In a study from MDACC, 2 cases of grade 3 lumbar plexopathy on reirradiation were observed with SBRT for recurrent spinal metastases in 59 patients (Garg et al., 2011).

Colleagues from Indiana University attempted to determine the tolerance of the brachial plexus to SBRT based on 36 patients with 37 apical primary lung cancer treated to a dose of 30–72 Gy in 3–4 fractions (Forquer et al., 2009). There were seven cases of grade 2–4 brachial plexopathy observed, and the cutoff dose was determined to be 26 Gy in 3–4 fractions. This result corroborates the constraint of 24 Gy in 3 fractions used in RTOG trials (Forquer et al., 2009). The 2-year brachial plexopathy rates were 46% and 8% when the maximum brachial plexus dose was more than 26 Gy and ≤ 26 Gy, respectively (Forquer et al., 2009). However, in this study, the subclavian/axillary vessels served as a surrogate for brachial plexus, instead of the full brachial plexus contour.

Colleagues from Sweden reported brachial plexopathy from the treatment of apical lung tumors with SBRT using a 3-fraction regimen (Lindberg et al., 2019). They analyzed 52 patients with 56 apical tumors and tracked the brachial plexus dose. With a median follow-up of 30 months, seven patients experienced maximum grade 2 (n = 3) or 3 (n = 4) brachial plexopathy after a median of 8.7 months. Median maximum BED_3 for the patients who developed brachial plexopathy was 381 Gy (range, 30–524) compared to maximum BED_3 of 34 Gy (range, 0.10–483) for those who did not develop brachial plexopathy. Using NTCP model, the authors suggested that the maximum BED_3 should be kept ≤130 Gy or 30 Gy in 3 fractions.

The constraints for nerves used by legacy RTOG are listed in Table 21.1.

Disclaimer: These dose constraints are intended to be used in RTOG trials and have not been thoroughly tested clinically. The authors do not assume responsibility for the use of these dose limits.

21.3.4 PAIN FLARE

Pain flare is a common side effect of SBRT for spinal metastases. Although it may not strictly be considered a late complication, given the common occurrence, it is covered in this chapter. In the recent years, high-quality data have emerged from high-volume centers such as Sunnybrook Health Science Centre, University of Toronto, and the M.D. Anderson Cancer Center, Texas (Chiang et al., 2013; Pan et al.,

Table 21.1 **Normal tissue constraints used by the Radiation Therapy Oncology Group (RTOG) trials**

NEURAL STRUCTURE	1-FRACTION	3-FRACTION	4-FRACTION	5-FRACTION
Brachial plexus	RTOG 0631 and 0915:17.5 Gy (< 0.03 cc or maximum)/ 14 Gy (< 3 cc)	RTOG 0236 and 0618: 24 Gy (maximum) RTOG 1021: 24 Gy (maximum)/ 20.4 Gy (< 3 cc)	RTOG 0915: 27.2 Gy (maximum)/ 23.6 Gy (< 3 cc)	RTOG 0813: 32 Gy (maximum)/ 30 Gy (< 3 cc)
Cauda equina	RTOG 0631: 16 Gy (< 0.03 cc)/ 14 Gy (< 5 cc)	Not available	Not available	Not available
Sacral plexus	RTOG 0631: 18 Gy (< 0.03 cc)/ 14.4 Gy (< 5 cc)	Not available	Not available	Not available

Source: Data extracted from www.rtog.org.

2014; Khan et al., 2015). Colleagues from University of Toronto conducted a prospective clinical trial to determine the incidence of pain flare after spine SBRT in steroid-naïve patients and identify predictive factors. A total of 41 patients were enrolled, and 18 patients were treated with 20–24 Gy in a single fraction, whereas 23 patients were treated with 24–35 Gy in 2–5 fractions. Pain flare was observed in 68.3% of patients, most commonly on day 1 after spinal SBRT. Significant predictive factors for pain flare included a higher Karnofsky performance status and tumor location in the cervical or lumbar region (Chiang et al., 2013). In those patients treated with dexamethasone, a significant decrease in pain scores over time were subsequently observed (Chiang et al., 2013). In a subsequent prospective observational study from University of Toronto, 47 patients were prophylactically treated with dexamethasone during spinal SBRT. The first cohort of 24 patients were treated with 4 mg of dexamethasone, and the second cohort of 23 patients with 8 mg of dexamethasone (Khan et al., 2015). The Brief Pain Inventory (BPI) was used to score pain and functional interference each day during SBRT and for 10 days afterwards. The total incidence of pain flare was 19%, and the incidence in the 4 and 8 mg cohorts were 25% and 13% (not significantly different), respectively (Khan et al., 2015). The 4-mg cohort had better profile in walking ability as well as relationships with others, compared to the 8-mg cohort. Compared to the earlier steroid-naïve cohort, the use of dexamethasone was associated with lower worst pain scores and improved general activity interference outcome (Khan et al., 2015). In a secondary analysis of the phase I/II trials of SBRT for spinal tumors from MDACC, the incidence of pain flare was 23%, and the median time to onset was 5 days. The only independent factor associated with pain flare was the number of fractions with a single-fraction regimen resulting in the highest rate (Pan et al., 2014). In that study, rigorous daily pain evaluation was not conducted and, therefore, it is difficult to compare the MDACC data to the University of Toronto study.

21.3.5 ESOPHAGEAL TOXICITY

The esophagus is located adjacent to the thoracic spinal column and is, therefore, susceptible to toxicity from SBRT to spinal tumors, especially when delivered in a single fraction. The most robust data on esophageal toxicity from single-fraction spinal SBRT came from MSKCC. Two-hundred and four spinal metastases abutting the esophagus in 182 patients were treated with single-dose SBRT to 24 Gy. The esophageal toxicity was scored using National Cancer Institute Common Toxicity Criteria for Adverse Events (version 4.0). The rates of acute and late esophageal toxicities were 15% and 12%, respectively (Cox et al., 2012). Grade 3 or higher acute or late toxicities occurred in 14 (6.8%) patients. For grade 3 or higher esophageal toxicities, the median splits were determined, and those for D2.5cm³ (the minimum dose to the 2.5 cc volume receiving the highest dose), V12 (the volume receiving at least 12 Gy), V15, V20, and V22 were 14 Gy, 3.78 cc, 1.87 cc, 0.11 cc, and 0 cc, respectively (Cox et al., 2012). The authors recommended the maximum point dose to be kept less than 22 Gy. Most notably, the seven patients who developed grade 4 or higher toxicities had either radiation recall reactions after chemotherapy with doxorubicin or gemcitabine or iatrogenic esophageal manipulation such as biopsy, dilatation, and stent placement (Cox et al., 2012).

Stephans et al. from CCF analyzed 52 patients treated with SBRT for liver or lung tumors who had a planning target volume within 2 cm of the esophagus, trying to determine the esophageal tolerance to SBRT (Stephans et al., 2014). The radiation dose given was 37.5–60 Gy in 3–10 fractions (median 50 Gy in 5 fractions). Two patients developed esophageal fistula and their maximum esophageal point doses were 51.5 and 52 Gy and 1-cc doses were 48.1 and 50 Gy, respectively. Interestingly, they both received adjuvant anti-angiogenic agents within 2 months of completing SBRT.

21.3.6 RADIATION-INDUCED MYOSITIS

Radiation-induced myositis is a rare and previously under-reported complication after SBRT. MRI usually shows increased T2-weighted signal intensity, patchy, irregular T1-weighted postcontrast enhancement, and muscle volume loss correlating spatially to the treatment field. It is important to distinguish between myositis and tumor recurrence. Colleagues from Memorial Sloan-Kettering Cancer Center (MSKCC) published the largest retrospective series of SBRT-related myositis in 667 patients with 891 spine lesions treated at MSKCC to either 24 Gy in 1 fraction or to 27 Gy in 3 fractions (Lockney et al., 2018). The 1-year rate of grade 1 or higher myositis rate was 1.9%, and the time to myofascial back pain was 1.4 months from SBRT, often preceding the imaging changes with a median time to development at 4.7 months. The only significant factor predicting the development of grade 1 or higher myositis was single-fraction SBRT. The areas with myositis received a median dose of 17.5 Gy (range, 9.2–21.6 Gy). Most of the cases occurred in lumbar spine treated with single-fraction SBRT. All patients in this study responded to treatment with analgesics and steroids.

21.3.7 STRATEGIES TO MITIGATE RISKS OF COMPLICATIONS FROM SPINAL IG-HSRT

To minimize the risks of serious complications from spinal SBRT or IG-HSRT, all relevant organs-at-risk (OARs) including the spinal cord, cauda equina, nerve plexuses/ roots, and esophagus must be contoured, and the dose constraint for each OAR must be respected (Sahgal et al., 2008; Lo et al., 2010; Foote et al., 2011). Fusion of the spinal MRI (axial T1 and T2 sequences) with the treatment planning CT should be performed to facilitate the accurate contouring of neural structures such as the spinal cord and cauda equina. The accuracy of the fusion must be verified before the image sets are used for delineation of the neural structures. Alternatively, a CT myelogram can be used for delineation of the spinal cord in patients with contraindications to MRI or when there are significant metallic artifacts as in postoperative cases. It is imperative that the window leveling is correct; otherwise, the cord contour could be drawn incorrectly, resulting in an inaccurate determination of cord dose which can potentially be detrimental (Figure 21.4). Ideally, the MRI or CT myelogram used for treatment planning should be obtained as close to the time of starting SBRT as possible because if there is tumor growth after a long-time lapse, the spinal cord position can change (Figure 21.5).

Another important aspect to consider in avoiding inaccurate determination of cord dose is the selection of an appropriate treatment planning algorithm, especially in the thoracic region. Okoye and colleagues from University Hospital Seidman Cancer Center, Case Western Reserve University, reviewed CyberKnife-based SBRT treatment plans in 37 patients with thoracic spinal tumors generated using Ray Tracing and Monte Carlo algorithms (Okoye et al., 2016). They found discrepancies in the coverage of the planning target volume (PTV) as well as the actual doses delivered to various organs-at-risk (OARs) including the spinal cord. In 14% of lesions, the actual dose delivered to the spinal cord determined with Monte Carlo calculation was found to be ≥ 5% (Okoye et al., 2016). These data underscore the importance of an optimal treatment planning algorithm especially in regions where there are tissues with substantially different electronic densities, such as the lungs. Interested readers are encouraged to visit: http://rpc.mdanderson.org/rpc/Services/Anthropomorphic_%20Phantoms/TPS%20-%20algorithm%20list%20updated.pdf.

Given the fact that the spinal cord is in very close proximity to the spinal clinical target volume (CTV) and the steep dose gradient between the spinal cord and spinal CTV, even a slight deviation in positioning can result in significant overdosing of the spinal cord, resulting in catastrophic complications such as radiation myelopathy (Wang et al., 2008). Therefore, robust immobilization is crucial. Colleagues from the

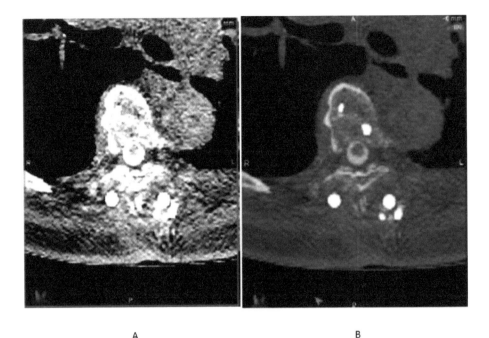

A   B

Figure 21.4 CT myelogram showing the incorrect window level (A) and the correct window level (B). The spinal cord contour is smaller in (A) compared to that in (B). Treatment planning based on the spinal cord contour in (A) will likely result in underestimation of the spinal cord dose.

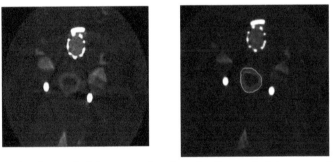

2 weeks after surgical decompression   6 weeks after surgical decompression

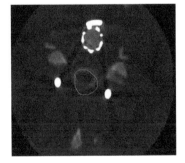

Residual tumor progressed, pushing the thecal sac and cord posteriorly

Figure 21.5 CT myelogram obtained 2 weeks after surgical compression is shown in the left upper panel; repeat CT myelogram 6 weeks after surgical decompression showed progressive epidural disease pushing the thecal sac and spinal cord posteriorly (right upper panel); lower panel shows an overlay of the thecal sac and spinal cord contours based on the 6-week CT myelogram on the 2-week myelogram showing the degree of displacement of these structures due to tumor growth.

University of Toronto have demonstrated that the BodyFIX (Elekta, Stockholm, Sweden) near-rigid body immobilization system is more robust in minimizing intrafraction motions than a simple Vac-Loc system, and it can limit the setup error to 2 mm (Li et al., 2012). Therefore, a dual vacuum system is recommended for the immobilization for spinal SBRT unless a CyberKnife system is used, as it can robotically track the spine in a near real-time fashion.

Despite the most robust immobilization system and advanced technology, intrafraction patient motion can occur especially when the treatment time is anticipated to be long. Colleagues from University of Toronto showed that there could be an intrafraction motion of 1.2 mm and 1 degree even with near-rigid body immobilization with BodyFIX, image guidance with kilovoltage cone-beam CT, and a robotic couch capable of adjusting shifts with six degrees of freedom for their LINAC-based SBRT treatments (Hyde et al., 2012). In order to attain this level of precision, they recommended an intrafraction repeat cone-beam CT every 20 minutes to check for any positional deviation. With the availability of newer technologies such as VMAT and the high-dose rate flattening filter-free feature, treatment time can potentially be reduced dramatically, and in that situation, intrafraction cone-beam CT will not be necessary. Physiologic spinal cord motion can also potentially create uncertainties in the estimation of true spinal cord dose from SBRT. Data from multiple institutions have showed that the spinal cord motion is typically less than 1 mm (Cai et al., 2007; Tseng et al., 2015). A recent study by Oztek et al. from University of Washington showed that the cord could excurse beyond 1 mm from the static position in 43% of patients, but there was no excursion beyond 1.5 mm. The average cord dose was more than 5% and more than 10% higher in 24% and 14% of the patients, respectively (Oztek et al., 2020). Although some spinal SBRT physicians from very experienced centers use the true spinal cord as the avoidance structure for inverse planning, it is prudent to create a planning-at-risk volume (PRV) which typically ranges from 1.5 to 2.0 mm. Alternatively, the thecal sac can be contoured as a surrogate for spinal cord. This practice serves to decrease the risk of RM resulting from potential errors that can lead to overdosing of the spinal cord (Sahgal et al., 2012; Sahgal, Weinberg et al., 2013).

Based on the analysis of radiation naïve and *reirradiated* patients who developed RM after spinal SBRT, Sahgal et al. has made recommendations on spinal cord constraints, and these have been discussed in detail in an earlier section of this chapter and in the radiation myelopathy chapter (Sahgal et al., 2012; Sahgal, Weinberg et al., 2013). When the recommendations are strictly followed, no RM has been observed at Sunnybrook Health Science Centre/University of Toronto in more than one decade (personal communication with one of the co-authors Arjun Sahgal). Alternatively, in the setting of reirradiation with prior conventional radiotherapy dose ≤ 45 Gy (1.8–2.0 Gy per fraction), authors of this chapter (including Simon S. Lo and Eric L. Chang) have not observed RM when spinal cord dose constraints of 10 Gy in 5 fractions or 9 Gy in 3 fractions are used.

Based on the studies available, certain risk factors have been identified to predispose patients to VCF after spinal SBRT. It is reasonable to avoid using a single fraction of ≥20 Gy, especially in patients with risk factors such as baseline VCF, lytic tumor, and spinal deformity, all of which are parameters of SINS (Cunha et al., 2012; Sahgal, Atenafu et al., 2013). Prophylactic kyphoplasty or vertebroplasty can be considered for patients with SINS of 13 or higher or pre-existing VCF before SBRT to reduce the risk of further fracture and relieve the mechanical pain, rendering SBRT more tolerable.

In order to spare the nerve plexuses and nerves from the damage from SBRT, these structures must be carefully and accurately contoured apart from respecting their tolerance to ablative radiation. MRI, especially T2 sequence, can be fused with treatment planning CT to facilitate accurate contouring of these structures. The sparing of these neural structures is particularly important at levels where the nerves roots or nerve plexuses are responsible for motor function of the extremities. Damage of nerve function at those levels results in loss of vital extremity function. Radiation Therapy Oncology Group (RTOG) has included a brachial plexus contouring atlas on their website, and interested readers are encouraged to access the document (www.rtog.org/CoreLab/ContouringAtlases/BrachialPlexus-ContouringAtlas.aspx). Another very useful and more comprehensive resource is the brachial plexus contouring guidelines based on both CT and MRI from Boston University (Truong et al. 2010).

Pain flare is a relatively common phenomenon after spinal SBRT. Investigators from Sunnybrook Health Science Centre, University of Toronto have reported in their prospective trial that approximately two out of three of patients developed pain flare during or after SBRT, and the symptoms can be effectively controlled with steroid therapy (Chiang et al., 2013). In another study, prophylactic dexamethasone with rapid taper has resulted in a much lower pain flare rate (Khan et al., 2015). There was no difference in pain flare rates between 4 mg and 8 mg per dose, but patients had better subsequent ability to walk and social relationships with others in the 4-mg cohort. It appears that dexamethasone 4 mg starting day 1, on days of SBRT, and 4 days post-SBRT, as used in the study, should be adequate for prophylaxis against pain flare. In the United States, a 6-day course with a Medrol (methylprednisolone) Dose Pack is very popular, and it can effectively control pain flare in most cases.

There can be considerable radiation dose exposure to the esophagus when the spinal segment to be treated with SBRT is located in the thoracic region. Therefore, it is imperative that the esophagus is contoured as an OAR, and the dose constraints have to be respected in order to avoid serious complications. Data from MSKCC showed that the maximum single dose tolerated was volume dependent, and it has been discussed previously in this chapter (Cox et al., 2012). They also identified risks factors for serious post-SBRT esophageal toxicities, namely, post-SBRT doxorubicin- or gemcitabine-based chemotherapy or surgical manipulation of the esophagus (Cox et al., 2012). These treatments or procedures should be avoided after substantial radiation dose exposure from thoracic spinal SBRT. The use of a multiple session regimen (2–5 fractions) may also decrease the risk of serious esophageal toxicities from SBRT if the dose constraint cannot be met in one fraction. Data from CCF has showed that anti-angiogenic agents should be avoided if the esophagus has been exposed to ablative doses of radiation from SBRT (Stephans et al., 2014). However, the study was based on lung tumors, and, therefore, the dose used was beyond the range used for spinal tumors.

21.4 CONCLUSION

Serious complications have been observed after HSRT for intracranial and spinal tumors. In most cases, there are risk factors which can be identified. With the accumulation of experience with intracranial and spinal HSRT, normal tissue constraints of various OARs are better understood. Given the sophistication of radiation technology, understanding of dose constraints, and advancement of clinical expertise, it is possible to deliver very potent doses of radiation to brain and spinal tumors without causing excessive toxicities.

REFERENCES

Boehling NS, Grosshans DR, Allen PK, McAleer MF, Burton AW, Azeem S, Rhines LD, Chang EL (2012) Vertebral compression fracture risk after stereotactic body radiotherapy for spinal metastases. *Journal of Neurosurgery Spine* 16:379–386.

Cai J, Sheng K, Sheehan JP, Benedict SH, Larner JM, Read PW (2007) Evaluation of thoracic spinal cord motion using dynamic MRI. *Radiotherapy and Oncology* 84:279–282.

Chen X, Gui C, Grimm J, Huang E, Kleinberg L, Lo L, Sciubba D, Khan M, Redmond KJ (2020) Normal tissue complication probability of vertebral compression fracture after stereotactic body radiotherapy for de novo spine metastasis. *Radiotherapy and Oncology* 150:142–149.

Chiang A, Zeng L, Zhang L, Lochray F, Korol R, Loblaw A, Chow E, Sahgal A (2013) Pain flare is a common adverse event in steroid-naive patients after spine stereotactic body radiation therapy: a prospective clinical trial. *International Journal of Radiation Oncology, Biology, Physics* 86:638–642.

Chun SG, Nedzi LA, Choe KS, Abdulrahman RE, Chen SA, Yordy JS, Timmerman RD, Kutz JW, Isaacson B (2014) A retrospective analysis of tumor volumetric responses to five-fraction stereotactic radiotherapy for paragangliomas of the head and neck (glomus tumors). *Stereotactic and Functional Neurosurgery* 92:153–159.

Clark BG, Souhami L, Pla C, Al-Amro AS, Bahary JP, Villemure JG, Caron JL, Olivier A, Podgorsak EB (1998) The integral biologically effective dose to predict brain stem toxicity of hypofractionated stereotactic radiotherapy. *International Journal of Radiation Oncology, Biology, Physics* 40:667–675.

Cohen-Inbar O, Xu Z, Schlesinger D, Vance ML, Sheehan JP (2015) Gamma Knife radiosurgery for medically and surgically refractory prolactinomas: long-term results. *Pituitary* 18:820–830.

Conti A, Pontoriero A, Midili F, Iati G, Siragusa C, Tomasello C, La Torre D, Cardali SM, Pergolizzi S, De Renzis C (2015) CyberKnife multisession stereotactic radiosurgery and hypofractionated stereotactic radiotherapy for perioptic meningiomas: intermediate-term results and radiobiological considerations. *SpringerPlus* 4:37.

Cox BW, Jackson A, Hunt M, Bilsky M, Yamada Y (2012) Esophageal toxicity from high-dose, single-fraction paraspinal stereotactic radiosurgery. *International Journal of Radiation Oncology, Biology, Physics* 83:e661–e667.

Cunha MV, Al-Omair A, Atenafu EG, Masucci GL, Letourneau D, Korol R, Yu E, Howard P, Lochray F, da Costa LB, Fehlings MG, Sahgal A (2012) Vertebral compression fracture (VCF) after spine stereotactic body radiation therapy (SBRT): analysis of predictive factors. *International Journal of Radiation Oncology, Biology, Physics* 84:e343–e349.

Faruqi S, Ruschin M, Soliman H, Myrehaug S, Zeng KL, Husain Z, Atenafu E, Tseng CL, Das S, Perry J, Maralani P, Heyn C, Mainprize T, Sahgal A (2020) Adverse radiation effect after hypofractionated stereotactic radiosurgery in 5 daily fractions for surgical cavities and intact brain metastases. *International Journal of Radiation Oncology, Biology, Physics* 106:772–779.

Flickinger JC, Kondziolka D, Lunsford LD, Kassam A, Phuong LK, Liscak R, Pollock B (2000) Development of a model to predict permanent symptomatic postradiosurgery injury for arteriovenous malformation patients. Arteriovenous malformation radiosurgery study group. *International Journal of Radiation Oncology, Biology, Physics* 46:1143–1148.

Flickinger JC, Kondziolka D, Niranjan A, Maitz A, Voynov G, Lunsford LD (2004) Acoustic neuroma radiosurgery with marginal tumor doses of 12 to 13 Gy. *International Journal of Radiation Oncology, Biology, Physics* 60:225–230.

Foote M, Letourneau D, Hyde D, Massicotte E, Rampersaud R, Fehlings M, Fisher C, Lewis S, Macchia NL, Yu E, Laperriere NJ, Sahgal A (2011) Technique for stereotactic body radiotherapy for spinal metastases. *Journal of Clinical Neuroscience* 18:276–279.

Forquer JA, Fakiris AJ, Timmerman RD, Lo SS, Perkins SM, McGarry RC, Johnstone PA (2009) Brachial plexopathy from stereotactic body radiotherapy in early-stage NSCLC: dose-limiting toxicity in apical tumor sites. *Radiotherapy and Oncology* 93:408–413.

Garg AK, Shiu AS, Yang J, Wang XS, Allen P, Brown BW, Grossman P, Frija EK, McAleer MF, Azeem S, Brown PD, Rhines LD, Chang EL (2012) Phase 1/2 trial of single-session stereotactic body radiotherapy for previously unirradiated spinal metastases. *Cancer* 118:5069–5077. Garg AK, Wang XS, Shiu AS, Allen P, Yang J, McAleer MF, Azeem S, Rhines LD, Chang EL (2011) Prospective evaluation of spinal reirradiation by using stereotactic body radiation therapy: The University of Texas MD Anderson Cancer Center experience. *Cancer* 117:3509–3516.

Graffeo CS, Link MJ, Stafford SL, Parney IF, Foote RL, Pollock BE (2019) Risk of internal carotid artery stenosis or occlusion after single-fraction radiosurgery for benign parasellar tumors. *Journal of Neurosurgery* 133:1388–1395.

Guss ZD, Batra S, Limb CJ, Li G, Sughrue ME, Redmond K, Rigamonti D, Parsa AT, Chang S, Kleinberg L, Lim M (2011) Radiosurgery of glomus jugulare tumors: a meta-analysis. *International Journal of Radiation Oncology, Biology, Physics* 81:e497–e502.

Huang L, Sun L, Wang W, Cui Z, Zhang Z, Li J, Wang Y, Wang J, Yu X, Ling Z, Qu B, Pan LS (2019) Therapeutic effect of hypofractionated stereotactic radiotherapy using CyberKnife for high volume cavernous sinus cavernous hemangiomas. *Technology in Cancer Research and Treatment* 18:1533033819876981.

Huang Z, Mayr NA, Yuh WT, Wang JZ, Lo SS (2013) Reirradiation with stereotactic body radiotherapy: analysis of human spinal cord tolerance using the generalized linear-quadratic model. *Future Oncology* 9:879–887.

Hyde D, Lochray F, Korol R, Davidson M, Wong CS, Ma L, Sahgal A (2012) Spine stereotactic body radiotherapy utilizing cone-beam CT image-guidance with a robotic couch: intrafraction motion analysis accounting for all six degrees of freedom. *International Journal of Radiation Oncology, Biology, Physics* 82:e555–e562.

Inoue HK, Sato H, Seto K, Torikai K, Suzuki Y, Saitoh J, Noda SE, Nakano T (2014) Five-fraction CyberKnife radiotherapy for large brain metastases in critical areas: impact on the surrounding brain volumes circumscribed with a single dose equivalent of 14 Gy (V14) to avoid radiation necrosis. *Journal of Radiation Research* 55:334–342.

Ivan ME, Sughrue ME, Clark AJ, Kane AJ, Aranda D, Barani IJ, Parsa AT (2011) A meta-analysis of tumor control rates and treatment-related morbidity for patients with glomus jugulare tumors. *Journal of Neurosurgery* 114:1299–1305.

Iwata H, Sato K, Tatewaki K, Yokota N, Inoue M, Baba Y, Shibamoto Y (2011) Hypofractionated stereotactic radiotherapy with CyberKnife for nonfunctioning pituitary adenoma: high local control with low toxicity. *Neuro-Oncology* 13:916–922.

Jawad MS, Fahim DK, Gerszten PC, Flickinger JC, Sahgal A, Grills IS, Sheehan J, Kersh R, Shin J, Oh K, Mantel F, Guckenberger M, Consortium obotESRR (2016) Vertebral compression fractures after stereotactic body radiation therapy: a large, multi-institutional, multinational evaluation. *Journal of Neurosurgery Spine* 24:928–936.

Kanesaka N, Mikami R, Nakayama H, Nogi S, Tajima Y, Nakajima N, Wada J, Miki T, Haraoka J, Okubo M, Sugahara S, Tokuuye K (2012) Preliminary results of fractionated stereotactic radiotherapy after cyst drainage for craniopharyngioma in adults. *International Journal of Radiation Oncology, Biology, Physics* 82:1356–1360.

Karam SD, Tai A, Strohl A, Steehler MK, Rashid A, Gagnon G, Harter KW, Jay AK, Collins SP, Kim JH, Jean W (2013) Frameless fractionated stereotactic radiosurgery for vestibular schwannomas: a single-institution experience. *Frontiers in Oncology* 3:121.

Khan L, Chiang A, Zhang L, Thibault I, Bedard G, Wong E, Loblaw A, Soliman H, Fehlings MG, Chow E, Sahgal A (2015) Prophylactic dexamethasone effectively reduces the incidence of pain flare following spine stereotactic body radiotherapy (SBRT): a prospective observational study. *Supportive Care in Cancer* 23:2937–2943.

Killory BD, Kresl JJ, Wait SD, Ponce FA, Porter R, White WL (2009) Hypofractionated CyberKnife radiosurgery for perichiasmatic pituitary adenomas: early results. *Neurosurgery* 64:A19–A25.

Kondziolka D, Lunsford LD, McLaughlin MR, Flickinger JC (1998) Long-term outcomes after radiosurgery for acoustic neuromas. *New England Journal of Medicine* 339:1426–1433.

Korytko T, Radivoyevitch T, Colussi V, Wessels BW, Pillai K, Maciunas RJ, Einstein DB (2006) 12 Gy Gamma Knife radiosurgical volume is a predictor for radiation necrosis in non-AVM intracranial tumors. *International Journal of Radiation Oncology, Biology, Physics* 64:419–424.

Leavitt JA, Stafford SL, Link MJ, Pollock BE (2013) Long-term evaluation of radiation-induced optic neuropathy after single-fraction stereotactic radiosurgery. *International Journal of Radiation Oncology, Biology, Physics* 87:524–527.

Li W, Sahgal A, Foote M, Millar BA, Jaffray DA, Letourneau D (2012) Impact of immobilization on intrafraction motion for spine stereotactic body radiotherapy using cone beam computed tomography. *International Journal of Radiation Oncology, Biology, Physics* 84:520–526.

Lindberg K, Grozman V, Lindberg S, Onjukka E, Lax I, Lewensohn R, Wersall P (2019) Radiation-induced brachial plexus toxicity after SBRT of apically located lung lesions. *Acta Oncologica* 58:1178–1186.

Lindvall P, Bergstrom P, Blomquist M, Bergenheim AT (2010) Radiation schedules in relation to obliteration and complications in hypofractionated conformal stereotactic radiotherapy of arteriovenous malformations. *Stereotactic and Functional Neurosurgery* 88:24–28.

Lo SS, Sahgal A, Wang JZ, Mayr NA, Sloan A, Mendel E, Chang EL (2010) Stereotactic body radiation therapy for spinal metastases. *Discovery Medicine* 9:289–296.

Lockney DT, Jia AY, Lis E, Lockney NA, Liu C, Hopkins B, Higginson DS, Yamada Y, Laufer I, Bilsky M, Schmitt AM (2018) Myositis following spine radiosurgery for metastatic disease: a case series. *Journal of Neurosurgery Spine* 28:416–421.

Mahadevan A, Floyd S, Wong E, Jeyapalan S, Groff M, Kasper E (2011) Stereotactic body radiotherapy reirradiation for recurrent epidural spinal metastases. *International Journal of Radiation Oncology, Biology, Physics* 81:1500–1505.

Marchetti M, Bianchi S, Pinzi V, Tramacere I, Fumagalli ML, Milanesi IM, Ferroli P, Franzini A, Saini M, DiMeco F, Fariselli L (2016) Multisession radiosurgery for sellar and parasellar benign meningiomas: long-term tumor growth control and visual outcome. *Neurosurgery* 78:638–646.

Mayo C, Martel MK, Marks LB, Flickinger J, Nam J, Kirkpatrick J (2010b) Radiation dose-volume effects of optic nerves and chiasm. *International Journal of Radiation Oncology, Biology, Physics* 76:S28–S35.

Mayo C, Yorke E, Merchant TE (2010a) Radiation associated brainstem injury. *International Journal of Radiation Oncology, Biology, Physics* 76:S36–S41.

Meijer OW, Vandertop WP, Baayen JC, Slotman BJ (2003) Single-fraction vs. fractionated linac-based stereotactic radiosurgery for vestibular schwannoma: a single-institution study. *International Journal of Radiation Oncology, Biology, Physics* 56:1390–1396.

Milano MT, Grimm J, Niemierko A, Soltys SG, Moiseenko V, Redmond KJ, Yorke E, Sahgal A, Xue J, Mahadevan A, Muacevic A, Marks LB, Kleinberg LR (2020) Single- and multifraction stereotactic radiosurgery dose/volume tolerances of the brain. *International Journal of Radiation Oncology, Biology, Physics*. (article in press)

Milano MT, Grimm J, Soltys SG, Yorke E, Moiseenko V, Tome WA, Sahgal A, Xue J, Ma L, Solberg TD, Kirkpatrick JP, Constine LS, Flickinger JC, Marks LB, El Naqa I (2018) Single- and multi-fraction stereotactic radiosurgery dose tolerances of the optic pathways. *International Journal of Radiation Oncology, Biology, Physics* (article in press).

Morimoto M, Yoshioka Y, Kotsuma T, Adachi K, Shiomi H, Suzuki O, Seo Y, Koizumi M, Kagawa N, Kinoshita M, Hashimoto N, Ogawa K (2013) Hypofractionated stereotactic radiation therapy in three to five fractions for vestibular schwannoma. *Japanese Journal of Clinical Oncology* 43:805–812.

Murphy ES, Suh JH (2011) Radiotherapy for vestibular schwannomas: a critical review. *International Journal of Radiation Oncology, Biology, Physics* 79:985–997.

Nguyen JH, Chen CJ, Lee CC, Yen CP, Xu Z, Schlesinger D, Sheehan JP (2014) Multisession Gamma Knife radiosurgery: a preliminary experience with a noninvasive, relocatable frame. *World Neurosurgery* 82:1256–1263.

Okoye CC, Patel RB, Hasan S, Podder T, Khouri A, Fabien J, Zhang Y, Dobbins D, Sohn JW, Yuan J, Yao M, Machtay M, Sloan AE, Miller J, Lo SS (2016) Comparison of ray tracing and monte Carlo calculation algorithms for thoracic spine lesions treated with CyberKnife-based stereotactic body radiation therapy. *Technology in Cancer Research and Treatment* 15:196–202.

Oztek MA, Mayr NA, Mossa-Basha M, Nyflot M, Sponseller PA, Wu W, Hofstetter CP, Saigal R, Bowen SR, Hippe DS, Yuh WTC, Stewart RD, Lo SS (2020) The dancing cord: inherent spinal cord motion and its effect on cord dose in spine stereotactic body radiation therapy. *Neurosurgery.* 87:1157–1166

Pan HY, Allen PK, Wang XS, Chang EL, Rhines LD, Tatsui CE, Amini B, Wang XA, Tannir NM, Brown PD, Ghia AJ (2014) Incidence and predictive factors of pain flare after spine stereotactic body radiation therapy: secondary analysis of phase 1/2 trials. *International Journal of Radiation Oncology, Biology, Physics* 90:870–876.

Park KJ, Kano H, Iyer A, Liu X, Tonetti DA, Lehocky C, Faramand A, Niranjan A, Flickinger JC, Kondziolka D, Lunsford LD (2018) Gamma Knife stereotactic radiosurgery for cavernous sinus meningioma: long-term follow-up in 200 patients. *Journal of Neurosurgery*:1–10.

Petit JH, Hudes RS, Chen TT, Eisenberg HM, Simard JM, Chin LS (2001) Reduced-dose radiosurgery for vestibular schwannomas. *Neurosurgery* 49:1299–1306; discussion 1306–1297.

Pollock BE, Link MJ, Leavitt JA, Stafford SL (2014) Dose-volume analysis of radiation-induced optic neuropathy after single-fraction stereotactic radiosurgery. *Neurosurgery* 75:456–460; discussion 460.

Puataweepong P, Dhanachai M, Hansasuta A, Dangprasert S, Swangsilpa T, Sitathanee C, Jiarpinitnun C, Vitoonpanich P, Yongvithisatid P (2016) The clinical outcome of hypofractionated stereotactic radiotherapy with CyberKnife robotic radiosurgery for perioptic pituitary adenoma. *Technology in Cancer Research and Treatment* 15:NP10–NP15.

Rose PS, Laufer I, Boland PJ, Hanover A, Bilsky MH, Yamada J, Lis E (2009) Risk of fracture after single fraction image-guided intensity-modulated radiation therapy to spinal metastases. *Journal of Clinical Oncology* 27:5075–5079.

Ruess D, Pohlmann L, Grau S, Hamisch C, Hoevels M, Treuer H, Baues C, Kocher M, Ruge M (2020) Outcome and toxicity analysis of single dose stereotactic radiosurgery in vestibular schwannoma based on the Koos grading system. *Scientific Reports* 10:9309.

Sahgal A, Atenafu EG, Chao S, Al-Omair A, Boehling N, Balagamwala EH, Cunha M, Thibault I, Angelov L, Brown P, Suh J, Rhines LD, Fehlings MG, Chang E (2013b) Vertebral compression fracture after spine stereotactic body radiotherapy: a multi-institutional analysis with a focus on radiation dose and the spinal instability neoplastic score. *Journal of Clinical Oncology* 31:3426–3431.

Sahgal A, Chang JH, Ma L, Marks LB, Milano MT, Medin P, Niemierko A, Soltys SG, Tome WA, Wong CS, Yorke E, Grimm J, Jackson A (2019) Spinal cord dose tolerance to stereotactic body radiation therapy. *International Journal of Radiation Oncology, Biology, Physics* S0360-3016(19)33862-3.

Sahgal A, Larson DA, Chang EL (2008) Stereotactic body radiosurgery for spinal metastases: a critical review. *International Journal of Radiation Oncology, Biology, Physics* 71:652–665.

Sahgal A, Ma L, Weinberg V, Gibbs IC, Chao S, Chang UK, Werner-Wasik M, Angelov L, Chang EL, Sohn MJ, Soltys SG, Letourneau D, Ryu S, Gerszten PC, Fowler J, Wong CS, Larson DA (2012) Reirradiation human spinal cord tolerance for stereotactic body radiotherapy. *International Journal of Radiation Oncology, Biology, Physics* 82:107–116.

Sahgal A, Weinberg V, Ma L, Chang E, Chao S, Muacevic A, Gorgulho A, Soltys S, Gerszten PC, Ryu S, Angelov L, Gibbs I, Wong CS, Larson DA (2013a) Probabilities of radiation myelopathy specific to stereotactic body radiation therapy to guide safe practice. *International Journal of Radiation Oncology, Biology, Physics* 85:341–347.

Sharma M, Meola A, Bellamkonda S, Jia X, Montgomery J, Chao ST, Suh JH, Angelov L, Barnett GH (2018) Long-term outcome following stereotactic radiosurgery for glomus jugulare tumors: a single institution experience of 20 years. *Neurosurgery* 83:1007–1014.Soderlund Diaz L, Hallqvist A (2020) LINAC-based stereotactic radiosurgery versus hypofractionated stereotactic radiotherapy delivered in 3 or 5 fractions for vestibular schwannomas: comparative assessment from a single institution. *Journal of Neuro-oncology* 147:351–359.

Song DY, Williams JA (1999) Fractionated stereotactic radiosurgery for treatment of acoustic neuromas. *Stereotactic and Functional Neurosurgery* 73:45–49.

Spiegelmann R, Cohen ZR, Nissim O, Alezra D, Pfeffer R (2010) Cavernous sinus meningiomas: a large LINAC radiosurgery series. *Journal of Neuro-oncology* 98:195–202.

Stafford SL, Pollock BE, Leavitt JA, Foote RL, Brown PD, Link MJ, Gorman DA, Schomberg PJ (2003) A study on the radiation tolerance of the optic nerves and chiasm after stereotactic radiosurgery. *International Journal of Radiation Oncology, Biology, Physics* 55:1177–1181.

Stephans KL, Djemil T, Diaconu C, Reddy CA, Xia P, Woody NM, Greskovich J, Makkar V, Videtic GM (2014) Esophageal dose tolerance to hypofractionated stereotactic body radiation therapy: risk factors for late toxicity. *International Journal of Radiation Oncology, Biology, Physics* 90:197–202.

Tishler RB, Loeffler JS, Lunsford LD, Duma C, Alexander E, 3rd, Kooy HM, Flickinger JC (1993) Tolerance of cranial nerves of the cavernous sinus to radiosurgery. *International Journal of Radiation Oncology, Biology, Physics* 27:215–221.

Truong MT, Nadgir RN, Hirsch AE, Subramaniam RM, Wang JW, Wu R, Khandekar M, Nawaz AO, Sakai (2010) Brachial plexus contouring with CT and MR imaging in radiation therapy planning for head and neck cancer. *Radiographics* 30(4):1095–1103.

Tsai JT, Lin JW, Lin CM, Chen YH, Ma HI, Jen YM, Ju DT (2013) Clinical evaluation of CyberKnife in the treatment of vestibular schwannomas. *BioMed Research International* 2013:297093.

Tseng CL, Sussman MS, Atenafu EG, Letourneau D, Ma L, Soliman H, Thibault I, Cho BC, Simeonov A, Yu E, Fehlings MG, Sahgal A (2015) Magnetic resonance imaging assessment of spinal cord and cauda equina motion in supine patients with spinal metastases planned for spine stereotactic body radiation therapy. *International Journal of Radiation Oncology, Biology, Physics* 91:995–1002.

Vivas EX, Wegner R, Conley G, Torok J, Heron DE, Kabolizadeh P, Burton S, Ozhasoglu C, Quinn A, Hirsch BE (2014) Treatment outcomes in patients treated with CyberKnife radiosurgery for vestibular schwannoma. *Otology & Neurotology* 35:162–170.

Wang H, Shiu A, Wang C, O'Daniel J, Mahajan A, Woo S, Liengsawangwong P, Mohan R, Chang EL (2008) Dosimetric effect of translational and rotational errors for patients undergoing image-guided stereotactic body radiotherapy for spinal metastases. *International Journal of Radiation Oncology, Biology, Physics* 71:1261–1271.

Wang X, Liu X, Mei G, Dai J, Pan L, Wang E (2012) Phase II study to assess the efficacy of hypofractionated stereotactic radiotherapy in patients with large cavernous sinus hemangiomas. *International Journal of Radiation Oncology, Biology, Physics* 83:e223–e230.

Wang X, Zhu H, Knisely J, Mei G, Liu X, Dai J, Mao Y, Pan L, Qin Z, Wang E (2018) Hypofractionated stereotactic radiosurgery: a new treatment strategy for giant cavernous sinus hemangiomas. *Journal of Neurosurgery* 128:60–67.

Witt TC (2003) Stereotactic radiosurgery for pituitary tumors. *Neurosurgery Focus* 14:e10.

22 MR-Guided Focused Ultrasound for Brain Tumors

Ying Meng, Christopher B. Pople, Suganth Suppiah, and Nir Lipsman

Contents

22.1 INTRODUCTION

Focused ultrasound (FUS) is an emerging approach for transcranial image-guided surgery, allowing the deposition of ultrasound energy onto discrete targets in the brain[1,2]. The range of ultrasound-induced biological effects relevant to neuro-oncology is broad, such as thermoablation, hyperthermia, blood–brain barrier (BBB) disruption, and histotripsy. Many such applications are showing promise in preclinical research and early clinical trials. Additionally, many of the putative mechanisms of FUS are likely to co-occur and may eventually be amenable to rational combined approaches targeting multiple aspects of cancer pathobiology. The mechanisms of FUS and the current state of preclinical and clinical research applicable to neuro-oncology will be discussed in this chapter.

22.2 FOCUSED ULTRASOUND

22.2.1 HIGH INTENSITY

Currently, the only Food and Drug Administration (FDA)-approved neurological applications of FUS are thermoablative for movement disorders in essential tremor (ET) and tremor-dominant Parkinson's disease. The approved device, ExAblate Neuro, utilizes a phased array of transducers operating at a frequency of 650 kHz.[1] High-intensity FUS achieves thermoablation by necrotic thermocoagulation of the tissue after heating to a temperature higher than 55°C, which occurs rapidly at approved clinical parameters.[2] FUS offers the benefit of a wide treatment envelope encompassing deep structures, and MRgFUS can achieve high accuracy with sub-millimeter precision by accounting for skull attenuation and intracranial obstacles to ultrasound paths using co-registered CT scans.[1,2]

High-intensity FUS is also being investigated for hyperthermia therapy involving prolonged sublesional temperatures for radiosensitization[3] and histotripsy for tissue ablation by mechanical forces.[4] In animal studies, mechanical forces induced by the effect of FUS on intravenously injected microbubbles can result in rapid liquification of the target tissue.[5] It is possible to precisely and rapidly ablate the target tissue without thermal injury to adjacent areas, improving the treatment envelope and treatment volume over thermoablation.[4,5] However, this approach cannot yet be readily achieved with the current clinical prototype.

22.2.2 LOW INTENSITY

In addition to thermoablation, a number of new applications for this technology are emerging. Low-intensity FUS is being investigated for blood–brain barrier (BBB) opening,[1] ultrasonic uncaging of therapeutics,[6] neuromodulation,[7] and liquid biopsy.[8] Additionally, a number of devices are currently being developed and investigated for ultrasound-mediated BBB opening. Such systems include the ExAblate Neuro 220 kHz (Insightec), NaviFUS, and SonoCloud (Carthera) devices. The former two systems are noninvasive, employing an external phased array of transducers with real-time and previous MRI for neuro-navigation, respectively. In contrast, SonoCloud is an implanted device, operating through a hole in the skull, which uses unfocused ultrasound and does not involve intraoperative image guidance. All of these devices disrupt the BBB by sonication after a microbubble-based ultrasound contrast agent is administered.

FUS BBB opening was first observed around the peripheries of high-intensity FUS lesions created by thermoablation.[9] Later research indicated that BBB opening was achievable in the absence of histological damage with low-intensity FUS.[10] It was determined that cavitation or the oscillation of gas bubbles was responsible for FUS BBB opening.[10] This led to the use of repurposing of existing intravenous (IV) microbubble ultrasound contrast agents as an artificial substrate for cavitation *in situ*, allowing more effective BBB opening.[2] In particular, the use of IV microbubbles allows BBB opening with sonication intensities as much as 100-fold lower.[11]

Investigations into the mechanisms of BBB opening identified two routes by which FUS can increase permeability: i) mechanical disruption of tight junctions between endothelial cells, ii) enhanced transcytosis across endothelial cells, possibly via caveolae (Figure 22.1).[12] The structural integrity and low permeability of the BBB appear to be restored after several hours (Figure 22.2).[12] Additionally, FUS BBB opening can also indirectly enhance drug delivery by a transient reduction in expression of efflux transporters such as P-glycoprotein, which has broad affinity for many chemotherapeutics.

Preliminary studies show FUS-induced BBB opening might also affect glymphatic function, such as with amyloid clearance in animal models[13] and release of CNS biomarkers for noninvasive liquid biopsy.[8,14] In combination with FUS, drug-loaded nanodroplets which are converted into microbubbles in an ultrasound-sensitive fashion can also be used for simultaneous, spatially localized drug uncaging with or without BBB opening.[6,15] This approach allows higher local concentrations of a therapeutic to be delivered to a target tissue with lower systemic exposure and potentially fewer side effects.

22.3 THERMOABLATION

Ablation of brain tumors with FUS is an area of active investigation,[16] and thermal ablation is common for many peripheral tumors.[17] MRgFUS has several attractive advantages for ablative procedures: (i) it is incisionless, having fewer risks than open surgery; (ii) it produces immediate lesions; (iii) it is image-guided, allowing intraoperative targeting; (iv) intraoperative MRI thermometry allows online target adjustment and treatment monitoring; and (v) treatment can be repeated indefinitely.[1,2] Additionally, radiation can contribute to tumoral immunosuppression, where FUS thermoablation may enhance tumor immunogenicity and induce an immune response.[18,19]

While the FDA has also approved multiple high-intensity FUS devices for thermal ablation body cancers such as prostate cancer,[20] brain and spinal tumors have remained technically challenging because of the skull's bony enclosure. High-intensity FUS utilizes a large number of ultrasound transducers to create controlled temperature elevations intracranially, with a small focal area of constructive interference.[2] Lesions created by MRgFUS are characterized by sharply demarcated coagulative necrosis.[2] However, the lesions are small and have not been successfully translated to brain tumor ablation.

Early trials using earlier iterations of the ExAblate device failed to achieve significant ablation as a result of device limitations on power.[21] Successful ablation has been partially performed in one centrally seated tumor, but systematic investigations, in larger clinical trials, have yet to be conducted.[22] Variation in location of intracranial tumors can affect the success of MRgFUS thermoablation, as centrally located targets reduce the angle of incidence for ultrasound paths, result in more usable transducers in the ultrasound array, and thus more effective energy is delivered to the target with less off-target skull heating.[2] In contrast, targets near the skull are more likely to lead to off-target heating due to ultrasound deflection from the angle of incidence, and fewer transducers are available, requiring more energy to delivered through a small area of skull bone. Emerging approaches intended to improve ablation of tumors are strictly investigational at the moment and will be discussed in detail later.

22.4 BBB OPENING FOR DRUG DELIVERY

The unique environment of the brain poses many challenges to therapeutic delivery, which are unique to cancers affecting the central nervous system (CNS). The BBB in particular is a major obstacle to the efficacy of currently available cancer therapeutics.[23] Capillaries in the brain have many unique features that pose significant challenges to drug delivery, including tight junctions between endothelial cells that limit paracellular diffusion of polar molecules and therapeutics larger than 400 Da, suppressed non-specific transcytosis, and drug efflux transporters which remove many chemotherapies from the brain.[12,24] FUS BBB disruption has been demonstrated to address all of these barriers by focally and reversibly disrupting tight junctions, enhancing transcytosis, and reducing expression of drug efflux transporters such as P-glycoprotein in preclinical models (Figure 21.1).[25–27] Additionally, many studies have demonstrated both enhanced therapeutic delivery and efficacy in animal models of brain tumors (Table 22.1).[12] This has prompted interest in employing FUS BBB opening for drug delivery in cancer such as glioblastoma multiforme (GBM), where prognosis is poor, and limited improvements have been made in recent years.[28]

Up to now clinical trials have employed FUS BBB opening for the delivery of common chemotherapies, primarily in patients with high-grade gliomas such as GBM. A trial using the transcranial phased

Table 22.1 Relative *in situ* increase in drug concentration after focused ultrasound blood–brain barrier disruption in a selection of animal studies. BPN, brain penetrating nanoparticles

THERAPEUTIC	FIRST AUTHOR, YEAR	APPROXIMATE RELATIVE CHANGE
NK-92 cells	Alkins, 2013[30]	2.3–10.5 ×
Interleukin-12	Chen, 2015[31]	2 ×
Carboplatin	Gorldwirt, 2016[32]	5.2 ×
Trastuzumab	Kinoshita, 2006[33]	2 ×
Carmustine	Liu, 2010[34]	2 ×
Temozolomide	Liu, 2014[35]	2.7 ×
Bevacizumab	Liu, 2016[36]	5.7 ×—56.7 ×
Liposomal paclitaxel	Shen, 2017[37]	2 ×
Cisplatin-loaded BPN	Timbie, 2017[38]	30 ×
Doxorubicin	Treat, 2007[39]	21 ×
Liposomal methotrexate	Wang, 2014[40]	9 ×
Adeno-associated virus	Wang, 2015[41]	6.5 ×
Cytarabine	Zeng, 2012[42]	4.4 ×

Source: Table adapted from Meng et al., 2018.[29]

array ExAblate device with Definity microbubbles for the delivery doxorubicin or temozolomide successfully demonstrated the safety and feasibility of this approach in five patients with high-grade gliomas (Figure 22.3).[43] In particular, no AEs of grade 3 or higher were reported, although surgical tissue collected 24 hours later did not contain high levels of temozolomide as assessed by mass spectrometry. In another trial, the safety and efficacy of the implanted SonoCloud device were demonstrated alongside SonoVue microbubbles in the delivery of carboplatin to 19 recurrent glioblastoma patients (Figure 22.4).[44] Grade 3 and 4 AEs were reported but deemed unrelated to the BBB opening procedure. In a comparison of patients with clear BBB opening to those with poor opening, progression-free survival was 4.1 versus 2.7 months. More recently, a trial of repeated BBB opening in six GBM patients successfully opened the BBB in 131 out of 145 trials in various locations across six treatment cycles.[45]

More clinical trials are currently recruiting or underway (NCT03744026, NCT03322813), hoping to treat larger numbers of patients with brain tumors including gliomas and metastases with the intent of enhancing delivery of cancer therapeutics. These trials aid in optimizing protocols for the management of neurooncological disease with FUS BBB opening and determining the safety and efficacy of this approach. Existing clinical trials have primarily tried to deliver small molecule therapeutics, but trials are underway to deliver larger therapeutics, for example, antibodies such as the HER2 inhibitor trastuzumab to intracranial HER2-positive breast cancer metastases (NCT03714243). The ability of FUS BBB opening to facilitate delivery of large therapeutics including biologics and cell-based therapies is of particular interest due to their very poor BBB permeability under normal circumstances and promising preclinical evidence.[12,46,47] If successful, this could improve the application of entire classes of therapeutics to the treatment of brain tumors.

Although most CNS applications of FUS have focused on the brain, efforts to extend FUS BBB opening to disruption of the blood–spinal cord barrier (BSCB) are well underway. Disruption of the BSCB with FUS and microbubbles has been demonstrated in rats, rabbits, and pigs.[48–51] The feasibility of this approach for drug delivery has also been demonstrated by focal transfection of spinal cord cells following intravenous infusion with an adenoviral vector.[52] More recently, reduced tumor growth was also observed in rats with leptomeningeal HER2+ tumors when FUS BSCB disruption was combined with trastuzumab.[53] Should ongoing FUS BBB opening research demonstrate benefit to patients with brain tumors, this research may facilitate rapid translation of these advances to patients with cancers affecting the spine.

22.4.1 DRUG CARRIER

The brain is served by a dense network of capillaries, with the brain parenchyma formed by a network of densely packed cells and extracellular matrix (ECM) that can impede drug mobility even after passage across the BBB. The ECM forms a charged mesh, with an approximate pore size of approximately 100 nm and a negative net charge, which filters particles by both physical and electrostatic interactions.[54–56] As such, particles which are charged and/or larger than the mesh size are greatly impeded by the ECM and may result in heterogenous drug delivery.[54,56] Modifications that neutralize surface charge, such as PEGylation, allow greater mobility for larger nanoparticles and may result in greater delivery and more homogenous distribution of drugs after delivery across the BBB.[54] However, these benefits must be balanced with impaired cellular interaction and uptake of PEGylated nanoparticles as well as the possibility that their cargo may be preferentially degraded.[57,58] Indeed, positively charged particles have a greater tendency to undergo uptake via electrostatic adsorption and trancytosis across the BBB.[59] Although non-specific transcytosis in this manner is believed to be upregulated by FUS BBB disruption and contribute to drug delivery, its relative importance in humans is yet to be determined.[60] Preclinical studies have nonetheless demonstrated that FUS BBB opening can enhance delivery and efficacy of PEGylated nanoparticles delivering intracellular gene therapy.[61] Furthermore, it has similarly been demonstrated that FUS hyperthermia can similarly enhance delivery and efficacy of PEGylated doxorubicin-containing liposomes in a mouse model of breast cancer brain metastases.[62] Thus, although the ECM can pose a significant barrier to therapeutic delivery to the brain even after

FUS BBB opening, rational design of drug delivery systems can overcome these challenges and may act synergistically with FUS to enhance efficacy.

22.5 OTHER APPLICATIONS OF MRGFUS

22.5.1 HYPERTHERMIA

An alternative approach to treating brain tumors by MRgFUS thermoablation or BBB opening is sublesional heating or hyperthermia therapy. This approach can enhance chemotherapy uptake and radiosensitivity, as well as tumor immunogenicity.[19,63] Hyperthermia therapy appears to inhibit a broad range of DNA repair and cell survival pathways and has radiosensitizing effects in a broad range of cancers.[64,65] Although clinical studies of ultrasound hyperthermia therapy have primarily been conducted in patients with systemic cancers,[3,66] one trial has been conducted in 1991, with GBM patients treated by sonications conducted through a cranial window.[67] Here, hyperthermia was feasible in most patients, but there was a substantial variation in the volume of tumor effectively treated due to technical limitations and tumor location. Modern FUS devices are noninvasive, with substantially improved treatment envelope, accuracy, precision, intraoperative monitoring, and control of ultrasound delivery.[2] Nonetheless, significant challenges remain in achieving sufficient temperature elevations in a large volume of tissue and treatment-limiting skull heating.

22.5.2 IMMUNOMODULATION

FUS BBB opening might also present a valuable tool for manipulating the immune microenvironment. In gliomas, the development of an immunosuppressive microenvironment is closely related to the progression to higher grades, increased tumor aggressiveness, and treatment resistance.[68] FUS BBB disruption in rodents induces an inflammatory response that includes features of endothelial cell activation including transient upregulation of cellular adhesion proteins that promote invasion by immune cells.[69,70] A number of further studies have found that FUS BBB opening can alter the tumor immune microenvironment in ways that oppose tumor immunosuppression, such as reducing the abundance immunosuppressive regulatory T cells and myeloid-derived suppressor cells, while also increasing the number of cytotoxic T cells and NK cells.[19] FUS BBB opening has also been used in preclinical studies for delivery of cytotoxic cell-based cancer therapies.[46]

22.5.3 LIQUID BIOPSY

A consequence of BBB disruption is the establishment of bidirectional exchange of substances in peripheral circulation and within the CNS. This introduces the possibility that FUS BBB opening could liberate biomarkers from the brain and thus facilitate the noninvasive liquid biopsy of otherwise unreachable CNS environments via the blood. Indeed, after FUS BBB opening in preclinical models, the concentrations of both xenograft tumor-derived nucleic acids[8] and CNS-derived proteins in peripheral circulation[14] are increased. Such findings are of particular interest, as successes seen in the sensitive detection of systemic cancers by liquid biopsy have not been effectively reproduced in cancers of the CNS.[71] The presence of the BBB has been proposed as a potential limiting factor in the systemic release of tumor-derived biomarkers, particularly nucleic acids on which the most successful diagnostics are based.[71] Liquid biopsy by FUS BBB opening also suggests the possibility of not only diagnostic applications, but also the combination of treatment with prognostic analyses, facilitating personalized medicine-based approaches.

22.5.4 SURGICAL MARKING

The use of FUS for marking tissues for later follow-up or treatment is an emerging concept. For example, dyes deposited by low-intensity FUS BBB opening could be used to demarcate areas for further surgical excision.[72,73] In extracranial studies, exploratory work using high-intensity MRgFUS to mark non-palpable tumors in *ex vivo* turkey breasts found superior surgical excision with a smaller area of tumor-positive margins compared to MR-guided needle-wire marking.[74] Interest is also growing in the combination of ultrasound and 5-ALA, a fluorescent photosensitizer preferentially taken up by many cancers and used to guide surgical resection, due to their combined cytotoxic effects.[75,76] Notably, an ongoing clinical trial is investigating the ability of FUS BBB opening immediately prior to surgery, to facilitate the marking of the tumor margins with 5-ALA in order to improve resection (NCT03322813). Although this approach has seen limited development, further progress could prove broadly applicable in improving surgical resection of tumors and also facilitate resection of progressive tumors after MRgFUS ablation.

22.5.5 HISTOTRIPSY

Significant limitations on high-intensity MRgFUS thermoablation of brain tumors remain, resulting from excessive skull heating during the treatment of large volumes and difficulties delivering sufficient energy to tumors. Histotripsy, or the rapid liquefaction of tissues by mechanical forces resulting from cavitation of gas bubbles, presents a possible solution to these challenges.[4,5] This approach has a number of benefits over traditional thermoablation, including lower energy requirements, higher accuracy, more sharp demarcation, more rapid lesion resolution, and potentially superior immunogenicity.[4,5,77,78] There is growing interest in using histrotripsy for both tissue ablation and clot dissolution.[78,79] FUS histotripsy ablation has been successfully accomplished in several animal models following craniectomy, without significant off-target effects.[78–80]

Furthermore, clot liquefication through *ex vivo* human skulls has also been successfully achieved.[81] Although sonication protocols for histotripsy have the ability to destroy unintended gas bubbles *in situ* prior to treatment, as well as the ability to generate them *de novo*,[5] the administration of exogenous microbubbles is of particular interest as a way to reduce the intensity threshold necessary for mechanical ablation in vivo.[82] Developments in FUS histotripsy have the potential to expand the treatment envelope of ablative FUS applications, particularly in the field of neuro-oncology where target volumes are often large and may appear in many anatomical locations currently difficult to target with conventional MRgFUS thermoablation.

22.6 CHALLENGES AND FUTURE DIRECTIONS

Despite the rapid progress of FUS therapies in recent years, clinical investigations in CNS tumors are still in their infancy with numerous trials and studies ongoing (Table 22.2). Significant barriers to effective clinical translation remain. Thermoablation of tumors with FDA-approved devices has seen limited success.[21,22] The treatment envelope of potential volume of treated tissue is limited by the absorption of ultrasonic energy at the skull bone and, until the treatment envelope is expanded, may limit the utility of this approach. Positive results in ongoing trials of ultrasound histotripsy devices for ablation of peripheral tumors do offer hope, however, for applying lesional FUS to intracranial tumors.[4] Additionally, FUS hyperthermia is beginning to enter clinical trials, with early reports suggesting the safety and feasibility of this approach in peripheral tumors.[3,66]

The future of FUS BBB opening depends critically on the ability to demonstrate improved efficacy over drug administration alone and practical challenges that can limit widespread adoption. This may depend on the properties of the drug or drug carrier administered, including its size, charge, and baseline BBB permeability. A handful of early clinical trials have been carried out in patients with oncological disorders, providing preliminary evidence that delivery of chemotherapeutics is safe and feasible in humans.[29–31,83] Pilot trials in neurodegenerative indications help to demonstrate similar safety and feasibility.[84,85] Further trials are planned or are underway for a range of neuro-oncological diseases including new or recurrent gliomas, as well as intracranial metastases (NCT03322813, NCT03744026, NCT03714243). The successful future development of these technologies may also open the door for further or complimentary applications, such as hyperthermia therapy, surgical marking, and liquid biopsy.

Despite these limitations, the rapidly expanding scope of indications and approaches utilizing ultrasound-based therapies suggests that the role of FUS in the neuro-oncology setting will continue to grow. Although many ultrasound therapies and their neuro-oncological applications are in the early stages, the range of potential applications and FUS mechanisms indicate an exciting future for this emerging technology.

Reproduced with permission from Mainprize et al., 2019[29] and Agessandro et al., 2019[86] under the Creative Commons License.

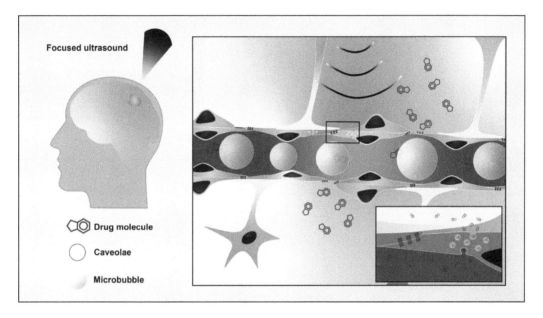

Figure 22.1 Focused ultrasound blood–brain barrier disruption. Mechanisms by which focused ultrasound in combination with intravascular microbubbles increase blood–brain barrier permeability, namely stable cavitation of microbubbles resulting in physical disruption of tight junctions between endothelial cells and increased transcytosis by caveolae in a magnified inset. Reproduced with permission from Meng et al., 2019.[12]

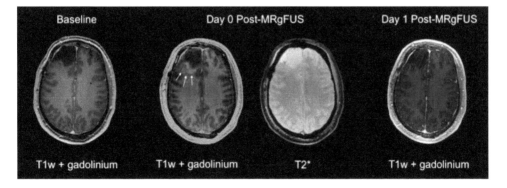

Figure 22.2 Transient disruption of the intact blood–brain barrier using an Exablate device and its subsequent resolution. T1-weighted contrast-enhanced and T2*-weighted MR images in a patient with a right frontal grade 3 anaplastic astrocytoma. From left to right: Intact blood–brain barrier at baseline (30 days pre-treatment), peritumoral blood–brain barrier disruption visualized by parenchymal contrast extravasation (immediately post-treatment), T2* image showing absence of microhemorrhages (immediately post-treatment), follow-up contrast-enhanced T1-weighted MR image showing resolution of BBB disruption (20 hours post-treatment). Reproduced with permission from Mainprize et al., 2019 under the Creative Commons License.[29]

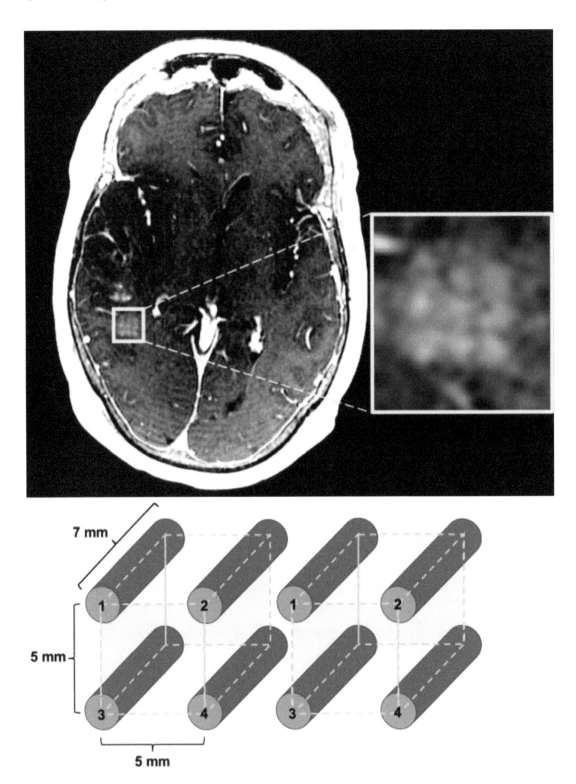

Figure 22.3 Top: Magnified view the gadolinium T1-weighted MRI of a patient shortly after MRgFUS induced BBB opening with the ExAblate device demonstrates contrast extravasation in a precise grid-like pattern of the sonication target. Bottom: A schematic diagram of the target volume showing the grid-like pattern in which the ultrasound targets are arranged by the ExAblate device.

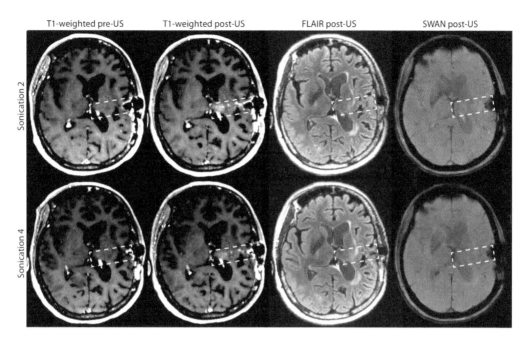

Figure 22.4 Disruption of the blood–brain barrier along the ultrasound path (observed within the white box) of an implanted Sonocloud device. Contrast-enhanced T1-weighted images show disruption of the blood–brain barrier without inflammation visible on FLAIR or bleeding visible on SWAN imaging post-treatment. Reproduced with permission from Carpentier et al., 2016.[69]

REFERENCES

1. Meng Y, Suppiah S, Mithani K, Solomon B, Schwartz ML, Lipsman N. Current and emerging brain applications of MR-guided focused ultrasound. *Journal of Therapeutic Ultrasound*. 2017;5:26. doi:10.1186/s40349-017-0105-z

2. Ghanouni P, Pauly KB, Elias WJ, et al. Transcranial MR-guided focused ultrasound: a review of the technology and neuro applications. *AJR American Journal of Roentgenology*. 2015;205(1):150–159. doi:10.2214/AJR.14.13632

3. Zhu L, Altman MB, Laszlo A, et al. Ultrasound hyperthermia technology for radiosensitization. *Ultrasound Med Biol*. 2019;45(5):1025–1043. doi:10.1016/j.ultrasmedbio.2018.12.007

4. Dubinsky TJ, Khokhlova TD, Khokhlova V, Schade GR. Histotripsy: the next generation of high-intensity focused ultrasound for focal prostate cancer therapy. *Journal of Ultrasound in Medicine*. 2020;39(6):1057–1067. doi:10.1002/jum.15191

5. Roberts WW. Development and translation of histotripsy: current status and future directions. *Curr Opin Urol*. 2014;24(1):104–110. doi:10.1097/MOU.0000000000000001

6. Lea-Banks H, O'Reilly MA, Hamani C, Hynynen K. Localized anesthesia of a specific brain region using ultrasound-responsive barbiturate nanodroplets. *Theranostics*. 2020;10(6):2849–2858. doi:10.7150/thno.41566

7. Blackmore J, Shrivastava S, Sallet J, Butler CR, Cleveland RO. Ultrasound neuromodulation: a review of results, mechanisms and safety. *Ultrasound in Medicine & Biology*. Published online May 2019:S0301562919300043. doi:10.1016/j.ultrasmedbio.2018.12.015

8. Zhu L, Cheng G, Ye D, et al. Focused ultrasound-enabled brain tumor liquid biopsy. *Sci Rep*. 2018;8(1):6553. doi:10.1038/s41598-018-24516-7

9. Patrick JT, Nolting MN, Goss SA, et al. Ultrasound and the blood–brain barrier. *Adv Exp Med Biol*. 1990;267:369–381. doi:10.1007/978-1-4684-5766-7_36

10. McDannold N, Vykhodtseva N, Raymond S, Jolesz FA, Hynynen K. MRI-guided targeted blood–brain barrier disruption with focused ultrasound: histological findings in rabbits. *Ultrasound in Medicine & Biology*. 2005;31(11):1527–1537. doi:10.1016/j.ultrasmedbio.2005.07.010

11. McDannold N, Arvanitis CD, Vykhodtseva N, Livingstone MS. Temporary disruption of the blood–brain barrier by use of ultrasound and microbubbles: safety and efficacy evaluation in rhesus macaques. *Cancer Res*. 2012;72(14):3652–3663. doi:10.1158/0008-5472.CAN-12-0128

12. Meng Y, Pople CB, Lea-Banks H, et al. Safety and efficacy of focused ultrasound induced blood–brain barrier opening, an integrative review of animal and human studies. *Journal of Controlled Release*. 2019;309:25–36. doi:10.1016/j.jconrel.2019.07.023

13. Leinenga G, Götz J. Scanning ultrasound removes amyloid-β and restores memory in an Alzheimer's disease mouse model. *Sci Transl Med*. 2015;7(278):278ra33–278ra33. doi:10.1126/scitranslmed.aaa2512

14. Pacia CP, Zhu L, Yang Y, et al. Feasibility and safety of focused ultrasound-enabled liquid biopsy in the brain of a porcine model. *Scientific Reports*. 2020;10(1):7449. doi:10.1038/s41598-020-64440-3

15. Chen CC, Sheeran PS, Wu S-Y, Olumolade OO, Dayton PA, Konofagou EE. Targeted drug delivery with focused ultrasound-induced blood–brain barrier opening using acoustically-activated nanodroplets. *Journal of Controlled Release*. 2013;172(3):795–804. doi:10.1016/j.jconrel.2013.09.025

16. Hersh DS, Kim AJ, Winkles JA, Eisenberg HM, Woodworth GF, Frenkel V. Emerging Applications of therapeutic ultrasound in neuro-oncology moving beyond tumor ablation. *Neurosurgery*. 2016;79(5):643–654. doi:10.1227/NEU.0000000000001399

17. Stransky-Heilkron N, Humbert-Mehier S, Allémann E. Combination treatments of tumors with thermoablation: principles and review of preclinical studies. *Journal of Drug Delivery Science and Technology*. 2012;22(5):435–446. doi:10.1016/S1773-2247(12)50070-5

18. Persa E, Balogh A, Sáfrány G, Lumniczky K. The effect of ionizing radiation on regulatory T cells in health and disease. *Cancer Letters*. 2015;368(2):252–261. doi:10.1016/j.canlet.2015.03.003

19. Cohen-Inbar O, Xu Z, Sheehan JP. Focused ultrasound-aided immunomodulation in glioblastoma multiforme: a therapeutic concept. *J Ther Ultrasound*. 2016;4. doi:10.1186/s40349-016-0046-y

20. Sundaram KM, Chang SS, Penson DF, Arora S. Therapeutic ultrasound and prostate cancer. *Semin Intervent Radiol*. 2017;34(2):187–200. doi:10.1055/s-0037-1602710

21. McDannold N, Clement GT, Black P, Jolesz F, Hynynen K. Transcranial magnetic resonance imaging-guided focused ultrasound surgery of brain tumors: initial findings in 3 patients. *Neurosurgery*. 2010;66(2):323–332; discussion 332. doi:10.1227/01.NEU.0000360379.95800.2F

22. Coluccia D, Fandino J, Schwyzer L, et al. First noninvasive thermal ablation of a brain tumor with MR-guided focused ultrasound. *J Ther Ultrasound*. 2014;2:17. doi:10.1186/2050-5736-2-17

23. Aldape K, Brindle KM, Chesler L, et al. Challenges to curing primary brain tumours. *Nat Rev Clin Oncol*. Published online February 7, 2019. doi:10.1038/s41571-019-0177-5

24. Andreone BJ, Chow BW, Tata A, et al. Blood–brain barrier permeability is regulated by lipid transport-dependent suppression of caveolae-mediated transcytosis. *Neuron*. 2017;94(3):581–594.e5. doi:10.1016/j.neuron.2017.03.043

25. Cho H, Lee H-Y, Han M, et al. Localized down-regulation of P-glycoprotein by focused ultrasound and microbubbles induced blood–brain barrier disruption in rat brain. *Scientific Reports*. 2016;6(1):31201. doi:10.1038/srep31201

26. Sheikov N, McDannold N, Sharma S, Hynynen K. Effect of focused ultrasound applied with an ultrasound contrast agent on the tight junctional integrity of the brain microvascular endothelium. *Ultrasound Med Biol*. 2008;34(7):1093–1104. doi:10.1016/j.ultrasmedbio.2007.12.015

27. Hynynen K, McDannold N, Vykhodtseva N, et al. Focal disruption of the blood–brain barrier due to 260-kHz ultrasound bursts: a method for molecular imaging and targeted drug delivery. *Journal of Neurosurgery*. 2006;105(3):445–454. doi:10.3171/jns.2006.105.3.445

28. Stupp R, Mason WP, van den Bent MJ, et al. Radiotherapy plus concomitant and adjuvant temozolomide for glioblastoma. *New England Journal of Medicine*. 2005;352(10):987–996. doi:10.1056/NEJMoa043330

29. Meng Y, Suppiah S, Surendrakumar S, Bigioni L, Lipsman N. Low-intensity MR-guided focused ultrasound mediated disruption of the blood–brain barrier for intracranial metastatic diseases. *Front Oncol*. 2018;8. doi:10.3389/fonc.2018.00338

30. Alkins R, Burgess A, Ganguly M, et al. Focused ultrasound delivers targeted immune cells to metastatic brain tumors. *Cancer Res*. 2013;73(6):1892–1899. doi:10.1158/0008-5472.CAN-12-2609

31. Chen P-Y, Hsieh H-Y, Huang C-Y, Lin C-Y, Wei K-C, Liu H-L. Focused ultrasound-induced blood–brain barrier opening to enhance interleukin-12 delivery for brain tumor immunotherapy: a preclinical feasibility study. *J Transl Med*. 2015;13. doi:10.1186/s12967-015-0451-y

32. Goldwirt L, Canney M, Horodyckid C, et al. Enhanced brain distribution of carboplatin in a primate model after blood–brain barrier disruption using an implantable ultrasound device. *Cancer Chemother Pharmacol*. 2016;77(1):211–216. doi:10.1007/s00280-015-2930-5

33. Kinoshita M, McDannold N, Jolesz FA, Hynynen K. Noninvasive localized delivery of Herceptin to the mouse brain by MRI-guided focused ultrasound-induced blood–brain barrier disruption. *Proc Natl Acad Sci USA*. 2006;103(31):11719–11723. doi:10.1073/pnas.0604318103

34. Liu H-L, Hua M-Y, Chen P-Y, et al. Blood–brain barrier disruption with focused ultrasound enhances delivery of chemotherapeutic drugs for glioblastoma treatment. *Radiology*. 2010;255(2):415–425. doi:10.1148/radiol.10090699

35. Liu H-L, Huang C-Y, Chen J-Y, Wang H-YJ, Chen P-Y, Wei K-C. Pharmacodynamic and therapeutic investigation of focused ultrasound-induced blood–brain barrier opening for enhanced temozolomide delivery in glioma treatment. *PLoS One.* 2014;9(12):e114311. doi:10.1371/journal.pone.0114311

36. Liu H-L, Hsu P-H, Lin C-Y, et al. Focused ultrasound enhances central nervous system delivery of bevacizumab for malignant glioma treatment. *Radiology.* 2016;281(1):99–108. doi:10.1148/radiol.2016152444

37. Shen Y, Pi Z, Yan F, et al. Enhanced delivery of paclitaxel liposomes using focused ultrasound with microbubbles for treating nude mice bearing intracranial glioblastoma xenografts. *Int J Nanomedicine.* 2017;12:5613–5629. doi:10.2147/IJN.S136401

38. Timbie KF, Afzal U, Date A, et al. MR image-guided delivery of cisplatin-loaded brain-penetrating nanoparticles to invasive glioma with focused ultrasound. *J Control Release.* 2017;263:120–131. doi:10.1016/j.jconrel.2017.03.017

39. Treat LH, McDannold N, Vykhodtseva N, Zhang Y, Tam K, Hynynen K. Targeted delivery of doxorubicin to the rat brain at therapeutic levels using MRI-guided focused ultrasound. *Int J Cancer.* 2007;121(4):901–907. doi:10.1002/ijc.22732

40. Wang X, Liu P, Yang W, et al. Microbubbles coupled to methotrexate-loaded liposomes for ultrasound-mediated delivery of methotrexate across the blood–brain barrier. *Int J Nanomedicine.* 2014;9:4899–4909. doi:10.2147/IJN.S69845

41. Wang S, Olumolade OO, Sun T, Samiotaki G, Konofagou EE. Noninvasive, neuron-specific gene therapy can be facilitated by focused ultrasound and recombinant adeno-associated virus. *Gene Therapy.* 2015;22(1):104–110. doi:10.1038/gt.2014.91

42. Zeng H-Q, Lü L, Wang F, Luo Y, Lou S-F. Focused ultrasound-induced blood–brain barrier disruption enhances the delivery of cytarabine to the rat brain. *Journal of Chemotherapy.* 2012;24(6):358–363. doi:10.1179/1973947812Y.0000000043

43. Mainprize T, Lipsman N, Huang Y, et al. Blood–brain barrier opening in primary brain tumors with non-invasive MR-guided focused ultrasound: A clinical safety and feasibility study. *Sci Rep.* 2019;9(1):321. doi:10.1038/s41598-018-36340-0

44. Idbaih A, Canney M, Belin L, et al. Safety and feasibility of repeated and transient blood–brain barrier disruption by pulsed ultrasound in patients with recurrent glioblastoma. *Clinical Cancer Research.* 2019;25(13):3793–3801. doi:10.1158/1078-0432.CCR-18-3643

45. Park SH, Kim MJ, Jung HH, et al. Safety and feasibility of multiple blood–brain barrier disruptions for the treatment of glioblastoma in patients undergoing standard adjuvant chemotherapy. *Journal of Neurosurgery.* 2020;1(aop):1–9. doi:10.3171/2019.10.JNS192206

46. Alkins R, Burgess A, Kerbel R, Wels WS, Hynynen K. Early treatment of HER2-amplified brain tumors with targeted NK-92 cells and focused ultrasound improves survival. *Neuro-Oncology.* 2016;18(7):974–981. doi:10.1093/neuonc/nov318

47. Park E-J, Zhang Y-Z, Vykhodtseva N, McDannold N. Ultrasound-mediated blood–brain/blood-tumor barrier disruption improves outcomes with trastuzumab in a breast cancer brain metastasis model. *J Control Release.* 2012;163(3):277–284. doi:10.1016/j.jconrel.2012.09.007

48. Wachsmuth J, Chopra R, Hynynen K. Feasibility of transient image-guided blood-spinal cord barrier disruption. *AIP Conference Proceedings.* 2009;1113(1):256–259. doi:10.1063/1.3131425

49. Montero A-S, Bielle F, Goldwirt L, et al. Ultrasound-induced blood–spinal cord barrier opening in rabbits. *Ultrasound in Medicine & Biology.* 2019;45(9):2417–2426. doi:10.1016/j.ultrasmedbio.2019.05.022

50. Fletcher S-MP, Choi M, Ogrodnik N, O'Reilly MA. A porcine model of transvertebral ultrasound and microbubble-mediated blood-spinal cord barrier opening. *Theranostics.* 2020;10(17):7758–7774. doi:10.7150/thno.46821

51. Payne AH, Hawryluk GW, Anzai Y, et al. Magnetic resonance imaging-guided focused ultrasound to increase localized blood-spinal cord barrier permeability. *Neural Regen Res.* 2017;12(12):2045–2049. doi:10.4103/1673-5374.221162

52. Weber-Adrian D, Thévenot E, O'Reilly MA, et al. Gene delivery to the spinal cord using MRI-guided focused ultrasound. *Gene Therapy.* 2015;22(7):568–577. doi:10.1038/gt.2015.25

53. O'Reilly MA, Chinnery T, Yee M-L, et al. Preliminary investigation of focused ultrasound-facilitated drug delivery for the treatment of leptomeningeal metastases. *Scientific Reports.* 2018;8(1):9013. doi:10.1038/s41598-018-27335-y

54. Nance EA, Woodworth GF, Sailor KA, et al. A dense poly(ethylene glycol) coating improves penetration of large polymeric nanoparticles within brain tissue. *Science Translational Medicine.* 2012;4(149):149ra119–149ra119. doi:10.1126/scitranslmed.3003594

55. Neeves KB, Sawyer AJ, Foley CP, Saltzman WM, Olbricht WL. Dilation and degradation of the brain extracellular matrix enhances penetration of infused polymer nanoparticles. *Brain Research.* 2007;1180:121–132. doi:10.1016/j.brainres.2007.08.050

56. Tomasetti L, Breunig M. Preventing obstructions of nanosized drug delivery systems by the extracellular matrix. *Advanced Healthcare Materials*. 2018;7(3):1700739. doi:10.1002/adhm.201700739

57. Mishra S, Webster P, Davis ME. PEGylation significantly affects cellular uptake and intracellular trafficking of non-viral gene delivery particles. *European Journal of Cell Biology*. 2004;83(3):97–111. doi:10.1078/0171-9335-00363

58. Remaut K, Lucas B, Braeckmans K, Demeester J, De Smedt SC. Pegylation of liposomes favours the endosomal degradation of the delivered phosphodiester oligonucleotides. *Journal of Controlled Release*. 2007;117(2):256–266. doi:10.1016/j.jconrel.2006.10.029

59. Georgieva JV, Hoekstra D, Zuhorn IS. Smuggling drugs into the brain: an overview of ligands targeting transcytosis for drug delivery across the blood–brain barrier. *Pharmaceutics*. 2014;6(4):557–583. doi:10.3390/pharmaceutics6040557

60. Burgess A, Nhan T, Moffatt C, Klibanov AL, Hynynen K. Analysis of focused ultrasound-induced blood–brain barrier permeability in a mouse model of Alzheimer's disease using two-photon microscopy. *Journal of Controlled Release*. 2014;192:243–248. doi:10.1016/j.jconrel.2014.07.051

61. Mead BP, Kim N, Miller GW, et al. Novel focused ultrasound gene therapy approach noninvasively restores dopaminergic neuron function in a rat Parkinson's disease model. *Nano Lett*. 2017;17(6):3533–3542. doi:10.1021/acs.nanolett.7b00616

62. Wu S-K, Chiang C-F, Hsu Y-H, et al. Short-time focused ultrasound hyperthermia enhances liposomal doxorubicin delivery and antitumor efficacy for brain metastasis of breast cancer. *Int J Nanomedicine*. 2014;9:4485–4494. doi:10.2147/IJN.S68347

63. Frey B, Weiss E-M, Rubner Y, et al. Old and new facts about hyperthermia-induced modulations of the immune system. *Int J Hyperthermia*. 2012;28(6):528–542. doi:10.3109/02656736.2012.677933

64. Oei AL, Vriend LEM, Crezee J, Franken NAP, Krawczyk PM. Effects of hyperthermia on DNA repair pathways: one treatment to inhibit them all. *Radiation Oncology*. 2015;10(1):165. doi:10.1186/s13014-015-0462-0

65. Man J, Shoemake JD, Ma T, et al. Hyperthermia sensitizes glioma stem-like cells to radiation by inhibiting AKT signaling. *Cancer Res*. 2015;75(8):1760–1769. doi:10.1158/0008-5472.CAN-14-3621

66. Gray MD, Lyon PC, Mannaris C, et al. Focused ultrasound hyperthermia for targeted drug release from thermosensitive liposomes: results from a phase I trial. *Radiology*. 2019;291(1):232–238. doi:10.1148/radiol.2018181445

67. Guthkelch AN, Carter LP, Cassady JR, et al. Treatment of malignant brain tumors with focused ultrasound hyperthermia and radiation: results of a phase I trial. *J Neurooncol*. 1991;10(3):271–284. doi:10.1007/bf00177540

68. Grabowski MM, Sankey EW, Ryan KJ, et al. Immune suppression in gliomas. *J Neurooncol*. Published online June 15, 2020. doi:10.1007/s11060-020-03483-y

69. Kovacs ZI, Kim S, Jikaria N, et al. Disrupting the blood–brain barrier by focused ultrasound induces sterile inflammation. *Proc Natl Acad Sci U S A*. 2017;114(1):E75–E84. doi:10.1073/pnas.1614777114

70. McMahon D, Bendayan R, Hynynen K. Acute effects of focused ultrasound-induced increases in blood–brain barrier permeability on rat microvascular transcriptome. *Scientific Reports*. 2017;7:45657.

71. Fontanilles M, Duran-Peña A, Idbaih A. Liquid biopsy in primary brain tumors: looking for stardust! *Current Neurology and Neuroscience Reports*. 2018;18(3):13. doi:10.1007/s11910-018-0820-z

72. Hipp E, Fan X, Partanen A, et al. Quantitative evaluation of internal marks made using MRgFUS as seen on MRI, CT, US, and digital color images—A pilot study. *Physica Medica*. 2014;30(8):941–946. doi:10.1016/j.ejmp.2014.04.007

73. Cox BF, Stewart F, Huang Z, Nathke IS, Cochran S. Ultrasound facilitated marking of gastrointestinal tissue with fluorescent material. In: *2016 IEEE International Ultrasonics Symposium (IUS)*; 2016:1–4. doi:10.1109/ULTSYM.2016.7728782

74. Schmitz AC, Bosch MAAJ van den, Rieke V, et al. 3.0-T MR-guided focused ultrasound for preoperative localization of nonpalpable breast lesions: an initial experimental *ex vivo* study. *Journal of Magnetic Resonance Imaging*. 2009;30(4):884–889. doi:10.1002/jmri.21896

75. Yoshida M, Kobayashi H, Terasaka S, et al. Sonodynamic therapy for malignant glioma using 220-kHz transcranial magnetic resonance imaging-guided focused ultrasound and 5-aminolevulinic acid. *Ultrasound in Medicine and Biology*. 2019;45(2):526–538. doi:10.1016/j.ultrasmedbio.2018.10.016

76. Wu S-K, Santos MA, Marcus SL, Hynynen K. MR-guided focused ultrasound facilitates sonodynamic therapy with 5-aminolevulinic acid in a rat glioma model. *Scientific Reports*. 2019;9(1):10465. doi:10.1038/s41598-019-46832-2

77. Schade GR, Wang Y-N, D'Andrea S, Hwang JH, Liles WC, Khokhlova TD. Boiling histotripsy ablation of renal cell carcinoma in the EKER rat promotes a systemic inflammatory response. *Ultrasound Med Biol*. 2019;45(1):137–147. doi:10.1016/j.ultrasmedbio.2018.09.006

78. Sukovich JR, Cain CA, Pandey AS, et al. In vivo histotripsy brain treatment. *Journal of Neurosurgery*. 2019;131(4):1331–1338. doi:10.3171/2018.4.JNS172652

79. Burgess A, Huang Y, Waspe AC, Ganguly M, Goertz DE, Hynynen K. High-intensity focused ultrasound (HIFU) for dissolution of clots in a rabbit model of embolic stroke. *PLoS One*. 2012;7(8). doi:10.1371/journal.pone.0042311

80. Gerhardson T, Sukovich JR, Chaudhary N, et al. Histotripsy clot liquefaction in a porcine intracerebral hemorrhage model. *Neurosurgery*. 2020;86(3):429–436. doi:10.1093/neuros/nyz089

81. Gerhardson T, Sukovich JR, Pandey AS, Hall TL, Cain CA, Xu Z. Effect of frequency and focal spacing on transcranial histotripsy clot liquefaction using electronic focal steering. *Ultrasound Med Biol*. 2017;43(10):2302–2317. doi:10.1016/j.ultrasmedbio.2017.06.010

82. Peng C, Sun T, Vykhodtseva N, et al. Intracranial non-thermal ablation mediated by transcranial focused ultrasound and phase-shift nanoemulsions. *Ultrasound in Medicine & Biology*. 2019;45(8):2104–2117. doi:10.1016/j.ultrasmedbio.2019.04.010

83. Carpentier A, Canney M, Vignot A, et al. Clinical trial of blood–brain barrier disruption by pulsed ultrasound. *Sci Transl Med*. 2016;8(343):343re2. doi:10.1126/scitranslmed.aaf6086

84. Lipsman N, Meng Y, Bethune AJ, et al. Blood–brain barrier opening in Alzheimer's disease using MR-guided focused ultrasound. *Nat Commun*. 2018;9(1):1–8. doi:10.1038/s41467-018-04529-6

85. Abrahao A, Meng Y, Llinas M, et al. First-in-human trial of blood–brain barrier opening in amyotrophic lateral sclerosis using MR-guided focused ultrasound. *Nature Communications*. 2019;10(1):4373. doi:10.1038/s41467-019-12426-9

86. Abrahao A, Meng Y, Llinas M, et al. First-in-human trial of blood–brain barrier opening in amyotrophic lateral sclerosis using MR-guided focused ultrasound. *Nat Commun*. 2019;10(1):1–9. doi:10.1038/s41467-019-12426-9

23 Laser Thermal Therapy for Brain Tumors

Josue Avecillas-Chasin and Gene H. Barnett

Contents

23.1 INTRODUCTION

Laser interstitial thermal therapy (LITT) is a minimally invasive therapy that has been used for intra-axial brain tumors including gliomas, metastases, and radiation necrosis (RN). This technology utilizes an optical fiber that emits photons that are absorbed by tissue chromophores, generating heat in the region of interest. Protein coagulation is produced when the tissue is exposed to the thermal energy above certain temperatures for a specific period of time (Thomsen, 1991). Therefore, this therapy is performed under imaging guidance in order to precisely locate the probe and ablate the area of interest under real-time magnetic resonance imaging (MRI) thermography. This MRI sequence provides the ability to approximate the extent of the lesion by measuring the change in temperature over time. This is important because the rapid increases in temperature could cause undesired effects including vaporization or carbonization. These effects may cause an increase in intracranial pressure and changes in the optical properties of the tissue preventing proper laser penetration. Other detrimental effects include an increase in thermal absorption, thermal damage to the device, and explosive tissue destruction (popcorn effect) (Kuroda, 2005; Schroeder *et al.*, 2014). One LITT system provides a flexible surgical approach to cytoreduce tumors using a guided side-firing probe, which can also be used for tumors with irregular shapes. LITT is particularly useful for tumors with difficult accessibility for open surgical approaches. The minimally invasive nature of this procedure makes it also suitable for patients with poor performance status, otherwise not eligible for open surgical approaches. Several reports have demonstrated safety and acceptable tumor control rates using LITT for intra-axial tumors. In this chapter, we will summarize key aspects of this approach and the outcomes in the setting of brain tumors.

23.2 TREATMENT

A preoperative volumetric post-gadolinium MRI is performed. In cases of non-enhancing tumors, a T2 or T2-FLAIR MRI volumetric sequences are also performed. Furthermore, functional MRI or tractography is often used in cases of tumors near eloquent areas. A navigation system is used to plan the trajectory of the laser probe and to visualize functional areas and white matter tracts of interest. The trajectory should be planned in a way that the laser probe is located in a position to maximize the ablation and minimize the damage to eloquent structures close to the tumor. Also, general principles of stereotaxis are applied including avoiding sulci, vessels, and ventricles (Figure 23.1). At our institution, we use an IMRIS (Winnipeg, Canada) intra-operative MRI system incorporating a 1.5 Tesla Siemens (Erlangen, Germany) magnet. The patient's head is secured in the desired position with the MRI-compatible immobilization system. Several fiducials are applied to the skull, and intraoperative MRI is acquired for the registration to the navigation system (or, in some cases, skull fiducials are applied after the patient is anesthetized and a volumetric CT scan obtained prior to the patient being taken to the operating room). Once the patient's head is registered to the preoperative plan, the entry point and the trajectory are defined, and the bur-hole or twist drill is aligned to the planned trajectory. In some cases, a tissue biopsy is performed along the same trajectory for diagnostic confirmation and molecular analyses. Depending on the number of trajectories, the stereotactic bolt(s) is/are placed in the bur-hole(s).

The NeuroBlate® System (Monteris Medical, Inc., Plymouth, MN) includes a robotically driven laser fiber and FUSION-S™ software for planning and treatment control. The NeuroBlate System uses a pulsed diode laser in the Nd: YAG range (1046 nm) with an output of 12W. This system uses gaseous CO_2 flowing through the tip of the fiber-optic probe as a cooling mechanism. This minimizes both vaporization of tissue at the probe–tissue interface and the adherence of the probe to the surrounding tissue, thus maximizing the volume of tissue treated. Gas cooling also improves the directionality of thermal energy for the side firing probe thereby allowing more conformal treatment. Thermal energy can also be delivered in a cylindrical or ellipsoid shape along the long axis of the fiber optic probe (Mohammadi and Schroeder, 2014). Once the laser probe is in place, the intraoperative MRI scan confirms the probe position. The intraoperative MRI

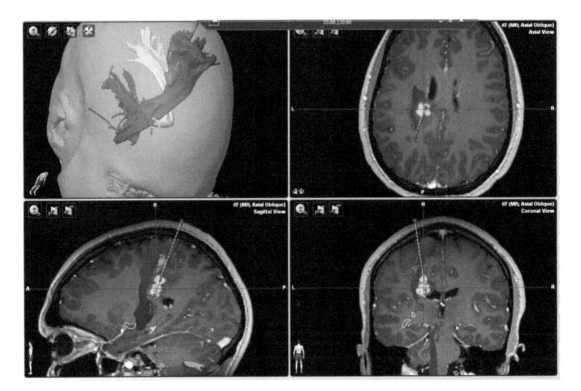

Figure 23.1 Navigation system view that indicates the preoperative planning for LITT treatment for a tumor near the corticospinal tract. The representations of the corticospinal tract (blue) and the optic radiations (yellow) are shown in relation to the tumor (red contour) The trajectory should be planned to avoid eloquent areas and maximize the ablation area.

is integrated with the FUSION-S software to contour the tumor and calculate the temperature of the surrounding brain areas in real time (Figure 23.2).

MR thermography provides real-time extent of the thermal energy delivered to the target tissue in the form of thermal damage threshold (TDT) lines that are monitored on the FUSION-S software during the procedure. A yellow line indicates the equivalent temperature exposures at 43°C for 2 minutes, a blue line at 43°C for 10 minutes, and a white line at 43°C for 60 minutes (Figure 23.3). The thermal properties

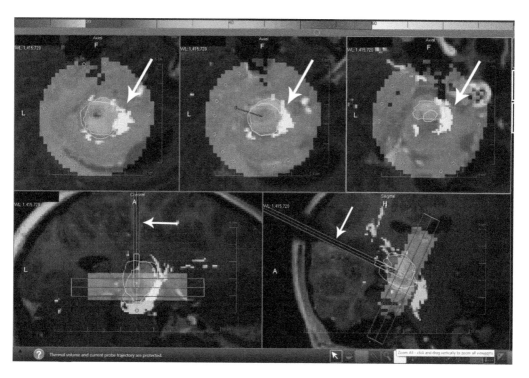

Figure 23.2 LITT workstation view with the tumor being ablated within the TDT lines. Eloquent white matter tracts can be included in the plan for more precise preservation (arrows in upper panels). The trajectory is optimized to maximize the ablation of the tumor (arrows in the lower panels).

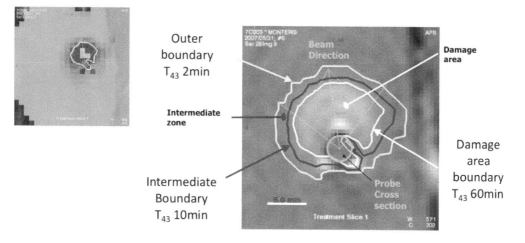

Figure 23.3 Illustration of the relation of the TDT lines with lesion being ablated. The yellow line indicates the equivalent temperature exposure at 43°C for 2 minutes (T_{43} 2min), the blue line at 43°C for 10 minutes (T_{43} 10min), and the white line at 43°C for 60 minutes (T_{43} 60min).

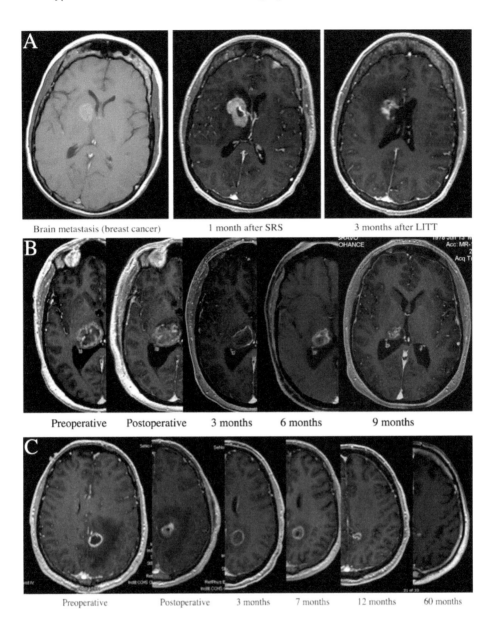

Figure 23.4 Imaging and clinical outcomes of patients treated with LITT. (A) Patient with metastatic breast cancer treated with stereotactic radiosurgery with tumor progression 1 month after treatment. The patient was treated with LITT and 3 months after treatment, there was a significant shrinkage of the tumor mass. (B) Patient with thalamic glioblastoma (GBM) treated with LITT; the T1 MRI shows a hyperintense appearance in the ablated area a few days after surgery. Three months after treatment, the images show hypointensity in the ablated area and a rim of contrast enhancement that 6 months after surgery decreased concentrically. (C) Patient with newly diagnosed posterior cingulate glioblastoma treated with LITT showed a complete response at 60 months after treatment with some residual enhancement.

of tissue being ablated and the temperature/energy output of the laser source determine the time required for the ablation, higher temperatures/energy output requiring shorter treatment times to ablate the same amount of tissue. The extent of tumor tissue damage by LITT is related to both the temperature to which the tissue is heated and the time for which the tissue is exposed to that temperature. For instance, with temperatures in the range between 50 to 80°C for less than 10-minute tumor, necrosis through protein denaturation is found in the tissue (Mohammadi and Schroeder, 2014).

23.3 IMAGING

The MRI changes after the procedure correlate with the histological effects of the ablation and will depend on the time of the MRI after the surgery. The area surrounding the LITT probe can be exposed to temperatures as high as 60°C. This results in coagulation necrosis, which contains damaged cell membranes and positive stains for markers of apoptosis in the core of the ablated area. In the next outer area, there is still permanent damage of the tumor tissue with thrombosed vessels and distended cell bodies, and it eventually undergoes liquefactive necrosis. In the end, the outermost area represents viable brain tissue with edema. In general, the areas of central liquefactive necrosis and interstitial edema are both associated with irreversible cell damage and would increase in size during the first 2 weeks after LITT. The appearance of the central zone on postoperative MRI consists of T1 hyperintensity and T2 hypointensity. This correlates with the damage of cellular membranes of the tumor tissue, hemoglobin degradation, and leakage from red blood cells. On the other hand, the peripheral zone demonstrates the typical appearance of vasogenic edema with T1 hypointense and T2 hyperintense areas. This area also correlates with necrotizing edema and is associated with blood–brain barrier (BBB) breakdown and therefore a rim of contrast enhancement. Immediately outside of these two areas, there is perilesional edema that contains viable nervous tissue. This perilesional edema usually resolves after 2 weeks to 2 months. Over time, the central core of the ablated area will be replaced with granulation tissue with subsequent lesion shrinkage. This is correlated with the T1 hyperintense area beginning to decrease in size and the T1 peripheral hypointense area increasing concentrically. Eventually, the ring-enhancement at the periphery will decrease with an 'eggshell' of residual enhancement that can be present years after the ablation (Mohammadi and Schroeder, 2014; Avecillas-Chasin *et al.*, 2020).

23.4 OUTCOMES

Several reports have demonstrated the utility of LITT in the treatment of newly diagnosed and recurrent gliomas. LITT offers a cytoreductive option for patients with recurrent HGG, who are not considered for open surgery either because of the performance status or co-morbidities. Furthermore, LITT provides additional advantages over open surgery including less pain, shorter hospital stay, and a smaller skin incision which makes early adjuvant treatment possible. Sloan et al. published the first human phase I study that used escalating doses of hyperthermia to assess the safety and efficacy of the procedure in patients with recurrent high-grade gliomas (Sloan *et al.*, 2013). The median overall survival (OS) was 10.5 months after LITT which was increased compared to historic controls by 3 to 9 months (Barker *et al.*, 1998; Friedman *et al.*, 2009). This study demonstrated that LITT was a feasible and safe treatment modality for recurrent HGG.

The standard treatment for patients with newly diagnosed HGG is maximal safe resection followed by chemoradiation. In some cases, this standard strategy cannot be carried out when eloquent or difficult-to-access areas are involved by the tumor because of an unacceptable risk of morbidity (Figure 23.4). In those cases, the standard of care includes biopsy followed by chemoradiation, and this is associated with a disadvantage in regards of tumor cytoreduction. Mohammadi et al. demonstrated an improved disease-specific OS and progression-free survival (PFS) in patients treated with upfront LITT followed by chemoradiation compared with a propensity score-matched control group based on age, gender, tumor location, and tumor volume. This control group, from other institutions not using LITT for this patient population, was treated with biopsy-only followed by chemoradiation. These authors also demonstrated that the extent of ablation is an independent predictor of disease-specific OS and PFS (Mohammadi *et al.*, 2019).

A recent metanalysis reported the use of LITT for newly diagnosed and recurrent high-grade gliomas (Ivan *et al.*, 2016). The results were similar to previous findings reported in the literature demonstrating benefit of LITT in terms of OS, PFS, as long as more than 95% of the tumor was ablated. LITT seems to be a reasonable option for patients with deep, inaccessible, or eloquent regions tumors. With this technique, these kind of tumors can be cytoreduced with minimal brain manipulation and with a rate of complications comparable to open surgery for accessible HGG (Mohammadi *et al.*, 2014; Barnett *et al.*, 2016). Similar to surgical series, the tumor should be ablated by at least 78% to 80% in order to achieve a meaningful survival benefit. Beaumont et al. demonstrated that treatment with LITT in patients with HGG in the corpus callosum has comparable median OS of 7 months when compared with surgical series with approximate 65% extent of resection. In this study, patients with larger tumors (>15 cm^3) were six times more likely to experience a complication, and based on this, the authors recommended a volume ceiling of 15 cm³ for LITT (Beaumont *et al.*, 2018). As previously demonstrated by others, this study also highlights the importance of tumor coverage for maximizing the survival benefit. This pooled evidence suggests that LITT is an effective treatment modality for newly diagnosed and recurrent HGG and that near-complete coverage of the tumor is as important as gross total resection in open surgery.

Currently, the management of radiation necrosis (RN) aims to either decrease the cerebral edema or surgically remove the area of necrosis (Vellayappan *et al.*, 2018). Bevacizumab has shown to be beneficial in the treatment of RN (Levin *et al.*, 2011). Rao et al. studied the role of LITT in brain metastasis/RN and included 16 patients with both recurrent metastatic tumors and RN. These authors reported a local control of 75.8%, median PFS of 37 weeks, and OS of 57% with a median follow-up of 24 weeks. The authors concluded that MRI-guided LITT is a well-tolerated procedure and may be effective in treating tumor recurrence/RN (Rao *et al.*, 2014). On the other hand, although stereotactic radiosurgery is an effective treatment in the management of brain metastases, there is a failure rate of approximately 15% (Flickinger *et al.*, 1994). Therefore, LITT therapy could be an option for metastatic lesions that progress after SRS. Banerjee et al. reviewed all the published studies on the utility of LITT in patients with recurrent metastasis and reported a median OS of 9.0–19.8 months with a PFS between 3.8 and 8.5 months and a complication rate of 12% to 16.7% (Banerjee *et al.*, 2015). These results suggest that LITT may be an effective treatment option in otherwise resistant metastatic lesions. A recently published multicenter clinical trial included 42 patients with radiographic progression after stereotactic radiosurgery treated with LITT for salvage treatment. This patient group included 19 biopsy-proven RN, 20 with recurrent tumor, and 3 with no diagnosis. The primary outcome was local PFS, and secondary outcomes were OS, LITT safety, neurocognitive function, and quality of life. LITT treatment in these patients was associated with PFS and OS rates of 74% and 72%, respectively, at 26 weeks. The results also showed stabilization of the KPS score, neurocognition, and quality of life over the survival time. Moreover, some patients were able to stop or reduce the steroid burden after surgery, and there was a 12% neurological complication rate in this group of patients. There was some advantage of the RN over recurrent tumor or tumor progression patients in terms of OS, which indicates that in cases of recurrence/progression patients would still need some additional treatment (Ahluwalia *et al.*, 2019).

23.5 COMPLICATIONS

LITT is a minimally invasive procedure with a lower complication rate compared to open surgery (Salem *et al.*, 2019). The most common complication for LITT is neurological deficit ranging from 0 to 29.4% for transient and 0 to 10% for permanent deficits. These are related to direct white matter damage from hyperthermia leading to permanent deficits and displacement or edema of white matter tracts leading to temporary deficits. Bleeding complications can also occur within the area of the tumor or in the trajectory. Refractory brain edema has also been associated with laser ablation of larger tumors (Salem *et al.*, 2019). Recent observations by our group suggest that these large lesions may need an immediate surgical debulking procedure (Avecillas-Chasin *et al.*, 2020). Another rare complication is pseudoaneurysm formation with rupture, and this seems to be associated with thermal damage to large- or mid-size cerebral arteries (Hawasli *et al.*, 2013). Careful preoperative planning using MRI angiography or tractography may increase the safety of the procedure. Other minor complications including infection or wound issues are less frequent than open surgical approaches because of the advantage of smaller skin.

23.6 INDICATIONS

- Newly diagnosed high-grade gliomas (HGG): Small, deep seated brain tumors (including butterfly gliomas) in which open surgery may have high complication risk, patient preference.
- Recurrent HGG: Small or nodular recurrence. For larger recurrences, LITT might have advantage over open surgery due to the minimally invasive nature and small size of incision in an irradiated scalp.
- If a total or near-total ablation is feasible with one or two trajectories regardless of newly diagnosed or recurrent HGG. Partial ablation seems to have limited benefit in terms of OS and PFS.

- For larger tumors, LITT may need to be combined with an immediate surgical debulking. However, although surgery is facilitated by the avascular nature of the tissue and may be performed through a minimal access craniotomy, this approach does lead to a longer surgical time for these combined procedures.
- Radiosurgery-resistant metastasis.
- Medically resistant RN, as a second line treatment option. Usually smaller than 40–50 cc in size.

23.7 FUTURE DIRECTIONS

In order to maximize the extent of ablation and minimize the thermal damage of eloquent brain tissue, it would be useful to have more planes of thermal monitoring than are currently available. Also, as there are several tissue factors that will influence the extent of ablation such as histological type of the tumor, blood vessels, or degree of malignancy, a preoperative means of predicting optimal probe placement and expected pattern of ablation should improve patient selection and outcomes. Furthermore, improvements in MRI coil design and developments in laser probe design would not only improve the flexibility of the treatment but may also expand the possibility to treat lesions with more complex shapes and locations. The use of tractography and fMRI are useful adjuncts to the LITT procedure to avoid damage of eloquent structures. Sharma et al. studied the extent of involvement of the corticospinal tract by the TDT lines used by the NeuroBlate System (Sharma *et al.*, 2016) and showed that even minimal involvement of the CST by the ablation area can cause motor deficit after ablation. Resting-state fMRI may provide more accurate delineation of eloquent cortex allowing this procedure to be done in or near speech areas.

Recent evidence suggests that the disruption of the blood–brain barrier (BBB) caused by the laser ablation can extend 1–2 cm outside of the tumor limits and may persist for up to 4 weeks after LITT. The authors state that this effect may enhance the delivery of chemotherapeutic agents that would be otherwise be hampered by the BBB (Leuthardt *et al.*, 2016). It has also been suggested that hyperthermia may sensitize tumor tissue to radiation therapy and therefore enhance the cytotoxic effects of radiation on the tumor tissue. The synergistic effect of laser ablation and radiation was found to impair the mechanisms of cell repair that translated into a consistent reduction in tumor size (Man *et al.*, 2015). Furthermore, clinical evidence has also shown the benefit of simultaneous hyperthermia and radiotherapy in terms of OS for patients with HGG (Sneed *et al.*, 1998; Rodriguez and Tatter, 2016).

CHECKLIST: KEY POINTS FOR CLINICAL PRACTICE

ACTIVITY	CONSIDERATIONS
Indications and patient factors	• Newly diagnosed and recurrent HGG: small, deep-seated, eloquent brain regions, not amenable for open surgery • Radiosurgery-resistant brain metastases • Medically refractory radiation necrosis • Previous scalp irradiation • High surgical risk
Tumor	• Need of tissue diagnosis or molecular analyses • Possibility to ablate ≥80% of the tumor. • Tumor size and shape: number of trajectories, additional surgical debulking, postoperative edema • Tumor located in or around eloquent brain areas, e.g., precentral gyrus, posterior limb of the internal capsule, optic radiations, etc. Need for additional imaging, e.g., fMRI, DTI • Proximity to the ventricular system. If invading ventricular wall, risk of CSF dissemination • Critical vascular structures, e.g., middle cerebral artery, anterior cerebral artery

(Continued)

ACTIVITY	CONSIDERATIONS
Planning	• Tumor contour and trajectory planning in the navigation system • Stereotaxis principles: if possible, avoid passing through eloquent brain areas, vessels, sulci, and ventricles, and minimize CSF egress • Intraoperative MRI with patient in the desired surgical position with fiducials applied • Intraoperative MRI registration with preoperative plan • Bur-hole or twist-drill aligned to the planned trajectory
Treatment delivery	• Thermal damage threshold (TDT) lines: Yellow line equivalent to temperature exposure at 43°C for at least 2 minutes, blue line at 43°C for at least 10 minutes, and white line at 43°C for 60 minutes • Monitor the temperature in the surrounding brain parenchyma

REFERENCES

Ahluwalia M, Barnett GH, Deng D, Tatter SB, Laxton AW, Mohammadi AM, et al. Laser Ablation After Stereotactic Radiosurgery: A Multicenter Prospective Study in Patients with Metastatic Brain Tumors and Radiation Necrosis. *Journal of Neurosurgery* 2019; 4;130(3):804–811.

Avecillas-Chasin JM, Atik A, Mohammadi AM, Barnett GH. Laser Thermal Therapy in the Management of High-Grade Gliomas. *International Journal of Hyperthermia* 2020; 37: 44–52.

Banerjee C, Snelling B, Berger MH, Shah A, Ivan ME, Komotar RJ. The Role of Magnetic Resonance-Guided Laser Ablation in Neurooncology. *British Journal of Neurosurgery* 2015; 29: 192–196.

Barker FG, Chang SM, Gutin PH, Malec MK, McDermott MW, Prados MD, et al. Survival and Functional Status after Resection of Recurrent Glioblastoma Multiforme. *Neurosurgery* 1998; 42: 709–719.

Barnett GH, Voigt JD, Alhuwalia MS. A Systematic Review and Meta-Analysis of Studies Examining the Use of Brain Laser Interstitial Thermal Therapy Versus Craniotomy for the Treatment of High-Grade Tumors in or near Areas of Eloquence: An Examination of the Extent of Resection and Major Comp. *Stereotact Funct Neurosurg* 2016; 94: 164–673.

Beaumont TL, Mohammadi AM, Kim AH, Barnett GH, Leuthardt EC. Magnetic Resonance Imaging-Guided Laser Interstitial Thermal Therapy for Glioblastoma of the Corpus Callosum. *Neurosurgery* 2018; 83: 556–565.

Flickinger JC, Kondziolka D, Dade Lunsford L, Coffey RJ, Goodman ML, Shaw EG, et al. A Multi-Institutional Experience with Stereotactic Radiosurgery for Solitary Brain Metastasis. *International Journal of Radiation Oncology* 1994; 28: 797–802.

Friedman HS, Prados MD, Wen PY, Mikkelsen T, Schiff D, Abrey LE, et al. Bevacizumab Alone and in Combination With Irinotecan in Recurrent Glioblastoma. *Journal of Clinical Oncology* 2009; 27: 4733–4740.

Hawasli AH, Bagade S, Shimony JS, Miller-Thomas M, Leuthardt EC. Magnetic Resonance Imaging-Guided Focused Laser Interstitial Thermal Therapy for Intracranial Lesions. *Neurosurgery* 2013; 73: 1007–1017.

Ivan ME, Mohammadi AM, De Deugd N, Reyes J, Rodriguez G, Shah A, et al. Laser Ablation of Newly Diagnosed Malignant Gliomas. *Neurosurgery* 2016; 79: S17–S23.

Kuroda K. Non-invasive MR Thermography Using the Water Proton Chemical Shift. *International Journal of Hyperthermia* 2005; 21: 547–560.

Leuthardt EC, Duan C, Kim MJ, Campian JL, Kim AH, Miller-Thomas MM, et al. Hyperthermic Laser Ablation of Recurrent Glioblastoma Leads to Temporary Disruption of the Peritumoral Blood–Brain Barrier. *PLoS One* 2016; 11: e0148613.

Levin VA, Bidaut L, Hou P, Kumar AJ, Wefel JS, Bekele BN, et al. Randomized Double-Blind Placebo-Controlled Trial of Bevacizumab Therapy for Radiation Necrosis of the Central Nervous System. *International Journal of Radiation Oncology, Biology, Physics* 2011; 79(5):1487–1495.

Man J, Shoemake JD, Ma T, Rizzo AE, Godley AR, Wu Q, et al. Hyperthermia Sensitizes Glioma Stem-Like Cells to Radiation by Inhibiting AKT Signaling. *Cancer Research* 2015; 75: 1760–1769.

Mohammadi AM, Hawasli AH, Rodriguez A, Schroeder JL, Laxton AW, Elson P, et al. The Role of Laser Interstitial Thermal Therapy in Enhancing Progression-Free Survival of Difficult-to-Access High-Grade Gliomas: A Multicenter Study. *Cancer Medicine* 2014; 3: 971–979.

Mohammadi AM, Schroeder JL. Laser Interstitial Thermal Therapy in Treatment of Brain Tumors—the Neuroblate System. *Expert Review of Medical Devices* 2014; 11(2):109–119.

Mohammadi AM, Sharma M, Beaumont TL, Juarez KO, Kemeny H, Dechant C, et al. Upfront Magnetic Resonance Imaging-Guided Stereotactic Laser-Ablation in Newly Diagnosed Glioblastoma: A Multicenter Review of Survival Outcomes Compared to a Matched Cohort of Biopsy-Only Patients. *Neurosurgery* 2019; 85: 762–772.

Rao MS, Hargreaves EL, Khan AJ, Haffty BG, Danish SF. Magnetic Resonance-Guided Laser Ablation Improves Local Control for Postradiosurgery Recurrence and/or Radiation Necrosis. *Neurosurgery* 2014; 74: 658–667.

Rodriguez A, Tatter SB. Laser Ablation of Recurrent Malignant Gliomas. *Neurosurgery* 2016; 79: S35–S39.

Salem U, Kumar VA, Madewell JE, Schomer DF, De Almeida Bastos DC, Zinn PO, et al. Neurosurgical Applications of MRI Guided Laser Interstitial Thermal Therapy (LITT). *Cancer Imaging* 2019; 19(1): Article 65: 1–13.

Schroeder JL, Missios S, Barnett GH, Mohammadi AM. Laser Interstitial Thermal Therapy as a Novel Treatment Modality for Brain Tumors in the Thalamus And Basal Ganglia. *Photonics and Lasers in Medicine* 2014; 3: 151–158.

Sharma M, Habboub G, Behbahani M, Silva D, Barnett GH, Mohammadi AM. Thermal Injury to Corticospinal Tracts and Postoperative Motor Deficits After Laser Interstitial Thermal Therapy. *Neurosurgery Focus* 2016; 41: E6.

Sloan AE, Ahluwalia MS, Valerio-Pascua J, Manjila S, Torchia MG, Jones SE, et al. Results of the Neuroblate System First-in-Humans Phase i Clinical Trial for Recurrent Glioblastoma. *Journal of Neurosurgery* 2013; 118: 1202–1219.

Sneed PK, Stauffer PR, McDermott MW, Diederich CJ, Lamborn KR, Prados MD, et al. Survival Benefit of Hyperthermia in a Prospective Randomized Trial of Brachytherapy Boost ± Hyperthermia for Glioblastoma Multiforme. *International Journal of Radiation Oncology* 1998; 40: 287–295.

Thomsen S. Pathologic Analysis of Photothermal and Photomechanical Effects of Laser-Tissue Interactions. *Photochem Photobiology* 1991; 53: 825–835.

Vellayappan B, Tan CL, Yong C, Khor LK, Koh WY, Yeo TT, et al. Diagnosis and Management of Radiation Necrosis in Patients with Brain Metastases. *Frontiers in Oncology* 2018; 8: Article 395: 1–9.

Updates on Laser Interstitial Therapy for Metastatic Tumors of the Spine

Christopher Alvarez-Breckenridge, Melissa Lo Presti, and Claudio Tatsui

Contents

24.1 INTRODUCTION

Tumor seeding along the spine is a common manifestation of metastatic disease, occurring in 40% of patients, and 10% of these patients ultimately present with symptomatic spinal cord compression.[1–4] Advancements in systemic treatment and radiation delivery methods resulted in improvements in local control and overall survival for many histology types. However, metastatic involvement of the spine still represents a significant source of morbidity, pain, and functional impairment for these patients.[5]

Management of epidural spinal cord compression has evolved in recent years. A randomized trial by Patchell et al.[6] compared circumferential decompression and stabilization followed by conventional external beam radiation therapy (cEBRT) to cEBRT alone in the management of symptomatic spinal cord compression. Patients in the surgical cohort experienced maintenance and significant improvement in rates of recovery for ambulation, pain control, urinary continence, and survival.

One of the most significant challenges in the management of spinal metastases is the efficacy of cEBRT to different tumor histologies based on their level of radiosensitivity.[7] Radiosensitive histologies include lymphoma, plasmacytoma, multiple myeloma, germ cell tumors, breast cancer, and prostate carcinomas. In response to cEBRT, these tumors have been shown to have 2-year local control rates of nearly 90%. In contrast, radioresistant malignancies such as non-small cell lung, thyroid, hepatocellular, colorectal, renal cell carcinoma (RCC), melanoma, and sarcomas exhibit 2-year local control approaching 30% following radiation therapy.[8]

The development of image-guided spinal stereotactic radiosurgery (SSRS) has enabled the delivery of highly conformal and tumoricidal doses of radiation as either a single treatment or hypofractionated regimen (two to five treatments) to the spine while limiting toxicity to adjacent normal tissue.[9–11] The biologically effective dose of radiation delivered with SSRS is estimated to be approximately three times greater than that with cEBRT. This leads to a cascade of downstream biologic effects including enhanced DNA damage within the tumor, endothelial disruption, and potentially enhanced immune environment with T-cell activation and pro-inflammatory cytokines.[12] SSRS has been shown to effectively treat canonically radioresistant tumors such as RCC with demonstrated 12-month local control rates of 85%.[13] SSRS delivers radiation to a contoured tumor volume with a steep dose gradient that spares surrounding tissues such as the spinal cord, nerves, and visceral organs and has been successfully used as a salvage treatment for patients with local recurrence after cEBRT.[9,14]

Despite the ability to achieve a steep falloff of radiation, there are still tumor dose limitations imparted by the radiation tolerance of surrounding structures such as the spinal cord.[15] While the rate of radiation-induced myelopathy has been reported at under 1% when the SSRS dose to the cord reached the maximum tolerated dose of 14Gy, treatment planning becomes more challenging with high-grade epidural spinal cord compression.[16,17] In these cases, the spinal cord or cauda equina radiation constraints requires adjustment to the prescribed treatment dose, potentially resulting in underdosing of the epidural tumor and poor local control. Lovelock et al.[18] corroborated this observation with the association of local failure when tumors received less than 15 Gy to any point in the treatment planning volume.

The role of surgery has evolved when combined with SSRS for the management of spinal metastasis. Rather than trying to maximize the extent of resection, surgery is now utilized to provide adequate distance between the tumor margins and spinal cord in order to achieve local control through the delivery of tumoricidal doses of radiation to residual tumor.[16] This approach, termed separation surgery, creates sufficient space between the residual epidural tumor and spinal cord to allow a safe margin for the delivery of a cytotoxic dose of radiation.[16,19–22] Separation surgery, cEBRT, and SSRS have subsequently integrated into a broader treatment algorithm that includes neurologic, oncologic, mechanical, and systemic factors.[16] As a result, the aim of surgery in the era of SSRS is (I) neurologic decompression, (II) spinal stabilization as indicated, and (III) to create separation between tumor and the spinal cord for adequate dosing to the tumor.

Patients with spinal metastasis are often frail, present with multiple medical comorbidities, and the impact of surgery must be considered terms of interruption of oncological management and quality of life.[23–25] As a result, the ideal surgical intervention that achieves local control for spinal metastases would limit delays in systemic treatment for the primary and disseminated tumor, facilitate prompt wound healing and functional recovery, and minimize post-operative pain and morbidity. In this setting, spinal laser interstitial thermal therapy (sLITT) has become an emerging treatment option for a select group of patients with metastatic disease to the spine. As a percutaneous, minimally invasive procedure, sLITT uses intraoperative magnetic resonance imaging (MRI) to facilitate laser probe placement approximately 5–6 mm from the dural margin.[26] Laser ablation is monitored in real time using MR-thermography to protect the spinal cord from thermal damage and facilitates intraoperative visualization of epidural tumor ablation.[26–28] This alternative to classical separation surgery[27] facilitates prompt administration of adjuvant radiotherapy at cytotoxic doses that cover the gross tumor volume, limited time off systemic therapy, limited treatment-associated morbidity, and the ability to incorporate percutaneous stabilization during the index or second-stage surgery for patients with spinal instability.[29] The indications, patient selection, technical considerations, and single institutions results of sLITT will be discussed later.

24.2 PATIENT SELECTION

We score the epidural tumor as proposed by Bilsky et al.[30] and have been selecting patients with high-grade spinal cord compression (grades 1c, 2, and 3) from tumors arising in the vertebral body, contained by the posterior longitudinal ligament located between T2 andT12, and without neurological deficits. Normal neurological exam is critical for sLITT candidates, as the ablated tumor does not retract immediately after the procedure. We have seen a 4- to 8-week interval between sLITT treatment and radiographic decompression of the spinal canal. As a result, patients presenting with acute or progressive neurological compromise from ESCC require more rapid decompression with separation surgery. The level and location of epidural involvement are additional factors to consider when evaluating candidacy for sLITT. As key functional motor nerve roots exit the neuroforamina from C3-T1 and the lumbar spine, avoidance of unintentional injury to these functional motor nerve roots of the cervical and lumbosacral plexus is paramount. In patients with ESCC at these levels, open surgical decompression for better direct visualization is favored. In contrast, patients with ESCC from T2–T12 who are asymptomatic or are associated thoracic radiculopathy secondary to foraminal tumor involvement are ideal candidates for laser ablation, as ablation of the foraminal tumor and sensory nerve roots can provide resolution of pain.[31]

Additional limitations to the use of sLITT include patients with contraindications to MRI as sLITT requires the use of MRI thermography. Similarly, instrumentation at the level of ablation creates metallic artifact that impairs the accuracy of MRI thermography and prevents the use of sLITT. In contrast, sLITT is particularly advantageous for patients who need to continue or rapidly resume systemic therapy as individuals in need for continued systemic therapy can safely be treated with sLITT without interruption in chemotherapy due to the procedure's minimally invasive nature and short recovery time.

Given its percutaneous nature, sLITT carries a significantly lower risk of wound-related complications in patients with recurrent tumors that failed prior cEBRT than an open salvage surgery. In cases of spinal instability, a percutaneous pedicle fixation with cement augmentation can be performed immediately after the laser ablation[29] or in a staged procedure. sLITT is also associated with minimal intraoperative blood loss and does not require a preoperative embolization for highly vascular tumors (i.e., renal cell carcinoma, thyroid carcinoma, hemangiopericytomas).

In summary, sLITT is best suited for patients who are asymptomatic or with mild sensory symptoms associated with T2-T12 ESCC (Bilsky 1c, 2, or 3) and confined by the posterior longitudinal ligament. The choice of sLITT is well suited for patients who are poor surgical candidates for traditional open surgical approaches but require decompression of the spinal cord prior to receiving SSRS, in patients who have failed SSRS and will be treated with salvage cEBRT, or as a stand-alone cytoreduction technique for patients who cannot be treated with additional radiation.[27,28] Additionally, sLITT is particularly well suited for patients who require minimal interruption or their systemic therapies. The ultimate goal of patient selection for sLITT is to optimize local control and provide palliation while minimizing the morbidity and mortality associated with more traditional, larger-scale oncologic spinal surgery.

24.3 TECHNICAL NOTES

sLITT offers real-time monitoring of thermal injury with the use of an intraoperative MRI (iMRI). Thus, the ability to perform sLITT may be limited by access to an iMRI or require coordination with the MRI department to perform the procedure. Workflow involves the patient being placed under general anesthesia and positioned prone with the upper extremities parallel to the body to afford surgical ergonomics and lack of interference with C-arm fluoroscopy, CT, or iMRI for localization and stereotactic laser fiber placement.[27,28,32,33] Skin fiducials are placed over the region of interest in a unique pattern to distinguish right-left and rostral-caudal and their positions marked [Figure 24.1A]. The surgical site is subsequently prepped

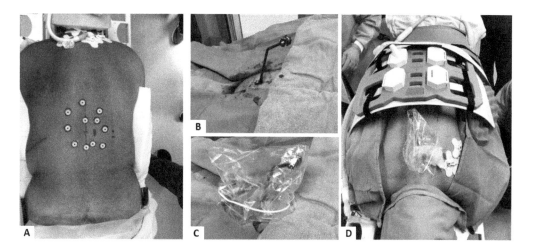

Figure 24.1 (A) The patient is intubated and positioned prone in the iMRI transfer board. Skin fiducials are applied in a configuration that is easy to identify superior/inferior and left/right. (B) An MRI-compatible spinous process clamp is applied and (C) covered with a sterile plastic bag. (D) The MRI coil is placed over the area of interest supported by a plastic cradle to avoid contact with the skin and prevent displacement of the fiducial markers.

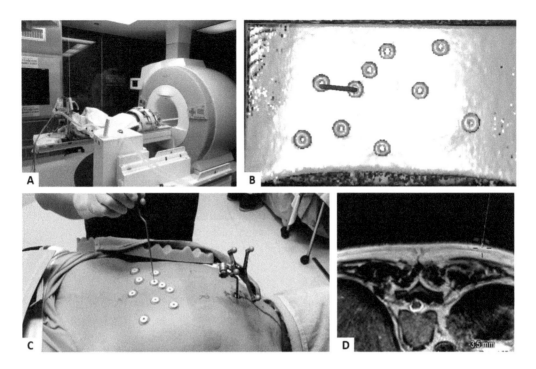

Figure 24.2 (A) Patient is positioned inside the MRI magnet and a T2 sequence of the region of interest is obtained. (B) A three-dimensional model is reconstruction in the navigation software. (C) A reference array is attached to the spinous process clamp and surface matching of the fiducial markers is performed. (D) Accuracy is verified by comparing surface landmarks and position of the probe inside the fiducial markers.

and draped. A small incision overlying a spinous process is made, and subperiosteal dissection is performed to reflect the soft-tissue away to apply a spinous process MRI-compatible titanium clamp (Medtronic Inc., Minneapolis, MN), which is covered with a sterile plastic bag [Figure 24.1B, 1C]. A Siemens body matrix coil secured to a plastic cradle is then placed over the region of interest, and the patient is positioned within the MRI scanner [Figure 24.1D]. A high-resolution T2-weighted image of the area spanning the fiducials and tumor is acquired and used for registration and navigation [Figure 24.2A]. After MRI, the patient is positioned safely outside of the MRI magnet, the sterile plastic bag over the spinous process clamp is removed, and a sterile reference array is attached to the clamp. The registration sequence is transferred to a Stealth S7 workstation (Medtronic, Minneapolis, MN), and registration is performed using point matching with the fiducials and a non-sterile navigation probe [Figure 24.2B, C, D]. Accuracy is confirmed anatomically, evaluating midline, fiducials, and skin surface for depth. Inline sagittal and axial reconstructed images are selected for navigation to reduce error.

Navigation allows for meticulous entry point and trajectory planning with easy identification of the spinal cord, tumor, and surrounding CSF spaces. Trajectories ideally place the laser fiber approximately 6 mm from the thecal sac; each fiber placed can achieve a 10 mm diameter of thermal injury. Thus, depending on the extent of disease in the rostral–caudal plane, multiple trajectories, 10 mm apart, may be required to achieve the desired ablation with no untreated segments between successive ablations. In our experience, up to nine trajectories have been used in a single patient. Similarly, bilateral trajectories may be used to completely treat ventral or lateral epidural disease.

After trajectory planning, the skin is marked at the entry point(s), unsterile fiducials are removed, and operative field is prepared and draped using standard sterile technique. Careful attention is paid to not displace the skin and the reference array during draping. A navigated Jamshidi needle (Medtronic Inc., Minneapolis, MN) is introduced, and navigation accuracy is again confirmed via easily identifiable landmarks. Small incisions are made at the entry sites, and the navigated Jamshidi needle is advanced to bone. The C-arm is used to confirm location and verify concordance of fluoroscopy with navigation [Figure 24.3]. Then the Jamshidi is advanced to target depth using navigation. A K-wire is introduced through the Jamshidi needle [Figure 24.4A] and exchanged with a 1.65-mm-diameter plastic catheter and stylet which is MRI-compatible and can accommodate the diameter of the laser within its lumen. This is repeated in succession for each trajectory [Figure 24.4B]. Once all of cannulas have been inserted, the reference array is removed, and sterile towels are placed to cover the skin, exposing the access cannulas to allow for placement and relocation of the laser fiber. Each individual access cannula is covered

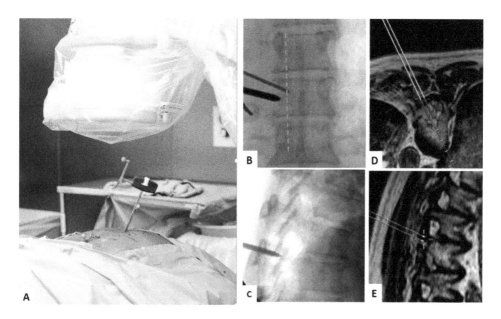

Figure 24.3 (A) Image guidance is used to advance a navigated Jamshidi needle, until it is docked in the lamina. (B) Anteroposterior and (C) lateral fluoroscopy images are obtained to compare the real location of the needle to the predicted position in the image-guidance system. (D) The image-guidance system allows for placement of an offset (red) which predicts the final position of the needle. (E) Comparison of the predicted location in the upper portion of the pedicle is compatible with the real location demonstrated by fluoroscopy, depicted in (C).

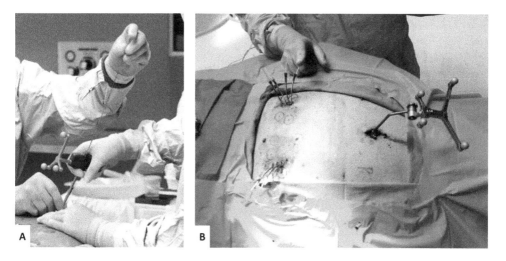

Figure 24.4 (A) A K-wire is inserted and the Jamshidi needle is removed, followed by insertion of a teflon access cannula over the K-wire. (B) Multiple access cannulas in tandem can be inserted in order to treat larger lesions, as we estimate each laser catheter can ablate a 10–12-mm radial zone.

with a sterile plastic bag [Figure 24.5A], and the non-sterile MRI coil is positioned in a way that gives easy access to the plastic bag covering the cannulas [Figure 24.5B]. The plastic bags are then removed with care to not contaminate the access cannulas. Sterile towels are placed over the MRI coil to drape it out, allowing sterile access to the cannulas. The patient is then transferred to the MRI scanner, and a trajectory localization scan is obtained to confirm the axial plane of each access cannula [Figure 24.5C].

A laser fiber, 980-nm diode encased, in a catheter is connected to a 15-W power source (Visualase, Medtronic Inc.). The laser fiber is introduced into one cannula and advanced to the targeted treatment depth (Fig 24.5D). It is then moved to each subsequent cannula after each cycle of ablation is completed. MR thermography is based on gradient-echo acquisition and used throughout the ablation process to monitor heat within the tissue. Proton resonance within the tissue is sensitive to temperature, and phase differences allow for temperature modeling within the exposed tissue. Three-millimeter slices are acquired every 5 to 6 seconds while the laser is activated [Figure 24.6]. The laser is deactivated when one of two

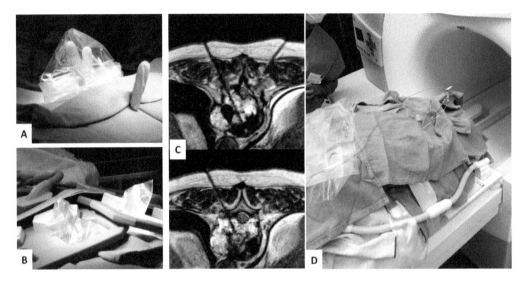

Figure 24.5 (A) Sterile towels are applied over the skin and the access cannulas are covered with glove fingers and a sterile plastic bag. (B) The MRI coil is positioned over the plastic bag, which is removed keeping the sterile glove fingers covering the access cannulas and the MRI coil is covered with sterile towels (not shown). (C) Patient is transported to the MRI magnet and a localization scan is performed to identify the exact axial plane of each access cannula (D) The laser catheter is inserted in each access cannula and the ablation is performed under MRI thermography guidance.

Figure 24.6 Intraoperative footage of the MRI thermography. Green boxes are monitoring the temperature in the tumor/dura-mater interface, which is set to an automatic stop at 48°C. The red boxes are monitoring the temperature around the laser probe, which is set to automatic shut off once the threshold of 90°C is reached.

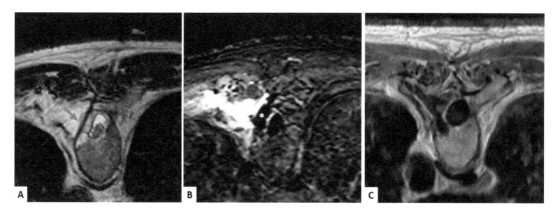

Figure 24.7 (A) Intra-operative axial T2 image demonstrating fibrous histiocytoma causing compression of the spinal cord. The laser catheter is positioned 5 mm from the dural edge (arrow). (B) Immediate post-operative axial T1 with contrast demonstrating absence of enhancement of the epidural component after treatment (asterisk). (C) 8 Weeks axial T1 image demonstrating near resolution of the epidural tumor and decompression of the spinal cord.

temperature thresholds are reached: an upper temperature limit of 48°C to 50°C at the boundary between dura and tumor and 90°C in the tissue adjacent to the laser fiber to prevent excessive heating of the tumor and tissue carbonization. As the thermal maps are degraded by motion, and the spine is vulnerable to respirophasic motion, ablation is performed at 2-minute intervals with breath holds, interrupted cyclically for ventilation. Typically, total ablation time is approximately 4 minutes at a single site. The laser fiber is manually advanced or withdrawn as needed to ensure ablation at all the intended areas along that trajectory.

After ablation is completed, laser fiber and cannulas are removed, and incisions are closed with an absorbable suture. MRI pre- and post-contrast T1-weighted imaging is acquired to visualize the extent of ablation by image subtraction from the contrasted and non-contrasted co-planar scans. On this subtraction image, the area of coagulative necrosis lacks contrast enhancement, appearing hypointense [Figure 24.7].

In patients with concomitant instability, stabilization can be performed during the same ablation procedure or as a separate staged surgery. Typically, percutaneous pedicle screw fixation with cement augmentation is performed the same day. After ablation, the patient is positioned at a safe distance of the MRI magnet, the spinous process clamp is reapplied, and the ink marks of the fiducials are re-registered, allowing image guidance. If registration is inaccurate after verification, fiducials can be replaced and new registration scans acquired, or fluoroscopy can be used to place screws with image guidance. Infrequently, stabilization can also be deferred to a separate day to accommodate CT navigation guidance.

24.4 OUTCOMES OF sLITT IN FACILITATING THE TREATMENT OF SPINE TUMORS

We have recently reported single institution outcomes of our first 120 sLITT cases.[34] Among this cohort, 110 patients had an intact neurologic exam (Frankel E), 8 patients presented with sensory radiculopathy, and 2 patients were unable to walk. Tumor involving the thoracic spine accounted for 107 cases, 5 cases involving the cervical spine, and 8 cases affecting the lumbar spine. The most commonly treated histologies were renal cell carcinoma (39%), non-small cell lung cancer (11%), thyroid cancer (8.3%), and hepatocellular carcinoma (5.8%). A total of 112 cases were associated with high-grade ESCC (1c, 2, and 3) with a median pre-treatment score of 2. The most common anatomical tumor location was the vertebral body (106 cases), followed by paraspinal/foraminal (9 cases), and posterior elements (5 cases). sLITT was followed by adjuvant radiotherapy in 87 cases of which 71 were treated with SSRS and 16 received cEBRT in the context in previously failed SSRS. sLITT was performed in 27% of patients presenting with mechanical back pain and/or spinal instability requiring stabilization. The median length of hospitalization was

2 days, median time to adjuvant radiotherapy was 6 days, and median time to post-operative systemic chemotherapy was 19 days.

Complications within 30 days after surgery were documented in 22 cases and included neurolgical,[9] medical,[6] and fracture[7] etiologies. Among the sLITT cases, four patients experienced functional decline from Frankel E to Frankel D. These cases were associated with motor root palsy[2], and two patients experienced spinothalamic dysfunction.[2] There were four additional patients who declined from Frankel E to Frankel C. Among these four patients, one patient was diagnosed with a cerebellar stroke and three patients declined due to tumor progression. A single patient presented with pre-existing cauda equina syndrome (Frankel C) and progressed to complete motor palsy (Frankel B). A significant correlation was attributed between sLITT involving the non-thoracic spine and neurologic complications. Utilizing multivariate logistic regression to evaluate for complications compared to the thoracic spine, sLITT involving the lumbar spine had an odds ratio of 15.4 and the cervical spine had an odds ratio of 17.1.

Both alone and in conjunction with SSRS, sLITT provides effective and durable local tumor control with minimal morbidity. The median overall survival among the 120 treated patients was 14 months. The median preoperative ESCC score was 2 and decreased to 1b after treatment and local tumor control rate as 81.7% at 1 year. Tumors arising from the vertebral body and contained by the posterior longitudinal ligament had the best progression-free survival (median PFS; not reached). This was followed by tumor arising from the dorsal elements (median PFS 10 months) and paraspinal region (median PFS 3 months). Adjuvant radiation (either SSRS or cEBRT) was associated with longer PFS compared to those who received sLITT alone (PFS note reached vs 14 months), and sLITT as a sole salvage treatment had an OR of 3.2 for failure. These findings were confirmed on multivariate analysis with epidural tumor derived from paraspinal tumor and sLITT as stand-alone therapy being independent predictors of failure (OR 6.3 and 3.3, respectively).

Pain scores have been reported with the visual analog scale (VAS) among 19 patients treated with sLITT.[28] The mean preoperative VAS score was 4.72 and decreased to 2.5 at 1 month and 3.25 at 3 months post-treatment. A similar significant improvement was observed in the EQ-5D index for quality of life. The median pre-operative score was 0.67 which improved at 3 months to 0.83. Among this same cohort of 19 patients, pre- and post-operative MRI was analyzed for radiographic response to sLITT. While immediate post-operative MRI did not demonstrate immediate spinal cord decompression, there was a significant reduction in the mean thickness of epidural tumor at 2 months post-treatment (8.0 mm to 6.4 mm).[28]

In order to further understand the treatment outcomes in patients treated with sLITT compared to classical separation surgery, our group performed a single institution, retrospective matched-group design study of 80 total patients with metastatic thoracic epidural spinal cord compression treated with either sLITT (40 patients) or open surgery (40 patients).[35] Among patients undergoing open surgery, there was an increase in the number of major and minor complications compared to the group treated with sLITT. Instrumentation was associated with 15% of patients treated with sLITT compared to 92.5% treated with open surgery. Differences were also observed in the degree of spinal cord decompression between groups. While 72.5% of patients treated with the sLITT had a reduction in ESCC, the open surgery group noted a reduction in 90% of cases. Regarding the degree of spinal cord decompression, 55% of patients treated with open surgery had a post-procedure Bilsky score of 0 compared to patients treated with sLITT who had a Bilsky score of 0 in 12.5% of cases. The median preoperative ESSC score in patients treated with sLITT was 2 which reduced to 1b after the procedure. In contrast, open surgery resulted in a median reduction from 2 to no compression. A number of differences were observed in additional intra-procedural and post-procedural variables in patients treated with sLITT compared to open surgery. These included lower blood loss (117 mL versus 1331 mL), mean hospital length of stay (3.4 days versus 9 days), shorter interval between procedure and initiation of radiotherapy (7.8 days versus 35.9 days), and the mean time to begin post-procedure systemic therapy (24.7 days versus 59 days). When comparing sLITT and open surgery, there was no difference between PFS, local control rates, or median overall survival.

24.5 DISCUSSION

sLITT represents an emerging minimally invasive treatment option for patients with ESCC that provides real-time monitoring of thermal injury and integrates into the NOMS framework for the evaluation and treatment of patients with ESCC outlined by Bilsky et al.[16,19,36] Through our single institution experience, we have obtained an emerging perspective for the optimal indications for sLITT, expected post-operative outcomes, technical nuances, and future lines of inquiry. Separation surgery, compared to en bloc resection, has become a foundational tool in the treatment of patients with ESCC.[19,20,28] Although associated with less morbidity than radical resection, separation surgery is associated with delays in systemic therapies that can be compounded by post-operative complications. sLITT represents a minimally invasive surgical alternative to traditional separation surgery[27,28,37,38] with a goal to achieve adequate spinal cord decompression to prevent neurologic decline and facilitate adjuvant radiation therapy. Importantly, sLITT and separation surgery have demonstrated similar levels of efficacy with 1-year local control rates of 81.7% and 93.7%–82%,[19–21] respectively. Additionally, in a retrospective matched-group design study comparing sLITT with open surgery, sLITT was found to be non-inferior to open surgical decompression when comparing local control rates, median overall survival, and the degree of decrease in ESCC Bilsky scores.[35]

Patient selection for this procedure is a critical factor in order to achieve post-operative success. Given the minimally invasive nature of the procedure, less intraoperative blood loss, shorter hospital stay, and fewer complications,[35] sLITT is an ideal procedure for patients who are poor operative candidates for a larger open separation surgery or require continuation of their systemic therapy without interruption.[27,28,35,39] However, our experience has demonstrated a significant difference in the dynamics of spinal cord decompression between sLITT and open separation surgery. sLITT is associated with an immediate loss of radiographic epidural enhancement and median decrease of 2 ESCC degrees that occurs over 4–8 weeks.[27,28,37,38] Treatment is also associated with the potential of transient post-operative edema and mass effect from the ablated epidural tumor.[27,31] As a result, patients presenting with neurologic dysfunction at the time of neurosurgical consultation are better suited for open separation surgery in order to achieve an immediate decompression of the spinal cord and avoid potential neurologic worsening post-ablation.

The location of tumor within the vertebral column is an additional consideration when deciding between sLITT and open separation surgery. Although initially attempted in the cervical and lumbar spine, procedures at these levels were associated with increased complication rates (17-fold and 15-fold, respectively)[34] due to an inability to adequately visualize and avoid traversing nerve roots. In contrast, tumor ablation from T2 to T12 that associated sensory nerves lead to resolution of radicular symptoms.[27] Additional limiting factors include tumor involving multiple neuroforamina and tumor arising from the paraspinal region as they are challenging to adequately cover with the laser treatment fields. This contrasts with tumor arising from the vertebral body and contained within the posterior longitudinal ligament which can be more adequately thermally ablated.

In summary, patient selection has been refined through our institutional experience of sLITT to include patients with high Bilsky grade ESCC (1c, 2, and 3), epidural tumor arising from the PLL, tumor involving T2–T12, and absence of neurologic deficits or myelopathy on exam. In contrast, sLITT is contraindicated in patients with cervical and lumbar disease due to increased risk of nerve root palsy, progressive weakness due to spinal cord compression, preexisting spinal hardware that creates metallic artifact impeding treatment planning, and inability to undergo MRI.

Our experience has provided additional insights into the broader treatment course and post-surgical expectations. Although most patients treated with sLITT did not require spinal instrumentation (74.2%),[34] the procedure can be readily adapted for at-risk individuals to include percutaneous fixation during the index procedure or as a second operation. Additionally, as emerging systemic therapies change the treatment landscape for patients with metastatic disease, the interaction between those agents with sLITT must be considered. For instance, sLITT has been associated with post-operative edema at the ablation site which may require the addition of steroids, a contraindication in patients receiving ongoing immune checkpoint blockade.

As the experience with sLITT continues to improve, it will be necessary to integrate technological advances to optimize the operative workflow and expand adoption of this technique to other centers. Ongoing work will also be needed to improve our understanding between sLITT and the broader evolving landscape of systemic oncologic therapies. While our current retrospective studies[34] suggest that sLITT is optimally suited for high-risk surgical, neurologically intact individuals with ESCC from radioresistant tumors, it will be critical to compare the rate of local control between sLITT, open separation surgery followed by SSRS, or SSRS alone in a prospective, randomized fashion. With this additional data, indications for sLITT will continue to integrate more clearly into pre-existing decision frameworks such as NOMS criteria.

REFERENCES

1. Bach, F., *et al.* Metastatic spinal cord compression. Occurrence, symptoms, clinical presentations and prognosis in 398 patients with spinal cord compression. *Acta Neurochir (Wien)* **107**, 37–43 (1990).
2. Laufer, I., *et al.* The NOMS framework: approach to the treatment of spinal metastatic tumors. *Oncologist* **18**, 744–751 (2013).
3. Barron, K.D., Hirano, A., Araki, S. & Terry, R.D. Experiences with metastatic neoplasms involving the spinal cord. *Neurology* **9**, 91–106 (1959).
4. Walsh, G.L., *et al.* Anterior approaches to the thoracic spine in patients with cancer: indications and results. *Ann Thorac Surg* **64**, 1611–1618 (1997).
5. Cole, J.S. & Patchell, R.A. Metastatic epidural spinal cord compression. *Lancet Neurology* **7**, 459–466 (2008).
6. Patchell, R.A., *et al.* Direct decompressive surgical resection in the treatment of spinal cord compression caused by metastatic cancer: a randomised trial. *Lancet* **366**, 643–648 (2005).
7. Maranzano, E. & Latini, P. Effectiveness of radiation therapy without surgery in metastatic spinal cord compression: final results from a prospective trial. *International Journal of Radiation Oncology, Biology, Physics* **32**, 959–967 (1995).
8. Mizumoto, M., *et al.* Radiotherapy for patients with metastases to the spinal column: a review of 603 patients at Shizuoka cancer center hospital. *International Journal of Radiation Oncology, Biology, Physics* **79**, 208–213 (2011).
9. Gerszten, P.C., Mendel, E. & Yamada, Y. Radiotherapy and radiosurgery for metastatic spine disease: what are the options, indications, and outcomes? *Spine (Phila Pa 1976)* **34**, S78–S92 (2009).
10. Yamada, Y., *et al.* High-dose, single-fraction image-guided intensity-modulated radiotherapy for metastatic spinal lesions. *International Journal of Radiation Oncology, Biology, Physics* **71**, 484–490 (2008).
11. Bishop, A.J., *et al.* Outcomes for spine stereotactic body radiation therapy and an analysis of predictors of local recurrence. *International Journal of Radiation Oncology, Biology, Physics* **92**, 1016–1026 (2015).
12. Greco, C., *et al.* Spinal metastases: from conventional fractionated radiotherapy to single-dose SBRT. *Rep Pract Oncol Radiother* **20**, 454–463 (2015).
13. Gerszten, P.C., *et al.* Stereotactic radiosurgery for spinal metastases from renal cell carcinoma. *Journal of Neurosurgery Spine* **3**, 288–295 (2005).
14. Sahgal, A., Larson, D.A. & Chang, E.L. Stereotactic body radiosurgery for spinal metastases: a critical review. *International Journal of Radiation Oncology, Biology, Physics* **71**, 652–665 (2008).
15. Chang, E.L., *et al.* Phase I/II study of stereotactic body radiotherapy for spinal metastasis and its pattern of failure. *Journal of Neurosurgery Spine* **7**, 151–160 (2007).
16. Bilsky, M. & Smith, M. Surgical approach to epidural spinal cord compression. *Hematol Oncol Clin North Am* **20**, 1307–1317 (2006).
17. Chang, J.H., *et al.* Stereotactic body radiotherapy for spinal metastases: what are the risks and how do we minimize them? *Spine (Phila Pa 1976)* **41 Suppl 20**, S238–S245 (2016).
18. Lovelock, D.M., *et al.* Correlation of local failure with measures of dose insufficiency in the high-dose single-fraction treatment of bony metastases. *International Journal of Radiation Oncology, Biology, Physics* **77**, 1282–1287 (2010).
19. Laufer, I., *et al.* Local disease control for spinal metastases following "separation surgery" and adjuvant hypofractionated or high-dose single-fraction stereotactic radiosurgery: outcome analysis in 186 patients. *Journal of Neurosurgery Spine* **18**, 207–214 (2013).
20. Moussazadeh, N., Laufer, I., Yamada, Y. & Bilsky, M.H. Separation surgery for spinal metastases: effect of spinal radiosurgery on surgical treatment goals. *Cancer Control* **21**, 168–174 (2014).
21. Bate, B.G., Khan, N.R., Kimball, B.Y., Gabrick, K. & Weaver, J. Stereotactic radiosurgery for spinal metastases with or without separation surgery. *Journal of Neurosurgery Spine* **22**, 409–415 (2015).
22. Tao, R., *et al.* Stereotactic body radiation therapy for spinal metastases in the postoperative setting: a secondary analysis of mature phase 1–2 trials. *International Journal of Radiation Oncology, Biology, Physics* **95**, 1405–1413 (2016).
23. Gerszten, P.C. Spine metastases: from radiotherapy, surgery, to radiosurgery. *Neurosurgery* **61 Suppl 1**, 16–25 (2014).
24. Vitaz, T.W., *et al.* Rotational and transpositional flaps for the treatment of spinal wound dehiscence and infections in patient populations with degenerative and oncological disease. *Journal of Neurosurgery* **100**, 46–51 (2004).

25. Zacharia, B.E., *et al.* Incidence and risk factors for preoperative deep venous thrombosis in 314 consecutive patients undergoing surgery for spinal metastasis. *Journal of Neurosurgery Spine* **27**, 189–197 (2017).

26. Moussazadeh, N., Evans, L.T., Grasu, R., Rhines, L.D. & Tatsui, C.E. Laser interstitial thermal therapy of the spine: technical aspects. *Neurosurgery Focus* **44**, V3 (2018).

27. Tatsui, C.E., *et al.* Utilization of laser interstitial thermotherapy guided by real-time thermal MRI as an alternative to separation surgery in the management of spinal metastasis. *Journal of Neurosurgery Spine* **23**, 400–411 (2015).

28. Tatsui, C.E., *et al.* Spinal laser interstitial thermal therapy: a novel alternative to surgery for metastatic epidural spinal cord compression. *Neurosurgery* **79 Suppl 1**, S73–S82 (2016).

29. Tatsui, C.E., *et al.* Percutaneous surgery for treatment of epidural spinal cord compression and spinal instability: technical note. *Neurosurgery Focus* **41**, E2 (2016).

30. Bilsky, M.H., *et al.* Reliability analysis of the epidural spinal cord compression scale. *Journal of Neurosurgery Spine* **13**, 324–328 (2010).

31. Thomas, J.G., *et al.* A novel use of the intraoperative MRI for metastatic spine tumors: laser interstitial thermal therapy for percutaneous treatment of epidural metastatic spine disease. *Neurosurg Clin N Am* **28**, 513–524 (2017).

32. Jimenez-Ruiz, F., Arnold, B., Tatsui, C.E. & Cata, J.P. Perioperative and anesthetic considerations for neurosurgical laser interstitial thermal therapy ablations. *Journal of Neurosurgery Anesthesiology* **30**, 10–17 (2018).

33. Boriani, S., Weinstein, J.N. & Biagini, R. Primary bone tumors of the spine. Terminology and surgical staging. *Spine (Phila Pa 1976)* **22**, 1036–1044 (1997).

34. Dhiego C. A. Bastos, R.A.V., Jeffrey I. Traylor, Amol Ghia, Jing Li, Marilou Oro, Andrew J. Bishop, Debra Yeboa, Behrang Amini, Vinodh A. Kumar, Ganesh Rao, Laurence D. Rhines, & Claudio E. Tatsui. Spinal laser interstitial thermal therapy: single-center experience and outcomes in the first 120 cases. *Journal of Neurosurgery: Spine* (2020).

35. de Almeida Bastos, D.C., *et al.* A comparison of spinal laser interstitial thermotherapy with open surgery for metastatic thoracic epidural spinal cord compression. *Journal of Neurosurgery Spine*, 1–9 (2020).

36. Bilsky, M.H., Laufer, I. & Burch, S. Shifting paradigms in the treatment of metastatic spine disease. *Spine (Phila Pa 1976)* **34**, S101–S107 (2009).

37. Tatsui, C.E., *et al.* Image guidance based on MRI for spinal interstitial laser thermotherapy: technical aspects and accuracy. *Journal of Neurosurgery Spine* **26**, 605–612 (2017).

38. Ghia, A.J., *et al.* The use of image guided laser interstitial thermotherapy to supplement spine stereotactic radiosurgery to manage metastatic epidural spinal cord compression: Proof of concept and dosimetric analysis. *Pract Radiat Oncol* **6**, e35–e38 (2016).

39. Vega, R.A., Ghia, A.J. & Tatsui, C.E. Percutaneous hybrid therapy for spinal metastatic disease: laser interstitial thermal therapy and spinal stereotactic radiosurgery. *Neurosurg Clin N Am* **31**, 211–219 (2020).

Stereotactic Radiotherapy and Intracranial Leptomeningeal Disease

Timothy K. Nguyen, Sten Myrehaug, Chia-Lin Tseng, Zain A. Husain, Jay Detsky, Sunit Das, Arjun Sahgal, and Hany Soliman

Contents

25.1 INTRODUCTION

Leptomeningeal disease (LMD) (otherwise known as leptomeningeal metastasis, neoplastic meningitis, or leptomeningeal carcinomatosis) describes the spread of tumor cells into the leptomeninges (comprising the arachnoid mater and pia mater) and the cerebrospinal fluid (CSF) of the subarachnoid space. LMD is a pattern of disease progression that carries a poor prognosis with a reported median survival of 1.6 months without treatment and 5.3 months with whole-brain radiotherapy (WBRT) at the time of diagnosis.[1] Approximately 5% to 15% of patients with metastatic solid or hematologic malignancies are diagnosed with LMD, and it is most commonly diagnosed in patients with breast cancer, lung cancer, gastrointestinal cancer, and melanoma.[2–4] Furthermore, the incidence of LMD may be rising with the continual emergence of improved and novel systemic therapies.[5] While these agents may enhance extracranial disease control in patients with metastatic malignancy, the blood–brain barrier (BBB) prevents most drug therapies from accessing the central nervous system (CNS), creating a potential "sanctuary site" for malignant cells.

Metastatic infiltration into the leptomeninges may occur by several means including hematogenous spread via the arachnoid vessels and choroid plexus, perivascular lymphatic spread, perineural spread, or direct invasion into CSF compartments from existing brain or skull metastases.[6] LMD is commonly diagnosed in association with existing or previous parenchymal brain metastases,[7] with one series reporting the coexistence of brain metastases and LMD in 70% of patients with a solid tumor histology.[3]

The presentation of LMD is variable, and often the signs and symptoms include non-specific neurologic deficits that may overlap with those caused by parenchymal brain metastases or treatment-related effects.

There should be elevated clinical suspicion in patients with symptoms suggesting the involvement of multiple neurologic sites and cranial nerve palsies (especially if symptoms are bilateral and if central spinal cord dysfunction is present).[5,8]

Therapeutic options and their effectiveness for patients with LMD are generally limited. Radiotherapy has long been an established treatment option and is primarily delivered as WBRT, focal spinal radiotherapy, or craniospinal irradiation (CSI) in carefully selected patients to palliate symptoms. Recently, there has been a shift in practice from WBRT to focal stereotactic radiosurgery or hypofractionated stereotactic radiotherapy (SRS/HSRT) for patients with limited parenchymal brain metastases. Accordingly, recent reports have suggested that focal SRS/HSRT may have a role in certain patients with limited LMD.

25.2 DIAGNOSTIC WORK-UP

Patients with intracranial LMD may present with a variety of non-specific neurologic signs and symptoms including headache, nausea, vomiting, cognitive changes, weakness, and cranial nerve dysfunction (Table 1). Given the variability in presentation, the history and physical exam alone are insufficient in establishing a diagnosis of LMD; however, they are still essential in identifying the need for further investigations. For the majority of intracranial LMD cases, a contrast-enhanced magnetic resonance imaging (MRI) scan of the head establishes the diagnosis with a reported specificity and sensitivity ranging from 75% to 90% and from 70% to 85%, respectively.[3,9,10] The presence of intracranial LMD necessitates a contrast-enhanced MRI scan of the brain and complete spine to rule out multifocal disease, given the risk to the whole neuroaxis. Sampling of cerebral spinal fluid (CSF) may also be considered, particularly in hematologic malignancies, and if imaging findings are equivocal. Repeat CSF sampling is sometimes required and can improve its sensitivity from 50% to 75%.[11,12] In many cases of LMD from solid tumor, characteristic imaging findings and a concordant clinical presentation are sufficient for establishing a diagnosis.

25.3 RADIOGRAPHIC PATTERNS OF LMD

25.3.1 CLASSIFICATION

Intracranial LMD is typically discovered on MRI with enhancement of cerebral sulci and cisterns, pial surfaces, cranial nerves (CN), cerebellar folliae, and/or the ependymal lining of lateral ventricles.[13–16] As reported in one series of 270 patients with cytology-proven LMD, the most common sites of enhancement identified are in the cerebellum, occipital lobe, and along the CN. Among patients with CN enhancement, CN VII/VIII, V, and III had the highest incidence of involvement and were observed in 20%, 10%, and 3% of cases, respectively.[16] The extent of disease on imaging is variable ranging from one or few focal sites to widespread diffuse involvement of the neuroaxis. Some patients may also present with associated hydrocephalus.

The existence of distinct radiographic patterns of leptomeningeal enhancement has long been recognized which is reflected in the distinct descriptors used in the literature, including linear, curvi-linear, bulky, and nodular. Recently, Turner et al. proposed an MRI-based classification system to clearly identify and standardize these descriptions.[17] Their report differentiates between local recurrence, distant intraparenchymal brain metastases, and two patterns of LMD: classical LMD (cLMD) and nodular LMD

Table 25.1 **Signs and symptoms of intracranial and/or spinal leptomeningeal disease**[11]

SYMPTOMS	SIGNS
Lower motor weakness (38%)	Reflex Asymmetry (71%)
Parasthesias (34%)	Weakness (60%)
Radicular pain (33%)	Mental status change (31%)
Headaches (33%)	Sensory loss (27%)
Back/neck pain (26%)	Cranial nerve III paresis (20%)
Diplopia (20%)	Cranial nerve VII paresis (17%)
Mental status change (17%)	

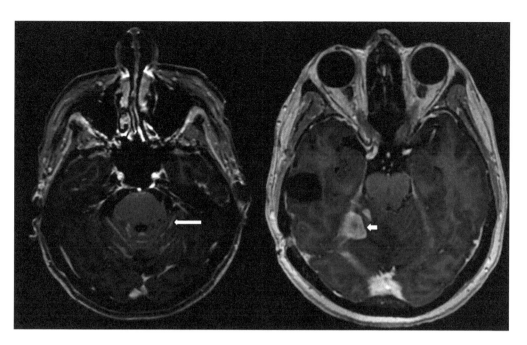

Figure 25.1 Examples of cLMD and nLMD on MRI head (axial images, T1-weighted post-contrast). (Left) An example of cLMD with the long arrow indicating linear enhancement along the cerebellar folia. (Right) An example of nLMD indicated by the short arrow.

(nLMD) (Figure 25.1). Local recurrence was defined as nodular enhancement within the resection cavity or within the 80% isodose line (IDL) of an intact metastasis treated with radiotherapy alone. This should be differentiated from distant intraparenchymal metastases, which are enhancing tumors deep to pial surfaces and outside of the 80% IDL of a previously treated metastasis. Imaging features consistent with cLMD include cranial nerve enhancement and curvilinear enhancement of cerebral sulci, cerebellar folia, cerebral cisterns, and brain surfaces. nLMD may involve similar intracranial sites such as cLMD, but the enhancing disease typically appears as bulky, mass-like, thickened, and/or irregular. In Turner's system, nLMD is an intracranial entity and differentiated from nodular-appearing LMD in the spine. In other reports, the term pachymeningeal seeding/disease has also been used to describe bulky, nodular metastatic disease involving the meninges.[18] Anatomically, pachymeningeal disease and nLMD are separate, with the former indicating the involvement of the dura mater and the latter referring to the involvement of the inner arachnoid and pia mater.[19] Practically, it can be challenging to distinguish whether nodular disease is involving solely the outer or inner meninges based on imaging alone; therefore, pachymeningeal disease and nLMD have been considered alternate descriptions of the same radiographic entity.[1,17]

25.3.2 IMPLICATIONS FOR SURVIVAL

In 1997, the impact of bulky metastatic CNS disease on survival was studied in a matched retrospective cohort study of 40 patients diagnosed with LMD based on lumbar puncture.[20] All patients received intraventricular chemotherapy and radiotherapy was delivered in 45% of patients without bulky disease and 100% of patients with bulky disease. They observed a significantly worse median survival in patients with bulky disease compared to those without (4 months versus 7 months; $p < 0.01$). It is noteworthy that in 15 of the 21 cases of bulky disease, the radiographic descriptions were more consistent with parenchymal brain metastases than nLMD described above. Furthermore, 16 of the 20 patients without bulky disease had normal MRI scans of the brain and spine, while the 4 with abnormal cranial findings had hydrocephalus without any comment of enhancing disease. The survival difference observed in this study was likely driven by the dramatic difference in radiographic disease burden as opposed to a survival difference associated with bulky LMD as a distinct radiographic entity. This study also highlights the importance of adopting a standardized system to classify LMD.

Under the modern classification system, several recent retrospective studies have examined the difference in survival between nLMD and cLMD with mixed findings. In a large retrospective series of 442 patients with 501 resected brain metastases treated with post-operative SRS, Shi et al. showed dramatically shorter survival in patients with cLMD compared with nLMD (2 months versus 11 months; $p < 0.01$) and also a higher incidence of neurologic death (67% versus 41%; $p = 0.02$). Likewise, Prabhu et al. reported on 129 patients who received salvage radiotherapy for LMD following initial treatment for parenchymal brain metastases.[21] In patients with nLMD, the median survival was much longer compared with patients with cLMD (12.5 months versus 4.4 months; $p < 0.001$). The presence of nLMD was also a significant predictor of OS on multivariable analysis (HR: 0.59; $p = 0.04$). Contrary to these studies, in a retrospective series of 235 patients treated with HSRT for intact or resected brain metastases, no difference in OS was observed between nLMD and cLMD.[1]

25.4 STANDARD PALLIATIVE TREATMENT OPTIONS

25.4.1 WHOLE-BRAIN RADIOTHERAPY

WBRT has long been a standard of care treatment option for patients with intracranial LMD. While high level evidence is lacking, WBRT has been associated with improved survival in several retrospective studies including patients with a wide range of solid tumor histologies.[22–25] Factors that may predict for improved survival include a favorable performance status, lack of presenting symptoms, absence of parenchymal brain metastases, and an increased interval of time between the primary cancer diagnosis and development of LMD.[22,23] Besides potentially extending life expectancy, WBRT can also provide symptomatic relief and/or stabilization. In one cohort, 51 patients with intracranial LMD received WBRT, and 84% experienced an improvement in presenting symptoms.

The dose fractionation for WBRT in the setting of LMD is the same as for patients with parenchymal metastasis: 20 Gy in 5 fractions (20 Gy/5) or 30 Gy in 10 fractions (30 Gy/10). In patients with a poor prognosis, 20 Gy/5 should be favored over 30 Gy/10. When placing the radiotherapy fields and shielding, it is essential that all brain tissue and CSF spaces (including the posterior orbital spaces, basal cisterns, and lamina cribrosa) are well covered. The acute toxicities associated with WBRT include fatigue, scalp dermatitis, alopecia, headache, nausea/vomiting, otitis media, xerostomia, and taste changes which may develop during radiotherapy or in the days after its completion. Late toxicities typically present months following treatment and include persistent fatigue, cognitive dysfunction, cerebrovascular effects, cataract development, and pituitary dysfunction.

25.4.2 CRANIOSPINAL RADIOTHERAPY

CSI is the delivery of radiotherapy to the entire neuroaxis and is traditionally a complex technique involving the use of WBRT lateral fields along with one to two spinal fields that are appropriately matched. At the junction where the spinal fields overlap, dose heterogeneity is present and has been traditionally addressed using multiple moving junctions to reduce hot and cold spots. CSI is infrequently prescribed in the context of LMD for many reasons. These include the palliative treatment intent, the absence of a demonstrated survival benefit, and the significant toxicity profile, such as myelosuppression which can hinder the timely administration of systemic therapy.[11] In select patients where CSI is considered, the use of intensity-modulated radiotherapy (IMRT) can provide more conformal treatment in an effort to minimize toxicity. In one retrospective series, 25 patients with LMD received CSI using an IMRT technique, predominantly with a dose of 36 Gy in 20 fractions.[26] The median overall survival was 19 weeks, and improved survival was associated with patients who were younger, with a favorable performance status, and who had improvement in their neurologic symptoms. Acute toxicities included grade I–II fatigue (84%), grade I–II nausea (36%), and grade III myelosuppression (32%). The use of proton CSI has also been examined in a small phase I trial of 24 patients with LMD from solid tumors. The trial used a 3 + 3 de-escalation design, starting at a dose of 30 Gy in 10 fractions and expanding the cohort if tolerated or de-escalating to 25 Gy in 10 fractions if sufficient cases of dose-limiting toxicity were observed.[27] In total, 20 patients were treated with 30 Gy in 10 fractions, and the authors concluded that this technique was safe to deliver. Two dose-limiting toxicities were reported and included grade 4 lymphopenia and grade 4 thrombocytopenia. CSI is an available treatment option for patients with LMD but is generally avoided and should only be considered selectively in patients where the potential benefits may outweigh the risks.

25.4.3 SYSTEMIC THERAPY

Chemotherapy agents may be delivered directly into CSF spaces of the neuroaxis using an intrathecal (IT) approach. This can be done through an indwelling intraventricular catheter or through repeated lumbar punctures. Common IT agents include methotrexate, cytarabine, and thiotepa; however, the optimal agent and dosing schedule for patients with LMD are not known.[6] Furthermore, there has not been clear evidence of a survival benefit with the addition of IT chemotherapy compared with standard treatment.[28,29] One of the limitations with IT chemotherapy is that drugs typically penetrate only 2–3 mm into spinal or brain tissue, which hinders its effectiveness for bulky, nodular LMD, and associated parenchymal brain metastases.[30] Furthermore, obstructions to CSF flow, due to tumor bulk for example, can reduce the distribution of drugs throughout the neuroaxis. In these cases, radiotherapy is indicated to relieve CSF obstruction and/or treat inadequately treated bulky disease.

Immune checkpoint inhibitors have found broad application in metastatic disease across multiple tumor histologies, and similarly its utility has been explored for patients with LMD. It has been suggested that innate and adaptive immune cells may permeate into the CSF and that interfaces in the choroid plexus may provide another access.[6,31,32] While immune checkpoint inhibitors have shown intracranial responses in patients with brain metastases, the evidence supporting their use in LMD are limited but promising. For example, the interim analysis of an ongoing phase II trial of pembrolizumab in LMD demonstrated a survival benefit when compared to historical controls and a toxicity profile that was similar to patients without LMD (NCT02886585). Several trials further examining this indication are underway (NCT03719768, NCT02939300, NCT03025256).

Similarly, targeted agents have also demonstrated CNS penetration and intracranial efficacy. For example, in the single arm phase II LANDSCAPE trial for patients with brain metastases from HER2+ breast cancer, the combination of lapatinib and capecitabine resulted in a CNS response in 65.9% of patients.[33] In addition, an exploratory subgroup analysis of the KAMILLA phase IIIb safety study demonstrated that trastuzumab emtansine (T-DM1) produced a radiographic response in 67% of nearly 400 patients with brain metastases.[34] These data are very promising and lead the way for studies examining a LMD-specific population.

25.5 RISK OF LMD FOLLOWING THE TREATMENT OF PARENCHYMAL BRAIN METASTASES

25.5.1 UPFRONT HSRT/SRS ALONE

For patients treated with upfront HSRT/SRS for intact brain metastases, the overall incidence of LMD following treatment ranges from 4.6% to 21%[35–37] with a cumulative 1-year incidence of 5.8% to 9%.[1,35,38] LMD presents a median of 7 to 8 months following completion of HSRT/SRS,[1,38] and most patients are symptomatic at the time of recurrence.[35] Both risk factors and protective factors for LMD have been identified in this patient population. For example, in a large study of 820 patients treated with upfront SRS, independent predictors of LMD included breast cancer histology and ≥ four brain metastases ($p < 0.05$).[38] Another report suggested that the use of targeted agents or immunotherapy may have a protective effect in reducing the risk of LMD (odds radio = 0.178; $p = 0.023$).[1]

25.5.2 SURGICAL RESECTION AND ADJUVANT HSRT/SRS

In patients who have undergone surgical resection of at least one brain metastasis, the reported rate of LMD ranges from 5% to 31%.[1,18,39–44] While one large retrospective study reported no difference in LMD rates between surgical and non-surgical patients,[45] multiple other studies that compared patients treated with surgical resection plus radiotherapy and radiotherapy alone have shown that surgical resection is associated with higher rates of LMD.[1,18,37,42] Pooling together cases of cLMD and nLMD, the incidence of

LMD has ranged from 5.2% to 20% in patients treated with radiotherapy alone compared with 16.9% to 31% for surgical patients.[1,18,42] In fact, a multivariable analysis completed by Ma et al. found that patients with previous resection of a brain metastasis were six times as likely to develop LMD than non-surgical patients (OR = 6.5; p = 0.01).[37]

Other factors associated with an increased risk of LMD in the surgical population include radiographic tumor features, tumor location, and surgical technique. Press et al. examined the impact of lesion-specific features on the risk of LMD after surgical resection in a retrospective review of 134 patients.[43] On multivariable analysis, they found that tumors with hemorrhagic features (HR: 2.34; p = 0.015) or cystic features (HR: 2.34; p = 0.013) were associated with an increased risk of LMD. Hemorrhagic features were defined as lesions with hyperintensity on T1-weighted pre-contrast MRI sequences and/or hypointensity on gradient-echo T2-weighted sequences, in addition to hyperdensity on non-contrast CT imaging. Cystic lesions had fluid-filled components that were hypointense on T1-weighted MRI sequences and hyperintense on T2-weighted sequences. Furthermore, at both 1 year and 2 years after radiotherapy, patients with hemorrhagic lesions had a higher risk of nLMD compared to patients without hemorrhagic features (11.2% versus 2.1% and 25.5% versus 6.5%, respectively). Tumor location may also impact LMD risk, which was demonstrated in one series where patients with surgical resection of a posterior fossa metastasis were at higher risk of LMD than patients with a resected supratentorial metastasis.[46] Suki et al. examined two separate cohorts of patients treated with surgical resection followed by radiotherapy: one with supratentorial metastases and the other with posterior fossa metastases.[39,47] In both cohorts, they reported an increased risk of LMD in patients who underwent piecemeal tumor resection compared with en bloc tumor resection (supratentorial metastases: 9% versus 3%; infratentorial metastases: 13.9% versus 5.7%). This finding is supported by another retrospective series which similarly found piecemeal tumor resections were associated with a higher LMD rate than en bloc resections (22% versus 5.7%; HR: 4.08; p < 0.01).[48]

With the characterization of nLMD as a distinct entity, recent studies have identified a correlation between nLMD and surgical resection. Cagney et al. reported striking differences in the rates of nLMD (which they described as pachymeningeal disease) between surgical and non-surgical patients. In those who underwent surgical resection, the incidence of nLMD was 11% compared with 0% in non-surgical patients. However, nLMD may not be exclusive to surgical patients as it was observed in both surgical and non-surgical patients in a series from Nguyen et al. In this series, the incidence of nLMD in the surgical subgroup was also higher at 19.5% compared with 8% in the non-surgical cohort. The increased incidence of nLMD in post-operative patients has led to the hypothesis that this pattern of recurrence is related to tumor spillage during surgery and/or microscopic residual disease that is inadequately controlled by current post-operative radiotherapy techniques.[18,49] These two studies with the addition of the previously discussed publication from Prabhu et al are the most recent reports on this topic, which stratify patients using the modern classification of nLMD and cLMD. For that reason, these are summarized in Table 25.2.

Following complete surgical resection of a brain metastasis, the risk of local recurrence is 43%–46% without any adjuvant therapy.[50,51] Historically, WBRT was the standard adjuvant therapy these patients received;[50] however, over the last several years, this approach has been superseded by surgical cavity-directed HSRT/SRS, given that this approach conferred no difference in OS, while avoiding the toxicities associated with WBRT.[52] With the identification of increased rates of LMD, particularly nLMD, after surgical resection of brain metastases, there has been a speculation whether this is a result of the transition from post-operative WBRT to focal techniques.[18,49] The impact of WBRT in reducing the risk of LMD remains unclear. Perhaps the strongest evidence is from a phase III randomized controlled trial from Brown et al. where patients received WBRT or cavity-directed SRS following surgical resection of a brain metastasis. In the analysis of secondary outcomes, no significant difference in LMD was observed at 12 months for patients receiving post-operative SRS compared with WBRT (7.2% versus 5.4%; p = 0.64).[52] The retrospective evidence for WBRT mitigating the risk of LMD remains more controversial. In one series, 650 patients with brain metastases were treated with SRS alone while 177 received SRS combined with upfront WBRT. On multivariable analysis, patients who received upfront WBRT had a significantly lower rate of LMD (HR = 2.24; p = 0.047).[38] Conversely, the retrospective series from Suki et al. previously discussed found that postoperative WBRT was not protective for LMD (HR = 1.1; p = 0.86).[39]

Table 25.2 A comparison of modern retrospective series that describe patients with recurrent intracranial nLMD/cLMD after initial treatment with radiotherapy +/– surgery

	CAGNEY ET AL.	PRABHU ET AL.	NGUYEN ET AL.
Compares resected versus intact metastases	Yes	No	Yes
Patients treated with surgery + SRS/HSRT	318	147	123
Patients treated with only SRS/HSRT for intact metastases	870	0	112
Overall rate of LMD	11.7%	100% (selected for patients with LMD)	19.2%
Rate of cLMD	Resected = 11% Intact = 7.8% ($p = 0.56$)	42.9%	Resected = 7% Intact = 11.6%
Rate of nLMD	Resected = 11.3% Intact = 0% ($p < 0.001$)	57.1%	Resected = 19.5% Intact = 8%
Survival	Median survival from diagnosis of nLMD = 11.1 months	Median survival for whole cohort = 5.4 months Patients who received cranial salvage RT ($n = 115$): nLMD versus cLMD (12.5 mo versus 4.4 mo, $p < 0.001$) Better survival with focal RT versus WBRT: 13 mo versus 6.6 mo, $p = 0.01$ Better survival with nLMD versus cLMD (8.2 mo versus 3.3 mo, $p < 0.001$)	Received salvage WBRT = 5.3 mo No salvage WBRT = 1.6 mo 6 mo OS: cLMD = 44% nLMD = 40% No difference in survival between cLMD and nLMD
Risk Factors for LMD (based on MVA)	nLMD = Prior radiotherapy	NR	Overall = surgical resection (OR = 2.3, CI 1.24–4.39, $p = 0.008$) Resected = radiosensitive tumor histology (OR = 2.35, 95%CI 1.04–5.35, $p = 0.041$)
Risk factors for poor survival (based on MVA)	NR	Patients who received cranial salvage RT = cLMD, melanoma histology	NR

(Continued)

Table 25.2 (*Continued*) **A comparison of modern retrospective series that describe patients with recurrent intracranial nLMD/cLMD after initial treatment with radiotherapy +/– surgery**

	CAGNEY ET AL.	PRABHU ET AL.	NGUYEN ET AL.
Salvage treatment	Most patients with nLMD received SRS/HSRT or WBRT. SRS/HSRT was associated with improved all-cause mortality rate versus WBRT: HR = 0.49; CI, 0.23–1.02, *p* = 0.06	Salvage treatment more common with nLMD versus cLMD: 94% versus 79.4%, *p* = 0.01 Patients receiving salvage treatment (*n* = 87.8%): 46.5% WBRT alone, 27.1% SRS, 10.1% craniospinal RT nLMD and cranial RT (*n* = 73): 47.9% received WBRT, 52.1% received focal RT cLMD and cranial RT (*n* = 42): 95.2% WBRT, 4.8% focal RT	75.6% patients with LMD received salvage WBRT

Abbreviations: SRS/HSRT, stereotactic radiosurgery or hypofractionated stereotactic radiotherapy; WBRT, whole-brain radiotherapy; cLMD, classical leptomeningeal disease; nLMD, nodular leptomeningeal disease; MVA, multivariable analysis; OR, odds ratio; CI, confidence interval.

25.5.3 NEOADJUVANT HSRT/SRS

Neoadjuvant HSRT/SRS is a promising strategy to reduce the risk of LMD following surgical resection of brain metastases based on several retrospective studies. Reported 1-year LC and median OS rates in these studies have ranged from 77.5% to 84.1%[53–55] and 10.5 months to 17.2 months,[53,54,56] respectively. The rates of LMD at 1 year are compelling, ranging from 3.2% to 6.3%,[53,54,56] which is a marked improvement from the 5% to 31% rates observed in the adjuvant HSRT/SRS studies. In one of the largest studies to date, 117 patients were treated with single-fraction preoperative SRS followed by resection. The median SRS dose was 15 Gy which was delivered a median of 2 days prior to surgery. The 2-year cumulative incidence of local recurrence and LMD were 25.1% and 4.3%, respectively.[53] A potential hypothesis is that neoadjuvant radiotherapy may impair malignant cells within the surgical cavity rendering them unable to proliferate and cause disease recurrence.[57] Randomized phase III trials are required to confirm these findings.

25.6 HSRT/SRS FOR LMD

The use of HSRT/SRS for highly selected patients with focal LMD may be a viable treatment option, although it is controversial, given a lack of supportive high-level clinical data. Apart from the limited evidence, there are several reasons why practitioners may feel uneasy with this treatment approach. These include the potential for LMD to present with disseminated disease, the risk of subclinical disease affecting the whole neuraxis, and a short life expectancy in many patients. As a result, WBRT has been and remains a standard of care radiotherapy option for most patients with intracranial LMD. With that said, international practice guidelines and limited small retrospective cohort studies suggest it may be reasonable to consider upfront focal cranial HSRT/SRS in certain patients with focal LMD. In a collaborative effort between the European Association of Neuro-Oncology (EANO) and the European Society for Medical Oncology (ESMO), clinical practice guidelines for patients with LMD were produced. Therein, the authors suggest that focal radiotherapy, including HSRT/SRT, may be considered in patients with nodular LMD or focal symptomatic sites.[7]

In support of these guidelines, a retrospective cohort of 465 patients with brain metastases identified 16 patients who received SRS to focal sites of LMD. Eleven patients were treated with upfront SRS, and five had previously received WBRT. In 13 of the 14 patients for which MRI follow-up was available, there was a partial response to SRS ($n = 8$) or stability ($n = 5$), and half of the patients developed distant recurrent LMD with a median time to progression of 7 months (range, 2–15 months).[58] The retrospective cohort from Prabhu et al. reported 115 patients who received salvage cranial radiotherapy for patients who developed LMD recurrence following surgery and adjuvant SRS. Patients who received focal RT had significantly longer median survival than patients treated with WBRT (12.5 months versus 4.4 months, $p < 0.001$). Furthermore, in patients with nLMD ($n = 73$), focal RT resulted in significantly longer median survival rates than WBRT once again (13 months versus 6.6 months, $p = 0.01$). While these data are still hypothesis-generating, they suggest that focal SRS/HSRT for certain patients (e.g., focal nLMD) may afford a reasonable intracranial progression-free survival rate and allow for the postponement or avoidance of WBRT and its associated toxicities in some patients. These data need to be validated in larger more robust studies, and there remains the challenge of appropriately identifying patients most suitable for this approach.

CHECKLIST: KEY POINTS FOR CLINICAL PRACTICE

✓	TOPIC	SUGGESTIONS
	Work-up	• Full neurologic review of systems and physical exam • Contrast-enhanced MR head and complete spine • Consider CSF in hematologic malignancies or solid tumor malignancy and uncertainties with radiographic control
	Radiographic patterns of LMD	• cLMD and nLMD are distinct radiographic patterns that should be differentiated from local recurrences and new parenchymal brain metastases • On contrast-enhanced MRI, cLMD typically appears as cranial nerve enhancement and/or curvilinear enhancement of the cerebral sulci, cerebellar folia, cerebral cisterns, and brain surfaces • On contrast-enhanced MRI, nLMD may appear as bulky, mass-like, thickened, and/or irregular enhancement • Focal nLMD may be associated with favorable outcomes
	WBRT versus SRS/HSRT	• The use of HSRT/SRS may be a reasonable approach for carefully selected patients with focal nLMD and a favorable prognosis and performance status

REFERENCES

1. Nguyen TK, Sahgal A, Detsky J, et al. Predictors of leptomeningeal disease following hypofractionated stereotactic radiotherapy for intact and resected brain metastases. *Neuro-Oncology.* 2020;22(1):84–93. doi:10.1093/neuonc/no z144.
2. Chamberlain MC. Comprehensive neuraxis imaging in leptomeningeal metastasis: A retrospective case series. *CNS Oncol.* 2013;2(2):121–128. doi:10.2217/cns.12.45.
3. Clarke JL, Perez HR, Jacks LM, Panageas KS, Deangelis LM. Leptomeningeal metastases in the MRI era. *Neurology.* 2010;74(18):1449–1454. doi:10.1212/WNL.0b013e3181dc1a69.
4. Buszek SM, Chung C. Radiotherapy in leptomeningeal disease: A systematic review of randomized and non-randomized trials. *Front Oncol.* 2019;9(November):1–15. doi:10.3389/fonc.2019.01224.
5. Wang N, Bertalan MS, Brastianos PK. Leptomeningeal metastasis from systemic cancer: Review and update on management. *Cancer.* 2018;124(1):21–35. doi:10.1002/cncr.30911.
6. Cheng H, Perez-soler R. Review leptomeningeal metastases in non-small-cell lung cancer. *Lancet Oncol.* 2018;19(1):e43–e55. doi:10.1016/S1470-2045(17)30689-7.
7. Le Rhun E, Weller M, Brandsma D, et al. EANO-ESMO clinical practice guidelines for diagnosis, treatment and follow-up of patients with leptomeningeal metastasis from solid tumours. *Ann Oncol.* 2017;28(Suppl 4):iv84–iv99. doi:10.1093/annonc/mdx221.

8. Chamberlain M, Soffietti R, Raizer J, et al. Leptomeningeal metastasis: A response assessment in neuro-oncology critical review of endpoints and response criteria of published randomized clinical trials. *Neuro-Oncology*. 2014;16(9):1176–1185. doi:10.1093/neuonc/nou089.

9. Pauls S, Fischer AC, Brambs HJ, Fetscher S, Höche W, Bommer M. Use of magnetic resonance imaging to detect neoplastic meningitis: Limited use in leukemia and lymphoma but convincing results in solid tumors. *Eur J Radiol*. 2012;81(5):974–978. doi:10.1016/j.ejrad.2011.02.020.

10. Weston CL, Glantz MJ, Connor JR. Detection of cancer cells in the cerebrospinal fluid: Current methods and future directions. *Fluids Barriers CNS*. 2011;8(1):14. doi:10.1186/2045-8118-8-14.

11. Wasserstrom WR, Glass JP, Posner JB. Diagnosis and treatment of leptomeningeal metastases from solid tumors: Experience with 90 patients. *Cancer*. 1982. doi:10.1002/1097-0142(19820215)49:4<759::AID-CNCR2820490427>3.0.CO;2-7.

12. Asselain B, Vincent-salomon A, Jouve M, Dieras V, Beuzeboc P. Meningeal carcinomatosis in patients with breast carcinoma. *Cancer*. 1996:1315–1323.

13. Chamberlain M, Junck L, Brandsma D, et al. Leptomeningeal metastases: A RANO proposal for response criteria. *Neuro-Oncology*. 2017;19(4):484–492. doi:10.1093/neuonc/now183.

14. Singh SK, Leeds NE, Ginsberg LE. MR imaging of leptomeningeal metastases: Comparison of three sequences. *Am J Neuroradiol*. 2002;23(5):817–821.

15. Freilich RJ, Krol G, Deangelis LM. Neuroimaging and cerebrospinal fluid cytology in the diagnosis of leptomeningeal metastasis. *Ann Neurol*. 1995. doi:10.1002/ana.410380111.

16. Debnam JM, Mayer RR, Chi TL, et al. Most common sites on MRI of intracranial neoplastic leptomeningeal disease. *J Clin Neurosci*. 2017;45:252–256. doi:10.1016/j.jocn.2017.07.020.

17. Turner BE, Prabhu RS, Burri SH, et al. Nodular leptomeningeal disease—a distinct pattern of recurrence after post-resection stereotactic radiosurgery for brain metastases: A multi-institutional study of inter-observer reliability. *International Journal of Radiation Oncology Biology Physics*. 2018;102(3):e363–e364.

18. Cagney DN, Lamba N, Sinha S, et al. Association of neurosurgical resection with development of pachymeningeal seeding in patients with brain metastases. *JAMA Oncol*. 2019. doi:10.1001/jamaoncol.2018.7204.

19. Smirniotopoulos J, Murphy F, Rushing E, Schroeder J. Patterns of contrast enhancement in the brain and meninges. *Radiographics*. 2007;27:525–552.

20. Chamberlain MC, Kormanik PA. Prognostic significance of coexistent bulky in patients with leptomeningeal metastases. *Arch Neurol*. 1997;54:1364–1368. doi:10.1017/CBO9781107415324.004.

21. Prabhu RS, Turner BE, Asher AL, et al. A multi-institutional analysis of presentation and outcomes for leptomeningeal disease recurrence after surgical resection and radiosurgery for brain metastases. *Neuro-Oncology*. 2019. doi:10.1080/03081080xxxxxxxxx.

22. Sakaguchi M, Maebayashi T, Aizawa T, Ishibashi N, Saito T. Patient outcomes of whole brain radiotherapy for brain metastases versus leptomeningeal metastases: A retrospective study. *Asia Pac J Clin Oncol*. 2017;13(5):e449–e457. doi:10.1111/ajco.12597.

23. Ozdemir Y, Yildirim BA, Topkan E. Whole brain radiotherapy in management of non-small-cell lung carcinoma associated leptomeningeal carcinomatosis: Evaluation of prognostic factors. *J Neurooncol*. 2016. doi:10.1007/s11060-016-2179-9.

24. Gani C, Müller AC, Eckert F, et al. Outcome after whole brain radiotherapy alone in intracranial leptomeningeal carcinomatosis from solid tumors. *Strahlentherapie und Onkol*. 2012. doi:10.1007/s00066-011-0025-8.

25. Xu Q, Chen X, Qian D, et al. Treatment and prognostic analysis of patients with leptomeningeal metastases from non-small cell lung cancer. *Thorac Cancer*. 2015. doi:10.1111/1759-7714.12188.

26. El Shafie RA, Böhm K, Weber D, et al. Outcome and prognostic factors following palliative craniospinal irradiation for leptomeningeal carcinomatosis. *Cancer Manag Res*. 2019;11:789–801. doi:10.2147/CMAR.S182154.

27. Yang TJ, Wijetunga NA, Yamada J, et al. Clinical trial of proton craniospinal irradiation for leptomeningeal metastases. *Neuro-Oncology*. 2020;(June):1–10. doi:10.1093/neuonc/noaa152.

28. Bokstein F, Lossos A, Siegal T. Leptomeningeal metastases from solid tumors: A comparison of two prospective series treated with and without intra-cerebrospinal fluid chemotherapy. *Cancer*. 1998;82(9):1756–1763. doi:10.1002/(SICI)1097-0142(19980501)82:9<1764::AID-CNCR24>3.0.CO;2-1.

29. Boogerd W, Van Den Bent MJ, Koehler PJ, et al. The relevance of intraventricular chemotherapy for leptomeningeal metastasis in breast cancer: A randomised study. *Eur J Cancer*. 2004;40(18):2726–2733. doi:10.1016/j.ejca.2004.08.012.

30. Chamberlain MC. Neoplastic meningitis. *Handb Neuro-Oncology Neuroimaging Second Ed*. 2016;23(15):63–77. doi:10.1016/B978-0-12-800945-1.00008-2.

31. O'Kane GM, Leighl NB. Are immune checkpoint blockade monoclonal antibodies active against CNS metastases from NSCLC?-current evidence and future perspectives. *Transl Lung Cancer Res*. 2016;5(6):628–636. doi:10.21037/tlcr.2016.09.05.

32. Berghoff AS, Ricken G, Wilhelm D, et al. Tumor infiltrating lymphocytes and PD-L1 expression in brain metastases of small cell lung cancer (SCLC). *J Neurooncol.* 2016;130(1):19–29. doi:10.1007/s11060-016-2216-8.

33. Bachelot T, Romieu G, Campone M, et al. Lapatinib plus capecitabine in patients with previously untreated brain metastases from HER2-positive metastatic breast cancer (LANDSCAPE): A single-group phase 2 study. *Lancet Oncol.* 2013;14(1):64–71. doi:10.1016/S1470-2045(12)70432-1.

34. Montemurro F, Ellis P, Delaloge S, et al. Safety and efficacy of trastuzumab emtansine (T-DM1) in 399 patients with central nervous system metastases: Exploratory subgroup analysis from the KAMILLA study. In: *San Antonio Breast Cancer Symposium.* San Antonio, TX and Philadelphia, PA: AACR; 2016.

35. Trifiletti DM, Romano KD, Xu Z, Reardon KA, Sheehan J. Leptomeningeal disease following stereotactic radiosurgery for brain metastases from breast cancer. *J Neurooncol.* 2015;124(3):421–427. doi:10.1007/s11060-015-1854-6.

36. Kaidar-Person O, Deal AM, Anders CK, et al. The incidence and predictive factors for leptomeningeal spread after stereotactic radiation for breast cancer brain metastases. *Breast J.* 2018;24(3):424–425. doi:10.1111/tbj.12919.

37. Ma R, Levy M, Gui B, et al. Risk of leptomeningeal carcinomatosis in patients with brain metastases treated with stereotactic radiosurgery. *J Neurooncol.* 2018;136(2):395–401. doi:10.1007/s11060-017-2666-7.

38. Jo K Il, Lim DH, Kim ST, et al. Leptomeningeal seeding in patients with brain metastases treated by Gamma Knife radiosurgery. *J Neurooncol.* 2012;109(2):293–299. doi:10.1007/s11060-012-0892-6.

39. Suki D, Hatiboglu MA, Patel AJ, et al. Comparative risk of leptomeningeal dissemination of cancer after surgery or stereotactic radiosurgery for a single supratentorial solid tumor metastasis. *Neurosurgery.* 2009;64(4):664–674. doi:10.1227/01.NEU.0000341535.53720.3E.

40. Atalar B, Modlin LA, Choi CYH, et al. Risk of leptomeningeal disease in patients treated with stereotactic radiosurgery targeting the postoperative resection cavity for brain metastases. *International Journal of Radiation Oncology, Biology, Physics.* 2013;87(4):713–718. doi:10.1016/j.ijrobp.2013.07.034.

41. Strauss I, Corn BW, Krishna V, et al. Patterns of failure after stereotactic radiosurgery of the resection cavity following surgical removal of brain metastases. *World Neurosurg.* 2015;84(6):1825–1831. doi:10.1016/j.wneu.2015.07.073.

42. Johnson MD, Avkshtol V, Baschnagel AM, et al. Surgical resection of brain metastases and the risk of leptomeningeal recurrence in patients treated with stereotactic radiosurgery. *International Journal of Radiation Oncology, Biology, Physics.* 2016;94(3):537–543. doi:10.1016/j.ijrobp.2015.11.022.

43. Press RH, Zhang C, Chowdhary M, et al. Hemorrhagic and cystic brain metastases are associated with an increased risk of leptomeningeal dissemination after surgical resection and adjuvant stereotactic radiosurgery. *Neurosurgery.* 2018:1–10. doi:10.1093/neuros/nyy436.

44. Keller A, Doré M, Cebula H, et al. Hypofractionated stereotactic radiation therapy to the resection bed for intracranial metastases. *International Journal of Radiation Oncology, Biology, Physics.* 2017;99(5):1179–1189. doi:10.1016/j.ijrobp.2017.08.014.

45. Huang AJ, Huang KE, Page BR, et al. Risk factors for leptomeningeal carcinomatosis in patients with brain metastases who have previously undergone stereotactic radiosurgery. *J Neurooncol.* 2014;120(1):163–169. doi:10.1007/s11060-014-1539-6.

46. Van Der Ree TC, Dippel DWJ, Avezaat CJJ, Sillevis Smitt PAE, Vecht CJ, Van Den Bent MJ. Leptomeningeal metastasis after surgical resection of brain metastases. *J Neurol Neurosurg Psychiatry.* 1999;66(2):225–227. doi:10.1136/jnnp.66.2.225.

47. Suki D, Abouassi H, Patel AJ, Sawaya R, Weinberg JS, Groves MD. Comparative risk of leptomeningeal disease after resection or stereotactic radiosurgery for solid tumor metastasis to the posterior fossa. *Journal of Neurosurgery.* 2008;108(2):248–257. doi:10.3171/JNS/2008/108/2/0248.

48. Ahn JH, Lee SH, Kim S, et al. Risk for leptomeningeal seeding after resection for brain metastases: Implication of tumor location with mode of resection: Clinical article. *Journal of Neurosurgery.* 2012. doi:10.3171/2012.1.JNS111560.

49. Vogelbaum MA, Yu HM. Nodular leptomeningeal disease after surgery for a brain metastasis—should we be concerned? *Neuro-Oncology.* 2019;21(8):959–960. doi:10.1093/neuonc/noz081.

50. Patchell RA, Tibbs PA, Regine WF, et al. Postoperative radiotherapy in the treatment of single metastases to the brain: A randomized trial. *J Am Med Assoc.* 1998;280(17):1485–1489. doi:10.1001/jama.280.17.1485.

51. Mahajan A, Ahmed S, McAleer MF, et al. Post-operative stereotactic radiosurgery versus observation for completely resected brain metastases: A single-centre, randomised, controlled, phase 3 trial. *Lancet Oncol.* 2017;18(8):1040–1048. doi:10.1016/S1470-2045(17)30414-X.

52. Brown PD, Ballman K V., Cerhan JH, et al. Postoperative stereotactic radiosurgery compared with whole brain radiotherapy for resected metastatic brain disease (NCCTG N107C/CEC·3): A multicentre, randomised, controlled, phase 3 trial. *Lancet Oncol.* 2017;18(8):1049–1060. doi:10.1016/S1470-2045(17)30441-2.

53. Prabhu RS, Miller KR, Asher AL, et al. Preoperative stereotactic radiosurgery before planned resection of brain metastases: Updated analysis of efficacy and toxicity of a novel treatment paradigm. *Journal of Neurosurgery.* 2018;131(5):1387–1394. doi:10.3171/2018.7.JNS181293.

54. Patel KR, Burri SH, Asher AL, et al. Comparing preoperative with postoperative stereotactic radiosurgery for resectable brain metastases: A multi-institutional analysis. *Neurosurgery.* 2016;79(2):279–285. doi:10.1227/NEU.0000000000001096.

55. Prabhu RS, Press RH, Patel KR, et al. Single-fraction stereotactic radiosurgery (SRS) alone versus surgical resection and SRS for large brain metastases: A multi-institutional analysis. *International Journal of Radiation Oncology, Biology, Physics.* 2017;99(2):459–467. doi:10.1016/j.ijrobp.2017.04.006.

56. Yamamoto M. When serendipity meets creativity. *J radiosurgery SBRT.* 2011.

57. Udovicich C, Phillips C, Kok DL, et al. Neoadjuvant stereotactic radiosurgery: A further evolution in the management of brain metastases. *Curr Oncol Rep.* 2019;21(8). doi:10.1007/s11912-019-0817-z.

58. Wolf A, Donahue B, Silverman JS, Chachoua A, Lee JK, Kondziolka D. Stereotactic radiosurgery for focal leptomeningeal disease in patients with brain metastases. *J Neurooncol.* 2017;134(1):139–143. doi:10.1007/s11060-017-2497-6.

26 Minimally Invasive Percutaneous Treatment in the Palliation of Painful Bone Metastases

Elizabeth David, Harley Meirovich, and Robert Koucheki

Contents

26.1 INTRODUCTION

Percutaneous vertebral augmentation in the form of vertebroplasty (VP) or kyphoplasty (KP) with or without radiofrequency ablation (RFA) is an accepted form of palliation in bone metastases and is often used as an adjunct to external beam radiotherapy. The advantage of this technique over others is that it provides rapid pain relief in a minimally invasive nature. Rapid pain relief is an important component in providing optimal palliative care since quality of life is crucial in this subset of patients. Although new lines of treatment have helped to increase the life expectancy of patients with bone metastases, patients are usually still in the advanced stages of their disease with a median survival reported to be less than 1 year (1). Given this, quality of life in this cohort becomes increasingly important. Bone metastases can be a significant source of disability and pain. Rapid relief of pain and maintaining mobility are essential goals when treating these patients. Some common side effects of bone metastases include fractures, instability, and considerable discomfort. In patients with metastatic bone lesions, pathological fractures can lead to pain crisis and further disability, both of which can adversely affect quality of life (2). Conservative strategies to manage these pathological fractures usually involve analgesics, often in the form of opioids, and bed rest; these can be effective but can decrease quality of life parameters for the patient. Progression in pain can lead to increased opioid utilization. This may create a cascade of negative side effects in patients, including physiological dependence, somnolence, nausea, and constipation. Often, radiotherapy is given concurrently, but this treatment requires time to achieve optimal efficacy; this waiting period can be difficult for the patient, especially if there is a pain flare post-radiation. Multiple studies have established that percutaneous vertebral and pelvic augmentation is a safe and effective form of palliation in this population (3–6). It can provide rapid pain relief and enhanced stability with minimal side effects; pain relief is often immediate, within 24–48 hours in most studies. Objective measures of reduction in pain intensity are often noted by a decrease in analgesic requirements. Patient satisfaction with the treatment in most studies exceeded 60%. A systematic review of vertebral augmentation involving VP or KP for cancer-related vertebral compression fractures in 4,235 patients (7) evaluated the safety and effectiveness of VP and KP; it demonstrated that these interventions rapidly reduce pain intensity, decrease the need for opioid pain medication, and improve mobility and performance status of the patient. VP and KP have low complication rates and do not require general anesthesia. These procedures are typically performed in an outpatient setting (8). Osteoplasty has also been effective in addressing pain and disability resulting from metastatic tumors in extraspinal locations including the bony pelvis (sacrum) and acetabulum (9).

26.2 VERTEBRAL AND PELVIC AUGMENTATION

Vertebral augmentation, also known as vertebroplasty (VP), is a minimally invasive, radiologically guided procedure that consists of percutaneous injection of surgical cement (polymethylmethacrylate or PMMA) into the vertebral body. The term "osteoplasty" is often used when cement injection is extraspinal. Vertebroplasty has been most heavily studied in the context of osteoporotic fractures. More recent studies, however, have consistently shown the benefit of vertebral augmentation when used in spinal bone metastases (10–12). A joint position statement supported by multiple interventional radiology and neurosurgical societies acknowledged the growing body of evidence supporting this procedure and stated that "percutaneous vertebroplasty remains medically appropriate therapy for the treatment of painful vertebral compression fractures refractory to nonoperative medical therapy and for vertebrae weakened by neoplasia when performed for medical indications outlined in the published standards" (3).

The etiology of pain resulting from metastatic bone lesions is complex; however, the mechanical compressions of periosteal nociceptors, combined with the cascade of cytokines and biochemical factors produced by neoplastic cells, are important factors (13). Cement injected into the vertebral body (VB) or pelvic bone (osteoplasty) provides structural stability to the treated bone and solidifies pathologic fractures, if present. The manner in which pain relief is obtained is thought to be from the closure of fracture clefts and enhanced stability, leading to reduced compression of periosteal nociceptors. Other mechanisms of pain relief have been hypothesized; one other possible mechanism is the destruction of nerve ending of pain fibers as a result of the exothermic reaction associated with the hardening of polymethylmethacrylate (PMMA). The cytotoxicity of PMMA may also be related to vascular or chemical effects (14). This cytotoxicity may also reduce the overall load of the pain inducing pro-inflammatory cytokines and bio-humoral factors produced by neoplastic cells and the surrounding tissue. A randomized controlled trial compared the injection of nonparticulate corticosteroids followed by vertebroplasty to vertebroplasty alone and showed further increase in early pain reduction when vertebroplasty was combined with corticosteroids up to 1 month post-procedure (13). Complication rates related to vertebroplasty vary greatly in the literature. Major complications are uncommon, especially with proper intraprocedural imaging. The complication rate per level treated ranges from 0 to 17%, whereas it is 0 to 33% per patient, as many patients have a few vertebral body levels treated (14). Most of these complications are minor. The majority of complications are due to cement leakage or extravasation outside of the target, either the vertebral body or pelvic bone. Most cement leaks are inconsequential (15, 16). Rare neurological complications have been described (17, 18). In one series, Cotton et al. (19) described 1 case out of 258, where the patient developed spinal cord compression requiring surgical decompression. Other rare complications include rib or pedicle fractures, pneumothorax, discitis, and pulmonary or fat emboli.

26.3 RADIOFREQUENCY ABLATION

Vertebral and pelvic augmentation have more recently been done following ablative therapy of the lesions treated. Ablative treatment in bone was first used in benign primary bone tumors, namely osteoid osteomas, more than 20 years ago. Percutaneous ablation has since become well established as the primary approach to treat osteoid osteomas (20, 21). More recently, ablative treatment has been used to cure painful bone metastases, usually in combination with percutaneous augmentation (vertebroplasty in the spine or osteoplasty in extraspinal sites). Radiofrequency ablation (RFA) is the most commonly used ablative modality that has been used in combination with osteoplasty; this combined procedure has been shown to be safe and effective in the treatment of painful bone metastases (22, 23). RFA is more commonly used when spinal fractures are associated with significant tumor mass (7). It remains unclear if there is a synergistic effect when RFA is combined with vertebroplasty or osteoplasty (24–26). Different factors that may contribute to the therapeutic effect of RFA prior to osteoplasty or vertebroplasty include the reduction of interleukins and TNF responsible for sensitization of nerve endings and the inhibition of osteoclasts. Direct destruction of sensible nerve endings within the periosteum may also play a role (22).

It has been anecdotally noted that the use of RFA prior to vertebroplasty or osteoplasty improves the control of cement injection (22, 27–30). RFA may diminish tumor bulk prior to cement deposition. It is postulated that ablation of the lesion changes the consistency of the tumor, which can shrink the tumor and destroy the "cohesion" of the tumor cells (22). This creation of a "thermal cavity" allows for a more controlled injection of cement. In a comparative study looking at various RFA systems, it was highlighted that the biomechanical stability of the metastatic spine was more significantly dependent on the distribution of cement rather than cement filling larger tumor-ablated volumes. The best spinal stability was achieved when an RFA system combined with vertebroplasty directed to yield a cement distribution that supported the posterior wall of the vertebral body (31).

Previous research has found that RFA produced an immediate reduction in pain scores, with a continued decrease in pain at various time points up to 3 months post-RFA (32–34), and RFA used prior to vertebroplasty resulted in fewer posterior and venous cement leaks compared to vertebroplasty alone (34). Furthermore, theoretically, the use of RFA prevents displacement of tumor cells by the cement into adjacent vascular spaces and therefore increases the safety of cement injection.

Previously, two randomized controlled trials have investigated the possibility of combining percutaneous vertebroplasty with interstitial implantation of Iodine-125 (^{125}I) seeds for metastatic spinal tumors. In these trials, combining ^{125}I seeds with vertebroplasty was shown to further decrease pain levels and improve the general quality of life compared to vertebroplasty alone; however, overall survival did not significantly increase (35, 36). In a retrospective study, the combination treatment of percutaneous vertebroplasty with RFA was compared with percutaneous vertebroplasty combined with ^{125}I seed implantation. RFA proved superior to ^{125}I seed implantations in terms of the overall pain relief and improvement of motion dysfunction (37).

One of the concerns when using RFA is the possibility of thermal injury to adjacent structures. This is particularly relevant in the spine due to proximity to the neural structures. Injury to adjacent structures has been described in the past (23). The insulative nature of cortical bone gives a measure of safety especially for structures in the epidural space. The risk of injury can be further reduced with bipolar design of the RFA probe. Most conventional RFA devices use a monopolar probe. Newer designs are bipolar in nature, limiting the deposition of energy outside the desired field. One such system has been used at our institution with good success (OsteoCool RF Ablation System, Medtronic, Inc., 99 Hereford Street, Brampton, Ontario, L6Y 0R3, Canada) (38, 39). The OsteoCool RF Ablation System uses an active tip in the ablation probe that is cooled using sterile water; this allows for better temperature control and can create larger thermal cavities while avoiding excessive heating. Improved cement distribution and vertebral stability have been achieved with bipolar probes (31).

26.4 PATIENT SELECTION

In order to achieve optimal and durable results with vertebral and pelvic augmentation, patient selection is crucial. The pain should localize to the region of pathology corresponding with imaging studies, and there should be a reasonable temporal relationship between the development of new pathology and onset of symptoms. Pain should be limited to one or few sites; in diffuse disease, it can be difficult to identify the source of maximal pain. Multiple levels can be treated in a staged fashion to optimize pain relief. However, patients with numerous lesions and diffuse rather than focal pain are better treated with systemic therapy or wide-field conventional radiation rather than a localized approach (40).

The degree and nature of the pain are also important factors in patient selection. Patients with mild pain may benefit from the use of analgesics alone, which are often effective. As for the nature of the pain, it is important to note that this intervention provides a little benefit in the treatment of nerve root or radicular pain. Anecdotally, we see that patients with nerve root or radicular pain who undergo vertebral or pelvic augmentation express that they have not experienced overall pain relief, as their focal pain may have subsided but their radiating pain persists, which confounds objective measures of pain scores evaluating the efficacy of PV or KP.

In the end, the structural integrity of weight-bearing or axial loading regions in the skeleton, namely lesions in the spine and peri-acetabular region, is another important consideration. Patients with lesions in these regions are therefore often good candidates for cement augmentation.

The procedure can be done concurrently with radiation therapy and can be performed either before or after radiotherapy. It can be done on an outpatient basis, under local anesthesia and conscious sedation, and in most centers does not require a general anesthetic. The minimally invasive nature of the procedure, combined with its ability to provide rapid pain relief, contributes to its utility in patients with bone metastases as this population requires preservation of quality of life and mobility, and they are often in the advanced stages of disease and may not tolerate a more aggressive approach.

26.5 ILLUSTRATIVE CASES

26.5.1 CASE 1

An 88-year-old male with metastatic prostate cancer with multiple lytic and sclerotic bone metastases recently developed a pathologic fracture of the left iliac wing (Figure 26.1). He was experiencing severe left pelvic pain (rated at 10 out of 10). Under conscious sedation and local anesthesia and under CT guidance, an 11-gauge bone needle was inserted into the left iliac wing. A 13-gauge bone drill was needed to get across sclerotic regions of the bone and across the pathologic fracture (Figure 26.2). Under fluoroscopic guidance, the RFA probe was inserted (Figure 26.3), and RFA was performed. Live monitoring of the temperature achieved as well as that of the power used and measured impedance was done as per routine (Figure 26.4). A temperature of 70°C was achieved and kept constant for 15 minutes. Once done, the RFA probe was removed, and cement (Kyphon—PMMA) injection was done through the same needle under fluoroscopic guidance (Figure 26.5). Post-procedure CT shows filling of the fracture cleft (Figure 26.6).

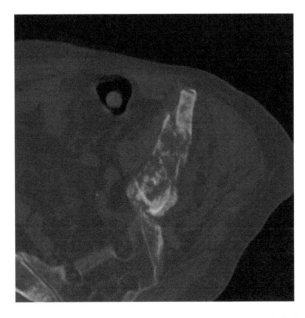

Figure 26.1 Pre-procedure CT shows multiple mixed lytic/sclerotic metastases in left iliac wing, with a subacute pathologic displaced fracture through the iliac wing.

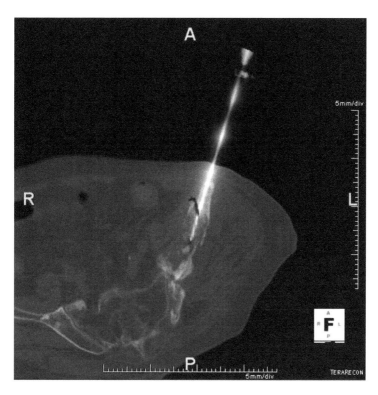

Figure 26.2 A bone needle is being inserted under CT guidance into the iliac bone via the iliac crest, and a bone drill is used for advancement across the pathologic fracture.

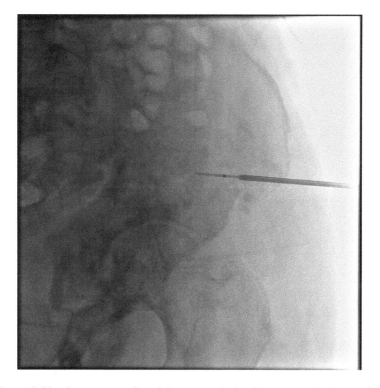

Figure 26.3 The bone drill has been removed, and the RFA probe has been inserted. The end of the active tip of probe is denoted by a radiopaque marker. With the bipolar design, no heating occurs posterior to this marker.

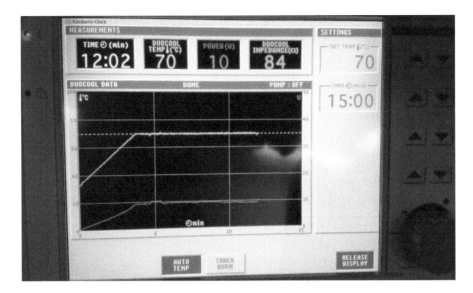

Figure 26.4 Live monitoring of the RF ablation, showing the measured temperature, power, and impedance used. Temperature of 70°C is achieved and then kept constant for 15 minutes. This achieves an ablation zone of 2 by 3 cm.

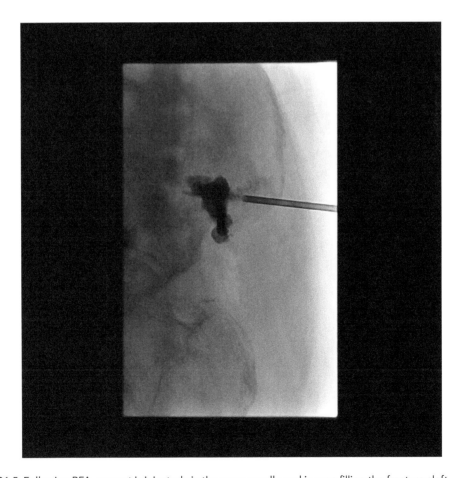

Figure 26.5 Following RFA, cement is injected via the same needle and is seen filling the fracture cleft.

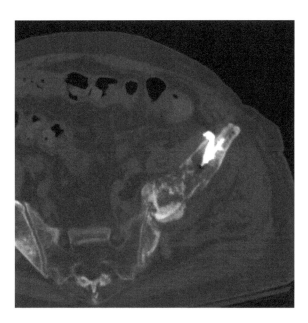

Figure 26.6 Post-procedure confirms good cement filling of the fracture cleft. A small amount of contrast extends through the cleft anterior to the bone of no consequence.

The patient had near-immediate pain relief (down to 4 out of 10). Follow-up CT over a year after the procedure showed interval stable appearance without progression of the fracture.

26.5.2 CASE 2

A 69-year-old male with metastatic prostate cancer was found to have a new painful lytic right acetabular lesion, at risk for fracture (Figure 26.7). Under conscious sedation and local anesthesia, a 13-gauge Cook M1M Osteo Site needle was inserted into the acetabular lesion by an anterior approach, using *XperCT* guide (Figure 26.8). A bipolar RFA probe was then used to perform a 12-minute ablation. This

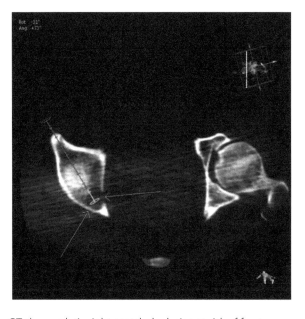

Figure 26.7 A preliminary CT shows a lytic right acetabular lesion at risk of fracture.

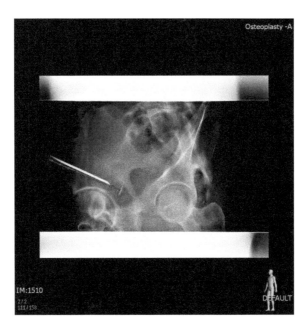

Figure 26.8 XperCT guidance of needle into the lytic lesion.

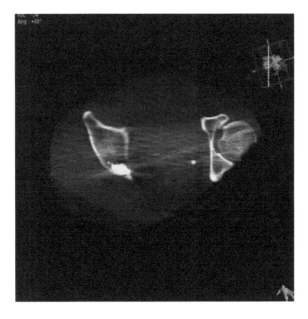

Figure 26.9. CT done post-cement delivery into the lesion

was followed by cement injection using the *Duro-Ject* system, achieving good filling of the lytic lesion (Figure 26.9). Though not immediate, the patient experienced improvement in right hip pain following the procedure.

26.5.3 CASE 3

An 83-year-old male with metastatic prostate cancer who developed a pathologic, painful metastasis at L2 was referred for treatment. RFA-assisted vertebroplasty was performed at L2, using a left unipedicular approach. Needle placement for vertebroplasties was done under biplanar fluoroscopy. Once in place, the RFA probe was inserted, and a 15-minute ablation took place. Following this, cement (PMMA) was injected. This patient had pain rated at 9.5 out of 10 prior to the procedure, which improved to 5 out of 10, at the end of the procedure (see Figures 26.10–26.12).

26.5.4 CASE 4

A 73-year-old male with metastatic renal cell carcinoma was found to have a painful lytic bone metastasis in the T12 vertebral body. Pre-procedure CT and MRI are shown in Figures 26.13 and 26.14. Pain was

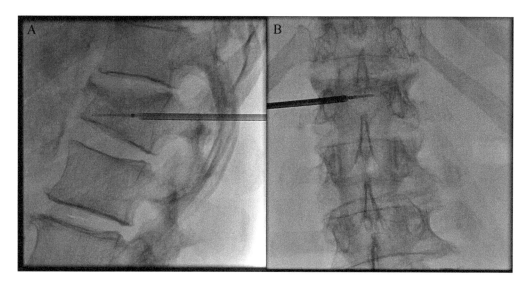

Figure 26.10 (A: lateral, B: frontal). A pathologic fracture is seen at L2. Needle placement was done under fluoroscopic guidance and using a unipedicular approach. The needle is placed just beyond the midline in order to optimize cement distribution. The RFA probe has been placed; its radiopaque marker has been uncovered by the outer needle. Ablation can now be performed.

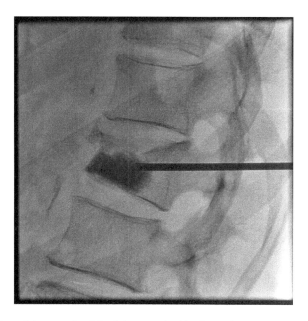

Figure 26.11 Cement filling of the anterior 2/3 of the vertebral body can be seen.

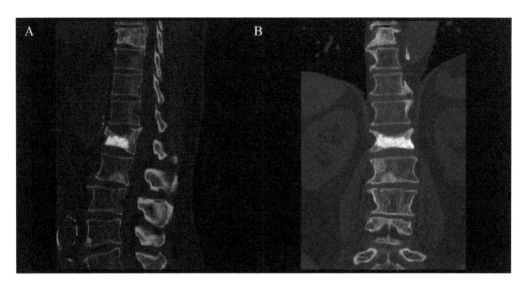

Figure 26.12 Post-procedure CT: (A) sagittal and (B) coronal. Good cross filling laterally can be seen without any leak.

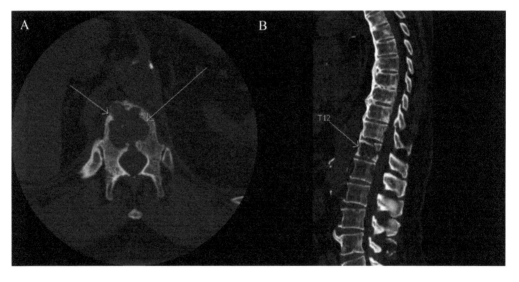

Figure 26.13 Pre-procedure CT: (A) axial and (B) sagittal. (B) A lytic lesion is seen at T12. There is breech of the posterior cortex and a superior endplate fracture posteriorly.

rated at 6 out of 10 prior to the procedure. An RFA-assisted vertebroplasty was done using a unipedicular access from the left pedicle (Figure 26.15). Good cement filling of the majority of the lesion was achieved as seen fluoroscopically (Figure 26.16) and on post-procedure CT (Figure 26.17) and MRI (Figure 26.18). Pain was alleviated down to 2.5 out of 10 following the procedure.

26.5.5 CASE 5

A 77-year-old male with multiple myeloma was admitted with a significant pain crisis due to multiple compression fractures in the lumbar spine. He was bed-bound and heavily dependent on narcotic medications. He required CADD pump during admission. He underwent conventional (no RFA used prior to cement injection) vertebroplasty at L1, L2, and L3 (Figure 26.19). He had tremendous pain relief almost

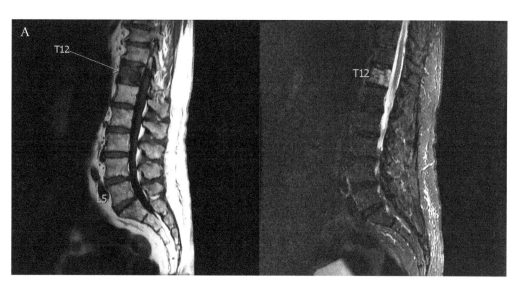

Figure 26.14 Pre-procedure MRI. (A) T1 sagittal and (B) STIR sagittal. A metastatic lesion replaces the normal bright fatty marrow by dark signal on T1. The lesion is hyperintense on STIR.

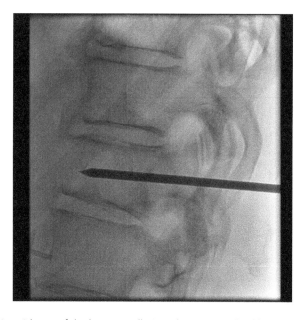

Figure 26.15 Fluoroscopic guidance of the bone needle into the T12 vertebral body. The lytic lesion is only faintly appreciated.

immediately following the procedure and was able to move around much better and required lesser doses of narcotics. On follow-up after18 days following the procedure, he only had mild residual lower back pain. On follow-up 6 weeks after the procedure, he had essentially no residual pain. Post-procedure CT (Figure 26.20) and MRI (Figure 26.21) are shown. Note this patient's diffuse disease, best appreciated on the MRI (Figure 26.21). This is a good example of a patient with diffuse disease but whose pain was more focal and localized to the treated levels. He, therefore, benefited from treatment of these levels.

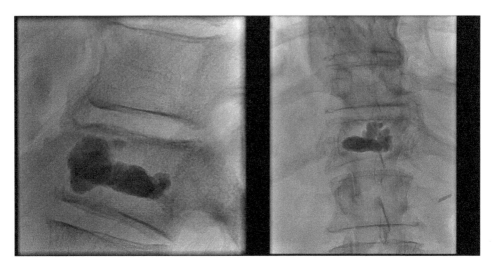

Figure 26.16 Following RF ablation, cement has been injected and fills the majority of the lesion.

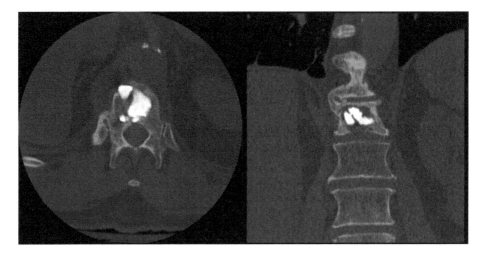

Figure 26.17 Post-procedure CT. Cement fills the majority of the lesion.

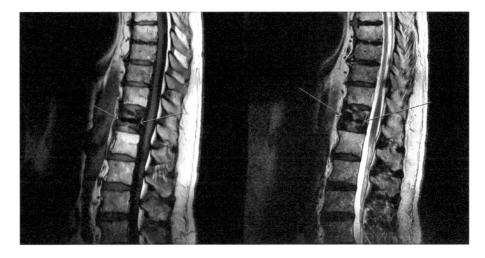

Figure 26.18 Post-procedure MRI. Cement (dark signal) is seen within the T12 vertebral body.

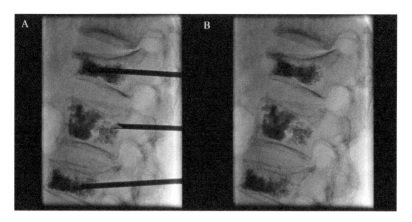

Figure 26.19 (A) Lateral fluoroscopic image shows needle placement into the L1, L2, and L3 vertebral bodies. The L2 and L3 needles have already been retracted. (B) Cement has been placed and the needles were removed.

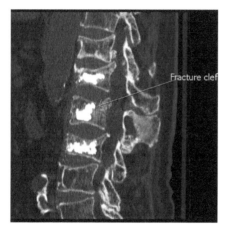

Figure 26.20 Post-procedure CT shows good filling of L1, L2, and L3 vertebral bodies, including into the superior end plate fracture at L2.

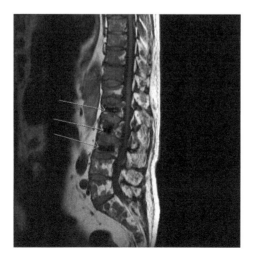

Figure 26.21 Post-procedure MRI (T1 sagittal) shows cement delivery into the compressed L1, L2, and L3 vertebral bodies. Note that the majority of the spine is infiltrated by tumor (low signal has replaced the majority of visualized bones), except for a few islands of normal residual fatty (bright) marrow, as seen in the posterior aspect of L4 and L5.

26.6 CONCLUSION

In summary, percutaneous vertebral and pelvic augmentation can provide rapid, durable pain relief without the need for major surgery and is an effective adjunct to radiotherapy in palliating bone metastases.

REFERENCES

1. Ellsworth S, Alcorn S, Hales R, McNutt T, DeWeese T, Smith T. Patterns of care among patients receiving radiation therapy for bone metastases at a large academic institution. *International Journal of Radiation Oncology, Biology, Physics*. 2014;89(5):1100–1105.

2. Piccioli A. CORR insights®: What factors are associated with quality of life, pain interference, anxiety, and depression in patients with metastatic bone disease? *Clinical Orthopedics and Related Research*. 2016;475(2): 508–510. doi:10.1007/s11999-016-5185-5.

3. Barr J, Jensen M, Hirsch J, McGraw J, Barr R, Brook A, et al. Position statement on percutaneous vertebral augmentation: A consensus statement developed by the Society of Interventional Radiology (SIR), American Association of Neurological Surgeons (AANS) and the Congress of Neurological Surgeons (CNS), American College of Radiology (ACR), American Society of Neuroradiology (ASNR), American Society of Spine Radiology (ASSR), Canadian Interventional Radiology Association (CIRA), and the Society of NeuroInterventional Surgery (SNIS). *Journal of Vascular and Interventional Radiology*. 2014;25(2):171–181.

4. Chi J, Gokaslan Z. Vertebroplasty and kyphoplasty for spinal metastases. *Current Opinion in Supportive and Palliative Care*. 2008;2(1):9–13.

5. Cai Z. A preliminary study of the safety and efficacy of radiofrequency ablation with percutaneous kyphoplasty for thoracolumbar vertebral metastatic tumor treatment. *Medical Science Monitor*. 2014;20:556–553.

6. Voormolen M, Lohle P, Lampmann L, van den Wildenberg W, Juttmann J, Diekerhof C, et al. Prospective clinical follow-up after percutaneous vertebroplasty in patients with painful osteoporotic vertebral compression fractures. *Journal of Vascular and Interventional Radiology*. 2006;17(8):1313–1320.

7. Health Quality Ontario. Vertebral augmentation involving vertebroplasty or kyphoplasty for cancer-related vertebral compression fractures: A systematic review. *Ont Health Technol Assess Ser* [Internet]. 2016 May;16(11):1–202. Available from: www.hqontario.ca/Evidence-to-Improve-Care/Journal-Ontario-Health-Technology-Assessment-Series.

8. Barragán-Campos H, Le Faou A, Rose M, et al. Percutaneous vertebroplasty in vertebral metastases from breast cancer: Interest in terms of pain relief and quality of life. *Interventional Neuroradiology*. 2014;20(5):591–602. doi:10.15274/inr-2014-10084.

9. Anselmetti G. Osteoplasty: Percutaneous bone cement injection beyond the spine. *Semin Intervent Radiol*. 2010;27(2):199–208. doi:10.1055/s-0030-1253518.

10. Kaemmerlen P, Thiesse P, Bouvard H, Biron P, Mornex F, Jonas P. Vertebroplastie percutanee dans le traitement des metastases: Technique et resultats. *Journal de Radiologie*. 1989;70(10):557–562.

11. Weill A, Chiras J, Simon J, Rose M, Sola-Martinez T, Enkaoua E. Spinal metastases: Indications for and results of percutaneous injection of acrylic surgical cement. *Radiology*. 1996;199(1):241–247.

12. Fourney D, Schomer D, Nader R, Chlan-Fourney J, Suki D, Ahrar K, et al. Percutaneous vertebroplasty and kyphoplasty for painful vertebral body fractures in cancer patients. *Journal of Neurosurgery*. 2003;98(1):21–30.

13. Basile A, Masala S, Banna G, et al. Intrasomatic injection of corticosteroid followed by vertebroplasty increases early pain relief rather than vertebroplasty alone in vertebral bone neoplasms: Preliminary experience. *Skeletal Radiology*. 2012;41(4):459–464. doi:10.1007/s00256-011-1300-6.

14. Halpin R, Bendok B, Liu J. Minimally invasive treatments for spinal metastases: Vertebroplasty, kyphoplasty, and radiofrequency ablation. *The Journal of Supportive Oncology*. 2004;2(4):339–355.

15. Zoarski G, Snow P, Olan W, Stallmeyer M, Dick B, Hebel J, et al. Percutaneous vertebroplasty for osteoporotic compression fractures: Quantitative prospective evaluation of long-term outcomes. *Journal of Vascular and Interventional Radiology*. 2002;13(2):139–148.

16. Cotten A, Dewatre F, Cortet B, Assaker R, Leblond D, Duquesnoy B, et al. Percutaneous vertebroplasty for osteolytic metastases and myeloma: Effects of the percentage of lesion filling and the leakage of methyl methacrylate at clinical follow-up. *Radiology*. 1996;200(2):525–530.

17. Lee B, Lee S, Yoo T. Paraplegia as a complication of percutaneous vertebroplasty with polymethylmethacrylate. *Spine*. 2002;27(19):419–422.

18. Harrington K. Major neurological complications following percutaneous vertebroplasty with polymethylmethacrylate: A case report. *The Journal of Bone and Joint Surgery American Volume*. 2001;83(7):1070–1073.

19. Cotten A, Boutry N, Cortet B, Assaker R, Demondion X, Leblond D, et al. Percutaneous vertebroplasty: State of the art. *RadioGraphics*. 1998;18(2):311–320.

20. Rosenthal D, Callstrom M. Critical review and state of the art in interventional oncology: Benign and metastatic disease involving bone. *Radiology*. 2012;262(3):765–780.

21. Rosenthal D, Alexander A, Rosenberg A, Springfield D. Ablation of osteoid osteomas with a percutaneously placed electrode: A new procedure. *Radiology*. 1992;183(1):29–33.

22. Hoffmann R, Jakobs T, Trumm C, Weber C, Helmberger T, Reiser M. Radiofrequency ablation in combination with osteoplasty in the treatment of painful metastatic bone disease. *Journal of Vascular and Interventional Radiology*. 2008;19(3):419–425.

23. Munk P, Rashid F, Heran M, Papirny M, Liu D, Malfair D, et al. Combined cementoplasty and radiofrequency ablation in the treatment of painful neoplastic lesions of bone. *Journal of Vascular and Interventional Radiology*. 2009;20(7):903–911.

24. Anselmetti G, Manca A, Ortega C, Grignani G, DeBernardi F, Regge D. Treatment of extraspinal painful bone metastases with percutaneous cementoplasty: A prospective study of 50 patients. *CardioVascular and Interventional Radiology*. 2008;31(6):1165–1173.

25. Kojima H, Tanigawa N, Kariya S, Komemushi A, Shomura Y, Sawada S. Clinical assessment of percutaneous radiofrequency ablation for painful metastatic bone tumors. *CardioVascular and Interventional Radiology*. 2006;29(6):1022–1026.

26. Proschek D. Prospective pilot-study of combined bipolar radiofrequency ablation and application of bone cement in bone metastases. *Anticancer Research*. 2009;29(7):2787–2792.

27. Halpin R, Bendok B, Sato K, Liu J, Patel J, Rosen S. Combination treatment of vertebral metastases using image-guided percutaneous radiofrequency ablation and vertebroplasty: A case report. *Surgical Neurology*. 2005;63(5):469–474.

28. Schaefer O, Lohrmann C, Herling M, Uhrmeister P, Langer M. Combined radiofrequency thermal ablation and percutaneous cementoplasty treatment of a pathologic fracture. *Journal of Vascular and Interventional Radiology*. 2002;13(10):1047–1050.

29. Georgy B. Bone cement deposition patterns with plasma-mediated radio-frequency ablation and cement augmentation for advanced metastatic spine lesions. *American Journal of Neuroradiology*. 2009;30(6):1197–1202.

30. Lane M, Le H, Lee S, Young C, Heran M, Badii M, et al. Combination radiofrequency ablation and cementoplasty for palliative treatment of painful neoplastic bone metastasis: Experience with 53 treated lesions in 36 patients. *Skeletal Radiology*. 2010;40(1):25–32.

31. Pezeshki P, Davidson S, Murphy K, et al. Comparison of the effect of two different bone-targeted radiofrequency ablation (RFA) systems alone and in combination with percutaneous vertebroplasty (PVP) on the biomechanical stability of the metastatic spine. *European Spine Journal* 2015;25(12):3990–3996. doi:10.1007/s00586-015-4057-0.

32. Bagla S, Sayed D, Smirniotopoulos J, et al. Multicenter prospective clinical series evaluating radiofrequency ablation in the treatment of painful spine metastases. *CardioVascular and Interventional Radiology*. 2016;39(9):1289–1297. doi:10.1007/s00270-016-1400-8.

33. Wallace AN, Huang AJ, Vaswani D, Chang RO, Jennings JW. Combination acetabular radiofrequency ablation and cementoplasty using a navigational radiofrequency ablation device and ultrahigh viscosity cement: Technical note. *Skeletal Radiology*. 2016;45(3):401–405. doi:10.1007/s00256-015-2263-9.

34. David E, Kaduri S, Yee A, et al. Initial single center experience: Radiofrequency ablation assisted vertebroplasty and osteoplasty using a bipolar device in the palliation of bone metastases. *Ann Palliat Med*. 2017 (January). doi:10.21037/apm.2016.12.02.

35. Yang Z, Yang D, Xie L, et al. Treatment of metastatic spinal tumors by percutaneous vertebroplasty versus percutaneous vertebroplasty combined with interstitial implantation of ^{125}I seeds. *Acta Radiol*. 2009;50(10):1142–1148. doi:10.3109/02841850903229133.

36. Yang Z, Tan J, Zhao R, et al. Clinical investigations on the spinal osteoblastic metastasis treated by combination of percutaneous vertebroplasty and ^{125}I seeds implantation versus radiotherapy. *Cancer Biotherapy and Radiopharmaceuticals*. 2013;28(1):58–64. doi:10.1089/cbr.2012.1204.

37. Lu C, Shao J, Wu Y, et al. Which combination treatment is better for spinal metastasis. *Am J Ther*. 2019;26(1):e38–e44. doi:10.1097/mjt.0000000000000449.

38. Gofeld M, Yee A, Whyne C, Akens M, Pezeshki P, Woo J. New palliative intervention for painful metastatic bone disease: The OsteoCool™ system. *European Cells and Materials*. 2012;23(3):11.

39. Woo J, Pezeshki P, Yee A, Whyne C, Akens M, Won E, et al. *Validation of a Novel Bone Tumor RF Ablation System: Physics and Animal Data*. Boston: Group de recherche interdisciplinaire sur les biomatériaux ostéo-articulaires injectable (GRIBOI); 2011.

40. Nicholas Kurup A, Callstrom M. Ablation of musculoskeletal metastases: Pain palliation, fracture risk reduction, and oligometastatic disease. *Techniques in Vascular and Interventional Radiology*. 2013;16(4):253–261.

Proton Stereotactic Radiosurgery for Brain Metastases

Kevin Oh

Contents

27.1 INTRODUCTION

Brain metastases are extremely common among oncology patients, with an estimated 170,000 new cases per year in the United States (Sperduto et al., 2010). While whole brain radiation therapy (WBRT) had been the mainstay of treatment for many decades, multiple randomized trials have demonstrated no benefit in overall survival when WBRT is added to either surgical resection or stereotactic radiosurgery (SRS) alone (Aoyama et al., 2006; Kocher et al., 2011; Patchell et al., 1998) for patients with one to four brain metastases. In this context, along with the availability of more effective CNS penetrant systemic therapy, SRS has become increasingly more common in the treatment for patients with brain metastases.

Proton therapy is the most common form of heavy-charged particle therapy and has dosimetric advantages over photon-based therapy in delivering equivalent therapeutic dose to the target while minimizing exit dose. Proton therapy is generated by initially ionizing hydrogen and accelerating the resultant protons within a cyclotron. When they reach a desired energy, they are ushered into a beam with a linear trajectory. Upon entering the body, the velocity of protons rapidly decreases due to Coulomb interactions with negatively charged tissues. As this occurs, the energy transferred exponentially increases until the particles come to rest. The "Bragg peak" is the region of maximal energy transfer. Because the range of the proton beam is energy dependent, modulation of energy can target a desired depth. By superimposing beams of varying energies, one can create a "spread out Bragg peak," which delivers the intended dose to a three-dimensional target. Brass apertures shape the dose according to the target shape, while a range compensators allow for more conformal shaping of the distal edge of the beam.

The dosimetric advantage of proton-based over photon-based radiation therapy is the minimization of both entrance and exit dose around the target and therefore the potential reduction of late toxicities related to low-medium doses of radiation. In cases of fractionated radiation therapy for primary brain tumors, the clinical implication is the theoretical reduction in the risk of late neurocognitive deficits, neuroendocrine changes, and second malignancy, which are believed to be associated with lower doses of radiation. In the context of proton radiosurgery, the minimization of low-medium dose penumbra may reduce the risk of late radiation changes including radionecrosis (Figure 27.1).

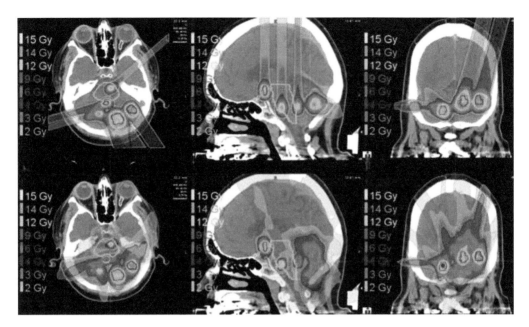

Figure 27.1 Comparison of radiosurgery plans delivered by proton therapy (passive scattering on Stereotactic Alignment for Radiosurgery (STAR) device, top row) versus LINAC (VMAT, bottom row).

27.2 INDICATIONS

Beginning in 1961, proton radiosurgery has been used at the Harvard Cyclotron Laboratory/Massachusetts General Hospital for a wide variety of benign and malignant primary brain tumors, brain metastases, and arteriovenous malformations (G. Harsh et al., 1999). Loeffler, Shih, and colleagues at the Massachusetts General Hospital have published series on the use of single-fraction proton therapy in a variety of benign tumors of the central nervous system including arteriovenous malformations (Barker et al., 2003; Hattangadi et al., 2012; Hattangadi-Gluth et al., 2014; Walcott et al., 2014), vestibular schwannomas (G. R. Harsh et al., 2002; Weber et al., 2003), pituitary adenomas (Petit et al., 2007), and meningiomas (Halasz et al., 2011). More recently, Atkins et al. published the experience of proton radiosurgery for brain metastases (Atkins et al., 2018).

27.3 TREATMENT

27.3.1 TREATMENT PLANNING

The steps of target volume and normal tissue contouring, immobilization, and treatment planning are analogous to those of LINAC -based radiosurgery. At the Massachusetts General Hospital, immobilization for either modality uses a modified Gill-Thomas-Cosman frame (Integra-Radionics Inc, Burlington, MA), which relies on an integrated custom dental mold. For those with insufficient dentition, a rigid thermoplastic mask is utilized. The CT simulation acquires thin cut (e.g., 1.25 mm) axial images with contrast. Registration with a recent MRI T1 gadolinium sequence is performed when available. To date, all proton radiosurgery has been planned and delivered using passive scattering.

27.3.2 TREATMENT DELIVERY

Over several decades, there have been several challenges in implementing proton therapy as simple fraction, high precision SRS. Most glaringly, there has never been a commercially available system for proton radiosurgery. Pioneered at the Massachusetts General Hospital, the first iteration of proton radiosurgery was delivered at the Harvard Cyclotron Laboratory in Cambridge, Massachusetts, beginning in 1961 using

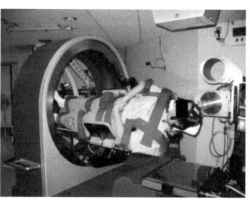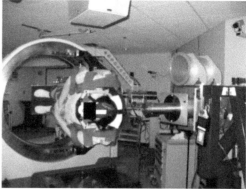

Figure 27.2 STAR device utilized at the Massachusetts General Hospital for proton radiosurgery.

an isocentric rotational alignment system while the patient was in a seated position. In 1993, Chapman and colleagues developed the STAR (Figure 27.2) device on which the patient was positioned supine and could move in five independent axes (three linear and two rotational) around a fixed beam (G. Harsh et al., 1999). At the current time, the primary difference between proton- and Linac-based radiosurgery is the strategy for image guidance. The current system for image guidance during proton SRS does not include cone-beam CT. Instead, three fiducial markers are placed in the outer table of the skull prior to CT simulation as reference coordinates.

27.4 OUTCOMES

In the largest published series of proton SRS for brain metastases, Atkins et al. reviewed the outcomes of 815 metastases treated in 370 patients treated with proton SRS between 1991 and 2016 at the Harvard Cyclotron Laboratory (Cambridge, Massachusetts) or the Francis H Burr Proton Therapy Center at Massachusetts General Hospital (Boston, Massachusetts) (Atkins et al., 2018). Immobilization was done using a modified Gill–Thomas–Cosman frame with dental fixation or thermoplastic mask with modified thermoplastic frame. Image guidance was performed using three fiducial markers surgically inserted into the outer table of the skull. The median dose delivered was 18 Gy RBE typically prescribed to the 90% isodose line. All therapy was either 160- or 230-MeV passively scattered beams. At a median follow-up of 9.2 months, the cumulative incidence of local failure at 6 and 12 months was 4.3% and 8.5%, respectively. At 6 and 12 months, the cumulative incidence of distant brain failure was 39.1% and 48.2%, and overall survival was 76% and 51.5%, respectively. Of note, most patients received prior radiation (55%) with the majority in the form of WBRT (45%). On multivariate analysis for local failure, only prior cranial radiation remained a significant predictor of reduced risk (HR 0.61, $p = 0.037$). On multivariate analysis for survival, melanoma histology was associated with higher risk of mortality (HR 1.83, $p < 0.006$), although most were treated in the era prior to immunotherapy and targeted therapies.

27.5 TOXICITY

Several groups have demonstrated the value of brain volume in the medium dose range (V8–V16) in predicting the risk of late radionecrosis (Blonigen et al., 2010; Minniti et al., 2011). For example, in an analysis of 206 patients treated with SRS for brain metastases, V10 and V12 were found on multivariate analysis to be most predictive for radionecrosis. Cases with V12 more than 8.5 cm³ carried a radionecrosis risk of more than 10% (Minniti et al., 2011). In their published series of proton radiosurgery for brain metastases, Atkins et al. included a planning exercise of 10 cases with three to four brain metastases modeled with either 6-MV photons using cones, dynamic conformal arcs (DCA), or volumetric modulated arc therapy (VMAT) versus passively scattered protons (Atkins et al., 2018). For three- and four-lesion cases, photon plans estimated a delivery of 10 Gy to normal brain tissue to additional 1 and 3 cm³, respectively, when compared to proton plans. In this the Massachusetts General Hospital (MGH) series of 370 patients, 65 required craniotomy after proton SRS. Of these, pathology confirmed radionecrosis in 26, or a crude incidence of 7%. The median time to radionecrosis was 12 months (range, 1.1 months–8.2 years), and the 12-month cumulative incidence of radionecrosis was 3.6%. On multivariate analysis, only target volume remained a significant predictor for pathologically confirmed radionecrosis (HR 1.13, $p < 0.0005$).

These numbers are generally consistent with the rates of radionecrosis reported in photon-based radiosurgery. For example, Kohutek et al. reported outcomes of 271 brain metastases treated with single-fraction Linac-based SRS. Seventy cases (26%) experienced radionecrosis diagnosed by either imaging (18%) or were pathologically confirmed (8%) (Kohutek et al., 2015). Siddiqui et al. reported a Gamma Knife (GK) series of 732 lesions treated in 198 patients. The estimated per lesion incidence of radionecrosis at 4 years was 6.8% (Siddiqui et al., 2020).

27.6 CONTROVERSIES

Cost and accessibility are the greatest limitations to the widespread implementation of proton radiosurgery. Although there are a sharply increasing number of proton centers of late, the techniques pioneered at the Massachusetts General Hospital utilize a dedicated eye-line (also used to treat cases of uveal melanoma) in order to achieve small-field dosimetry, which is not available within most proton centers.

With respect to outcomes, the local control and rate of radionecrosis reported by Atkins et al. are similar but not necessarily superior to more commonly used Linac-based radiosurgery (Atkins et al., 2018). While the integral low/medium doses achieved with proton radiosurgery may be superior dosimetrically (Figure 27.1), there is no robust clinical data on preservation of neurocognition and quality of life.

Uncertainty in biological modeling of dose is a recognized challenge in the implementation of proton therapy, including radiosurgery. The relative biological effectiveness (RBE) value of proton therapy is the ratio of the dose of Cobalt-60 required to produce the same clinical/biological effect as proton therapy. For this reason, proton radiation doses are expressed as "Cobalt Gray Equivalents" (CGE), as opposed to "Gray" (Gy). The reported RBE values for proton are in the range of 0.9 to 1.3, but the most widely used coefficient is 1.1 based on in vitro and animal studies performed in the early 1970s (Paganetti et al., 2002). However, RBE is heavily dependent on biological characteristics (such as the α/β tissue), dose per fraction, and linear energy transfer (LET), which is the average amount of energy imparted to the local medium per unit length and approximates a particle's ionization track structure. The RBE has been modeled to increase with the depth ranging from 1.1 at its entrance to 1.74 at the distal falloff. For example, this uncertainty could result in unexpectedly high toxicities if the modeled RBE for proton therapy was lower than the actual value. With respect to proton radiosurgery, this uncertainty of RBE may be mitigated by an apparent decrease as dose per fraction increases (Paganetti, 2014).

For these reasons, it is recommended that proton radiosurgery be delivered only in high-volume proton centers with robust clinical and treatment planning experience.

27.7 FUTURE DIRECTIONS

While pencil beam scanning (PBS) is now in widespread use for the delivery of proton therapy for fractionated tumors of the central nervous system, its use in proton radiosurgery largely remains in development with the hope of improving dose conformality and homogeneity.

McAuley et al. has also investigated potential improvements in dose distribution of proton radiosurgery with a triplet of quadrupole permanent magnets placed immediately upstream from tissue entry in order to focus protons. Their work demonstrated that this system could be useful for proton radiosurgery with lower entrance dose and increased dose delivery efficiency which could possibly result in shorter treatment times (McAuley et al., 2018, 2019).

Ultimately, widespread adoption of proton radiosurgery will be dependent on improvements in financial requirements, accessibility, and dosimetric reliability.

CHECKLIST: KEY POINTS FOR CLINICAL PRACTICE

✓	ACTIVITY	SOME CONSIDERATIONS
	Patient selection	*Is the patient appropriate for proton intracranial SRS?* Institutional policies critical to patient selection Is proton SRS being delivered at an experienced center?
	Simulation	*Immobilization* Noninvasive stereotactic head frame (often using custom dental mouthpiece) • Invasive stereotactic head frame with pins • Aquaplast mask *Imaging* • CT scan with contrast (1.25 mm) • MRI with gadolinium enhancement (1 mm) • Verified CT to MRI fusion over area of tumor involvement
	Treatment planning	*Contours* • GTV should be contrast-defined gross disease as defined by CT and MRI • CTV expansion is not used for SRS • PTV expansion should be judiciously used. Consider 1 mm expansion if • using non-rigid immobilization *Treatment planning* • Passive scattering currently in use • Pencil beam scanning in development *Dose* Similar to photon-based SRS. Target coverage typically to 90% *Dose constraints* Similar to photon-based SRS
	Treatment delivery	*Imaging* • kV matching to 3 fiducial markers placed in outer table of calvarium • Treatment couch with five to six degrees of freedom to correct translation and • Rotational shifts if needed

REFERENCES

Aoyama, H., Shirato, H., Tago, M., Nakagawa, K., Toyoda, T., Hatano, K., . . . Kobashi, G. (2006). Stereotactic radiosurgery plus whole-brain radiation therapy vs stereotactic radiosurgery alone for treatment of brain metastases: A randomized controlled trial. *JAMA, 295*(21), 2483–2491. doi:10.1001/jama.295.21.2483

Atkins, K. M., Pashtan, I. M., Bussière, M. R., Kang, K. H., Niemierko, A., Daly, J. E., . . . Shih, H. A. (2018). Proton stereotactic radiosurgery for brain metastases: A single-institution analysis of 370 patients. *International Journal of Radiation Oncology, Biology, Physics, 101*(4), 820–829. doi:10.1016/j.ijrobp.2018.03.056

Barker, F. G., Butler, W. E., Lyons, S., Cascio, E., Ogilvy, C. S., Loeffler, J. S., & Chapman, P. H. (2003). Dose-volume prediction of radiation-related complications after proton beam radiosurgery for cerebral arteriovenous malformations. *Journal of Neurosurgery, 99*(2), 254–263. doi:10.3171/jns.2003.99.2.0254

Blonigen, B. J., Steinmetz, R. D., Levin, L., Lamba, M. A., Warnick, R. E., & Breneman, J. C. (2010). Irradiated volume as a predictor of brain radionecrosis after linear accelerator stereotactic radiosurgery. *International Journal of Radiation Oncology, Biology, Physics, 77*(4), 996–1001. doi:10.1016/j.ijrobp.2009.06.006

Halasz, L. M., Bussière, M. R., Dennis, E. R., Niemierko, A., Chapman, P. H., Loeffler, J. S., & Shih, H. A. (2011). Proton stereotactic radiosurgery for the treatment of benign meningiomas. *International Journal of Radiation Oncology, Biology, Physics, 81*(5), 1428–1435. doi:10.1016/j.ijrobp.2010.07.1991

Harsh, G., Loeffler, J. S., Thornton, A., Smith, A., Bussiere, M., & Chapman, P. H. (1999). Stereotactic proton radiosurgery. *Neurosurg Clin N Am, 10*(2), 243–256.

Harsh, G. R., Thornton, A. F., Chapman, P. H., Bussiere, M. R., Rabinov, J. D., & Loeffler, J. S. (2002). Proton beam stereotactic radiosurgery of vestibular schwannomas. *International Journal of Radiation Oncology, Biology, Physics, 54*(1), 35–44. doi:10.1016/s0360-3016(02)02910-3

Hattangadi, J. A., Chapman, P. H., Bussière, M. R., Niemierko, A., Ogilvy, C. S., Rowell, A., . . . Shih, H. A. (2012). Planned two-fraction proton beam stereotactic radiosurgery for high-risk inoperable cerebral arteriovenous malformations. *International Journal of Radiation Oncology, Biology, Physics, 83*(2), 533–541. doi:10.1016/j.ijrobp.2011.08.003

Hattangadi-Gluth, J. A., Chapman, P. H., Kim, D., Niemierko, A., Bussière, M. R., Stringham, A., . . . Shih, H. A. (2014). Single-fraction proton beam stereotactic radiosurgery for cerebral arteriovenous malformations. *International Journal of Radiation Oncology, Biology, Physics, 89*(2), 338–346. doi:10.1016/j.ijrobp.2014.02.030

Kocher, M., Soffietti, R., Abacioglu, U., Villà, S., Fauchon, F., Baumert, B. G., . . . Mueller, R. P. (2011). Adjuvant whole-brain radiotherapy versus observation after radiosurgery or surgical resection of one to three cerebral metastases: Results of the EORTC 22952–26001 study. *Journal of Clinical Oncology, 29*(2), 134–141. doi:10.1200/JCO.2010.30.1655

Kohutek, Z. A., Yamada, Y., Chan, T. A., Brennan, C. W., Tabar, V., Gutin, P. H., . . . Beal, K. (2015). Long-term risk of radionecrosis and imaging changes after stereotactic radiosurgery for brain metastases. *J Neurooncol, 125*(1), 149–156. doi:10.1007/s11060-015-1881-3

McAuley, G. A., Heczko, S. L., Nguyen, T. T., Slater, J. M., Slater, J. D., & Wroe, A. J. (2018). Monte Carlo evaluation of magnetically focused proton beams for radiosurgery. *Physics in Medicine and Biology, 63*(5), 055010. doi:10.1088/1361-6560/aaaa92

McAuley, G. A., Teran, A. V., McGee, P. Q., Nguyen, T. T., Slater, J. M., Slater, J. D., & Wroe, A. J. (2019). Experimental validation of magnetically focused proton beams for radiosurgery. *Physics in Medicine and Biology, 64*(11), 115024. doi:10.1088/1361-6560/ab0db1

Minniti, G., Clarke, E., Lanzetta, G., Osti, M. F., Trasimeni, G., Bozzao, A., . . . Enrici, R. M. (2011). Stereotactic radiosurgery for brain metastases: Analysis of outcome and risk of brain radionecrosis. *Radiat Oncol, 6*, 48. doi:10.1186/1748-717X-6-48

Paganetti, H. (2014). Relative biological effectiveness (RBE) values for proton beam therapy. Variations as a function of biological endpoint, dose, and linear energy transfer. *Physics in Medicine and Biology, 59*(22), R419–R472. doi:10.1088/0031-9155/59/22/R419

Paganetti, H., Niemierko, A., Ancukiewicz, M., Gerweck, L. E., Goitein, M., Loeffler, J. S., & Suit, H. D. (2002). Relative biological effectiveness (RBE) values for proton beam therapy. *International Journal of Radiation Oncology, Biology, Physics, 53*(2), 407–421. doi:10.1016/s0360-3016(02)02754-2

Patchell, R. A., Tibbs, P. A., Regine, W. F., Dempsey, R. J., Mohiuddin, M., Kryscio, R. J., . . . Young, B. (1998). Postoperative radiotherapy in the treatment of single metastases to the brain: A randomized trial. *JAMA, 280*(17), 1485–1489. doi:10.1001/jama.280.17.1485

Petit, J. H., Biller, B. M., Coen, J. J., Swearingen, B., Ancukiewicz, M., Bussiere, M., . . . Loeffler, J. S. (2007). Proton stereotactic radiosurgery in management of persistent acromegaly. *Endocr Pract, 13*(7), 726–734. doi:10.4158/EP.13.7.726

Siddiqui, Z. A., Squires, B. S., Johnson, M. D., Baschnagel, A. M., Chen, P. Y., Krauss, D. J., . . . Grills, I. S. (2020). Predictors of radiation necrosis in long-term survivors after Gamma Knife stereotactic radiosurgery for brain metastases. *Neurooncol Pract, 7*(4), 400–408. doi:10.1093/nop/npz067

Sperduto, P. W., Chao, S. T., Sneed, P. K., Luo, X., Suh, J., Roberge, D., . . . Mehta, M. (2010). Diagnosis-specific prognostic factors, indexes, and treatment outcomes for patients with newly diagnosed brain metastases: A multi-institutional analysis of 4,259 patients. *International Journal of Radiation Oncology, Biology, Physics, 77*(3), 655–661. doi:10.1016/j.ijrobp.2009.08.025

Walcott, B. P., Hattangadi-Gluth, J. A., Stapleton, C. J., Ogilvy, C. S., Chapman, P. H., & Loeffler, J. S. (2014). Proton beam stereotactic radiosurgery for pediatric cerebral arteriovenous malformations. *Neurosurgery, 74*(4), 367–373; discussion 374. doi:10.1227/NEU.0000000000000294

Weber, D. C., Chan, A. W., Bussière, M. R., Harsh, G. R., Ancukiewicz, M., Barker, F. G., . . . Loeffler, J. S. (2003). Proton beam radiosurgery for vestibular schwannoma: Tumor control and cranial nerve toxicity. *Neurosurgery, 53*(3), 577–586; discussion 586–578. doi:10.1227/01.neu.0000079369.59219.c0

28

The Zap-X: A Novel 3 Megavolt Linear Accelerator for Dedicated Intracranial Stereotactic Radiosurgery

Georg A. Weidlich and John R. Adler

Contents

28.1 INTRODUCTION

This radiosurgical system is deemed self-shielded in that radiation leakage values outside the system are below dose limits for the public as stipulated by the National Council on Radiation Protection (NCRP) [1]. Importantly, the patient undergoing treatment is shielded from target leakage radiation 50 times more effectively than required by the International Electrotechnical Commission (IEC) [2]. As a result, the goal of self-shielding can be achieved under all but the most exceptional conditions. The Zap-X is depicted schematically in Figure 28.1 and has been described in more detail previously [3–5].

Accurate therapeutic beam positioning is accomplished through the dual axes, independent rotations of the accelerator, and precise movements of a robotic patient table. The patient is supported on a moveable treatment table that extends outside the treatment spherical shield. The Zap-X accomplishes precise three-dimensional (3D) patient registration by means of an integrated planar kilovolt (kV) imaging system that rotates around the MV isocenter. Pairs of non-coaxial x-ray images and image-to-image correlation are utilized to determine the location of the patient's anatomy with respect to the machine isocenter, both prior to and during radiosurgical treatment.

The specific array of the Zap-X's various structural elements provide shielding comparable to the walls, ceiling, and floor of a radiotherapy vault [1]. Most components needed to produce the therapeutic beam such as the radiofrequency power source, waveguide system, beam-triggering electronics, and a dedicated beam stop are mounted on or integrated into the primary spherical supporting structure. Furthermore, the patient, who is positioned supine, is enclosed by additional scatter shielding consisting of a rotatable iron shell and a shielded, pneumatically elevated door at the foot of the treatment table. By being mounted onto a shielded treatment sphere with dual axes of independent rotation, the treatment beam from the linear accelerator can be isocentrically positioned across a solid angle of just over 2π, as necessitated for high-quality cranial SRS. Figure 28.2 shows a rendering of the Zap-X system in operation.

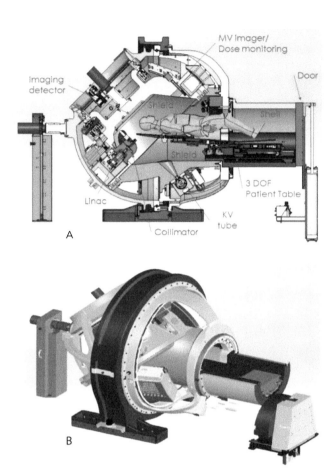

Figure 28.1 Cross-sectional (A) and room's eye view (B) of Zap-X. MV, megavoltage; DOF, degree of freedom; KV, kilovoltage.

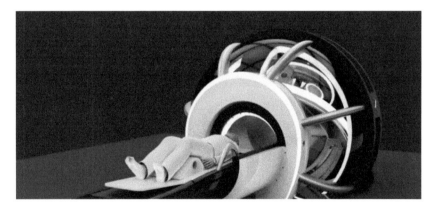

Figure 28.2 Room's eye view of Zap-X with patient in the loaded position.

The unique collimator design is critical to the overall performance of the Zap-X system. Consisting of a shielded tungsten wheel oriented with its rotational axis perpendicular to the beam's central axis, this device minimizes radiation leakage while enabling rapid changes in beam aperture. As shown in Figure 28.3, beam selection is accomplished by rotating the wheel within its tungsten shield.

A dedicated Treatment Planning System for the Zap-X system, especially with its unique collimator, enables better [6] target coverage and dose conformity while also sparing critical structures and

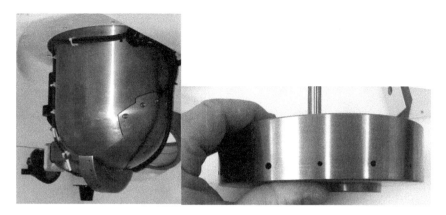

Figure 28.3a Collimator housing   **Figure 28.3b** Collimator cax view

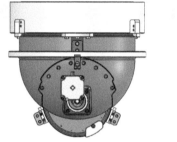

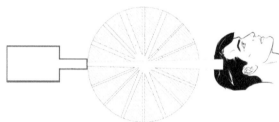

Figure 28.3c Collimator schematic drawing   **Figure 28.3d** Collimator selection for treatment delivery

maximizing the dose falloff outside the target volume. A recently released treatment-planning module utilizes both forward and inverse planning methodologies.

28.2 SYSTEM CHARACTERIZATION

The Zap-X system was characterized with a focus on:

1. System shielding
2. the energy of the linear accelerator, thereby enabling both the nominal energy and accelerating potential to be derived
3. the focal spot size and penumbra
4. the linac collimator leakage
5. a comprehensive radiation survey outside the shielded sphere and housing of the Zap-X with a derivation of the expected annual dose
6. beam performance specifications
7. beam data acquisition and treatment planning system verification; and
8. a comprehensive Quality Assurance Program.

28.2.1 SYSTEM SHIELDING DESIGN

The Zap-X was designed to incorporate shielding from the primary beam, leakage, and patient scatter radiation. The self-shielding designation was specified to achieve a public annual dose limit at 1 m distance from the periphery of the system while limiting patient irradiation from leakage to less than 0.1% of the primary dose. Monte Carlo Simulation models were applied to achieve this desired effect as described by Rogers et al. [7]. The linac energy, the maximum dose rate, materials used for collimation, and the system geometry were all considered when applying the model. The maximum dose for the general public in uncontrolled areas as required by the National Council on Radiation Protection and Measurement (NCRP) is 1 millisievert (mSv)/year [4], which is equivalent to 100 milliroentgens (mR)/year. For a total of 2,000 work hours per year, the Zap-X will result in 0.05 mR exposure to an operator in any one hour.

The workload and "Beam On" time are calculated by use of the number of patients (Np) per week, the average treatment time (T min/patient), the qualification and calibration time to not exceed a certain duration (Tqc) per week, the total workload (W = Np × T + Tqc hours per week), the utilization (U), add a safety factor (SF), and number of weeks (Nw) of operation in 1 year. The equivalent radiation on time per year is calculated by applying the following formula:

$$T_{Ann} = U \times W \times SF \times Nw \text{ (h/year)}$$

To maximize the clearance of the accelerator and collimator assembly rotating around the patient's head, tungsten was used to minimize the thickness for a given shielding effect. Other high-density shielding materials, such as lead and iron, are also utilized in the Zap-X, with iron being the main material blocking secondary scatter. Lead and tungsten are more selectively used.

28.2.2 LINEAR ACCELERATOR ENERGY CHARACTERIZATION

The linear accelerator is a key component of the Zap-X System as it produces the therapeutic photon radiation used for patient treatment. As a very common depth for intracranial radiosurgery is approximately 5 cm, the 3.0 MV photon beam with a 66% PDD at this depth appears well-matched to such targets. The photon radiation is produced by the bremsstrahlung interaction of the accelerated electrons with the tungsten target. The Linac beam energy was evaluated by means of a depth dose scan in a PTW 3D water

R100 [mm]	R80 [mm]	R50 [mm]	Ds [%]	D100 [%]	QI	NAP [MV]	SSD [cm]	Field Size [cm x cm]	Curve Type
7.98	32.62	78.52	75.68	40.29	0.4307	2.56	45.0	2.5 x 2.5	PDD

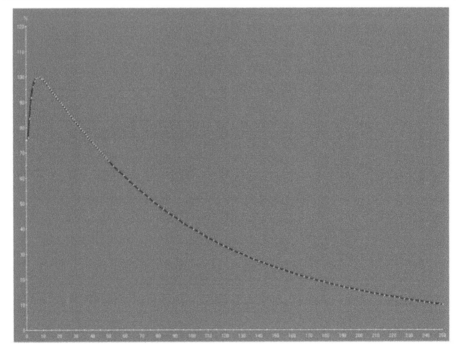

Figure 28.4 Measured percent depth dose of 40.29% for the Zap-X 3.0 MV photon beam. R, reading; Ds, surface dose; D100, dose at 100 mm depth; QI, quality index; NAP (MV), nominal accelerating potential (megavoltage); SSD, source surface distance; PDD, percent depth dose.

phantom (Physikalisch-Technische Werkstaetten, Freiburg, Germany) at 100 cm SSD (source-to-surface distance) and reference data from the *British Journal of Radiology*, Supplement 17 [8]. The resulting photon bremsstrahlung PDD is shown in Figure 28.4.

28.2.3 FOCAL SPOT SIZE AND BEAM PENUMBRA

The focal spot size has a direct impact on the geometric penumbra of the beam and a small penumbra will allow the creation of steep dose gradients at the target to critical anatomical structure interface.

The focal spot size was measured with a device consisting of thin tungsten (W) leaves alternating with thin paper spacers. A 7.5 mm collimator was selected, and exposures were made with Gafchromic™ film (Ashland Advanced Materials, Bridgewater, NJ) to record the radiographic transmission pattern, both in the plane of the collimator wheel and in the orientation perpendicular to the collimator wheel. Nine line-pairs were identified which translates to a 1.8-mm focal spot size. The focal spot size can be considered circular.

The beam penumbra was measured for all collimator sizes of 4, 5, 7.5, 10.0, 12.5, 15, 20, and 25 mm at a 5 cm and 10 cm depth at SSD of 45 cm. The beam penumbra was defined as the distance between the 80% and 20% intensity points. Figure 28.5 shows the beam profile for the 25 mm collimator which is the calibration field size. Table 28.1 shows the penumbra values for all collimator sizes at 5.0 cm and 10 cm depth.

The Zap-X collimator was designed to achieve rapid dose falloff at the beam edge, thereby minimizing penumbra. The Zap-X system has a penumbra between 1.77 mm and 2.69 mm at treatment depth. This means that the Zap-X system is designed to deliver more of the radiation to the intended target and reduces radiation reaching off-target areas of the brain. Furthermore, the short SAD reduces beam spread, which also helps reduce off-target radiation. Low collimator leakage contributes to the improved delivery of complex yet highly conformal radiosurgical treatments to oddly shaped targets, which, by nature, require both many beams and MUs.

28.2.4 LINAC COLLIMATOR PERFORMANCE AND LEAKAGE RADIATION

The Zap-X system has a Source-to-Axis Distance (SAD) of 45 cm (Figure 28.1). Given the short SAD, an isocentric treatment geometry, and the fact that all imaging and shielding components rotate with the beam, the resulting treatment sphere is relatively compact [9].

A comprehensive radiation survey of the Zap-X collimator was performed with two sets of measurements, one in the patient isocenter plane perpendicular to the beam central axis (cax) and the other set alongside the linac at 1 m distance from the cax. Stations are equidistantly spaced at 20-cm intervals. We used a large-volume leakage chamber from a Radcal (model #ADDM+, S/N 47–0657, Radcal Corporation, Monrovia, CA) and a survey meter (Fluke Biomedical Model 451 BRYR, S/N 0000003284, Fluke Biomedical, Everett, WA) to measure radiation leakage. Figure 28.6 shows the location of the measurement stations.

In the patient plane, a maximum leakage was measured at station #8 with a value of 10.4 mR or 0.00104% of the primary beam intensity. Along the linear accelerator, a maximum leakage was measured at station #3 with a value of 6.8 mR or 0.00068% of the primary beam intensity. As the IEC standards require an average of 0.1% of leakage radiation in the patient plane and along the linear accelerator [5], Zap-X exceeds the requirements of this standard by two orders of magnitude.

Collimator leakage represents radiation that will be absorbed by the patient in areas other than the target volume. As the leakage radiation is expected to be absorbed by a large portion of the patient's body

Table 28.1 Penumbra Values in mm for all Collimator Sizes at 5.0 cm and 10.0 cm depth. The higher values (from both scan directions) are recorded.

COLLIMATOR SIZE	4 MM	5 MM	7.5 MM	10 MM	12.5 MM	15 MM	20 MM	25 MM
Left penumbra 5 cm	1.77	1.88	1.94	2.00	2.06	2.11	2.19	2.27
Left penumbra 10 cm	2.00	2.12	2.21	2.29	2.35	2.43	2.54	2.64
Right penumbra 5 cm	1.78	1.85	1.93	1.99	2.08	2.11	2.19	2.27
Right penumbra 10 cm	2.02	2.11	2.19	2.28	2.39	2.44	2.53	2.69

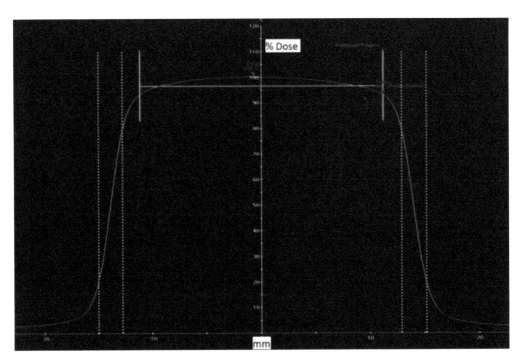

Figure 28.5 Measured beam profile of 25.0 mm collimator at 5.0 cm depth, with indication of symmetry and penumbra.

distant from the treatment volume, the whole-body dose from collimator leakage in patients treated with the Zap-X system is expected to be very low.

28.2.5 RADIATION SURVEY AND RADIATION SAFETY

The goals of this self-shielded system are to provide shielding to personnel outside a 1-meter (m) safety zone from the perimeter of the Zap-X system to levels that are acceptable to the public (1 millisievert [mSv]/year; NRC and NCRP [4]) and, therefore, provide all or most required shielding typically present in the facility shielding. The required treatment sphere wall thicknesses were determined according to the Radiation Protection guidelines defined in NCRP reports 49, 51, 116, and 151 [4].

Instantaneous exposure rates were determined for 14 equidistant stations around the perimeter of accessible areas of the Zap-X at 1 m distance, using a Victoreen Model 451 survey meter (Fluke Biomedical, Everett, Washington, United States of America). Figure 28.7 illustrates the position of measurement locations, as viewed from above.

Standard heavy operating conditions assume that a total of 2,250 treatments per year will be delivered for a very busy operation, all with a single fraction. It is assumed that a spherical head (20 cm diameter) will be treated with the target in the center of the sphere at 10 cm depth. For the target, the prescription is 2,000 centiGray (cGy) to 80% isodose, collimator size 25 mm (output factor (OF) 1.0), and a depth of 10 cm (tissue-phantom-ratio [TPR] ~0.4). The resulting Monitor Units (MUs) per treatment are 6,250 MU. With the given patient distribution, a total of 56,250 MU will be delivered per day or 1.406×10^7 MU per year. This translates into 156.25 hours/year of "Rad On" time; therefore, we will have a utilization factor of 156.25/2,000 = 0.0781. The shielding should be designed for 500 mrem/year or 0.25 mrem/hour. Applying the utilization factor, the allowable instantaneous exposure rate is 3.205 mrem/hr at 1 m.

The results of the exposure rate-based radiation survey are shown in Table 28.2 relating to Figure 28.7 that shows the measurement locations. The maximum projected annual dose of 0.936 mSv was determined at the foot end of the table with an exposure rate value of 3.0 mR/hour. All projected annual dose values were derived to be below 1.0 mSv which is equivalent to the Annual Public Dose limit.

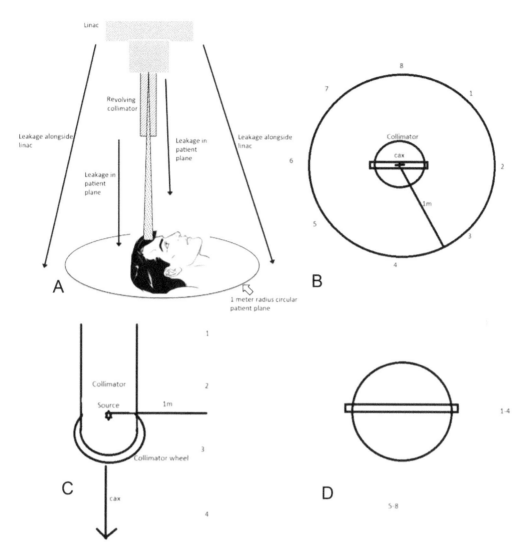

Figure 28.6 Orientation of Linac during treatment (A), Location of measurement stations in the patient plane (B), along the linear accelerator (C), and view along the beam cax (D).

Therefore, the Zap-X system, under the specified workload, would allow unrestricted access for non-radiation workers outside a 1-m system perimeter and to all areas on the floor above the ZAP treatment room, and no additional shielding would be needed in the walls of the treatment room.

Only under the most extreme operating conditions (with more than 50 patients being treated per week), would the self-shielding characteristics not hold true. In that case, a reassignment of the area in the immediate vicinity of the Zap-X to occupationally exposed access with 5.0 mSv/year would allow the patient census to be quintupled and remain self-shielded.

Several benefits are seen to accrue from the reported findings: For a heavy workload, all necessary shielding required for operating personnel and the public is integrated as part of the system design, removing the need for construction of a dedicated radiation bunker, and the technical burden of building a dedicated treatment room is removed from the end user institution. Since the shielding requirements for the Zap-X were carefully studied, optimized, and do not depend on a local designer, expected radiation levels outside the treatment sphere are known to meet regulations and safety standards. During clinical implementation, the total required time for facility preparation is decreased by approximately 6 to 12

Table 28.2 **Results of Radiation Survey in Units of mR/hr; Beam pos: beam position; mR/hr: milliroentgens per hour; U: use factor; D/C: duty cycle; Ann Dose (mSv): annual dose in millisieverts; G: gantry**

LOCATION	DESCRIPTION OF STATIONS	BEAM POS.	mR/hr	U	D/C	ANN DOSE (mSv)
1	Foot End Table Sheild	G=45	3.0	0.2	0.078	0.936
2	Right Side Main Gantry	G=45	0.09	0.2	0.078	0.028
3	Left Side Main Gaintry	G=90	0.1	0.2	0.078	0.031
4	Head End of System	G=45	0.36	0.2	0.078	0.112
5	Table Right	G=90	0.45	0.2	0.078	0.140
6	Table - Orbit Right	G=90	0.37	0.2	0.078	0.115
7	Right Gantry	G=0	0.37	0.2	0.078	0.059
8	Right Gantry	G=270	1.72	0.2	0.078	0.537
9	Right - Head	G=45	2.70	0.2	0.078	0.542
10	Left - Head	G=45	2.80	0.2	0.078	0.874
11	Left Gantry	G=45	1.70	0.2	0.078	0.530
12	Left Gantry	G=0	0.57	0.2	0.078	0.178
13	Table - Orbit Left	G=0	0.84	0.2	0.078	0.262
14	Table Left	G=0	0.4	0.2	0.078	0.125
15	Control Console	G=0	0.11	0.2	0.078	0.034

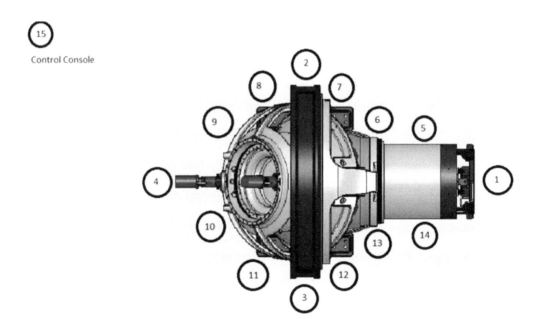

Figure 28.7 Description of measurement locations for Zap-X radiation survey.

months as no treatment bunker construction will be necessary, and total construction costs will be reduced significantly.

28.2.6 BEAM PERFORMANCE SPECIFICATIONS

Zap-X beam performance specifications and system acceptance testing criteria were established based on clinical requirements and system capabilities and were established as follows:

1. Photon beam energy: Single photon beam energy of 3.0 MV; Percent depth dose = 40% ± 2% for 2.5 cm circular field size at 45 cm SAD
2. Dose rate: 1,500 MU/min at 45 cm SAD; 1 MU = 1 cGy at SAD = 45 cm, 2.5 cm field size, at depth of dose maximum; linearity < ±3%; reproducibility of ± 2 MU or 2%, whichever is greater; Orientation dose stability of ± 2 MU or 3%, whichever is greater.
3. Beam symmetry shall not exceed ± 2%.
4. The leakage radiation to the patient plane and outside the patient plane shall be less than 0.1%.
5. The collimator leakage transmission shall be less than 0.1%.

28.2.7 BEAM DATA ACQUISITION AND TREATMENT PLANNING SYSTEM QUALIFICATION

A complete set of beam data is acquired with the PTW 3D water phantom. Percent Depth Dose (PDD) scans for all available collimator sizes and to a depth of 20 cm, Off-Center Ratios (OCR) for all collimator sizes at 0.7 cm (D_{max}), 2.5 cm, 5.0 cm, 10 cm, and 20 cm, as well as collimator output factors at D_{max} and SAD = 45.0 cm for all collimator sizes of 4.0, 5.0, 7.5, 10.0, 12.5, 15.0, 20.0, and 25.0 mm. Sample data sets for the PDD are shown in Figure 28.8, while OCR values are shown in Figure 28.9.

Output factors are measured with the PTW model 31022 pinpoint ionization chamber and the PTW model TN60019 diamond detector (Physikalisch-Technische Werkstaetten, Freiburg, Germany).

Because the Zap-X has a unique system configuration and the confined available space, the recommended water phantom is a PTW MP3XS 3D. A dedicated supporting frame was designed to hold the water phantom. The 25 mm collimator is the reference collimator for absolute dose calibration. The 3D PTW water phantom is shown in Figure 28.10.

The functionality and accuracy of the treatment planning system were verified by independently calculating expected dose in various scenarios: single beam SSD setup (Figure 28.11a), single beam source axis distance (SAD) setup (Figure 28.11b), and two beam set, isocentric setup (Figure 28.12). The calculated expected dose was computed using the following formula:

$$D(x) = 1 \text{ MU} \times 1 \text{ cGy/MU} \times OCR \text{ (coll, } d_{eff}, OAD) \times TPR \text{ (coll, } d_{eff}) \times OF \text{ (coll)} \times (450/SAD)^2$$

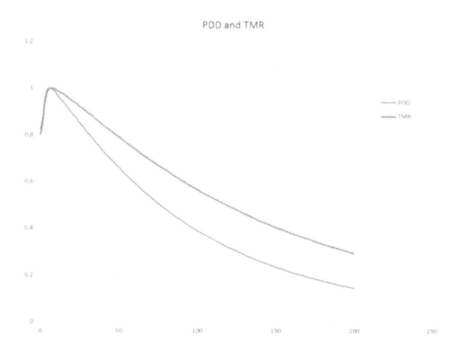

Figure 28.8 Percent depth dose (PDD) and tissue-maximum ratio (TMR) for the 3 MV Zap-X beam.

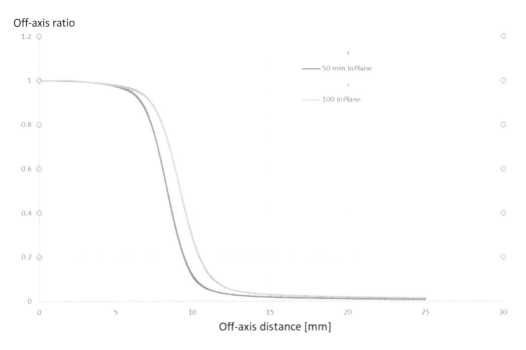

Figure 28.9 Off-center-ratios for the 3 MV Zap-X beam in-plane and cross-plane at 50 mm and 100 mm depth.

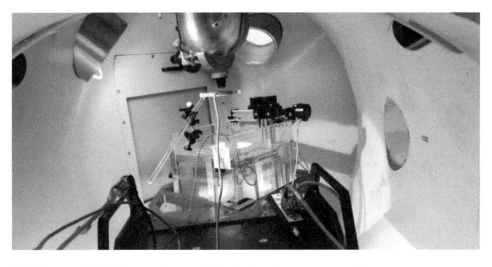

Figure 28.10 Setup of 3D PTW water phantom inside the Zap-X treatment capsule.

The single isocenter conformity indices for the prescription isodose lines 80%, 70%, 60%, 50%, 40%, and 30% were calculated in relation to the target volume and are shown in Table 28.3.

28.2.8 QUALITY ASSURANCE PROGRAM

A complete set of principal quality assurance procedures was developed and implemented following AAPM guidelines [10, 11], including daily output verification, weekly beam performance verification, monthly output calibration, and an annual calibration for verification of baseline data established during the commissioning of the system. Figure 28.13 shows Zap-X-specific quality assurance tools.

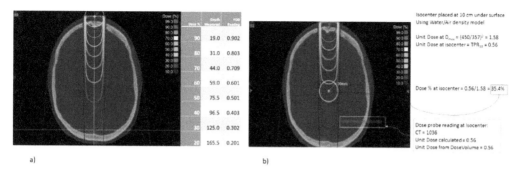

Figure 28.11 Verification of a) PDD and b) TMR values on the cax.

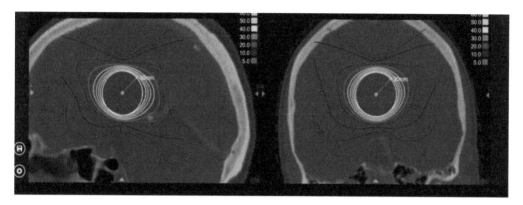

Figure 28.12 Verification of treatment plan with full beam set.

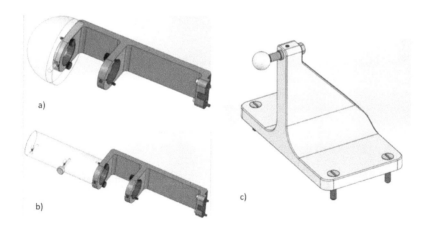

Figure 28.13 Zap-X quality assurance tools used for (a) absolute dose calibration, (b) ionization ratio measurement, and (c) radiation isocenter verification.

To verify the overall system accuracy, a weekly end-to-end test will be performed that includes computed tomography (CT) imaging, contouring, treatment planning, treatment delivery, and treatment evaluation.

28.3 CONCLUSION

No standard currently exists for an external beam-based dedicated intracranial stereotactic radiosurgery (SRS) system at the treatment energy used by the Zap-X. As a result, the above-presented analysis and

Table 28.3 **Conformity Indices for Prescription Isodose Line, 80%, 70%, 60%, 50%, 40%, and 30% Isodoses; Normalized to the 80% isodose volume.**

Prescription Iodose Line [%]	Volume covered (cm³)	Conformality index
80	13.38	1.0
70	16.51	1.23
60	19.95	1.49
50	24.57	1.84
40	32.30	2.41
30	48.73	3.64

cm³: cubic centimeter

results can serve as a set of baseline parameters for any new class of specialized radiation treatment devices using comparable x-ray energies. Additional parameters, such as reproducibility of parameters for subsequent production systems and individual long-term stability, remain to be explored and determined.

Experimental analysis described here-in demonstrated that a radiosurgical system can be self-shielded, such that it produces radiation exposure levels deemed safe to the public by the NCRP standards while operating under a full clinical workload. The Zap-X has been shown to be a self-shielded radiosurgery system without the need for a radiation treatment bunker at most clinical user sites.

Ultimately, the Zap-X system was found to meet safety, accuracy, and performance requirements widely accepted in the radiation oncology and radiosurgery communities. In its earliest clinical experience, the Zap-X appears to meet the practical clinical requirements of a typical radiosurgical facility.

REFERENCES

1. Weidlich GA, Schneider M, Adler JR. Self-shielding analysis of the Zap-X system. *Cureus*. 2017, 9:1917. DOI: 10.7759/cureus.1917
2. Weidlich GA., Bodduluri M, Achkire Y, et al. Characterization of a novel 3 megavolt linear accelerator for dedicated intracranial stereotactic radiosurgery. *Cureus*. March 19, 2019, 11(3):e4275. DOI: 10.7759/cureus.4275
3. Jenkins CH, Kahn R, Weidlich GA, Adler JR. Radiosurgical treatment verification using removable megavoltage radiation detectors. *Cureus*. 2017, 9:1889. DOI: 10.7759/cureus.1889
4. National Council on Radiation Protection and Measurement—NCRP reports (#49, 51, 116, 151). (2014). Accessed: November 12, 2017: www.ncrppublications.org/reports.
5. International Electrotechnical Commission. Medical electrical equipment—Part 2–1: Particular requirements for the basic safety and essential performance of electron accelerators in the range 1 MeV to 50 MeV. IEC 60601–2–1:2009. *International Electrotechnical Commission*. 2009, 3:132.3.4.
6. Adler JR, Schweikard A, Achkire Y, Blanck O, Bodduluri M, Ma L, Zhang H. Treatment planning for self-shielded radiosurgery. *Cureus*. 2017, 9:1663. DOI: 10.7759/cureus.1663
7. Rogers DW, Faddegon BA, Ding GX, et al. BEAM: A Monte Carlo code to simulate radiotherapy treatment units. *Medical Physics*. 1995, 22:503–524. DOI: 10.1118/1.597552
8. LaRiviere PD. The quality of high-energy X-ray beams. *Br J Radiol*. 1989, 62:473–481. DOI: 10.1259/0007-1285-62-737-473
9. Weidlich GA, Schneider MB, Adler JR. Characterization of a novel revolving radiation collimator. *Cureus*. 2017, 9:2146. DOI: 10.7759/cureus.2146
10. Benedict SH, Yenice KM, Followill D, et al. Stereotactic body radiation therapy: The report of AAPM task group 101. *Medical Physics*. 2010, 37:4078–4101. DOI: 10.1118/1.3438081
11. Almond PR, Biggs PJ, Coursey BM, et al. AAPM's TG-51 protocol for clinical reference dosimetry of high-energy photon and electron beams. *Medical Physics*. 1999, 26:1847–1870. DOI: 10.1118/1.598691

Immunotherapy and Radiosurgery for Brain and Spine Metastases

Diana A. Roth O'Brien, Horia Vulpe, and Tony J.C. Wang

Contents

29.1 INTRODUCTION

Brain metastases (BrM) are a significant cause of morbidity and mortality, occurring in approximately 10% to 30% of cancer patients at one point in the course of their disease (Loeffler, 2016). Historically, BrM portended a poor prognosis, with survival on the order of months (Johnson and Young, 1996; Gaspar et al., 1997; Sampson et al., 1998; Schouten et al., 2002; Barnholtz-Sloan et al., 2004; Kohler et al., 2011; Nieder et al., 2011; Cagney et al., 2017). With improved diagnostic technologies, cancer-directed therapies, and supportive care, improvements in overall survival (OS) have been accompanied by an increase in the incidence of BrM (Loeffler, 2016). Traditionally, cytotoxic chemotherapy has played only a minor role in the treatment of BrM, as many chemotherapeutic agents have limited penetration across the blood–brain barrier (BBB), though exceptions exist. Recent advances in targeted therapies have further shown that novel tyrosine kinase inhibitors cross the BBB and can provide local control in a subset of lung cancer patients harboring epidermal growth factor receptor (EGFR) mutations (Khandekar et al., 2018). Immunotherapy has revolutionized the care of cancer patients in the last decade, but the ability of T cells to effectively find and treat cancer across the BBB continues to be actively studied. In this chapter, we review the literature on the interplay between immunotherapy and stereotactic radiosurgery (SRS) for brain and spine metastases.

29.1.1 BACKGROUND

The immune system plays an important role in the management of malignancy. Antigen-presenting cells (APCs) can recognize the overexpression of normally occurring antigens, mutated self-antigens, and foreign antigens, such as viral proteins, and subsequently drive cytotoxic T cells to kill malignant cells (Hanahan and Weinberg, 2011; Fridman et al., 2012; Platten et al., 2016). Regulatory T cells (Tregs) function to limit immune overreaction by secreting inhibitory cytokines such as interleukin 10 (IL-10) and transforming growth factor (TGF) beta (Schmidt et al., 2012). Tregs also express cytotoxic T-lymphocyte-associated protein 4 (CTLA-4), which downregulates APCs through competition with CD28 (Weber, 2010; Francisco et al., 2010; Herbst et al., 2014; Baumeister et al., 2016). Furthermore, programmed cell death protein 1

(PD-1) is expressed on the surface of lymphocytes and binds with the ligand PD-L1 to induce cytotoxic T cell apoptosis, downregulate apoptosis of Tregs, and increase IL-10.

Tumor cells have been known to avoid immune detection and destruction by downregulating the immune response. Tumor cells have decreased expression of tumor antigens to reduce antigen presentation and express PD-L1 to interfere with the action of cytotoxic T lymphocytes (CTLs) and recruitment of Tregs (Francisco et al., 2010; Wherry, 2011; Chen and Han, 2015; Oleinika et al., 2013). Sites of metastatic disease are associated with the production of immunosuppressive cytokines and activation of immune checkpoints, principally through CTLA-4, PD-1, and PD-L1, which enhance self-tolerance and suppress the function and proliferation of T cells (Weber, 2010; Herbst et al., 2014).

Immune checkpoint inhibitors (ICI) have revolutionized systemic therapy for metastatic melanoma, non-small cell lung cancer (NSCLC), renal cell carcinoma (RCC), and other malignancies, with improved disease control and prolonged OS (Hodi et al., 2010, 2016; Chuk et al., 2017; Robert et al., 2015; Larkin et al., 2015; Gettinger et al., 2016; Reck et al., 2016; McDermott et al., 2015; Brahmer et al., 2015; Rittmeyer et al., 2017; Herbst et al., 2016; Borghaei et al., 2015; Paz-Ares et al., 2018; Gandhi et al., 2018; Sharma et al., 2011; Dart, 2018).

Historically, the central nervous system (CNS) has been considered an immune-privileged site, as the BBB, formed by astroglial and endothelial cells, regulates permeability into the CNS (Han and Suk, 2005). Healthy brain parenchyma has only rare lymphocytes and the absence of a large draining lymphatic system. Macrophages within the CNS have limited major histocompatibility complex (MHC) expression, making them ineffective APCs (Ransohoff et al., 2003). However, the notion of the CNS as an immune-privileged site has been re-evaluated over time. There is evidence that the immune system can be effective within the brain if a sufficient systemic immune response is triggered (Carson et al., 2006; Louveau et al., 2015; Dunn and Okada, 2015). There is evidence that T cells can enter the CNS via the choroid plexuses, without passing through the BBB, and antigens injected directly in the CNS do travel through the subarachnoid space to the cervical and retropharyngeal lymph nodes, where they interact with peripheral immune cells (Ransohoff et al., 2003; Laman et al., 2013; Ransohoff and Engelhardt, 2012). BrM are also associated with an inflammatory microenvironment that can increase the permeability of the BBB, allowing greater infiltration of immune cells and antigen detection within the CNS (Carson et al., 2006; Berghoff and Preusser, 2015; Berghoff et al., 2015). A study of over 100 BrM patients revealed that nearly all samples contained CD3+ and CD8+ tumor-infiltrating lymphocytes (TILs), which generally correlate with improved clinical outcomes across a variety of histologies (Fridman et al., 2012; Berghoff and Preusser, 2015; Berghoff et al., 2015; Gooden et al., 2011). While these findings are promising, the CNS remains a relatively immune deficient site, and the movement of T cells and APCs into and out of the CNS remains more restricted than within other tissues (Louveau et al., 2015; Berghoff and Preusser, 2015).

29.1.2 INTERPLAY BETWEEN IMMUNOTHERAPY AND RADIOTHERAPY

The term abscopal effect refers to the rare observation that, on occasion, radiotherapy (RT) to one metastatic site of disease causes regression at distant sites, well outside of the treated area (Hodge et al., 2008; Postow et al., 2012). Historically, RT has been considered immunosuppressive (Piotrowski et al., 2018). Conventional RT with large fields irradiates a substantial amount of lymph node tissue, as well as circulating lymphocytes that are particularly radiosensitive, with a propensity for apoptotic death. This can induce transient lymphopenia, which has been associated with worse clinical outcomes (Crocenzi et al., 2016; Grossman et al., 2015). However, there is increasing evidence that intratumoral CD8+ T cells are more radioresistant than their lymphoid counterparts and can survive clinically relevant radiation, maintaining their function (Arina et al., 2019).

In parallel, RT can also interact synergistically with the immune system. First, RT may induce tumor cell necrosis, resulting in the release of tumor antigens, which can then be recognized by APCs and presented to cytotoxic T cells. Second, RT has been shown to enhance antigen presentation and to promote maturation of APCs (Fonteneau et al., 2002; Lugade et al., 2005; Sharabi, Lim et al., 2015; Sharabi, Nirschl et al., 2015). Third, RT promotes the release of cytokines, such as tumor necrosis factor (TNF) alpha, interferon (IFN) gamma, and chemokine C-X-C motif ligand (CXCL) 16, which alter the inflammatory microenvironment and increase infiltration of activated T cells across the BBB (Sharabi, Lim et al.,

2015; Sharabi, Nirschl et al., 2015; Demaria et al., 2015; Escorcia et al., 2017; Lee et al., 2009; Frey et al., 2014; Sistigu et al., 2014). Fourth, RT can result in priming of CD8 T cells through IFN-gamma production and increased tumor cell MHC I and Fas expression (Hodge et al., 2008; Tang et al., 2014; Garnett et al., 2004; Reits et al., 2006; Grass et al., 2016; Ngwa et al., 2018; Dovedi et al., 2014; Weichselbaum et al., 2017). Fifth, RT can induce immunogenic cell death (Kroemer et al., 2013).

The mechanism by which stereotactic body radiation therapy (SBRT)/SRS treatments result in tumor cell kill may differ from conventional, long-course RT. Hypofractionation, or the use of high dose per fraction, has been demonstrated to induce endothelial apoptosis and vascular damage not seen with conventional RT (Fuks and Kolesnick, 2005; Suit and Willers, 2003). Furthermore, SBRT/SRS plans reduce the volume of irradiation to circulating lymphocytes and lymph node basins, thereby minimizing the immunosuppressive effects of RT (Yovino et al., 2013). Given these key distinctions, radiosurgery may interact with systemic therapy, especially immunotherapy, to ultimately overcome the immunosuppressive tumor microenvironment, increase immune response, induce the abscopal effect, and recruit the immune system for greater anti-tumor effects (Hodge et al., 2008; Tang et al., 2014; Reynders et al., 2015; Popp et al., 2016; Filippi et al., 2016). Increased PD-L1 expression has been observed following RT, suggesting the potential for increased efficacy of anti-PD-1 or anti-PD-L1 therapy, and there is preclinical evidence to support this notion of a synergistic effect between ICI and RT (Hodge et al., 2008; Tang et al., 2014; Dovedi et al., 2014; Demaria et al., 2005).

29.1.3 TIMING OF IMMUNOTHERAPY AND RADIOTHERAPY

The ideal timing of the delivery of ICI relative to RT to induce a maximal tumor response remains an area of uncertainty. If SRS is delivered after ICI, APCs and effector lymphocytes are in circulation when SRS causes tumor cell kill and antigen release. However, SRS may also kill newly recruited lymphocytes leading to an immunosuppressive effect. Conversely, when SRS is delivered prior to ICI, it increases circulating tumor antigens and potentially also the permeability of the BBB in anticipation of immune cell activity (Kalbasi et al., 2013). Several preclinical trials suggest greatest efficacy when ICI is delivered concurrently or before RT. In a murine colorectal cancer model, mice were irradiated with 5 fractions of 2 Gy with a PD-L1 inhibitor. Disease outcomes, including OS, were maximized when ICI were given concurrent with RT, on day 1 or 5. If ICI were delivered 7 days after the completion of RT, no survival benefit was seen relative to RT alone. The efficacy of RT and ICI correlated with PD-1 levels following irradiation. At 24 hours post-RT, both CD4+ and CD8+ cells had increased levels of PD-1, and by 7 days, levels had returned to baseline (Dovedi et al., 2014). In another mouse model, CTLA-4 antibody was delivered 1 week before, 1 day after or 5 days after high dose, single-fraction RT (20 Gy in 1 fraction). The best outcomes were for ICI delivered prior to RT. When CTLA-4 inhibition was administered prior to RT, all tumors demonstrated a complete response (CR), and median OS was not reached. When ICI were delivered after RT, only half of tumors underwent CR. If ICI were given 1 day after RT, median survival was 92 days, compared to only 53 days when delivered 5 days after RT (Young et al., 2016). In a third study by Dewan et al., mammary carcinoma cells were injected into a primary and secondary site in a murine model. Mice were then treated with CTLA-4 inhibition alone, RT to the primary site alone (20 Gy in 1, 24 Gy in 3, or 30 Gy in 5 fractions), or both. Combined CTLA-inhibition and RT was most effective when the drug was delivered concurrently with RT or 2 days prior, compared to 2 days after RT (Dewan et al., 2009).

Several clinical studies support the findings of the above-mentioned murine models with regard to administration of ICI and RT concurrently. The definition of concurrent treatment (in clinical studies) varies between a few weeks to several months, and, given the long half-life and protracted activity of some immune modulators, the optimal timing remains to be elucidated. However, with respect to sequential treatments, most clinical studies favor the delivery of RT *before*, not after, ICI, as some pre-clinical murine studies would suggest. In clinical practice, the ideal timing of RT and ICI may depend upon a variety of factors, including RT dose and fractionation, type of ICI employed, and tumor histology. In addition to the timing of RT, fraction size may play a determining role. There is preclinical and emerging clinical evidence that high-dose RT in few fractions may be most effective (Dovedi et al., 2014; Demaria and Formenti, 2012). In an early murine model, a single fraction of 14 to 25 Gy resulted in increased T cell priming and a decrease in primary tumor and metastatic sites of disease, compared to fractionated RT (Lee et al., 2009). In another mouse melanoma model, single-fraction doses of 7.5 to 15 Gy resulted in a maximal tumor control, while a single fraction of 5 Gy did not induce immunostimulation. Fractionated regimens were also evaluated, specifically 15 Gy in 2, 3, or 5 fractions. Here, both 2- and 3-fraction regimens were superior to single-fraction treatment, and immune response and tumor control were maximized when 2 fractions were used (Schaue et al., 2012). In the study by Dewan et al., mammary carcinoma cells were injected into a primary and secondary site and treated with CTLA-4 inhibition and 20 Gy in 1, 24 Gy in 3, or 30 Gy in 5 fractions. Interestingly, single-fraction RT resulted in tumor response at the primary site alone, while hypofractionated RT in 3 or 5 fractions caused tumor response at the secondary site as well. The most marked response was seen with 3 fractions of 8 Gy (Dewan et al., 2009).

29.2 IMMUNOTHERAPY FOR BRAIN METASTASES

As of May 2019, more than 28,000 patients had been enrolled in 361 clinical trials of immunotherapy and RT on www.ClinicalTrials.gov, with some reporting durable disease control after RT and ICI in the metastatic setting (www.ClinicalTrials.gov; Brix et al., 2017; Shaverdian et al., 2017; Twyman-Saint Victor et al., 2015; Chandra et al., 2015). Unfortunately, patients with BrM are often excluded from such trials. A review of www.ClinicalTrials.gov reveals that patients with untreated NSCLC BrM were eligible for enrollment in only approximately one quarter of available trials (www.ClinicalTrials.gov; McCoach et al., 2016). As a result, evidence regarding efficacy and safety in this patient population is limited (Sampson et al., 1998; Silk et al., 2013; Flanigan et al., 2011). Prospective trials of immunotherapy that report intracranial response rates are discussed in the following section and their results summarized in Table 29.1.

The majority of trials regarding immunotherapy in patients with BrM have employed ipilimumab, a monoclonal antibody against CTLA-4 in patients with metastatic melanoma. An American expanded-access program of 165 melanoma BrM reported a 1-year OS of 20%, while a similar program conducted in Italy found an objective response rate (ORR) of 12% (Heller et al., 2011; Queirolo et al., 2014). In the phase II NIBIT-M1 study, Di Giacomo et al. evaluated 20 advanced melanoma patients treated with ipilimumab and fotemustine of which 7 had prior RT and found a 40% overall CNS response rate with 10% CNS CR (Di Giacomo et al., 2012, 2015). A prospective phase II trial from the University of Washington evaluated ipilimumab in 72 patients with melanoma BrM, with or without prior whole-brain RT or SRS, and reported an ORR of 5% and 16% and local control (LC) of 10% and 24% for symptomatic and asymptomatic patients, respectively (Margolin et al., 2012).

The evidence for monoclonal antibodies targeting PD-1, such as nivolumab and pembrolizumab, is somewhat more limited. An expanded-access program of metastatic RCC patients managed with nivolumab reported 19% CNS response in 32 patients with BrM (De Giorgi et al., 2019). A larger program in Italy included 38 patients with NSCLC BrM and showed disease control in 47% of patients (Bidoli et al., 2016). And a series of five asymptomatic NSCLC BrM patients managed with nivolumab found response in three patients, with one CR, one partial response, and one stable disease (Dudnik et al., 2016). In a single institution phase II trial, patients with asymptomatic, untreated, or progressive BrM from NSCLC or melanoma were managed with pembrolizumab, a PD-1 inhibitor, at a dose of 10 mg/kg every 2 weeks. PD-L1 expression of ≥1% was required for the NSCLC cohort. ORR was noted in 29% of NSCLC and 22% of melanoma patients. Median OS was 8.9 months (Goldberg et al., 2018).

The highest CNS response rates have been attained with the combination of ipilimumab and nivolumab. The phase II CheckMate-204 trial evaluated melanoma patients with asymptomatic intracranial disease, with at least one untreated BrM treated with combination ipilimumab (3 mg/kg) and nivolumab (1 mg/kg) every 3 weeks for four cycles, followed by maintenance nivolumab (3 mg/kg) every 2 weeks. To date, 94 patients had a 52% overall CNS response with 26% achieving CR (Tawbi et al., 2017). Discontinuation of immunotherapy on account of neurological toxicity was required in only 5% of patients (Tawbi et al., 2018). The phase II ABC trial included treatment-naïve patients with melanoma BrM subdivided into Cohort A (asymptomatic, treated with ipilimumab and nivolumab followed by maintenance nivolumab), Cohort B (asymptomatic, treated with nivolumab alone), and Cohort C (symptomatic/progressive disease or leptomeningeal spread, treated with nivolumab alone). The CNS response rate in 66 patients was 46%, 20%, and 6% in Cohorts A, B, and C, respectively (Long et al., 2017).

Table 29.1 Studies of immunotherapy in patients with metastatic disease reporting central nervous system outcomes

STUDY		PHASE	HISTOLOGY	ICI	N		ARMS	OUTCOMES		
								CNS ORR	CNS PFS	PFS
Margolin et al., 2012	NCT00623766	II	Melanoma	Ipilimumab	72	51	1. Asymptomatic, no corticosteroids	1.16%	1.5 mos	1.4 mos
						21	2. Symptomatic, corticosteroid use	2.5%	1.2 mos	1.2 mos
Di Giacomo et al., 2012	NIBIT-M1	II	Melanoma	Ipilimumab plus fotemustine	20		Asymptomatic	40%	3.0 mos	4.5 mos
Goldberg et al., 2016	NCT02085070	II	Melanoma, NSCLC	Pembrolizumab	36	18	1. Melanoma	1.22%	NR	NR
						18	2. NSCLC	2.33%		
Tawbi et al., 2018	CheckMate-204	II	Melanoma	Ipilimumab and nivolumab	94		Nivolumab plus ipilimumab → nivolumab maintenance	55%	64%	61%
Long et al., 2018	ABC	II	Melanoma	Ipilimumab and nivolumab	66	36	1. Asymptomatic treatment naïve, nivolumab plus ipilimumab → nivolumab maintenance	46%	53%	50%
						27	2. Asymptomatic treatment naïve, nivolumab	20%	20%	29%
						16	3. Symptomatic or failed previous local therapy, nivolumab	6%	13%	0%

Taken together, the existing evidence published to date suggests that both anti-PD-1 and anti-CTLA-4 agents, including ipilimumab, nivolumab, pembrolizumab, and the combination of ipilimumab and nivolumab, all have demonstrated intracranial activity for patients with BrM from melanoma, NSCLC, and RCC. Larger, phase II–III studies of immunotherapy are ongoing (NCT02460068, NCT02681549, NCT03175432) including studies of newer ICIs such as indoleamine (2,3)-dioxygenase (IDO) and others (www.ClinicalTrials.gov).

29.3 IMMUNOTHERAPY AND STEREOTACTIC RADIOSURGERY FOR BRAIN METASTASES

In recent years, numerous investigators have explored the combination of immunotherapy and SRS in the management of patients with BrM in an effort to improve rates of local and distant intracranial control. Some, but not all, trials suggest improved outcomes with the combination of SRS and ICI, especially in melanoma. A summary of trials evaluating ICI and SRS is provided in Table 29.2.

An early publication retrospectively reviewed 77 melanoma patients managed with SRS and ipilimumab or SRS alone and reported an improved 2-year OS rate of 47% and 20%, respectively (Knisely et al., 2012). A similar study from University of Michigan reported improved median OS in 70 patients treated with SRS and ipilimumab compared to SRS alone (18.3 versus 5.3 months) (Silk et al., 2013). A retrospective study on SRS in a subset of 26 patients with 73 BrM enrolled in two prospective trials of nivolumab (anti PD-1) reported LC at 12 months of 84% and median OS of 12.0 months, comparing favorably to historical controls (Ahmed et al., 2016). A similar study of 51 patients with 167 BrM treated with SRS plus anti-CTLA-4 or anti-PD-1 showed improved LC for the combination, compared to SRS alone (Yusuf et al., 2017). A retrospective study compared 121 patients treated with SRS with 48 patients treated with SRS plus ICI within 3 months (anti-CTLA-4 or anti-PD-1 therapy), and reported an improvement in 1-year distant brain control (60% versus 12%) and LC (85% vs 66%), favoring the SRS and ICI group. In a sub-group analysis in this study, SRS with both anti-PD-1 (nivolumab or pembrolizumab) and anti-CTLA-4 (ipilimumab) ($n = 14$) resulted in superior 6-month distant intracranial control compared to either drug alone ($n = 34$; 84% versus 50%; $p = 0.016$) (Acharaya et al., 2017).

The question of ICI timing with SRS was studied in several retrospective cohorts including 46 patients with 113 BrM treated with SRS. Delivery of SRS before or during ipilimumbab resulted in improved regional control and OS when compared with SRS delivered after ICI. There was a transient increase in BrM diameter of more than 150% in half the patients who received SRS during or before ICI, but in only 13% in those who were treated after ICI, possibly caused by the interplay between SRS and the immunomodulatory effects of impilimumab. Five of 11 lesions resected for suspected progression in fact revealed only complete necrosis (Kiess et al., 2015). A large study from Yale with 75 melanoma patients and 560 BrM further supports that combination SRS and immunotherapy (within 4 weeks) results in better response, with smaller lesion volume compared to treatments delivered non-concurrently (Qian et al., 2016). A study at the University of Virginia found improved OS (59% versus 33%) and LC (54.4% versus 16.5%) when SRS was delivered before or concurrently with ICI, compared to SRS after ICI (Cohen-Inbar et al., 2017). In the end, a multi-institutional retrospective study of 99 metastatic melanoma patients managed with ipilimumab and SRS showed that delivery of SRS in less than 5.5 months after ICI resulted in longer duration of intracranial response compared to patients completing SRS at more than 5.5 months (8.1 versus 3.6 months), with a similar OS (An et al., 2017).

An Australian study investigated differences between anti-CTLA4, anti-PD-1, BRAF, and MEK inhibitors, when combined with SRS in 108 melanoma metastases. Median OS was improved with SRS plus anti-PD1 compared to anti-CTLA4 (20.4 months vs 7.5 months), as was median brain control (12.7 months versus 7.5 months) (Choong et al., 2017). Further studies would help to determine the relative efficacy of these agents when combined with SRS.

While the majority of trials published to date regarding ICI and SRS have evaluated melanoma patients with BrM, a handful of studies have reported on patients with other primary histologies. In a study of SRS and anti-PD1 in 37 NSCLC patients with 85 BrM, patients treated concurrently had improved 1-year OS (87%) compared to those receiving SRS before ICI (70%) or SRS after ICI (0%) (Schapira et al.,

Table 29.2 Retrospective studies of combined immune checkpoint inhibition and stereotactic radiosurgery

STUDY	N		HISTOLOGY	ICI	SRS MEDIAN DOSE (RANGE) (GY)	ARMS	OS (MEDIAN OR %)	P	1Y LC (%)	P	RN INCIDENCE	CONCLUSION
Knisely et al., 2012	77	50	Melanoma	Ipilimumab	NR	SRS	4.9 mos	0.044	NR		NR	Improved OS with ICI and SRS
		27				ICI and SRS	21.3 mos					No difference in OS by timing of ICI and SRS
Mathew et al., 2013	58	33	Melanoma	Ipilimumab	20 (15–20)	SRS	45%	0.18	63	0.55	0	No difference in OS with ICI
		25				ICI and SRS	56%		65			
Silk et al., 2013	70	37	Melanoma	Ipilimumab	NR (14–24)	WBRT or SRS	5.3 mos	0.005	NR			Improved OS with ICI and SRS
		33				ICI and WBRT or SRS	18.3 mos					No difference in OS by timing of ICI and SRS
Kiess et al., 2015	46	19	Melanoma	Ipilimumab	21 (15–24)	SRS → ICI	56%	0.008	87	0.21	0.003	Improved OS and regional control with concurrent ICI and SRS
		15				SRS = ICI	65%		100			
		12				ICI → SRS	50%		89			
Patel et al., 2017	54	34	Melanoma	Ipilimumab	NR	SRS	39%	0.84	92	0.4	21%	No difference in OS or LC with ICI

(Continued)

Table 29.2 (Continued) Retrospective studies of combined immune checkpoint inhibition and stereotactic radiosurgery

STUDY	N	HISTOLOGY	ICI	SRS MEDIAN DOSE (RANGE) (GY)	ARMS	OS (MEDIAN OR %)	P	1Y LC (%)	P	RN INCIDENCE	CONCLUSION
	20				SRS and ICI	37%		71		30%	Trend toward increased RN with ICI and SRS
Qian et al., 2015	75	Melanoma	CTLA-4 or PD-1	20 (12–24)	Non-concurrent SRS and ICI	9 mos	0.07	NR			Trend toward improved OS with concurrent ICI and SRS
	33				Concurrent SRS and ICI	19.1 mos					
	20				Both concurrent and non-concurrent SRS and ICI						
Tazi et al., 2015	31	Melanoma	Ipilimumab	NR	ICI (no BM at baseline)	33.1 mos	0.9	NR			No difference in OS with ICI and SRS
	10				ICI and SRS (BM at baseline)	29.3 mos					
Cohen-Inbar et al., 2017	46	32	Melanoma	Ipilimumab	20 (14–22)	SRS → ICI and SRS = ICI	13.8 mos	0.12	54	0.005	31	Improved LC and LRFS with SRS before or concurrent with ICI

(Continued)

Table 29.2 (Continued) Retrospective studies of combined immune checkpoint inhibition and stereotactic radiosurgery

STUDY	N	HISTOLOGY	ICI	SRS MEDIAN DOSE (RANGE) (GY)	ARMS	OS (MEDIAN OR %)	P	1Y LC (%)	P	RN INCIDENCE	CONCLUSION
	14				ICI→SRS	6.4 mos		17		7	Increased RN with SRS before or concurrent with ICI
Yusuf et al., 2017	12	Melanoma	CTLA-4 or PD-1	18 (13–24)	SRS = ICI	7.4 mos	0.21	NR		2	Improved lesion regression, LC and distant intracranial control with concurrent SRS and ICI compared to SRS alone.
	6				SRS	7.1 mos				0	
Acharya et al., 2017	72	Melanoma	CTLA-4 and/ or PD-1	20 (11–24)	SRS	31%	0.02	66%	0.04	NR	Improved LC and distant intracranial control with SRS and ICI. Improved OS with SRS and ICI on UVA.
	121										

(Continued)

Table 29.2 (Continued) Retrospective studies of combined immune checkpoint inhibition and stereotactic radiosurgery

STUDY	N	HISTOLOGY	ICI	SRS MEDIAN DOSE (RANGE) (GY)	ARMS	OS (MEDIAN OR %)	P	1Y LC (%)	P	RN INCIDENCE	CONCLUSION
	48				SRS and ICI	58%		85%			
Chen et al., 2018	260	Melanoma, NSCLC, RCC	CTLA-4 and/ or PD-1	20 (16–25)	SRS	12.9 mos	SS	82% LRFS		NR	Improved OS, intracranial disease control with concurrent SRS and ICI
	181										
	51				Non-concurrent SRS and ICI	14.5 mos		9% LRFS			No difference in toxicity with concurrent SRS and ICI
	28				SRS = ICI	24.7 mos		88% LRFS			
Schapira et al., 2018	37	NSCLC	PD-1	NR	SRS = ICI	87%	0.008	100%	0.02	1	Improved OS and LC with concurrent SRS and ICI
					SRS→ICI	70%		72%		2	
					ICI→SRS	0%					

2018). A study by Ahmed et al. of 16 NSCLC patients with 49 BrM treated with SRS and nivolumab or durvalumab found favorable 6-month distant brain control of 48%, 1-year OS of 40%, and LC of 96%. There was an effect of SRS timing, with 6-month distant brain control of 57% if SRS was delivered during or prior to ICI, compared to 0% for SRS delivered after ICI (Ahmed et al., 2017). A meta-analysis of 17 studies across three countries encompassing more than 500 BrM patients with a variety of primary cancers reported a 1-year OS of 65% when ICI and SRS were used concurrently, compared to 52% for non-concurrent administration ($p < 0.001$) (Lehrer et al., 2019). Further, a recent large retrospective study of SRS and ICI in 260 patients with 623 BrM from NSCLC, melanoma, or RCC showed improved OS for SRS delivered concurrently within 2 weeks of ICI (24.7 months), compared to non-concurrent SRS and ICI (14.5 months) or SRS alone (12.9 months) (Chen et al., 2018).

Not all studies found a benefit of combining ICI and SRS, though most of these studies evaluated fewer patients. In a retrospective study from New York University, SRS was administered to 58 patients with metastatic melanoma including 25 patients who also received ipilimumab (3 mg/kg) every 3 weeks. The authors reported no significant difference in LC, freedom from intracranial disease, OS, or toxicity (Mathew et al., 2013). A study from Emory on 54 melanoma patients treated with SRS or SRS and ipilimumab similarly did not report a statistically significant difference in outcomes between groups, and other, smaller, studies have also reported no differences in outcomes (Patel et al., 2017; Tazi et al., 2015; Liniker et al., 2016).

There are numerous ongoing trials evaluating the efficacy and safety of immunotherapy and SRS in the treatment of BrM, as well as the optimal timing and dosing regimen. Such trials include NCT01454102, NCT 01703507, NCT02662725, NCT02107755, NCT02097732, NCT02858869, NCT02716948, NCT02696993, NCT02886585, NCT02978404, and NCT03340129 (www.ClinicalTrials.gov). Table 29.3 provides a summary of these trials.

29.4 ADVERSE EFFECTS OF IMMUNOTHERAPY IN THE BRAIN

When compared to traditional chemotherapeutic agents, immunotherapy drugs have a unique complement of adverse effects (AEs). Key to their mechanism of action, these agents upregulate the immune system, causing inflammatory AEs (Postow et al., 2018). The mechanisms underpinning these inflammatory or autoimmune reactions are not entirely understood. ICI may cause inflammation in tissues that express the targets being inhibited, such as CTLA-4 expression in normal pituitary tissue that may account for the hypophysitis rarely seen with CTLA-4 inhibitors (Iwama et al., 2014). There is potential for cross-reactivity of normal and malignant tissues, as in the case of vitiligo induced by immunotherapy treatment (Byrne and Fischer, 2017). It has also been posited that ICI may elevate levels of pre-existing antibodies and alter the cytokine environment (Osorio et al., 2017; Esfahani and Miller, 2017).

Side effects vary, but most ICI are associated with potential for fever, chills, lethargy, maculopapular rash, pruritis, hypophysitis, thyroiditis, and adrenal insufficiency. Of particular relevance to the patient population being considered in this chapter, the most commonly observed CNS AE is headache, but other reported neurological toxicities include hypophysitis, encephalitis, demyelinating polyneuropathy, encephalomyelitis, confusion, edema, and seizures (Margolin et al., 2012; Weber et al., 2015; Gauvain et al., 2018).

Anti-CTLA-4 agents are associated with diarrhea, colitis with ulceration, and elevations in liver function tests (LFTs). Neuropathy, nephritis, Guillain–Barre, myasthenia gravis, sarcoid, and thrombocytopenia are rarely observed (Osorio et al., 2017; Weber et al., 2015; Abdel-Rahman et al., 2016; Naidoo et al., 2017; Morganstein et al., 2017). In a prospective phase II trial, ipilimumab was also associated with two cases of grade 4 confusion and three instances of intracranial edema (Gauvain et al., 2018). Anti-PD-1 antibodies are associated with diarrhea, colitis with ulceration, pneumonitis, and abnormal LFTs. Neuropathy, Guillain–Barre, myasthenia gravis, and nephritis are also rarely seen (Osorio et al., 2017; Weber et al., 2015; Abdel-Rahman et al., 2016; Naidoo et al., 2017; Morganstein et al., 2017). PD-L1 inhibitors can lead to diarrhea and ulcerative colitis, LFT elevation, pneumonitis, and anemia (Weber et al., 2015). In the prospective CheckMate-204 trial that combined nivolumab and ipilimumab for melanoma BrM, CNS toxicities were observed in 8% of patients. In half of these patients, immunotherapy had to be stopped as a result of toxicity (Tawbi et al., 2017).

Table 29.3 Ongoing clinical trials of immune checkpoint inhibition and stereotactic radiosurgery

NCT	STUDY	START	PHASE	ENROLLMENT	HISTOLOGY	ICI	TREATMENT	STATUS	PRIMARY ENDPOINT
01454102	CheckMate 012	2011	I	NA	NSCLC	Nivolumab	One arm with BM treated with nivolumab	Active, not recruiting	AEs
01703507		2012	I	17	Melanoma	Ipilimumab	Ipilimumab and WBRT	Completed	Maximum tolerated dose
							Ipilimumab and SRS		
02662725		2012	II	73	Melanoma	Ipilimumab	Ipilimumab and SRS	Completed	OS
02107755		2014	II	8	Melanoma	Ipilimumab	Ipilimumab and SRS	Active, not recruiting	PFS
02097732		2014	II	4	Melanoma	Ipilimumab	Ipilimumab and SRS	Active, not recruiting	LC
02858869		2016	I	30	Melanoma and NSCLC	Pembrolizumab	Pembrolizumab and 3 doses of SRS	Recruiting	Dose limiting toxicity
02716948		2016	I	90	Melanoma	Nivolumab	Nivolumab and SRS for spine and brain metastases	Active, not recruiting	AEs
02696993		2016	I/II	88	NSCLC	Ipilimumab and nivolumab	Nivolumab and SRS	Recruiting	Dose limiting toxicity; Intracranial PFS
							Nivolumab and WBRT		
							Nivolumab and ipilimumab and SRS		
							Nivolumab and ipilimumab and WBRT		
02886585		2016	II	102	Multiple	Pembrolizumab	Untreated BM, progressive, neoplastic meningitis, metastasis from melanoma. One cohort and SRS.	Recruiting	ORR; OS; Extracranial ORR

(Continued)

Table 29.3 (Continued) Ongoing clinical trials of immune checkpoint inhibition and stereotactic radiosurgery

NCT	STUDY	START	PHASE	ENROLLMENT	HISTOLOGY	ICI	TREATMENT	STATUS	PRIMARY ENDPOINT
02978404		2017	II	26	NSCLC, SCLC, Melanoma, RCC	Nivolumab	Nivolumab and SRS	Active, not recruiting	Intracranial PFS
04047602	RADREMI	2019	N/A	42	Multiple	ICI	ICI and reduced dose SRS	Recruiting	Symptomatic RN
04042220		2019	N/A	200	Melanoma and NSCLC	ICI	SRS and ICI	Recruiting	OS
							SRS and ICI and steroids		
							SRS		
03340129	ABC-X	2019	II	218	Melanoma	Ipilimumab and nivolumab	Nivolumab and ipilimumab → nivolumab	Recruiting	Neurologic specific cause of death
							Nivolumab and ipilimumab → nivolumab and SRS		
							Nivolumab and ipilimumab → nivolumab and WBRT		
03955198	SILK BM	2020	II	100	NSCLC	Durvalumab	Fractionated SRS	Recruiting	Time to intracranial progression
							Fractionated SRS and ICI		

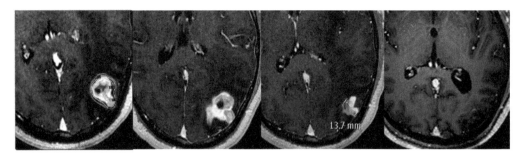

Figure 29.1 A patient with a renal cell carcinoma metastasis treated with 25 Gy in 5 fractions using the Gamma Knife Icon. After two subsequent cycles of a PDL-1 inhibitor, the patient developed headache, and MRI showed increased enhancement and tumor size. Clinically, this was thought to represent pseudoprogression, and after 2 weeks of dexamethasone, the tumor regressed. After resuming immunotherapy, the patient achieved a complete response approximately 2 months later.

29.4.1 ADVERSE EFFECTS OF IMMUNOTHERAPY AND STEREOTACTIC RADIOSURGERY IN THE BRAIN

Existing toxicity data is largely based upon retrospective studies. Potential CNS toxicities after SRS and ICI include hemorrhage, cerebral edema, seizures, and radionecrosis (RN) (Silk et al., 2013; Ahmed et al., 2016; Kiess et al., 2015; Patel et al., 2017; Tazi et al., 2015). A meta-analysis of 17 studies included 534 patients treated with SRS and ICI and found an overall rate of RN of 5% (Lehrer et al., 2019). Pseudoprogression, a phenomenon occasionally observed in tumors treated with immunotherapy, occurs when there is initial apparent tumor growth, followed by tumor regression. Pseudoprogression can be distinguished from true disease progression through biopsy or serial imaging (Reckamp 2018; Chiou and Burotto, 2015).

The study by Chen et al. including melanoma, NSCLC, and RCC patients managed with a variety of systemic agents found no increased risk of immune-related or neurological toxicities, including RN, with the combination if ICI and SRS compared to SRS alone (Chen et al., 2018).

However, several other studies suggest that combining ICI and SRS may increase rates of RN. Kiess et al. reported a trend toward increased overall CNS toxicities with SRS and ipilimumab, and Patel et al. observed a trend toward increased RN of 30% for ICI plus SRS versus 21% with SRS alone ($p = 0.078$) (Kiess et al., 2015; Patel et al., 2017). Cohen-Inbar et al. also reported a trend toward increased rates of RN when SRS was delivered prior to or concurrent with ICI compared to SRS after ICI (19% vs 10%, $p = 0.066$) (Cohen-Inbar et al., 2017). Martin et al. reviewed 480 patients with melanoma, NSCLC, or RCC who received SRS, including 115 treated with ICI, and found an increased risk of symptomatic RN with combination therapy (hazard ratio (HR) 2.56; $p = 0.004$) that was more pronounced for the melanoma subgroup (HR 4.02; $p = 0.03$) (Martin et al., 2018). A rare dose-finding prospective phase I study accrued 16 melanoma patients with BrM treated with ipilimumab and WBRT or ipilimumab and SRS. There was no dose-limiting toxicity, and no patients experienced grade 4–5 AEs, with no cases of RN. However, the WBRT arm was closed early due to limited accrual (Williams et al., 2017). A recent large single-institution review of 1118 BrM from Wake Forest University found an increase rate of AEs (defined as pseudoprogression or RN) in the combination SRS and ICI group compared to SRS alone (5% versus 2%). The increased toxicity was limited to grade 2 events, with no grade 5 events in the cohort. Prescription dose and V12Gy were associated with increased AEs in patients receiving ICI (Helis et al., 2020).

29.5 IMMUNOTHERAPY AND STEREOTACTIC BODY RADIOTHERAPY FOR SPINAL METASTASES

Much like BrM, spinal metastases are a significant source of morbidity and mortality for cancer patients. Spinal metastases can cause pain, vertebral compression fractures, and cord compression, which can result in deficits of motor and sensory function, as well as urinary, and bowel incontinence (Vellayappan et al.,

2018). Spinal SBRT to intact tumors or administered postoperatively delivers high doses per fraction, with a steep dose falloff outside of the target region, sparing adjacent organs at risk (Vellayappan et al., 2018; Seung et al., 2013; Sahgal et al., 2012; Hall et al., 2011). SBRT allows for better LC, with delivery of higher biologically effective dose, which may be especially beneficial for radioresistant pathologies. Numerous clinical trials have established the safety and efficacy of spinal SBRT, with reported rates of LC of 80%–95% at 1 year (Tseng et al., 2018).

Over the past decade, retrospective evidence has started to emerge regarding the safety profile and efficacy of the combination of SRS and ICI for BrM, as well as the optimal techniques for combining these modalities. However, there is little published literature regarding the efficacy of combining ICI and spinal SBRT. A case report from Wake Forest describes a patient with alveolar soft part sarcoma who achieved a CR after spine SBRT, followed by pembrolizumab to a bulky L3 lesion (Cramer et al., 2019). However, prospective studies of SBRT in oligometastatic disease have generally precluded concurrent administration of ICI, and studies to support a synergistic effect in the spine are greatly needed. There are similarly scarce data on toxicity to guide risk–benefit discussion. Luke et al. published the first prospective study of the combination of pembrolizumab and SBRT to sites of metastatic disease, including 11 patients with spinal metastases treated with 30 Gy in 3 fractions. Pembrolizumab was delivered within 1 week of completion of RT. Dose-limiting toxicity was encountered in 1 out 11 spine patients, with one case of grade 3 colitis, but no reported instances of radiation myelopathy (Luke et al., 2018). Several other case reports have described spinal toxicity in patients treated with ICI alone, including radiation myelitis and meningo-radicular neuritis (Blackmon et al., 2016; Bompaire et al., 2012; Chang et al., 2018). Whether spinal SBRT increases the risk of myelopathy or not in patients undergoing ICI treatment remains to be determined. An active phase I trial, NCT02716948, is evaluating the combination of SRS and nivolumab for melanoma patients with brain or spinal metastases with safety as the primary endpoint (www.ClinicalTrials.gov). Until more data emerge on the toxicity profile of combination SRS/SBRT and ICI, we consider it prudent to interrupt ICI for 1 week before and after treatment, where possible. Ultimately, this is an individualized multi-disciplinary decision based on the patient's burden of disease and expected toxicity.

29.6 FUTURE DIRECTIONS

Attaining an abscopal response by irradiating a single site of disease remains challenging, and there is increasing doubt surrounding this strategy. A recent phase II study randomized 62 head and neck cancer patients to nivolumab with or without SBRT to a single lesion and found no appreciable abscopal effect (McBride, 2020). Similarly, the study by Theelen et al. in advanced NSCLC recently failed to demonstrate a clinical benefit from adding SBRT to a single site of oligometastasis followed by pembrolizumab (Theelen et al., 2019). It may be that metastases are heterogeneous in their immunologic signature, thereby escaping immune recognition, or that an unwelcoming microenvironment prevents equal penetration of T cells at all metastatic sites. It is also possible that radiation-induced cytoreduction at more than one site is important to facilitate the penetration and effect of ICI. Irradiating all metastatic sites could circumvent heterogeneity in immune recognition, decrease the burden of disease to improve penetration of systemic agents, and possibly increase T cell infiltration by modulating the microenvironment at every tumor location (Gutiontov et al., 2020). This cytoreductive approach is an area of active research. The recently published single-arm phase 2 study by Bauml et al. included 45 patients treated with pembrolizumab after local therapy (SBRT or surgery) to all NSCLC oligometastatic sites and found favorable PFS of 19.1 months compared to a historical control of 6.6 months (Bauml et al., 2019). Future studies are eagerly awaited. Outside of the oligometastatic setting, SRS/SBRT is further being investigated as a systemic line of therapy for polymetastatic disease (10, 20 metastases, or more) with a potentially acceptable side-effect profile compared to next-line chemotherapy (Palma et al., 2020; Patel et al., 2020). In the end, assessing treatment response after immunotherapy will be of paramount importance in the era of ICI to help understand solid tumor behavior after ICI, assess immune-related AEs, and distinguish between true and pseudo-progression (Flavell et al., 2020). Newer criteria have been devised for this purpose, including the immune-related Response Criteria (iRC), iRECIST, and immune-modified RECIST criteria (imRECIST) (Wolchok et al., 2009; Hodi et al., 2016; Seymour et al., 2019; Hodi et al., 2018).

REFERENCES

Abdel-Rahman O, ElHalawani H, Fouad M. (2016) Risk of endocrine complications in cancer patients treated with immune check point inhibitors: a meta-analysis. *Future Oncol Lond Engl* 12:413–425.

Acharya S, Mahmood M, Mullen D, et al. (2017) Distant intracranial failure in melanoma brain metastases treated with stereotactic radiosurgery in the era of immunotherapy and targeted agents. *Adv Radiat Oncol* 2(4):572–580.

Ahmed KA, Kim S, Arrington J, et al. (2017) Outcomes targeting the PD-1/PD-L1 axis in conjunction with stereotactic radiation for patients with non-small cell lung cancer brain metastases. *Journal of Neurooncology* 133(2):1–8.

Ahmed KA, Stallworth DG, Kim Y, et al. (2016) Clinical outcomes of melanoma brain metastases treated with stereotactic radiation and anti-PD-1 therapy. *Ann Oncol* 27:434–441.

An Y, Jiang W, Kim BYS, et al. (2017) Stereotactic radiosurgery of early melanoma brain metastases after initiation of anti-CTLA-4 treatment is associated with improved intracranial control. *Radiotherapy and Oncology* 125:80–88.

Arina A, Beckett M, Fernandez C, et al. (2019) Tumor-reprogrammed resident T cells resist radiation to control tumors. *Nat Commun* 10(1):3959.

Barnholtz-Sloan JS, Sloan AE, Davis FG, et al. (2004) Incidence proportions of brain metastases in patients diagnosed (1973 to 2001) in the metropolitan Detroit cancer surveillance system. *Journal of Clinical Oncology* 22:2865–2872.

Baumeister SH, Freeman GJ, Dranoff G, Sharpe AH. (2016) Coinhibitory pathways in immunotherapy for cancer. *Annu Rev Immunol* 34:539–573.

Bauml JM, Mick R, Ciunci C, et al. (2019) Pembrolizumab after completion of locally ablative therapy for oligometastatic non-small cell lung cancer: a phase 2 trial. *JAMA Oncology* 5(9):1283–1290.

Berghoff AS, Fuchs E, Ricken G, et al. (2015) Density of tumor-infiltrating lymphocytes correlates with extent of brain edema and overall survival time in patients with brain metastases. *Oncoimmunology* 5(1):e1057388.

Berghoff AS, Preusser M. (2015) The inflammatory microenvironment in brain metastases: Potential treatment target? *Chin Clin Oncol* 4:21.

Bidoli P, Chiari R, Catino A, et al. (2016) Efficacy and safety data from patients with advanced squamous NSCLC and brain metastases participating in the nivolumab expanded access programme (EAP) in Italy. *Ann Oncol* 27:1228.

Blackmon JT, Viator TM, Conry RM. (2016) Central nervous system toxicities of anti-cancer immune checkpoint blockade. *J Neurol Neuromedicine* 1(4):39–45.

Bompaire F, Mateus C, Taillia H, et al. (2012) Severe meningo-radiculo-neuritis associated with ipilimumab. *Invest New Drug* 30(6):2407–2410.

Borghaei H, Paz-Ares L, Horn L, et al. (2015) Nivolumab versus docetaxel in advanced nonsquamous non-small-cell lung cancer. *New England Journal of Medicine* 373(17):1627–1639.

Brahmer J, Reckamp KL, Baas P, et al. (2015) Nivolumab versus docetaxel in advanced squamous-cell non-small-cell lung cancer. *New England Journal of Medicine* 373:123–135.

Brix N, Tiefenthaller A, Anders H, Belka C, Lauber K. (2017) Abscopal, immunological effects of radiotherapy: narrowing the gap between clinical and preclinical experiences. *Immunol Rev* 280(1):249–279.

Cagney DN, Martin AM, Catalano PJ, et al. (2017) Incidence and prognosis of patients with brain metastases at diagnosis of systemic malignancy: a population-based study. *Neuro-Oncology* 19:1511–1521.

Carson MJ, Doose JM, Melchior B, Schmid CD, Ploix CC. (2006) CNS immune privilege: hiding in plain sight. *Immunol Rev* 213:48–65.

Chandra RA, Wilhite TJ, Balboni TA, et al. (2015) A systematic evaluation of abscopal responses following radiotherapy in patients with metastatic melanoma treated with ipilimumab. *Oncoimmunology* 4(11):e1046028.

Chang VA, Simpson DR, Daniels GA, Piccioni DE. (2018) Infliximab for treatment-refractory transverse myelitis following immune therapy and radiation. *J Immunother Cancer* 6:153.

Chen A, Douglass J, Kleinberg L, et al. (2018) Concurrent immune checkpoint inhibitors and stereotactic radiosurgery for brain metastases in non-small cell lung cancer, melanoma, and renal cell carcinoma. *International Journal of Radiation Oncology, Biology, Physics* 100(4):916–925.

Chen L, Han X. (2015) Anti-PD-1/PD-L1 therapy of human cancer: past, present, and future. *J Clin Invest* 125(9):3384–3391.

Chiou VL, Burotto M. (2015) Pseudoprogression and immune-related response in solid tumors. *Journal of Clinical Oncology* 33:3541–3543.

Choong ES, Lo S, Drummond M, et al. (2017) Survival of patients with melanoma brain metastasis treated with stereotactic radiosurgery and active systemic drug therapies. *Eur J Cancer* 75:169–178.

Chuk MK, Chang JT, Theoret MR, et al. (2017) FDA approval summary: accelerated approval of pembrolizumab for second-line treatment of metastatic melanoma. *Clinical Cancer Research* 23:5666–5670.

ClinicalTrials.gov [Internet]. Bethesda (MD): National Library of Medicine (US). Feb 29, 2000 [cited August 22, 2020]; Available from: https://clinicaltrials.gov/ct2/home.

Cohen-Inbar O, Shih HH, Xu Z, Schlesinger D, Sheehan JP. (2017) The effect of timing of stereotactic radiosurgery treatment of melanoma brain metastases treated with ipilimumab. *Journal of Neurosurgery* 127(5):1007–1014.

Cramer CK, Ververs JD, Jones FS, Hsu W, Chan, MD, Quintero Wolfe S. (2019) Will immunotherapy change the role of spine radiosurgery in high-grade epidural disease? A case report and a call for an update of current treatment algorithms. *J Radiosurg SBRT* 6(2):153–156.

Crocenzi T, Cottam B, Newell P, et al. (2016) A hypofractionated radiation regimen avoids the lymphopenia associated with neoadjuvant chemoradiation therapy of borderline resectable and locally advanced pancreatic adenocarcinoma. *J Immunother Cancer* 4:45.

Dart A. (2018) New targets for cancer immunotherapy. *Nature Reviews Cancer* 18:667.

De Giorgi U, Carteni G, Giannarelli D, et al. (2019) Safety and efficacy of nivolumab for metastatic renal cell carcinoma: real-world results from an expanded access programme. *BJU Int* 123(1):98–105.

Demaria S, Formenti SC. (2012) Radiation as an immunological adjuvant: current evidence on dose and fractionation. *Front Oncol* 2:153.

Demaria S, Golden EB, Formenti SC. (2015) Role of local radiation therapy in cancer immunotherapy. *JAMA Oncology* 1:1325–1332.

Demaria S, Kawashima N, Yang AM, et al. (2005) Immune-mediated inhibition of metastases after treatment with local radiation and CTLA-4 blockade in a mouse model of breast cancer. *Clinical Cancer Research* 11(2):728–734.

Dewan MZ, Galloway AE, Kawashima N, et al. (2009) Fractionated but not single-dose radiotherapy induces an immune-mediated abscopal effect when combined with anti—CTLA-4 antibody. *Clinical Cancer Research* 15(17):5379–5388.

Di Giacomo AM, Ascierto PA, et al. (2015) Three-year follow-up of advanced melanoma patients who received ipilimumab plus fotemustine in the Italian network for tumor biotherapy (NIBIT)-M1 phase II study. *Ann Oncol Off J Eur Soc Med Oncol* 26:798–803.

Di Giacomo AM, Ascierto PA, Pilla L, et al. (2012) Ipilimumab and fotemustine in patients with advanced melanoma (NIBIT-M1): an open-label, single-arm phase 2 trial. *Lancet Oncology* 13:879–886.

Dovedi SJ, Adlard AL, Lipowska-Bhalla G, et al. (2014) Acquired resistance to fractionated radiotherapy can be overcome by concurrent PD-L1 blockade. *Cancer Research* 74:5458–5468.

Dudnik E, Yust-Katz S, Nechushtan H, et al. (2016) Intracranial response to nivolumab in NSCLC patients with untreated or progressing CNS metastases. *Lung Cancer* 98(Suppl 4):114–117.

Dunn GP, Okada H. (2015) Principles of immunology and its nuances in the central nervous system. *Neuro-Oncology* 17(Suppl 7):3–8, vii.

Escorcia FE, Postow MA, Barker CA. (2017) Radiotherapy and immune checkpoint blockade for melanoma: a promising combinatorial strategy in need of further investigation. *Cancer J* 23:32–39.

Esfahani K, Miller WH. (2017) Reversal of autoimmune toxicity and loss of tumor response by interleukin-17 blockade. *New England Journal of Medicine* 376:1989–1991.

Filippi AR, Fava P, Badellino S, Astrua C, Ricardi U, Quaglino P. (2016) Radiotherapy and immune checkpoints inhibitors for advanced melanoma. *Radiotherapy and Oncology* 120:1–12.

Flanigan JC, Jilaveanu LB, Faries M et al. (2011) Melanoma brain metastases: is it time to reassess the bias? *Curr Probl Cancer* 35:200–210.

Flavell RR, Evans MJ, Villanueva-Meyer JE, Yom SS. (2020) Understanding response to immunotherapy using standard of care and experimental imaging approaches. *International Journal of Radiation Oncology, Biology, Physics* 108(1):242–257.

Fonteneau JF, Larsson M, Bhardwaj N. (2002) Interactions between dead cells and dendritic cells in the induction of antiviral CTL responses. *Curr Opin Immunol* 14:471–477.

Francisco LM, Sage PT, Sharpe AH. (2010) The PD-1 pathway in tolerance and autoimmunity. *Immunol Rev* 236:219–242.

Frey B, Rubner Y, Kulzer L, et al. (2014) Antitumor immune responses induced by ionizing irradiation and further immune stimulation. *Cancer Immunol Immunother* 63:29–36.

Fridman WH, Pages F, Sautes-Fridman C, Galon J. (2012) The immune contexture in human tumours: impact on clinical outcome. *Nature Reviews Cancer* 12:298–306.

Fuks Z, Kolesnick R. (2005) Engaging the vascular component of the tumor response. *Cancer Cell* 8:89–91.

Gandhi L, Rodriguez-Abreu D, Gadgeel S, et al. (2018) Pembrolizumab plus chemotherapy in metastatic non-small-cell lung cancer. *New England Journal of Medicine* 378(22):2078–2092.

Garnett CT, Palena C, Chakraborty M, Tsang KY, Schlom J, Hodge JW. (2004) Sublethal irradiation of human tumor cells modulates phenotype resulting in enhanced killing by cytotoxic T lymphocytes. *Cancer Research* 64(21):7985–7994.

Gaspar L, Scott C, Rotman M, et al. (1997) Recursive partitioning analysis (RPA) of prognostic factors in three radiation therapy oncology group (RTOG) brain metastases trials. *International Journal of Radiation Oncology, Biology, Physics* 37(4):745–751.

Gauvain C, Vauléon E, Chouaid C, et al. (2018) Intracerebral efficacy and tolerance of nivolumab in non-small-cell lung cancer patients with brain metastases. *Lung Cancer Amst Neth* 116:62–66.

Gettinger S, Rizvi NA, Chow LQ, et al. (2016) Nivolumab monotherapy for first-line treatment of advanced non-small-cell lung cancer. *Journal of Clinical Oncology* 34:2980–2997.

Goldberg S, Mahajan A, Herbst R, et al. (2018) Durability of brain metastasis response and overall survival in patients with non-small cell lung cancer (NSCLC) treated with pembrolizumab. In: ASCO Annual Meeting.

Gooden MJM, de Bock GH, Leffers N, Daemen T, Nijman HW. (2011) The prognostic influence of tumour-infiltrating lymphocytes in cancer: a systematic review with meta-analysis. *Br J Cancer* 105(1):93–103.

Grass GD, Krishna N, Kim S. (2016) The immune mechanisms of abscopal effect in radiation therapy. *Curr Probl Cancer* 40:10–24.

Grossman SA, Ellsworth S, Campian J, et al. (2015) Survival in patients with severe lymphopenia following treatment with radiation and chemotherapy for newly diagnosed solid tumors. *J Natl Compr Canc Netw* 13(10):1225–1231.

Gutiontov SI, Pitroda SP, Chmura SJ, Arina A, Weichselbaum RR. (2020) Cytoreduction and the optimization of immune checkpoint inhibition with radiation therapy. *International Journal of Radiation Oncology, Biology, Physics* 108(1):17–26.

Hall WA, Stapleford LJ, Hadjipanayis CG, Curran WJ, Crocker I, Shu HK. (2011) Stereotactic body radiosurgery for spinal metastatic disease: an evidence-based review. *Int J Surg Oncol* 2011:979214.

Han HS, Suk K. (2005) The function and integrity of the neurovascular unit rests upon the integration of the vascular and inflammatory cell systems. *Curr Neurovasc Res* 2(5):409–423.

Hanahan D, Weinberg RA. (2011) Hallmarks of cancer: the next generation. *Cell* 144:646–674.

Helis CA, Hughes RT, Glenn CW, et al. (2020) Predictors of adverse radiation effect in brain metastasis patients treated with stereotactic radiosurgery and immune checkpoint inhibitor therapy. *International Journal of Radiation Oncology, Biology, Physics* 108(1):295–303.

Heller KN, Pavlick AC, Hodi FS, et al. (2011) Safety and survival analysis of ipilimumab therapy in patients with stable asymptomatic brain metastases. *Journal of Clinical Oncology* 29:8581.

Herbst RS, Baas P, Kim D-W, et al. (2016) Pembrolizumab versus docetaxel for previously treated, PD-L1-positive, advanced non-small-cell lung cancer (KEYNOTE-010): a randomised controlled trial. *Lancet* 387(10027):1540–1550.

Herbst RS, Soria JC, Kowanetz M, et al. (2014) Predictive correlates of response to the anti-PD-L1 antibody MPDL3280A in cancer patients. *Nature* 515:563–567.

Hodge JW, Guha C, Neefjes J, Gulley JL. (2008) Synergizing radiation therapy and immunotherapy for curing incurable cancers. Opportunities and challenges. *Oncology* 22:1064–1070.

Hodi FS, Ballinger M, Lyons B, et al. (2018) Immune-modified response evaluation criteria in solid tumors (imRECIST): refining guidelines to assess the clinical benefit of cancer immunotherapy. *Journal of Clinical Oncology* 36(9):850–858.

Hodi FS, Chesney J, Pavlick AC, et al. (2016) Combined nivolumab and ipilimumab versus ipilimumab alone in patients with advanced melanoma: 2-Year overall survival outcomes in a multicentre, randomised, controlled, phase 2 trial. *Lancet Oncology* 17: 1558–1568.

Hodi FS, Hwu WJ, Kefford R, et al. (2016) Evaluation of immune-related response criteria and recist v1.1 in patients with advanced melanoma treated with pembrolizumab. *Journal of Clinical Oncology* 34(13):1510–1517.

Hodi FS, O'Day SJ, McDermott DF, et al. (2010) Improved survival with ipilimumab in patients with metastatic melanoma. *New England Journal of Medicine* 363:711–723.

Iwama S, Armi H. (2017) Clinical practice and mechanism of endocrinological adverse events associated with immune checkpoint inhibitors. *Nihon Rinsho Meneki Gakkai Kaishi* 40(2):90–94.

Johnson JD, Young B. (1996) Demographics of brain metastasis. *Neurosurg Clin N Am* 7:337–344.

Kalbasi A, June CH, Haas N, Vapiwala N. (2013) Radiation and immunotherapy: a synergistic combination. *J Clin Investig* 123:2756–2763.

Khandekar MJ, Piotrowska Z, Willers H, Sequist LV. (2018) Role of epidermal growth factor receptor (EGFR) inhibitors and radiation in the management of brain metastases from EGFR mutant lung cancers. *Oncologist* 23(9):1054–1062.

Kiess AP, Wolchok JD, Barker CA, et al. (2015) Stereotactic radiosurgery for melanoma brain metastases in patients receiving ipilimumab: safety profile and efficacy of combined treatment. *International Journal of Radiation Oncology, Biology, Physics* 92:368–375.

Knisely JP, Yu JB, Flanigan J, Sznol M, Kluger HM, Chiang VL. (2012) Radiosurgery for melanoma brain metastases in the ipilimumab era and the possibility of longer survival. *Journal of Neurosurgery* 117:227–233.

Kohler BA, Ward E, McCarthy BJ, et al. (2011) Annual report to the nation on the status of cancer, 1975–2007, featuring tumors of the brain and other nervous system. *J Natl Cancer Inst* 103(9):714–736.

Kroemer G, Galluzzi L, Kepp O, Zitvogel L. (2013) Immunogenic cell death in cancer therapy. *Annu Rev Immunol* 31:51–72.

Laman JD, Weller RO. (2013) Drainage of cells and soluble antigen from the CNS to regional lymph nodes. *J Neuroimmune Pharmacol* 8:840–856.

Larkin J, Hodi FS, Wolchok JD. (2015) Combined nivolumab and ipilimumab or monotherapy in untreated melanoma. *New England Journal of Medicine* 373:1270–1271.

Lee Y, Auh SL, Wang Y, Burnette B, et al. (2009) Therapeutic effects of ablative radiation on local tumor require CD8+ T cells: changing strategies for cancer treatment. *Blood* 114(3):589–595.

Lehrer EJ, Peterson J, Brown PD, et al. (2019) Treatment of brain metastases with stereotactic radiosurgery and immune checkpoint inhibitors: an international meta-analysis of individual patient data. *Radiotherapy and Oncology* 130:104–112.

Liniker E, Menzies AM, Kong BY, et al. (2016) Activity and safety of radiotherapy with anti-PD-1 drug therapy in patients with metastatic melanoma. *Oncoimmunology* 5:e1214788.

Loeffler JS. (2016) Epidemiology, clinical manifestations, and diagnosis of brain metastases. In: Wen PY, editor. *UpToDate*. Waltham, MA: UpToDate. [Accessed on August 22, 2020].

Long GV, Atkinson V, Lo S, et al. (2018) Combination nivolumab and ipilimumab or nivolumab alone in melanoma brain metastases: a multicentre randomised phase 2 study. *Lancet Oncology* 19:672–681.

Louveau A, Harris TH, Kipnis J. (2015) Revisiting the mechanisms of CNS immune privilege. *Trends Immunol* 36(10):569–577.

Lugade AA, Moran JP, Gerber SA, et al. (2005) Local radiation therapy of B16 melanoma tumors increases the generation of tumor antigen-specific effector cells that traffic to the tumor. *J Immunol* 174:7516–7523.

Luke JJ, Lemons JM, Karrison TG, et al. (2018) Safety and clinical activity of pembrolizumab and multisite stereotactic body radiotherapy in patients with advanced solid tumors. *Journal of Clinical Oncology* 36(16):1611–1618.

Margolin K, Ernstoff MS, Hamid O, et al. (2012) Ipilimumab in patients with melanoma and brain metastases: an open-label, phase 2 trial. *Lancet Oncology* 13:459–465.

Martin AM, Cagney DN, Catalano PJ, et al. (2018) Immunotherapy and symptomatic radiation necrosis in patients with brain metastases treated with stereotactic radiation. *JAMA Oncology* 4(8):1123–1124.

Mathew M, Tam M, Ott PA, et al. (2013) Ipilimumab in melanoma with limited brain metastases treated with stereotactic radiosurgery. *Melanoma Res* 23:191–195.

McBride S, Sherman E, Tsai CJ, et al. (2020) Randomized phase II trial of nivolumab with stereotactic body radiotherapy versus nivolumab alone in metastatic head and neck squamous cell carcinoma. *Journal of Clinical Oncology*.

McCoach CE, Berge EM, Lu X, Baron AE, Camidge DR. (2016) A brief report of the status of central nervous system metastasis enrollment criteria for advanced non—small cell lung cancer clinical trials: a review of the clinicaltrials.gov trial registry. *J Thorac Oncol* 11(3):407–413.

McDermott DF, Drake CG, Sznol M, et al. (2015) Survival, durable response, and long-term safety in patients with previously treated advanced renal cell carcinoma receiving nivolumab. *Journal of Clinical Oncology* 33:2013–2020.

Morganstein DL, Lai Z, Spain L, et al. (2017) Thyroid abnormalities following the use of cytotoxic T-lymphocyte antigen-4 and programmed death receptor protein-1 inhibitors in the treatment of melanoma. *Clin Endocrinol (Oxf)* 86:614–620.

Naidoo J, Wang X, Woo KM, et al. (2017) Pneumonitis in patients treated with anti-programmed death-1/programmed death ligand 1 therapy. *J Clin Oncol Off J Am Soc Clin Oncol* 35:709–717.

Ngwa W, Irabor OC, Schoenfeld JD, Hesser J, Demaria S, Formenti SC. (2018) Using immunotherapy to boost the abscopal effect. *Nature Reviews Cancer* 18:313–322.

Nieder C, Spanne O, Mehta MP, Grosu AL, Geinitz H. (2011) Presentation, patterns of care, and survival in patients with brain metastases: what has changed in the last 20 years? *Cancer* 117(11):2505–2512.

Oleinika K, Nibbs RJ, Graham GJ, Fraser AR. (2013) Suppression, subversion and escape: the role of regulatory T cells in cancer progression. *Clin Exp Immunol* 171:36–45.

Osorio JC, Ni A, Chaft JE, et al. (2017) Antibody-mediated thyroid dysfunction during T-cell checkpoint blockade in patients with non-small-cell lung cancer. *Ann Oncol Off J Eur Soc Med Oncol* 28:583–589.

Palma DA, Bauman GS, Rodrigues GB. (2020) Beyond oligometastases. *International Journal of Radiation Oncology, Biology, Physics* 107(2):253–256.

Patel KR, Shoukat S, Oliver DE, et al. (2017) Ipilimumab and stereotactic radiosurgery versus stereotactic radiosurgery alone for newly diagnosed melanoma brain metastases. *Am J Clin Oncol* 40(5):444–450.

Patel RR, Verma V, Barsoumian H, et al. (2020) Use of multi-site radiation therapy as systemic therapy: a new treatment approach personalized by patient immune status. *International Journal of Radiation Oncology, Biology, Physics* S0360–3016(20):34114–34116.

Paz-Ares L, Luft A, Vicente D, et al. (2018) Pembrolizumab plus chemotherapy for squamous non-small-cell lung cancer. *New England Journal of Medicine* 379(21):2040–2051.

Piotrowski AF, Nirschl TR, Velarde E, et al. (2018) Molecular mechanisms of Treg-mediated T cell suppression systemic depletion of lymphocytes following focal radiation to the brain in a murine model. *Oncoimmunology* 7:e1445951.

Platten M, Bunse L, Wick W, Bunse T. (2016) Concepts in glioma immunotherapy. *Cancer Immunol Immunother* 65(10):1269–1275.

Popp I, Grosu AL, Niedermann G, Duda DG. (2016) Immune modulation by hypofractionated stereotactic radiation therapy: therapeutic implications. *Radiotherapy and Oncology* 120:185–194.

Postow MA, Callahan MK, Barker CA, et al. (2012) Immunologic correlates of the abscopal effect in a patient with melanoma. *New England Journal of Medicine* 366(10):925–931.

Postow MA, Sidlow R, Hellman MD. (2018) Immune-related adverse effects associated with immune checkpoint blockade. *New England Journal of Medicine*378:158–168.

Qian JM, Yu JB, Kluger HM, Chiang VLS. (2016) Timing and type of immune checkpoint therapy affect the early radiographic response of melanoma brain metastases to stereotactic radiosurgery. *Cancer* 122:3051–3058.

Queirolo P, Spagnolo F, Ascierto PA, et al. (2014) Efficacy and safety of ipilimumab in patients with advanced melanoma and brain metastases. *J Neuro-Oncol* 118:109–116.

Ransohoff RM, Engelhardt B. (2012) The anatomical and cellular basis of immune surveillance in the central nervous system. *Nature Reviews Immunology* 12(9):623–635.

Ransohoff RM, Kivisakk P, Kidd G. (2003) Three or more routes for leukocyte migration into the central nervous system. *Nature Reviews Immunology* 3(7):569–581.

Reck M, Rodríguez-Abreu D, Robinson AG, et al. (2016) Pembrolizumab versus chemotherapy for PD-L1-positive non-small-cell lung cancer. *New England Journal of Medicine* 375:1823–1833.

Reckamp KL. (2018) Real-world pseudoprogression: an uncommon phenomenon. *J Thorac Oncol*13:880–882.

Reits EA, Hodge JW, Herberts CA, et al. (2006) Radiation modulates the peptide repertoire, enhances MHC class I expression, and induces successful antitumor immunotherapy. *J Exp Med* 203: 1259–1271.

Reynders K, Illidge T, Siva S, Chang JY, De Ruysscher D. (2015) The abscopal effect of local radiotherapy: using immunotherapy to make a rare event clinically relevant. *Cancer Treat Rev* 41:503–510.

Rittmeyer A, Barlesi F, Waterkamp D, et al. (2017) Atezolizumab versus docetaxel in patients with previously treated non-small-cell lung cancer (OAK): A phase 3, open-label, multicentre randomised controlled trial. *Lancet* 389:255–265.

Robert C, Schachter J, Long GV, et al. (2015) Pembrolizumab versus ipilimumab in advanced melanoma. *New England Journal of Medicine* 372:2521–2532.

Sahgal A, Roberge D, Schellenberg D, et al. (2012) The Canadian association of radiation oncology-stereotactic body radiotherapy task force. Canadian association of radiation oncology scope of practice guidelines for lung, liver and spine stereotactic body radiotherapy. *Clinical Oncology (R Coll Radiol)* 24(9):629–639.

Sampson JH, Carter JH, Friedman AH, et al. (1998) Demographics, prognosis, and therapy in 702 patients with brain metastases from malignant melanoma. *Journal of Neurosurgery* 88:11–20.

Schapira E, Hubbeling H, Yeap BY et al. (2018) Improved overall survival and locoregional disease control with concurrent PD-1 pathway inhibitors and stereotactic radiosurgery for lung cancer patients with brain metastases. *International Journal of Radiation Oncology, Biology, Physics* 101(3):624–629.

Schaue D, Ratikan JA, Iwamoto KS, McBride WH (2012). Maximizing tumor immunity with fractionated radiation. *International Journal of Radiation Oncology, Biology, Physics* 83:1306–1310.

Schmidt A, Oberle N, Krammer PH. (2012) Molecular mechanisms of Treg-mediated T cell suppression. *Front Immunology* 3:51.

Schouten LJ, Rutten J, Huveneers HAM, Twijnstra A. (2002) Incidence of brain metastases in a cohort of patients with carcinoma of the breast, colon, kidney, and lung and melanoma. *Cancer* 94(10):2698–2705.

Seung SK, Larson DA, Galvin JM, et al. (2013) American college of radiology (ACR) and American society for radiation oncology (ASTRO) practice guideline for the performance of stereotactic radiosurgery (SRS). *Am J Clin Oncol* 36(3):310–315.

Seymour L, Bogaerts J, Perrone A, et al. (2017) iRECIST: guidelines for response criteria for use in trials testing immunotherapeutics [published correction appears in Lancet Oncol. 2019 May;20(5):e242]. *Lancet Oncology* 18(3):e143–152.

Sharabi AB, Lim M, DeWeese TL, et al. (2015) Radiation and checkpoint blockade immunotherapy: radiosensitisation and potential mechanisms of synergy. *Lancet Oncology* 16:e498–e509.

Sharabi AB, Nirschl CJ, Kochel CM, et al. (2015) Stereotactic radiation therapy augments antigen-specific PD-1-mediated antitumor immune responses via cross-presentation of tumor antigen. *Cancer Immunology Research* 3:345–355.

Sharma P, Wagner K, Wolchok JD, Allison JP. (2011) Novel cancer immunotherapy agents with survival benefit: recent successes and next steps. *Nature Reviews Cancer* 11:805–812.

Shaverdian N, Lisberg AE, Bornazyan K, et al. (2017) Previous radiotherapy and the clinical activity and toxicity of pembrolizumab in the treatment of non-small-cell lung cancer: a secondary analysis of the KEYNOTE-001 phase 1 trial. *Lancet* 18(7):895–903.

Silk AW, Bassetti MF, West BT, Tsien CI, Lao CD. (2013) Ipilimumab and radiation therapy for melanoma brain metastases. *Cancer Medicine* 2:899–906.

Sistigu A, Yamazaki T, Vacchelli E, et al. (2014) Cancer cell-autonomous contribution of type I interferon signaling to the efficacy of chemotherapy. *Nat Med* 20:1301–1309.

Suit HD, Willers H. (2003) Comment on "Tumor response to radiotherapy regulated by endothelial cell apoptosis" (I). *Science* 302:1894.

Tang C, Wang X, Soh H, et al. (2014) Combining radiation and immunotherapy: a new systemic therapy for solid tumors? *Cancer Immunology Research* 2:831–838.

Tawbi HA-H, Forsyth PAJ, Algazi AP, et al. (2017) Efficacy and safety of nivolumab (NIVO) plus ipilimumab (IPI) in patients with melanoma (MEL) metastatic to the brain: Results of the phase II study CheckMate 204. *Journal of Clinical Oncology* 35:9507–9517.

Tawbi HA, Forsyth PA, Algazi A, et al. (2018) Combined nivolumab and ipilimumab in melanoma metastatic to the brain. *New England Journal of Medicine* 379:722–730.

Tazi K, Hathaway A, Chiuzan C, Shirai K. (2015) Survival of melanoma patients with brain metastases treated with ipilimumab and stereotactic radiosurgery. *Cancer Medicine* 4:1–6.

Theelen WSME, Peulen HMU, Lalezari F, et al. (2019) Effect of pembrolizumab after stereotactic body radiotherapy vs pembrolizumab alone on tumor response in patients with advanced non-small cell lung cancer: results of the PEMBRO-RT phase 2 randomized clinical trial. *JAMA Oncology* 5(9):1276–1282.

Tseng CL, Soliman H, Myrehaug S, et al. (2018) Imaging-based outcomes for 24 Gy in 2 daily fractions for patients with de novo spinal metastases treated with spine stereotactic body radiation therapy (SBRT). *International Journal of Radiation Oncology, Biology, Physics* 102(3):499–507.

Twyman-Saint Victor C, Rech AJ, Maity A, et al. (2015) Radiation and dual checkpoint blockade activate non-redundant immune mechanisms in cancer. *Nature* 520(7547):373–377.

Vellayappan BA, Chao ST, Foote M, et al. (2018) The evolution and rise of stereotactic body radiotherapy (SBRT) for spinal metastases. *Expert Rev Anticancer Ther* 18(9):887–900.

Weber J. (2010) Immune checkpoint proteins: a new therapeutic paradigm for cancer 2014; preclinical background: CTLA-4 and PD-1 blockade. *Semin Oncol* 37:430–439.

Weber JS, Yang JC, Atkins MB, Disis M. (2015) Toxicities of immunotherapy for the practitioner. *Journal of Clinical Oncology* 33(18):2092–2099.

Weichselbaum RR, Liang H, Deng L, Fu YX. (2017) Radiotherapy and immunotherapy: a beneficial liaison? *Nat Rev Clin Oncol* 14:365–379.

Wherry EJ. (2011) T cell exhaustion. *Nature Immunology* 12(6):492–499.

Williams NL, Wuthrick EJ, Kim H, et al. (2017) Phase 1 study of ipilimumab combined with whole brain radiation therapy or radiosurgery for melanoma patients with brain metastases. *International Journal of Radiation Oncology, Biology, Physics* 99:22–30.

Wolchok JD, Hoos A, O'Day S, et al. (2009) Guidelines for the evaluation of immune therapy activity in solid tumors: immune-related response criteria. *Clinical Cancer Reserch* 15(23):7412–7420.

Young KH, Baird JR, Savage T, et al. (2016) Optimizing timing of immunotherapy improves control of tumors by hypofractionated radiation therapy. *PLoS One* 11:e0157164.

Yovino S, Kleinberg L, Grossman SA, et al. (2013) The etiology of treatment-related lymphopenia in patients with malignant gliomas: modeling radiation dose to circulating lymphocytes explains clinical observations and suggests methods of modifying the impact of radiation on immune cells. *Cancer Investigation* 31:140–144.

Yusuf MB, Amsbaugh MJ, Burton E, Chesney J, Woo S. (2017) Peri-SRS Administration of immune checkpoint therapy for melanoma metastatic to the brain: investigating efficacy and the effects of relative treatment timing on lesion response. *World Neurosurgery* 100:632–640.

30

24 Gy in Two Daily Spine SBRT Fractions as Developed by the Sunnybrook Health Sciences Centre

Ahmed Abugharib, Arjun Sahgal, K. Liang Zeng, Sten Myrehaug, Hany Soliman, Mark Ruschin, Arman Sarfehnia, Pejman Maralani, Chris Heyn, Jeremie Larouche, Victor Yang, Zain A. Husain, Jay Detsky, and Chia-Lin Tseng

Contents

30.1 INTRODUCTION

Approximately 40% of cancer patients will develop spinal metastases (1), which can present with axial pain, vertebral compression fracture (VCF), and radiculopathy, with 5 to 10% having extra osseous extension in the form of epidural disease. The latter is critical, given that progression to metastatic epidural spinal cord compression (MESCC) can lead to a significant impairment in quality of life resulting from severe pain and devastating neurologic deficits (2). Moreover, patients may require surgical intervention (3), which has its own consequences with respect to morbidity, mortality, and delays in subsequent oncologic care.

The application of stereotactic body radiotherapy (SBRT) to spinal metastases is now an established treatment option in appropriately selected patients as an alternative to conventional palliative radiation. Reported spine SBRT series have used a range of dose-fractionation schemes (4), which presents a significant challenge when comparing outcomes and adverse events. The Canadian Cancer Trials Group (CCTG) Symptom Control (SC)-24 study (NCT02512965) is a multicenter phase 3 randomized control trial (RCT) comparing 24 Gy in 2 SBRT fractions to 20 Gy in 5 conventional external beam radiation therapy (EBRT) fractions (5), with preliminary positive findings in favor of SBRT discussed subsequently in this chapter. The 24 Gy in 2-fraction regimen evaluated in this RCT trial was developed at the University of Toronto as a middle ground between single-fraction stereotactic radiosurgery (SRS), typically at a dose of 24 Gy, which has been associated with high rates of VCF (6), and more prolonged 3 to 5 fractionation regimens. Experience with this fractionation schedule has been widely published, and this approach has been adopted globally by several centers (7–18). This chapter will detail the spine SBRT technique developed by the Sunnybrook Health Sciences Centre (SHSC) of the University of Toronto and summarize the spine SBRT literature specific to the 24 Gy in 2 daily fractions regimen.

30.2 INDICATIONS

The appropriate management of spinal metastases is complex and requires multidisciplinary assessment. In general, suitability for spine SBRT depends on a number of patient and disease factors including the presence or absence of pain, neurological status, spinal stability as determined using the Spinal Instability Neoplastic Score (SINS) (19), the presence and grade of epidural disease (using the Bilsky criteria) (20), tumor histology, available systemic therapeutic options, the overall metastatic burden, performance status, and life expectancy.

Although there are prediction models for spinal metastases (21), there are no refined models that are histology specific as compared to the brain metastases for which the graded prognostic assessment (GPA) can be used to estimate survival (22–24). One common theme emerging from the spine metastases literature is that performance status is critical to the assessment, and patients presenting with an ECOG more than 2 (KPS <70) and neurologic deficits are among those with shorter survival (12, 25). In a recent publication specific to patients treated with spine SBRT at the SHSC, prognostic factors such as non-breast/prostate primary cancer, ECOG performance status ≥2, polymetastatic disease, painful lesions, and paraspinal disease were identified as predictors of surviving less than 3 months after spine SBRT (26).

With respect to surgical interventions, referral to a spine surgeon should be considered in the setting of frank spinal instability based on a SINS more than 12, for patients with potential instability (SINS 7 to 12), and/or for patients with symptomatic high-grade epidural disease (27, 28). Increasingly, cement augmentation in conjunction with local tumor ablation procedures are being applied to patients with a baseline VCF as a prophylactic/therapeutic intervention, which is a pattern of care that is likely going to become more prevalent as we improve our understanding of the complexity related to these patients with respect to optimizing long-term pain and stability (29–31). Well-established framework and criteria have been reported in the literature to guide clinical decision-making (28, 32, 33), and the neurologic, oncologic, mechanical, and systemic (NOMS) decision framework is a good starting point to build a decision-making tree for emerging practices (34). Following surgery, the use of postoperative spine SBRT is emerging as a means to consolidate the surgical intent and optimize local control and remains an area of active investigation (15, 27, 28, 35).

30.3 TECHNICAL SPECIFICATIONS

30.3.1 RADIATION SIMULATION

Spine SBRT is a highly precise radiation therapy technique that delivers high doses of radiation to metastatic spinal lesions while sparing the surrounding organs-at-risk (OARs), the most critical of which are the spinal cord and cauda equina nerve roots. Therefore, near-rigid body immobilization and adequate visualization of the thecal sac and spinal cord are of utmost importance.

In cervical and upper thoracic spine metastases (from C1 caudally to approximately T4/T5), a head and shoulder thermoplastic mask is used for immobilization. For lower thoracic and lumbosacral lesions, patients at the SHSC are immobilized using a vacuum-assisted body cushion known as the BodyFIX (Elekta AB, Stockholm, Sweden) (36).

The importance of near-rigid body immobilization was reported by Li et al. (36). The authors analyzed the cone-beam computed tomography (CBCT) setup accuracy for 102 spinal metastatic lesions in 84 patients who were immobilized with thermoplastic S-frame (SF) masks for the cervical and upper thoracic vertebrae (18 spinal lesions) and a semi-rigid vacuum body fixation (60 lesions) or evacuated cushion (24 lesions) for the thoracic and lumbar vertebrae. It was concluded that a 3-mm margin for OARs and clinical target volumes (CTVs) was necessary to encompass total setup error for SF and the evacuated cushion device, while a 2-mm margin was sufficient for semi-rigid vacuum body fixation during spine SBRT. Importantly, this analysis pre-dated the incorporation of a robotic couch top that allows for fine translation and rotational corrections.

The simulation CT is acquired with 1-mm slice thickness. The simulation MRI sequences consist of axial thin-slice volumetric T1 and T2 sequences (1.5–2 mm slice thickness) to provide accurate visualization and aid delineation of both the spinal target and OARs, specifically the spinal cord and thecal sac at/below the conus. The MRI acquisition includes the affected segment and at least one vertebra above and below this level. Due to the short height of cervical spine vertebrae, two levels above and below are typically needed. In the postoperative setting, both the preoperative MRI and planning MRI are used for accurate delineation of the target volumes. The pre-operative images are often not volumetric and cannot be fused within the treatment planning software, but they are critical to guide planning as the extent of pre-operative epidural disease in the cranio-caudal axis can guide the selection of what levels are needed for the inclusion in the post-operative simulation MRI acquisition. Chan et al. reported in their patterns of failure analyses that the preoperative extent of epidural disease determined patterns of failure as opposed to the post-operative MRI findings alone (37).

These MRI sequences are then fused to the CT simulation dataset. In the event of spinal hardware, especially in the postoperative setting which may cause significant ferromagnetic artifacts obscuring the spinal cord, a computed tomography (CT) myelogram may be required (35). If a myelogram is performed, which is infrequent in our experience at the SHSC, the myelogram dye is injected and the patient taken to simulation for what is effectively a radiation simulation CT myelogram. MRI is still required for target visualization and contouring. In addition, there are instances where the myelogram is insufficient in allowing for accurate delineation of the spinal cord, and, in these situations, we have observed that the MRI can be reliable for delineation. Visualization depends on the location of the hardware, the degree of artifact, and the amount of image distortion. Therefore, even in the presence of surgical hardware, it is not a foregone conclusion that a myelogram is needed. If the spinal cord or thecal sac cannot be properly visualized and contoured, conventional radiation should be used instead of SBRT.

30.3.2 TARGET VOLUME AND ORGAN-AT-RISK CONTOURING

Gross tumor volume (GTV) is defined as the gross disease within the spinal segment, including any epidural and paravertebral extension, which can be readily seen on the MRI and CT images (38). The T1 non-gadolinium sequence is reliable to determine the intra-osseous disease extent and epidural disease, and the T2 can enable better delineation of the paraspinal disease. At the SHSC, no gadolinium is used, and, if needed in a post-operative situation, then a pre-gad T1 can be acquired. In our practice, almost all contouring, including the central nervous system structures, is completed based on the T1 non-gadolinium volumetric treatment planning MRI acquisition.

The delineation of the CTV is per published international consensus guidelines which are based on an anatomic classification system; there are now several guidelines to enable accurate, reproducible, and anatomically driven contouring for intact metastases of the cervical, thoracic, and lumbar spine, and post-operative spine and sacrum (27, 39, 40). These guidelines importantly do not address paraspinal or epidural margins. At the SHSC and in the SC24 protocol, a 5-mm anatomically respectful margin is added to paraspinal soft tissue extension to encompass potential microscopic paraspinal disease extension. Furthermore, in the presence of epidural disease, our standard practice is to apply a 5-mm margin in the cranio-caudal direction within the spinal canal (excluding the cord and/or thecal sac as an anatomically respectful margin) as no boundaries to spread exist within this axis (41). The same principles are applied to the reirradiation setting as well.

For the planning target volume (PTV), a margin of 2 mm is typically added beyond the CTV. This margin is based on our published experience of our technique and will vary according to what technology is being used (42, 43). The PTV is not modified for overlap into adjacent OAR as the coverage is dictated by the dose limits to the OAR and not the PTV. This ensures consistency in reporting of outcomes. For treatment planning purposes, the OAR is subtracted from the PTV to create a planning structure for optimization called a PTV-EVAL.

Precise contouring of the OAR is critical due to the proximity of the high-dose gradients to the spinal cord, cauda equina, nerve roots, gastrointestinal tract, kidneys, and other organs. As there may be significant variability in the delineation of OAR by different physicians (44), it is strongly recommended for practicing radiation oncologists to follow established guidelines and protocols (45, 46). Critically, the spinal cord and/or thecal sac is delineated based on the T1 and/or T2 axial MRI acquisition fused to the planning CT and/or CT myelogram as applicable. A planning organ-at-risk volume (PRV) is applied to the spinal cord contour, and the dose limit is applied to the spinal cord PRV. In the event the cauda equina is contoured for spinal segments below the spinal cord level, or if at the transition phase of the cord to cauda equina which typically begins mid T12-L1, then no PRV is applied. The rationale for this approach originates from early work establishing spinal cord tolerance constraints for spine SBRT based on the thecal sac and not the true spinal cord. This was primarily due to limitations in fusion software, which necessitated the use of the thecal sac as the reliable OAR to which dose limits were defined. (47–49). Therefore, the thecal sac naturally incorporates a PRV to the true cord and, for true cauda equina, the omission of the PRV inherently makes an accommodation for what generally is accepted as a greater tolerance of the nerve roots versus spinal cord tissue itself.

With respect to nerve root contouring, it is only at the level of the brachial plexus and lumbar-sacral nerve plexus that we contour the exiting nerve roots. Our practice has been to limit the dose to the nerve roots to no hot spots (>5% of the prescribed dose) as long as the nerve roots are not within the CTV. There is as yet no consensus on nerve root tolerance. Our experience in those patients surviving 3 years or greater showed the development of plexopathy at a rate of 0.74%, 1.5%, 2.2%, and 5.1% at 1, 2, 3, and 5 years, respectively, with most of the cases occurring after receiving more than one course of radiotherapy (including repeat SBRT) to the same site (26).

30.3.3 TREATMENT PLANNING AND DELIVERY

SBRT treatment entails delivery of high doses of radiation to the target volumes, so strict adherence to the OAR dose constraints is essential to minimize the risk of toxicities.

For the spinal cord, in the setting of *de novo* irradiation, the maximum point dose to the spinal cord PRV and/or thecal sac (below the level of the conus medullaris) has been 17 Gy in 2 daily fractions (47). In cases associated with epidural involvement, the treating clinician may consider increasing the constraint to 18.7 Gy to maximize dose coverage within the epidural space. In cases of reirradiation of spinal metastases, the maximum point doses to the spinal cord PRV and/or thecal sac is 12.2 Gy, and the treating physician may consider giving doses in 14.6 Gy in the presence of epidural disease (48, 50, 51). Tables 30.1 and 30.2 summarize our institutional protocol dose–volume constraints for spine SBRT using 24 Gy in 2 daily fractions.

In order to optimize the coverage of the epidural space, where failure is most common, the dose delivered to the spinal cord PRV and/or thecal sac is typically maximized to the maximum point dose constraint (8, 15, 18, 35). For example, in 2-fraction spine SBRT, the cord PRV or thecal sac point dose is maximized up to 17 Gy with the secondary objective of maximizing dose prescription coverage to the PTV. In the SC-24 protocol, it is stipulated that the maximum point dose must be 17 Gy with an allowed deviation of only -5% with 0% over 17 Gy, which ensures that the plan is optimized for the dose gradient at the CTV–spinal cord interface (5).

Step-and-shoot intensity modulated radiation therapy (IMRT) with more than 6 fields or volumetric modulated arc therapy (VMAT) are typically used in the planning of spine SBRT (52). The aim is to achieve coverage of the PTV by at least 80% of the prescribed dose (e.g., PTV Dmin ≥80% dose or V80% dose = 100% volume) and, to achieve this, 80% of the CTV should get 95 to 100% of the prescribed dose (53). Figures 30.1 and 30.2 illustrate two spine SBRT cases treated with 24 Gy in 2 fractions.

During treatment delivery, spine SBRT requires kilovoltage CBCT for image guidance. The initial CBCT is done prior to the treatment and is overlaid with the treatment planning CT to verify the patient

Table 30.1 **Dose–volume constraints for** *de novo* **irradiation of spinal metastases with SBRT using 24 Gy in 2 daily fractions**

MAXIMUM POINT DOSES WITH NO PRIOR IRRADIATION AND NO EPIDURAL DISEASE (GY)	
CORDprv, thecal sac, cauda equina	17 Gy
Nerve roots	No hot spots, i.e., no dose > 105% of prescribed dose in the contoured nerve roots
Esophagus	22 Gy
Trachea, bronchus, small airways	22 Gy
Heart	26 Gy
Colon, stomach, duodenum	22 Gy
Dose–volume constraints with no prior irradiation and no epidural disease (Gy)	
Each lung	V20 < 3–5%, V10 < 10%, V5 < 35%, mean < 5 Gy
Liver	As low as reasonably achievable, mean < 8 Gy
Each kidney	Mean < 6 Gy, V18 Gy < 67% (combined V18 Gy < 35%)
Maximum point doses with no prior irradiation with epidural disease (Gy)	
CORDprv, thecal sac, cauda equina	18.7 Gy

Table 30.2 **Dose–volume constraints for reirradiation of spinal metastases with SBRT using 24 Gy in 2 daily fractions**

MAXIMUM POINT DOSES (GY) IN THE EVENT OF NO EPIDURAL DISEASE AND PRIOR SPINAL CORD/THECAL SAC EQD2 EXPOSURE RANGING FROM 30 GY$_2$ TO 42.8 GY$_2$	
CORDprv, thecal sac, cauda equina	12.2 Gy
Nerve roots	No hot spots, i.e., no dose> 105% of prescribed dose in the contoured nerve roots
Esophagus	18 Gy
Trachea, bronchus, small airways	18 Gy
Heart	20 Gy
Colon, stomach, duodenum	18 Gy
Dose–volume constraints (Gy) in the event of no epidural disease and prior spinal cord/thecal sac EQD2 exposure ranging from 30 Gy$_2$ to 42.8 Gy$_2$	
Each lung	V20 < 3–5%, V10 < 10%, V5 < 35%, mean < 5 Gy
Liver	Mean < 8 Gy
Each kidney	Mean < 6 Gy, V18 Gy < 67% (combined V18 Gy < 35%)
Maximum point doses (Gy) in the event of no epidural disease and prior spinal cord/thecal sac EQD2exposure of 50Gy$_2$ as upper limit	
CORDprv, thecal sac, cauda equina	11 Gy
Maximum point doses (Gy) in the event of epidural disease and prior spinal cord/thecal sac EQD2 exposure ranging from 30 Gy$_2$ to 42.8 Gy$_2$	
CORDprv, thecal sac, cauda equina	14.6 Gy
Maximum point doses (Gy) in the event epidural disease and prior spinal cord/thecal sac EQD2 exposure of 50 Gy$_2$ as upper limit	
CORDprv, thecal sac, cauda equina	11 Gy

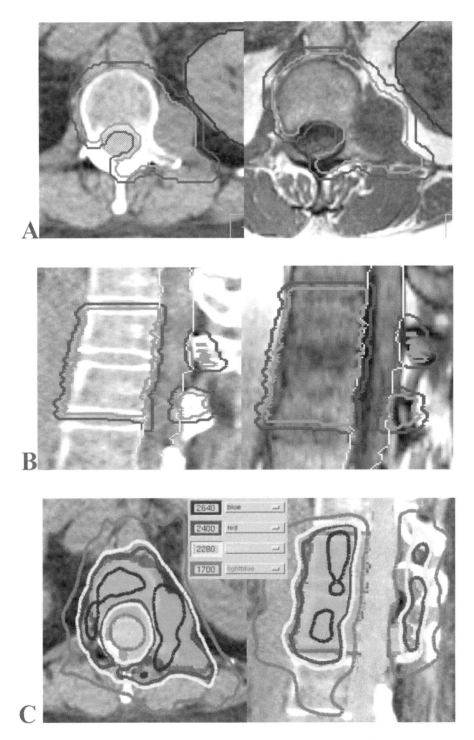

Figure 30.1 Fifty-three-year-old female patient with metastatic uterine sarcoma to T12, L1 vertebrae with left paraspinal soft tissue mass. (A) Axial CT and axial T1 MRI showing OARs (thecal sac, kidneys, bowel, liver), CTV expansion by 5-mm margin within soft tissue while respecting the anatomic boundaries such as the peritoneal reflection (green line) and PTV (red line). (B) Sagittal CT and sagittal T1 MRI showing OARs (thecal sac), CTV (green line), and PTV (red line). (C) Axial and sagittal CT showing isodose lines with CTV (green color wash), and PTV (red color wash).

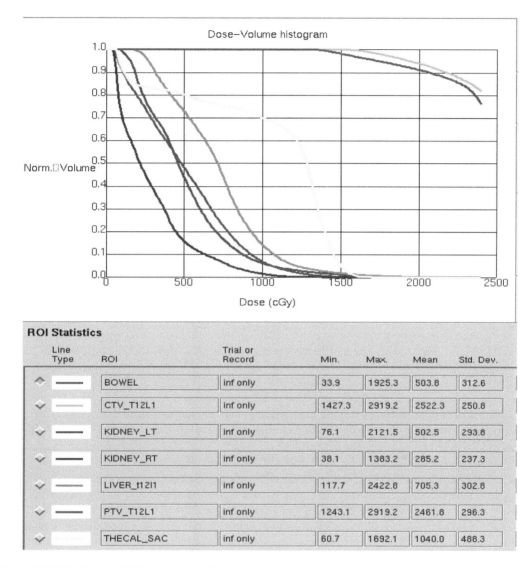

ROI Statistics

Line Type	ROI	Trial or Record	Min.	Max.	Mean	Std. Dev.
—	BOWEL	inf only	33.9	1925.3	503.8	312.6
—	CTV_T12L1	inf only	1427.3	2919.2	2522.3	250.8
—	KIDNEY_LT	inf only	76.1	2121.5	502.5	293.8
—	KIDNEY_RT	inf only	38.1	1383.2	285.2	237.3
—	LIVER_t12l1	inf only	117.7	2422.8	705.3	302.8
—	PTV_T12L1	inf only	1243.1	2919.2	2461.8	296.3
—	THECAL_SAC	inf only	60.7	1692.1	1040.0	488.3

Figure 30.1 (Continued) (D) Dose–volume histogram.

position through matching of the bony anatomy and contours (CTV, PTV, and spinal cord). A robotic couch top with 6 degrees-of-freedom is used to allow for correction of translational and rotational errors (42). If the treatment time is excessively long (according to center protocol), or if there is suspected patient movement, intrafractional verification imaging may be done. The tolerance for repositioning is 1 mm and 1 degree at our institution. At present, the role of MR Linac technology is unknown for spine SBRT, given the potential physical limitation of the MRI bore to accommodate a semi-rigid vacuum body fixation device, the geometrical limitations in the beam angle arrangement, the fixed collimator angle, and the lack of rotational corrections. The dosimetric feasibility of spine SBRT using the Viewray MRIdian MR LINAC system (ViewRay, Inc., Oakwood Village, Ohio), which employs a relatively low magnetic field of 0.35T, has been studied (54, 55); however, similar studies have not yet been undertaken on a higher field system such as the 1.5T Elekta Unity. On the Elekta Unity, feasibility has been demonstrated for the generation of clinically acceptable stereotactic radiation plans for single brain metastases (56). Future work is needed to confirm feasibility of spine SBRT in a 1.5T MR LINAC system, including a comprehensive evaluation of geometrical distortions if an MRI-only approach is to be used.

30.3.4 TECHNICAL EVALUATION

Hyde et al. (42) reported the technical evaluation of our approach. This study was conducted on 42 patients for whom a CBCT was acquired during the initial setup process and prior to the treatment to assess for residual shifts from the initial position. A third and fourth CBCT was performed during and after the treatment, respectively, to quantify intrafractionalmotion. In some cases due to excessively long treatment times, given the technology at the time, two intra-fraction and post-treatment CBCTs were obtained if the overall treatment time was to exceed 60 minutes per fraction. For the 48 spinal metastatic lesions included in the study, a total of 106 fractions and 307 image registrations were analyzed. The analyses determined that following the initial corrected CBCT setup, 90% and 97% of shifts were observed within 1 mm and 1 degree, respectively. Moreover, based on a 1-mm and 1 degree correction threshold, the target was localized within 1.2 mm and 0.9 degree with 95% confidence. Therefore, safe spine SBRT treatment can be achieved through near-rigid body immobilization and strict adherence to the repositioning thresholds, and these data provided the foundation for the 2-mm PTV. Since that evaluation, technologies such as VMAT and flattening-filter-free (FFF) linac delivery have emerged and reduced the treatment time to below 20 minutes in total. As a result, we no longer perform an intra-fraction CBCT or a post-treatment CBCT as our technique is established.

With respect to including multiple spinal segments in the target volume and the impact on PTV, Chang et al. (43) reviewed our experience of intrafractional motion during SBRT treatment of 44 single vertebral metastases (SVM) and 21 multiple vertebral metastases (MVM) planned with a single isocenter. In this study, the authors demonstrated that although the intrafraction translational errors were smaller with SVM than for MVM ($P = .0019$), there was no difference in rotational errors. Moreover, the calculated PRV and PTV margins (PRV was 0.8 mm for SVM and 1.2 mm for MVM; PTV was 1.4 mm for SVM and 1.9 mm for MVM) were smaller than that used clinically (PRV: 1.5 mm; PTV: 2.0 mm). Thus, based on this analysis, a 2-mm margin is safe in SBRT up to three contiguous vertebral segments; however, consideration for a larger PTV margin may be appropriate when treating multiple (four or greater) contiguous vertebral metastases, which constituted only a small portion of the analyzed cohort (four vertebral bodies in two patients and five vertebral bodies in one patient).

Spinal cord motion was evaluated by Tseng et al. (57), who assessed the change in the position of critical neural tissues (spinal cord and cauda equina nerve roots) using dynamic axial and sagittal MRI in 65 patients with metastatic lesions planned for spine SBRT. It was observed that during spine SBRT treatment, there was relatively minor oscillatory spinal cord motion; however, a PRV margin was still required to account for the bulk shifts associated with gross patient motion. More recently, the Washington University group reported their cord motion analysis, where dynamic balanced fast field echo (BFFE) MRI was performed on 21 patients treated with spine SBRT. The authors reported that inherent spinal cord motion during SBRT can lead to an increase in the maximum radiation dose delivered to the cord even if a 1-mm PRV margin was used. The study confirmed that a PRV margin of at least 1.5 to 2 mm is required during spine SBRT to account for such motion (58). These data combined provide support to the standard practice of a 1.5-mm spinal cord PRV margin.

A dosimetric analysis was recently reported by Lee et al. (59). In this planning study, the intent was to compare step and shoot IMRT and the two-arc VMAT plans created by Pinnacle v9.2 (VMAT-P and IMRT-P) and Monacov5.10 (IMRT-M and VMAT-M) treatment planning systems (TPS). The study demonstrated that both Pinnacle and Monaco plans showed comparable dose–volume statistics for the PTV but the monitor units were higher for VMAT-P as compared to VMAT-M. The authors concluded that it was technically feasible to generate clinically acceptable plans on both TPS as all planning objectives were met. The impact of arc discretization in VMAT planning has also been assessed by Marchand et al. (60), who compared target coverage, OAR sparing, and delivery performance in both of VMAT and IMRT plans done for spine SBRT. In this study, 15 spine IMRT plans were included, and then corresponding VMAT plans were generated. Both techniques were able to achieve good coverage of the targets while meeting dose constraints of the OARs; however, VMAT plans had better delivery efficiency compared with IMRT plans, reducing treatment time by 53% and number of delivered MUs by 23% on average.

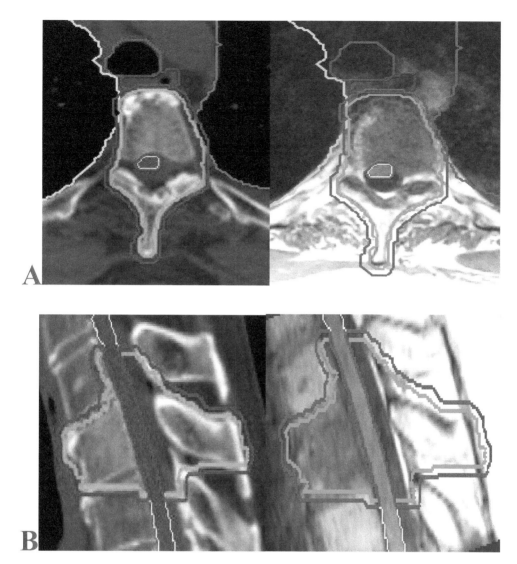

Figure 30.2 Fifty-five-year-old male patient with metastatic squamous cell carcinoma of the base of the tongue, reirradiation toT4 vertebral body, bilateral pedicles, and posterior element, with soft tissue extension into the anterior and left lateral epidural space. Cranially, the epidural disease tracked into the T3 vertebral segment. (A) Axial CT and axial T1 MRI showing OARs (spinal cord, esophagus, trachea, lungs), CTV which is donut shaped due to epidural, bilateral pedicles and posterior element involvement (green line) and PTV (red line). (B) Sagittal CT and sagittal T1 MRI showing OARs (spinal cord), CTV expansion by 5-mm margin in the cranio-caudal direction within the spinal canal to reduce the risk of marginal epidural failure (green line), and PTV (red line).

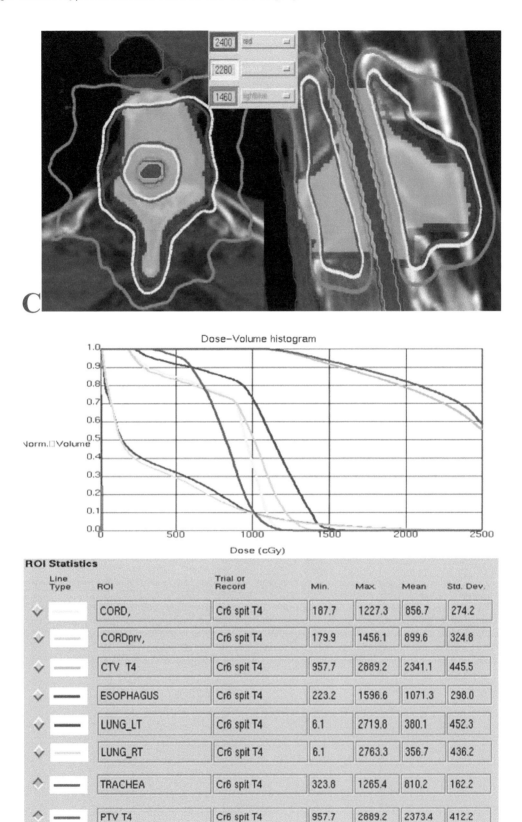

ROI Statistics

Line Type	ROI	Trial or Record	Min.	Max.	Mean	Std. Dev.
◇	CORD,	Cr6 spit T4	187.7	1227.3	856.7	274.2
◇	CORDprv,	Cr6 spit T4	179.9	1456.1	899.6	324.8
◇	CTV T4	Cr6 spit T4	957.7	2889.2	2341.1	445.5
◇	ESOPHAGUS	Cr6 spit T4	223.2	1596.6	1071.3	298.0
◇	LUNG_LT	Cr6 spit T4	6.1	2719.8	380.1	452.3
◇	LUNG_RT	Cr6 spit T4	6.1	2763.3	356.7	436.2
◇	TRACHEA	Cr6 spit T4	323.8	1265.4	810.2	162.2
◇	PTV T4	Cr6 spit T4	957.7	2889.2	2373.4	412.2

Figure 30.2 (Continued). (C) Axial and sagittal CT showing isodose lines with PRV (blue line), CTV (green color wash), and PTV (red color wash). (D) Dose–volume histogram.

30.4 A SUMMARY OF THE OUTCOMES LITERATURE FOR 24 GY IN 2 DAILY FRACTIONS

30.4.1 *DE NOVO* TREATMENT

Recently, the Canadian Cancer Trials Group-Symptom Control study (CCTG-SC-24), NCT02512965) results were published in abstract form (61). This was a phase II/III multicenter randomized trial in which cancer patients with painful spinal metastatic lesions and a pain score of at least ≥ 2 according to the Brief Pain Inventory (BPI) were randomized to receive either 24 Gy in 2 SBRT fractions or 20 Gy in 5 conventional EBRT fractions (CRT). The study included patients with ≤ three spinal metastases with no prior therapeutic radiation nor surgery to the target study spinal segment(s). The study demonstrated that SBRT to painful spinal lesions is associated with improved complete pain relief rates compared to CRT at 3 and 6 months. The complete response rates to pain in the SBRT versus CRT arm at 3 months were 36% versus 14%, $p < 0.001$, respectively, and at 6 months, 33% versus 16%, $p = 0.004$, respectively. There was no significant difference in radiation site progression-free survival (RSS PFS) between the two treatment techniques ($p = 0.42$). The VCF rates were 17% after CRT and 11% after SBRT, and no patient progressed to epidural spinal cord compression following SBRT, while two patients developed tumor-associated cord compression after CRT. Grade 2+ adverse events were observed in 12% and 11% in the CRT and SBRT arms, respectively, with no grade 5 events.

The largest single institutional experience in spine SBRT using 24 Gy in two daily fractions was published in 2018 by Tseng et al. (8). This study included 145 patients with 279 *de novo* spinal metastases treated between 2009 and 2015. The rates of local failure at 1 year and 2 years were 9.7% and 17.6%, respectively. The combination of intra- and extra-osseous progression within the epidural space and/or paraspinal tissues was the most common pattern of failure. In this study, no significant difference was observed in the rates of local failure between the radioresistant histologies and other histologies, and the only factor that predicted for local failure was the presence of epidural disease. In this selected population, the overall survival (OS) rates at 1 year and 2 years were 73.1% and 60.7%, respectively, and the presence of epidural disease, certain primary histologies (lung and renal cell), and diffuse metastases was prognostic for poor OS.

Zeng et al. (7) published the results of treating 52 patients with 93 spinal metastatic lesions in the cervical and sacral vertebrae with SBRT, most of which were treated with 24 Gy in 2 fractions (80.4% of cervical lesions and 73% of the sacral lesions, respectively). The 1- and 2-year cumulative rates of local control were 94.5% and 92.7% and 86.5% and 78.7% for the cervical and sacral cohorts, respectively. Median OS was 16.3 months in the cervical group and 28.5 months in the sacrum group.

Similar results were shown by a multicenter Australian study. In this study, a median dose fractionation of 24 Gy in 2 fractions was used to treat 72 spinal segments in 60 patients. The 1-year and 2-year rates of local control were comparable to those from the previous study at 92% and 86%, respectively. However, this study demonstrated better OS (90% and 76% at 1 and 2 years, respectively), which is likely reflective of the inclusion of only oligometastatic patients with up to three sites of metastases (9).

A prior Australian single-institution study presented the clinical outcomes of 34 patients treated with spine SBRT. In this cohort of patients, the local control was 86%, with the epidural space being the most common site of recurrence. In patients with solitary metastatic disease (14 patients), the overall survival was 87% at a median follow-up of 9.3 months (62).

Thibault et al. (10) analyzed the data of 37 renal cell cancer patients treated with SBRT for 71 spinal lesions, 85% of which received radiation for the first time. Within this cohort, 37% were treated with a dose of 20–24 Gy in 2 fractions. The rates of OS and local control at 1 year were 64% and 83%, respectively.

30.4.2 REIRRADIATION

Detsky et al. (12) reported on treatment outcomes for 43 patients with 83 spinal segments who received salvage spine SBRT following the failure of prior radiation (re-treat SBRT in 5 patients and prior EBRT in 38 patients). In this cohort, the local control rates were 86% and 81% at 1 and 2 years, respectively, while the overall survival rate was 53% at 1 year and 26% at 2 years. The presence of extensive paraspinal disease was associated with a higher incidence of local failure.

A Japanese group published their experience in which 134 lesions in 131 patients were treated with SBRT using 24 Gy in 2 fractions; however, in this cohort, 61.2% of the patients were previously irradiated with conventional EBRT, and 33.6% had surgical decompression of epidural disease prior to SBRT. The median follow-up after SBRT was 9 months, and the rate of local control at 1 year was 72.3%. On multivariable analysis, both colorectal primary and prior radiation were independent predictors of poorer local control. Notably, 79.5% of patients with painful metastatic lesions had pain relief, and the 1-year pain progression-free rate was 61.7% (63). Furthermore, the same group analyzed 66 patients who received reirradiation for painful bone metastases (most lesions were in spine, and 77% received 24 Gy in 2 fractions), showing a pain response in 57 patients (86%) and a 1-year pain failure-free rate of 55%(14). More recently, they published an update of their data in which they showed the outcomes of SBRT reirradiation of 133 spinal metastatic lesions in 123 patients. The median follow-up in this study was 12 months, and the 1-year overall survival and local control rates were 60.6% and 74.2%, respectively. They did not find an association between reirradiation and toxicity (11).

30.4.3 POSTOPERATIVE SBRT

One of the early postoperative spine SBRT studies was reported by Al-Omair et al. (18), who analyzed the treatment outcomes in 80 patients with spinal metastatic lesions treated with postoperative SBRT at University of Toronto. In this cohort, the median dose fractionation was 20 Gy in 2 fractions, and the median follow-up was 8.3 months. The OS and local control at 1 year were 64% and 84%, respectively. A radiation dose of 18–26 Gy in 1 or 2 fractions and no or low-grade (0 or 1) postoperative epidural disease were identified as predictors of improved local control.

Thibault et al. (17) published the treatment outcomes of 40 patients with 56 metastatic spinal lesions who received a salvage second SBRT course to the same spinal level. In this study, 8.9% of this cohort received 24–26 Gy in 2 fractions, and the rates of OS and local control at 1 year were 48% and 81%, respectively. Absence of baseline paraspinal disease was a predictor for better local control, while longer time interval between the first and second SBRT courses was predicted for better OS.

A Japanese group reviewed the treatment outcomes of 28 patients with metastatic epidural spinal cord compression who received postoperative SBRT reirradiation using 24 Gy in 2 fractions. The mean interval since the most recent course of radiation was 16 months, and the median follow-up after SBRT was 13 months. In this cohort of patients, the rate of local control at 1 year was 70% (16).

In 2019, Alghamdi et al. (15) retrospectively reviewed the data for 47 patients who received postoperative SBRT to 83 spinal metastatic lesions. The OS and local control rates at 1 year were 55% and 83%, respectively. On multivariable analysis, lower postoperative grade of epidural disease and longer time interval between the radiotherapy courses were associated with better local control.

Table 30.3 summarizes the studies reporting 24 Gy in two spine SBRT in the *de novo*, reirradiation, and postoperative settings.

30.5 TOXICITY

The toxicity profile of spine SBRT is generally favorable, and acute toxicities are mild. The most feared late complication of radiation-induced myelopathy can be serious and debilitating, but fortunately, the occurrence of this complication is extremely rare. A more common late effect of spine SBRT is VCF which is discussed in this section.

30.5.1 PAIN FLARE

Although one of the main indications of treating spinal metastatic lesions with SBRT is to provide a durable pain response, there may be a transient increase in pain during or shortly after the completion of treatment, known as pain flare, which is also observed after CRT. This increase in pain occurs following SBRT treatment and can affect up to 70% of patients, especially those with metastases affecting the cervical and lumbar vertebrae. Chiang et al. (64) assessed the incidence of pain flare in 41 patients who received spine SBRT, of whom 58.5% were treated with 24 to 35 Gy/2 to 5 fractions. Any pain flare event was observed in 68.3% of the patients, most commonly on day 1 after SBRT. These results were compared with data from a prospective

Table 30.3 Summary of findings from studies reporting spine SBRT using 24 Gy in 2 fractions

STUDY	NO. OF PATIENTS	NO. OF SEGMENTS RECEIVING 24 GY IN 2 FRACTIONS / TOTAL NO. OF SEGMENTS	TYPE OF SBRT	LOCAL CONTROL AT 1 YEAR	OVERALL SURVIVAL AT 1 YEAR	VCF AT 1 YEAR	PAIN CONTROL
De novo							
CCTG SC-24 (61)	114	114	De novo	In the SBRT arm 92% at 3 months and 75% at 6 months	NR in abstract	11% in SBRT arm	Complete response to pain in the SBRT arm Was 36% at 3 months and 33% at 6 months
Zeng et al. (7)	52	72/93	De novo 66.7% Reirradiation 20.4% Postoperative 12.9%	Cervical 94.5% Sacrum 86.5%	Cervical 53% Sacrum 72%	Cervical 0% Sacrum 5.4%	--
Tseng et al. (8)	145	279/279	De novo	90.3%	73.1%	8.5%	--
Chang et al. (9)	60	30/72	De novo	92%	90%	6.7%	--
Finnigan et al. (62)	34	--	De novo 65% Reirradiation 39% Postoperative 22%	86%	--	22%	--
Thibault et al. (10)	37	26/71	De novo 85% Reirradiation 15% Postoperative 14%	83%	64.1%	18%	--
Reirradiation							
Ito et al. (11)	123	133/133	Reirradiation	74.2%	60.6%	13.8%	Any pain response (complete + partial) Was 75% at 3 months and 64% at 6 months
Detsky et al. (12)	43	26/83	Reirradiation	86%	53%	4%	--

(Continued)

Table 30.3 (*Continued*) Summary of findings from studies reporting spine SBRT using 24 Gy in 2 fractions

STUDY	NO. OF PATIENTS	NO. OF SEGMENTS RECEIVING 24 GY IN 2 FRACTIONS / TOTAL NO. OF SEGMENTS	TYPE OF SBRT	LOCAL CONTROL AT 1 YEAR	OVERALL SURVIVAL AT 1 YEAR	VCF AT 1 YEAR	PAIN CONTROL
Ito et al. (63)	131	134/134	Reirradiation 61.2% Postoperative 33.6%	72.3%	65 %	11.9%	79.5% achieved any pain response, 50 % achieved complete pain response
Ogawa et al. (14)	66	51/66	Reirradiation	--	--	7.6%	86% achieved any pain response, 52 % achieved complete pain response
Postoperative							
Alghamdi et al. (15)	47	51/83	Postoperative	83%	55%	3.6%	--
Ito et al. (16)	28	28/28	Postoperative reirradiation	70%	63%	10.7%	--
Thibault et al. (17)	40	5/56	Postoperative reirradiation 66.1% Reirradiation only 33.9%	80.6%	48%	0% in reirradiation only group	--
Al-Omair et al. (18)	80	--	Postoperative	84%	64%	11.3%	--

observational study which included 47 patients who received spine SBRT but were treated with prophylactic oral dexamethasone 1 hour before and for 4 days following SBRT. The incidence of pain flare was found to be significantly reduced with prophylactic oral dexamethasone (68 versus 19%, $p < 0.0001$). Moreover, patients who received dexamethasone had lower pain scores and better general activity interference outcomes when compared with the steroid naïve cohort (65). Although other groups have reported a lower pain flare event rate, the patients in the aforementioned studies were prospectively studied with the intent to determine pain flare.

30.5.2 VERTEBRAL COMPRESSION FRACTURE (VCF)

The use of single-fraction SBRT for the treatment of spinal metastases is associated with a high risk of VCF. This was first reported in a study from the Memorial Sloan-Kettering Cancer Center, where 71 spinal segments in 62 patients were treated with single-fraction SBRT (18–24 Gy), and VCF was observed in 39% of the treated segments (6). On the other hand, the 24 Gy in 2 fractionation regimen has reported a much lower incidence of VCF, where the cumulative risk of VCF was shown to be 8.5% and 13.8% at 1 year and 2 years, respectively, by Tseng et al. (8). Similarly, the Australian group reported a 6.7% risk of VCF in their series (9). In the CCTG-SC24 randomized trial, VCF rates were 17% after CRT and 11% after SBRT, and the difference between the two arms was not significantly much. This reflects that spine SBRT dose is only one component of fracture risk as discussed in Chapter 19.

The landmark multi-institutional study of 252 patients with 410 spinal metastatic lesions treated with spine SBRT showed that the median time to VCF was 2.46 months and 65% occurred within the first 4 months following SBRT. Furthermore, it identified increasing dose per fraction, baseline VCF, lytic tumor, and spinal deformity as significant predictors of VCF (66). In 2018, the largest and most comprehensive systematic literature review of VCF following spine SBRT treatment was reported by Faruqi et al. (67). The review included 11 studies reporting the risk of VCF following spine SBRT. The risk of VCF was 13.9% following SBRT in 2911 spinal segments. Surgical intervention was required in 37% of VCF, most commonly a cement augmentation procedure. On multivariable analysis, lytic lesions, the presence of VCF prior to SBRT, higher dose per fraction, spinal deformity, older age, and an involvement of more than 40% of vertebral body disease were predictors of incidence or progression of VCF following spine SBRT.

30.5.3 MYELOPATHY

Myelopathy is a late toxicity secondary to overdosing of the spinal cord. It is rarely seen with conventional fractionation. In the modern era of spine SBRT, the risk of myelopathy is extremely rare when proper techniques are adopted (68).

The spinal cord tolerance while using high dose per fraction is different from that of conventional treatment, and this should be taken into consideration while treating spinal metastases with SBRT. Also, strict precaution should be followed to minimize intrafraction motion as any change in position can be associated with unintentional overdosing of the spinal cord and consequently can increase the probability of myelopathy (69).

Sahgal et al. (47) analyzed the dose–volume histogram (DVH) results of nine patients who developed myelopathy following spine SBRT and compared them to a cohort of 66 patients treated with spine SBRT without radiation myelopathy. This study demonstrated the first logistic regression model yielding estimates for the probability of human radiation myelopathy specific to SBRT. Point maximum dose constraints were established specific to 2 fractions SBRT, and a thecal sac point maximum dose of 17 Gy was associated with less than 5% probability of developing post-treatment myelopathy.

In the context of SBRT reirradiation, Sahgal et al. (48) reviewed the DVHs for the thecal sac of 5 patients who developed radiation myelopathy following spine SBRT reirradiation and for 14 patients who did not develop such toxicity. This study concluded that it is safe to deliver reirradiation SBRT with thecal sac point maximum normalized biologically equivalent dose (nBED) of 20 to 25 $Gy_{2/2}$ on the condition that the total point maximum nBED does not exceed approximately 70 $Gy_{2/2}$ and that the SBRT thecal sac point maximum nBED comprises no more than approximately 50% of the total nBED.

Recently, the Hypofractionation Treatment Effects in the Clinic (HyTEC) reported its evaluation of spinal cord dose tolerance during SBRT. This study extracted the existing published dose/volume data and used dose–response models, including Sahgal and Katsoulakis–Gibbs models, to estimate the upper and lower limits of spinal cord tolerance. The authors concluded that *de novo* spine SBRT using a 2-fraction schedule is associated with 1% to 5% risk of radiation myelopathy for doses between 17 Gy and 19.3 Gy. Consequently, this supports the safety of our practice at SHSC in cases associated with epidural disease, where a maximum dose of 18.7 Gy in 2 fractions is allowed. For reirradiation, the available data is insufficient; therefore, the recommendation follows that established by the original report by Sahgal et al. (48). Similarly, HyTEC recommended to allow a minimum of 5 months between the two courses and to limit the cord PRV/thecal sac D_{max} to 12.2 Gy. The same dose constraints apply for *de novo* and reirradiation in the postoperative setting (50).

30.6 CONTROVERSIES AND FUTURE DIRECTIONS

30.6.1 CONVENTIONAL FRACTIONATION VERSUS SBRT

Traditionally, EBRT with conventional fractionation has been the standard of care for treatment of cancer patients with spinal metastases. However, local control rates have been suboptimal with conventional radiotherapy in the setting of bulky lesions or radio-resistant histologies (70, 71). The emergence of SBRT enables the delivery of high biologically effective dose to the spinal metastases without jeopardizing the surrounding OARs, which may translate into improved pain response and local control. Specific to 2-fraction spine SBRT, the results from the CCTG SC-24 study (NCT02512965), as described previously in the chapter, has a provided the first level 1 evidence of improved complete pain response at both 3 and 6 months after treatment as compared to CRT (61). Further trials establishing the superiority of SBRT are needed in the reirradiation and post-operative scenarios, and eventually will be required in histology and molecular genetics specific studies.

30.6.2 DIFFERENT SBRT FRACTIONATION SCHEDULES

Currently, there is no consensus as to the SBRT dose-fractionation scheme that is optimal for spine metastases. Several dose-fractionation regimens are in use which include 16 to 24 Gy in 1 fraction, 24 Gy in 2 fractions, 24 to 27 Gy in 3 fractions, and 30 to 35 Gy in 5 fractions. No high-level evidence exists to support the use of one fractionation over another. Some published reports have suggested improved local control with single-fraction SBRT compared with a multi-fraction approach (72), but this may come at the expense of an increased risk of VCF (8). Ultimately, the question of optimal fractionation schedule must be addressed in a prospective randomized fashion. In this regard, a randomized clinical trial (NCT01223248) is currently awaiting publication, comparing 24 Gy in 1 fraction to 27 Gy in 3 fractions delivered to sites of cancer metastases (not specific to spine). At SHSC, we have escalated our dose to 28 Gy in 2 fractions and will report data in the years to come with the intent to determine if local control can be further improved with this more aggressive regimen.

CHECKLIST: KEY POINTS FOR CLINICAL PRACTICE

✓	ACTIVITY	SOME CONSIDERATIONS
	Patient selection	• Spinal metastatic lesions • 24 Gy in 2 fractions • *De novo*, reirradiation, postoperative
	Simulation	• Immobilization using head and shoulder thermoplastic mask for cervical and upper thoracic spine metastases, while using vacuum-assisted body cushion for lower thoracic and lumbosacral lesions • CT simulation with 1-mm slice thickness • Axial thin-slice volumetric T1 and T2 MRI of the affected segment at least one vertebra above and below • Computed tomography (CT) myelogram in certain cases • Verifying the accuracy of the CT–MRI fusion

(Continued)

✓	ACTIVITY	SOME CONSIDERATIONS
	Target volume and organ-at-risk contouring	• Gross tumor volume (GTV) • Clinical target volume (CTV) in accordance with published guidelines with the stipulation of a 5-mm margin beyond paraspinal and/or epidural soft tissue extension while respecting the anatomic boundaries • Planning target volume (PTV) of 2 mm • Spinal cord and/or thecal sac contoured per simulation MRI and/or CT myelogram if applicable • Planning organ-at-risk volume (PRV) of 1.5 mm for spinal cord
	Treatment planning	• Strict adherence to the OAR dose constraints • For *de novo* and postoperative cases, maximum point dose to the spinal cord PRV and/or thecal sac is 17 Gy in the absence of epidural disease and 18.7 Gy in the presence of epidural disease • For reirradiation, the maximum point doses to the spinal cord PRV and/or thecal sac is 12.2 Gy and 14.6 Gy in the absence and presence of epidural disease, respectively • Step-and-shoot IMRT with > 6 fields or VMAT • The aim is to cover at least 80% of CTV with 100% of prescription dose (24 Gy), although it is acknowledged depending on the spinal segment(s) and proximity to OARs that a CTV Dmin ≥ 80% may not always be achievable
	Treatment delivery	• CBCT for image guidance • A 6 degrees-of-freedom robotic couch top • Consider intrafractional CBCT for long treatments (> 20 minutes) • The tolerance for repositioning set to at a minimum of 1 mm and 1 degree.

REFERENCES

1. Chang EL, Shiu AS. Spinal metastases: fractionated radiation therapy perspective. In: Chin LS, Regine WF, editors. *Principles and Practice of Stereotactic Radiosurgery*. New York, NY: Springer; 2008. pp. 455–458.
2. Klimo P, Jr., Thompson CJ, Kestle JRW, Schmidt MH. A meta-analysis of surgery versus conventional radiotherapy for the treatment of metastatic spinal epidural disease. *Neuro-Oncology* 2005;7(1):64–76.
3. Fehlings MG, Nater A, Tetreault L, Kopjar B, Arnold P, Dekutoski M, et al. Survival and clinical outcomes in surgically treated patients with metastatic epidural spinal cord compression: results of the prospective multicenter AOSpine study. *Journal of Clinical Oncology: Official Journal of the American Society of Clinical Oncology* 2016;34(3):268–276.
4. Husain ZA, Sahgal A, De Salles A, Funaro M, Glover J, Hayashi M, et al. Stereotactic body radiotherapy for de novo spinal metastases: systematic review. *Journal of Neurosurgery Spine* 2017;27(3):295–302.
5. Sahgal A, Myrehaug S, Dennis K, Liu M, Chow E, Wong R, et al. A randomized phase II/III study comparing stereotactic body radiotherapy (SBRT) versus conventional palliative radiotherapy (CRT) for patients with spinal metastases (NCT02512965). *Journal of Clinical Oncology* 2017;35Suppl 15:TPS10129-TPS.
6. Rose PS, Laufer I, Boland PJ, Hanover A, Bilsky MH, Yamada J, et al. Risk of fracture after single fraction image-guided intensity-modulated radiation therapy to spinal metastases. *Journal of Clinical Oncology: Official Journal of the American Society of Clinical Oncology* 2009;27(30):5075–5079.
7. Zeng KL, Myrehaug S, Soliman H, Tseng CL, Atenafu EG, Campbell M, et al. Stereotactic body radiotherapy for spinal metastases at the extreme ends of the spine: imaging-based outcomes for cervical and sacral metastases. *Neurosurgery* 2019;85(5):605–612.
8. Tseng CL, Soliman H, Myrehaug S, Lee YK, Ruschin M, Atenafu EG, et al. Imaging-based outcomes for 24 Gy in 2 daily fractions for patients with de novo spinal metastases treated with spine stereotactic body radiation therapy (SBRT). *International Journal of Radiation Oncology, Biology, Physics* 2018;102(3):499–507.
9. Chang JH, Gandhidasan S, Finnigan R, Whalley D, Nair R, Herschtal A, et al. Stereotactic ablative body radiotherapy for the treatment of spinal oligometastases. *Clinical Oncology (Royal College of Radiologists (Great Britain))* 2017;29(7):e119–e125.
10. Thibault I, Al-Omair A, Masucci GL, Masson-Côté L, Lochray F, Korol R, et al. Spine stereotactic body radiotherapy for renal cell cancer spinal metastases: analysis of outcomes and risk of vertebral compression fracture. *Journal of Neurosurgery Spine* 2014;21(5):711–718.

11. Ito K, Ogawa H, Nakajima Y. Efficacy and toxicity of reirradiation spine stereotactic body radiotherapy with respect to irradiation dose history. *Japanese Journal of Clinical Oncology* 2021 Feb;51(2):264–270.

12. Detsky JS, Nguyen TK, Lee Y, Atenafu E, Maralani P, Husain Z, et al. Mature imaging-based outcomes supporting local control for complex reirradiation salvage spine stereotactic body radiotherapy. *Neurosurgery* 2020 Sep 15;87(4):816–822.

13. Ito K, Ogawa H, Shimizuguchi T, Nihei K, Furuya T, Tanaka H, et al. Stereotactic body radiotherapy for spinal metastases: clinical experience in 134 cases from a single Japanese institution. *Technology in Cancer Research & Treatment* 2018;17:1533033818806472.

14. Ogawa H, Ito K, Shimizuguchi T, Furuya T, Nihei K, Karasawa K. Reirradiation for painful bone metastases using stereotactic body radiotherapy. *Acta Oncologica (Stockholm, Sweden)* 2018;57(12):1700–1704.

15. Alghamdi M, Sahgal A, Soliman H, Myrehaug S, Yang VXD, Das S, et al. Postoperative stereotactic body radiotherapy for spinal metastases and the impact of epidural disease grade. *Neurosurgery* 2019;85(6):E1111–E1118.

16. Ito K, Nihei K, Shimizuguchi T, Ogawa H, Furuya T, Sugita S, et al. Postoperative reirradiation using stereotactic body radiotherapy for metastatic epidural spinal cord compression. *Journal of Neurosurgery Spine* 2018;29(3):332–338.

17. Thibault I, Campbell M, Tseng CL, Atenafu EG, Letourneau D, Yu E, et al. Salvage stereotactic body radiotherapy (SBRT) following in-field failure of initial SBRT for spinal metastases. *International Journal of Radiation Oncology, Biology, Physics* 2015;93(2):353–360.

18. Al-Omair A, Masucci L, Masson-Cote L, Campbell M, Atenafu EG, Parent A, et al. Surgical resection of epidural disease improves local control following postoperative spine stereotactic body radiotherapy. *Neuro-Oncology* 2013;15(10):1413–1419.

19. Fisher CG, DiPaola CP, Ryken TC, Bilsky MH, Shaffrey CI, Berven SH, et al. A novel classification system for spinal instability in neoplastic disease: an evidence-based approach and expert consensus from the spine oncology study group. *Spine* 2010;35(22):E1221–E1229.

20. Bilsky MH, Laufer I, Fourney DR, Groff M, Schmidt MH, Varga PP, et al. Reliability analysis of the epidural spinal cord compression scale. *Journal of Neurosurgery Spine* 2010;13(3):324–328.

21. Chao ST, Koyfman SA, Woody N, Angelov L, Soeder SL, Reddy CA, et al. Recursive partitioning analysis index is predictive for overall survival in patients undergoing spine stereotactic body radiation therapy for spinal metastases. *International Journal of Radiation Oncology, Biology, Physics* 2012;82(5):1738–1743.

22. Sperduto PW, Kased N, Roberge D, Xu Z, Shanley R, Luo X, et al. Summary report on the graded prognostic assessment: an accurate and facile diagnosis-specific tool to estimate survival for patients with brain metastases. *Journal of Clinical Oncology: Official Journal of the American Society of Clinical Oncology* 2012;30(4):419–425.

23. Jensen G, Tang C, Hess KR, Bishop AJ, Pan HY, Li J, et al. Internal validation of the prognostic index for spine metastasis (PRISM) for stratifying survival in patients treated with spinal stereotactic radiosurgery. *Journal of Radiosurgery and SBRT* 2017;5(1):25–34.

24. Tang C, Hess K, Bishop AJ, Pan HY, Christensen EN, Yang JN, et al. Creation of a prognostic index for spine metastasis to stratify survival in patients treated with spinal stereotactic radiosurgery: secondary analysis of mature prospective trials. *International Journal of Radiation Oncology, Biology, Physics* 2015;93(1):118–125.

25. Verlaan JJ, Choi D, Versteeg A, Albert T, Arts M, Balabaud L, et al. Characteristics of Patients who survived < 3 months or > 2 years after surgery for spinal metastases: can we avoid inappropriate patient selection? *Journal of Clinical Oncology: Official Journal of the American Society of Clinical Oncology* 2016;34(25):3054–3061.

26. Zeng KL, Sahgal A, Tseng CL, Myrehaug S, Soliman H. Prognostic factors associated with surviving less than 3 months vs. greater than 3 years specific to spine SBRT and late adverse events. *Neurosurgery.* 2021 Jan 20:nyaa583

27. Redmond KJ, Robertson S, Lo SS, Soltys SG, Ryu S, McNutt T, et al. Consensus contouring guidelines for postoperative stereotactic body radiation therapy for metastatic solid tumor malignancies to the spine. *International Journal of Radiation Oncology, Biology, Physics* 2017;97(1):64–74.

28. Redmond KJ, Lo SS, Fisher C, Sahgal A. Postoperative stereotactic body radiation therapy (SBRT) for spine metastases: a critical review to guide practice. *International Journal of Radiation Oncology, Biology, Physics* 2016;95(5):1414–1428.

29. Wardak Z, Bland R, Ahn C, Xie XJ, Chason D, Morrill K, et al. A phase 2 clinical trial of SABR followed by immediate vertebroplasty for spine metastases. *International Journal of Radiation Oncology, Biology, Physics* 2019;104(1):83–89.

30. Fisher C, Ali Z, Detsky J, Sahgal A, David E, Kunz M, et al. Photodynamic therapy for the treatment of vertebral metastases: a phase I clinical trial. *Clinical Cancer Research* 2019;25(19):5766–5776.

31. Steverink JG, Willems SM, Philippens MEP, Kasperts N, Eppinga WSC, Versteeg AL, et al. Early tissue effects of stereotactic body radiation therapy for spinal metastases. *International Journal of Radiation Oncology Biology Physics* 2018;100(5):1254–1258.

32. Tseng C-L, Eppinga W, Charest-Morin R, Soliman H, Myrehaug S, Maralani PJ, et al. Spine stereotactic body radiotherapy: indications, outcomes, and points of caution. *Global Spine Journal.* 2017;7(2):179–197.

33. Jabbari S, Gerszten PC, Ruschin M, Larson DA, Lo SS, Sahgal A. Stereotactic body radiotherapy for spinal metastases: practice guidelines, outcomes, and risks. *Cancer Journal (Sudbury, Mass)* 2016;22(4):280–289.

34. Laufer I, Rubin DG, Lis E, Cox BW, Stubblefield MD, Yamada Y, et al. The NOMS framework: approach to the treatment of spinal metastatic tumors. *The Oncologist* 2013;18(6):744–751.

35. Sahgal A, Bilsky M, Chang EL, Ma L, Yamada Y, Rhines LD, et al. Stereotactic body radiotherapy for spinal metastases: current status, with a focus on its application in the postoperative patient. *Journal of Neurosurgery Spine* 2011;14(2):151–166.

36. Li W, Sahgal A, Foote M, Millar BA, Jaffray DA, Letourneau D. Impact of immobilization on intrafraction motion for spine stereotactic body radiotherapy using cone beam computed tomography. *International Journal of Radiation Oncology, Biology, Physics* 2012;84(2):520–526.

37. Chan MW, Thibault I, Atenafu EG, Yu E, John Cho BC, Letourneau D, et al. Patterns of epidural progression following postoperative spine stereotactic body radiotherapy: implications for clinical target volume delineation. *Journal of Neurosurgery Spine* 2016;24(4):652–659.

38. Burnet NG, Thomas SJ, Burton KE, Jefferies SJ. Defining the tumor and target volumes for radiotherapy. *Cancer Imaging* 2004;4(2):153–161.

39. Dunne EM, Sahgal A, Lo SS, Bergman A, Kosztyla R, Dea N, et al. International consensus recommendations for target volume delineation specific to sacral metastases and spinal stereotactic body radiation therapy (SBRT). *Radiotherapy and Oncology: Journal of the European Society for Therapeutic Radiology and Oncology* 2020;145:21–29.

40. Cox BW, Spratt DE, Lovelock M, Bilsky MH, Lis E, Ryu S, et al. International spine radiosurgery consortium consensus guidelines for target volume definition in spinal stereotactic radiosurgery. *International Journal of Radiation Oncology, Biology, Physics* 2012;83(5):e597–e605.

41. Tseng C-L, Eppinga WSC, Charest-Morin R, Soliman H, Myrehaug S, Maralani PJ, et al. Spine stereotactic body radiotherapy: indications, outcomes, and points of caution. *Global Spine Journal.* 2017;7:179–197.

42. Hyde D, Lochray F, Korol R, Davidson M, Wong CS, Ma L, et al. Spine stereotactic body radiotherapy utilizing cone-beam CT image-guidance with a robotic couch: intrafraction motion analysis accounting for all six degrees of freedom. *International Journal of Radiation Oncology, Biology, Physics* 2012;82(3):e555–e562.

43. Chang JH, Sangha A, Hyde D, Soliman H, Myrehaug S, Ruschin M, et al. Positional accuracy of treating multiple versus single vertebral metastases with stereotactic body radiotherapy. *Technology in Cancer Research & Treatment* 2017;16(2):231–237.

44. Kong FM, Ritter T, Quint DJ, Senan S, Gaspar LE, Komaki RU, et al. Consideration of dose limits for organs at risk of thoracic radiotherapy: atlas for lung, proximal bronchial tree, esophagus, spinal cord, ribs, and brachial plexus. *International Journal of Radiation Oncology, Biology, Physics* 2011;81(5):1442–1457.

45. Mir R, Kelly SM, Xiao Y, Moore A, Clark CH, Clementel E, et al. Organ at risk delineation for radiation therapy clinical trials: global harmonization group consensus guidelines. *Radiotherapy and Oncology: Journal of the European Society for Therapeutic Radiology and Oncology* 2020;150:30–39.

46. Thibault I, Chang EL, Sheehan J, Ahluwalia MS, Guckenberger M, Sohn MJ, et al. Response assessment after stereotactic body radiotherapy for spinal metastasis: a report from the SPIne response assessment in Neuro-Oncology (SPINO) group. *The Lancet Oncology* 2015;16(16):e595–e603.

47. Sahgal A, Weinberg V, Ma L, Chang E, Chao S, Muacevic A, et al. Probabilities of radiation myelopathy specific to stereotactic body radiation therapy to guide safe practice. *International Journal of Radiation Oncology, Biology, Physics* 2013;85(2):341–347.

48. Sahgal A, Ma L, Weinberg V, Gibbs IC, Chao S, Chang UK, et al. Reirradiation human spinal cord tolerance for stereotactic body radiotherapy. *International Journal of Radiation Oncology, Biology, Physics* 2012;82(1):107–116.

49. Sahgal A, Ma L, Gibbs I, Gerszten PC, Ryu S, Soltys S, et al. Spinal cord tolerance for stereotactic body radiotherapy. *International Journal of Radiation Oncology, Biology, Physics* 2010;77(2):548–553.

50. Sahgal A, Chang JH, Ma L, Marks LB, Milano MT, Medin P, et al. Spinal cord dose tolerance to stereotactic body radiation therapy. *International Journal of Radiation Oncology, Biology, Physics* 2019 Oct 10:S0360-3016(19)33862-3.

51. Hashmi A, Guckenberger M, Kersh R, Gerszten PC, Mantel F, Grills IS, et al. Reirradiation stereotactic body radiotherapy for spinal metastases: a multi-institutional outcome analysis. *Journal of Neurosurgery Spine* 2016;25(5):646–653.

52. Sangha A, Korol R, Sahgal A. Stereotactic body radiotherapy for the treatment of spinal metastases: an overview of the university of Toronto, Sunnybrook Health Sciences Odette Cancer Centre, Technique. *Journal of Medical Imaging and Radiation Sciences* 2013;44(3):126–133.

53. Sahgal A. NCIC CTG Protocol Number: SC.24, A phase II randomized feasibility study comparing stereotactic body radiotherapy (SBRT) versus conventional palliative radiotherapy (CRT) for patients with spinal metastases. 2015.

54. Yadav P, Musunuru HB, Witt JS, Bassetti M, Bayouth J, Baschnagel AM. Dosimetric study for spine stereo-tactic body radiation therapy: magnetic resonance guided linear accelerator versus volumetric modulated arc therapy. *Radiology and Oncology* 2019;53(3):362–368.

55. Redler G, Stevens T, Cammin J, Malin M, Green O, Mutic S, et al. Dosimetric Feasibility of utilizing the ViewRay magnetic resonance guided linac system for image-guided spine stereotactic body radiation therapy. *Cureus* 2019;11(12):e6364.

56. Tseng CL, Eppinga W, Seravalli E, Hackett S, Brand E, Ruschin M, et al. Dosimetric feasibility of the hybrid Magnetic Resonance Imaging (MRI)-linac System (MRL) for brain metastases: the impact of the mag-netic field. *Radiotherapy and Oncology: Journal of the European Society for Therapeutic Radiology and Oncology* 2017;125(2):273–279.

57. Tseng CL, Sussman MS, Atenafu EG, Letourneau D, Ma L, Soliman H, et al. Magnetic resonance imag-ing assessment of spinal cord and cauda equina motion in supine patients with spinal metastases planned for spine stereotactic body radiation therapy. *International Journal of Radiation Oncology, Biology, Physics* 2015;91(5):995–1002.

58. Oztek MA, Mayr NA, Mossa-Basha M, Nyflot M, Sponseller PA, Wu W, et al. The dancing cord: inherent spinal cord motion and its effect on cord dose in spine stereotactic body radiation therapy. *Neurosurgery* 2020 Dec;87(6):1157–1166

59. Lee YK, Munawar I, Mashouf S, Sahgal A, Ruschin M. Dosimetric comparison of two treatment planning systems for spine SBRT. *Medical Dosimetry: Official Journal of the American Association of Medical Dosimetrists* 2020;45(1):77–84.

60. Marchand EL, Sahgal A, Zhang TJ, Millar BA, Sharpe M, Moseley D, et al. Treatment planning and delivery evaluation of volumetric modulated arc therapy for stereotactic body radiotherapy of spinal tumours: impact of arc discretization in planning system. *Technology in Cancer Research & Treatment* 2012;11(6):599–606.

61. Sahgal A, Myrehaug S, Siva S, Masucci GL, Foote M, Brundage MD. CCTG SC.24/TROG 17.06: a random-ized phase II/III study comparing 24Gy in 2 stereotactic body radiotherapy (SBRT) fractions versus 20gy in 5 conventional palliative radiotherapy (CRT) fractions for patients with painful spinal metastases. *International Journal of Radiation Oncology, Biology, Physics* 2020; ASTRO's 62nd Annual Meeting (October 23–29, 2020) Late-breaking Abstracts.

62. Finnigan R, Burmeister B, Barry T, Jones K, Boyd J, Pullar A, et al. Technique and early clinical outcomes for spinal and paraspinal tumors treated with stereotactic body radiotherapy. *Journal of Clinical Neuroscience: Official Journal of the Neurosurgical Society of Australasia* 2015;22(8):1258–1263.

63. Ito K, Ogawa H, Shimizuguchi T, Nihei K, Furuya T, Tanaka H, et al. Stereotactic body radiotherapy for spinal metastasis: clinical experience in 134 cases from a single Japanese institution. *Technology in Cancer Research & Treatment* 2018;17:1533033818806472.

64. Chiang A, Zeng L, Zhang L, Lochray F, Korol R, Loblaw A, et al. Pain flare is a common adverse event in steroid-naïve patients after spine stereotactic body radiation therapy: a prospective clinical trial. *International Journal of Radiation Oncology, Biology, Physics* 2013;86(4):638–642.

65. Khan L, Chiang A, Zhang L, Thibault I, Bedard G, Wong E, et al. Prophylactic dexamethasone effectively reduces the incidence of pain flare following spine stereotactic body radiotherapy (SBRT): a prospective obser-vational study. *Supportive Care in Cancer: Official Journal of the Multinational Association of Supportive Care in Cancer* 2015;23(10):2937–2943.

66. Sahgal A, Atenafu EG, Chao S, Al-Omair A, Boehling N, Balagamwala EH, et al. Vertebral compression fracture after spine stereotactic body radiotherapy: a multi-institutional analysis with a focus on radiation dose and the spinal instability neoplastic score. *Journal of Clinical Oncology: Official Journal of the American Society of Clinical Oncology* 2013;31(27):3426–3431.

67. Faruqi S, Tseng CL, Whyne C, Alghamdi M, Wilson J, Myrehaug S, et al. Vertebral compression fracture after spine stereotactic body radiation therapy: a review of the pathophysiology and risk factors. *Neurosurgery* 2018;83(3):314–322.

68. Arrillaga-Romany I, Monje M, Wen PY. Chapter 15 — neurologic complications of oncologic therapy. In: Newton HB, editor. *Handbook of Neuro-Oncology Neuroimaging* (2nd Edition). San Diego, CA: Academic Press; 2016. pp. 125–142.

69. Chuang C, Sahgal A, Lee L, Larson D, Huang K, Petti P, et al. Effects of residual target motion for image-tracked spine radiosurgery. *Medical Physics* 2007;34(11):4484–4490.

70. Glicksman RM, Tjong MC, Neves-Junior WFP, Spratt DE, Chua KLM, Mansouri A, et al. Stereotactic abla-tive radiotherapy for the management of spinal metastases: a review. *JAMA Oncology* 2020;6(4):567–577.

71. Mizumoto M, Harada H, Asakura H, Hashimoto T, Furutani K, Hashii H, et al. Radiotherapy for patients with metastases to the spinal column: a review of 603 patients at Shizuoka cancer center hospital. *International Journal of Radiation Oncology, Biology, Physics* 2011;79(1):208–213.

72. Folkert MR, Bilsky MH, Tom AK, Oh JH, Alektiar KM, Laufer I, et al. Outcomes and toxicity for hypofrac-tionated and single-fraction image-guided stereotactic radiosurgery for sarcomas metastasizing to the spine. *International Journal of Radiation Oncology, Biology, Physics* 2014;88(5):1085–1091.

31 MRI for Spinal Metastases and Response Determination

Pejman Jabehdar Maralani, Aimee Chan, and Jay Detsky

Contents

31.1 INTRODUCTION

Following the lung and liver, bone is the most common site of metastatic disease; and within the bony skeleton, the spine is the most common site of metastasis[1]. Due to the complex anatomy of the spine, imaging—especially cross-sectional orthogonal planes—is essential for the proper diagnosis of metastases and their delineation for radiation planning. This chapter will outline the use of imaging, in particular magnetic resonance imaging (MRI) as the primary modality for (1) the diagnosis of spinal metastases, (2) spine stereotactic body radiotherapy (SBRT) planning, and (3) response assessment after treatment.

31.2 DIAGNOSIS OF SPINAL METASTASES

31.2.1 SPINE STAGING

The need to stage the spine in the setting of metastatic cancer is directed by symptoms (pain, weakness, neurologic decline) or due to the suspicion of spinal metastases on a CT scan, bone scan, PET scan, or other imaging modalities. When considering aggressive upfront therapy such as surgery or SBRT to manage spinal metastases, an MRI of the entire spine should be obtained covering all cervical, thoracic, lumbar, and sacral levels. This is used to rule out multi-level pathology and may alter the patient's classification from oligo-metastatic to polymetastatic disease. Sagittal, non-contrast T1- and T2- weighted images covering the entire spine are typically obtained, and axial T2-weighted images should be obtained at the levels where disease is known or suspected to evaluate for the presence and extent of epidural and paraspinal disease (Table 31.1).

Table 31.1 Evaluation of the extent of epidural and paraspinal diseases

APPLICATION IN CONTEXT OF SPINAL METASTASES CHARACTERIZATION	COMMONLY USED MRI SEQUENCE
Detection of metastases	Sagittal T1WI and T2WI with fat suppression (i.e., STIR)
Detection of VCF	Sagittal T1WI and T2WI
Delineation of epidural disease	Axial T2WI
Assessment of leptomeningeal disease	Axial and sagittal T1WI with and without gadolinium-based contrast agents
SBRT planning	Volumetric T1WI and T2WI

According to the Spine Response Assessment in Neuro-Oncology (SPINO) group,[2,3] the eligibility for spine SBRT requires an MRI to determine the Spinal Instability Neoplastic Score (SINS)[4] and the Bilsky Epidural Spinal Cord Compression (ESCC) scale.[3] While the majority of components required for the SINS can be obtained from MRI, CT remains the gold standard for differentiating osteolytic from osteosclerotic bone metastases which is an important factor in the SINS and helps predict the risk of vertebral compression fracture (VCF).[5] The utilities of various MRI sequences are outlined later in this chapter.

31.2.2 IMAGING PROTOCOLS FOR ASSESSMENT

Owing to its high spatial and contrast resolution, conventional MRI is the preferred modality for the diagnosis and management of spinal metastases.[2,6] Compared to CT, which has a sensitivity, specificity, and diagnostic accuracy of 66.2, 98.5, and 88.8% respectively, MRI has a sensitivity, specificity, and diagnostic accuracy of 98.5, 98.9, and 98.7%, respectively, for the detection of spinal metastases.[7] The MRI signal is generated by exploiting the differences between water and fat composition with regards to bone marrow. Tissues with abundant water and fat content—such as bone marrow—are ideal tissues for MRI assessment. The marrow intensity is described as hypo-, iso-, or hyper-intense relative to the intervertebral disks or skeletal muscles, and different MRI sequences have been developed for appropriate assessment.

31.2.2.1 T1-Weighted Images (T1WI)

Fat, which is present in high amounts in the marrow of adult vertebral bodies, appears hyperintense on T1WI.[8–10] Comparatively, marrow-replacing tumors appear hypointense relative to normal marrow. The intervertebral disks serve as internal standard; signal from the vertebral body marrow should not be more hypointense than the intervertebral disks. Following this, the T1WI has a high sensitivity and specificity, ranging from 90 to 100% in diagnosing abnormal marrow.[11]

31.2.2.2 T2-Weighted Images (T2WI)

T2WI are sensitive to water/fluid content which can be present inside the spinal lesion, as most primary tumors have higher water content than marrow. Axial T2WI is ideal for the delineation of epidural disease. However, as turbo (fast) spin-echo sequences are typically used in the clinical setting to shorten overall scan times, both fluids and fat will appear hyperintense. To suppress this effect and increase lesion conspicuity, techniques such as short tau inversion recovery (STIR),[12] spectrally selective fat suppression,[13] or DIXON-based fat suppression[14] are used to suppress the signal from marrow fat.

31.2.2.3 Gadolinium-Based Contrast Agent (GBCA)-Enhanced MRI

A combination of T1WI and STIR is most effective when assessing spinal metastases, especially lesions that are marrow-replacing, using the sagittal plane. GBCA-enhanced MRI is not routinely used; however, if it is, then the scan should be obtained with fat suppression techniques to null the signal from the T1 hyperintense marrow. Use of fat-suppressed, contrast-enhanced MRI can also help to delineate extraosseous tumoral extensions, identify invasion into the epidural space or paraspinal tissues,[2] and differentiate postoperative or residual tumor from fluid collection.[15] To assess for leptomeningeal spread, fat suppression is not needed for GBCA-enhanced MRI.

Clinical context is vital for appropriate assessment when using T1WI and STIR. For example, in circumstances where there is limited water content in the tumor, as in heavily sclerotic metastases from prostate cancer or when methyl methacrylate—a commonly used material in vertebral body augmentation—signal may be limited or void.[16] As MRI is sensitive to fat, chemotherapy regimens such as granulocyte colony-stimulating factor can also influence the marrow signal[17,18] by diffusely increasing red marrow component resulting in diffuse hypointensity on T1WI. Last, acute-to-subacute vertebral compression fractures that have extensive edema may mimic metastases;[19] proper assessment may require use of other modalities in conjunction with MRI.

31.2.2.4 Diffusion-Weighted Imaging (DWI)

DWI signals are produced from the free Brownian motion of water molecules in tissue. Its intensity decreases when diffusion of water is limited, as in areas of high cellularity, but increases in areas where diffusion can occur more freely, as in the extracellular space, in neoplasms that have responded to therapy, or in perilesional edema.[20] This relationship appears to be biphasic and dependent on the extent of bone marrow replacement.[20] Therefore, DWI can provide quantitative information without the need for exogenous contrast agents. Recently, it has shown promise in differentiating benign from metastatic spinal lesions[21] and in the evaluation of therapy effectiveness;[22,23] however, these still require further validation. Drawbacks to DWI include its susceptibility to artifacts at tissue interfaces in addition to those produced from surgical hardware, which are common in patients with spinal metastases.

31.2.2.4.1 Intravoxel Incoherent Motion (IVIM) Imaging

When DWI are acquired with several low ($\leq$200 s/mm^2) and high ($\geq$1500 s/mm^2) b-values,[24,25] it becomes possible to separate diffusion and perfusion signals; this is known as IVIM imaging.[26] Biexponential modelling produces several maps, including the apparent diffusion coefficient (ADC), which takes into account both diffusion and perfusion effects; the perfusion fraction (f), which represents the water perfusing in the microcirculation; the tissue diffusion coefficient (D), and the pseudodiffusion coefficient (D^*) which represents the perfusion-related diffusion. Further validation is still required, but some studies have demonstrated that different IVIM parameters can be used to differentiate malignant spinal tumors from traumatic compression fractures[27] and pathologic VCFs from nodular hyperplastic hematopoietic bone marrow.[28]

31.2.2.5 Perfusion-Weighted Imaging/Dynamic Contrast-Enhanced (DCE) Imaging

DCE is a T1WI perfusion technique that relies on GBCA kinetics to provide quantitative measures of vascularization and permeability differences between normal and neoplastic tissues. It can be used to differentiate vascular from non-vascular metastases,[29] benign from malignant lesions in vertebral bone marrow,[30] and non-pathologic from pathologic VCFs.[31] DCE parameters can be affected by many factors, including older age, where decreased perfusion is observed; spine location, with more perfusion in the upper compared to lower lumbar spine; and increasing marrow fat, which leads to decreased perfusion.[32] In addition, there is a lack of standardization for image acquisition and post-processing of DCE in spine with respect of different centres and different vendors and field strengths. Therefore, more work is needed to integrate DCE into routine practice.

31.2.2.6 In-Phase and Out-of-Phase (IP/OP) Imaging

This technique, also known as "chemical shift" MR imaging, utilizes the slight differences in resonance frequencies between fat and water molecules and is used primarily to differentiate benign or abnormal lesions from normal bone marrow. When protons from fat and water are in-phase, they both contribute to T1 hyperintensity. However, when they are out-of-phase, the signal cancels out and results in a decreased T1 signal intensity.[33,34] In normal marrow, the average decrease ranges from 11% to 93%, with an average of 58.5%.[33] Comparatively, in marrow-replacing diseases, there is no difference between IP/OP images, with reported changes ranging between 0.5% and 8.7%.[33] A signal decrease of 20% or more from in-phase to out-of-phase images has been recommended for detecting marrow-replacing lesions on 1.5 T and 3 T scanners.[33,35,36] Due to the replacement of bone marrow by fat after radiation, there is limited application of IP/OP imaging post-radiation.

31.2.2.7 Scanning Patients with Metallic Hardware

A number of patients may require spinal hardware placement, particularly after surgical debulking and radiation therapy. Metallic hardware creates magnetic field inhomogeneities, resulting in artifacts in the surrounding tissue. In these cases, the composition, size, type of pulse sequences applied, and sequence parameters can all influence the resultant images.[37,38]

Spinal hardware made of stainless steel, which is ferromagnetic, creates marked local magnetic field inhomogeneities during the MR scan resulting in geometric distortions known as "susceptibility artifacts." These artifacts appear as areas of signal loss around the hardware, with larger implants creating larger areas of signal loss, making tumor delineation challenging. Comparatively, titanium implants are non-ferromagnetic and produce less severe susceptibility artifacts.[39] If the long axis of a metallic hardware is perpendicular to the long axis of the magnet bore (z axis), the artifact is minimal; when it is perpendicular to z axis, the artifact is maximal.[40,41] However, for those with spinal hardware, limited options are available as the spine is already aligned with the z axis of the scanner, and therefore, the pedicular screws will remain perpendicular to direction of the main magnetic field.

Low magnetic field strength[40,42] or using broader receiver bandwidth on a high magnetic field strength scanner[43] can reduce artifacts or distortion effects. Reducing voxel size can reduce the apparent size of the signal loss[44] around the spinal hardware and can be achieved by decreasing the field of view, increasing the matrix size, or minimizing slice thickness.[44] However, these may also result in decreased signal-to-noise ratio. Increasing the echo train length can also reduce artifacts but results in increased scan time.[45]

31.3 IMAGING FOR RADIATION PLANNING

Spine SBRT requires strict planning guidelines to safely deliver ablative doses to tumor while minimizing risk of toxicity, specifically by limiting the dose to the nearby spinal cord or thecal sac, collectively referred to as the critical neural tissue. MRI is required in order to delineate the clinical target volume (CTV) and the organs at risk (OAR). A 1.5-Tesla (T) or 3-T MRI can be used for treatment planning, with improved signal-to-noise ratio with 3 T but at a cost of increased susceptibility artifact in the post-operative setting and longer relaxation times leading to longer scan times.[46]

CT simulation is still required for Linac-based SBRT with an ideal slice thickness of 1 mm with a maximum of 3 mm. Cross-sectional (axial) volumetric T1WI and T2WI with a slice thickness of 1–3 mm, without any gap between slices, should be obtained with the patient approximating the SBRT treatment position.[2] Figure 31.1 shows axial CT, T1WI, and T2WI images used for SBRT treatment planning. The MRI should span a minimum of one vertebral body above and below the target levels; this may need to be expanded if epidural or paraspinal disease extends superior or inferior to the target levels.[47,48] This ensures accurate contouring of the critical neural tissue anywhere the SBRT fields/arcs may pass through. Careful image registration between each MRI and CT dataset needs to be performed and independently verified. When surgical hardware obscures visualization of the spinal cord on MRI, a CT myelogram may be performed; however, volumetric T1- and T2-weighted MRI should still be obtained for identifying residual metastatic disease post-operatively. Preoperative MRI scans can be used to help identify levels and regions to be included in the CTV, but thin slice axial images are typically not obtained prior to surgery.

T1WI is typically used to identify all bony regions involved with disease in order to guide target volume definitions according to the accepted guidelines.[47] T2WI is best suited to identify epidural and paraspinal disease. TThe Bilsky ESCC scale[49] should be noted, as the presence of epidural disease guides dose constraints to the critical neural tissue.[50] Caution is warranted when considering SBRT for Bilsky grade 2 disease (cord compression with visible cerebral spinal fluid) while SBRT is contraindicated for Bilsky grade 3 disease (cord compression with no visible cerebral spinal fluid).[51] The use of post-contrast T1-weighted images are reserved for specific scenarios at the discretion of the treating physician. The most common scenario where post-contrast images are used is in the setting of extensive paraspinal disease with possible invasion into adjacent muscle to aid in CTV contouring.

31.4 RESPONSE ASSESSMENT

31.4.1 CURRENT GUIDELINES

The SPINO guidelines defines response assessment after SBRT for spinal metastases.[2] The previous Response Evaluation Criteria in Solid Tumors (RECIST) V1.1 for radiographic response assessment has limited applicability as comparisons of tumor diameter may be inaccurate on follow-up imaging, where non-volumetric, 3–4 mm slices are commonly used. Lytic or mixed lesions are considered nonmeasurable in RECIST unless they have soft tissue (extra-osseous) disease over 1 cm, while sclerotic lesions that are also considered are nonmeasurable, further reducing its applicability. Previously, no guidelines addressed the MRI parameters for follow-up, nor their frequency. To address this, the consensus guidelines from SPINO were developed to help standardize definitions of response to SBRT. CT alone does not provide enough soft tissue contrast to reliably evaluate the response of epidural or paraspinal disease. Therefore, response to SBRT should be assessed with an MRI every 2 to 3 months for the first 12 to 18 months after treatment and every 6 to 12 months subsequently. Any new pain or neurologic deficits may require earlier imaging re-assessment.[2]

According to SPINO, the post-SBRT MRI should be reviewed by both a neuro-radiologist and radiation oncologist with expertise in spine SBRT. Each vertebral level should be evaluated independently even if multiple contiguous levels were treated concurrently with SBRT. Local control is defined as the absence of progression on two consecutive MRI scans at least 8 weeks apart. Local progression is defined as any of the following: gross unequivocal increase in tumor volume or linear dimension; any new or progressive tumor within the epidural space; and neurological deterioration attributable to pre-existing epidural disease with equivocal increased epidural dimensions on MRI.[2] An example of progression following SBRT is shown in Figure 31.2. Pseudoprogression can be seen after spine SBRT[52,53] with an apparent increase in bone-confined tumor dimensions typically on T1-weighted images. A repeat MRI in 4 to 8 weeks should be performed to differentiate pseudoprogression from true progression. A biopsy can be considered when ongoing imaging investigations are equivocal. Pseudoprogression is further discussed in Section 31.4.4.

Variable patterns of T1 and T2 signal changes can be seen after SBRT:[54] the most common pattern for stable (non-progressing) lesions is an increased but heterogeneous T2-signal. For lesions that demonstrate local failure, the volume of T2-signal changes expands with no change in T2 signal intensity.[54] DCE imaging for response assessment is considered investigational, but preliminary reports indicate that an increase in blood plasma volume (V_p) is associated with tumor progression.[55] Positron emission tomography (PET) may have a role in the post-operative setting when surgical hardware makes response assessment with MRI difficult.[56]

In the end, the SPINO guidelines also recommend the Brief Pain Inventory as the preferred method for evaluating pain response.[2] The International Consensus Pain Response Endpoints (ICPRE) should be used to define overall response rates specific to bone metastases with the benefit of incorporating opioid analgesic use into pain response assessment.[57] The primary pain response to spine radiation should be evaluated 3 months after treatment.

31.4.2 IMAGING FOR RESPONSE ASSESSMENT—KEY FACTORS

In a routine MRI follow-up of the post-SBRT spine, attention to the following items is recommended: (1) change in the size of the target lesion, extension of the lesion into new sectors of the vertebral body, and presence of new metastasises;[58] (2) epidural extension of tumor and whether or not it has changed; (3) paraspinal tumor if present and whether or not it has changed; and (4) interval development of vertebral compression fractures (detailed further in section 31.4.3).

It should be noted that tumor progression into the bone, epidural space, and paraspinal tissues occur at different frequencies, with progression within the bone and/or epidural space being the most common forms of treatment failure after SBRT.[58,59] In a study with 70 patients, of whom 26 had local disease recurrence, Chan et al. state that 73% of cases had a component of in-field recurrence for bone and paraspinal tissues.[58] Therefore, an accurate depiction of paraspinal and epidural disease is needed at baseline (prior to intervention; Figure 31.2a–c) in order to later assess treatment response and define progression (Figure 31.2d–g).

31.4.3 VERTEBRAL COMPRESSION FRACTURE (VCF)

Patients with spinal metastases, in particular those with lytic lesions, are at risk of malignant (pathologic) VCF. Patients with metastatic disease may also present with pain from a VCF whose underlying etiology (malignant versus a benign osteoporotic-related VCF) is unclear. The use of palliative conventional external beam radiation to the spine is associated with an approximate 5% risk of subacute or late radiation-induced VCF.[60] Spine SBRT has a higher risk of VCF with a meta-analysis reporting a crude rate of 13.9%.[5] After spinal radiation,

VCF may be due to tumor progression, radiation effects, or a combination of both. Due to this complexity, the management of VCF relies on appropriate imaging and expert interpretation of the resulting images.

The imaging work-up for a patient suspected of having a malignant VCF should include both MRI and CT scans. MRI is the primary modality for characterizing VCF. MRI features differentiating benign from malignant VCF are summarized by Mauch et al.[61] Morphologically, benign VCF generally demonstrates normal posterior element signal, retropulsed bony fragments, and the presence of additional benign VCFs at other levels, while malignant VCF is suggested by abnormal signal in the posterior elements, epidural or paravertebral soft tissue masses, expanded posterior vertebral contour, and the presence of metastases at other vertebral levels. The signal intensity (SI) on T1 and T2 imaging also helps differentiate benign from malignant VCF. Neoplastic bone marrow infiltration and/or replacement related to malignant VCF results in diffuse low signal on T1WI, while bone marrow is preserved in osteoporosis yielding preserved high T1 and intermediate T2 SI. However, in the acute phase, benign VCF may demonstrate low signal on T1WI.[62] Generally, post-contrast T1-weighted imaging is not required to evaluate VCF. Theoretically, malignant VCF should demonstrate restricted diffusion in bone (due to the presence of tumor cells) that could be detected with DWI; however, DWI has not been shown to have any advantage over routine non-contrast MRI in this scenario.[63] In the acute or subacute setting, differentiation between benign and malignant VCF may not be possible as extensive edema from fracture can be similar to metastatic disease involving almost all of the vertebral body unless there is clear evidence of extra-osseous mass. In these cases, follow-up imaging is advised as marrow signal abnormality secondary to fracture subsides over time but marrow changes due to neoplasm stays the same or progresses.

Similar to morphological MRI, features on CT that are frequently found in benign VCF include anterolateral or posterior cortex VB fracture, retropulsed bone, fracture lines within cancellous bone, sharp fracture lines without cortical destruction, and diffuse thin paraspinal soft tissue thickening.[64] Malignant VCF on CT commonly demonstrates bony destruction and epidural or paraspinal soft tissue masses.[64] The combination of features from both MRI and CT yields the highest accuracy for predicting the etiology of VCF.[65] Data supporting the use of nuclear medicine (PET, SPECT) for VCF is sparse.

31.4.4 PSEUDOPROGRESSION FOLLOWING SPINE SBRT

Similar to the phenomena observed in patients with glioma following radiotherapy, osseous pseudoprogression has been defined by SPINO as an imaging-based, transient increase in apparent tumor size following SBRT. This increase mimics disease progression; however, in pseudoprogression, the enlargement subsequently stabilizes or resolves without any further intervention (Figure 31.3). Currently, only a limited number of studies have investigated this phenomenon.[52,53,66–68] Due to considerable number of different primary histologies included in these studies, it is unclear whether the timeframe or incidence of spinal pseudoprogression is affected by the underlying biology of the primary neoplasm.

Case series have reported timeframes of tumor enlargement range from 3 weeks[52] to 3 years[66] post-SBRT. Studies with larger cohorts of mixed histologies reported tumor enlargement occurring, on average, 4 to 5 months post-SBRT.[67,68] In one study where the study population was limited to those with either *de novo* spinal metastases arising from primary prostate or renal cell carcinoma, the average timeframe to tumor enlargement was similar to those previously reported.[53] However, when analyzed separately, patients with prostate primaries demonstrated tumor enlargement later (4 versus 6 months post-SBRT), but it also resolved quicker (9 to 12 months vs 6 to 18 months post-SBRT), suggesting that lytic- vs sclerotic- type diseases have a profound effect on the timeframe of pseudoprogression.

Incidence of pseudoprogression also seems to vary in the available literature. Amini et al.[67] and Bahig et al.[68] reported an overall incidence of 14% to 18%; however, Maralani et al.[53] reported 37%. Further validation is required to define a threshold of change, and definition that is based on primary histology and timeline.

31.5 CONCLUSION

MRI is the primary imaging modality for the diagnosis of spinal metastases due to its high sensitivity and specificity. MRI is also critical for SBRT planning and response determination. Standard T1WI and T2WI may be complemented by more advanced MRI techniques in specific instances. Due to the complex nature of spine metastases, appropriate imaging sequences are necessary to accurately interpret the underlying

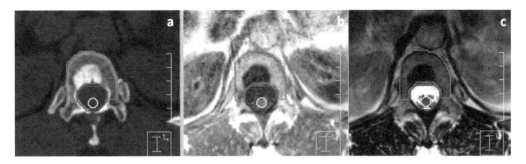

Figure 31.1 A sclerotic bone metastasis from breast cancer in the posterior vertebral body of T12 planned for spine SBRT. Images used for radiation planning include (a) axial CT, (b) axial T1WI MRI, and (c) axial T2WI MRI. The CTV is outlined in blue, the spinal cord in yellow, and the thecal sac in purple.

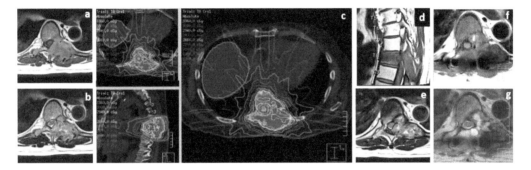

Figure 31.2 A spinal metastasis at T8 from renal cell carcinoma involving the spinous process, left posterior elements, and left transverse process with a large left paraspinal component as shown on the (a) T1WI MRI and (b) T2WI MRI. Spine SBRT was delivered to 28 Gy in two fractions (c). After an initial pain response, the patient had increasing pain 3 months later and an MRI (sagittal T1WI in d, axial T2WI in e) demonstrated a decreased paraspinal component but worsening left and posterior epidural disease. After decompression, an axial T1WI (f) and T2WI (g) was used for salvage SBRT planning.

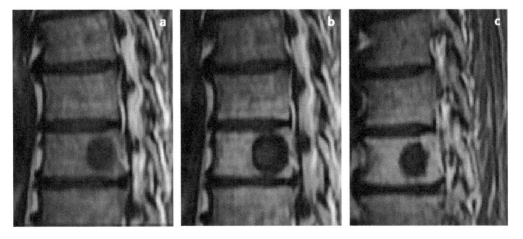

Figure 31.3 Sagittal T1WI (a) of a spine metastasis at baseline. On 3-month post-SBRT MRI the mass demonstrates interval increase in size (b) while at 6-month follow-up, the size of the mass has decreased without any other interventions (c) demonstrating pseudoprogression phenomenon.

biology. Further research is needed to better understand MRI signal changes after SBRT in order to differentiate treatment effect from tumor progression.

REFERENCES

1. Aaron AD. The management of cancer metastatic to bone. *JAMA* 1994;272(15):1206–1209.
2. Thibault I, Chang EL, Sheehan J, et al. Response assessment after stereotactic body radiotherapy for spinal metastasis: a report from the SPIne response assessment in neuro-oncology (SPINO) group. *The Lancet Oncology* 2015;16(16):e595–e603.
3. Laufer I, Lo SS, Chang EL, et al. Population description and clinical response assessment for spinal metastases: part 2 of the SPIne response assessment in neuro-oncology (SPINO) group report. *Neuro-Oncology* 2018;20(9):1215–1224.
4. Fourney DR, Frangou EM, Ryken TC, et al. Spinal instability neoplastic score: an analysis of reliability and validity from the spine oncology study group. *Journal of Clinical Oncology: Official Journal of the American Society of Clinical Oncology* 2011;29(22):3072–3077.
5. Faruqi S, Tseng CL, Whyne C, et al. Vertebral compression fracture after spine stereotactic body radiation therapy: a review of the pathophysiology and risk factors. *Neurosurgery* 2018;83(3):314–322.
6. Jabehdar Maralani P, Lo SS, Redmond K, et al. Spinal metastases: multimodality imaging in diagnosis and stereotactic body radiation therapy planning. *Future Oncology* 2017;13(1):77–91.
7. Buhmann Kirchhoff S, Becker C, Duerr HR, Reiser M, Baur-Melnyk A. Detection of osseous metastases of the spine: comparison of high resolution multi-detector-CT with MRI. *European Journal of Radiology* 2009;69(3):567–573.
8. Chakraborty D, Bhattacharya A, Mete UK, Mittal BR. Comparison of 18F fluoride PET/CT and 99mTc-MDP bone scan in the detection of skeletal metastases in urinary bladder carcinoma. *Clinical Nuclear Medicine* 2013;38(8):616–621.
9. Even-Sapir E, Metser U, Mishani E, Lievshitz G, Lerman H, Leibovitch I. The detection of bone metastases in patients with high-risk prostate cancer: 99mTc-MDP Planar bone scintigraphy, single- and multi-field-of-view SPECT, 18F-fluoride PET, and 18F-fluoride PET/CT. *Journal of Nuclear Medicine: Official Publication, Society of Nuclear Medicine* 2006;47(2):287–297.
10. Kruger S, Buck AK, Mottaghy FM, et al. Detection of bone metastases in patients with lung cancer: 99mTc-MDP planar bone scintigraphy, 18F-fluoride PET or 18F-FDG PET/CT. *European Journal of Nuclear Medicine and Molecular Imaging* 2009;36(11):1807–1812.
11. Carroll KW, Feller JF, Tirman PF. Useful internal standards for distinguishing infiltrative marrow pathology from hematopoietic marrow at MRI. *Journal of Magnetic Resonance Imaging: JMRI* 1997;7(2):394–398.
12. Krinsky G, Rofsky NM, Weinreb JC. Nonspecificity of short inversion time inversion recovery (STIR) as a technique of fat suppression: pitfalls in image interpretation. *AJR American Journal of Roentgenology* 1996;166(3):523–526.
13. Haase A, Frahm J, Hänicke W, Matthaei D. 1H NMR chemical shift selective (CHESS) imaging. *Physics in Medicine and Biology* 1985;30(4):341–344.
14. Dixon WT. Simple proton spectroscopic imaging. *Radiology* 1984;153(1):189–194.
15. Georgy BA, Hesselink JR. Evaluation of fat suppression in contrast-enhanced MR of neoplastic and inflammatory spine disease. *AJNR American Journal of Neuroradiology* 1994;15(3):409–417.
16. Schmidt GP, Schoenberg SO, Reiser MF, Baur-Melnyk A. Whole-body MR imaging of bone marrow. *European Journal of Radiology* 2005;55(1):33–40.
17. Ciray I, Lindman H, Astrom GK, Wanders A, Bergh J, Ahlstrom HK. Effect of granulocyte colony-stimulating factor (G-CSF)-supported chemotherapy on MR imaging of normal red bone marrow in breast cancer patients with focal bone metastases. *Acta Radiologica (Stockholm, Sweden: 1987)* 2003;44(5):472–484.
18. Carmona R, Pritz J, Bydder M, et al. Fat composition changes in bone marrow during chemotherapy and radiation therapy. *International Journal of Radiation Oncology, Biology, Physics* 2014;90(1):155–163.
19. Baker LL, Goodman SB, Perkash I, Lane B, Enzmann DR. Benign versus pathologic compression fractures of vertebral bodies: assessment with conventional spin-echo, chemical-shift, and STIR MR imaging. *Radiology* 1990;174(2):495–502.
20. Khoo MM, Tyler PA, Saifuddin A, Padhani AR. Diffusion-weighted imaging (DWI) in musculoskeletal MRI: a critical review. *Skeletal Radiology* 2011;40(6):665–681.
21. Zhou XJ, Leeds NE, McKinnon GC, Kumar AJ. Characterization of benign and metastatic vertebral compression fractures with quantitative diffusion MR imaging. *AJNR American Journal of Neuroradiology* 2002;23(1):165–170.
22. Byun WM, Shin SO, Chang Y, Lee SJ, Finsterbusch J, Frahm J. Diffusion-weighted MR imaging of metastatic disease of the spine: assessment of response to therapy. *AJNR American Journal of Neuroradiology* 2002;23(6):906–912.

23. Pui MH, Mitha A, Rae WI, Corr P. Diffusion-weighted magnetic resonance imaging of spinal infection and malignancy. *Journal of Neuroimaging* 2005;15(2):164–170.
24. Iima M, Le Bihan D. Clinical intravoxel incoherent motion and diffusion MR imaging: past, present, and future. *Radiology* 2016;278(1):13–32.
25. Iima M, Yano K, Kataoka M, et al. Quantitative non-gaussian diffusion and intravoxel incoherent motion magnetic resonance imaging: differentiation of malignant and benign breast lesions. *Investigative Radiology* 2015;50(4):205–211.
26. Le Bihan D, Breton E, Lallemand D, Aubin ML, Vignaud J, Laval-Jeantet M. Separation of diffusion and perfusion in intravoxel incoherent motion MR imaging. *Radiology* 1988;168(2):497–505.
27. Chen Y, Yu Q, La Tegola L, et al. Intravoxel incoherent motion MR imaging for differentiating malignant lesions in spine: a pilot study. *European Journal of Radiology* 2019;120:108672.
28. Park S, Kwack KS, Chung NS, Hwang J, Lee HY, Kim JH. Intravoxel incoherent motion diffusion-weighted magnetic resonance imaging of focal vertebral bone marrow lesions: initial experience of the differentiation of nodular hyperplastic hematopoietic bone marrow from malignant lesions. *Skeletal Radiology* 2017;46(5):675–683.
29. Khadem NR, Karimi S, Peck KK, et al. Characterizing hypervascular and hypovascular metastases and normal bone marrow of the spine using dynamic contrast-enhanced MR imaging. *AJNR American Journal of Neuroradiology* 2012;33(11):2178–2185.
30. Moulopoulos LA, Maris TG, Papanikolaou N, Panagi G, Vlahos L, Dimopoulos MA. Detection of malignant bone marrow involvement with dynamic contrast-enhanced magnetic resonance imaging. *Annals of Oncology: Official Journal of the European Society for Medical Oncology/ESMO* 2003;14(1):152–158.
31. Arevalo-Perez J, Peck KK, Lyo JK, Holodny AI, Lis E, Karimi S. Differentiating benign from malignant vertebral fractures using T1 -weighted dynamic contrast-enhanced MRI. *Journal of Magnetic Resonance Imaging: JMRI* 2015;42(4):1039–1047.
32. Savvopoulou V, Maris TG, Vlahos L, Moulopoulos LA. Differences in perfusion parameters between upper and lower lumbar vertebral segments with dynamic contrast-enhanced MRI (DCE MRI). *European Radiology* 2008;18(9):1876–1883.
33. Zajick DC, Jr., Morrison WB, Schweitzer ME, Parellada JA, Carrino JA. Benign and malignant processes: normal values and differentiation with chemical shift MR imaging in vertebral marrow. *Radiology* 2005;237(2):590–596.
34. Zampa V, Cosottini M, Michelassi C, Ortori S, Bruschini L, Bartolozzi C. Value of opposed-phase gradient-echo technique in distinguishing between benign and malignant vertebral lesions. *European Radiology* 2002;12(7):1811–1818.
35. Disler DG, McCauley TR, Ratner LM, Kesack CD, Cooper JA. In-phase and out-of-phase MR imaging of bone marrow: prediction of neoplasia based on the detection of coexistent fat and water. *AJR American Journal of Roentgenology* 1997;169(5):1439–1447.
36. Del Grande F, Subhawong T, Flammang A, Fayad LM. Chemical shift imaging at 3 Tesla: effect of echo time on assessing bone marrow abnormalities. *Skeletal Radiology* 2014;43(8):1139–1147.
37. Stradiotti P, Curti A, Castellazzi G, Zerbi A. Metal-related artifacts in instrumented spine. Techniques for reducing artifacts in CT and MRI: state of the art. *European Spine Journal: Official Publication of the European Spine Society, the European Spinal Deformity Society, and the European Section of the Cervical Spine Research Society* 2009;18 Suppl 1:102–108.
38. Lee MJ, Kim S, Lee SA, et al. Overcoming artifacts from metallic orthopedic implants at high-field-strength MR imaging and multi-detector CT. *Radiographics: A Review Publication of the Radiological Society of North America, Inc* 2007;27(3):791–803.
39. Ganapathi M, Joseph G, Savage R, Jones AR, Timms B, Lyons K. MRI susceptibility artefacts related to scaphoid screws: the effect of screw type, screw orientation and imaging parameters. *Journal of Hand Surgery (Edinburgh, Scotland)* 2002;27(2):165–170.
40. Guermazi A, Miaux Y, Zaim S, Peterfy CG, White D, Genant HK. Metallic artefacts in MR imaging: effects of main field orientation and strength. *Clinical Radiology* 2003;58(4):322–328.
41. Harris CA, White LM. Metal artifact reduction in musculoskeletal magnetic resonance imaging. *The Orthopedic Clinics of North America* 2006;37(3):349–359, vi.
42. Farahani K, Sinha U, Sinha S, Chiu LC, Lufkin RB. Effect of field strength on susceptibility artifacts in magnetic resonance imaging. *Computerized Medical Imaging and Graphics :The Official Journal of the Computerized Medical Imaging Society* 1990;14(6):409–413.
43. Hargreaves BA, Worters PW, Pauly KB, Pauly JM, Koch KM, Gold GE. Metal-induced artifacts in MRI. *AJR American Journal of Roentgenology* 2011;197(3):547–555.
44. Törmänen J, Tervonen O, Koivula A, Junila J, Suramo I. Image technique optimization in MR imaging of a titanium alloy joint prosthesis. *Journal of Magnetic Resonance Imaging: JMRI* 1996;6(5):805–811.
45. White LM, Buckwalter KA. Technical considerations: CT and MR imaging in the postoperative orthopedic patient. *Seminars in Musculoskeletal Radiology* 2002;6(1):5–17.

46. Dahele M, Zindler JD, Sanchez E, et al. Imaging for stereotactic spine radiotherapy: clinical considerations. *International Journal of Radiation Oncology, Biology, Physics* 2011;81(2):321–330.

47. Cox BW, Spratt DE, Lovelock M, et al. International spine radiosurgery consortium consensus guidelines for target volume definition in spinal stereotactic radiosurgery. *International Journal of Radiation Oncology, Biology, Physics* 2012;83(5):e597–e605.

48. Tseng CL, Soliman H, Myrehaug S, et al. Imaging-based outcomes for 24 Gy in 2 daily fractions for patients with de novo spinal metastases treated with spine stereotactic body radiation therapy (SBRT). *International Journal of Radiation Oncology, Biology, Physics* 2018;102(3):499–507.

49. Bilsky MH, Laufer I, Fourney DR, et al. Reliability analysis of the epidural spinal cord compression scale. *Journal of Neurosurgery Spine* 2010;13(3):324–328.

50. Detsky JS, Nguyen TK, Lee Y, et al. Mature imaging-based outcomes supporting local control for complex reirradiation salvage spine stereotactic body radiotherapy. *Neurosurgery* 2020;87(4):816–822.

51. Tseng CL, Eppinga W, Charest-Morin R, et al. Spine stereotactic body radiotherapy: indications, outcomes, and points of caution. *Global Spine Journal* 2017;7(2):179–197.

52. Taylor DR, Weaver JA. Tumor pseudoprogression of spinal metastasis after radiosurgery: a novel concept and case reports. *Journal of Neurosurgery Spine* 2015;22(5):534–539.

53. Jabehdar Maralani P, Winger K, Symons S, et al. Incidence and time of onset of osseous pseudoprogression in patients with metastatic spine disease from renal cell or prostate carcinoma after treatment with stereotactic body radiation therapy. *Neurosurgery* 2019;84(3):647–654.

54. Hwang YJ, Sohn MJ, Lee BH, et al. Radiosurgery for metastatic spinal tumors: follow-up MR findings. *AJNR American Journal of Neuroradiology* 2012;33(2):382–387.

55. Chu S, Karimi S, Peck KK, et al. Measurement of blood perfusion in spinal metastases with dynamic contrast-enhanced magnetic resonance imaging: evaluation of tumor response to radiation therapy. *Spine* 2013;38(22):E1418–E1424.

56. Gwak HS, Youn SM, Chang U, et al. Usefulness of (18)F-fluorodeoxyglucose PET for radiosurgery planning and response monitoring in patients with recurrent spinal metastasis. *Minim Invasive Neurosurgery* 2006;49(3):127–134.

57. Chow E, Hoskin P, Mitera G, et al. Update of the international consensus on palliative radiotherapy endpoints for future clinical trials in bone metastases. *International Journal of Radiation Oncology, Biology, Physics* 2012;82(5):1730–1737.

58. Chan MW, Thibault I, Atenafu EG, et al. Patterns of epidural progression following postoperative spine stereotactic body radiotherapy: implications for clinical target volume delineation. *Journal of Neurosurgery Spine* 2016;24(4):652–659.

59. Chang EL, Shiu AS, Mendel E, et al. Phase I/II study of stereotactic body radiotherapy for spinal metastasis and its pattern of failure. *Journal of Neurosurgery Spine* 2007;7(2):151–160.

60. Chow E, Harris K, Fan G, Tsao M, Sze WM. Palliative radiotherapy trials for bone metastases: a systematic review. *Journal of Clinical Oncology: Official Journal of the American Society of Clinical Oncology* 2007;25(11):1423–1436.

61. Mauch JT, Carr CM, Cloft H, Diehn FE. Review of the imaging features of benign osteoporotic and malignant vertebral compression fractures. *AJNR American Journal of Neuroradiology* 2018;39(9):1584–1592.

62. Yamato M, Nishimura G, Kuramochi E, Saiki N, Fujioka M. MR appearance at different ages of osteoporotic compression fractures of the vertebrae. *Radiation Medicine* 1998;16(5):329–334.

63. Castillo M, Arbelaez A, Smith JK, Fisher LL. Diffusion-weighted MR imaging offers no advantage over routine noncontrast MR imaging in the detection of vertebral metastases. *AJNR American Journal of Neuroradiology* 2000;21(5):948–953.

64. Kubota T, Yamada K, Ito H, Kizu O, Nishimura T. High-resolution imaging of the spine using multidetector-row computed tomography: differentiation between benign and malignant vertebral compression fractures. *Journal of Computer Assisted Tomography* 2005;29(5):712–719.

65. Yuzawa Y, Ebara S, Kamimura M, et al. Magnetic resonance and computed tomography-based scoring system for the differential diagnosis of vertebral fractures caused by osteoporosis and malignant tumors. *Journal of Orthopaedic Science: Official Journal of the Japanese Orthopaedic Association* 2005;10(4):345–352.

66. Al-Omair A, Smith R, Kiehl TR, et al. Radiation-induced vertebral compression fracture following spine stereotactic radiosurgery: clinicopathological correlation. *Journal of Neurosurgery Spine* 2013;18(5):430–435.

67. Amini B, Beaman CB, Madewell JE, et al. Osseous pseudoprogression in vertebral bodies treated with stereotactic radiosurgery: a secondary analysis of prospective phase I/II clinical trials. *AJNR American Journal of Neuroradiology* 2016;37(2):387–392.

68. Bahig H, Simard D, Letourneau L, et al. A study of pseudoprogression after spine stereotactic body radiation therapy. *International Journal of Radiation Oncology, Biology, Physics* 2016;96(4):848–856.

32 SBRT for Metastatic Disease to the Sacrum

Emma M. Dunne, M. Liu, S.S. Lo, A.M. Bergman, R. Kosztyla, N. Dea,
K. Liang Zeng, and Arjun Sahgal

Contents

32.1 INTRODUCTION

The sacrum is an anatomically complex region composed of five progressively smaller vertebrae, S1 to S5. Unlike the rest of the 'mobile' spine (cervical, thoracic, and lumbar regions) where vertebral bodies are separated by intervertebral disc structures, each sacral vertebra is separated by the remnants of previous intervertebral spaces represented as transverse lines of fibrocartilage on complete fusion on the ventral aspect of the sacrum (1–3). In addition to affording the spine flexibility, the intervertebral disc structures are also thought to prevent direct metastatic spread between vertebrae due to factors such as avascularization and high intra-disc pressure. Given the contiguous nature of the sacral vertebrae, it is unlikely the sacrum benefits from similar mechanical or biochemical factors to prevent spread, though there is scant information in the literature on this area (4–7).

The sacrum is located deep within the pelvis adjacent to organs such as the rectum, bladder, and ureter in addition to neural and vascular structures. Although metastatic disease to the sacrum is rare, estimated at only 5% in recent series, if untreated, progression can result in significant functional cost to the patient (1, 8–10). Charest-Morin et al. conducted the largest retrospective study to date on the treatment of symptomatic sacral metastases and was the first study to report on health-related quality of life (HRQOL) outcomes for such a cohort (11). Although the sample size was small (n = 23), the results showed that treatment with surgery and/or radiotherapy significantly improved pain and quality of life in both groups suggesting treatment is beneficial for suitable patients.

Due to the complexity of sacral anatomy however, surgical treatment can be challenging with potentially high morbidity. Increasingly, stereotactic body radiation therapy (SBRT) is being used to treat metastatic disease to the sacrum, including within clinical trials. Understandably, given the relatively rarity of metastatic disease at this site, outcome data on the effectiveness of SBRT is sparse. One recent dedicated series of imaged-based SBRT outcomes for sacral metastases reported a cumulative local control of 86.5% and 78.7% at 1 and 2 years, respectively, with a median time to local recurrence of 4.7 months (range, 0.4–54.6 months). In this study, the authors made specific reference to the complexities associated with planning and treating disease in the sacrum. This was partly thought to be due to larger treatment volumes as a result of bulkier disease (12). However, it is not uncommon for the disease in the sacrum to be more advanced at diagnosis due to the subtleties of presenting symptoms.

It is challenging to gain clinical experience treating the sacrum with SBRT, given the relatively low incidence of disease at this site coupled with the anatomical heterogeneity of the sacrum. In 2020, Dunne et al. published consensus recommendations for target delineation when treating metastatic disease in the sacrum in an effort to improve uniformity of practice internationally (13). Standardization of contouring across institutions is crucial to inform the treatment approach and allow meaningful comparison of outcome metrics and pattern of failure analysis going forward.

Recognising the evolving role of SBRT in the treatment of metastatic disease to the sacrum, this chapter aims to provide a practical guide on the management of disease at this site, summarising current guidelines to support this practice and highlighting the many challenges that arise when treating this area.

32.2 TREATMENT PLANNING OVERVIEW

32.2.1 PATIENT SELECTION

All patients with metastatic disease to the spine require careful selection when considering treatment with SBRT. Various survival models have been created to guide decision-making such as the Neurological Oncologic Mechanical and Systemic (NOMS) framework and treatment algorithms developed by the International Spine Oncology Consortium (14,15). No specific framework or guidance exists for deciding treatment in patients with metastatic disease to the sacrum, which is understandable given the rarity of presentation. Nonetheless these patients are still included in clinical trials and single institution studies. However the heterogeneity of this patient population due to the uniqueness of the sacral anatomy, the increased risk of toxicity at this site, and the complexity of the planning process mean careful patient selection is paramount. In the absence of high-level evidence, it is the authors' opinion that it is practical and reasonable to extrapolate from the decision frameworks mentioned above when deciding on who best to treat. However, it is strongly recommended that a multidisciplinary discussion takes place with radiation oncologists, medical oncologists and spine surgeons to determine who would most benefit from SBRT or, indeed, a combined approach to treatment. Until further data is available, we would suggest adhering to the following recommendations as detailed in Table 32.1.

32.3 TECHNICAL REQUIREMENTS FOR SPINE SBRT

32.3.1 Immobilization

Treating with SBRT requires all patients to be immobilized in the supine position and have the ability to lie immobile in this position for up to 60 minutes. If the patient's breathing status or pain control is such that lying supine is not possible for this length of time, then the patient will not be eligible for treatment.

As with any spine SBRT treatment, the recommended practice for delivery of SBRT to the sacrum is a validated immobilization system and that can limit intrafraction motion to less than 2–3 mm (depending on planning organ at risk volume (PRV) and planning target volume (PTV) margins). This may be achieved in several different ways, including whole-body vacuum bag systems (e.g. ELEKTA BodyFIX) or custom

Table 32.1 **Clinical factors with ideal requirements**

CLINICAL FACTOR	IDEAL REQUIREMENTS
Clinical situation	• Painful or radiologically progressive sacral metastasis with or without previous radiotherapy • *De novo* oligometastatic disease • Limited extra spinal systemic disease
Performance status	• KPS >50
Life expectancy	• ≥3 months
SINS score	• 0–6 (stable) • 7–12 (indeterminate and potentially unstable spine)* *Acceptable in select cases but should be evaluated by a neurosurgeon first prior to considering treatment
Imaging	• Spine MRI < 8 weeks prior to CT planning simulation or CT myelogram
Spinal segments treated	• ≤3 contiguous or non-contiguous spinal segments to be included in PTV • No more than two spine sites requiring treatment at the same time
Immobilization	• Able to tolerate body fix immobilization • Patient able to maintain stable supine position and tolerate immobilization for SBRT treatment for up to 60 minutes* *The patient should be considered ineligible for treatment if unable to lie comfortably supine due to poor pain control or breathing difficulties

vacuum cushion systems—with or without additional strap/compression devices (16,17). Some groups utilize simpler immobilization systems for the pelvic region and rely on intrafraction monitoring and rapid treatment delivery (18). When treating with CyberKnife, the patient is placed in a vacuum cushion or cradle, and X-Sight Spine is used for near real-time tracking to facilitate the adjustment of intrafractional positional deviations. Li et al. analyzed three different immobilization devices for spine SBRT (evacuated cushion, BodyFIX and thermoplastic S frame Mask) and found that residual setup errors were similar across all three immobilization systems. They did, however, find that the BodyFIX resulted in the least amount of intrafraction motion. However, it is worth noting that at the time of the study, they did not have a robotic couch that could correct for rotational offsets so any rotations ≤2° were not corrected for (17).

32.3.2 IMAGING

Computed tomography (CT) simulation scan thickness must be ≤1.25 mm. It is imperative that the extent of scanning include at least the entire sacrum and the full extent of normal structures adjacent to the disease. Adjacent organs at risk (OARs) are likely to receive some dose and therefore need to be accounted for during planning. Suggested scanning levels would be from the level of the first lumbar vertebra to below the pelvis. A volumetric magnetic resonance image (MRI) scan for co-registration with the CT simulation scan can be obtained from diagnostic imaging, ideally with the patient in the simulated position or in the immobilization device. The preferred MRI sequences to visualise spinal metastases are volumetric T1-weighted (with and without gadolinium) and T2-weighted axial non-contrast-enhanced sequences. Specifically, axial T1 three-dimensional (3D) fast field echo (FFE) and axial T2 3D turbo field echo (TFE) sequences are used (voxels: 1 × 1 × 2 mm for both sequences). If CT artefacts are present due to metallic implants, the artefact must be contoured and assigned the electron density of water. Using 'bone windows' when contouring can also help minimise the effect of the artefact. The metallic implant additionally must be contoured and assigned the appropriate electron density in the treatment planning system if possible. In those patients with titanium fixation devices in place, a CT myelogram (performed immediately prior to CT planning) is often superior to an MRI for delineation of the thecal sac. It is important to remember, however, that a myelogram is not useful to aid delineation caudal to the termination of the thecal sac, which typically occurs at the level of S1 or S2 (though this may vary in each individual patient).

32.3.3 TREATMENT PLANNING

Sacral PTVs can be bulky and oddly shaped. In the pelvic region, factors that can contribute to initial setup reproducibility include variations in the pelvic tilt. This may be influenced by patient factors such as differences in gluteal muscle contractions, patient comfort, stress or superior–inferior positioning in the immobilization device (19). To correct for possible angular rotations (i.e. tilt) of the sacral area, a 6 degree-of-freedom (DoF) couch should be considered for pre-positioning (20).When planning a sacral target for treatment with SBRT, step and shoot intensity modulated radiation therapy (IMRT) may be preferable to volumetric modulated arc therapy (VMAT) due to the ability to rotate the collimator for each beam allowing alignment with the long axis of the sacrum. This ultimately results in a steeper dose gradient. Furthermore, in the presence of metallic implants resulting in dose scatter, step and shoot IMRT can decrease the weight of the beams going through the metal.

It is worth appreciating that non-random intrafraction target motion is a function of treatment time; therefore, the length of treatment is an important consideration. A study by Ma et al. demonstrated that the average time needed to maintain a translation of 1 mm or rotation of 1 degree was 7.1 minutes in the lumbar-sacrum ($n = 24$) compared to 5.5 minutes and 5.9 minutes in cervical ($n = 20$) and thoracic ($n = 20$) locations, respectively. It is therefore important that when choosing patients for treatment, they are able to maintain their position and any pain appropriately managed to prevent any elongation of the treatment time (21).

For long treatment times (e.g. >5-minute beam-on), periodic interventions to assess and correct any intrafraction motion should be considered. This can be achieved by intrafraction cone-beam CT (CBCT), dual-orthogonal X-ray systems (e.g. ExacTrac), or surface-guided radiotherapy (SGRT) systems (e.g. Varian AlignRT) (19, 22, 23).

32.4 TARGET AND CRITICAL NEURAL ELEMENT DELINEATION

32.4.1 TARGET DELINEATION IN THE SACRUM

The single largest uncertainty throughout the planning process is the variability in contouring of the clinical target volume (CTV). The use of consensus contouring guidelines is a recognised way of decreasing uncertainty and standardising practice, though of course it is not possible to fully eliminate variation completely. Published consensus guidelines on CTV delineation when treating metastatic disease to the spine in the *de novo* and post-operative setting have increased the uniformity of practice across the field (24, 25). In 2020, consensus recommendations by Dunne et al. for target delineation in the sacrum were published (13). Similar to guidelines published for the *de novo* spine, an anatomic compartment approach is applied when delineating the CTV in the sacrum. In essence, this involves contouring the entirety of the compartment in which the gross tumor volume (GTV) is centered and the 'prophylactic' contouring of the entirety of the compartment(s) adjacent to the GTV to cover for potential microscopic extension. There are certain nuances to this general approach; for example if disease within the vertebral body is small and discrete, the CTV can be confined to just the vertebral body without prophylactically including the adjacent compartment(s) as part of the CTV volume. Treating the alae also takes some consideration; at the level of S1–S2 (and occasionally S3), fusion (or ossification) lines can be identified separating each ala into an anterior and posterior compartment. These ossification lines can be used to limit the overall size of the CTV should the volume require the *prophylactic* inclusion of the ala. Inferiorly in the sacrum (S3–S5), the alae are smaller and have developed from only one ossification centre; therefore, the entirety of that compartment would be included if appropriate. It is important to note, however, that ossification lines are not thought to be barriers to spread; therefore, if the GTV involves any portion of the ala, it is not advisable to use these lines to limit the CTV volume, and the authors would advise that the entirety of the ala (anterior and posterior compartments) is included in the treatment volume. With bone-only disease, extra-osseous expansion of the CTV is not necessary. The PTV is typically ≤ 3 mm and is institution-specific depending on the type of immobilization and image-guidance technique used.

It should be noted that the recommendations for CTV delineation in the sacrum do not address the management of patient in the post-operative or reirradiation setting or in those cases of epidural or paraspinal disease. The recommendations are specific to disease confined within the bone only. Clinical expertise needs to be applied in such circumstances taking into account the intent of treatment and the ability to deliver an ablative dose while respecting adjacent normal tissue constraints. The development of the sacral guidelines is described in more detail in the next section.

32.4.2 DEVELOPMENT OF THE SACRAL CLASSIFICATION SYSTEM FOR CTV DELINEATION

When developing international consensus guidelines for target volume delineation in spine SBRT, the International Spine Radiosurgery Consortium (ISRC) developed an anatomic description of contours using a modification of the Weinstein–Boriani–Biagini (WBB) surgical staging system (24, 26). The ISRC adapted and simplified the WBB system, dividing the spine in to 6 sectors instead of the original 12, to aid uniformity of contouring. This classification system was subsequently used to guide contouring in postoperative spine SBRT (25).

Although the development of the sacrum is remarkably complex, there are many aspects that are similar to the embryological development of the rest of the thoraco-lumbar spine. The sacrum classification system (SCS) to guide CTV delineation was modeled on the ISRS classification system but adapted to consider the unique embryological and anatomical development of the sacrum. The SCS uses eight sectors, four of which (the vertebral body, spinous process and left and right laminae) are shared with the ISRC classification system. The final four sectors divide up the alae or lateral surfaces of the sacrum, unique to this part of the spine (Figure 32.1).

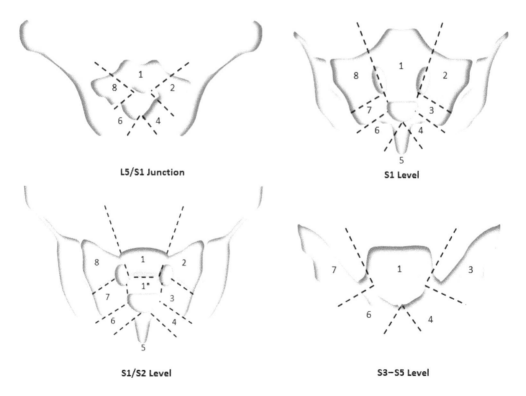

Figure 32.1 Stereotactic Radiosurgery Sacral Classification System. 1. Vertebral body [1* represents the subsequent caudal vertebral body (S2 in this illustration)]; 2. Left anterior ala; 3. Left posterior ala; 4. Left lamina; 5. Spinous process; 6. Right lamina; 7. Right posterior ala; 8. Right anterior ala (13).

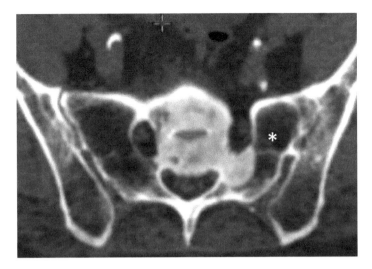

Figure 32.2 Axial CT image at the level of S1/S2 * Ossification line separating the right anterior and posterior ala. CT, computed tomography.

Each ala forms from the fusion of the two ossification centres, the costal elements and the neural arches, which comprise the anterior and posterior portions of the ala, respectively. The division between these ossification centres can be seen as calcified lines on helical CT scan or as low-signal intensity relative to cartilage on MR imaging (3, 27). The SCS uses this ossification line to divide the each ala in to its anterior and posterior segment (Figure 32.2).

The WBB surgical staging system was created to facilitate effective communication of results from en bloc vertebrectomy across institutions in order to encourage more consistent surgical planning. Similarly, the SCS was developed as a framework for decision making when treating metastatic disease to the sacrum both in clinical trials and in general practice. However, unlike the ISRC anatomic classification system for the *de novo* spine, these recommendations are not based on pattern of failure data. Given the rarity of sacral metastatic disease and the inherent anatomic heterogeneity of the sacrum, achieving comprehensive pattern of failure data specific to each sacral vertebra would be a challenging if not an impossible undertaking. The uniformity of approach the SCS facilitates allows the standardization of pattern of failure analysis, not previously available, going forward and will allow results to be reported in a consistent and reproducible way.

Strength, however, does lie in the similarity of these internationally peer-reviewed recommendations to those previously published by the ISRC. However, the SCS will need to be validated with retrospective or preferably prospective pattern of failure analysis in order to strengthen its applicability and adapted for clinical practice accordingly.

The published recommendations for target volume definition of the sacrum in spinal stereotactic body radiation therapy/stereotactic radiosurgery (SBRT/SRS) are detailed in Table 32.2.

32.4.3 THECAL SAC DELINEATION

Within the sacrum, the thecal sac (TS) is the major dose-limiting organ for spinal SBRT. The proximity of the TS to the high-dose gradient will impact delivered dose to the PTV and dictate how much the adjacent OAR is spared. Precise TS contouring is a prerequisite for accurate assessment of delivered dose to this critical neural structure (CNS). However, in the absence of a defined gold standard or reference contour, accurate delineation of this structure is challenging. Uncertainty defining the TS and the availability of appropriate imaging to assist in delineation can impact the verity of the delineated contour. It is essential that the TS is defined on MR images; therefore, the first step to guide contouring is the accurate fusion of thin-sliced volumetric T1-weighted and T2-weighted axial non-contrast-enhanced MRI sequences to the

Table 32.2 Recommendations for target volume definition of the sacrum in spinal stereotactic body radiation therapy/stereotactic radiosurgery (SBRT/SRS) (13)

GTV INVOLVEMENT	SACRUM ANATOMIC MAP CLASSIFICATION	SACRUM BONY CTV RECOMMENDATION	CTV DESCRIPTION
Any portion of the VB	1	1	Entire VB
Lateralised within the VB (S1–S2)	1	1, 2, 3	Entire VB and the ipsilateral ala. When contouring the ala, use the ossification line if visible to limit the extent of the CTV. The superior and inferior extent of the CTV is determined by the superior and inferior extent of the adjacent VB
Lateralised within the VB (S3–S5)	1	1, 3	Entire VB and the ipsilateral posterior ala. The superior and inferior extent of the CTV is determined by the superior and inferior extent of the adjacent VB
Diffusely involves the VB (S1–S2)	1	1, 2, 3, 7, 8	Entire VB and bilateral alae. When contouring the ala, use the ossification line if visible to limit the extent of the CTV. The superior and inferior extent of the CTV is determined by the superior and inferior extent of the adjacent VB
Diffusely involves the VB (S3–S5)	1	1, 3, 7	Entire VB, bilateral posterior ala. The superior and inferior extent of the CTV is determined by the superior and inferior extent of the adjacent VB
GTV involves VB and unilateral ala (S1–S2)	1, 2, 3	1, 2, 3, 4	Entire VB, ipsilateral ala, and ipsilateral lamina. The superior and inferior extent of the CTV is determined by the superior and inferior extent of the adjacent VB
GTV involves VB and unilateral ala (S3–S5)	1, 3	1, 3, 4	Entire VB, ipsilateral posterior ala, and ipsilateral lamina. The superior and inferior extent of the CTV is determined by the superior and inferior extent of the adjacent VB
GTV involves VB and bilateral ala (S1–S2)	1, 2, 3, 7, 8	1, 2, 3, 4, 6, 7, 8	Entire VB, bilateral alae, and bilateral laminae. The superior and inferior extent of the CTV is determined by the superior and inferior extent of the adjacent VB

(*Continued*)

Table 32.2 *(Continued)* **Recommendations for target volume definition of the sacrum in spinal stereotactic body radiation therapy/stereotactic radiosurgery (SBRT/SRS) (13)**

GTV INVOLVEMENT	SACRUM ANATOMIC MAP CLASSIFICATION	SACRUM BONY CTV RECOMMENDATION	CTV DESCRIPTION
GTV involves VB and bilateral ala (S3–S5)	1, 3, 7	1, 3, 4, 6, 7	Entire VB, bilateral posterior alae, and bilateral laminae. The superior and inferior extent of the CTV is determined by the superior and inferior extent of the adjacent VB
GTV involves the unilateral ala (S1–S2)	2, 3	2, 3, ± 1	Entire ipsilateral ala ± the entire adjacent VB. The superior and inferior extent of the CTV is determined by the superior and inferior extent of the adjacent VB†
GTV involves unilateral lamina	4	4, 5, ± 1	Ipsilateral lamina, spinous process ± VB
GTV involves bilateral laminae	4, 6	4, 5, 6, ± 1	Bilateral laminae, spinous process ± VB
GTV involves spinous process	5	4, 5, 6	Spinous process and bilateral laminae

†Except if the GTV involves any part of the S1 ala (see text). *Abbreviations:* VB, vertebral body.

treatment planning CT. Accuracy of fusion is paramount to minimize the level of uncertainty delineating this structure.

The TS is contoured as a surrogate for the cauda equina. Knowledge of the TS anatomy is key. Although the TS typically terminates at the level of S2, we would currently advise contouring the structure up to the inferior aspect of the S5 vertebral body as nerve rootlets are still passing within the bony canal distal to the TS termination. Without knowledge of their specific dose tolerance, it would be advisable to include them in the overall contour. Unlike the spinal cord, no PRV is added to the TS contour for planning. We also advise contouring the individual lumbo-sacral nerve roots for the lumbo sacral plexus as per the guidelines developed by Yi et al (28). Dose tolerance to nerve roots and the sacral plexus is an area of further research.

Thecal sac dose constraints vary between institutions which is unsurprising, given the paucity of data in the literature to guide practice. Current spine SBRT trial protocols have set the TS dose constraint at 16 Gy in 1 fraction, D5 cc ≤ 14 Gy (RTOG 0631) and D_{max} ≤17 Gy in 2 fractions (National Cancer Institute of Canada (NCIC) Canadian Clinical Trials Group (CCTG) Symptom Control (SC)-24 (SC-24) trial [NCT02512965]) and the sacral nerve roots (S1–S5) at 18 Gy in 1 fraction, D5 cc ≤14.4 Gy (RTOG 0631) and D_{max} of 26 Gy in 2 fractions: (NCIC CCTG SC.24) (29,30). The maximum point dose to the TS should be strictly adhered to and treatment planning adjusted so dose constraints are met within approximately 5%. As such it is acceptable to compromise PTV coverage to ensure this occurs. In the setting of reirradiation, the previous dose fractionation schedule, irradiation volume and time from previous radiotherapy need to be considered when considering the dose constraints for the TS and other OARs. The minimal time to reirradiation should be at least 5 months. Comprehensive modern dose limits to the spinal cord have recently been published by the HyTEC group and can be extrapolated to the TS as a surrogate for the spinal cord (31).

32.5 DOSE/FRACTIONATION

There remains no consensus as to the optimal dose/fractional schedule to use when treating metastatic disease to the spine in the *de novo* setting. As yet, there is no level 1 evidence to inform practice;

however, a phase III multicentre randomised study comparing 27 Gy in 3 fractions in 3 consecutive days with 24 Gy in a single fraction (NCT01223248) is due to complete in 2021 (32). Although this trial is not specific for metastatic disease to the spine, with disease to lymph nodes, soft tissue and bone also included among the numbers, it is hoped that the results may improve consistency in the treatment schedule used across the field. It is unlikely, however, that this trial will give any specific outcomes on metastatic disease to the sacrum, given the very small numbers of cases likely to be recruited. The study by Zeng et al, conducted at the Sunnybrook Health Sciences Centre of the University of Toronto, looked at imaged-based SBRT outcomes for sacral metastases. They used a median total dose of 24 Gy (range, 24–30 Gy) in 2 fractions (range, 2–5). In this study, the median percent volume of the CTV receiving 80% and 90% of the prescribed dose (V80 and V90) was 96% (range, 75%–100%) and 93% (range, 60%–100%), respectively. Outcomes were favourable with a cumulative local control of 86.5% and 78.7% at 1 and 2 years, respectively, and a median time to local recurrence of 4.7 months (12).

Typically disease to the sacrum is treated with the same dose/fractionation regimen as prescribed for the rest of the mobile spine. For complex large volume cases in the sacrum, it is reasonable to use lower doses such as 30 Gy in 4 fractions while adhering to published CNT constraints (12, 31, 33). However, recognising the inherent limitations of published studies in the literature, with the inclusion of few (if any) cases of metastatic disease to the sacrum within their data set, dose/fraction and CNT recommendations are based on clinical experience. The authors recommend that the individual physician use their own clinical judgement and expertise, taking into account factors particular to their own practice, when deciding on the appropriate schedule to use until more comprehensive data is available.

32.6 PATTERN OF FAILURE ANALYSIS

Pattern of failure studies have helped inform target volume delineation when treating metastatic disease to the spine with SBRT both in the *de novo* and post-operative setting (24, 25, 34–36). In these studies, very few (if any) cases of metastatic disease to the sacrum were included among the numbers, so unsurprisingly no outcomes specific to the sacrum have been reported on. Given the anatomic heterogeneity of the sacrum and the rarity of presentation at this site, achieving comprehensive failure data on sites of recurrence following SBRT would be a challenging undertaking. The one existing study with data on sacral recurrence following SBRT recorded eight sites of failure in a cohort of 22 patients and 37 treated sacral segments (12). The most common site of sacral failure was the progression within the vertebral body together with the paraspinal tissue (4/8, 50 %). This was followed by isolated epidural progression in two out of eight cases (25 %), failure in the paraspinal tissue alone in one case (12.5%) and a combination of epidural and vertebral body failure in one case (12.5%). Certainly more data is needed to help inform practice. The hope is that the consensus contouring recommendations to guide CTV delineation in the sacrum will help standardise target delineation at this site and will be used as a baseline for meaningful pattern of failure analysis hereon in (13).

32.7 TOXICITY

32.7.1 VERTEBRAL COMPRESSION FRACTURE

The most common late toxicity as a result of SBRT to the spine is radiation-induced vertebral compression fracture (VCF). This can be a debilitating consequence of radiation treatment resulting in pain, destabilization of the spine or neurological compromise. In certain cases, surgical intervention is required to manage this condition. Rates of VCF following SBRT have been reported as high as 40% in certain series, and it is thought that treatment with higher doses per fraction may increase the rate of this condition (37–40). None of these series however, specifically mentions the risk of VCF in cases of metastatic disease to the sacrum following SBRT. Referring back again to the Zeng et al. study, two cases of sacral VCF (out of 37 segments treated) at 4.3 months and 19.5 months were observed. However, both patients likely had unstable spine lesions prior to treatment as baseline SINS scores were 8 and 10 (12).

Table 32.3 **SBRT organ at risk (OAR) constraints for the sacral plexus (30)**

FRACTION NUMBER	2	3	4	5	8
Sacral Plexus	$D_{max} \leq 26$ Gy (+ nerve roots)	$D_{max} \leq 24$ Gy V22.5 Gy $\leq$ 5 cc	$D_{max} \leq 29$ Gy V27 Gy ≤ 5 cc	$D_{max} \leq 32$ Gy V30 Gy ≤ 5 cc	$D_{max} \leq 39$ Gy V36 Gy ≤ 5 cc

32.7.2 RADIATION-INDUCED LUMBOSACRAL PLEXOPATHY

The most commonly reported neurological complication of pelvic radiation is radiation-induced lumbosacral plexopathy (RILSP), a condition resulting in pain, paraesthesia, hypaesthesia, numbness, diminished reflexes or lower extremity weakness. RILSP is thought to occur as a result of late fibrotic changes which can cause progressive demyelination. The full pathophysiological mechanism of RILSP is still not fully understood however. MRI is a useful tool to differentiate RILSP from tumor involvement of the plexus; however, the presence of myokymic discharges on electomyography (EMG) is thought to be pathognomonic for RILSP, distinguishing it from neoplastic plexopathy (41–43). Due to limited studies in the literature, little is known about the effect of dose on the development of this condition, particularly in the context of SBRT. Extrapolating from studies on brachial plexopathy following treatment with conventional fractionation for breast cancer, it is likely that both fraction size and total dose increase the frequency of this condition (44). In a study specific to treatment with SBRT, Forquer et al. advised keeping the brachial plexus maximum dose to less than 26 Gy in 3 or 4 fractions (45). In the limited literature on sacral plexopathy, one single institution study reported 4 cases of sacral plexopathy out of 2140 treated with conventional external beam and intracavity radiation for cervical and endometrial cancer. The estimated dose to the plexus in this study was between 70 and 79 Gy. Other case series have additionally suggested that the tolerance of the lumbosacral plexus decreases to between 50 and 60 Gy in those receiving concomitant chemotherapy with radiation treatment (46–48). In the Zeng et al. study, only one case of lumbar–sacral plexopathy (out of 37 treated sacral segments) was recorded at 7.8 months following (*de novo*) treatment at a dose of 24 Gy in 2 fractions (EQD2/2 84Gy; EQD2/3 72 Gy) (12). The paucity of data in this area means that there exists no convincing dose threshold above which the incidence of RILSP is known to increase. Furthermore, the impact of an ablative dose of radiotherapy when treating with SBRT on the incidence of plexopathy is still relatively unexplored. Until further data is available, we would suggest adhering to dose tolerance as suggested by the Phase 3 NCIC CCTG SC.24 trial and Task Group 101 of the American Association of Physicists in Medicine (AAPM) (TG101) (Table 32.3) (30). Certainly, faithful and consistent contouring using published guidelines by Yi et al. (28) and collection of toxicity data going forward, including the long-term toxicity data from the SC24 trial, will improve our knowledge on this area.

32.8 CHALLENGES AND FUTURE DIRECTIONS

The optimal management of metastatic disease to the sacrum is complex; however, increasingly, SBRT is being used both in general practice and within the context of clinical trials with variable consistency in approach across institutions. As clinicians treating this area, it is imperative we understand and try and address areas of uncertainty associated with treating this anatomically complex region. Until level 1 randomised evidence is available to understand the clinical indications, patient selection and clinical outcomes, a multidisciplinary approach to treatment, with input from radiation oncologists, spine surgeons, medical oncologists and radiologists, remains paramount. The publication of international consensus contouring recommendations is a critical first step in achieving some standardization within this field. Further work however needs to be done on understanding the appropriate patients to treat, technical difficulties associated with delivering an ablative dose to this area and dose tolerance of adjacent normal tissues to ensure we can optimise treatment and outcomes for this challenging group of patients.

CHECKLIST: KEY POINTS FOR CLINICAL PRACTICE

ACTIVITY	SOME CONSIDERATIONS
Patient Selection	• Patient able to maintain stable position and tolerate immobilization for SBRT treatment • Spine MRI (<8 weeks prior to CT planning simulation) or CT myelogram* • Painful or radiologically progressive sacral metastasis with or without previous RT • *De novo* oligometastatic disease • Limited extra spinal systemic disease • No more than 2–3 consecutive sacral segments involved by tumor to be included in PTV • No more than 2 spine sites requiring treatment at the same time • Stable spinal column—calculate Spinal Instability Neoplastic Score (SINS score) • KPS >50
Simulation	*Immobilization* • Near rigid immobilization system capable of maintaining intrafraction motion <2–3 mm. For example a vacuum cushion with immobilization straps or whole-body vacuum bag system (e.g. BodyFIX) *Imaging* • Planning CT performed w/o contrast ≤1.25 mm slices • Volumetric T1 and T2 MRI for fusion with planning CT scan • Axial T2 for the evaluation of epidural disease • CT myelogram to evaluate epidural disease in postoperative patients with spinal hardware* **CT myelogram is not useful to delineate the thecal sac caudal to its termination at the level of S1/S2*
Treatment Planning	*Contours* • Utilize the consensus contouring guidelines by Dunne et al. (13) • PTV = CTV + 1–3 mm, minus thecal sac as per institutional preferences and standards • Dose prescribed to PTV, goal ≥80% of PTV and ≥90% of CTV covered by prescribed isodose line • Planning priority (1) thecal sac and (2) PTV coverage
Prescription Dose and Fractionation	• 24 Gy in 2 fractions daily • 30 Gy in 4 fractions daily • 35 Gy in 5 fractions daily

REFERENCES

1. Cheng JS, Song JK. Anatomy of the sacrum. *Neurosurgery Focus* 2003;15:1–4.
2. Cardoso HFV, Pereira V, Rios L. Chronology of fusion of the primary and secondary ossification centers in the human sacrum and age estimation in child and adolescent skeletons. *American Journal of Physical Anthropology*. 2014;153:214–225.
3. Broome DR, Hayman LA, Herrick RC, et al. Postnatal maturation of the sacrum and coccyx: MR imaging, helical CT, and conventional radiography. *American Journal of Roentgenology* 1998;170:1061.
4. Asdourian PL, Weidenbaum M, DeWald RL, Hammerberg KW, Ramsey RG. The pattern of vertebral involvement in metastatic vertebral breast cancer. *Clinical Orthopedics and Related Research*. 1990;250:164–170.

5. Fujita T, Ueda Y, Kawahara N, Baba H, Tomita K. Local spread of metastatic vertebral tumors: a histologic study. *Spine* 1997;22:1905–1912.

6. Langer R, Brem H, Falterman K, Klein M, Folkman J. Isolations of a cartilage factor that inhibits tumor neo-vascularization. *Science* 1976;193:70–72.

7. Park J, Lee J, Cho S, Park E, Daniel Riew K. A biochemical mechanism for resistance of intervertebral discs to metastatic cancer: Fas ligand produced by disc cells induces apoptotic cell death of cancer cells. *European Spine Journal.* 2007;16:1319–1324.

8. Feldenzer JA, McGauley JL, McGillicuddy JE. Sacral and presacral tumors: problems in diagnosis and management. *Neurosurgery* 1989;25:884.

9. Zhang Z, Hua Y, Li G, et al. Preliminary proposal for surgical classification of sacral tumors. *Journal of Neurosurgery Spine* 2010;13:651.

10. Fourney DR, Rhines LD, Hentschel SJ, et al. En bloc resection of primary sacral tumors: classification of surgical approaches and outcome. *Journal of Neurosurgery Spine* 2005;3:111–122.

11. Charest-Morin R, Fisher CG, Versteeg AL, et al. Clinical presentation, management and outcomes of sacral metastases: a multicenter, retrospective cohort study. *Annals of Translational Medicine.* 2019;7(10):214. doi:10.21037/atm.2019.04.88.

12. Zeng K, Myrehaug S, Soliman H, et al. Stereotactic body radiotherapy for spinal metastases at the extreme ends of the spine: imaging-based outcomes for cervical and sacral metastases. Neurosurgery Published Online Ahead of Print 2018 Aug 29. doi:10.1093/neuros/nyy393.

13. Dunne EM, Sahgal A, Lo SS, Bergman A, Kosztyla R, Dea N, et al. International consensus recommendations for target volume delineation specific to sacral metastases and spinal stereotactic body radiation therapy (SBRT).*Radiotherapy and Oncology* 2019 Dec 21;145:21–29. doi:10.1016/j.radonc.2019.11.026. [Epub ahead of print].

14. Laufer I, Rubin DG, Lis E, Cox BW, Stubblefield MD, Yamada Y, et al. The NOMS framework: approach to the treatment of spinal metastatic tumors. *The Oncologist* 2013 Jun;18(6):744–751.

15. Spratt DE, Beeler WH, de Moraes FY, Rhines LD, Gemmete JJ, Chaudhary N, et al. An integrated multidisciplinary algorithm for the management of spinal metastases: an international spine oncology consortium report. *Lancet Oncology* 2017 Dec 1;18(12):e720.

16. Ma L, Sahgal A, Hossain S, et al. Nonrandom intrafraction target motions and general strategy for correction of spine stereotactic body radiotherapy. *International Journal of Radiation Oncology* 2009;75(4):1261–1265. doi:10.1016/j.ijrobp.2009.04.027.

17. Li W, Sahgal A, Foote M, Millar B, Jaffray DA, Letourneau D. Impact of immobilization on intrafraction motion for spine stereotactic body radiotherapy using cone beam computed tomography. *International Journal of Radiation Oncology, Biology, Physics* 2012;84(2):520–526.

18. Dahele M, Slotman B, Verbakel W. Stereotactic body radiotherapy for spine and bony pelvis using flattening filter free volumetric modulated arc therapy, 6D cone-beam CT and simple positioning techniques: treatment time and patient stability. *Acta Oncologica*2016;55(6):795–798. doi:10.3109/0284186X.2015.1119885.

19. Leong B, Padilla L. Impact of use of optical surface imaging on initial patient setup for stereotactic body radiotherapy treatments. *Journal of Applied Clinical Medical Physics* 2019;20(12):149–158. doi:10.1002/acm2.12779.

20. Redmond KJ, Lo SS, Fisher C, Sahgal A. Postoperative stereotactic body radiation therapy (SBRT) for spine metastases: a critical review to guide practice. *International Journal of Radiation Oncology, Biology, Physics* 2016;95(5):1414–1428. doi:10.1016/j.ijrobp.2016.03.027.

21. Ma L, Sahgal A, Hossain S, et al. Nonrandom intrafraction target motions and general strategy for correction of spine stereotactic body radiotherapy. *International Journal of Radiation Oncology, Biology, Physics* 2009;75(4):1261–1265. doi:10.1016/j.ijrobp.2009.04.027.

22. Bertholet J, Knopf A, Eiben B, et al. Real-time intrafraction motion monitoring in external beam radiotherapy. *Physics in Medicine and Biology* 2019;64(15):15TR01. Published 2019 Aug 7. doi:10.1088/1361-6560/ab2ba8.

23. Heinzerling JH, Hampton CJ, Robinson M, Bright M, Moeller BJ, Ruiz J, Prabhu R, Burri SH, and Foster RD. Use of surface-guided radiation therapy in combination with IGRT for setup and intrafraction motion monitoring during stereotactic body radiation therapy treatments of the lung and abdomen. *Journal of Applied Clinical Medical Physics* 2020;21:48–55. doi:10.1002/acm2.12852.

24. Cox BW, Spratt DE, Lovelock M, Bilsky MH, Lis E, Ryu S, et al. International spine radiosurgery consortium consensus guidelines for target volume definition in spinal stereotactic radiosurgery. *International Journal of Radiation Oncology, Biology, Physics* 2012;83(5):e597–e605.

25. Redmond KJ, Robertson S, Lo SS, Soltys SG, Ryu S, McNutt T, et al. Consensus contouring guidelines for postoperative stereotactic body radiation therapy for metastatic solid tumor malignancies to the spine. *International Journal of Radiation Oncology, Biology, Physics* 2017 Jan;97(1):64–74.

26. Boriani S, Weinstein JN, Biagini R. Primary bone tumors of the spine: terminology and surgical staging. *Spine* 1997;22:1036–1044.

27. Whelan MA, Gold RP. Computed tomography of the sacrum: 1. Normal anatomy. *American Journal of Roentgenology* 1982;139:1183–1190.

28. Yi, Sun K. et al. Development of a standardized method for contouring the lumbosacral plexus: a preliminary dosimetric analysis of this organ at risk among 15 patients treated with intensity-modulated radiotherapy for lower gastrointestinal cancers and the incidence of radiation-induced lumbosacral plexopathy. *IJROBP* 2012;84(2):376–382.

29. Canadian Cancer Trials Group SC24. Study comparing stereotactic body radiotherapy vs Conventional Palliative Radiotherapy (CRT) for Spinal metastases. In:*ClinicalTrials.gov NLM Identifier:NCT02512965*. Bethesda, MD: National Library of Medicine; Web site. clinicaltrials.gov/ct2/show/NCT02512965.

30. Schaub SK, Tseng YD, Chang EL, et al. Strategies to mitigate toxicities from stereotactic body radiation therapy for spine metastases. *Neurosurgery* 2019;85(6):729–740. doi:10.1093/neuros/nyz213.

31. Sahgal A, Chang JH, Ma L, Marks LB, Milano MT, Medin P et al. Spinal cord dose tolerance to stereotactic body radiation therapy. *International Journal of Radiation Oncology, Biology, Physics* 2019 Oct 10. doi:10.1016/j.ijrobp.2019.09.038. [Epub ahead of print].

32. A Phase III randomized study comparing two dosing schedules for hypofractionated image guided radiation therapy in patients with metastatic cancer. National Library of Medicine (US). 2017. [Internet]. 2019 Dec 09 [cited 2019 Dec 09]. Available from: https://clinicaltrials.gov/show/NCT01223248.

33. Sahgal A, Weinberg V, Ma L, Chang E, Chao S, Muacevic A, et al. Probabilities of radiation myelopathy specific to stereotactic body radiation therapy to guide safe practice. *International Journal of Radiation Oncology, Biology, Physics* 2013 Feb 1;85(2):341.

34. Patel VB, Wegner RE, Heron DE, Flickinger JC, Gerszten P, Burton SA. Comparison of Whole versus partial vertebral body stereotactic body radiation therapy for spinal metastases. *Technology in Cancer Research & Treatment* 2012 Apr;11(2):105–115.

35. Ryu S, Rock J, Rosenblum M, Kim JH. Patterns of failure after single-dose radiosurgery for spinal metastasis. *Journal of Neurosurgery* 2004 Nov;101 Suppl 3:402–405.

36. Eric L Chang, Almon S Shiu, Ehud Mendel, Leni A Mathews, Anita Mahajan, Pamela K Allen, et al. Phase I/II study of stereotactic body radiotherapy for spinal metastasis and its pattern of failure. *Journal of Neurosurgery Spine* 2007 Aug 1;7(2):151–160.

37. Rose PS, Laufer I, Boland PJ, Hanover A, Bilsky et al. Risk of fracture after single fraction image-guided intensity-modulated radiation therapy to spinal metastases. *Journal of Clinical Oncology* 2009;27:5075–5079.

38. Cunha MV, Al-Omair A, Atenafu EG, Masucci GL, Letourneau D, Korol R, et al. Vertebral compression fracture (VCF) after spine stereotactic body radiation therapy (SBRT): analysis of predictive factors. *International Journal of Radiation Oncology, Biology, Physics* 2012;84:e343–e349.

39. Boehling NS, Grosshans DR, Allen PK, McAleer MF, Burton AW, Azeem S, et al. Vertebral compression fracture risk after stereotactic body radiotherapy for spinal metastases. *Journal of Neurosurgery Spine* 2012;16:379–386.

40. Sahgal A, Whyne CM, Ma L, Larson DA, Fehlings MG. Vertebral compression fracture after stereotactic body radiotherapy for spinal metastases. *Lancet Oncology* 2013 Jul 1;14(8):e320.

41. Harper CM, Thomas JE, Cascino TL, Litchy WJ. Distinction between neoplastic and radiation-induced brachial plexopathy, with emphasis on the role of EMG. *Neurology* 1989;39:502–506.

42. Delanian S, Lefaix JL, Pradat PF. Radiation-induced neuropathy in cancer survivors. *Radiotherapy and Oncology* 2012;105:273–383.

43. Dropcho EJ. Neurotoxicity of radiation therapy. *Neurology Clinical* 2010;28:217–234.

44. Olsen NK, Pfeiffer P, Johannsen L, et al. Radiation-induced brachial plexopathy: neurological follow-up in 161 recurrence-free breast cancer patients. *International Journal of Radiation Oncology, Biology, Physics* 1993;26:43e–49e.

45. Forquer JA, Fakiris AJ, Timmerman RD, et al. Brachial plexopathy from stereotactic body radiotherapy in early-stage NSCLC: dose-limiting toxicity in apical tumor sites. *Radiotherapy and Oncology: Journal of the European Society for Therapeutic Radiology and Oncology* 2009 Dec;93(3):408–413. doi:10.1016/j.radonc.2009.04.018.

46. Georgiou A, Grigsby PW, Perez CA. Radiation induced lumbosacral plexopathy in gynecologic tumors: clinical findings and dosimetric analysis. *International Journal of Radiation Oncology, Biology, Physics* 1993;26:479e–482e.

47. Badin S, Iqbal A, Sikder M, et al. Persistent pain in anal cancer survivors. *Journal of Cancer Survivorship*. 2008;2:79e83.

48. Frykholm GJ, Sintorn K, Montelius A, et al. Acute lumbosacral plexopathy during and after preoperative radiotherapy of rectaladeno carcinoma. *Radiotherapy and Oncology* 1996;38:121e–130e.

33 Radiomics and Its Role in Predicting Radiotherapy Response and Outcome in Brain Metastasis

Seyed Ali Jalalifar, Hany Soliman, Arjun Sahgal, and Ali Sadeghi-Naini

Contents

33.1 BRAIN METASTASIS

Brain metastases are the most common malignancy of the central nervous system, occurring in about 10% to 30% of all cancer patients [1]. The incidence of brain metastasis is increasing because of the improvement in control of systemic disease, better radiologic detection, and prolonged survival of cancer patients [2]. The estimated incidence of new brain metastases in the United States is between 7 and 14 per 100,000 persons based on population studies. The prevalence of metastases to the brain is expected to continue to increase in the future [2]. Autopsy and clinical data reveal that over 100,000 patients develop brain metastases annually in the United States [2].

Brain metastasis may present as single or multiple tumors. Clinical records indicate that approximately 29% of brain metastases arise as a single tumor, 35% as two to three tumors, and 36% as more than three tumors [3]. In the past, different subgroups of brain metastases suffered from similar poor prognosis but with advances in technology and therapeutic options, including stereotactic radiosurgery (SRS), the one-size-fits-all treatment paradigm has been changed to choosing appropriate therapy based on brain metastases subgroups and expected survival [4]. As the treatment outcomes are vastly different, an essential step in clinical trials is to classify patients based on measurable prognostic scores. Recursive partitioning analysis (RPA), a statistical method for multivariant analysis, was one the first methods used to prognosticate patients into separate classes based on age, performance status, control of primary, and extent of extracranial disease [5–7]. A major development at the time but overly simplistic, RPA is now replaced by more sophisticated classification methods such as diagnosis-specific graded prognostic assessment (DS-GPA) [8]. Proposed by Sperduto et al. [4], significant prognostic factors were used to define the DS-GPA with a GPA of 4.00 and 0.0 for the best and worst prognosis, respectively. The prognosis factors vary based on the diagnostic information regarding the primary cancer. For example, lung cancer prognostic factors included the Karnofsky performance score (an index which classifies patients based on their functional impairment [9]), age, presence of extracranial metastases, and number of brain metastases, in addition to the presence of epidermal growth factor receptor (EGFR) and anaplastic lymphoma kinase (ALK) mutations in patients with adenocarcinoma [10]. Similarly, for gastrointestinal cancers, prognostic factors included the Karnofsky performance score, age, extracranial metastases, and the number of brain metastases [11].

A population-based study by Cagney et al. has provided generalizable estimates of the incidence and prognosis for patients with brain metastases in the United States [12]. Out of 1,302,166 patients diagnosed with nonhematologic malignancies originating outside of the CNS between 2010 and 2013, 26,430 patients were identified with brain metastasis. Patients diagnosed with small cell lung cancer (SCLC) (16%) and lung adenocarcinoma (14%) had an incidence proportion of more than 10%. Patients with prostate cancer, bronchioloalveolar carcinoma, and breast cancer as the primary cancer had the longest median survivals (12, 10, and 10 months, respectively). Several studies by Sperduto et al. on patients diagnosed with brain metastasis have also reported overall median survival times for non-small cell lung cancer (NSCLC), SCLC, melanoma cancer, and breast cancer as 12, 5, 10, and 16 months respectively, where higher DS-GPA scores were associated with longer survival time, for example. as long as 4 years for NSCLC [10, 13, 14].

33.2 TREATMENT OPTIONS

Early diagnosis and precise treatment of the brain metastasis may lead to the reduction of brain symptoms and may enhance the quality of life and survival for patients [2]. Many factors, including the origin of cancer, associated symptoms, and size/number of metastases, are considered in planning a treatment strategy for the patient. Radiation therapy, chemotherapy, and surgery are the main treatment options for the management of metastatic brain tumors. Available options for radiotherapy include whole-brain radiation therapy (WBRT), hypofractionated stereotactic radiotherapy (SRT), and single-fraction stereotactic radiosurgery (SRS).

Important factors in deciding whether a patient should proceed with surgical resection of brain metastasis include the accessibility and size of the tumor, its relative proximity to eloquent brain areas, the degree of mass effect, patient's age, and the presence of other extracranial diseases [15]. Surgical resection is mainly recommended for patients with single brain metastasis in an accessible location when the tumor is large (>3–4 cm) and/or causing a considerable mass effect [16]. Neurosurgical intervention is also used in case a pathological diagnosis is required [17]. In addition, surgery is preferred in the presence of significant peri-tumoral edema to relieve and potentially reverse the associated neurological complications [18]. In most cases, surgical resection of the tumor is combined with adjuvant radiation therapy. For the treatment of a single metastatic tumor, a combination of surgical resection and postoperative radiotherapy has been shown to outperform radiotherapy alone [19]. In a randomized trial, Patchell et al. investigated whether postoperative radiotherapy can increase the survival and improve the neurological control of the disease [19]. Ninety-five patients participated in this study, and the outcome demonstrated that the recurrence of metastasis anywhere in the brain was less frequent in the radiotherapy group compared to the observation group (18% in radiotherapy group compared to 70% in the observation group). In a more recent randomized trial, Mahajan et al. investigated whether SRS after the surgical resection of brain metastasis improves the time to local recurrence [20]. One hundred thirty-two patients were randomly assigned after surgery to the observation (n = 68) or cavity SRS (n = 64). The findings of the study indicate that in patients who had complete resection of one, two, or three brain metastases (the majority of the study population had one metastasis), adjuvant SRS significantly decreased local recurrence compared to the observation cohort. A multicenter randomized controlled phase 3 trial by Brown et al. studied the efficacy of postoperative SRS compared to WBRT for resected brain metastasis [21]. One hundred ninety-four adult patients from 48 institutions with at least one resected brain metastasis of less than 5 cm in maximum diameter were randomly assigned to either postoperative SRS (n = 98) or WBRT (n = 96). Cognitive-deterioration-free survival was longer in patients assigned to SRS (median = 3.7) compared to WBRT (median = 3). The median survival rate was 12.2 months for the SRS and 11.6 months for WBRT cohorts, respectively.

For WBRT, the radiation dose is delivered to the whole brain in 5 to 20 fractions over a period of 1 to 4 weeks. Common fractionation schedules include 20/5, 30/10, 37.5/15, and 40/20. Conventionally, WBRT has been the treatment of choice for patients with multiple brain metastases [22]. WBRT, however, is associated with adverse side effects which typically include mild fatigue, mild dermatitis, temporary alopecia, otitis media, or externa [22]. Multiple clinical trials have shown cognitive deterioration after WBRT [21, 23, 24]. Apart from the detrimental effects of WBRT, some studies have also shown no benefit with respect to survival. Most recently, the QUARTZ trial in patients with non-small-cell lung cancer, brain metastasis, and poor performance status observed no survival advantage following WBRT versus best supportive care (median survival of 9.2 weeks for patients who receive optimal supportive care plus WBRT and 8.5 weeks for patients who only receive optimal supportive care) [25].

There has been a gradual shift in the past two decades away from WBRT to stereotactic radiosurgery (SRS), particularly in patients with limited brain metastases. SRS delivers a focused ablative radiation treatment with sub-millimeter precision to the tumor localized in three dimensions in a single fraction. This treatment modality is usually used as the sole treatment for patients who have less than four metastatic brain tumors [26]. SRS is demonstrated to have lower toxic effects over the non-tumor areas within the brain compared to WBRT [27]. However, several studies have shown that using WBRT reduces the risk of development of new brain metastases compared to SRS [27].

In some cases, due to the large size or location of the tumor applying the prescribed radiation dose in a single fraction is not recommended. In such cases, high-dose radiation is frequently delivered in very few fractions using the same precisely targeted method. This approach is known as hypofractioned stereotactic radiation therapy (SRT). Because of the number of SRT fractions that typically range from 3 to 5, a thermoplastic mask, or a more sophisticated noninvasive mask system, is used as the immobilization device, and the precision of therapy is coupled to image guidance when using such systems as compared to an invasive frame.

It should be noted that despite the evidence against WBRT, this treatment is still recommended in certain situations such as when patients present with innumerable metastases and/or diffuse leptomeningeal or pachymeningeal disease.

33.3 RESPONSE EVALUATION

In an effort to provide more standardization in clinical trials, the Response Assessment in Neuro-Oncology-Brain Metastases (RANO-BM) group has presented standard criteria for evaluating response of brain metastases to treatment [28]. The RANO-BM criteria are based on changes in the longest diameter of the target lesion to define its outcome after treatment as complete response, partial response, stable disease, or progressive disease (local failure). Complete response occurs when no target lesion remains. Partial response happens when there is more than a 30% decrease in the largest diameter of tumor compared to baseline. Stable disease occurs when there is less than a 30% decrease but no greater than a 20% increase in the longest diameter of the tumor, compared to baseline. If there is an increase in the longest diameter of the tumor more than 20%, this is considered a progressive disease.

Although SRS is usually well tolerated [29], there is a risk of radiation-related damage to the brain parenchyma. Adverse radiation effect (ARE) may result in a temporary or stable enlargement of lesion on serial imaging. Tumor progression and ARE usually have similar appearance on standard imaging [30]. In the case of patients who have been treated with SRS or immunotherapy-based approaches and there has been radiographical evidence of enlargement of target and non-target lesions, the RANO-BM group acknowledges that the enlargement does not necessarily represent tumor progression. If radiographical evidence of progression exists, but clinical evidence indicates that the radiological changes are due to radiation effect (instead of tumor progression), additional evidence is needed to distinguish between true progression and treatment effect, and the standard MRI alone is insufficient. For more details on the application of RANO-BM in different scenarios, the readers are encouraged to refer to the original paper [28].

While local response to stereotactic radiation treatment is evaluated based on changes in tumor size using serial imaging, it may take months before a local response is evident on follow-up images. Also, early changes in tumor size are not always correlated with long-term local control. Early prediction of local failure can potentially permit effective treatment adjustments and improved treatment outcome. In the following sections, the use of biomarkers in general, and radiomics in particular, will be discussed for therapy response prediction and monitoring.

33.4 RESPONSE PREDICTION

Despite ongoing improvements in cancer therapeutics, tumor response is variable despite similar cancer therapy. There are many patient-related factors contributing to therapy response including genetics, age, nutrition, health status, environmental exposure, and epigenetic factors [31]. Predicting efficacy of a treatment for an individual patient before, or early after, therapy initiation could lead to improvements in prognosis and clinical outcome.

Biological markers are any measurable indicators of the severity or presence of a disease. Biomarkers are formally defined as "a characteristic that is objectively measured and evaluated as an indicator of normal biological processes, pathogenic processes, or biological responses to a therapeutic intervention." [32] Different types of biomarkers can potentially provide beneficial diagnostic and prognostic information for therapy response prediction and, in combination with available therapeutic options, can pave the way for the paradigm of personalized medicine.

The methodologies for cancer therapy response prediction could be either invasive or noninvasive depending on the type of the applied biomarkers. Important invasive biomarkers of cancer therapy response include those obtained via biopsy or blood samples, including the histopathologic, molecular, genomic, and proteomic data, or circulating tumor DNA (ctDNA) [33–35]. Noninvasive biomarkers are often derived from images or raw data acquired using different modalities such as positron emission tomography (PET), magnetic resonance imaging (MRI), computed tomography (CT), diffused optical imaging (DOI), and ultrasound [36]. With the advancements in imaging technologies and improvements in standard equipment and protocols, development of methodologies to extract, standardize, and mine noninvasive biomarkers for cancer characterization and therapy outcome prediction has become the focus of many works in the literature [37–40].

33.4.1 INVASIVE BIOMARKERS OF THERAPY RESPONSE

Invasive response prediction involves the acquisition of biomarkers in an invasive manner. Invasive predictive biomarkers could be obtained through quantification of specific proteins or DNAs in the blood, identifying genetic signatures, immunohistochemistry, histopathology on tumor biopsy samples, etc. An example of invasive biomarkers for response prediction is the expression of the human epidermal growth factor receptor 2 (HER2/neu) receptor in breast cancer. HER2/neu is a member of the epidermal growth factor receptor (EGFR) family of transmembrane receptors. HER2/neu overexpression has been identified in 10 to34% of invasive breast cancers [41]. The expression of HER2/neu is assessed in breast tumors as a biomarker to administer anti-HER2-targeted drugs such as trastuzumab [42]. As another example, the expression of estrogen and progesterone receptors plays an important role in mediating the effects of therapeutic agents and serves as a biomarker for responsiveness to endocrine therapy in breast cancer [43].

It is now well-known that non-small cell lung cancer is not a single entity but is, in fact, various pathologies with unique molecular signature [44]. Since there are multiple subtypes of NSCLC, careful screening of predictive and prognostic biomarkers has been shown beneficial for the optimal management of NSCLC [45]. For example, in patients diagnosed with metastatic non-small cell lung cancer, patients with tumors harboring either EGFR exon 19 or 21 gene mutations or the echinoderm microtubule-associated protein-like 4-anaplastic lymphoma kinase (EML4-ALK) fusion protein can be offered oral targeted therapies against EGFR (gefitinib or erlotinib) or ALK (crizotinib), respectively, which are more effective than cytotoxic chemotherapy [45, 46].

In prostate cancer, the prostate-specific antigen (PSA) level in blood is considered as a diagnostic biomarker [47]. Blood and urine samples are also analyzed for predictive biomarkers [48]. Circulating tumor cells (CTC), tumor-educated platelets, and cell-free nucleic acid released from tumor in the bloodstream are examples of such biomarkers [48]. Among markers in blood, increased concentration of cell-free DNA (cfDNA) and increased number of CTC are associated with a worse prognosis of prostate cancer [49].

Urinary markers are either DNA-based, RNA-based, or protein-based markers [50]. Some examples are prostatic acid phosphatase which is a protein-based prognostic urinary biomarker associated with biochemical relapse after prostatectomy [51] and prostate cancer antigen 3 (PCA3) which is an RNA-based urinary predictive biomarker of tumor volume and positive surgical margin in prostatectomy specimens [52, 53].

Recent studies have suggested that the expression of glucose-regulated protein 94 (GRP94) and fibroblast growth factor-inducible 14 (FN14) proteins in breast cancer can be predictive biomarkers of developing brain metastasis [54]. A study by Darlix et al. has demonstrated that the serum level of MMP-9 and HER2-ECD can be considered as a predictive biomarker of brain metastasis in patients with breast cancer [55]. In lung cancer, the EGFR and Kirsten rat sarcoma viral oncogene homolog (KRAS) mutation at codon 12 and different chromosomal imbalances have been demonstrated to be correlated with the development of brain metastasis [56].

One limitation associated with invasive biomarkers is that they are acquired through, more or less, invasive approaches that in some cases are challenging or even technically not feasible. Further, processing technologies required for the genomic and proteomic biomarkers are not always accessible. Another limitation is that these biomarkers may not be robust as they may not represent the entirety of the tumor due to the heterogeneity and high degree of genomic diversity within the tumor [57].

33.4.2 QUANTITATIVE IMAGING BIOMARKERS

A quantitative imaging biomarker can be defined as "an objective characteristic derived from an in vivo image measured on a ratio or interval scale as indicators of normal biological processes, pathogenic processes, or a response to a therapeutic intervention [58]." The tumor volume measured based on volumetric CT or MRI could be considered as an imaging biomarker for response monitoring, for example, to describe a patient response to therapy. Tissue radioactivity concentration in positron emission tomography (PET) scans is another example of imaging biomarkers [58].

Different quantitative imaging biomarkers can be utilized for the diagnosis of abnormalities and tissue characterization, such as differentiation of benign versus malignant lesions in breast [59] or low- and high-grade malignancies in brain using various textural biomarkers derived from MRI [60]. As another example, lesion-to-normal background ratio in single proton emission computed tomography (SPECT) and PET images are useful in differentiating tumor recurrence from radiation necrosis [61]. A study by Hutter et al. shows that textural biomarkers derived from the proton magnetic resonance spectroscopy (MRS) can be utilized to distinguish between lesion and bacterial abscess in brain [62]. Research on MRS is exploring the possibility of differentiating tumor histology noninvasively. In a study by Tran et al., histopathological findings were predicted based on the characteristic proton spectra for five common adult supratentorial brain tumors: low-grade and anaplastic astrocytomas, glioblastoma multiforme (GBM), meningioma, and metastasis [63].

Other imaging modalities such as ultrasound have also been investigated for quantitative imaging of tissue characteristics. Breast lesion characterization using quantitative ultrasound (QUS) parametric maps has been studied in [64]. Specifically, QUS spectral-analysis techniques on raw radiofrequency (RF) data were applied with a sliding window analysis to generate parametric maps of mid-band fit (MBF), spectral slope (SS), spectral intercept (SI), spacing among scatterers (SAS), average scatterer diameter (ASD), and average acoustic concentration (AAC), in addition to the attenuation coefficient estimate (ACE) parameter. Textural analysis was subsequently applied on each QUS parametric map to derive imaging biomarkers such as mean, contrast, correlation, energy, and homogeneity. Using a hybrid biomarker, benign versus malignant lesions have been distinguished with a sensitivity of 96% and specificity of 84%. In another study, the efficacy of QUS spectral parametric imaging has been demonstrated to characterize the extent of the prostate cancer noninvasively [65]. The findings of that study demonstrate that the mid-band fit and 0-MHz intercept parametric images can show the presence of the disease with a good correlation to the whole-mount histopathology images (Figure 33.1).

Quantitative imaging biomarkers have also shown promise in prediction of response to anti-cancer therapies. Such predictive models of therapy response usually apply machine learning (ML) methods to differentiate between different types of response, for example, complete response, partial response, and no response, using imaging biomarkers.

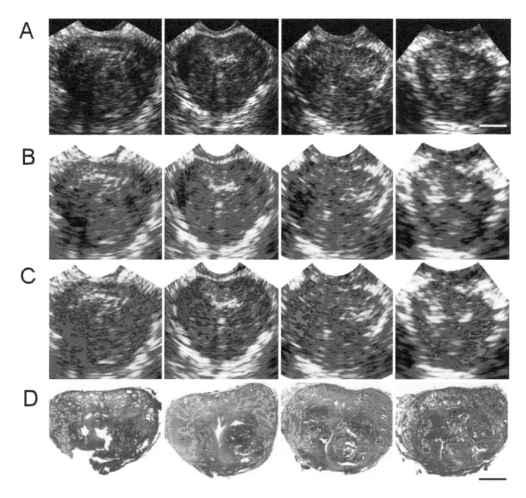

Figure 33.1 QUS spectral parametric imaging to characterize the extent of prostate cancer [65]. (A) Ultrasound B-mode image. (B) The corresponding MBF parametric image. (C) The MBF parametric images with identified areas of putative disease segmented over the images. (D) The whole-mount histopathology slides of the prostatectomy specimens. Areas outlined in green and orange indicate tumor and hyperplastic areas of abnormality, respectively.

A study by Tadayyon et al. demonstrated that QUS parameters can be used to predict tumor response to neoadjuvant chemotherapy (NAC) in patients with locally advanced breast cancer (LABC) early after therapy initiation [66]. In that study, QUS parameters including MBF, SS, SI, SAS, ACE, ASD, and AAC were derived from tumor regions using the ultrasound RF data and processed through a feature selection procedure. The selected features were applied with a binary classifier for chemotherapy response prediction where an accuracy of 70 9% and 80 5% were obtained at weeks 1 and 4 after the start of treatment, respectively. Another recent study demonstrates that changes in cancerous cell nuclei and tumor microstructure in response to chemotherapy in breast cancer patients are initially evident as alternations in textural characteristics of QUS spectral parametric maps, followed by consequent changes in their mean values [67]. On this basis, by incorporating textural biomarkers of QUS spectral parametric maps, their model could differentiate between responding and non-responding patients to chemotherapy with up to 100% sensitivity and 93% specificity at week 1 after treatment initiation. Other imaging modalities such as diffuse optical imaging (DOI) have also shown promise in quantifying heterogeneous changes in tumor metabolism in response to chemotherapy. In responding and non-responding breast cancer patients, different changes in DOI-based textural and mean-value parameters have been observed after the start of chemotherapy [68]. The cross-validated sensitivities and specificities of the predictive models that applied these quantitative biomarkers for response prediction ranged between 80% and 100% at week 1 and 4 after the start of the treatment [68].

The efficacy of new imaging methods such as chemical exchange saturation transfer (CEST) has recently been evaluated for therapy outcome prediction. Desmond et al. investigated quantitative CEST metrics for predicting local response in brain metastases treated with stereotactic radiosurgery (SRS) [69]. CEST spectra were collected from 25 patients, and the amide proton transfer-weighted images as well as the maps of the amplitude and width of Lorentzian-shaped CEST peaks and the relaxation-compensated AREX metric were constructed. The findings of the paper show that CEST metrics, particularly the nuclear overhauser effect (NOE) peak amplitude, can predict changes in tumor volume 1 month after SRS. Mehrabian et al. have studied the cellular and metabolic characteristics of glioblastoma over the course of standard 6-week chemoradiation treatment with CEST-MRI [70]. The results showed that changes in magnetization transfer ratio (MTR) of NOE at day 14 after the treatment compared to the baseline were significantly different between the non-progressors and progressors. In another study, Mehrabian et al. have investigated the CEST metrics to distinguish between the true progression and radiation necrosis in brain metastasis [71]. The findings of that study demonstrate that the CEST parameters including MTRs of NOE and amide are capable of differentiating radiation necrosis from tumor progression.

Kapdia et al. have evaluated intravoxel incoherent motion (IVIM) imaging against dynamic contrast-enhanced MRI (DCE-MRI) to measure the blood volume within the brain metastasis [72]. Using the data acquired from 20 tumors, they concluded that blood volumes measured with IVIM and DCE-MRI are not equivalent. Mehrabian et al. designed a study to evaluate whether changes in metastatic brain tumors after SRS can be detected with quantitative MRI early after treatment [73]. A three-water-compartment tissue model consisting of intracellular (I), extracellular-extravascular (E), and vascular (V) was used to assess the intra-extracellular water exchange rate constant (kIE), efflux rate constant (kep), and water compartment volume fractions ($M_{0,I}$, $M_{0,E}$, $M_{0,V}$). The results show that early changes in kIE are highly correlated with long-term tumor response.

Quantitative imaging methodologies have been shown effective in a variety of clinical applications and specially in cancer therapeutics. Research is ongoing to develop new methods of data acquisition, biomarker extraction, and multi-modal analysis to be used in robust models for various diagnostic, tissue characterization, and prognostic applications.

33.4.3 RADIOMICS

Radiomics is an emerging translational field of research concerned with the high-throughput mining of high-dimensional medical imaging data to discover quantitative diagnostic and prognostic biomarkers [74]. The quantitative features extracted from the medical imaging data for biomarker discovery in radiomic analysis are usually categorized as [75]:

- **Morphological features** that quantify the geometry and shape of the region of interest, for example, tumor. Some examples of morphological features include volume, maximum diameter, surface area, elongation, and sphericity.
- **First-order statistical features** that describe the voxel intensities without considering the spatial relationship between them, for example mean intensity, standard deviation, energy, and entropy,
- **Second-order statistical features** also known as textural features are obtained by calculating the statistical inter-relationships between the intensity of neighboring voxels. One example of such features are the ones derived from gray-level co-occurrence matrix (GLCM), gray-level run length matrix (GRLM), and gray-level size zone matrix (GLSZM) [76].
- **Higher-order statistic features** that are statistical features obtained after applying a transformation on the image, for example features extracted from textural parametric images.

In radiomic analysis, hundreds of quantitative features are extracted from an image and its transformed versions within different regions of interests including the intra- and peri-tumoral regions. Data mining is performed in conjunction with advanced statistical analyses on these high-dimensional data under the hypothesis that an appropriate low-dimensional combination of these features, possibly along with clinical data, can represent crucial tissue properties, useful for diagnosis, prognosis, or treatment planning of individual patients [75]. Since many of the extracted radiomic features are redundant, initial steps of the radiomic analysis usually include dimensionality reduction and feature selection to identify the best single and hybrid features as candidates for subsequent rigorous evaluations.

After identifying the most informative radiomic features, for example imaging biomarkers, the next step is to analyze their relationship with diagnosis, prognosis, or therapy outcome. In this step, machine learning (ML) plays a major role as it is a powerful tool for data analytics in classification problems such as those mentioned above. A number of common ML methods include Random Forest [77], Support Vector Machines [78], and K-nearest neighbor model [79]. Random Forest is one of the most common classification techniques which involves training a combination of trees on randomly sampled vectors. Random Forest is resistant to overfitting and has the power to handle a large dataset with high dimensionality [77]. Support vector machines try to find the best hyperplane which segregates members of different classes as best as possible [80]. K-nearest neighbor assigns a class to a sample based on the class of the nearest samples in previously available data [81].

33.4.3.1 Radiomics and Tumor Biology

Recent studies demonstrate correlations between the radiomic signature of tumors and their phenotypes and genomic and proteomic profiles. Aerts et al. have presented a radiomic analysis of 440 features quantifying the intensity, shape, and texture of tumors in CT imaging data of 1,019 patients with lung or head-and-neck cancer [82]. The results of that study suggest that radiomics are capable of identifying a general prognostic phenotype existing in both lung and head-and-neck cancer and that these features have prognostic power in independent datasets of lung and head-and-neck cancer patients. Extracted radiomic features from CT images of the lung and head and neck tumors also show correlations between radiomic features and underlying gene-expression patterns [82].

The interest to study the link between the imaging phenotypes and the tumor genetic profiles has led to the emergence of a new field of research known as radiogenomics [83]. Diehn et al. investigated the association of GBM neuroimaging features with microarray DNA data in order to map the gene expression within the tumor noninvasively [84]. The volumetric imaging features have been shown associated with PERIOSTIN expression, a gene linked with decreased survival, shorter time of recurrence, and the mesenchymal GBM subtype [85]. Tumor shape [86, 87] and tumor-associated edema on T2-FLAIR [88] have also been reported as possible imaging biomarkers of molecular subtype in GBM. Panth et al. have studied the casual relationship between genetic changes and radiomic features of doxycycline inducible GADD34 tumor cells [89]. Radiomic analysis was performed on CT images of the tumors grown in mice. The results demonstrate that gene induction and/or irradiation are translated into significant changes in radiomics features [89].

Considering the potential association of the radiomic features with tissue/tumor phenotypes and their genomic and proteomic profiles, several studies have explored the potential of radiomic features in conjunction with ML methods to differentiate between different tissue types and to predict or monitor the outcome of cancer-targeting therapies. Parekh et al. developed an MRI radiomic framework to describe the breast tumor biology. Their proposed model could differentiate between the benign versus malignant lesions with a sensitivity of 0.93 and a specificity of 0.85 [90]. Bickelhaupt et al. conducted a study to assess radiomics as a tool to classify suspicious lesions found in screening X-ray mammography into benign and malignant [91]. Seven hundred and twenty-five radiomic features were extracted from contrast agent-free diffusion MRI for each patient and out of those, 11 features were selected for classification. Their model could achieve an area under curve of 84% using a Lasso-supervised machine-learning classifier. A study by Morales et al. demonstrates that the peri- and intra-tumoral radiomic features extracted from low-dose CT screening images can identify aggressive early-stage lung cancers and be used to predict patient survival [92]. Statistical root mean square, one of the two radiomic features applied to stratify patients into different risk-groups, was shown correlated with the occurrence of the genes LOC285403 and FOXF2, with the latter associated with poor outcomes in non-small cell lung cancer [92]. The second identified feature was the neighborhood gray tone difference matrix (NGTDM) busyness that is a measure of the speed of intensity change from a pixel to its neighbor [92].

33.5 APPLICATION OF RADIOMICS IN CANCER MANAGEMENT

In the previous section, a formal definition of radiomics was presented, and its link to the tumor genetic profile was studied. In this section, the application of radiomics in cancer diagnostic, treatment, and response will be discussed. In particular, we will discuss the application of radiomics in brain metastasis diagnosis and prognosis.

Multiple studies show that therapy response of patients can be predicted prior to or early after the treatment using imaging biomarkers derived from radiomic analysis. Predicting the outcome of standard treatments in the early days of diagnosis is of extreme importance because it is a key to the personalized medicine paradigm. Response monitoring using radiomics is also highly desirable and can facilitate radiomics-guided therapies [93].

Radiomics has recently been investigated for cancer diagnosis, therapy response/outcome prediction, and monitoring in various types of malignancies [94–96] for outcome assessment and prediction in many fields of research. Numerous studies have shown a high potential of radiomics methods to be adapted for lung cancer management. In a study by Huynh et al., a combination of two conventional tumor features (volume and diameter) and two CT radiomic features has been demonstrated to be correlated to the overall survival of stage I–II NSCLC patients treated with SBRT [97]. The study also shows that, unlike any of the conventional and clinical parameters, one radiomic feature (wavelet-LLH-stats-range) has a significant prognostic power for prediction of distant metastasis (concordance index = 0.67, q-value <0.1). Li et al. have investigated the association of radiomic features extracted from pre-treatment planning CT scans with the overall survival (OS), recurrence-free survival (RFS), and loco-regional recurrence-free survival (LR-RFS) of NSCLC patients treated with stereotactic body radiotherapy [98]. The results show that the Harrell's C-index is improved by adding the radiomic features to the clinical models for survival prediction. Harrell's C-index is a measure of goodness of fit for models that produce risk score [99]. It is commonly used to evaluate risk models in survival analysis where data may be censored.

Li et al. have studied the efficacy of contrast-enhanced and non-contrast CT textural features for patient risk stratification beyond conventional prognostic factors (CPFs) in stage III NSCLC [100]. Their findings show that the pre-treatment tumor texture can provide prognostic information beyond that obtained from the CPFs. Bak et al. studied the prognostic power of biomarkers reflecting longitudinal change of radiomic features. A model for predicting survival was developed with multivariate cox regression. The areas under the receiver operating characteristic (ROC) curve (AUC) obtained for the predictive models were 0.948 and 0.862 for the outcomes determined at 1 and 3 years of follow-up, respectively. Their study suggests that longitudinal change of radiomic tumor features may serve as prognostic biomarkers in patients with advanced NSCLC [101].

Radiomics have also shown promise in breast cancer diagnosis, monitoring its progression, and assessment of its response to treatment [102]. Several research studies have investigated the effectiveness and reliability of radiomics in distinguishing breast malignancies from benign lesions or normal parenchyma [102–104]. Generally, studies show that incorporating radiomic features to standard radiological workflow can improve the diagnostic power of breast imaging [105]. Among radiomic features extracted from breast MRI, the intra-lesion entropy in describing the randomness and uncertainty among the pixels have been shown to have a considerable diagnostic power to detect malignancies. Specifically, the MRI entropy values are significantly higher in malignant tumors compared to the benign lesions due to the higher levels of intra-lesion heterogeneity in malignancies [90]. In a study by Whitney et al. on a large clinical breast MRI dataset, a set of two radiomic features (irregularity and entropy) extracted from dynamic contrast-enhanced MRI could significantly improve the ability to distinguish between benign lesions and luminal A breast cancers, compared to using maximum linear size alone. [106]. Two other studies have shown that MRI-based radiomic models are able to predict breast cancer metastasis to axillary lymph nodes [107] [108], implying the clinical utility of such models in diagnostic and prognostic evaluation of breast cancer patients. Another study by Guo et al. demonstrates high correlation of radiomic features extracted from ultrasound images of invasive ductal carcinoma (IDC) tumors with sentinel lymph node metastases [109].

A relevant study has shown the potential of deep-learning methodologies that derive quantitative feature maps from the mammography images automatically in predicting the likelihood of breast cancer development up to 5 years in advance [110]. The developed model has been reported to have a higher prediction accuracy (AUC = 0.7) compared to the Tyrer-Cuzick model (AUC = 0.62) [111]. The Tyrer-Cuzick model estimates the likelihood of a woman developing breast cancer in 10 years by calculating the likelihood of carrying the breast cancer gene 1 (BRCA1) or BRCA2 mutations.

A challenge for characterizing malignancies such as GBM using biopsy specimens alone is the heterogeneity within such tumors. Because of the intra-tumor heterogeneity, a local specimen often cannot reflect the characteristics of other parts of the tumor accurately [112]. Radiomics using volumetric MRI can provide key insight into critical features of glioblastoma, the most common astrocytic primary brain malignancy, with an annual incidence of 2 to 3 cases per 100,000 adults in North America and Europe [113], [114]. In a recent study, three distinct subtypes of GBM with different clinical and molecular characteristics (isocitrate dehydrogenase-1 (IDH1), O6-methylguanine–DNA methyltransferase, epidermal growth factor receptor variant III (EGFRvIII), and transcriptomic subtype composition) were revealed using MRI radiomic signatures, demonstrating the prognostic value of these features [115]. In another study, Lu et al. have proposed a model to classify GBM from lower-grade gliomas using MRI radiomics and a three-level machine learning model [116]. Their model could classify IDH and 1p/19q status of gliomas with AUC of 0.92% and accuracy of 87.7%. Li et al. have studied the effect of image standardization parameters on reproducibility and prognostic performance of a fully automatic radiomics model for predicting the overall survival in patients diagnosed with GBM [117]. They have showed that the voxel size, quantization method, and gray level may have an influence on reproducibility and prognostic performance of radiomic features for GBM overall survival prediction. Beig et al. have proposed a method to create survival risk-score using radiomic features of the tumor habitats on standard MRI to predict progression-free survival (PFS) in GBM and also obtain a biological basis for these prognostic radiomic features [118]. In that study, a radiogenomic analysis has revealed associations of MRI radiomic features with signaling pathways for cell differentiation, cell adhesion, and angiogenesis, which contribute to chemo-resistance in GBM [102].

33.5.1 POTENTIAL ROLE OF RADIOMICS IN MANAGEMENT OF BRAIN METASTASIS

Given the abundance of MRI and CT data acquired as part of standard of care for brain metastasis patients, radiomics has a good potential to be adopted in clinics to provide an efficient and noninvasive method of characterizing metastatic brain tumors and predicting the outcome of their treatment. Similar to primary brain tumors, brain metastases are heterogeneous within the tumor. Further, the genetic profiles within the surgically biopsied brain metastasis may differ from other distant metastases in the brain [119]. For example, these unbiopsied regions may demonstrate clinically relevant sensitivity to targeted therapies [120]. In such cases, radiomics can be adapted as a clinically viable option, offering insight on potentially unique targeted drugs for the residual unresected tumor segments [121].

Up to 20% of metastatic brain tumors progress despite radiation treatment, and it may take up to a couple of months until this progression is evident on follow-up imaging [122]. Karami et al. have recently proposed an MRI-based radiomic framework for early prediction of treatment outcome in brain metastasis patients treated with hypofractionated stereotactic radiation therapy (SRT) [122]. Quantitative MRI (qMRI) biomarkers were constructed through a multistep feature extraction/reduction/selection framework and later were fed to an SVM classifier to predict the outcome in terms of local control (LC) or local failure (LF). Figure 33.2 shows the scheme of their radiomic-based outcome prediction framework. Among various geometrical and textural features extracted within the tumor and edema regions and the corresponding margins from contrast-enhanced T1w and T2-FLAIR images, the optimal qMRI biomarkers with most prognostic power were the ones extracted from edema, tumor margin, and lesion margin rather than the tumor itself. Figure 33.3 shows the spatial variations in five selected MRI radiomic features acquired at the baseline and the first follow-up for representative LC and LF tumors. The qMRI biomarker consisting of these five features could predict the outcome of LF with an AUC of 0.79 and a cross-validated sensitivity and specificity of 81% and 79%, respectively. Risk assessment analyses demonstrated that the patients whose tumors were predicted as LC had a significantly better rate of survival, compared to those predicted as LF.

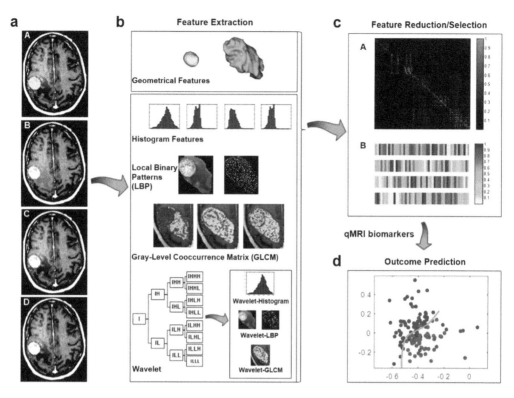

Figure 33.2 Scheme of the radiomics-based outcome prediction framework in [122]. (a) The binary masks including the tumor delineated by expert oncologists (A), edema segmented semi-automatically (B), tumor-margin (C), and the lesion-margin (D). (b) Extracting the geometrical and textural features from T1w and T2-FLAIR images within the binary masks. (c) Correlation-based feature reduction (A) and multi-step feature selection (B). (d) Outcome prediction (LC versus LF) using the SVM classifier.

A recent study by Mouraviev et al. has investigated whether MRI radiomic features provide any additional value to clinical variables for predicting local control in brain metastasis following SRS [123]. Four hundred and forty radiomic features were extracted from the tumor core and the peri-tumoral regions on pre-treatment contrast-enhanced T1w and T2-FLAIR images. Top radiomic features were selected based on resampled random forest feature importance. The results show that the addition of any one of the top 10 radiomic features to the set of clinical features increases the area under the ROC curve considerably.

In 30% of patients with metastatic brain tumors, the diagnosis of brain metastasis is before the diagnosis of the primary site of cancer [124]. For these patients, fast identification of the primary site of cancer is crucial for therapy planning and prognosis [125]. Kniep et al. have recently investigated the feasibility of brain tumor histology classification in patients with unknown primary lesion at the time of diagnosis using MRI radiomic features and a multiclass machine learning approach [126]. The dataset used in this research included metastases originated from breast cancer, small cell lung cancer, NSCLC, gastrointestinal cancer, and melanoma. A total of 1,423 quantitative image features and basic clinical parameters were evaluated using Random Forest machine learning algorithm. Among those features, 59 were ranked as most important features by Gini impurity measures [127], including 29 first-order features and 20 texture features. The areas under the ROC curve of the five-class problem ranged between 0.64 (for non-small cell lung cancer) and 0.82 (for melanoma). The prediction performance of the classifier was superior to the radiologist readings.

Chen et al. developed and evaluated an unenhanced CT-based radiomic model to predict brain metastasis (BM) in patients with category T1 lung adenocarcinoma [128]. A total of 89 eligible patients with category T1 lung adenocarcinoma were enrolled and classified as patients with BM ($n = 35$) or patients without

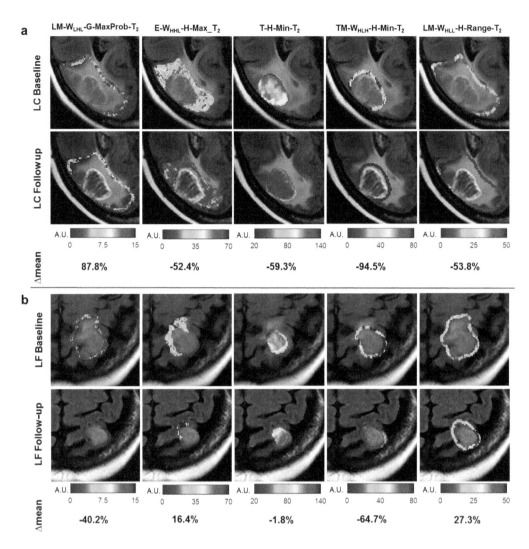

Figure 33.3 Representative parametric maps of the features in the optimal qMRI biomarker for the LC/LF outcome in [122]. The parametric maps show the spatial variations in the features derived from the MR images acquired at the baseline and the first follow-up for representative tumors with LC (a) and LF (b) outcomes. The mean relative change from the baseline at the first follow-up (Δ mean) is given for each feature.

BM ($n = 54$). A total of 1160 quantitative radiomic features were extracted from unenhanced thoracic CT images of each patient. Three prediction models including a clinical model, a radiomics model, and a hybrid (clinical plus radiomics) model were investigated. A 10-fold cross-validation was used to evaluate the prediction performance of the models. In terms of predictive performance for the hybrid model, a sensitivity and specificity of 82.9% and 83.3% was obtained, respectively, compared to a sensitivity and specificity of 82.9% and 57.4% for the clinical model and 80.0% and 81.5% for the radiomic model, respectively.

Although patients diagnosed with melanoma brain metastasis suffer from poor prognosis, recent advances in treatments with immune checkpoint inhibitors (ICIs) have resulted in durable responses in some patients [129]. In a recent study, Bhatia et al. have demonstrated that high-order MRI radiomic features in melanoma brain metastasis treated with ICIs can stratify patients in terms of overall survival [130]. Specifically, the Laplacian of Gaussian edge features could best differentiate the variation in outcome with a p-value of 0.001.

Contrast-enhanced MRI often remains inconclusive for differentiating between recurrent brain metastasis and radiation injury [131]. To tackle this problem, Lohmann et al. have proposed a combined

FET–PET/MRI radiomic model [132]. Forty-two textural features were calculated on both original and filtered CE-MRI and summed FET–PET images using GLCM, NGLDM, and GLRLM methods. After feature selection, logistic regression models were developed using a maximum of five features to avoid overfitting. The models were constructed for each imaging modality separately and for the combined FET–PET/MRI features. The diagnostic accuracy of the hybrid model in differentiating the radiation injury from recurrent brain metastasis was 89% with a sensitivity of 85% and a specificity of 96%. For the model trained on FET–PET, the accuracy was 83% with a sensitivity of 88% and specificity of 75%, while for the model trained on CE-MRI the accuracy, sensitivity, and specificity were 81%, 67%, 90%, respectively.

In this chapter, the applications of radiomics in diagnosis, prognosis, and treatment planning of cancer in general and brain metastases in particular were discussed. There is a large body of research available supporting the efficacy of radiomics in cancer therapeutics. Advancements in imaging technologies and data-driven algorithms creates opportunities to develop and assess standardized radiomic-based platforms in neuro-oncology that can facilitate effective personalized treatment for patients diagnosed with brain metastasis.

33.5.2 FINAL REMARKS

Radiomics have shown promising potential in improving cancer diagnosis and prognosis alone or in combination with other clinical information. In the case of brain metastasis, tumor biopsies usually cannot characterize the entirety of tumor due to intra-tumor heterogeneity. Radiomics could potentially provide invaluable information to characterize a tumor in terms of histology, phenotype, genomic profile, and radiation-resistance. Further, radiomics can potentially be adapted to predict treatment outcomes and differentiate between tumor progression and treatment effects.

Having said that, there are some challenges that must be addressed before the integration of radiomics in clinical decision-making. Many radiomic-based models do not provide any probability for their prediction, whereas likelihood plays an important role in the clinic. For example, characterizing a tumor with 55% likelihood of EGFR amplification results in different clinical implications compared to a tumor with 99% likelihood of EGFR amplification [111]. Another challenge is that the radiomic features should be biologically interpretable, since a purely black-box approach is not adoptable in clinic for diagnosis, prognosis, and altering standard treatments [133]. Radiomics shows promising results and strong link with underlaying tissue characteristics, but further work needs to be done to improve its interpretability and demonstrate its robust performance on multi-institutional data acquired from larger cohorts of patients.

REFERENCES

[1]   P. Y. Wen, P. M. Black, and J. S. Loeffler, "Metastatic brain cancer," in *Cancer: Principles and Practice of Oncology*, V. DeVita, S. Hellman, and S. A. Rosenberg, Eds. Lippincott Williams & Wilkins, Philadelphia, 2001. p. 2655–2670.

[2]   B. D. Fox, V. J. Cheung, A. J. Patel, D. Suki, and G. Rao, "Epidemiology of metastatic brain tumors," *Neurosurg. Clin. N. Am.*, vol. 22, no. 1, pp. 1–6, Jan. 2011, doi: 10.1016/j.nec.2010.08.007.

[3]   C. Nieder, O. Spanne, M. P. Mehta, A. L. Grosu, and H. Geinitz, "Presentation, patterns of care, and survival in patients with brain metastases," *Cancer*, vol. 117, no. 11, pp. 2505–2512, Jun. 2011, doi: 10.1002/cncr.25707.

[4]   P. W. Sperduto et al., "Summary report on the graded prognostic assessment: an accurate and facile diagnosis-specific tool to estimate survival for patients with brain metastases," *J. Clin. Oncol.*, vol. 30, no. 4, pp. 419–425, Feb. 2012, doi: 10.1200/JCO.2011.38.0527.

[5]   E. Y. Saito et al., "Whole brain radiation therapy in management of brain metastasis: results and prognostic factors," *Radiat. Oncol.*, vol. 1, p. 20, Jun. 2006, doi: 10.1186/1748-717X-1-20.

[6]   C. Nieder and M. P. Mehta, "Prognostic indices for brain metastases—usefulness and challenges," *Radiat. Oncol.*, vol. 4, no. 1, p. 10, Dec. 2009, doi: 10.1186/1748-717X-4-10.

[7]   C. Nieder, U. Nestle, B. Motaref, K. Walter, M. Niewald, and K. Schnabel, "Prognostic factors in brain metastases: should patients be selected for aggressive treatment according to recursive partitioning analysis (RPA) classes?" *Int. J. Radiat. Oncol.*, vol. 46, no. 2, pp. 297–302, Jan. 2000, doi: 10.1016/S0360-3016(99)00416-2.

[8]   H. Soliman, S. Das, D. A. Larson, and A. Sahgal, "Stereotactic radiosurgery (SRS) in the modern management of patients with brain metastases," *Oncotarget*, vol. 7, no. 11, pp. 12318–12330, Mar. 2016, doi: 10.18632/oncotarget.7131.

[9]   J. W. Yates, B. Chalmer, and F. P. McKegney, "Evaluation of patients with advanced cancer using the karnofsky performance status," *Cancer*, vol. 45, no. 8, pp. 2220–2224, Apr. 1980, doi: 10.1002/1097-0142(19800415)45:8<2220::AID-CNCR2820450835>3.0.CO;2-Q.

[10]   P. W. Sperduto *et al.*, "Estimating survival in patients with lung cancer and brain metastases," *JAMA Oncol.*, vol. 3, no. 6, p. 827, Jun. 2017, doi: 10.1001/jamaoncol.2016.3834.

[11]   P. W. Sperduto *et al.*, "Estimating survival in patients with gastrointestinal cancers and brain metastases: an update of the graded prognostic assessment for gastrointestinal cancers (GI-GPA)," *Clin. Transl. Radiat. Oncol.*, vol. 18, pp. 39–45, Sep. 2019, doi: 10.1016/j.ctro.2019.06.007.

[12]   D. N. Cagney *et al.*, "Incidence and prognosis of patients with brain metastases at diagnosis of systemic malignancy: a population-based study," *Neuro. Oncol.*, vol. 19, no. 11, pp. 1511–1521, Oct. 2017, doi: 10.1093/neuonc/nox077.

[13]   P. W. Sperduto *et al.*, "Estimating survival in melanoma patients with brain metastases: an update of the graded prognostic assessment for melanoma using molecular markers (Melanoma-molGPA)," *Int. J. Radiat. Oncol.*, vol. 99, no. 4, pp. 812–816, Nov. 2017, doi: 10.1016/j.ijrobp.2017.06.2454.

[14]   P. W. Sperduto *et al.*, "Beyond an updated graded prognostic assessment (Breast GPA): a prognostic index and trends in treatment and survival in breast cancer brain metastases from 1985 to today," *Int. J. Radiat. Oncol.*, vol. 107, no. 2, pp. 334–343, Jun. 2020, doi: 10.1016/j.ijrobp.2020.01.051.

[15]   T. K. Owonikoko *et al.*, "Current approaches to the treatment of metastatic brain tumours," vol. 11, no. 4, pp. 203–222, 2014, doi: 10.1038/nrclinonc.2014.25.Current.

[16]   C. M. Carapella, N. Gorgoglione, and P. A. Oppido, "The role of surgical resection in patients with brain metastases," *Curr. Opin. Oncol.*, vol. 30, no. 6, pp. 390–395, 2018, doi: 10.1097/CCO.0000000000000484.

[17]   S. Campos *et al.*, "Brain metastasis from an unknown primary, or primary brain tumour? A diagnostic dilemma," *Curr. Oncol.*, vol. 16, no. 1, pp. 62–66, Jan. 2009 [Online]. Available: www.ncbi.nlm.nih.gov/pubmed/19229374.

[18]   E. W. Sankey *et al.*, "Operative and peri-operative considerations in the management of brain metastasis," *Cancer Med.*, vol. 8, no. 16, pp. 6809–6831, Nov. 2019, doi: 10.1002/cam4.2577.

[19]   R. A. Patchell *et al.*, "Postoperative radiotherapy in the treatment of single metastases to the brain," *JAMA*, vol. 280, no. 17, Nov. 1998, doi: 10.1001/jama.280.17.1485.

[20]   A. Mahajan *et al.*, "Post-operative stereotactic radiosurgery versus observation for completely resected brain metastases: a single-centre, randomised, controlled, phase 3 trial," *Lancet Oncol.*, vol. 18, no. 8, pp. 1040–1048, Aug. 2017, doi: 10.1016/S1470-2045(17)30414-X.

[21]   P. D. Brown *et al.*, "Postoperative stereotactic radiosurgery compared with whole brain radiotherapy for resected metastatic brain disease (NCCTG N107C/CEC·3): a multicentre, randomised, controlled, phase 3 trial," *Lancet Oncol.*, vol. 18, no. 8, pp. 1049–1060, Aug. 2017, doi: 10.1016/S1470-2045(17)30441-2.

[22]   P. D. Brown, M. S. Ahluwalia, O. H. Khan, A. L. Asher, J. S. Wefel, and V. Gondi, "Whole-brain radiotherapy for brain metastases: evolution or revolution?" *J. Clin. Oncol.*, vol. 36, no. 5, pp. 483–491, 2018, doi: 10.1200/JCO.2017.75.9589.

[23]   P. D. Brown *et al.*, "Effect of radiosurgery alone vs radiosurgery with whole brain radiation therapy on cognitive function in patients with 1 to 3 brain metastases," *JAMA*, vol. 316, no. 4, p. 401, Jul. 2016, doi: 10.1001/jama.2016.9839.

[24]   E. L. Chang *et al.*, "Neurocognition in patients with brain metastases treated with radiosurgery or radiosurgery plus whole-brain irradiation: a randomised controlled trial," *Lancet Oncol.*, vol. 10, no. 11, pp. 1037–1044, Nov. 2009, doi: 10.1016/S1470-2045(09)70263-3.

[25]   P. Mulvenna *et al.*, "Dexamethasone and supportive care with or without whole brain radiotherapy in treating patients with non-small cell lung cancer with brain metastases unsuitable for resection or stereotactic radiotherapy (QUARTZ): results from a phase 3, non-inferiority," *Lancet*, vol. 388, no. 10055, pp. 2004–2014, Oct. 2016, doi: 10.1016/S0140-6736(16)30825-X.

[26]   M. O'Beirn *et al.*, "The expanding role of radiosurgery for brain metastases," *Medicines*, vol. 5, no. 3, p. 90, Aug. 2018, doi: 10.3390/medicines5030090.

[27]   P. Venkatesan, "SRS is non-inferior to WBRT for brain metastases," *Lancet Oncol.*, vol. 19, no. 8, p. e386, Aug. 2018, doi: 10.1016/S1470-2045(18)30490-X.

[28]   N. U. Lin *et al.*, "Response assessment criteria for brain metastases: proposal from the RANO group," *Lancet Oncol.*, vol. 16, no. 6, pp. e270–e278, Jun. 2015, doi: 10.1016/S1470-2045(15)70057-4.

[29]   M. Werner-Wasik *et al.*, "Immediate side effects of stereotactic radiotherapy and radiosurgery," *Int. J. Radiat. Oncol.*, vol. 43, no. 2, pp. 299–304, Jan. 1999, doi: 10.1016/S0360-3016(98)00410-6.

[30]   P. K. Sneed *et al.*, "Adverse radiation effect after stereotactic radiosurgery for brain metastases: incidence, time course, and risk factors," *J. Neurosurg.*, vol. 123, no. 2, pp. 373–386, Aug. 2015, doi: 10.3171/2014.10.JNS141610.

[31]   F. R. Vogenberg, C. Isaacson Barash, and M. Pursel, "Personalized medicine: part 1: evolution and development into theranostics," *Pharmacy and Therapeutics*, vol. 35, no. 10, pp. 560–576, Oct. 2010.

[32] "DDT Glossary," Available: www.fda.gov/drugs/drug-development-tool-ddt-qualification-programs/ddt-glossary (accessed Mar. 22, 2020).

[33] A. Boire *et al.*, "Liquid biopsy in central nervous system metastases: a RANO review and proposals for clinical applications," *Neuro. Oncol.*, vol. 21, no. 5, pp. 571–584, 2019, doi: 10.1093/neuonc/noz012.

[34] F. Xiao *et al.*, "Cerebrospinal fluid biomarkers for brain tumor detection: clinical roles and current progress," *Am. J. Transl. Res.*, vol. 12, no. 4, pp. 1379–1396, 2020, [Online]. Available: www.ncbi.nlm.nih.gov/pubmed/32355549.

[35] F. Calabrese *et al.*, "Are there new biomarkers in tissue and liquid biopsies for the early detection of non-small cell lung cancer?" *J. Clin. Med.*, vol. 8, no. 3, Mar. 2019, doi: 10.3390/jcm8030414.

[36] W. B. Pope, "Brain metastases: neuroimaging," *Handb. Clin. Neurol.*, vol. 149, pp. 89–112, 2018, doi: 10.1016/B978-0-12-811161-1.00007-4.

[37] K. Brindle, "New approaches for imaging tumour responses to treatment," *Nat. Rev. Cancer*, vol. 8, no. 2, pp. 94–107, Feb. 2008, doi: 10.1038/nrc2289.

[38] A. Sadeghi-Naini *et al.*, "Imaging innovations for cancer therapy response monitoring," *Imaging Med.*, vol. 4, no. 3, pp. 311–327, Jun. 2012, doi: 10.2217/iim.12.23.

[39] H. Mehrabian, J. Detsky, H. Soliman, A. Sahgal, and G. J. Stanisz, "Advanced magnetic resonance imaging techniques in management of brain metastases," *Front. Oncol.*, vol. 9, Jun. 2019, doi: 10.3389/fonc.2019.00440.

[40] J. P. B. O'Connor *et al.*, "Imaging biomarker roadmap for cancer studies," *Nat. Rev. Clin. Oncol.*, vol. 14, no. 3, pp. 169–186, Mar. 2017, doi: 10.1038/nrclinonc.2016.162.

[41] J. S. Ross *et al.*, "The HER-2/ neu gene and protein in breast cancer 2003: biomarker and target of therapy," *Oncologist*, vol. 8, no. 4, pp. 307–325, Aug. 2003, doi: 10.1634/theoncologist.8-4-307.

[42] S. Paik *et al.*, "Real-world performance of HER2 testing--national surgical adjuvant breast and bowel project experience," *JNCI J. Natl. Cancer Inst.*, vol. 94, no. 11, pp. 852–854, Jun. 2002, doi: 10.1093/jnci/94.11.852.

[43] C. Williams and C.-Y. Lin, "Oestrogen receptors in breast cancer: basic mechanisms and clinical implications," *Ecancermedicalscience*, vol. 7, p. 370, Nov. 2013, doi: 10.3332/ecancer.2013.370.

[44] H. Greulich, "The genomics of lung adenocarcinoma: opportunities for targeted therapies," *Genes Cancer*, vol. 1, no. 12, pp. 1200–1210, Dec. 2010, doi: 10.1177/1947601911407324.

[45] B. A. Chan and B. G. M. Hughes, "Targeted therapy for non-small cell lung cancer: current standards and the promise of the future," *Transl. Lung Cancer Res.*, vol. 4, no. 1, pp. 36–54, Feb. 2015, doi: 10.3978/j.issn.2218-6751.2014.05.01.

[46] R. Rosell, M. Taron, N. Reguart, D. Isla, and T. Moran, "Epidermal growth factor receptor activation: how exon 19 and 21 mutations changed our understanding of the pathway," *Clin. Cancer Res.*, vol. 12, no. 24, pp. 7222–7231, Dec. 2006, doi: 10.1158/1078-0432.CCR-06-0627.

[47] J. Tkac *et al.*, "Prostate-specific antigen glycoprofiling as diagnostic and prognostic biomarker of prostate cancer," *Interface Focus*, vol. 9, no. 2, p. 20180077, Apr. 2019, doi: 10.1098/rsfs.2018.0077.

[48] E. Boerrigter, L. N. Groen, N. P. Van Erp, G. W. Verhaegh, and J. A. Schalken, "Clinical utility of emerging biomarkers in prostate cancer liquid biopsies," *Expert Rev. Mol. Diagn.*, vol. 20, no. 2, pp. 219–230, Feb. 2020, doi: 10.1080/14737159.2019.1675515.

[49] N. Terada, S. Akamatsu, T. Kobayashi, T. Inoue, O. Ogawa, and E. S. Antonarakis, "Prognostic and predictive biomarkers in prostate cancer: latest evidence and clinical implications," *Ther. Adv. Med. Oncol.*, vol. 9, no. 8, pp. 565–573, Aug. 2017, doi: 10.1177/1758834017719215.

[50] K. Fujita and N. Nonomura, "Urinary biomarkers of prostate cancer," *Int. J. Urol.*, vol. 25, no. 9, pp. 770–779, Sep. 2018, doi: 10.1111/iju.13734.

[51] T. Khalid *et al.*, "Urinary volatile organic compounds for the detection of prostate cancer," *PLoS One*, vol. 10, no. 11, p. e0143283, Nov. 2015, doi: 10.1371/journal.pone.0143283.

[52] E. J. Whitman *et al.*, "PCA3 score before radical prostatectomy predicts extracapsular extension and tumor volume," *J. Urol.*, vol. 180, no. 5, pp. 1975–1979, Nov. 2008, doi: 10.1016/j.juro.2008.07.060.

[53] X. Durand *et al.*, "The value of urinary prostate cancer gene 3 (PCA3) scores in predicting pathological features at radical prostatectomy," *BJU Int.*, vol. 110, no. 1, pp. 43–49, Jul. 2012, doi: 10.1111/j.1464-410X.2011.10682.x.

[54] A. Martínez-Aranda *et al.*, "FN14 and GRP94 expression are prognostic/predictive biomarkers of brain metastasis outcome that open up new therapeutic strategies," *Oncotarget*, vol. 6, no. 42, Dec. 2015, doi: 10.18632/oncotarget.5471.

[55] A. Darlix *et al.*, "Serum NSE, MMP-9 and HER2 extracellular domain are associated with brain metastases in metastatic breast cancer patients: predictive biomarkers for brain metastases?" *Int. J. Cancer*, vol. 139, no. 10, pp. 2299–2311, Nov. 2016, doi: 10.1002/ijc.30290.

[56] T. G. Whitsett *et al.*, "Molecular determinants of lung cancer metastasis to the central nervous system," *Transl. Lung Cancer Res.*, vol. 2, no. 4, pp. 273–283, Aug. 2013, doi: 10.3978/j.issn.2218-6751.2013.03.12.

[57] K. Cyll *et al.*, "Tumour heterogeneity poses a significant challenge to cancer biomarker research," *Br. J. Cancer*, vol. 117, no. 3, pp. 367–375, Jul. 2017, doi: 10.1038/bjc.2017.171.

[58] L. G. Kessler *et al.*, "The emerging science of quantitative imaging biomarkers terminology and definitions for scientific studies and regulatory submissions," *Stat. Methods Med. Res.*, vol. 24, no. 1, pp. 9–26, Feb. 2015, doi: 10.1177/0962280214537333.

[59] A. Ahmed, P. Gibbs, M. Pickles, and L. Turnbull, "Texture analysis in assessment and prediction of chemotherapy response in breast cancer," *J. Magn. Reson. Imaging*, vol. 38, no. 1, pp. 89–101, Jul. 2013, doi: 10.1002/jmri.23971.

[60] C. J. Rose *et al.*, "Quantifying spatial heterogeneity in dynamic contrast-enhanced MRI parameter maps," *Magn. Reson. Med.*, vol. 62, no. 2, pp. 488–499, Aug. 2009, doi: 10.1002/mrm.22003.

[61] N. Verma, M. C. Cowperthwaite, M. G. Burnett, and M. K. Markey, "Differentiating tumor recurrence from treatment necrosis: a review of neuro-oncologic imaging strategies," *Neuro. Oncol.*, vol. 15, no. 5, pp. 515–534, May 2013, doi: 10.1093/neuonc/nos307.

[62] A. Hutter, K. E. Schwetye, A. J. Bierhals, and R. C. McKinstry, "Brain neoplasms: epidemiology, diagnosis, and prospects for cost-effective imaging," *Neuroimaging Clin. N. Am.*, vol. 13, no. 2, pp. 237–250, May 2003, doi: 10.1016/S1052-5149(03)00016-9.

[63] M. C. Preul, Z. Caramanos, R. Leblanc, J. G. Villemure, and D. L. Arnold, "Using pattern analysis of in vivo proton MRSI data to improve the diagnosis and surgical management of patients with brain tumors," *NMR Biomed.*, vol. 11, no. 4–5, pp. 192–200, Jun. 1998, doi: 10.1002/(SICI)1099-1492(199806/08)11:4/5 < 192::AID-NBM535 > 3.0.CO;2-3.

[64] A. Sadeghi-Naini *et al.*, "Breast-lesion characterization using textural features of quantitative ultrasound parametric maps," *Sci. Rep.*, vol. 7, no. 1, p. 13638, Dec. 2017, doi: 10.1038/s41598-017-13977-x.

[65] A. Sadeghi-Naini *et al.*, "Quantitative ultrasound spectroscopic imaging for characterization of disease extent in prostate cancer patients," *Transl. Oncol.*, vol. 8, no. 1, pp. 25–34, Feb. 2015, doi: 10.1016/j.tranon.2014.11.005.

[66] H. Tadayyon *et al.*, "Quantitative ultrasound assessment of breast tumor response to chemotherapy using a multi-parameter approach," *Oncotarget*, vol. 7, no. 29, Jul. 2016, doi: 10.18632/oncotarget.8862.

[67] A. Sadeghi-Naini *et al.*, "Early prediction of therapy responses and outcomes in breast cancer patients using quantitative ultrasound spectral texture," *Oncotarget*, vol. 5, no. 11, Jun. 2014, doi: 10.18632/oncotarget.1950.

[68] A. Sadeghi-Naini *et al.*, "Early detection of chemotherapy-refractory patients by monitoring textural alterations in diffuse optical spectroscopic images," *Med. Phys.*, vol. 42, no. 11, pp. 6130–6146, Nov. 2015, doi: 10.1118/1.4931603.

[69] K. L. Desmond *et al.*, "Chemical exchange saturation transfer for predicting response to stereotactic radiosurgery in human brain metastasis," *Magn. Reson. Med.*, vol. 78, no. 3, pp. 1110–1120, Sep. 2017, doi: 10.1002/mrm.26470.

[70] H. Mehrabian, S. Myrehaug, H. Soliman, A. Sahgal, and G. J. Stanisz, "Evaluation of glioblastoma response to therapy with chemical exchange saturation transfer," *Int. J. Radiat. Oncol. Biol. Phys.*, vol. 101, no. 3, pp. 713–723, 2018, doi: 10.1016/j.ijrobp.2018.03.057.

[71] H. Mehrabian, K. L. Desmond, H. Soliman, A. Sahgal, and G. J. Stanisz, "Differentiation between radiation necrosis and tumor progression using chemical exchange saturation transfer," *Clin. Cancer Res.*, vol. 23, no. 14, pp. 3667–3675, Jul. 2017, doi: 10.1158/1078-0432.CCR-16-2265.

[72] A. Kapadia *et al.*, "Temporal evolution of perfusion parameters in brain metastases treated with stereotactic radiosurgery: comparison of intravoxel incoherent motion and dynamic contrast enhanced MRI," *J. Neurooncol.*, vol. 135, no. 1, pp. 119–127, Oct. 2017, doi: 10.1007/s11060-017-2556-z.

[73] H. Mehrabian *et al.*, "Water exchange rate constant as a biomarker of treatment efficacy in patients with brain metastases undergoing stereotactic radiosurgery," *Int. J. Radiat. Oncol. Biol. Phys.*, vol. 98, no. 1, pp. 47–55, 2017, doi: 10.1016/j.ijrobp.2017.01.016.

[74] R. J. Gillies, P. E. Kinahan, and H. Hricak, "Radiomics: images are more than pictures, they are data," *Radiology*, vol. 278, no. 2, pp. 563–577, Feb. 2016, doi: 10.1148/radiol.2015151169.

[75] S. Rizzo *et al.*, "Radiomics: the facts and the challenges of image analysis," *Eur. Radiol. Exp.*, vol. 2, no. 1, p. 36, Dec. 2018, doi: 10.1186/s41747-018-0068-z.

[76] M. M. Galloway, "Texture analysis using gray level run lengths," *Comput. Graph. Image Process.*, vol. 4, no. 2, pp. 172–179, Jun. 1975, doi: 10.1016/S0146-664X(75)80008-6.

[77] L. Breiman, "Random forests," *Mach. Learn.*, vol. 45, no. 1, pp. 5–32, Oct. 2001, doi: 10.1023/A:1010933404324.

[78] Y. Shi, Y. Tian, G. Kou, Y. Peng, and J. Li, "Support vector machines for classification problems," *Adv. Inf. Knowl. Process.*, vol. 24, no. 9780857295033, pp. 3–13, 2011, doi: 10.1007/978-0-85729-504-0_1.

[79] G. Guo, H. Wang, D. Bell, Y. Bi, K. Greer, "KNN Model-Based Approach in Classification," in *On the Move to Meaningful Internet Systems 2003: CoopIS, DOA, and ODBASE. OTM 2003. Lecture Notes in Computer Science*, vol 2888, R. Meersman, Z. Tari, and D. C. Schmidt, Eds., Springer, Berlin, Heidelberg. pp. 986–996. https://doi.org/10.1007/978-3-540-39964-3_62.

[80] C. Cortes and V. Vapnik, "Support-vector networks," *Mach. Learn.*, vol. 20, no. 3, pp. 273–297, Sep. 1995, doi: 10.1007/BF00994018.

[81] L. Peterson, "K-nearest neighbor," *Scholarpedia*, vol. 4, no. 2, p. 1883, 2009, doi: 10.4249/scholarpedia.1883.

[82] H. J. W. L. Aerts *et al.*, "Decoding tumour phenotype by noninvasive imaging using a quantitative radiomics approach," *Nat. Commun.*, vol. 5, no. 1, p. 4006, Sep. 2014, doi: 10.1038/ncomms5006.

[83] Z. Bodalal, S. Trebeschi, T. D. L. Nguyen-Kim, W. Schats, and R. Beets-Tan, "Radiogenomics: bridging imaging and genomics," *Abdom. Radiol.*, vol. 44, no. 6, pp. 1960–1984, Jun. 2019, doi: 10.1007/s00261-019-02028-w.

[84] M. Diehn *et al.*, "Identification of noninvasive imaging surrogates for brain tumor gene-expression modules," *Proc. Natl. Acad. Sci.*, vol. 105, no. 13, pp. 5213–5218, Apr. 2008, doi: 10.1073/pnas.0801279105.

[85] P. O. Zinn *et al.*, "Radiogenomic mapping of edema/cellular invasion MRI-phenotypes in glioblastoma multiforme," *PLoS One*, vol. 6, no. 10, p. e25451, Oct. 2011, doi: 10.1371/journal.pone.0025451.

[86] N. M. Czarnek, K. Clark, K. B. Peters, L. M. Collins, and M. A. Mazurowski, "Radiogenomics of glioblastoma: a pilot multi-institutional study to investigate a relationship between tumor shape features and tumor molecular subtype," Mar. 2016, p. 97850V, doi: 10.1117/12.2217084.

[87] M. A. Mazurowski, K. Clark, N. M. Czarnek, P. Shamsesfandabadi, K. B. Peters, and A. Saha, "Radiogenomics of lower-grade glioma: algorithmically-assessed tumor shape is associated with tumor genomic subtypes and patient outcomes in a multi-institutional study with the cancer genome atlas data," *J. Neurooncol.*, vol. 133, no. 1, pp. 27–35, May 2017, doi: 10.1007/s11060-017-2420-1.

[88] P. Grossmann, D. A. Gutman, W. D. Dunn, C. A. Holder, and H. J. W. L. Aerts, "Imaging-genomics reveals driving pathways of MRI derived volumetric tumor phenotype features in glioblastoma," *BMC Cancer*, vol. 16, no. 1, p. 611, Dec. 2016, doi: 10.1186/s12885-016-2659-5.

[89] K. M. Panth *et al.*, "Is there a causal relationship between genetic changes and radiomics-based image features? An in vivo preclinical experiment with doxycycline inducible GADD34 tumor cells," *Radiother. Oncol.*, vol. 116, no. 3, pp. 462–466, Sep. 2015, doi: 10.1016/j.radonc.2015.06.013.

[90] V. S. Parekh and M. A. Jacobs, "Integrated radiomic framework for breast cancer and tumor biology using advanced machine learning and multiparametric MRI," *npj Breast Cancer*, vol. 3, no. 1, p. 43, Dec. 2017, doi: 10.1038/s41523-017-0045-3.

[91] S. Bickelhaupt *et al.*, "Prediction of malignancy by a radiomic signature from contrast agent-free diffusion MRI in suspicious breast lesions found on screening mammography," *J. Magn. Reson. Imaging*, vol. 46, no. 2, pp. 604–616, Aug. 2017, doi: 10.1002/jmri.25606.

[92] J. P. Morales *et al.*, "Peritumoral and intratumoral radiomic features identify aggressive screen-detected early-stage lung cancers," *J. Thorac. Oncol.*, vol. 14, no. 11, p. S1130, Nov. 2019, doi: 10.1016/j.jtho.2019.09.030.

[93] L. Shi *et al.*, "Radiomics for response and outcome assessment for non-small cell lung cancer," *Technol. Cancer Res. Treat.*, vol. 17, p. 153303381878278, Jan. 2018, doi: 10.1177/1533033818782788.

[94] H. Arimura, M. Soufi, K. Ninomiya, H. Kamezawa, and M. Yamada, "Potentials of radiomics for cancer diagnosis and treatment in comparison with computer-aided diagnosis," *Radiol. Phys. Technol.*, vol. 11, no. 4, pp. 365–374, Dec. 2018, doi: 10.1007/s12194-018-0486-x.

[95] L. E. Court, A. Rao, and S. Krishnan, "Radiomics in cancer diagnosis, cancer staging, and prediction of response to treatment," *Transl. Cancer Res.*, vol. 5, no. 4, pp. 337–339, Aug. 2016, doi: 10.21037/tcr.2016.07.14.

[96] R. T. H. M. Larue, G. Defraene, D. De Ruysscher, P. Lambin, and W. van Elmpt, "Quantitative radiomics studies for tissue characterization: a review of technology and methodological procedures," *Br. J. Radiol.*, vol. 90, no. 1070, p. 20160665, Feb. 2017, doi: 10.1259/bjr.20160665.

[97] E. Huynh *et al.*, "CT-based radiomic analysis of stereotactic body radiation therapy patients with lung cancer," *Radiother. Oncol.*, vol. 120, no. 2, pp. 258–266, Aug. 2016, doi: 10.1016/j.radonc.2016.05.024.

[98] Q. Li *et al.*, "Imaging features from pretreatment <scp>CT</scp> scans are associated with clinical outcomes in nonsmall-cell lung cancer patients treated with stereotactic body radiotherapy," *Med. Phys.*, vol. 44, no. 8, pp. 4341–4349, Aug. 2017, doi: 10.1002/mp.12309.

[99] F. E. Harrell, "Evaluating the yield of medical tests," *JAMA J. Am. Med. Assoc.*, vol. 247, no. 18, p. 2543, May 1982, doi: 10.1001/jama.1982.03320430047030.

[100] D. V. Fried *et al.*, "Prognostic value and reproducibility of pretreatment CT texture features in stage III non-small cell lung cancer," *Int. J. Radiat. Oncol.*, vol. 90, no. 4, pp. 834–842, Nov. 2014, doi: 10.1016/j.ijrobp.2014.07.020.

[101] S. H. Bak, H. Park, I. Sohn, S. H. Lee, M.-J. Ahn, and H. Y. Lee, "Prognostic impact of longitudinal monitoring of radiomic features in patients with advanced non-small cell lung cancer," *Sci. Rep.*, vol. 9, no. 1, p. 8730, Dec. 2019, doi: 10.1038/s41598-019-45117-y.

[102] A. S. Tagliafico, M. Piana, D. Schenone, R. Lai, A. M. Massone, and N. Houssami, "Overview of radiomics in breast cancer diagnosis and prognostication," *The Breast*, vol. 49, pp. 74–80, Feb. 2020, doi: 10.1016/j.breast.2019.10.018.

[103] N. C. D'Amico *et al.*, "A machine learning approach for differentiating malignant from benign enhancing foci on breast MRI," *Eur. Radiol. Exp.*, vol. 4, no. 1, p. 5, Dec. 2020, doi: 10.1186/s41747-019-0131-4.

[104] A. Conti, A. Duggento, I. Indovina, M. Guerrisi, and N. Toschi, "Radiomics in breast cancer classification and prediction," *Semin. Cancer Biol.*, May 2020, doi: 10.1016/j.semcancer.2020.04.002.

[105] P. Crivelli, R. E. Ledda, N. Parascandolo, A. Fara, D. Soro, and M. Conti, "A new challenge for radiologists: radiomics in breast cancer," *Biomed Res. Int.*, vol. 2018, pp. 1–10, Oct. 2018, doi: 10.1155/2018/6120703.

[106] H. M. Whitney *et al.*, "Additive benefit of radiomics over size alone in the distinction between benign lesions and luminal a cancers on a large clinical breast MRI dataset," *Acad. Radiol.*, vol. 26, no. 2, pp. 202–209, Feb. 2019, doi: 10.1016/j.acra.2018.04.019.

[107] Y. Dong *et al.*, "Preoperative prediction of sentinel lymph node metastasis in breast cancer based on radiomics of T2-weighted fat-suppression and diffusion-weighted MRI," *Eur. Radiol.*, vol. 28, no. 2, pp. 582–591, Feb. 2018, doi: 10.1007/s00330-017-5005-7.

[108] L. Han *et al.*, "Radiomic nomogram for prediction of axillary lymph node metastasis in breast cancer," *Eur. Radiol.*, vol. 29, no. 7, pp. 3820–3829, Jul. 2019, doi: 10.1007/s00330-018-5981-2.

[109] Y. Guo *et al.*, "Radiomics analysis on ultrasound for prediction of biologic behavior in breast invasive ductal carcinoma," *Clin. Breast Cancer*, vol. 18, no. 3, pp. e335–e344, Jun. 2018, doi: 10.1016/j.clbc.2017.08.002.

[110] A. Yala, C. Lehman, T. Schuster, T. Portnoi, and R. Barzilay, "A deep learning mammography-based model for improved breast cancer risk prediction," *Radiology*, vol. 292, no. 1, pp. 60–66, Jul. 2019, doi: 10.1148/radiol.2019182716.

[111] D. O. Himes, A. E. Root, A. Gammon, and K. E. Luthy, "Breast cancer risk assessment: calculating lifetime risk using the tyrer-cuzick model," *J. Nurse Pract.*, vol. 12, no. 9, pp. 581–592, Oct. 2016, doi: 10.1016/j.nurpra.2016.07.027.

[112] M. Snuderl *et al.*, "Mosaic amplification of multiple receptor tyrosine kinase genes in glioblastoma," *Cancer Cell*, vol. 20, no. 6, pp. 810–817, Dec. 2011, doi: 10.1016/j.ccr.2011.11.005.

[113] T. Koca *et al.*, "Comparison of linear accelerator and helical tomotherapy plans for glioblastoma multiforme patients," *Asian Pacific J. Cancer Prev.*, vol. 15, no. 18, pp. 7811–7816, Oct. 2014, doi: 10.7314/APJCP.2014.15.18.7811.

[114] Q. T. Ostrom, H. Gittleman, G. Truitt, A. Boscia, C. Kruchko, and J. S. Barnholtz-Sloan, "CBTRUS statistical report: primary brain and other central nervous system tumors diagnosed in the united states in 2011–2015," *Neuro. Oncol.*, vol. 20, no. suppl_4, pp. iv1–iv86, Oct. 2018, doi: 10.1093/neuonc/noy131.

[115] S. Rathore *et al.*, "Radiomic MRI signature reveals three distinct subtypes of glioblastoma with different clinical and molecular characteristics, offering prognostic value beyond IDH1," *Sci. Rep.*, vol. 8, no. 1, p. 5087, Dec. 2018, doi: 10.1038/s41598-018-22739-2.

[116] C.-F. Lu *et al.*, "Machine learning—based radiomics for molecular subtyping of gliomas," *Clin. Cancer Res.*, vol. 24, no. 18, pp. 4429–4436, Sep. 2018, doi: 10.1158/1078-0432.CCR-17-3445.

[117] Q. Li *et al.*, "A fully-automatic multiparametric radiomics model: towards reproducible and prognostic imaging signature for prediction of overall survival in glioblastoma multiforme," *Sci. Rep.*, vol. 7, no. 1, p. 14331, Dec. 2017, doi: 10.1038/s41598-017-14753-7.

[118] N. Beig *et al.*, "Radiogenomic-based survival risk stratification of tumor habitat on Gd-T1w MRI Is associated with biological processes in glioblastoma," *Clin. Cancer Res.*, vol. 26, no. 8, pp. 1866–1876, Apr. 2020, doi: 10.1158/1078-0432.CCR-19-2556.

[119] P. K. Brastianos *et al.*, "Genomic characterization of brain metastases reveals branched evolution and potential therapeutic targets," *Cancer Discov.*, vol. 5, no. 11, pp. 1164–1177, Nov. 2015, doi: 10.1158/2159-8290.CD-15-0369.

[120] L. S. Hu and K. R. Swanson, "Roadmap for the clinical integration of radiomics in neuro-oncology," *Neuro. Oncol.*, Mar. 2020, doi: 10.1093/neuonc/noaa078.

[121] L. S. Hu *et al.*, "Radiogenomics to characterize regional genetic heterogeneity in glioblastoma," *Neuro. Oncol.*, vol. 19, no. 1, pp. 128–137, Jan. 2017, doi: 10.1093/neuonc/now135.

[122] E. Karami *et al.*, "Quantitative MRI biomarkers of stereotactic radiotherapy outcome in brain metastasis," *Sci. Rep.*, vol. 9, no. 1, p. 19830, Dec. 2019, doi: 10.1038/s41598-019-56185-5.

[123] A. Mouraviev *et al.*, "Use of radiomics for the prediction of local control of brain metastases after stereotactic radiosurgery," *Neuro. Oncol.*, Jan. 2020, doi: 10.1093/neuonc/noaa007.

[124] N. E. Gilhus, M. P. Barnes, and M. Brainin, *European Handbook of Neurological Management*. Oxford, UK: Wiley-Blackwell, 2010.

[125] R. Soffietti *et al.*, "Diagnosis and treatment of brain metastases from solid tumors: guidelines from the European Association of Neuro-Oncology (EANO)," *Neuro. Oncol.*, vol. 19, no. 2, pp. 162–174, Feb. 2017, doi: 10.1093/neuonc/now241.

[126] H. C. Kniep *et al.*, "Radiomics of brain MRI: utility in prediction of metastatic tumor type," *Radiology*, vol. 290, no. 2, pp. 479–487, Feb. 2019, doi: 10.1148/radiol.2018180946.

[127] G. Louppe, L. Wehenkel, A. Sutera, and P. Geurts, "Understanding variable importances in forests of random-ized trees," in *Advances in Neural Information Processing Systems*, C. J. C. Burges, L. Bottou, M. Welling, Z. Ghahramani, and K. Q. Weinberger, Eds. Curran Associates, Inc., Red Hook, NY, United States, 2013, vol. 26, pp. 431–439.

[128] A. Chen *et al.*, "CT-based radiomics model for predicting brain metastasis in category T1 lung adenocarci-noma," *Am. J. Roentgenol.*, vol. 213, no. 1, pp. 134–139, Jul. 2019, doi: 10.2214/AJR.18.20591.

[129] A. S. Berghoff, V. A. Venur, M. Preusser, and M. S. Ahluwalia, "Immune checkpoint inhibitors in brain metastases: from biology to treatment," *American Society of Clinical Oncology Educational Book*, vol. 36, pp. e116–e122, May 2016, doi: 10.1200/EDBK_100005.

[130] A. Bhatia *et al.*, "MRI radiomic features are associated with survival in melanoma brain metastases treated with immune checkpoint inhibitors," *Neuro. Oncol.*, Oct. 2019, doi: 10.1093/neuonc/noz141.

[131] A. J. Walker *et al.*, "Postradiation imaging changes in the CNS: how can we differentiate between treatment effect and disease progression?" *Futur. Oncol.*, vol. 10, no. 7, pp. 1277–1297, May 2014, doi: 10.2217/fon.13.271.

[132] P. Lohmann *et al.*, "Combined FET PET/MRI radiomics differentiates radiation injury from recurrent brain metastasis," *NeuroImage Clin.*, vol. 20, pp. 537–542, 2018, doi: 10.1016/j.nicl.2018.08.024.

[133] W. Yang *et al.*, "Sex differences in GBM revealed by analysis of patient imaging, transcriptome, and survival data," *Sci. Transl. Med.*, vol. 11, no. 473, p. eaao5253, Jan. 2019, doi: 10.1126/scitranslmed.aao5253.

Index

Note: Numbers in **bold** indicate a table. Numbers in *italics* indicate a figure of the corresponding page.